Applied Innovation and Technology Management

Series Editors

Tuğrul U. Daim
Dept of Engineering & Technology Mgmt
Portland State University
Portland, OR, USA

Marina Dabić
Faculty of Economics & Business
University of Zagreb
Zagreb, Croatia

More information about this series at http://www.springer.com/series/16548

Tuğrul U. Daim
Editor

Roadmapping Future

Technologies, Products and Services

Springer

Editor
Tuğrul U. Daim
Department of Engineering and Technology Management
Portland State University
Portland, OR, USA

ISSN 2662-9402 ISSN 2662-9410 (electronic)
Applied Innovation and Technology Management
ISBN 978-3-030-50504-2 ISBN 978-3-030-50502-8 (eBook)
https://doi.org/10.1007/978-3-030-50502-8

This Springer imprint is published by the registered company Springer Nature Switzerland AG
The registered company address is: Gewerbestrasse 11, 6330 Cham, Switzerland

This book is dedicated to my students who made this volume possible. We want this book to remind us of Roland Richards, whom we lost past year, and his brilliance

VANPORT
COLLEGE

Contents

Part I
Technology Roadmap Analysis

Chapter 1
Technology Roadmapping Maturity Assessment: A Case Study in Energy Sector

Chih-Jen Yu and Tuğrul U. Daim

1.1 Technology Planning

Whether technology roadmapping (TRM) is success or failure from a perspective of maturity assessment, it may depend on how the TRM is viewed and what criteria or metrics are used for the evaluation. Therefore, before providing the cases, it seems necessary to conduct brief overview on criteria or metrics used for evaluating the success level of TRM. After reviewing key criteria or metrics, this section of chapter will cover some TRM case studies for illustrating how maturity have some influence on their consequences of success or failure.

Criteria or metrics used for evaluating the success level of TRM

UK MOD analyzed the characteristics of good roadmapping practice as follows (MOD 2006) (pp. 8–9):

- Clear title, contact detail, and date of publishing or the latest update should be included.
- The format of three layer inherent in the roadmap should be compatible to the architecture/guidance.
- Symbols and colors should be denoted and explained.
- Funding status of R&D program should be clearly shown.
- Time scale should be marked to show the short-term and long-term goals and tendency.
- Capability should be linked to equipment, systems, or platforms.
- Research projects should be linked to the equipment programs.
- Key milestones and decision points should be listed.
- Technology Readiness Levels and availability should be coded and indicated.

C.-J. Yu · T. U. Daim (✉)
Department of Engineering and Technology Management, Portland State University, Portland, OR, USA
e-mail: tugrul.u.daim@pdx.edu

T. U. Daim (ed.), *Roadmapping Future*, Applied Innovation and Technology Management, https://doi.org/10.1007/978-3-030-50502-8_1

- Supporting texts such as scope, aim, definition, options, trade-offs, and risk should be added for complementing the roadmaps.

Gerdsri et al. (2009) proposed some measures for success in different stages of TRM implementation. For initiation stage of TRM, the key measures for success include acceptance by key stakeholders and development of customized process. During development stage, the measures for success incorporate quality of content presented in a roadmap and knowledge sharing among different groups of participants. For the integration stage, successful TRM should be characterized by the linkage between a roadmap and corporate strategic plan and continuation of the TRM implementation (Gerdsri et al. 2009).

Jeffrey et al. (2013) reviewed some success factors for TRM and proposed a set of metrics for assessing the success level of a multi-organization TRM. Part of these metrics identified are derived from the literature of traditional TRM. The tradition TRM here means that the author is a single organization, the intended audience of the TRM is the same organization, and the purpose of the TRM is to act as a strategic planning. The other parts of the metrics are based on whether its objectives have been translated to actions or policies by the target organization, as shown in the table below (Table 1.1) (Jeffrey et al. 2013).

McDowall (2012) proposed four criteria for evaluating the TRM for system innovation. The first one is "credibility", meaning the roadmap should indicate a view of the future that is credible and persuasive. To be credible and persuasive, McDowall argued that it required the roadmap to be developed on the basis of reasonable assumptions and methods, relevant expertise, the commitment of key actors, and embedding a vision of technological change under the context of social, political, market, and cultural aspects of envisaged transition. The second criterion is "desirability," referring to the degree of the future pathway defensible as a good choice for society. The third criterion is "utility," implying the degree of helping the advancement of the innovation system. The fourth criterion "adaptability" refers to how roadmap process be consistent with reflexive, adaptive management (McDowall 2012). These criteria appear to correspond to the four perspectives proposed in my Comp. Exam. proposal as shown below (Table 1.2).

A background paper on TRM published by United Nations Framework Convention on Climate Change (UNFCCC) outlined six substantive elements in the TRM for review their degree of quality (UNFCCC 2013) (Table 1.3).

Oliveira and Fleury (2015) proposed three indicators to improve the roadmap. This first one is gaps of layer content, which aims to examine if the roadmap contains sufficient information on the content of each layer. For a traditional TRM, the detailed information needed for the layer of market/business, product/service, or technology/resource may include market trends, user needs, product/system, technology readiness level, funding opportunities, etc. The second indicator is gaps of layer integration. This mainly focuses on the relationship among all the information contained in each layer. Without integrated efforts, the elements or information in each layer may not connect to each other and clarify a clear path for the future target. The third indicator is gaps of forecasting. This generally involves in using

Table 1.1 Metrics used for accessing the success level of a multi-organization TRM (Jeffrey et al. 2013)

Type	Metric	How metric is assessed
Metrics assessing the architecture of the TRM and how it was prepared	1. Author	Scored depending on the reputation of the author and who they selected to be a part of the TRM process (this is a traditional success factor for compiling a roadmap)
	2. Target audience	Scored based on how well the roadmap addresses its entire target audience
	3. Roadmap message, effectiveness of delivery	Analyses a roadmap's message and how well it is delivered, taking into account format consistency and language
	4. Are the stakeholders adequately addressed?	Measures how well, and how evenly, the stakeholders relevant to the roadmap are addressed
	5. Ease of use – method used	Measures how easy to follow the roadmap is for readers from a range of backgrounds
Metrics assessing the results of the TRM and whether it has achieved its objectives	6. Status of suggested policies	Scored based on whether the roadmap's suggested policies have been implemented or are in the process of being implemented
	7. Citations and references	Scored based on the number of times the roadmap has been cited (highest weighting for citations by another roadmap or by government)
	8. Technology	Scored based on whether the roadmap's technology recommendations have been, or are in the process of being developed
	9. Supply chain	Scored based on whether the roadmap's supply chain recommendations have been or are in the process of being implemented

appropriate scenario planning or forecasting techniques to help predict the future market trend or technology evolution. The questions used for checking these three gaps are listed below (Oliveira and Fleury 2015) (Table 1.4).

In summary, the criteria and success factors involve various aspects and perspectives on how TRM can be successfully implemented for guiding organization's future technology development.

Oliveira and Fleury (2015) discussed three roadmaps by using three indicators for identifying the gaps of the TRM (Oliveira and Fleury 2015). The first roadmap was about renewable fuel, a case study of biobutanol in Brazil (Natalense and Zouain 2013). This roadmap has been developed by using interviews instead of workshop for collecting information. In terms of the layer integration, it can be obviously noticed that there is no connection or linkage among the elements in the neighboring layers. There is only a technology push arrow pointing toward the future timeline, which lacks clear explanation about the layer integration. The author suggested to adopt linking grips or Program Evaluation and Review Technique - Critical Path Method (PERT-CPM) for improvements (Oliveira and Fleury 2015).

Table 1.2 Criteria for roadmap evaluation (McDowall 2012)

Criteria	Key questions	Category/perspectives
Credibility	Is the roadmap based on sound analysis?	Methodological approach
	Does the roadmap draw on the right breadth of expertise?	Knowledge & expertise
	Has the roadmap secured the participation and commitment of key actors in the innovation system?	Organizational support
	Does the roadmap adequately address the political, social, and economic aspects of the transition?	Strategic alignment
Desirability	Does the transition meet social goals established through democratic institutions?	Strategic alignment
	Does the roadmap give a clear account of the justification for the proposed pathway, with transparency in aims, process, and who took part?	Methodological approach, organizational support
	Is the roadmap process inclusive and participatory?	Organizational support
Utility	Does the roadmap effectively articulate a path forward that can enable alignment around common goals?	Strategic alignment
	Is the roadmapping approach appropriate for the stage of innovation system maturity?	Methodological approach
Adaptability	Does the roadmapping process involve periodic reviews, updates and learning?	Methodological approach
	Is the roadmapping process embedded in a broader institutional structure that enables reflexivity and learning?	Methodological approach, knowledge & expertise

Adapted from McDowall (2012)

Table 1.3 Six substantive TRM elements and evaluation questions (UNFCCC 2013) (p. 18–19)

Substantive TRM elements	Evaluation questions
Process description	Is the roadmapping process and methodology described in enough detail to be sure of the roadmaps validity and reliability?
Stakeholders specified	Are the stakeholders (participants in workshops, experts consulted, etc.) specified explicitly?
Quantifiable targets	Are targets quantified and developed in a vision, and are they measurable?
Actions assigned	Are actions assigned to specific individuals or organizations?
Visual representation	Does the TRM contain a visual representation containing the three TRM perspectives (Trends and drivers; Applications, products, services, or other tangible systems; Technologies or other capabilities and resources) along a timeline?
Plan for update	Does the TRM set out a plan for a future update or show any evidence of being updated?

Table 1.4 Three indicators for improving TRM (Oliveira and Fleury 2015)

Indicators	Questions
Gaps of layer content	What information should be presented in each layer?
Gaps of layer integration	How does information relate to each other across layers?
Gaps of forecasting	What information should be presented for the long-term/vision?

The second roadmap to be investigated was Wind Energy roadmap developed by SETIS (Strategic Energy Technologies Information System)-European Commission (European Commission n.d.). According to the comments from Oliveira and Fleury (2015), the roadmap appears to have no layer integration and a gap of forecasting. Like the previous roadmap, there is not any indication about how the elements in neighboring layers are connected or linked. This makes the information in the roadmap look unorganized and discrete. Another drawback is on the gap of forecasting. There are no business drivers and long-term vision identified, which may confuse the stakeholders and make the technology development difficult to be well justifiable. In addition, the timeframe of 10 years seems to be too short, as far as a wind technology is concerned. It has been suggested that this roadmap could be enhanced by utilizing the scenario planning, blue ocean strategy, and open innovation to enhance the future planning efforts (Oliveira and Fleury 2015).

Third roadmap was ICT roadmap developed by National Institute of Information and Communications Technology (NICT) in Japan. The roadmap mainly shows three layers including R&D, Practical use, and Traffic in a timeframe of 40 years (NICT n.d.). The layer integration is visually noticed on the linkage between the R&D and the practical use. However, it appears to lack information that supports the market and business layer, meaning the justification for the technology development targets seems insufficient. Also, the whole roadmap lacks a detailed explanation on its TRM process. Therefore, there is no supporting message to support the content of layer and forecasting (Oliveira and Fleury 2015).

Jeffrey et al. (2013) evaluated four roadmaps for their level of success by using the 9-metrics mentioned in previous section. First roadmap is UKERC (UK Energy Research Centre) Marine Renewable Energy Technology Roadmap. The second roadmap is Marine Energy Roadmap published by Forum for Renewable Energy Development in Scotland's (FREDS) Marine Energy Group's (MEG). The third roadmap is IEA (International Energy Agency) Technology Roadmap: Wind Energy. The fourth roadmap is European Wind Energy Association (EWEA) oceans of opportunity: harnessing Europe's largest domestic energy resource. As it is shown below, all the four cases did not reach the highest level of success. The performance on each individual metric can be used for identifying the strength and weakness. It may be inferred that the better performance/maturity, the more chances of enabling a successful TRM implementation.

The success or failure of TRM can be judged by various success factors. The higher level of complying the success factors, the more probability of implementing a successful TRM program. On the contrary, the lower level of achieving the success criteria, the more chances of leading the TRM into a failure result.

1.1.1 Overview of Strategic Technology Planning

Strategic Technology Planning was defined as "The process of determining which technologies not yet adopted will have a strategic impact on the company" (Fenn et al. 2003). A report published from Gartner provided a Strategic Technology Planning Process called STREET meaning Scope, Track, Rank, Evaluate, Evangelize, and Transfer. Initially, this process began by understanding corporate objective, industry direction, and business bottlenecks by means of competitive analysis, visioning, and scenario building. Later, new technology opportunities need to be scanned, ranked, evaluated to confirm the benefit to the enterprise and the operationally applicability. Next, the efforts need to be evangelized or directed to bring the technology into production and be proactive to overcome any possible organizational resistance. Finally, the required knowledge and responsibility need to be transferred to the responsible people to take charge of the whole developmental initiatives (Fenn et al. 2003).

There are some success factors involving in conducting strategic technology planning, which include senior management commitment, networking across business units, balancing short- and long-term objectives, balancing staffing mix, planning the route to deployment, making prioritization explicit, associating technologies with business need, preparing to say no, starting transfer early, and keeping an open mind as to the best solution (Fenn et al. 2003). Other key factors pertaining to strategic technology planning are briefly described below (Table 1.5).

1.1.2 Overview of Strategic Technology Planning Tools

Suharto and Daim (2013) reviewed methods and tools applied in strategic technology planning during the period of 1970 through 2010. Based on their literature review, they found that the tools and methods in strategic technology planning can be divided into three categories including market analysis approach, technology analysis approach, and combined approaches of market and technology (Suharto and Daim 2013). The tools or methods included in each category are listed and briefly described in the tables as follows (Tables 1.6, 1.7, and 1.8):

Table 1.5 Key factors pertaining to strategic technology planning

Authors	Key factors	Applications/sectors	References
Fenn et al. (2003)	Senior management commitment, networking across business units, balancing short- and long-term objectives, balancing staffing mix, planning the route to deployment, making prioritization explicit, associating technologies with business need, preparing to say no, starting transfer early, and keeping an open mind as to the best solution	From the standpoint of advanced technology development	Fenn et al. (2003)
Haselkorn et al. (2007)	Use of a standardized process with clear roles and objectives, authority of committee, utilization of a criteria-based review structure, use of specific metrics, provision of ongoing communication, integration of departmental function and process, supportive culture, proactive approach in seeking emerging technology, monitoring of technology implementation	From the perspective of medical technology	Haselkorn et al. (2007)
Bai and Lee (2003)	CEO/CIO relationship, task coordination, stakeholder interaction, stakeholder involvement, IS maturity, centralization of organizational structure, quality of ISSP process	In the context of IS/IT strategic planning process	Bai and Lee (2003)
Lee and Bai (2003)	Group interaction, knowledge management, organizational learning, change management	From the perspective of organizational mechanisms for successful IS/IT strategic planning	Lee and Bai (2003)
Aron (2010)	Clarifying the purpose and scope in light of the business's vision, mission, and value proposition; what will make the business win; what capabilities are needed; planning for uncertainty, the required flexibility; determining how best to create a group-level strategy	From the viewpoint of Strategic Planning Key Initiative for IT leaders	Aron (2010)
Mittenthal (2002)	A clear and comprehensive grasp of external opportunities and challenges, a realistic and comprehensive assessment of the organization's strengths and limitations, an inclusive approach, an empowered planning committee, involvement of senior leadership, sharing of responsibility by board and staff members, learning from best practices, clear priorities and an implementation plan, patience, a commitment to change	From a perspective of nonprofit and foundation leaders	Mittenthal (2002)

(continued)

Table 1.5 (continued)

Authors	Key factors	Applications/sectors	References
Nauda and Hall (1991)	A well-defined process for strategic technology planning; systematically define customer needs, assess competitive posture, identify new technologies for future growth, forecast availability, monitor progress of critical technologies, identify high-level allocation of resources; be integrated into strategic business planning process; managing technology investment effectively and efficiently; make decision within dynamic framework	From a viewpoint of developing roadmaps for competitive advantage	Nauda and Hall (1991)

Table 1.6 The market analysis approaches applied in strategic technology planning

Category	Tools or methods	Brief description	References
Market analysis approach	Experience curve	A curve showing the percentage reduction in cost occurs, after the cumulative production is doubled	Krawiec et al. (1980), Ghemawat (1985)
	Porter's five forces	Analysis on the five forces (threat of new entrants, bargaining power of buyers, bargaining power of suppliers, threat of substitute products or services, and rivalry among existing competitors) shape industry competition	Porter (2008)
	SWOT analysis	Conducting analysis on Strength, Weakness, Opportunities, and Threats for an organization or industry	Phaal et al. (2005), Yuan (2013), Bull et al. (2016), Menon et al. (1999)
	Concept visioning	Aligning an external opportunity of market demand with an internal driver	Passey et al. (2006), Latham (1995)
	Priority and level of investment	Aimed to create technology investment options and determine relative technology expenditure compared to competitors	Larsson (2005), Christensen (1998)
	Timing concept	To decide a proper timing or time horizon for technology development or R&D investment	Christensen (2002), Grienitz and Ley (2007), Christensen (1996)
	Product portfolio (BCG matrix)	A market share/market growth matrix for facilitating product portfolio strategies	Day (1977), Hedley (1977)

Adapted from (Suharto and Daim 2013)

Table 1.7 The technology analysis approaches applied in strategic technology planning

Category	Tools or methods	Brief description	References
Technology analysis approach	Bibliometric	A method using counts of publications, patents, citations etc. to develop science and technology performance indicators	Kostoff et al. (2005), Kostoff and Schaller (2001), Kostoff et al. (2004)
	Technology acquisition	Forming a proper strategy on acquiring technology with consideration on patents, licensing agreements, technology transfer, or technological alliances	Christensen (2002), Clarke (1987), Caviggioli et al. (2017)
	Organizing of technological assets	Effectively sharing, transferring, expanding, or organizing technological assets to meet the short-term and long-term R&D goals	Christensen (1998)
	Technology integration	Integrating corporate technology into the corporate and business unit strategy	Edler et al. (2002)
	Soft system methodology	An action research method utilizing models to structure, observe, or solve a problem by investigating different or conflicting objectives, needs, purposes, interests, values, or perspectives	Checkland (2000)
	Database tomography	A textual database analysis system including algorithms for extracting multi-word phrase frequencies and performing proximity analyses	Kostoff et al. (1998)
	Technology development envelope	A method used for identifying the optimum technology path by using the combination of Delphi method and Hierarchical Decision Modeling (HDM)	Gerdsri (2007), Kockan et al. (2009)
	Patent analysis	A toolset involving in using data mining techniques to analyze patent information for identifying technology capacity or performance of R&D activity	Yoon and Lee (2008), Lee et al. (2008), Ernst and Soll (2003), Ernst (2003)
	AHP	A comprehensive framework using multi-criterion decision-making and pairwise comparison judgement for developing preference or priorities	Gerdsri and Kocaoglu (2007), Kocaoglu (1983)

Adapted from (Suharto and Daim 2013)

Table 1.8 The combined approaches applied in strategic technology planning

Category	Tools or methods	Brief description	References
Combined approach	Change management	Involving awareness of change, desire to make a change, knowledge about how to change, ability to implement new skills, and reinforcement to retain the change	Gerdsri et al. (2008)
	Synergy-oriented technology strategy	Guiding technology diversification to promote technology integration and supplementing existing technologies in innovative process	Christensen (2002)
	Technology acquisition strategy	Acquiring companies or units with needed R&D capacity for business growth or globalization	Christensen (2002), Christensen (1998)
	Scenario planning	A technique to analyze alternative future conditions, evaluate the impacts and develop firm strategies	Pagani (2009), Drew (2006)
	Cost of innovation	Used for measuring the level of R&D investment necessary to generate inventions in a particular industry or industry segment	Bigwood (2000)
	Axiomatic design approach	A methodology using mapping methods to systematically transform the customer needs into functional requirements, design parameters, and process variables	Koc and Mutu (2006)
	Innovation matrix	A planning technique showing the relationship between technology uncertainty and required availability of product feasibility and commitment to resource allocation	Groenveld (2007)
	Technology audit	An inventory of the technologies possessed or utilized by a firm	Martino (1994)
	Integrated portfolio	A portfolio approach integrating market and technology portfolios to support market-oriented R&D planning	Ernst et al. (2004)

Adapted from (Suharto and Daim 2013)

1.1.3 Other Strategic Technology Planning Tools

Apart from the tools identified above, there are some other tools such as strategic planning tools, which can be perceived as a subset of strategic technology planning tools. These tools aim to enhance the strategic guidance and the intention to foresee the future business environment and technological prospect. For example, Trainer (2004) identified top ten strategic planning tools including SWOT analysis, TOW (Turning Opportunities and Weaknesses into Strength) analysis, Nominal group technique, Affinity diagrams, SMART (Specific, Measurable, Achievable or attainable, Results-oriented, and Time-bound) language, Responsibility matrix, Flowcharting, Cause-and-effect diagrams, Presentation of quantitative data, and

Goal attainment team (Trainer 2004). Phaal et al. (2006) discussed two tools used for supporting strategic planning and can be utilized with technology roadmapping (TRM). The first tool is "Portfolio Matrix," which can be used for supporting prioritization, selection of research project, or strategic options. The second tool, "Linked Analysis Grids," is an orthogonal structure that can be used to link one set of domain/parameter to another (Phaal et al. 2006a). Ervural et al. (2018) proposed an ANP (Analytic Network Process) and fuzzy TOPSIS (Technique for Order Performance by Similarity to Ideal Solution) –based SWOT analysis for Turkey's energy planning (Cayir Ervural et al. 2018).

Strategic technology planning involves the combination of strategic planning and technology planning efforts. Each tool has been designed or utilized to facilitate the planning activities. However, the focus may need to be on effective integrating the usage of tools, so as to enhance the implementation of the strategic technology plan.

1.1.4 The Tools Identified from Public Documents

The public documents here are perceived as the documents published by government agencies, nonprofit organizations, or private companies/organizations with an intent for making the documents accessible to the public. The documents were mainly collected from the internet or PSU library database. The main key words used to search for the public documents including "Strategic Energy Technology Plan," "Strategy and Energy and Technology and Plan," "Strategic Planning and Energy and Technology," "Strategic Planning and Energy," "Strategy and Energy and R&D and Plan," "Strategy and Energy and Plan," "Department of Energy and Strategic Plan," "Department of Energy and R&D and Plan." In addition to using key words in search for the relevant documents, specific websites have been explored as follows:

- Department of Energy (DOE) Policies and Regulation website (https://www.energy.gov/cio/office-chief-information-officer/policy-and-regulations/doe-policies)
- DOE Directives website (https://www.directives.doe.gov/directives-browse#c10=&b_start=0)
- DOE Develop an Energy Plan website (https://www.energy.gov/eere/slsc/develop-energy-plan)

These DOE websites have been explored to search for relevant documents related to strategic energy technology planning or strategic energy system/resource planning.

After collecting the documents and reports from the internet or PSU library database, the documents can be categorized into three degrees of relevance. The first degree is that the title of the document matches the strategic energy technology plan exactly. The second degree of relevance means that the content of the document is strategic energy technology planning oriented, but the titles did not show the exact wording. For examples, some research and development plans published by DOE, or some strategic plans published by the technology organization/companies in the energy sector. The third degree of relevance is that the main intent of the document is not targeted on the strategic energy technology; however, the content contains some energy related technology planning or suggestion for guiding future development. The scope of the documents collected is not limited to the United States. Some of the documents were published from other countries. The selected public documents with strategic energy technology planning tools are listed and briefly introduced in the table below (Table 1.9).

1.1.5 The Tools Identified from Interviews with Responsible Energy System Planners

Four experts from three different organizations was interviewed via either telephone or face-to-face approach. The meeting notes are listed in Appendices 1–3. (Table 1.10):

1.1.5.1 Summary of the Tools Identified

Expert A identified five major tools used for energy technology/system planning. These tools include Load Forecasting, Resource Planning, Transmission Planning Non-Wire Transmission, and Integrated Resource Planning (IRP). For technology forecasting or demand forecasting tools, Production Cost Model and Probabilistic Model are used for transmission planning and contingency planning. He also said that BPA is now technology roadmapping (TRM) oriented. In addition to TRM, BPA also used portfolio management such as P3M3 (Portfolio, Program, and Project Maturity Management Model) for planning and managing their technology development programs and projects. Other tools used for strategic planning purposes include SWOT, Scenario Analysis, and Long-term trend tracking analysis. In addition to the technology roadmaps, BPA publishes strategic plan and is working on integrated resource and transmission plan documents to provide their future direction on power product and service.

Expert B said that PGE use "Monte Carlo Simulation" method to conduct a flexible capacity analysis, use "Economic Model" to evaluate the options of energy resources, and employ "Scenario Planning" together with "Simulation Model" for analyzing future demand and energy resources. He also said that their planning and decision on future energy is sometimes attributed to the regulation. For example, the 70% of renewable energy is the target that PGE need to strive for. As far as the strategic planning documents are concerned, PGE publishes strategic plan on outlining the future business and customer-related targets and issues. The Integrated Resource Plan addresses the planning for future energy portfolio mix.

Table 1.9 Selected public documents and the strategic energy technology planning tools

Organization	Title of the document	Purpose of the document	Date	Selected key tools
BPA (Bonneville Power Administration 2017)	2016–2021 Energy Efficiency Action Plan	Serves at BPA's roadmap for achieving energy efficiency savings from 2016 to 2021 for partially responding the demand forecast outlined in the 7th Power Plan	March, 2017	New technology assessment, cost saving assessment, program retirement assessment
BPA (Bonneville Power Administration 2018)	2018–2023 Strategic Plan	Describe the actions to be more competitive and responsive to customer needs, to enable industry change, and to deliver public responsibility	Jan., 2018	Strategic planning
California Public Utilities Commission (California Public Utilities Commission 2008)	California's Long-Term Energy Efficiency Strategic Plan	To set forth a roadmap for energy efficiency in California through the year 2020 and beyond and to articulate near-term, mid-term, and long-term strategies for achieving the goals	Sept., 2008	Strategic planning
European Union (European Commission 2017)	The Strategic Energy Technology (SET) Plan	To provide research & innovation goals/ achievements and implementation plan towards "Clean Energy for all Europeans"	2017	Strategic planning
Illinois Power Agency (Illinois Power Agency 2017)	Long-Term Renewable Resources Procurement Plan	To address how the agency will undertake a variety of programs and procurements for fulfilling the requirements of Illinois Renewable Portfolio Standard	Sep. 29, 2017	Energy resource planning
Northwest Energy Efficiency Alliance (NEEA) (Northwest Energy Efficiency Alliance 2014)	Strategic Plan 2015–2019	To provide and document vision, mission, principles and values, strategic goals and key strategies for northwest stakeholders to guide their transformation towards a sustainable future	July 8, 2014	Strategic planning process
International Renewable Energy Agency (IRENA) (European Commission 2018)	Renewable Energy Prospects for the European Union	To identify cost-effective renewable energy options for all EU member states to accelerate the deployment of renewables toward 2030	Feb., 2018	Energy technology forecasting

(continued)

Table 1.9 (continued)

Organization	Title of the document	Purpose of the document	Date	Selected key tools
International Renewable Energy Agency (IRENA) (International Renewable Energy Agency 2017)	Renewable Energy Outlook Thailand	To evaluate power generation thermal use and bioenergy, and identify key challenges and actions needed to meet or exceed the country's target of 30% renewables in the energy mix by 2036	2017	Renewable readiness assessment, renewable energy roadmap analysis
NWCouncil (NWPCC) (Crow 2016)	7th Power Plan	To provide energy demand forecasting and resource planning for Northwest region	2016	Energy demand forecasting, energy market price forecasting, energy resource strategy planning
PGE (PGE 2016)	Integrated Resource Plan	To offer detailed technical analysis of PGE's strategy of supplying the electric power to meet customers need	November 2016	Scenario analysis, scoring metrics, portfolio analysis, market assessment, need assessment

Table 1.10 The list of energy system planner interviewed

Name	Jot title, department/ division	Organization	Types of interview	Date of interview
A	Policy Strategist, Strategic Planning, Corporate Strategy	BPA (Bonneville Power Administration)	Telephone interview	March 16, 2018, during 2:00–2:18 pm
B	Director, Retail Technology Development	PGE (Portland General Electric)	Telephone interview	March 17, 2018, during 2:00–2:25 pm;
C	Project Analyst, Power Division	NWPCC (Northwest Power and Conservation Council)	Panel interview (face to face) at 851 SW Sixth, Suite 1100, Portland, OR	March 19, 2018, during 03:00–4:05 pm
D	Power System Analyst, Power Division	NWPCC (Northwest Power and Conservation Council)	Panel interview (face to face) at 851 SW Sixth, Suite 1100, Portland, OR	March 19, 2018, during 03:00–4:05 pm

Experts C and D mentioned that NWPCC use some models and tools in council plan development, including Energy 2020, Fuel Price Forecasting Model, AURORA$^{XMP®}$ Electricity Market Model, GENESYS, and Regional Portfolio Model (RPM). The Energy 2020, developed by Systematic Solutions, Inc., has been customized for and by the NWPCC to be used for forecasting the hourly electricity demand. AURORA$^{XMP®}$ Electricity Market Model is a production cost model developed by EPIS, Inc. and used for forecasting hourly wholesale electricity market prices. The Fuel Price Forecasting Model, GENESYS, and Regional Portfolio Model are developed by NWPCC and used to convert assumptions about fuel commodity process to regional wholesale prices, to perform hourly simulation of resource, and to identify low-cost and low-risk resource strategies under some future assumptions respectively.

Proper technology roadmapping is really an indication of maturity in strategic technology planning. The reasons are described as follows:

From the perspective of planning process, TRM has played a key role in strategic technology planning process. Phaal et al. (2005) discussed the relationship between technology roadmapping and strategic planning. The roadmap was perceived as a discrete step, where the output of the strategic planning process was captured and displayed for guiding the implementation of the follow-on technology development projects (Phaal et al. 2005).

From the perspective of benefits, TRM itself or combined with other tools providing benefits for strategic technology planning. Bray and Garcia (1997) claimed that technology roadmapping has been regarded as a form of technology planning to help organizations to handle the increasing competitive environment. TRM can help develop consensus about the need and required technology, provide a mechanism for forecasting the areas of technology development, provide a framework for coordination among companies or industries, and facilitate technology investment decisions (Garcia and Bray 1997).

Lizaso and Reger (2004) proposed that technology roadmapping can be linked with scenario technique as an approach for strategic technology planning. The proposed method consists of six steps including roadmapping preparation, system analysis, scenario projection, scenario building, time assessment, and roadmapping. This comprehensive method aimed to facilitate the planning of technology development and deployment of new and existing technologies and applications (Lizaso and Reger 2004).

Groenveld (2007) provided benefits of technology roadmapping (Groenveld 2007):

1. Establishment of a shared product-technology strategy
2. Set an interlinked approach for long-range product and technology planning and vision building
3. Stimulation of learning and improvement of cross-function communication
4. Improvement of time-to-market and time-to-money, thereby achieving a better competitive edge
5. Supporting working in the process way

UNFCC (2013) provided some advantages of using TRM as follows (UNFCCC 2013) (p. 41):

1. The structured depiction of trends, objectives, and actions provide a highly synthesized view of strategy, which is beneficial for communication among stakeholders.
2. TRM aligns technology and product strategy and link technological consideration to policy.
3. TRM has been identified as being pro-active to articulate the future that should be created.
4. TRM is characterized by a strong consensus, shared vision, and agreed actions, which is reinforced by the effective visual representation of a multilayered time-based diagram.
5. The strong consensus-building element of roadmaps enables the enhancement of cooperation in international partnership among public and private stakeholders.
6. The flexibility of TRM allows the method to be applied in different circumstances with some adaption to suit particular context or to be integrated with other tools such as scenario planning and portfolio management.

From the perspective of integration, TRM integrates key elements in strategic technology planning. TRM integrated vision and business driver with technology characteristics, and others. This is a visual effect and good for communication, monitoring, open discussion, feedback, and implementation.

1.1.6 Overview of NW Power and Conservation Council (NWPCC)

The Council was founded by Congress in 1980 when the Northwest Power Act was passed (United States Congress 1980). The Act gives Council the authority to develop and maintain a regional power plan and a fish and wildlife program for balancing the environment and meeting the energy need in northwest region including the state of Oregon, Washington, Idaho, and Montana (Crow 2016). The Council is located at Portland Oregon and has been structured to operate in central office and offices in these four states. There are council members appointed by their governors to serve in the Council. Several committees including executive, fish and wildlife, power, and public affair were established to provide professional opinions on the on-going regional plan or program development (NWCOUNCIL n.d.-a) (Tables 1.11 and 1.12).

One of the main responsibilities of the Council is to develop a 20-year electric power plan aiming to acquire adequate and reliable energy at the least economic cost and reduce environmental impact to the northwest. This power plan is updated

Table 1.11 NWPCC organization structure description

Organization	Position/member	Job description	Notes
Council organization	Council member	Set policy and provide overall leadership for Council activities	Two members from each of Idaho, Montana, Oregon, and Washington state are appointed by their governors
State office organization	Staff/Analyst	Provide technical review and assistance to Council members and participate in designing and developing public involvement programs	Each state (Oregon, Washington, Idaho, and Montana) maintain its own office
Central office organization	Council's Executive Director	Responsible for coordinating with eight-member Council, supervising the central office staff, administering the contracts, and overseeing the day-to-day operations of the council	The Central Office is in Portland Oregon
	Director, Manager, Analyst, Coordinator Assistant	Perform managerial or analytical tasks in Administrative, Power, Fish and Wildlife, Legal and Public Affair division	

Table 1.12 The advisory committee in energy division

Advisory committee	Key function
Regional Technical Forum	To develop standards to verity and evaluate energy efficiency savings
Regional Technical Forum (RTF) Policy Advisory Committee	To advise the Council regarding policy and governance issues of the RTF
Conservation Resources Advisory Committee	To advise the Council regarding the formulating and reviewing policy and program alternatives to effectively develop the region's cost-effective conservation potential
Demand Forecast Advisory Committee	To assist in review of the demand forecasting tools, input assumptions and forecast results

every 5 years (Crow 2016). Now the 7th power plan was adopted in Feb 2016 and 8th power plan is under developing (NWCOUNCIL n.d.-b).

1.1.6.1 Overview of CMMI

In view of the schedule, budget, productivity, and quality problems of managing software process, Paulk et al. (1993) proposes Capability Maturity Model (CMM) as a software maturity framework for improving organizational processes of developing and maintaining software. This model contains five maturity levels including initial, repeatable, defined, managed, and optimizing. The initial level is characterized as ad hoc and few defined processes. The repeatable level embraces some basic project management process for tacking and functionality. The defined level demonstrates documented, standardized, and integrated management and engineering activities involving in the software process. For the managed level, quantitative measures are applied for software process and product quality. The optimizing level incorporates continuous process improvement by quantitative feedback and adopting innovation and technology (Paulk et al. 1993). In terms of the usages, the CMM is regarded as a framework that can be used for laying out a path of improvements for software organizations to increase their capability. It is expected to be used by various assessments teams, evaluation groups, or managers and technical staffs for identifying strength and weaknesses, the risks involving in selecting contractors, and software process improvement program (Paulk et al. 1993).

The Software CMM has later replaced by CMM Integration (CMMI) in 1997, which integrates System Engineering Capability, Software Engineering and Integrated Product Development in a single model (Paulk 2009). After some version changes, the CMMI series of standards have been developed to include CMMI for Acquisition, Development, and Services, which are targeted on improving processes for acquiring, developing, or providing better product and services, respectively (Software Engineering Institute 2010a; Software Engineering Institute 2010b; Software Engineering Institute 2010c).

1.1.6.2 Overview of Energy Management Maturity Model

The increased awareness of sustainability and environmental impact has raised organizational attention in energy efficiency, energy consumption, and cost reduction. An energy management system containing critical elements to establish an energy policy and objective is aimed to monitor energy performance and facilitate the efforts for continuous improvements (Introna et al. 2014). With the need to develop a robust energy management system, energy management maturity model has been developed to measure the degree of maturity on the energy management performance (Introna et al. 2014; Finnerty et al. 2017; Antunes 2014; Leritz et al. 2014; Northwest Energy Efficiency Alliance (NEEA) 2013; Böttcher and Müller 2016; Ates and Durakbasa 2012; Ngai et al. 2013). These efforts for developing energy management maturity model correspond to the requirement of ISO 50001 for monitoring and controlling energy management performance (Gopalakrishnan et al. 2014).

The structure of energy management maturity model generally consists of various levels to represent the degree of maturity. It also includes some key dimensions, perspectives, attributes, or criteria that are relevant to the success of energy management. For example, Introna et al. (2014) proposed an Energy Management Maturity Model, which was structured with five levels including Initial, Occasional, Planning, Managerial, and Optimal. The model also consists of five dimensions including awareness, knowledge and skills, methodological approach, energy performance management and information system, organization structure, and strategy and alignment, which serve as key criteria for measuring the maturity of energy management performance (Introna et al. 2014). As shown in Table 1.13, the characteristics for the dimensions are defined in each level of maturity, whereby the design of the survey questionnaires can be based upon.

For the CMMI, it is clear that this maturity model is mainly used for software industry to access and monitor their capabilities. The NWPCC is not a software or R&D-oriented organization. Hence, the CMMI did not fit to the nature of the NWPCC. However, if NWPCC would like to develop a more sophisticated software program, the CMMI may be considered for adoption.

For the Energy Management Maturity Model, it is relevant to the energy industry. Therefore, it looks suitable to be applied in NWPCC. However, the maturity model is designed specifically to correspond the ISO 50001 requirement. It focuses on the companies or organizations, which consume lots of energy and would like to monitor the performance of meeting the sustainability objectives. What NWPCC needs is to evaluate the performance or maturity of their planning efforts on developing the 7th Power Plan and other analytical modelling tasks.

Table 1.13 Levels and dimensions of maturity in EMMM adapted from (Introna et al. 2014) (p. 114)

	Dimensions of maturity				
Level	Awareness, knowledge, and skills	Methodological approach	Energy performance management and I.S.	Organizational structure	Strategy and alignment
5: Optimal	Optimized	Optimized and in use	Optimized and in use	Optimized and in use	Optimized and in use
4: Managerial	Highly advanced	Energy management system in use	Improved, stabilized, and in use	Improved, stabilized, and in use	Complete alignment
3: Planning	Significant progress	Project approach in use	Standardized and in use	Project organization	Significant progress (shared targets)
2: Occasional	Basic	Occasional identification of intervention	Basic	Designation of a person responsible for energy	Policy definition and awareness campaign
1: Initial	Scattered	Nonexistent	Nonexistent	Scattered (nonexistent)	Nonexistent

1.2 Technology Planning Maturity Assessment Model (Fig. 1.1)

Based on the literature review and the conceptual input-process-output viewpoint, the key factors in each perspective have been combined or extracted to be selected as key criteria for assessing the maturity of the strategic technology planning. The proposed preliminary maturity model incorporating the perspectives of organization, methodology, knowledge, and documents are briefly described as follows:

- For organization perspective, there are five criteria including Executive's Commitments, Teaming & Staffing, Communication & Collaboration, Stakeholder Involvement, and Resource Allocation.
- For the methodology perspective, there are five criteria including Technology Roadmapping, Technology Forecasting, Technology Assessment, Data Collection & Management, and Review, Update & Change Management.
- For the knowledge perspective, there are five criteria including Guideline & Handbook, Instruction & Training, Skill/Competency, Lesson Learned & Dissemination, and Expert Engagement.
- For the document's perspective, there are five criteria including Environmental Scan & Technology Prospect, Mission, Vision, and Capability Gaps, Goals & Objectives with quantitative metrics, Strategy & Action/Implementation Plan, and Interconnection & Alignment.

The preliminary maturity model for strategic technology planning is shown below. The detailed description of each perspective and criteria will be explained (Fig. 1.2).

As shown in the preliminary maturity model for Strategic Technology Planning, there are four perspectives including organization, methodology, knowledge, and documents. For each perspective, there are five criteria contributing to the maturity assessment of strategic technology planning. It is proposed to use hierarchical decision model (HDM) combined with desirability curve analysis to be used for assessing the maturity of the strategic technology planning.

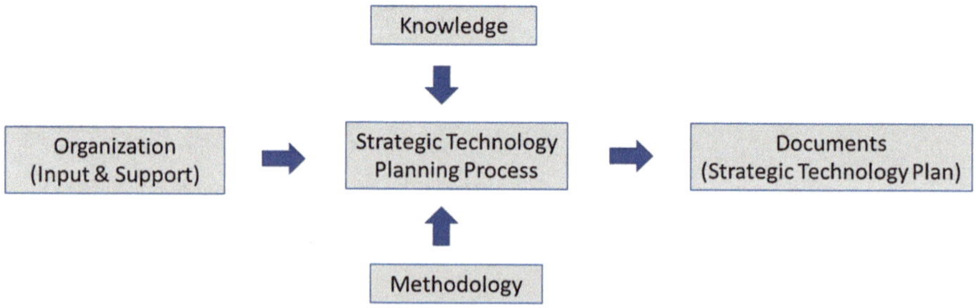

Fig. 1.1 Strategic technology planning from input-process-output view

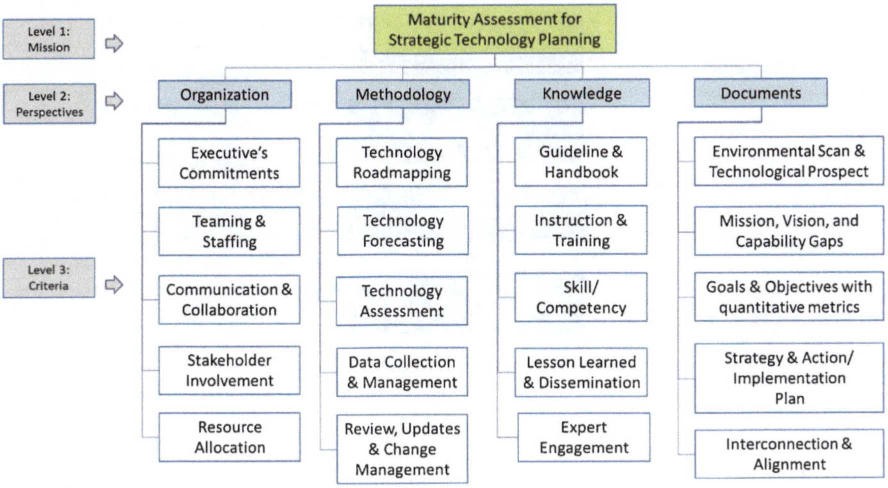

Fig. 1.2 Preliminary maturity model for strategic technology planning

Table 1.14 List of expert panels

Expert panel	Tasks	Tools	Numbers of experts
1	Validation of perspectives	Qualtrics	8
2	Validation of criteria	Qualtrics	8
3	Validation of desirability metrics	Qualtrics	8
4	Quantification of perspectives	HDM online software	8
5	Quantification of criteria	HDM online software	8
6	Quantification of desirability metrics	Qualtrics	8

Based on this initial research planning, there are six expert panels required to be set up. As shown in the Table 1.14, first, second, and third expert panels are designed for validating the perspective, criteria, and desirability metrics, respectively. The fourth, fifth, and sixth expert panels are formed for quantifying the perspectives, criteria, and desirability metrics correspondingly.

Ideally, the expert panels will need to be formed by selecting the subject matter experts pertaining to strategic technology planning. To demonstrate the application of this model, experts were chosen among the PhD students at Portland State University's Technology Management Doctoral Program.

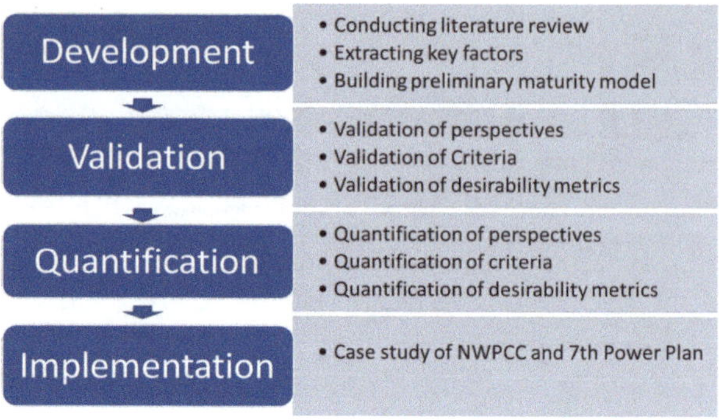

Fig. 1.3 HDM research approach for assessing maturity of strategic technology planning

For this research, a hierarchical decision model (HDM) combined with desirability curve is proposed to be used for assessing the maturity of strategic technology planning.

The research includes the following phases:

1. Model development: Building a maturity assessment model for Strategic Technology Planning
2. Model validation: Validating the perspectives, criteria, and desirability metrics included in the maturity model
3. Model quantification: Quantifying perspective, criteria, and desirability metrics for application of maturity assessment
4. Model implementation: Using case study to illustrate how the maturity model can be applied (Fig. 1.3)

The detailed steps for applying HDM research steps are listed as follows (Table 1.15):

For purpose of calculating maturity score, the Technology Value approach developed by Dr. Gerdsri and Dr. Kocaoglu is proposed to be modified to Maturity Score (M_S) and then used for analysis (Gerdsri and Kocaoglu 2004).

The maturity score can be calculated by using the equations below.

$$S_{n,jn}^{M} = \sum_{N}^{n=1} \sum_{Jn}^{jn=1} \left(P_n^M\right)\left(C_{n,jn}^P\right)$$

Where:

- $S_{n,jn}^{M}$ = Relative value of the J_n^{th} criteria under the nth perspective with respect to the maturity score
- P_n^M = Relative priority of the nth perspective with respect to the maturity score, $n = 1,2,3\ldots N$
- $C_{n,jn}^P$ = Relative contribution of the J_n^{th} criteria under the nth perspective, $j_n = 1,2,3\ldots N$

Table 1.15 Research steps for maturity assessment on strategic technology planning

Phase number	Activities/steps	Notes
Phase I: Hierarchical model development	1. Conducting literature review	
	2. Extracting and combing key factors	
	3. Building preliminary maturity model	
Phase II: Validation of model, criteria, and metrics	1. Design and apply Research Instrument 1 (RI 1) for validating perspective	RI 1 is depicted in Appendix 4
	2. Design and apply Research Instrument 2 (RI2) for validating criteria	RI 2 is depicted in Appendix 5
	3. Design and apply Research Instrument 3 (RI3) for validating desirability metrics	RI 3 is depicted in Appendix 6
Phase III: Quantification of perspective, criteria, and desirability metrics	1. Design and apply Research Instrument 4 (RI 4) for quantifying perspective (conducting pairwise comparison on the 4 perspectives)	RI 4 is depicted in Appendix 7
	2. Obtain the weight of perspectives	
	3. Design and apply Research Instrument 5 (RI5) for quantifying criteria (conducting pairwise comparison on the 5 criteria in each perspective)	RI 5 is depicted in Appendix 8
	4. Obtain the weight of criteria	
	5. Design and apply the Research Instrument 6 (RI 6) for quantifying desirability metrics (assigning desirability value on 5 metrics associated with each of the 20 criteria)	RI 6 is depicted in Appendix 9
	6. Obtain the maturity assessment value for each criterion	
	7. Calculate the Maturity Scores M_S	
Phase IV: Implementation of maturity model	1. Conduct case application/analysis on NWPCC and 7th Power Plan	
	2. Identify the strength and weakness on NWPCC and its 7th Power Plan	

$$\text{Maturity Score}\left(M_S\right) = \sum_{N}^{n=1} \sum_{Jn}^{jn=1} \left(S_{n,jn}^{M}\right)\left(D_{n,jn}\right)$$

Where:

- $S_{n,jn}^{M}$ = Relative value of the J_n^{th} criteria under the nth perspective with respect to the maturity score
- $D_{n,jn}$ = Desirability value of the performance measure corresponding to the J_n^{th} criteria under the *n*th perspective

There are four perspectives and twenty criteria, which have been identified from the literature. The description is provided in the following tables with references (Tables 1.16, 1.17, 1.18, 1.19, and 1.20).

This section will include the description of the desirability metrics and the results of applying the Research Instruments 1–6 on validating and quantifying the maturity model for strategic technology planning.

Table 1.16 The description of perspectives

Perspectives	Description	References
Organization	The category of factors that are related to organizational inputs and support for facilitating the implementation of strategic technology planning	Haselkorn et al. (2007), Bai and Lee (2003), Lee and Bai (2003), Irfan et al. (2017)
Methodology	The category of factors pertaining to technical, analytical or managerial methods, approaches, tools, and techniques that are utilized during strategic technology planning process	Bryson (1988), US DOE (2012), Whalen (2007), Vishnevskiy et al. (2016)
Knowledge	The category of factors which are related to knowledge assets and management for enabling the implementation of strategic technology planning process	Lee and Bai (2003), Irfan et al. (2017), Vishnevskiy et al. (2016), Chen (1999) Phaal et al. (2006b)
Documents	The category of key factors or elements that are included in the content of the strategic technology plan for guiding future technology direction	Garcia and Bray (1997), United Nations (n.d.), US Department of Energy (2013), Hervás Soriano and Mulatero (2011)

Table 1.17 The description of criteria for organization perspective

Criteria for organization perspective	Description	References
Executives' Commitments	The promises or assurance from high executives or senior managers, which support the implementation of strategic technology planning	Fenn et al. (2003), Mittenthal (2002), Nauda and Hall (1991), Bryson (1988)
Teaming & Staffing	The team forming and responsibility assignment within or beyond the organization structure, which are structured specifically for strategic technology planning	Fenn et al. (2003), Mittenthal (2002), Bryson (1988), US Department of Energy (2013)
Communication & Collaboration	The communication and collaboration efforts across different departments within or beyond the whole organization, aiming to facilitate strategic technology planning	Lee and Bai (2003), Irfan et al. (2017), Vishnevskiy et al. (2016), US Department of Energy (2013)
Stakeholder Involvement	The proper arrangement, interaction, and involvement with various stakeholders for the sake of successfully managing strategic technology planning	Bai and Lee (2003), Mittenthal (2002), Irfan et al. (2017), Bryson (1988), United Nations (n.d.), US Department of Energy (2013)
Resource Allocation	The financial or other organizational resources that are allocated to the usage for strategic technology planning	Nauda and Hall (1991), Irfan et al. (2017), United Nations (n.d.), US Department of Energy (2013), Vishnevskiy et al. (2016)

Table 1.18 The description of criteria for methodology perspective

Criteria for methodology perspective	Description	References
Technology Roadmapping	The approach serves the purpose of identifying driver, capability gaps, technology characteristics, R&D program initiatives, and the linkages by means of an integrated and timeline-based technology roadmap diagram	Phaal et al. (2005), Whalen (2007), Lichtenthaler (2008), McCarthy et al. (2001), Phaal and Muller (2009) Nauda and Hall (1991), Garcia and Bray (1997), Phaal et al. (2006b), Bray and Garcia (1997), Phaal et al. (2004)
Technology Forecasting	The method is intended to provide prediction and tendency of the future target technology by means of mathematical models, expert judgements, or other scientific approaches	US DOE (2012), Daim et al. (2005), Linstone (1974), Lamb et al. (2010)
Technology Assessment	The approach deals with identifying the needed technology and selecting the prioritized technology alternatives	Fenn et al. (2003), Nauda and Hall (1991), US DOE (2012), Vishnevskiy et al. (2016) Mohaghegh and Shirazi (2017), Tran and Daim (2008) Iskin and Daim (2014)
Data Collection & Management	The efforts and capability of managing the relevant technological data, including storage, analysis, updates, and protection	US DOE (2012), United Nations (n.d.), US Department of Energy (2013)
Review & Updates & Change Management	The monitoring and control mechanism including review, updates, and change management for assuring the robust process and adaptively responding to the dynamically changing environment	Lee and Bai (2003), Nauda and Hall (1991), Garcia and Bray (1997), Irfan et al. (2017), Chen (1999)

1.2.1 Description of the Model Elements

Apart from the description of each perspective and criteria mentioned in the answers to question 2 (d), all the desirability metrics are summarized in the following tables for clearly showing how the metrics can be used for maturity assessment of strategic technology planning.

For each criterion, there are five metrics associated with five levels of maturity showing the various degrees of the maturity of conducting strategy technology planning (Tables 1.21, 1.22, 1.23, 1.24, 1.25, 1.26, 1.27, 1.28, 1.29, 1.30, 1.31, 1.32, 1.33, 1.34, 1.35, 1.36, 1.37, 1.38, 1.39, and 1.40).

Table 1.19 The description of criteria for knowledge perspective

Criteria for knowledge perspective	Description	References
Guideline & Handbook	The knowledge assets such as guidebook, handbook, or manual in document form used for guiding organization to conduct strategic technology planning	United Nations (n.d.), Phaal and Muller (2009), Phaal et al. (2004)
Instruction & Training	The arrangement of instructional seminar or training programs for team or relevant personnel to build up, enhance, or transfer the knowledge of strategic technology planning	Irfan et al. (2017), Whalen (2007), Hervás Soriano and Mulatero (2011)
Skill/Competency	The skill or competency level of team members who are engaged in conducting strategic technology planning	Phaal et al. (2005), Irfan et al. (2017), McCarthy et al. (2001), Phaal and Muller (2009)
Lesson-Learned & Dissemination	The collection and organizing the lesson-learned information and disseminating these critical knowledge assets to personnel involving in strategic technology planning	Fenn et al. (2003), Lee and Bai (2003), Mittenthal (2002)
Expert Involvement	The arrangement and interaction with subject matter experts for providing professional knowledge, experience, and insights to assist the implementation of strategic technology planning	Vishnevskiy et al. (2016), McCarthy et al. (2001), Phaal and Muller (2009)

1.2.2 Validation of Perspectives

By applying the RI 1, the results of validating the perspectives are listed in Table 1.41. All the perspectives have been agreed by the experts.

1.2.3 Validation of Criteria

By applying the RI 2, the results of validating the criteria are listed in the Table 1.42. All the criteria have been agreed by the experts.

1. Validation of Desirability Metrics

By applying the RI 3, the results of validating the desirability metrics are listed in the following tables. All the metrics have been agreed by the experts, except for the third metric "the Partial team members and structure with clear responsibility and authority have been specified" for the criterion of "Teaming and Staffing." Since the rate of agreement (87.5%) is still above the 75%, the metric is kept for the following analysis (Tables 1.43, 1.44, 1.45, 1.46, 1.47, 1.48, 1.49, 1.50, 1.51, 1.52, 1.53, 1.54, 1.55, 1.56, 1.57, 1.58, 1.59, 1.60, 1.61, and 1.62).

Table 1.20 The description of criteria for documents perspective

Criteria for documents perspective	Description	References
Environmental Scan & Technological Prospect	The external business environment analysis with a focus on technological prospect, whereby technology opportunities, trend, or challenge can be identified	Nauda and Hall (1991), Irfan et al. (2017), Bryson (1988), United Nations (n.d.) European Commission (2017), Harrison (2010)
Mission, Vision, Capability Gap	The statement of what the organization does, the ultimate purpose that the organization strive for, and the current inability to achieve the expected capability requirement	Aron (2010), Mittenthal (2002), United Nations (n.d.), US Department of Energy (2013) Harrison (2010), Gates (2010)
Goals & Objectives with quantitative metrics	The statement of what the organization wants to achieve and the specific and measurable target performance they want to accomplish	Fenn et al. (2003), Mittenthal (2002), Nauda and Hall (1991), United Nations (n.d.) European Commission (2017), US Department of Energy (2013), Gates (2010)
Strategy & Action/ Implementation Plan	The statement of how the organization wants to do for achieving the goals and objectives, as well as the detailed planning efforts listed in a document form	Aron (2010), Mittenthal (2002), Irfan et al. (2017), US DOE (2012) Garcia and Bray (1997), US Department of Energy (2013), Gates (2010), European Commission (2010)
Interconnection & Alignment	The consistency, alignment, or inter-connection among Mission, Vision, Capability Gap, Goals, Objectives, Metrics, Strategy, and Implementation Plan, which are expected to enable synergy toward successful strategic technology planning	Phaal et al. (2005), European Commission (2017), US DOE (2012), Phaal and Muller (2009)

1.2.4 *Quantification of Perspectives*

By applying RI 4, the results for quantifying the perspectives are listed in the Table 1.63. It appears that the inconsistency and disagreement are less than 0.1, which is acceptable for proceeding to the following research steps.

Table 1.21 Desirability questions and metrics for "Executives' Commitments" factor

Desirability question for "Executives' Commitments"	Metrics
What is the degree of the Executives' Commitments for supporting strategic technology planning in your organization?	1. No commitments or supports from high executives, senior managers or middle level managers
	2. There are few informal commitments and supports from middle level managers. The commitments are not documented
	3. Commitments and supports from middle managers have been documented and demonstrated
	4. Commitments and supports from middle and senior managers have been documented and demonstrated
	5. Commitments and supports from high executives, senior managers and middle managers across the organization has been fully documented and demonstrated in a continuous manner

Table 1.22 Desirability questions and metrics for "Teaming & Staffing" factor

Desirability question for "Teaming & Staffing"	Metrics
What is the degree of the Teaming & Staffing for supporting strategic technology planning in your organization?	1. No specific teaming or staffing for strategic technology planning
	2. Very few team members have been specified. The responsibility and authority are not clearly assigned
	3. Partial team members and structure with clear responsibility and authority have been specified
	4. Most of the dedicated team and staff members across the organization have been specified with clear responsibility and authority
	5. All required team & staff members have been specified and assigned in Responsibility Assignment Matrix (RAM) showing clear responsibility and authority

1.2.5 Quantification of Criteria

By applying RI 5, the results of quantifying the criteria are listed in the following tables. It appears that the all the inconsistency and disagreement are less than 0.1, which is acceptable for proceeding to the following research steps (Tables 1.64, 1.65, 1.66, and 1.67).

Table 1.23 Desirability questions and metrics for "Communication & Collaboration" factor

Desirability question for "Communication & Collaboration"	Metrics
What is the degree of the Communication & Collaboration for supporting strategic technology planning in your organization?	1. No specific planning or interactions for departmental communication & collaboration
	2. No formal planning efforts. The communication & collaboration within the team members has been partially demonstrated
	3. The planning has been done. The communication and collaboration within the team members has been mostly demonstrated
	4. The required communication and collaboration efforts across the organization have been documented and demonstrated
	5. The required communication and collaboration efforts have been fully documented, demonstrated, and updated across and beyond the organization in a continuous manner

Table 1.24 Desirability questions and metrics for "Stakeholder Involvement" factor

Desirability question for "Stakeholder Involvement"	Metrics
What is the degree of the Stakeholder Involvement for supporting strategic technology planning?	1. No specific stakeholder involvement planning and interaction efforts
	2. No formal planning efforts. Stakeholder involvement have been partially demonstrated within the organization
	3. The planning has been done. Stakeholder involvement have been mostly demonstrated within the organization
	4. The required stakeholder involvement efforts across the organization have been documented and demonstrated
	5. The required stakeholder involvement efforts have been fully documented, demonstrated, and updated across and beyond the organization in a continuous manner

1.2.6 Quantification of Desirability Metrics

By applying RI 6, the results of quantifying desirability metrics are listed and depicted in the following tables and figures (Tables 1.68, 1.69, 1.70, 1.71, 1.72, 1.73, 1.74, 1.75, 1.76, 1.77, 1.78, 1.79, 1.80, 1.81, 1.82, 1.83, 1.84, 1.85, 1.86, and 1.87; Figs. 1.4, 1.5, 1.6, 1.7, 1.8, 1.9, 1.10, 1.11, 1.12, 1.13, 1.14, 1.15, 1.16, 1.17, 1.18, 1.19, 1.20, 1.21, 1.22, and 1.23).

Table 1.25 Desirability questions and metrics for "Resource Allocation" factor

Desirability question for "Resource Allocation"	Metrics
What is the degree of the Resource Allocation (budget, equipment, and facility) for supporting strategic technology planning in your organization?	1. No specific resource allocation planning and implementation efforts
	2. No formal resource allocation planning. Budget, equipment, and facility resources have been partially committed and demonstrated
	3. Formal resource allocation planning has been done. Budget, equipment, and facility resources have been mostly committed and demonstrated
	4. Formal resource allocation planning has been done. Budget, equipment, and facility resources have been committed, demonstrated, monitored, and controlled
	5. The required resource allocation efforts have been fully documented, allocated, demonstrated, monitored, controlled, and continuously updated across the organization

Table 1.26 Desirability questions and metrics for "Technology Roadmapping" factor

Desirability question for "Technology Roadmapping"	Metrics
What is the degree of utilizing Technology Roadmapping (TRM) approach for conducting strategic technology planning in your organization?	1. TRM has not been adopted
	2. TRM has been adopted on an ad hoc basis
	3. TRM has been adopted; however, the TRM process is not consistent
	4. A handbook or manual for conducting TRM has been developed. The implementation of TRM has been mostly demonstrated
	5. A handbook or manual for conducting TRM has been developed. The implementation of TRM has been fully demonstrated, monitored, controlled, and continuously updated across the organization

1.3 Case of NWPCC

NWPCC 7th Power plan is a comprehensive planning document containing 20 chapters (442 pages) and appendices (565 pages), which add up to 1007 pages. The development process for this Power Plan is depicted in Fig. 1.24 (Tables 1.88 and 1.89).

The application of strategic technology planning maturity model on NWPCC and 7th Power Plan is based on the information collected from the NWPCC website, as well as the information and insights gathered from interview with two staff members currently serving in NWPCC. The interview meeting notes are attached in

Table 1.27 Desirability questions and metrics for "Technology Forecasting" factor

Desirability question for "Technology Forecasting"	Metrics
What is the degree of utilizing Technology Forecasting tools for conducting strategic technology planning in your organization?	1. No specific Technology Forecasting tools have been used
	2. Technology Forecasting tools have been adopted on an ad hoc basis
	3. Technology Forecasting tools have been adopted and used for part of the target technology
	4. Technology Forecasting tools have been consistently adopted and used for all the required target technology
	5. All the required technology forecasting tools have been extensively adopted, reviewed, and updated across the organization with a manner of continuous improvement

Table 1.28 Desirability questions and metrics for "Technology Assessment" factor

Desirability question for "Technology Assessment"	Metrics
What is the degree of utilizing Technology Assessment for conducting strategic technology planning in your organization?	1. No specific Technology Assessment approaches have been used
	2. Technology Assessment has been used on an ad hoc basis
	3. Technology Assessment approaches have been adopted; however, the process is not consistent
	4. Technology Assessment approaches have been consistently adopted across the organization
	5. All the required Technology Assessment approaches have been extensively adopted, demonstrated, reviewed, and continuously updated across the organization with a manner of continuous improvement

Appendix. The level of maturity for each criterion was given, based on all the data collected and my personal best judgement within this very limited timeframe. The maturity assessment for NWPCC and 7th Power Plan are as follows:

1.3.1 Organization Perspective

For Executive's Commitment, NWPCC appear to achieve level 5 maturity, which means that the commitments and supports from high executives, senior managers, and middle managers across the organization has been fully documented and demonstrated in a continuous manner. The staffs in NWPCC replied me that they

Table 1.29 Desirability questions and metrics for "Data Collection & Management" factor

Desirability question for "Data Collection & Management"	Metrics
What is the degree of utilizing Data Collection & Management for supporting strategic technology planning in your organization?	1. No specific Data Collection & Management planning and implementation have been conducted
	2. No formal Data Collection & Management planning efforts. Data Collection for target technology has been conducted on an ad hoc basis
	3. Data Collection & Management planning have been done. Partial Data Collection & Management for target technology have been demonstrated
	4. Data Collection & Management planning and implementation have been done for most of the target technology
	5. All the required Data Collection and Management have been extensively conducted, demonstrated, reviewed, and continuously updated across the organization with a manner of continuous improvement

Table 1.30 Desirability questions and metrics for "Review, Update & Change Management" factor

Desirability question for "Review, Update, & Change Management"	Metrics
What is the degree of utilizing the Review, Update & Change Management for managing strategic technology planning process in your organization?	1. No specific reviews and updates (frequency of updating plan or process = 0, within 3 years)
	2. The review and updates has been done on an ad hoc basis (frequency of updating plan or process $\leq$1, within 3 years)
	3. The reviews and updates have been partially demonstrated (1 <frequency of updating plan or process $\leq$2, within 3 years)
	4. The reviews and updates have been conducted at least annually (3 frequency of updating plan or process, within 3 years)
	5. The reviews and updates have been conducted at least annually across the organization with benchmarking efforts for continuous improvement (3 $\leq$ frequency of updating plan or process, within 3 years)

received strong commitment and support from their directors and executives. Many documents and reports needed to be submitted to and approved by the director and managers, which shows their commitment and support in document form.

For Teaming & Staffing, NWPCC central office consists of executive, administrative, fish and wildlife, power, public affairs, and legal division. Seventh Power Plan is the responsibility of power division which includes the positions such as Energy Policy Analyst, Power Systems Analyst, Economic Analyst, Project Analyst, Technical Forum Manager and other managers, or Senior Analyst. These

Table 1.31 Desirability questions and metrics for "Guideline & Handbook" factor

Desirability question for "Guideline & Handbook"	Metrics
What is the degree of developing and issuing guidelines or handbooks to guide strategic technology planning in your organization?	1. No guidelines. No handbook/manual
	2. Guidelines have been issued. No handbook/manual
	3. Guidelines and handbook have been developed and issued; however, they are only adopted by the dedicated team
	4. Both guidelines and handbook have been developed, issued, and adopted across the organization; however, there is no update for a long time (e.g., more than 2 years)
	5. The required guidelines and handbook have been developed, issued, adopted, and regularly updated in a continuous improvement manner (reviewed and updated within 2 years)

Table 1.32 Desirability questions and metrics for "Instruction & Training" factor

Desirability question for "Instruction & Training"	Metrics
What is the degree of arranging Instruction & Training to support strategic technology planning in your organization?	1. No instructional materials. No training programs
	2. Instructional materials have been developed. There is no training program
	3. Instructional materials have been developed. The training programs have been arranged on an ad hoc basis ($1 \leq$ training frequency ≤ 2, within 3 years)
	4. Instructional materials have been developed. The training programs have been arranged at least annually ($3 \leq$ training frequency, within 3 years); however, they have not been updated regularly (updates frequency <2, within 3 years)
	5. The required instructional materials have been developed. The training programs have been arranged at least annually ($3 \leq$ training frequency, within 3 years). They have been updated regularly in a continuous improvement manner ($3 \leq$ updates frequency, within 3 years)

professional staffs and analysts need to work with several advisory committees for assuring the quality and correctness of their analytical results and planning efforts. For example, the Demand Forecast Advisory Committee has been formed by electric and gas public and private utilities, as well as experts in regional economy and state energy offices. Their teaming and staffing are demonstrated in Charter document and list of members on their website (https://www.nwcouncil.org/energy/df/home/). With the dedicated power division in charge of the development of power plan together with the assistance or consultation from the advisory committees, NWPCC also achieves level 5 maturity on this criterion, meaning all required team

Table 1.33 Desirability questions and metrics for "Skill & Competency" factor

Desirability question for "Skill & Competency"	Metrics
What is the degree of managing Skill & Competency of dedicated team members for effectively conducting strategic technology planning in your organization?	1. No managing efforts for skill & competency
	2. Draft plan for the required skill & competency (types and levels) are under development
	3. Planning has been completed. The monitoring and controlling of skill and competency for team members have been partially demonstrated (percentage of team members ≤50%)
	4. Planning has been completed. The monitoring and controlling of the skill and competency for team members have been mostly demonstrated (50% < percentage of team members <100%)
	5. Planning has been completed. The monitoring and controlling of the skill and competency for team members have been fully demonstrated with annual review and updates (percentage of team members = 100%)

Table 1.34 Desirability questions and metrics for "Lesson Learned & Dissemination" factor

Desirability question for "Lesson-Learned & Dissemination"	Metrics
What is the degree of managing Lesson Learned & Dissemination for facilitating the implementation of strategic technology planning in your organization?	1. No managing efforts for Lesson Learned & Dissemination
	2. Draft plan is under development
	3. Planning has been completed. The implementation (collection and dissemination) has been partially demonstrated (percentage of Lesson Learned ≤50%)
	4. Planning has been completed. The implementation (collection and dissemination) has been mostly demonstrated (50% < percentage of Lesson Learned <100%)
	5. Planning has been completed. The implementation (collection and dissemination) has been fully demonstrated with annual review and updates (percentage of Lesson Learned = 100%)

& staff members have been specified in Responsibility Assignment matrix (RAM) showing clear responsibility and authority.

For Communication & Collaboration, NWPCC staffs shows strong attitude to actively communicate within and beyond the council. Since part of the mission of NWPCC is to ensure an affordable and reliable energy system with public participation. They generally try their best to be transparent to the public. They frequently announce council meeting, issue Newsletters, and publish meeting minutes to

Table 1.35 Desirability questions and metrics for "Expert Engagements" factor

Desirability question for "Expert Engagements"	Metrics
What is the degree of leveraging the Expert Engagements for assisting strategic technology planning in your organization?	1. No specific managing efforts for Expert Engagements
	2. Draft plan is under development
	3. Planning has been completed. The implementation (selection and invitation) has been partially demonstrated (percentage of required experts ≤50%)
	4. Planning has been completed. The implementation (selection and invitation) has been mostly demonstrated (50% < percentage of required experts <100%)
	5. Planning has been completed. The implementation (selection and invitation) has been fully demonstrated with annual review and updates (percentage of required experts = 100%)

Table 1.36 Desirability questions and metrics for "Environmental Scan & Technological Prospect" factor

Desirability question for "Environmental Scan & Technological Prospect"	Metrics
What is the degree of completeness of Environmental Scan & Technological Prospect being indicated in the strategic technology plan?	1. No SWOT analysis. No technology trends analysis
	2. Either SWOT or technology trends analysis has been demonstrated
	3. SWOT analysis has been included. Technology trends analysis has been partially demonstrated (percentage of the required technology ≤50%)
	4. SWOT analysis has been included. Technology trends analysis has been mostly demonstrated (50% < percentage of the required technology <100%)
	5. All the required SWOT analysis and technology trends analysis have been included (percentage of the required technology = 100%)

inform the public. They also demonstrated some joint efforts in collaborating with other divisions or committees to accomplish their planning or analytical tasks. For example, the Regional Technical Forum (RTF) cooperates with RTF Policy Advisory Committee to develop guidelines, identify significant non-energy impacts of efficiency measures (NWPCC 2016). In fact, members including representatives from public or private utilities and experts form state agencies or the interested parties have laid out a good structure to facilitate these collaboration efforts. Therefore, NWPCC also achieves level 5 maturity on this criterion, meaning that the required communication & collaboration efforts have been fully documented, demonstrated, and updated across and beyond the organization in continuous manner.

Table 1.37 Desirability questions and metrics for "Mission, Vision, and Capability Gaps" factor

Desirability question for "Mission, Vision, and Capability Gaps"	Metrics
What is the degree of completeness of Mission, Vision, and Capability Gaps being indicated in the strategic technology plan?	1. No Mission. No Vision. No Capability Gaps
	2. Mission and Vision have been included. No Capability Gaps
	3. Mission and Vision have been included. Capability Gaps has been partially demonstrated (percentage of the required capability gaps ≤50%)
	4. Mission and Vision have been included. Capability Gaps has been mostly demonstrated (50% <percentage of the required capability gaps <100%)
	5. All the required Mission, Vision, and Capability Gaps have been included (percentage of the required capability gaps = 100%)

Table 1.38 Desirability questions and metrics for "Goals & Objectives with Quantitative Metrics" factor

Desirability question for "Goals & Objectives with quantitative metrics"	Metrics
What is the degree of completeness of Goals & Objectives with quantitative metrics being indicated in the strategic technology plan?	1. No goals. No objectives. No quantitative metrics
	2. Goals or objectives have been included. No quantitative metrics
	3. Goals and objectives have been included. The quantitative metrics have been partially demonstrated (percentage of the required metrics ≤50%)
	4. Goals and objectives have been included. The quantitative metrics have been mostly demonstrated (50% < percentage of the required metrics <100%)
	5. All the required goals, objectives, and quantitative metrics have been included (percentage of the required metrics=100%)

Stakeholders for NWPCC may include public utilities, investor-owned utilities, state agencies, interested parties, etc. As mentioned before, NWPCC tried to be transparent in most of their work and results. They would like to receive comments and feedback from all the stakeholders and public, so that they can keep updating their planning and analytical results. The Council is formed by the law and has responsibility to inform all the stakeholders to participate in the process of developing the power plan. This way will enable to reach consensus among all various parties or agencies, so that the probability of successfully implementing the power plan will be significantly increased. Therefore, NWPCC achieves level 5 maturity in this criterion, meaning the required stakeholders involvement efforts have been fully

Table 1.39 Desirability questions and metrics for "Strategy & Action/Implementation Plan" factor

Desirability question for "Strategy & Action/Implementation Plan"	Metrics
What is the degree of completeness of Strategy & Action/Implementation being indicated in the strategic technology plan?	1. No strategy. No action/implementation plan. No prioritization of technology alternatives
	2. Strategy has been included. No action/ implementation plan. No prioritization of technology alternatives
	3. Strategy has been included. Action/implementation plan have been developed; however, it only covers partial target technology (percentage of the required technology ≤50%). No prioritization of technology alternatives
	4. Strategy has been included. Action/implementation plan have been developed and cover most target technology (50% < percentage of the required technology <100%). Prioritization of technology alternatives has been demonstrated
	5. All the required strategy & action/implementation plan have been included (percentage of the required technology = 100%). Prioritization of technology alternatives has been demonstrated

Table 1.40 Desirability questions and metrics for "Interconnection & Alignment" factor

Desirability question for "Interconnection & Alignment"	Metrics
What is the degree of compliance of Interconnection & Alignment among the elements of the strategic technology plan?	1. No connection and alignment among the elements of strategic technology plan
	2. Environmental scan (SWOT) is connected with capability gaps. Other elements are not connected or aligned
	3. Environmental scan (SWOT) is connected with capability gaps. Capability gaps are connected with goals & objectives. The quantitative metrics are not connected or aligned with the objectives. The remaining elements are not connected either
	4. Environmental scan (SWOT) is connected with capability gaps. Capability gaps are connected with goals & objectives. The quantitative metrics are aligned with objectives. The strategy & action/implementation plan are not connected with goals & objectives
	5. All the required elements in strategic technology plan are connected and aligned

documented, demonstrated, and updated across and beyond the organization in a continuous manner.

In terms of Resource Allocation, NWPCC receives the funding resource from Bonneville Power Administration (BPA) to carry out its functions and responsibility, according to Northwest Power Act. The Act also establishes a formula to

Table 1.41 The results of validating the perspectives

Perspectives	Numbers of agreed responses	Numbers of disagreed responses	Total responses	Percentage of agreed responses
Organization	8	0	8	100
Methodology	8	0	8	100
Knowledge	8	0	8	100
Documents	8	0	8	100

Table 1.42 The results of validating the criteria

Perspectives	Criteria	Numbers of agreed responses	Numbers of disagreed responses	Total responses	Percentage of agreed responses
Organization	Executive's Commitments	8	0	8	100
	Teaming & Staffing	8	0	8	100
	Communication & Collaboration	8	0	8	100
	Stakeholder Involvement	8	0	8	100
	Resource Allocation	8	0	8	100
Methodology	Technology Roadmapping	8	0	8	100
	Technology Forecasting	8	0	8	100
	Technology Assessment	8	0	8	100
	Data Collection & Management	8	0	8	100
	Review, Update, and Change Management	8	0	8	100
Knowledge	Guideline & Handbook	8	0	8	100
	Instruction & Training	8	0	8	100
	Skill/Competence	8	0	8	100
	Lesson Learned & Dissemination	8	0	8	100
	Expert Engagement	8	0	8	100
Documents	Environmental Scan & Technological Prospect	8	0	8	100
	Mission, Vision, and Capability Gaps	8	0	8	100
	Goals & Objectives with quantitative metrics	8	0	8	100
	Strategy & Action/ Implementation Plan	8	0	8	100
	Interconnection & Alignment	8	0	8	100

Table 1.43 The results of validating the metrics for "Executive's Commitments"

Metrics	Numbers of agreed responses	Numbers of disagreed responses	Total responses	Percentage of agreed responses
1. No commitments or supports from high executives, senior managers, or middle-level managers	8	0	8	100
2. There are few informal commitments and supports from middle-level managers. The commitments are not documented	8	0	8	100
3. Commitments and supports from middle managers have been documented and demonstrated	8	0	8	100
4. Commitments and supports from middle and senior managers have been documented and demonstrated	8	0	8	100
5. Commitments and supports from high executives, senior managers, and middle managers across the organization has been fully documented and demonstrated in a continuous manner	8	0	8	100

Table 1.44 The results of validating the metrics for "Teaming & Staffing"

Metrics	Numbers of agreed responses	Numbers of disagreed responses	Total responses	Percentage of agreed responses
1. No specific teaming or staffing for strategic technology planning	8	0	8	100
2. Very few team members have been specified. The responsibility and authority are not clearly assigned	8	0	8	100
3. Partial team members and structure with clear responsibility and authority have been specified	7	1	8	87.5
4. Most of the dedicated team and staff members across the organization have been specified with clear responsibility and authority	8	0	8	100
5. All required team & staff members have been specified and assigned in Responsibility Assignment Matrix (RAM) showing clear responsibility and authority	8	0	8	100

Table 1.45 The results of validating the metrics for "Communication & Collaboration"

Metrics	Numbers of agreed responses	Numbers of disagreed responses	Total responses	Percentage of agreed responses
1. No specific planning or interactions for departmental communication & collaboration	8	0	8	100
2. No formal planning efforts. The communication & collaboration within the team members has been partially demonstrated	8	0	8	100
3. The planning has been done. The communication & collaboration within the team members has been mostly demonstrated	8	0	8	100
4. The required communication & collaboration efforts across the organization have been documented and demonstrated	8	0	8	100
5. The required communication & collaboration efforts have been fully documented, demonstrated, and updated across and beyond the organization in a continuous manner	8	0	8	100

Table 1.46 The results of validating the metrics for "Stakeholder Involvement"

Metrics	Numbers of agreed responses	Numbers of disagreed responses	Total responses	Percentage of agreed responses
1. No specific stakeholder involvement planning, and interaction efforts	8	0	8	100
2. No formal planning efforts. Stakeholder involvement have been partially demonstrated within the organization	8	0	8	100
3. The planning has been done. Stakeholder involvement have been mostly demonstrated within the organization	8	0	8	100
4. The required stakeholder involvement efforts across the organization have been documented and demonstrated	8	0	8	100
5. The required stakeholder involvement efforts have been fully documented, demonstrated, and updated across and beyond the organization in a continuous manner	8	0	8	100

Table 1.47 The results of validating the metrics for "Resource Allocation"

Metrics	Numbers of agreed responses	Numbers of disagreed responses	Total responses	Percentage of agreed responses
1. No specific resource allocation planning and implementation efforts	8	0	8	100
2. No formal resource allocation planning. Budget, equipment, and facility resources have been partially committed and demonstrated	8	0	8	100
3. Formal resource allocation planning has been done. Budget, equipment, and facility resources have been mostly committed and demonstrated	8	0	8	100
4. Formal resource allocation planning has been done. Budget, equipment, and facility resources have been committed, demonstrated, monitored, and controlled	8	0	8	100
5. The required resource allocation efforts have been fully documented, allocated, demonstrated, monitored, controlled, and continuously updated across the organization	8	0	8	100

Table 1.48 The results of validating the metrics for "Technology Roadmapping"

Metrics	Numbers of agreed responses	Numbers of disagreed responses	Total responses	Percentage of agreed responses
1. TRM has not been adopted	8	0	8	100
2. TRM has been adopted on an ad hoc basis	8	0	8	100
3. TRM has been adopted, however the TRM process is not consistent	8	0	8	100
4. A handbook or manual for conducting TRM has been developed. The implementation of TRM has been mostly demonstrated	8	0	8	100
5. A handbook or manual for conducting TRM has been developed. The implementation of TRM has been fully demonstrated, monitored, controlled, and continuously updated across the organization	8	0	8	100

Table 1.49 The results of validating the metrics for "Technology Forecasting"

Metrics	Numbers of agreed responses	Numbers of disagreed responses	Total responses	Percentage of agreed responses
1. No specific Technology Forecasting tools have been used	8	0	8	100
2. Technology Forecasting tools have been adopted on an ad hoc basis	8	0	8	100
3. Technology Forecasting tools have been adopted and used for part of the target technology	8	0	8	100
4. Technology Forecasting tools have been consistently adopted and used for all the required target technology	8	0	8	100
5. All the required technology forecasting tools have been extensively adopted, reviewed, and updated across the organization with a manner of continuous improvement	8	0	8	100

Table 1.50 The results of validating the metrics for "Technology Assessment"

Metrics	Numbers of agreed responses	Numbers of disagreed responses	Total responses	Percentage of agreed responses
1. No specific Technology Assessment approaches have been used	8	0	8	100
2. Technology Assessment has been used on an ad hoc basis	8	0	8	100
3. Technology Assessment approaches have been adopted; however, the process is not consistent	8	0	8	100
4. Technology Assessment approaches have been consistently adopted across the organization	8	0	8	100
5. All the required Technology Assessment approaches have been extensively adopted, demonstrated, reviewed, and continuously updated across the organization with a manner of continuous improvement	8	0	8	100

Table 1.51 The results of validating the metrics for "Data Collection & Management"

Metrics	Numbers of agreed responses	Numbers of disagreed responses	Total responses	Percentage of agreed responses
1. No specific Data Collection & Management planning and implementation have been conducted	8	0	8	100
2. No formal Data Collection & Management planning efforts. Data Collection for target technology has been conducted on an ad hoc basis	8	0	8	100
3. Data Collection & Management planning have been done. Partial Data Collection & Management for target technology have been demonstrated	8	0	8	100
4. Data Collection & Management planning and implementation have been done for most of the target technology	8	0	8	100
5. All the required Data Collection and Management have been extensively conducted, demonstrated, reviewed, and continuously updated across the organization with a manner of continuous improvement	8	0	8	100

Table 1.52 The results of validating the metrics for "Review, Update, and Change Management"

Metrics	Numbers of agreed responses	Numbers of disagreed responses	Total responses	Percentage of agreed responses
1. No specific reviews and updates (frequency of updating plan or process = 0, within 3 years)	8	0	8	100
2. The review and updates has been done on an ad hoc basis (frequency of updating plan or process ≤ 1, within 3 years)	8	0	8	100
3. The reviews and updates have been partially demonstrated ($1 <$ frequency of updating plan or process ≤ 2, within 3 years)	8	0	8	100
4. The reviews and updates have been conducted at least annually ($3 \leq$ frequency of updating plan or process, within 3 years)	8	0	8	100
5. The reviews and updates have been conducted at least annually across the organization with benchmarking efforts for continuous improvement ($3 \leq$ frequency of updating plan or process, within 3 years)	8	0	8	100

Table 1.53 The results of validating the metrics for "Guideline & Handbook"

Metrics	Numbers of agreed responses	Numbers of disagreed responses	Total responses	Percentage of agreed responses
1. No guidelines. No handbook/ manual	8	0	8	100
2. Guidelines have been issued. No handbook/manual	8	0	8	100
3. Guidelines and handbook have been developed and issued; however, they are only adopted by the dedicated team	8	0	8	100
4. Both guidelines and handbook have been developed, issued and adopted across the organization; however, there is no updates for a long time (e.g., more than 2 years)	8	0	8	100
5. The required guidelines and handbook have been developed, issued, adopted and regularly updated in a continuous improvement manner (reviewed and updated within 2 years)	8	0	8	100

determine a funding limitation, which may involve negotiation and agreement between NWPCC and BPA. Nevertheless, the budget has been committed and has been increased from 11,624,000 (FY 2018) to 11,914,000 (FY 2019) (NWPCC n.d.). For other equipment or facility resource, NWPCC staffs and analysts engage in lots of computer modeling, forecasting, and simulation to perform their analytical or planning tasks. The information system such as networking system, complex computer models, and extensive databases for various analysis has been listed in their budget plan. The required expenditure has been decomposed to several categories to make sure that the budget will be able to support their intended work requirements and meet the performance target. Based on the explanation and justification, NWPCC achieves level 5 of maturity for this criterion, implying that the required resource allocation efforts have been fully documented, allocated, demonstrated, monitored, controlled, and continuously updated across the organization.

1.3.2 Methodology Perspective

In terms of technology roadmapping (TRM), it is found that NWPCC is not an R&D-oriented organization; instead, it is a government agency responsible for coordinating and developing an energy resource plan to guide other organizations striving to meet the energy efficiency or energy saving targets. As far as the technology is concerned, there are some technology development suggestions outlined in

Table 1.54 The results of validating the metrics for "Instruction & Training"

Metrics	Numbers of agreed responses	Numbers of disagreed responses	Total responses	Percentage of agreed responses
1. No instructional materials. No training programs	8	0	8	100
2. Instructional materials have been developed. There is no training program	8	0	8	100
3. Instructional materials have been developed. The training programs have been arranged on an ad hoc basis ($1 \leq$ training frequency ≤ 2, within 3 years)	8	0	8	100
4. Instructional materials have been developed. The training programs have been arranged at least annually ($3 \leq$ training frequency, within 3 years). However, they have not been updated regularly (updates frequency <2, within 3 years)	8	0	8	100
5. The required instructional materials have been developed. The training programs have been arranged at least annually ($3 \leq$ training frequency, within 3 years). They have been updated regularly in a continuous improvement manner ($3 \leq$ updates frequency, within 3 years)	8	0	8	100

the Chapter 4: Action Plan. For example, it is suggested to develop a regional work plan to provide adequate focus on emerging technologies to help ensure adoption. This suggestion has been denoted to associate with BPA, NEEA, Utilities, National Labs, and Energy Trust of Oregon. In addition, several emerging technologies including Solar PV, Energy/Battery Storage, Enhanced Geothermal Systems (EGS), Offshore wind, Wave and tidal energy, Small modular reactors, Ocean energy, Natural gas-fired technology, Lighting technology, and Variable Refrigerant Flow (VRF) technology have been called for monitoring, tracking, exploration, or enhancement. The staffs in NWCCC replied me that there was an IT project using TRM approach to guide the system development program. Based on the above analysis, NWCCC seems to achieve level 2 on this criterion, implying that the TRM has been adopted on an ad hoc basis.

There are lots of energy demand and price forecasting efforts and results included in the 7th Power Plan (Part 2, Chapter 7–11 and Appendices B–E). The tools utilized for their energy forecast include the following (NWPCC 2014):

- Energy 2020: An open-source model used to forecast the hourly demand for electricity

Table 1.55 The results of validating the metrics for "Skill/Competency"

Metrics	Numbers of agreed responses	Numbers of disagreed responses	Total responses	Percentage of agreed responses
1. No managing efforts for skill & competency	8	0	8	100
2. Draft plan for the required skill & competency (types and levels) are under development	8	0	8	100
3. Planning has been completed. The monitoring & controlling of skill & competency for team members have been partially demonstrated (percentage of team members ≤50%)	8	0	8	100
4. Planning has been completed. The monitoring & controlling of the skill & competency for team members have been mostly demonstrated (50% < percentage of team members <100%)	8	0	8	100
5. Planning has been completed. The monitoring & controlling of the skill & competency for team members have been fully demonstrated with annual review and updates (percentage of team members = 100%)	8	0	8	100

Table 1.56 The results of validating the metrics for "Lesson Learned & Dissemination"

Metrics	Numbers of agreed responses	Numbers of disagreed responses	Total responses	Percentage of agreed responses
1. No managing efforts for Lesson Learned & Dissemination	8	0	8	100
2. Draft plan is under development	8	0	8	100
3. Planning has been completed. The implementation (collection and dissemination) has been partially demonstrated (percentage of Lesson Learned ≤50%)	8	0	8	100
4. Planning has been completed. The implementation (collection and dissemination) has been mostly demonstrated (50% < percentage of Lesson Learned <100%)	8	0	8	100
5. Planning has been completed. The implementation (collection and dissemination) has been fully demonstrated with annual review and updates (percentage of Lesson Learned = 100%)	8	0	8	100

Table 1.57 The results of validating the metrics for "Expert Engagement"

Metrics	Numbers of agreed responses	Numbers of disagreed responses	Total responses	Percentage of agreed responses
1. No specific managing efforts for Expert Engagements	8	0	8	100
2. Draft plan is under development	8	0	8	100
3. Planning has been completed. The implementation (selection and invitation) has been partially demonstrated (percentage of required experts ≤50%)	8	0	8	100
4. Planning has been completed. The implementation (selection and invitation) has been mostly demonstrated (50% < percentage of required experts <100%)	8	0	8	100
5. Planning has been completed. The implementation (selection and invitation) has been fully demonstrated with annual review and updates (percentage of required experts = 100%)	8	0	8	100

Table 1.58 The results of validating the metrics for "Environmental Scan & Technological Prospect"

Metrics	Numbers of agreed responses	Numbers of disagreed responses	Total responses	Percentage of agreed responses
1. No SWOT analysis. No technology trends analysis	8	0	8	100
2. Either SWOT or technology trends analysis has been demonstrated	8	0	8	100
3. SWOT analysis has been included. Technology trends analysis has been partially demonstrated (percentage of the required technology ≤50%)	8	0	8	100
4. SWOT analysis has been included. Technology trends analysis has been mostly demonstrated (50% < percentage of the required technology <100%)	8	0	8	100
5. All the required SWOT analysis and technology trends analysis have been included (percentage of the required technology = 100%)	8	0	8	100

Table 1.59 The results of validating the metrics for "Mission, Vision, and Capability Gaps"

Metrics	Numbers of agreed responses	Numbers of disagreed responses	Total responses	Percentage of agreed responses
1. No Mission. No Vision. No Capability Gaps	8	0	8	100
2. Mission and Vision have been included. No Capability Gaps	8	0	8	100
3. Mission and Vision have been included. Capability Gaps has been partially demonstrated (percentage of the required capability gaps ≤50%)	8	0	8	100
4. Mission and Vision have been included. Capability Gaps have been mostly demonstrated (50% < percentage of the required capability gaps <100%)	8	0	8	100
5. All the required Mission, Vision, and Capability Gaps have been included (percentage of the required capability gaps = 100%)	8	0	8	100

Table 1.60 The results of validating the metrics for "Goals & Objectives with Quantitative Metrics"

Metrics	Numbers of agreed responses	Numbers of disagreed responses	Total responses	Percentage of agreed responses
1. No goals. No objectives. No quantitative metrics	8	0	8	100
2. Goals or objectives have been included. No quantitative metrics	8	0	8	100
3. Goals and objectives have been included. The quantitative metrics have been partially demonstrated (percentage of the required metrics ≤50%)	8	0	8	100
4. Goals and objectives have been included. The quantitative metrics have been mostly demonstrated (50% < percentage of the required metrics <100%)	8	0	8	100
5. All the required goals, objectives, and quantitative metrics have been included (percentage of the required metrics = 100%)	8	0	8	100

Table 1.61 The results of validating the metrics for "Strategy & Action/Implementation Plan"

Metrics	Numbers of agreed responses	Numbers of disagreed responses	Total responses	Percentage of agreed responses
1. No strategy. No action/ implementation plan. No prioritization of technology alternatives	8	0	8	100
2. Strategy has been included. No action/implementation plan. No prioritization of technology alternatives	8	0	8	100
3. Strategy has been included. Action/ implementation plan have been developed; however, it only covers partial target technology (percentage of the required technology ≤50%). No prioritization of technology alternatives	8	0	8	100
4. Strategy has been included. Action/ implementation plan have been developed and cover most target technology (50% < percentage of the required technology <100%) Prioritization of technology alternatives has been demonstrated	8	0	8	100
5. All the required strategy & action/ implementation plan have been included (percentage of the required technology = 100%) Prioritization of technology alternatives has been demonstrated	8	0	8	100

- AURORA Electricity Market Model: A model developed by EPIS and used to forecast hourly wholesale electricity market prices in western US area

It appears that these tools are not developed and used for forecasting the technology advancement. Rather, they mainly are used for energy demand or price forecast. Regardless, these tools did share and utilize some common forecasting algorithm or methods to conduct complicated prediction under many assumptions or scenarios. Therefore, if we customize the criterion into "Energy Forecasting," then NWPCC seems to achieve the level 4 maturity level, meaning that "Energy" forecasting tools have been adopted and used for all the required target energy resources.

The main purpose of using Technology Assessment is to evaluate the impact of the technology onto the organization and try to prioritize the technology alternatives based on the level of cost effectiveness or relative importance among the relevant decision attributes. For the NWPCC, the attention has been directed to the issues of energy resource shortage and the environmental sustainability. Therefore, the focus is on the assessment of the energy need, the load-resource balance, the energy adequacy, and the priority of energy resource with the least cost and risk. To meet this objective, NWPCC has been using two assessment tools (GENESYS and RPM) to

Table 1.62 The results of validating the metrics for "Interconnection & Alignment"

Metrics	Numbers of agreed responses	Numbers of disagreed responses	Total responses	Percentage of agreed responses
1. No connection and alignment among the elements of strategic technology plan	8	0	8	100
2. Environmental scan (SWOT) is connected with capability gaps. Other elements are not connected or aligned	8	0	8	100
3. Environmental scan (SWOT) is connected with capability gaps. Capability gaps are connected with goals & objectives. The quantitative metrics are not connected or aligned with the objectives. The remaining elements are not connected either	8	0	8	100
4. Environmental scan (SWOT) is connected with capability gaps. Capability gaps are connected with goals & objectives. The quantitative metrics are aligned with objectives. The strategy & action/implementation plan are not connected with goals & objectives	8	0	8	100
5. All the required elements in strategic technology plan are connected and aligned	8	0	8	100

Table 1.63 The results for "Quantification of Perspectives"

Expert	Pairwise comparison for perspectives				
	Organization	Methodology	Knowledge	Documents	Inconsistency
1	0.32	0.21	0.27	0.19	0
2	0.27	0.21	0.32	0.2	0
3	0.28	0.26	0.25	0.22	0.01
4	0.25	0.38	0.21	0.16	0.01
5	0.24	0.3	0.3	0.15	0
6	0.26	0.14	0.36	0.25	0.01
7	0.26	0.21	0.34	0.19	0.01
8	0.33	0.23	0.26	0.18	0
Mean	0.28	0.24	0.29	0.19	
		Disagreement			0.042

Table 1.64 The results for "Pairwise Comparison for Criteria in Organization Perspective"

Expert	Pairwise comparison for criteria in organization perspective					
	Executive's Commitments	Teaming & Staffing	Communication & Collaboration	Stakeholder Involvement	Resource Allocation	Inconsistency
1	0.24	0.17	0.14	0.25	0.19	0
2	0.2	0.19	0.26	0.2	0.15	0
3	0.22	0.2	0.2	0.14	0.24	0
4	0.29	0.14	0.24	0.19	0.14	0.01
5	0.25	0.2	0.24	0.2	0.13	0
6	0.21	0.14	0.2	0.23	0.22	0
7	0.16	0.25	0.2	0.16	0.22	0
8	0.28	0.19	0.21	0.12	0.21	0
Mean	0.23	0.19	0.21	0.19	0.19	
			Disagreement			0.037

Table 1.65 The results for "Pairwise Comparison for Criteria in Methodology Perspective"

Expert	Pairwise comparison for criteria in methodology perspective					
	Technology Roadmapping	Technology Forecasting	Technology Assessment	Data Collection & Management	Review, Update, & Change Management	Inconsistency
1	0.17	0.19	0.21	0.27	0.15	0
2	0.19	0.17	0.23	0.2	0.21	0
3	0.18	0.19	0.22	0.21	0.2	0.01
4	0.31	0.14	0.14	0.22	0.18	0.01
5	0.24	0.19	0.19	0.19	0.19	0
6	0.22	0.22	0.22	0.18	0.15	0
7	0.22	0.16	0.25	0.19	0.18	0
8	0.27	0.15	0.2	0.22	0.17	0
Mean	0.23	0.18	0.21	0.21	0.18	
			Disagreement			0.029

perform need assessment and identify the optimal resource strategy/portfolio, which are briefly descripted below (NWPCC 2014):

- GENESYS (Generation Evaluation System): It was developed by Council and used to perform hourly chronological simulation of Northwest's resources.
- Regional Portfolio Model (RPM): It was developed by Council and used to identify low-cost and low-risk resource strategies under uncertain future conditions and policies.

Based on the above observation and explanation, NWPCC appears to reach the level 4 maturity on this criterion, if we make some customization on the maturity

Table 1.66 The results for "Pairwise Comparison for Criteria in Knowledge Perspective"

Expert	Pairwise comparison for criteria in knowledge perspective					
	Guideline & Handbook	Instruction & Training	Skill/ Competency	Lesson Learned & Dissemination	Expert Engagement	Inconsistency
1	0.14	0.18	0.26	0.17	0.25	0
2	0.14	0.19	0.22	0.21	0.24	0
3	0.16	0.22	0.21	0.17	0.25	0
4	0.22	0.29	0.15	0.19	0.16	0.01
5	0.14	0.22	0.28	0.22	0.14	0
6	0.28	0.15	0.2	0.27	0.1	0.01
7	0.16	0.2	0.24	0.22	0.18	0.01
8	0.28	0.2	0.19	0.15	0.18	0
Mean	0.19	0.21	0.22	0.2	0.19	
				Disagreement		0.043

Table 1.67 The results for "Pairwise Comparison for Criteria in Documents Perspective"

Expert	Pairwise comparison for criteria in documents perspective					
	Environmental Scan & Technological Prospect	Mission, Vision, and Capability Gaps	Goals & Objectives with quantitative metrics	Strategy & Action/ Implementation Plan	Interconnection & Alignment	Inconsistency
1	0.13	0.25	0.23	0.22	0.16	0
2	0.15	0.18	0.25	0.23	0.18	0
3	0.17	0.2	0.21	0.24	0.17	0
4	0.13	0.15	0.23	0.31	0.17	0.01
5	0.17	0.21	0.21	0.21	0.21	0
6	0.2	0.2	0.12	0.26	0.23	0.01
7	0.16	0.28	0.18	0.23	0.14	0.02
8	0.18	0.21	0.24	0.2	0.18	0
Mean	0.16	0.21	0.21	0.24	0.18	
				Disagreement		0.029

model and switch "Technology Assessment" into "Energy Resource Assessment." Namely, in NWPCC, the Energy Resource Assessment approaches have been perceived as consistently adopted across the organization.

For the data collection and management, NWPCC considers its power plan as extremely data and model intensive. It requires significant efforts to maintain data on electricity demand, energy prices, energy efficiency, etc. Therefore, NWPCC urges the utilities and relevant organization to collaborate on collecting or improving their required energy data such as demand response, regional operating reserve planning, and industry sale. With these on-going efforts of improving data collection, it can be inferred that NWPCC achieves level 4 maturity on this criterion,

Table 1.68 The desirability quantification results for "Executive's Commitments"

Expert	Executive's Commitments				
	1. No commitments or supports from high executives, senior managers, or middle-level managers	2. There are few informal commitments and supports from middle-level managers. The commitments are not documented	3. Commitments and supports from middle managers have been documented and demonstrated	4. Commitments and supports from middle and senior managers have been documented and demonstrated	5. Commitments and supports from high executives, senior managers, and middle managers across the organization have been fully documented and demonstrated in a continuous manner
1	0	12	40	68	100
2	0	10	35	71	100
3	0	14	41	69	100
4	0	20	60	80	100
5	0	16	40	91	100
6	0	10	30	75	100
7	0	23	24	34	100
8	0	50	70	85	100
Mean	0.00	19.38	42.50	71.63	100.00

Table 1.69 The desirability quantification results for "Teaming & Staffing"

Expert	Teaming & Staffing				
	1. No specific teaming or staffing for strategic technology planning	2. Very few team members have been specified. The responsibility and authority are not clearly assigned	3. Partial team members and structure with clear responsibility and authority have been specified	4. Most of the dedicated team and staff members across the organization have been specified with clear responsibility and authority	5. All required team & staff members have been specified and assigned in Responsibility Assignment Matrix (RAM) showing clear responsibility and authority
1	0	1	54	58	100
2	5	12	23	72	100
3	0	3	29	60	100
4	0	20	40	70	100
5	0	5	50	100	100
6	0	10	25	75	100
7	0	16	50	65	100
8	0	40	70	90	100
Mean	1.25	13.38	42.63	73.75	100.00

Table 1.70 The desirability quantification results for "Communication & Collaboration"

Expert	Communication & Collaboration				
	1. No specific planning or interactions for departmental communication & collaboration	2. No formal planning efforts. The communication & collaboration within the team members has been partially demonstrated	3. The planning has been done. The communication & collaboration within the team members has been mostly demonstrated	4. The required communication & collaboration efforts across the organization have been documented and demonstrated	5. The required communication & collaboration efforts have been fully documented, demonstrated, and updated across and beyond the organization in a continuous manner
1	0	2	49	65	100
2	0	15	40	85	100
3	0	6	32	70	100
4	0	25	60	70	100
5	0	9	60	81	100
6	0	8	42	80	100
7	0	27	30	52	100
8	0	30	60	75	100
Mean	0.00	15.25	46.63	72.25	100.00

Table 1.71 The desirability quantification results for "Stakeholder Involvement"

Expert	Stakeholder Involvement				
	1. No specific stakeholder involvement planning and interaction efforts	2. No formal planning efforts. Stakeholder involvement have been partially demonstrated within the organization	3. The planning has been done. Stakeholder involvement have been mostly demonstrated within the organization	4. The required stakeholder involvement efforts across the organization have been documented and demonstrated	5. The required stakeholder involvement efforts have been fully documented, demonstrated, and updated across and beyond the organization in a continuous manner
1	0	1	66	70	100
2	0	15	40	82	100
3	0	7	42	70	100
4	0	25	60	80	100
5	0	39	81	95	100
6	0	10	30	65	100
7	0	15	37	61	100
8	0	10	65	85	100
Mean	0.00	15.25	52.63	76.00	100.00

Table 1.72 The desirability quantification results for "Resource Allocation"

Expert	Resource Allocation				
	1. No specific resource allocation planning and implementation efforts	2. No formal resource allocation planning. Budget, equipment, and facility resources have been partially committed and demonstrated	3. Formal resource allocation planning has been done. Budget, equipment, and facility resources have been mostly committed and demonstrated	4. Formal resource allocation planning has been done. Budget, equipment, and facility resources have been committed, demonstrated, monitored and controlled	5. The required resource allocation efforts have been fully documented, allocated, demonstrated, monitored, controlled, and continuously updated across the organization
1	0	4	62	72	100
2	0	10	30	80	100
3	0	10	39	71	100
4	0	25	50	70	100
5	0	29	54	84	100
6	0	10	43	80	100
7	0	23	29	33	100
8	0	5	30	85	100
Mean	0.00	14.50	42.13	71.88	100.00

Table 1.73 The desirability quantification results for "Technology Roadmapping"

Expert	Technology Roadmapping				
	1. TRM has not been adopted	2. TRM has been adopted on an ad hoc basis	3. TRM has been adopted; however, the TRM process is not consistent	4. A handbook or manual for conducting TRM has been developed. The implementation of TRM has been mostly demonstrated	5. A handbook or manual for conducting TRM has been developed. The implementation of TRM has been fully demonstrated, monitored, controlled, and continuously updated across the organization
1	0	16	49	76	100
2	0	11	34	79	100
3	0	30	30	79	100
4	0	20	30	75	100
5	0	50	73	85	100
6	0	14	41	83	100
7	0	15	33	56	100
8	0	40	50	85	100
Mean	0.00	24.50	42.50	77.25	100.00

Table 1.74 The desirability quantification results for "Technology Forecasting"

Expert	Technology Forecasting				
	1. No specific Technology Forecasting tools have been used	2. Technology Forecasting tools have been adopted on an ad hoc basis	3. Technology Forecasting tools have been adopted and used for part of the target technology	4. Technology Forecasting tools have been consistently adopted and used for all the required target technology	5. All the required technology forecasting tools have been extensively adopted, reviewed, and updated across the organization with a manner of continuous improvement
1	0	5	9	95	100
2	0	10	30	77	100
3	0	13	39	81	100
4	0	15	35	75	100
5	0	50	71	87	100
6	0	17	49	86	100
7	0	8	12	42	100
8	0	50	50	90	100
Mean	0.00	21.00	36.88	79.13	100.00

Table 1.75 The desirability quantification results for "Technology Assessment"

Expert	Technology Assessment				
	1. No specific Technology Assessment approaches have been used	2. Technology Assessment has been used on an ad hoc basis	3. Technology Assessment approaches have been adopted; however, the process is not consistent	4. Technology Assessment approaches have been consistently adopted across the organization	5. All the required Technology Assessment approaches have been extensively adopted, demonstrated, reviewed, and continuously updated across the organization with a manner of continuous improvement
1	0	55	56	85	100
2	0	10	30	82	100
3	0	16	37	74	100
4	0	20	35	60	100
5	0	50	71	87	100
6	0	17	43	75	100
7	0	15	17	53	100
8	0	50	55	85	100
Mean	0.00	29.13	43.00	75.13	100.00

Table 1.76 The desirability quantification results for "Data Collection & Management"

	Data Collection & Management				
Expert	1. No specific Data Collection & Management planning and implementation have been conducted	2. No formal Data Collection & Management planning efforts. Data Collection for target technology has been conducted on an ad hoc basis	3. Data Collection & Management planning have been done. Partial Data Collection & Management for target technology have been demonstrated	4. Data Collection & Management planning and implementation have been done for most of the target technology	5. All the required Data Collection and Management have been extensively conducted, demonstrated, reviewed, and continuously updated across the organization with a manner of continuous improvement
1	0	25	54	85	100
2	0	10	30	82	100
3	0	17	50	72	100
4	0	15	30	60	100
5	0	50	68	88	100
6	0	23	61	83	100
7	0	15	44	59	100
8	0	25	35	90	100
Mean	0.00	22.50	46.50	77.38	100.00

meaning data collection & management planning and implementation have been done for most of the target energy resource.

For the Review, Update, & Change Management, NWPCC has published seven Power Plans since 1983. According to the Northwest Power Act, the Power Plan will need be updated at least every 5 years. In addition, NWPCC organizes a mid-term assessment to review the progress on the 7th Power Plan implementation. The agenda extends to cover Markets and Demand Comparison & Updates, Conservation Updates, Demand Response Updates, Generating Resource Updates, and Resource Adequacy Updates (NWPCC 2018). It can be inferred that NWPCC achieves the level 4 maturity on this criterion, implying that the reviews and updates have been conducted at least annually.

1.3.3 Knowledge Perspective

For the Guideline & Handbook, it has been found that the content of the Power Plan has been somewhat specified by the Power Act. For example, an energy conservation program, recommendation of research and development, a methodology for

Table 1.77 The desirability quantification results for "Review, Update, & Change Management"

	Review, Update & Change Management				
Expert	1. No specific reviews and updates (frequency of updating plan or process = 0, within 3 years)	2. The review and updates has been done on an ad hoc basis (frequency of updating plan or process ≤1, within 3 years)	3.The reviews and updates have been partially demonstrated (1 < frequency of updating plan or process ≤2, within 3 years)	4. The reviews and updates have been conducted at least annually (3 ≤ frequency of updating plan or process, within 3 years)	5. The reviews and updates have been conducted at least annually across the organization with benchmarking efforts for continuous improvement (3 ≤ frequency of updating plan or process, within 3 years)
1	0	42	44	51	100
2	0	10	31	81	100
3	0	16	26	63	100
4	0	30	50	71	100
5	0	19	50	77	100
6	0	20	44	70	100
7	0	29	50	44	100
8	0	50	55	80	100
Mean	0.00	27.00	43.75	67.13	100.00

determining quantifiable environmental costs and benefits, an electricity demand forecast of at least 20 years, a forecast of power resource, an analysis of electricity reserve, etc. (NWPCC 2016). This can be considered the major guideline for developing the content of the plan. Some other specific guideline such as roadmap for the assessment of energy efficiency measures was issued by Regional Technical Forum (RTF) to illustrate how RTF select, develop, and maintain methods for assessing cost and benefits of energy efficiency measures (Regional Technical Forum 2015). Although a comprehensive handbook has not been found yet, NWPCC did draft an overview presentation on the process of developing the Power Plan with an illustrated video posted on its website. Based on the above justification, it may be inferred that NWPCC nearly achieves level 3 maturity on this criterion, meaning that the guideline and handbook have been partially developed, issued, and has been adopted by the dedicated team.

In terms of Instruction and Training, the presentation material and video about the process of developing the Power Plan can be used for training and learning purposes. As for the tools and models used for their forecasting and analytics, NWPCC has posted relevant information on the website for public access. For example, there is lot of information about implementing Regional Portfolio Model (RPM). The public can get access to the detailed documentation and online model for exploring some limited functionality of the tool. NWPCC appears to reach the level 4 maturity

Table 1.78 The desirability quantification results for "Guideline & Handbook"

	Guideline & Handbook				
Expert	1. No guidelines. No handbook/ manual	2. Guidelines have been issued. No handbook/manual	3. Guidelines and handbook have been developed and issued; however, they are only adopted by the dedicated team	4. Both guidelines and handbook have been developed, issued, and adopted across the organization, however, there is no updates for a long time (e.g., more than 2 years)	5. The required guidelines and handbook have been developed, issued, adopted, and regularly updated in a continuous improvement manner (reviewed and updated within 2 years)
1	0	5	36	47	100
2	0	10	40	82	100
3	0	20	35	48	100
4	0	20	35	65	100
5	0	31	51	91	100
6	0	10	34	72	100
7	0	18	28	52	100
8	0	10	50	70	100
Mean	0.00	15.50	38.63	65.88	100.00

on this criterion, meaning that instructional materials have been mostly developed and training has been arranged at least annually.

For the Skill and Competency, NWPCC introduced their staffs and analysts by posting their qualification and work experience information on their website. It seems that the skill set needed for the position has been considered and evaluated during the employment process. The extensive forecasting and planning results in the Power Plan may partially reflect the excellent performance of these skillful planners and analysts. NWPCC seems to reach to level 4 maturity on this criterion, implying that the morning & controlling of the skill & competency for team members has been mostly demonstrated.

In terms of Lesson Learned, NWPCC collects, analyzes, and responds to the comments about their draft Power Plan and put this response document in the website for public access. They try to improve the way they develop the plan by modifying or updating some analytical process. NWPCC appears to reach level 4 maturity on this criterion, implying the planning and dissemination of lesson learned have been mostly demonstrated.

For the Expert Engagement, we can observe that there are experts appointed as members in these advisory committees. Many experts specializing in energy resource planning have participated in the regular meetings and contributed their

Table 1.79 The desirability quantification results for "Instruction & Training"

Expert	Instruction & Training				
	1. No instructional materials. No training programs	2. Instructional materials have been developed. There is no training program	3. Instructional materials have been developed. The training programs have been arranged on an ad hoc basis (1 ≤ training frequency ≤2, within 3 years)	4. Instructional materials have been developed. The training programs have been arranged at least annually (3 ≤ training frequency, within 3 years). However, they have not been updated regularly (updates frequency <2, within 3 years)	5. The required instructional materials have been developed. The training programs have been arranged at least annually (3 ≤ training frequency, within 3 years). They have been updated regularly in a continuous improvement manner (3 ≤ updates frequency, within 3 years)
1	0	2	13	65	100
2	0	10	30	82	100
3	0	14	27	42	100
4	0	20	50	80	100
5	0	40	55	78	100
6	0	17	41	74	100
7	0	14	16	20	100
8	0	15	75	85	100
Mean	0.00	16.50	38.38	65.75	100.00

insight to the development of the power plan. The Charter shows all the names, affiliation, and certain timeline of service. NWPCC seems to achieve level 5 maturity on this criterion, meaning the selection and invitation of the experts has been fully demonstrated with annual review and updates.

1.3.4 Documents Perspective

In terms of Environmental Scan & Technological Prospect, Chapter 2 (State of the Northwest Power System) of the Power Plan addresses environment scan and energy status in that region. For example, the issues of the regional economic conditions, electricity demand, natural gas market and prices, emissions regulations and impacts, etc. have been analyzed. If we switch technology prospect into energy resource prospect, NWPCC appears to achieve level 4 of maturity, meaning that the trend analysis of energy resource has been mostly demonstrated.

Table 1.80 The desirability quantification results for "Skill/Competency"

Expert	Skill/competency 1. No managing efforts for skill & competency	2. Draft plan for the required skill & competency (types and levels) are under development	3. Planning has been completed. The monitoring & controlling of skill & competency for team members have been partially demonstrated (percentage of team members ≤ 50%)	4. Planning has been completed. The monitoring & controlling of the skill & competency for team members have been mostly demonstrated (50% < percentage of team members <100%)	5. Planning has been completed. The monitoring & controlling of the skill & competency for team members have been fully demonstrated with annual review and updates (percentage of team members = 100%)
1	0	24	55	80	100
2	0	10	31	85	100
3	0	10	31	61	100
4	0	25	60	75	100
5	0	30	59	82	100
6	0	14	42	75	100
7	0	29	45	45	100
8	0	10	50	85	100
Mean	0.00	19.00	46.63	73.50	100.00

For the remaining criteria, mission and vision have been clearly articulated in the website and annual report. The energy shortage or challenge has been analyzed in the Chapter 2 of the Power Plan. The goals, objectives, and metrics have been included and explained in Chapters 3 and 4. The entire Chapter 4 addresses the action plan, whereas Chapter 3 describes the resource strategy. All elements including the supporting or supplementing report listed in the Appendix appear to link each other for cross-reference or mutual justification. Therefore, NWPCC appears to reach level 5 maturity on these criteria, meaning that all required elements in the Power Plan have been included and connected with each other.

Based on the above analysis, the value of maturity assessment is shown below (Tables 1.90 and 1.91):

The total sum of maturity scores for NWPCC and its 7th Power Plan is 81.25, which is considered a very good maturity performance. By looking at the individual maturity for each criterion, it is found that NWPCC performs very well in Executive's Commitments, Teaming & Staffing, Communication & Collaboration, Stakeholder Involvement, Resource Allocation, Expert Judgement, Mission, Vision, and Capability Gaps, Goals & Objectives with quantitative metrics, Strategy & Action/ Implementation and Interconnection & Alignment. One of the main reasons why

Table 1.81 The desirability quantification results for "Lesson Learned & Dissemination"

	Lesson Learned & Dissemination				
Expert	1. No managing efforts for Lesson Learned & Dissemination	2. Draft plan is under development	3. Planning has been completed. The implementation (collection and dissemination) has been partially demonstrated (percentage of Lesson Learned ≤50%)	4. Planning has been completed. The implementation (collection and dissemination) has been mostly demonstrated (50% < percentage of Lesson Learned <100%)	5. Planning has been completed. The implementation (collection and dissemination) has been fully demonstrated with annual review and updates (percentage of Lesson Learned = 100%)
1	0	7	45	51	100
2	0	10	28	82	100
3	0	7	38	72	100
4	0	25	45	75	100
5	0	22	70	88	100
6	0	9	38	70	100
7	0	14	35	56	100
8	0	10	50	80	100
Mean	0.00	13.00	43.63	71.75	100.00

Table 1.82 The desirability quantification results for "Expert Engagement"

	Expert Engagement				
Expert	1. No specific managing efforts for Expert Engagements	2. Draft plan is under development	3. Planning has been completed. The implementation (selection and invitation) has been partially demonstrated (percentage of required experts ≤50%)	4. Planning has been completed. The implementation (selection and invitation) has been mostly demonstrated (50% < percentage of required experts <100%)	5. Planning has been completed. The implementation (selection and invitation) has been fully demonstrated with annual review and updates (percentage of required experts = 100%)
1	0	18	46	48	100
2	0	11	25	82	100
3	0	12	34	60	100
4	0	20	40	65	100
5	0	28	60	82	100
6	0	8	30	73	100
7	11	18	14	100	51
8	0	10	40	70	100
Mean	2.75	15.63	36.13	72.50	93.88

Table 1.83 The desirability quantification results for "Environmental Scan & Technological Prospect"

Expert	Environmental Scan & Technological Prospect				
	1. No SWOT analysis. No technology trends analysis	2. Either SWOT or technology trends analysis has been demonstrated	3. SWOT analysis has been included. Technology trends analysis has been partially demonstrated (percentage of the required technology ≤50%)	4. SWOT analysis has been included. Technology trends analysis has been mostly demonstrated (50% < percentage of the required technology <100%)	5. All the required SWOT analysis and technology trends analysis have been included (percentage of the required technology = 100%)
1	0	4	45	50	100
2	0	11	28	81	100
3	0	21	39	75	100
4	0	25	50	80	100
5	0	57	71	95	100
6	0	17	42	72	100
7	0	11	30	100	45
8	0	20	50	65	100
Mean	0.00	20.75	44.38	77.25	93.13

Table 1.84 The desirability quantification results for "Mission, Vision, and Capability Gaps"

Expert	Mission, Vision, and Capability Gaps				
	1. No Mission. No Vision. No Capability Gaps	2. Mission and Vision have been included. No Capability Gaps	3. Mission and Vision have been included. Capability Gaps has been partially demonstrated (percentage of the required capability gaps ≤50%)	4. Mission and Vision have been included. Capability Gaps has been mostly demonstrated (50% < percentage of the required capability gaps <100%)	5. All the required Mission, Vision, and Capability Gaps have been included (percentage of the required capability gaps = 100%)
1	0	23	24	49	100
2	0	12	31	82	100
3	0	3	40	56	100
4	0	30	60	75	100
5	0	50	65	84	100
6	0	23	54	82	100
7	0	17	36	100	60
8	0	5	30	50	100
Mean	0.00	20.38	42.50	72.25	95.00

Table 1.85 The desirability quantification results for "Goals & Objectives with Quantitative Metrics"

Expert	Goals & Objectives with quantitative metrics				
	1. No goals. No objectives. No quantitative metrics	2. Goals or objectives have been included. No quantitative metrics	3. Goals and objectives have been included. The quantitative metrics have been partially demonstrated (percentage of the required metrics ≤50%)	4. Goals and objectives have been included. The quantitative metrics have been mostly demonstrated (50% < percentage of the required metrics <100%)	5. All the required goals, objectives, and quantitative metrics have been included (percentage of the required metrics = 100%)
1	0	25	51	53	100
2	0	10	26	90	100
3	0	19	48	74	100
4	0	20	40	71	100
5	0	50	73	89	100
6	0	21	50	75	100
7	0	19	39	100	43
8	0	25	40	65	100
Mean	0.00	23.63	45.88	77.13	92.88

NWPCC performs so well in these criteria may be attributed to the enactment and enforcement of Northwest Power Act in 1980. Specific sections or issues are specified to be included in the content of Power Plan. However, NWPCC may need to think about extending their utilization of methods or tools. Especially, in the Chapter 4 Action Plan, there are many technology development or enhancement suggestions, which were only briefly described. If NWPCC can adopt more technology roadmapping approaches, the suggestions or guidance for other utilities will be more specific and understandable.

The model has been developed to comprehensively include the relevant key factors. However, too many factors resulting in too many pairwise comparisons may deter the participation of the expert. In addition, using wordings with great clarity will minimize the risk of a wrong interpretation (Estep 2016).

The proposed Maturity Model is developed by reviewing strategic technology planning related literature. For other areas of applications, specific modification is deemed necessary. For example, the NWPCC's 7th Power Plan is not a pure strategic technology plan. Rather, it is a Strategic Energy (Power) Resource Plan. However, due to its extended scope of content, the plan incorporates some technology elements for future development, enhancement, or exploration. Therefore, it may be necessary to customize my model a little bit to suit the uniqueness of the Power Plan. For example, NWPCC has done lots of sophisticated modeling on forecasting energy demand or future market price. Hence, the criterion "Technology

Table 1.86 The desirability quantification results for “Strategy & Action/Implementation Plan”

Expert	Strategy & Action/Implementation Plan				
	1. No strategy. No action/ implementation plan. No prioritization of technology alternatives	2. Strategy has been included. No action/ implementation plan. No prioritization of technology alternatives	3. Strategy has been included. Action/ implementation plan have been developed; however, it only covers partial target technology (percentage of the required technology ≤50%). No prioritization of technology alternatives	4. Strategy has been included. Action/ implementation plan have been developed and cover most target technology (50% < percentage of required technology <100%). Prioritization of tech. alt. has been demonstrated	5. All the required strategy & action/implementation plan have been included (percentage of the required technology = 100%). Prioritization of technology alternatives has been demonstrated
1	0	37	41	46	100
2	0	12	27	85	100
3	0	30	48	65	100
4	0	20	50	70	100
5	0	50	64	85	100
6	0	14	45	76	100
7	0	19	52	100	35
8	0	15	50	70	100
Mean	0.00	24.63	47.13	74.63	91.88

Table 1.87 The desirability quantification results for "Interconnection & Alignment"

	Interconnection & Alignment				
Expert	1. No connection and alignment among the elements of strategic technology plan	2. Environmental scan (SWOT) is connected with capability gaps. Other elements are not connected or aligned	3. Environmental scan (SWOT) is connected with capability gaps. Capability gaps are connected with goals & objectives. The quantitative metrics are not connected or aligned with the objectives. The remaining elements are not connected either	4. Environmental scan (SWOT) is connected with capability gaps. Capability gaps are connected with goals & objectives. The quantitative metrics are aligned with objectives. The strategy & action/ implementation plan are not connected with goals & objectives	5. All the required elements in strategic technology plan are connected and aligned
1	0	28	60	89	100
2	0	10	29	82	100
3	0	10	39	62	100
4	0	40	60	80	100
5	0	50	70	87	100
6	0	17	48	79	100
7	0	32	53	69	100
8	0	10	30	50	100
Mean	0.00	24.63	48.63	74.75	100.00

Forecasting" is suggested to be replaced by "Energy Forecasting" during my evaluation. They also have done lots of simulation modelling to identify the optimal energy resource strategy under about 800 different scenarios. Therefore, "Technology Assessment" has been suggested to be switched to "Energy Resource Assessment" to comply the characteristics of Power Plan. In other words, the maturity model may need to be tailored for meeting specific applications.

Expert engagements are considered significantly critical for the research outcomes, because they involve subjective judgements (Lingga 2016). Therefore, the selection of experts needs to be conducted in a scientific and systematic manner.

As technology roadmapping (TRM) can be seen as a form of technology planning (Bray and Garcia 1997), some TRM maturity models might be modified or customized for assessing the maturity of strategic technology planning.

There are few maturity models proposed to assess TRM maturity. Petrick (2008) proposes Roadmapping Maturity Model, which is based on the discussions with over 150 executives during 4 years' period of time and contain six-stage maturity levels. This model begins with "Level 1: Ad Hoc" referring to the inconsistency of the RM application and lack of linkage between layers. Level 2 is characterized by

Fig. 1.4 Desirability curve for Executive's Commitments

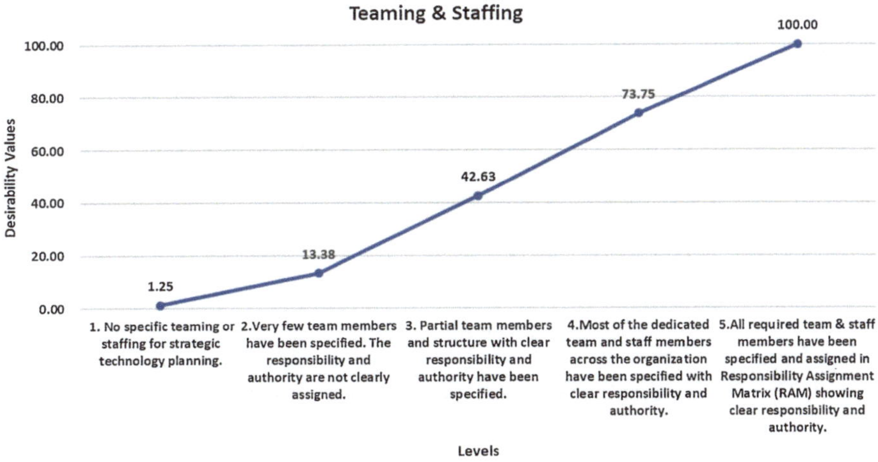

Fig. 1.5 Desirability curve for Teaming & Staffing

documented practices, available training, some communication efforts, and stand-alone activity. Level 3 maturity should demonstrate the linkage with new product development and technology project selection. Level 4 is expected to adopt scenario analysis and involve participation across the organization. Level 5 aims to link with portfolio analysis and risk analysis. It also should be able to leverage formalized inputs form suppliers & customer in a collaborated manner. Level 6 – the highest level of maturity – represents the cross-organization effectiveness and situational awareness across supply network enhanced via excellent roadmap efforts (Petrick 2008).

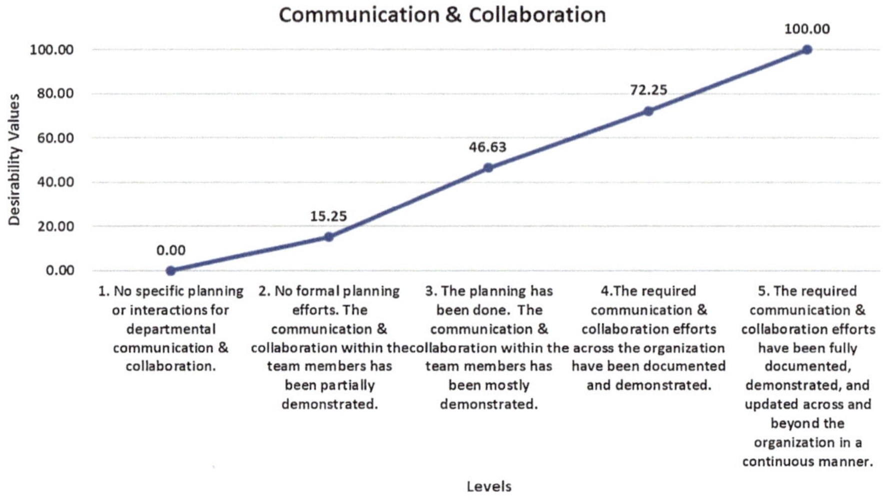

Fig. 1.6 Desirability curve for Communication & Collaboration

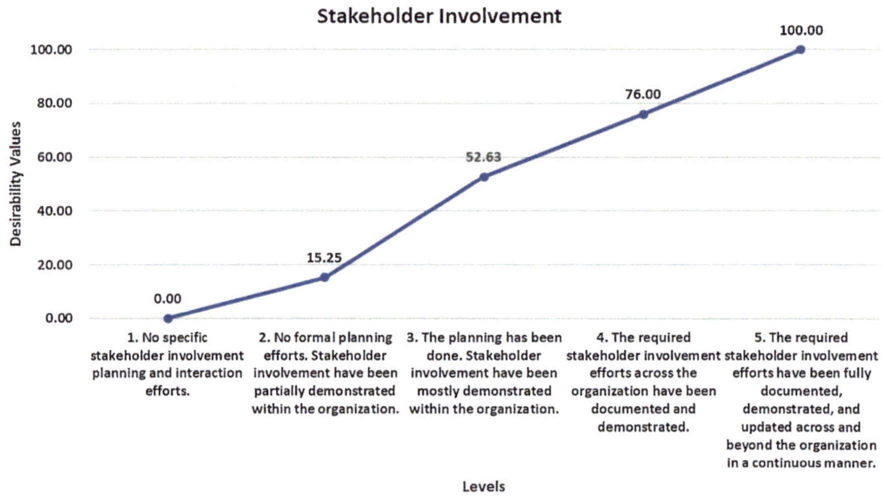

Fig. 1.7 Desirability curve for Stakeholder Involvement

Phaal et al. (2005) discusses a three-stage maturity/influence model proposed by Kappel (2001) (Phaal et al. 2005; Kappel 2001). Level 1 is denoted as understanding the strategic position with some metrics such as forecast accuracy or clarity of priorities. Level 2 is intended to show the maturity of persuading an external audience for the enhancement of communication and influence. The possible measures used in level 2 may include the degree of aligned priorities and decisions. Level 3 refers to a synchronized mechanism for ongoing coordination between labs and business units. A possible measure for this level is how long the group maintains

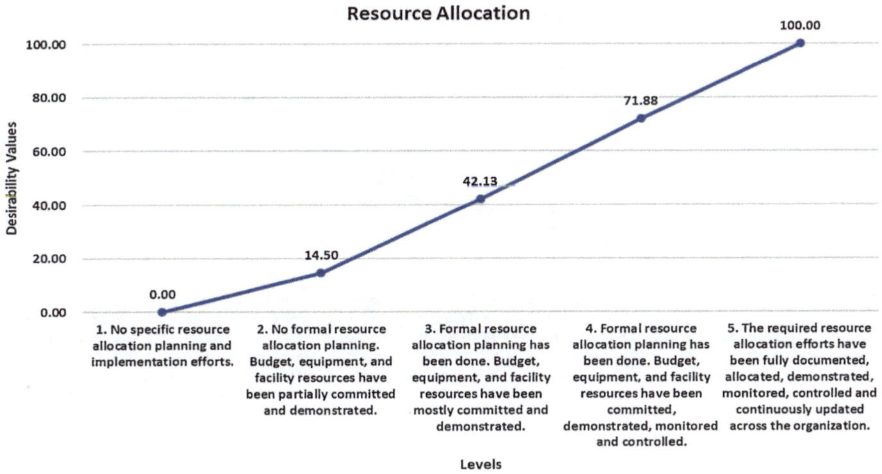

Fig. 1.8 Desirability curve for Resource Allocation

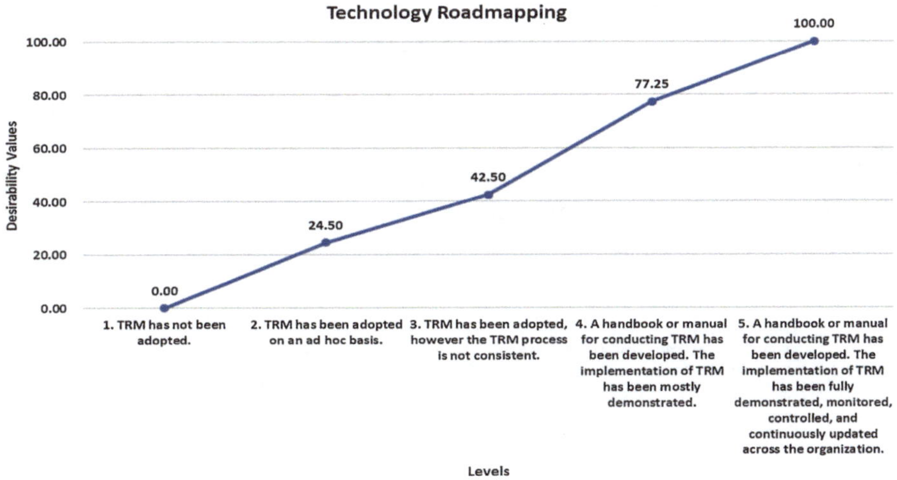

Fig. 1.9 Desirability curve for Technology Roadmapping

coordination. There are some challenges involving using these three degrees maturity/influence model, such as resistance to change, fitness of decentralized authority structure, and so on (Kappel 2001). Phaal et al. (2005) also briefly discuss a four-stage maturity model proposed by "Alignent," which defines the maturity levels as Organizing, Proactive, Collaborative, and Comprehensive, as shown in Table 1.3 (Phaal et al. 2005) (Table 1.92).

UK MOD (Ministry of Defense) addressed three level of maturity used for characterizing the TRM. For level 1, the roadmap is regarded as communicable and

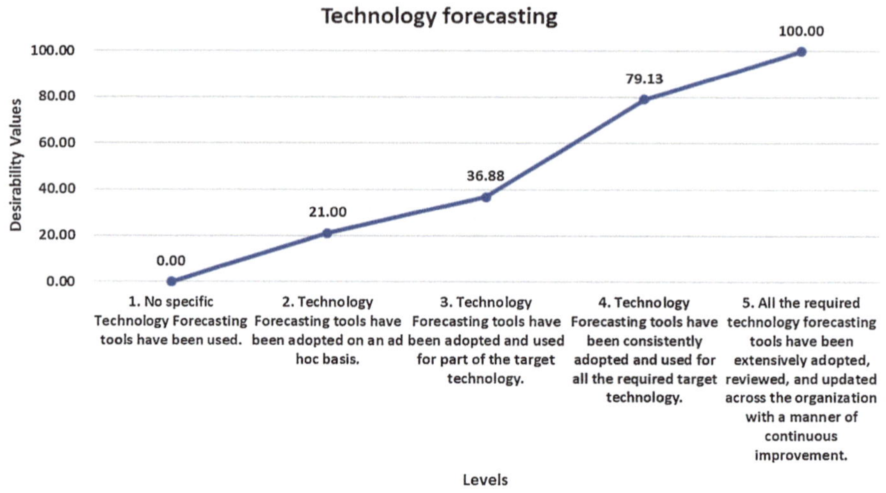

Fig. 1.10 Desirability curve for Technology Forecasting

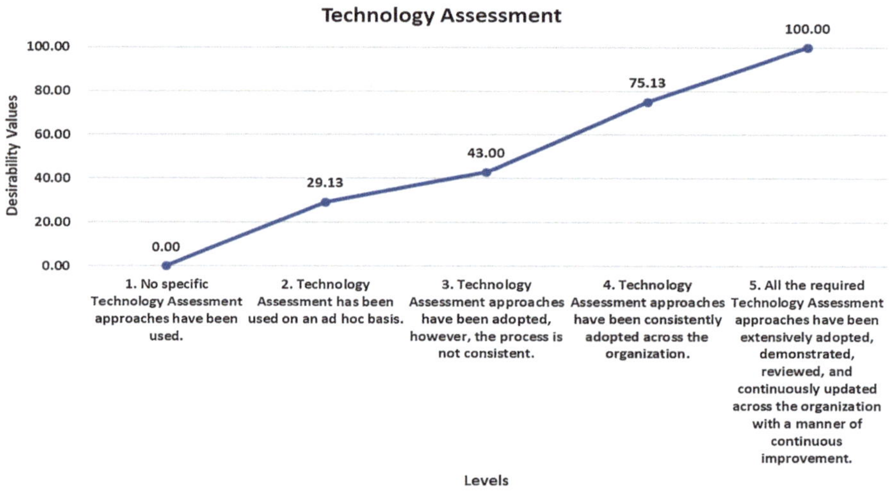

Fig. 1.11 Desirability curve for Technology Assessment

understandable. Level 2 implies that the roadmaps should reach a quality standard for facilitating the decision-making or resulting in a positive change. Level 3 roadmaps are perceived to be able to support the synchronization and alignment across the organization (MOD 2006) (p. 3). It seems that the maturity model addressed by UK MOD is the same as that of Kappel.

These TRM maturity models are expected to be used for checking the robustness of the TRM process and the quality of the roadmap. This concept of maturity can be applied to strategic technology planning. Namely, there is also a need to check the

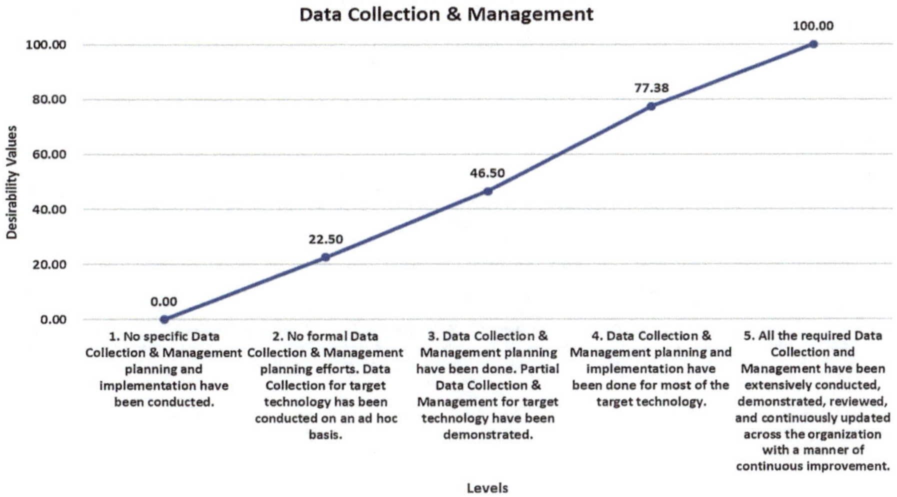

Fig. 1.12 Desirability curve for Data Collection & Management

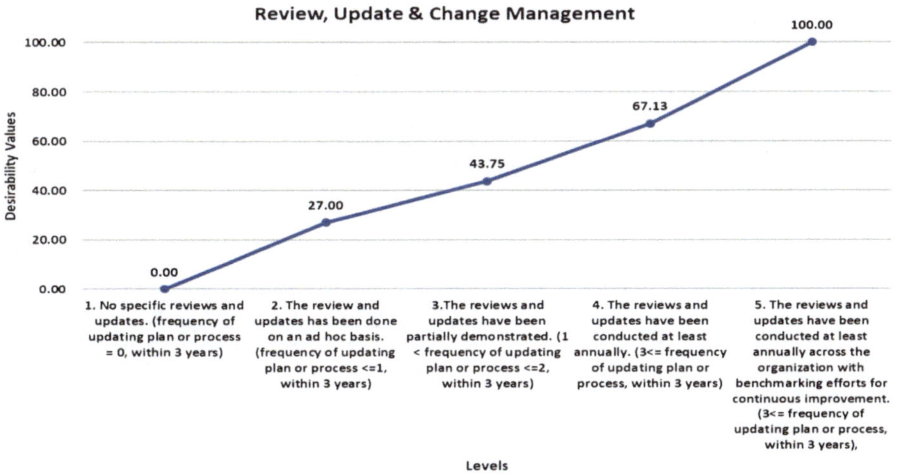

Fig. 1.13 Desirability curve for Review, Update, & Change Management

robustness of the strategic technology planning process and the quality of the strategic technology plan. Both the roadmaps and strategic technology plan need to be carefully developed to remain its quality and robustness, so that the organization can leverages its usefulness.

These maturity models tend to categorize different levels of the maturity, which might be served as an effective tool for organizations to conduct self-assessment for continuous improvement. However, these models seem to lack implementing steps

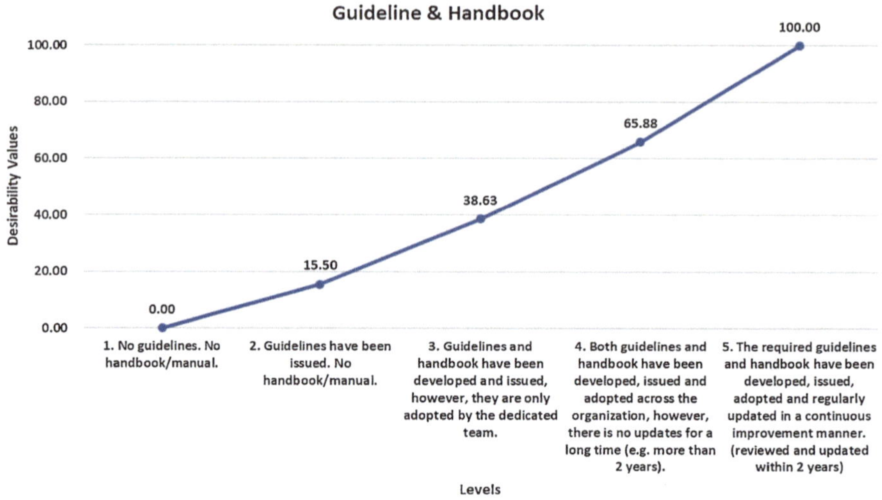

Fig. 1.14 Desirability curve for Guideline & Handbook

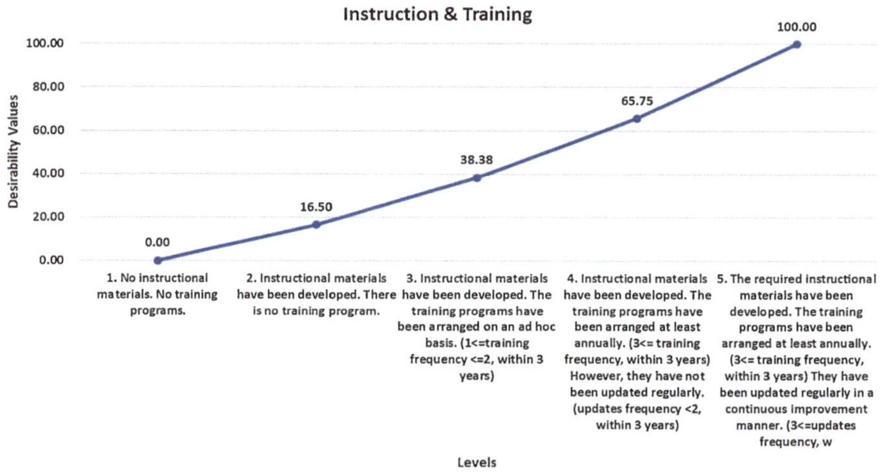

Fig. 1.15 Desirability curve for Instruction & Training

for assessing the maturity. The required attributes or criteria for maturity assessment are still unclear or insufficient. The table below summarizes strength and weakness of these maturity models (Table 1.93).

The HDM approach has been chosen to serve the purpose of assessment due to its ease of use and extensive applications. HDM was developed by Dr. Kocaoglu and has been regarded as one of the effective Multiple Criteria Decision Making (MCDM) approaches (Kocaoglu 1983). HDM approach is deemed as similar to Saaty's Analytic Hierarchy Process (AHP). However, the difference is that HDM utilizes Constant-Sum approach and AHP uses Eigenvectors for calculating the

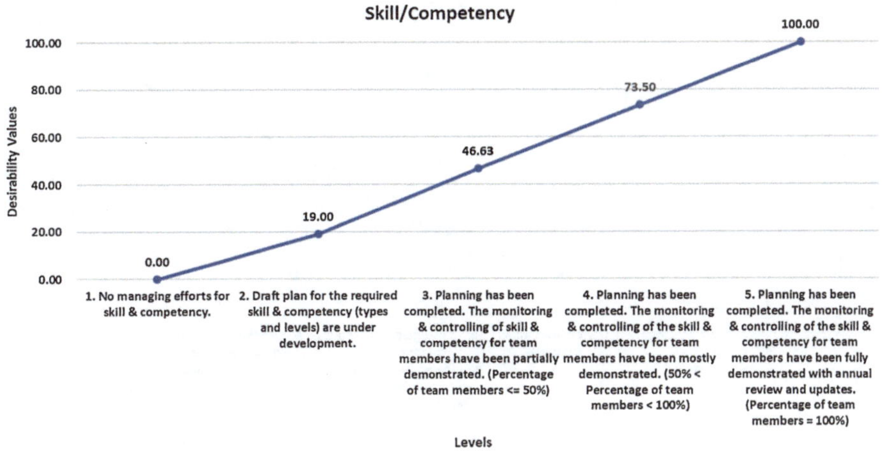

Fig. 1.16 Desirability curve for Skill/Competency

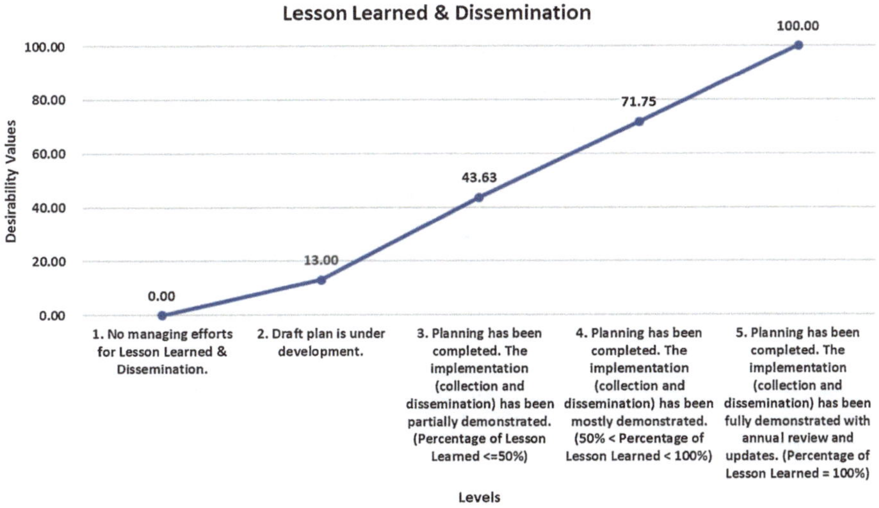

Fig. 1.17 Desirability curve for Lesson Learned & Dissemination

pairwise comparison between each criterion (Kocaoglu 1983). This HDM approach has been applied in many decision-making settings or program planning/evaluation cases (Iskin and Daim 2016; Eastham et al. 2014; Lingga and Kocaoglu 2013; Rahimi and Koosawangsri 2013; Saatchi et al. 2013).

Combining with "Desirability Curve" method, HDM has been applied in evaluating technology transfer proposals, measuring the performance of science and research center, assessing sustainability performance, and calculating the degree of innovation performance via innovativeness index (Estep 2016; Gibson 2016; Tekin 2014; Phan 2013).

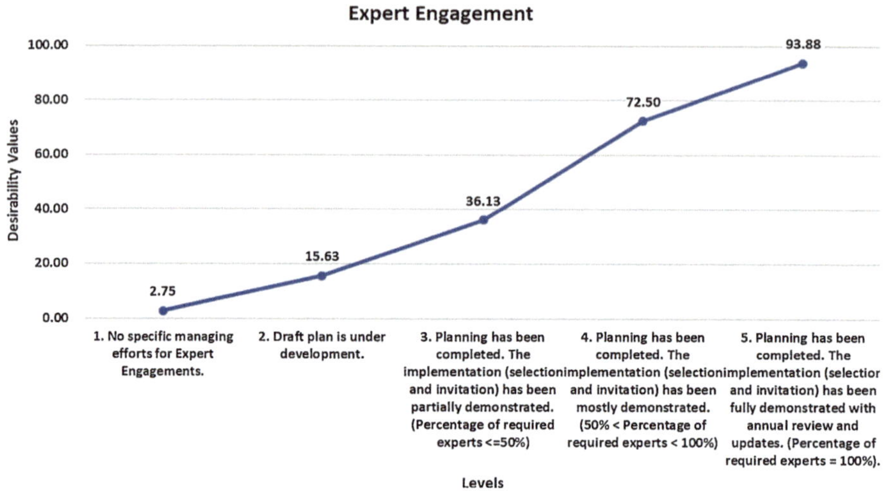

Fig. 1.18 Desirability curve for Expert Engagement

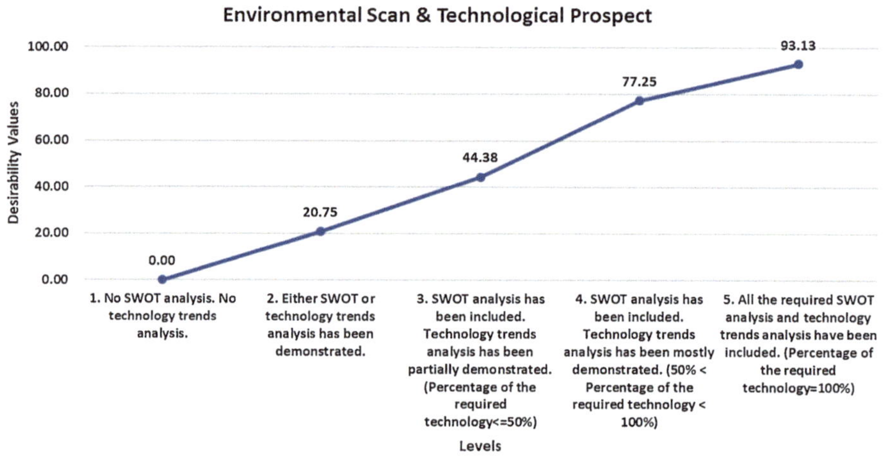

Fig. 1.19 Desirability curve for Environmental Scan & Technological Prospect

In view of its extensive applications in decision-making, program evaluation, and performance measurement, HDM is perceived as a feasible and suitable approach to be used for assessing the maturity of either TRM or strategic technology planning.

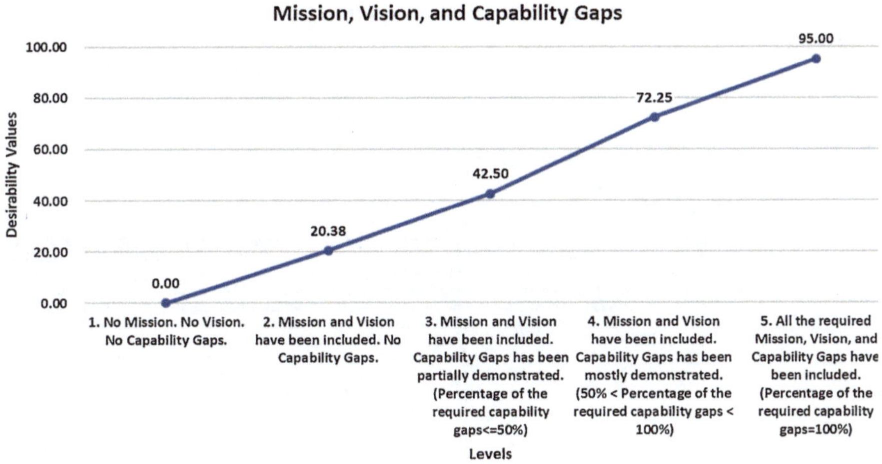

Fig. 1.20 Desirability curve for Mission, Vision, and Capability Gaps

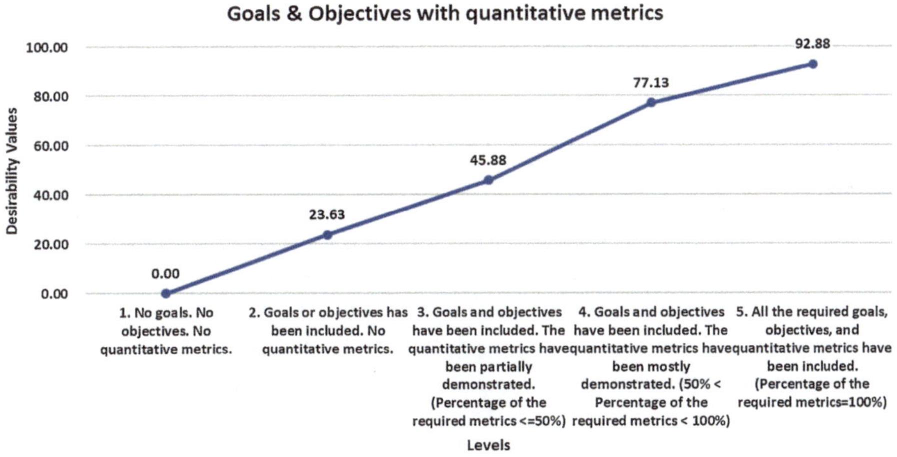

Fig. 1.21 Desirability curve for Goals & Objectives with Quantitative Metrics

1.3.4.1 Issues of Forming an Expert Panel

Expert panels are often arranged when specialized knowledge is required. The intention is to gather the experts with needed expertise to discuss or debate and come up with various course of actions for decision-making (Department of Sustainability and Environment 2005) (p. 36). An effective expert panel depends upon how the experts be selected and how to manage the potential issues carefully. The issues of expert panels are briefly discussed from the perspectives of government agency and research evaluation as follows:

Fig. 1.22 Desirability curve for Strategy & Action/Implementation Plan

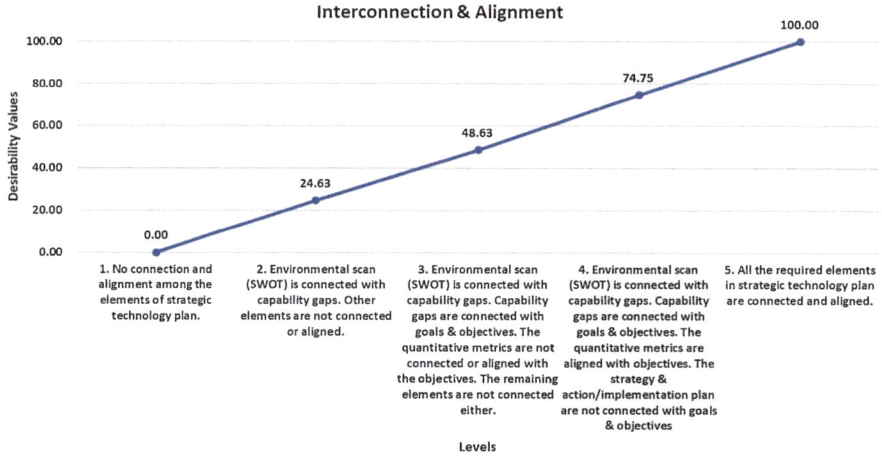

Fig. 1.23 Desirability curve for Interconnection & Alignment

The issues regarding use or forming of Expert Panel from a government agency's perspective are briefly summarized below (Federal Highway Administration US Department of Transportation 2002):

1. Relationship of the expert panel to other organizations and authorities.

   In a government types of project or contract setting, the relationship between the implementation agency and the expert panel is critical. The implementation agency should be cautious about facilitating the panel's work. A third party is considered preferable.
2. How the expert panel is defined?

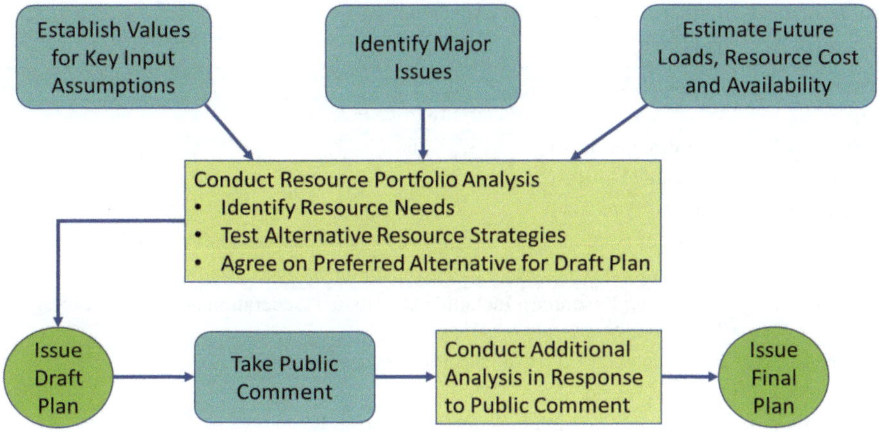

Fig. 1.24 Power plan development process (NWPCC 2014)

Table 1.88 The outline of NWPCC's 7th Power Plan

Part No.	Chapter No.	Title
	Chapter 1	Executive Summary
	Chapter 2	State of the System
Part 1: Resource Strategy and Action Plan	Chapter 3	Resource Strategy
	Chapter 4	Action Plan
	Chapter 5	Bonneville's Loads and Resources
	Chapter 6	Power Act Requirements and the Power Plan
Part 2: Demand and Price Forecasts, Existing Resources, and System Needs	Chapter 7	Electricity Demand Forecast
	Chapter 8	Electricity and Fuel Price Forecasts
	Chapter 9	Existing Resources and Retirements
	Chapter 10	Operating and Planning Reserves
	Chapter 11	System Needs Assessment
Part 3: New Resource Potential	Chapter 12	Conservation Resources
	Chapter 13	Generating Resources
	Chapter 14	Demand Response Resources
Part 4: Developing a Resource Strategy	Chapter 15	Analysis of Alternative Resource Strategies
	Chapter 16	Analysis of Cost Effective Reserves and Reliability
	Chapter 17	Model Conservation Standards and Surcharge Policy
Part 5: Other Plan Elements	Chapter 18	Coordinating with Regional Transmission Planning
	Chapter 19	Environmental Methodology and Due Consideration for Environmental Quality and Fish and Wildlife
	Chapter 20	Fish and Wildlife Program

Table 1.89 The outline of 7th Power Plan Appendix

Appendix No.	Title
Appendix A	Financial Assumptions
Appendix B	Wholesale and Retail Electricity Price Forecast
Appendix C	Fuel Price Forecast
Appendix D	Economic Forecast
Appendix E	Demand Forecast
Appendix F	Impact of Federal Standards on Loads and Conservation Potential
Appendix G	Conservation Resources and Direct Application Renewables
Appendix H	Generating Resources, Including Distributed Generation and Energy Storage Technologies
Appendix I	Environmental Effects of Electric Power Production
Appendix J	Demand Response
Appendix K	Reserve and Reliability Assessment Methods
Appendix L	Regional Portfolio Model
Appendix M	Climate Change Impacts to Loads and Resources
Appendix N	Direct Use of Natural Gas
Appendix O	Initial Scenario Analysis
Appendix P	Glossary

The panel size, ground rules, the level of detail on the quantitative analysis, motivation, or incentive are the issues that need to be considered.

3. The process by which the expert panel does its work.

   The issue of decision by consensus or consent needs to be taken care of. The consensus requires unanimity, whereas the consent can be satisfied by any decision rules. Whether or not opening the panel meeting to public will have some impact on expert's decision.
4. Analytical decisions of the expert panel.

   The decision made by panel may have some influence on themselves. This is an issue that panel sponsors need to consider before forming the expert panel. Other issues may include the credibility of data source and the scope of what the expert panel can question or review.

From a viewpoint of evaluating research, Langfeldt (2004) evaluated six expert panels involving research cases in the field of natural science, technology/engineering, social sciences, humanities, natural science, and multidisciplinary. It was found that there are two types of biases including scholarly/professional bias and non-professional/personal bias. For each type of bias, there are two categories of biases including cognitive constraints and interests. Therefore, the bias can be summarized as the constraints of a professional platform, research interests, general or personal cognitive constraints, and personal interests, as illustrated in Table 1.94.

Table 1.90 The maturity level and corresponding value assigned to NWPCC

Perspective	Criteria	Maturity level assigned to NWPCC	The Value from maturity assessment on NWPCC (V_C)
Organization	Executive's Commitments	5	100
	Teaming & Staffing	5	100
	Communication & Collaboration	5	100
	Stakeholder Involvement	5	100
	Resource Allocation	5	100
Methodology	Technology Roadmapping	2	25
	Technology Forecasting (Energy Forecasting)	4	79
	Technology Assessment (Energy Resource Assessment)	4	75
	Data Collection & Management	4	77
	Review, Update, & Change Management	4	67
Knowledge	Guideline & Handbook	3	39
	Instruction & Training	4	66
	Skill/Competency	4	74
	Lesson Learned & Dissemination	4	72
	Expert Engagement	5	94
Documents	Environmental Scan & Technology Prospect	4	77
	Mission, Vision, and Capability Gaps	5	95
	Goals & Objectives with quantitative metrics	5	93
	Strategy & Action/ Implementation Plan	5	92
	Interconnection & Alignment	5	100

1.3.4.2 Issues of Selecting the Experts

In view of the conflict-of-interest policy, policy makers or government agency need to carefully select the experts as their advisory committee members. Rowe et al. (2013) provided some guiding principles for advisory eligibility as follows:

1. The eligibility criteria for the required knowledge and experience are required.
2. Financial interests should be fully disclosed.
3. Potential biases such as political, ideological, religious, philosophical, and financial should be considered.
4. Various backgrounds such as different geographic area and ethnic group should be encouraged to participate.

Table 1.91 The maturity score for NWPCC

Perspective	Weight of Per.	Criteria	Weight of Cri.	Value of Desir.	Maturity score	Sum of maturity score for each perspective
Organization	0.28	Executive's Commitments	0.23	100	6.44	28.28
		Teaming & Staffing	0.19	100	5.32	
		Communication & Collaboration	0.21	100	5.88	
		Stakeholder Involvement	0.19	100	5.32	
		Resource Allocation	0.19	100	5.32	
Methodology	0.24	Technology Roadmapping	0.23	25	1.35	15.36
		Technology Forecasting (Energy Forecasting)	0.18	79	3.42	
		Technology Assessment (Energy Resource Assessment)	0.21	75	3.79	
		Data Collection & Management	0.21	77	3.90	
		Review, Update, & Change Management	0.18	67	2.90	
Knowledge	0.29	Guideline & Handbook	0.19	39	2.13	20.16
		Instruction & Training	0.21	66	4.00	
		Skill/Competency	0.22	74	4.69	
		Lesson Learned & Dissemination	0.2	72	4.16	
		Expert Engagement	0.19	94	5.17	
Documents	0.19	Environmental Scan & Technology Prospect	0.16	77	2.35	17.45
		Mission, Vision, and Capability Gaps	0.21	95	3.79	
		Goals & Objectives with quantitative metrics	0.21	93	3.71	
		Strategy & Action/ Implementation Plan	0.24	92	4.19	
		Interconnection & Alignment	0.18	100	3.42	
Total sum of maturity scores						81.25

5. Eligibility should be extended to various sectors including academia, professional societies, government agencies, and private sector institutions.

For the inconsistency in HDM approach, Kocaoglu (1987) explained, "Variances of W_j for each j is a measure of inconsistency of the subjective responses" (Kocaoglu 1987) (p. 13). Should we encounter a higher variance, discrepancies should be explained, and the new measurements need to be taken again. The variance may

Table 1.92 Alignent maturity model (Phaal et al. 2005) (p. 108)

Maturity level 1 Organizing	Maturity level 2 Proactive	Maturity level 3 Collaborative	Maturity level 4 Comprehensive
Roadmaps are often done by individuals to clarify their own vision May utilize different types of roadmaps, but roadmaps are not interconnected Roadmaps are not shared widely or distributed beyond small workgroups Tools include a variety of desktop software applications, and typically roadmaps do not share a consistent look	Roadmapping becomes a group effort Roadmaps are updated at regular planning intervals Roadmap data is more commonly used in gate reviews and other business decisions Approved roadmaps are shared across functions Consistent roadmap structures begin to form Common approaches to creating roadmaps are adopted	Participants from cross-functional areas contribute planning information to roadmaps Linkages between roadmaps create agreements between participants Shared planning elements are commonly used to create cross-functional roadmaps Environmental roadmaps include customer requirements, competitor plans, industry roadmaps, academic research, etc. Technology roadmaps can include supplier planning data	Roadmaps are shared across business units High-level "whole enterprise" roadmaps emerge to provide "the big picture" Roadmapping practice is tightly coupled with other company processes Roadmap data is used to support improved decision making Real-time changes in roadmap data trigger off-cycle evaluation of the plan of record Supplier roadmapping process includes procurement department, engineers and all key suppliers

include ordinal and cardinal inconsistency (Estep 2016) (p. 75–78). If an expert's preference is A > B and B > C, he or she should then express the preference of A > C. An ordinal inconsistency will incur when the expert replies the preference of A < C. If an expert's preference is that A is two times better than B and B is two times better than C, then the logical reasoning should be A is four times better than C. If the response did not match the reasoning, there is a cardinal inconsistency occurred.

The way the inconsistency is measured is to calculate the differences/variance among the relative values of the elements calculated in the n! orientations (Estep 2016) (p. 76–77). The mathematical equations are given below:

Let

r_{ij} = relative value of the *i*th element in the *j*th orientation for an expert
$\overline{r_i}$ = relative value of the *i*th element in the *j*th orientation for an expert

$$\frac{1}{n!}\sum_{j=1}^{n!} r_{ij}$$

Inconsistency in the relative value of the *i*th element is:

$$\sqrt{\frac{1}{n!}\sum_{n!}^{j=1}\left(\overline{r_i}-r_{ij}\right)^2}$$

Table 1.93 Strength and weakness of the maturity models

Maturity model	Maturity levels	Strength	Weakness
Petrick's Roadmapping maturity model	There are six levels: Ad hoc, consistent, integrated, advanced, company optimizing, network optimizing	There are more levels of maturity, which enables the assessment more precisely	Lack of comprehensive attributes or criteria for conducting the maturity assessment Each level involves different attributes, which is not consistent Lack of implementation steps
Kappel's Roadmapping influence model	There are three levels: understand, persuade, synchronize	Measures have been clearly specified in the corresponding level	Lack of comprehensive attributes or criteria for conducting the maturity assessment The proposed measures ought to be included in each level, not just for its corresponding level Lack of implementation steps
Alignent's maturity model	There are 4 levels: organizing, proactive, collaborative, comprehensive	Using same attributes or criteria across all levels Good justification for each maturity level	Lack of implementation steps

Table 1.94 Categories of bias in research evaluation (Langfeldt 2004) (p. 58)

	Cognitive constraints	Interests
Scholarly/professional bias	(a) The constraints of a professional platform Preconceptions of good and valuable research Selective perceptions = Looking through the glasses' of your school/scholarly viewpoint/ profession	(b) Research interests Taking effects on economic and political standing of the field/ research area into consideration. Nepotism = helping heirs or other colleagues because of school/ scholarly viewpoint or research topic
Non-professional/personal bias	(c) General or personal cognitive constraints Sub-optimal thoroughness and information seeking Selective perceptions = Disregarding information due to routines/ limited capacity for handling information	(d) Personal interests Taking effects on personal situation or situation of friends, partners or competitors into consideration Nepotism = Helping colleagues because of friendship

Table 1.95 The potential organizations to find experts from

Organization	Strategic Energy Technology Plan or the equivalent planning documents
BPA (Bonneville Power Administration)	Bonneville Power Administration Power Services Technology Roadmap Bonneville Power Administration 2016–2021 Energy Efficiency Action Plan
California Energy Commission Efficiency Division	CA Energy Efficiency Strategic Plan: Codes and Standards Action Plan 2012–2015
DOE (US Department of Energy) Office of Science	The US Department of Energy's Ten-Year-Plans for the Office of Science National Laboratories
DOE (US Department of Energy) Office of Electricity Delivery & Energy Reliability	Smart Grid Research & Development Multi-Year Program Plan (MYPP) 2010–2014 Power Electronics Research and Development Program Plan April 2011
DOE (US Department of Energy) Office of Energy Efficiency and Renewable Energy	2016–2020 STRATEGIC PLAN and Implementing Framework Guide To Community Energy Strategic Planning March 2013 Solid-State Lighting R&D Plan
DOE (US Department of Energy) Science and Energy Plan Executive Steering Committee	Fiscal Year 2016 Science and Energy Plan September 2015
DOE (US Department of Energy) Advanced Research Projects Agency	Advanced Research Projects Agency–Energy Annual Report for FY 2016
Energy Trust of Oregon	2015–2019 Strategic Plan
European Commission's Directorates-General for Research and Innovation	The Strategic Energy Technology (SET) Plan 2017
Fermi National Accelerator Laboratory	Science and Technology Strategy for the Future: Fermilab's Strategic Plan (Updated May 2017)
IAEA (International Atomic Energy Agency)	IAEA Tools and Methodologies for Energy System Planning and Nuclear Energy System Assessments 2009
IEA (International Energy Agency)	Technology Roadmap: Energy Storage 2014
Illinois Power Agency	Long-Term Renewable Resources Procurement Plan 2017
IRENA (International Renewable Energy Agency)	Renewable Energy Prospects for the European Union 2018
NASEO (National Association of State Energy Officials)	NASEO's State Energy Planning Guidelines: Guidance for States in Developing Comprehensive Energy Plans and Policy Recommendations 2018
NEEA (Northwest Energy Efficiency Alliance)	Strategic Plan: 2015–2019 July 8, 2014
NREL (National Renewable Energy Laboratory)	Guam Strategic Energy Plan 2013
New York State Smart Grid Consortium	Strategic Smart Grid Vision and Technical Plan Report 2009

(continued)

Table 1.95 (continued)

Organization	Strategic Energy Technology Plan or the equivalent planning documents
NW Natural	2014 Integrated Resource Plan
NWPCC (Northwest Power and Conservation Council)	7th Power Plan 2016
PacifiCorp	2017 INTEGRATED RESOURCE PLAN
PGE (Portland General Electric)	Integrated Resource Plan 2016
PSE (Puget Sound Energy	2017 PSE Integrated Resource Plan
Sandia National Laboratories	FY16–FY20 Strategic Plan
Southwestern Power Administration	Strategic Plan March 2013
Tacoma Power	2015 Integrated Resource Plan
West Virginia Energy	Energy Plan 2013–2017
Western Area Power Administration	Strategic Roadmap 2024: Powering the Energy Frontier October 2016

Table 1.96 The potential experts identified from Compendex database

Author	Affiliation	Degree	Betweenness
Jordand, D	E2V Technol., Chelmsford, United Kingdom	17	128
Kruger, M	Inst. of Tech. Thermodynamics, German Aerosp. Center, Stuttgart, Germany	15	85.059
Giuliano, S	Inst. of Solar Res., German Aerosp. Center, Stuttgart, Germany	17	49
Puppe, M	Inst. of Solar Res., German Aerosp. Center, Stuttgart, Germany	17	49
Yonghua Song	Dept. of Electr. Eng., Tsinghua Univ., Beijing, China	13	45
Rochelle, G	McKetta Dept. of Chem. Eng., Univ. of Texas at Austin, Austin, TX, United States	22	44
Shah,N	Dept. of Chem. Eng., Imperial Coll. London, London, United Kingdom	22	44
Koritarov, V	NA	13	40
Fulli, G	Inst. for Energy & Transp., Eur. Comm., Petten, Netherlands	10	24
Catalao, J	Inst. de Eng. de Sist. e Comput., Tecnol. e Cienc., Univ. of Porto, Porto, Portugal	9	20
Szklo, A	Energy Planning Program, Cidade Univ., Rio de Janeiro, Brazil	11	19
Bak-Jensen, B	Dept. of Energy Technol., Aalborg Univ., Aalborg, Denmark	8	15
Kit Po Wong	Sch. of Electr., Electron., & Comput. Eng., Univ. of Western Australia, Perth, WA, Australia	8	12.143
Zhao Xu	Dept. of Electr. Eng., Hong Kong Polytech. Univ., Kowloon, China	8	12.143

(continued)

Table 1.96 (continued)

Author	Affiliation	Degree	Betweenness
Ke Meng	Centre for Intell. Electr. Networks, Univ. of Newcastle, Newcastle, NSW, Australia	7	10.625
Yan Xu	Centre for Intell. Electr. Networks, Univ. of Newcastle, Newcastle, NSW, Australia	7	10.625
Van der Burgt, J	KEMA Nederland BV, Arnhem, Netherlands	7	10
Meilhan, J	CEA Leti-MINATEC, Grenoble, France	10	9
Simoens, F	CEA Leti-MINATEC, Grenoble, France	10	9
Svendsen, H	SINTEF Energy Res., Trondheim, Norway	6	9
Bansal, R	Sch. of Inf. Technol. & Electr. Eng., Univ. of Queensland, St. Lucia, QLD, Australia	5	8.5
Prica, M	Dept. of Electr. & Comput. Eng., Carnegie Mellon Univ., Pittsburgh, PA, United States	6	8
Liu, J	Harvard Univ., Boston, MA, United States	5	6
Wei-Jen Lee	Energy Syst. Res. Center, Univ. of Texas at Arlington, Arlington, TX, United States	5	6
Srivastava, A	Dept. of Electr. & Comput. Eng., Mississippi State Univ., Starkville, MS, United States	4	5
Jorgensen, B	Center for Smart Energy Solutions, Univ. of Southern Denmark, Odense, Denmark	4	4
Hanwei Liang	Collaborative Innovation Center on Forecast & Evaluation of Meteorol. Disaster, Nanjing Univ. of Inf. Sci. & Technol., Nanjing, China	11	3.75
Jingzheng Ren	Collaborative Innovation Center on Forecast & Evaluation of Meteorol. Disaster, Nanjing Univ. of Inf. Sci. & Technol., Nanjing, China	11	3.75
Liang Dong	Collaborative Innovation Center on Forecast & Evaluation of Meteorol. Disaster, Nanjing Univ. of Inf. Sci. & Technol., Nanjing, China	11	3.75
Zhiqiu Gao	Collaborative Innovation Center on Forecast & Evaluation of Meteorol. Disaster, Nanjing Univ. of Inf. Sci. & Technol., Nanjing, China	11	3.75

The inconsistency of the expert in providing relative values for the n element can be calculated below:

$$\text{Inconsistency} = \frac{1}{n}\sum_{n}^{i=1}\sqrt{\frac{1}{n!}\sum_{j=1}^{n!}\left(\overline{r_i} - r_{ij}\right)^2}$$

The value of the inconsistency has been recommended to be less than 10% for claiming a consistent judgement (Gibson 2016; Chan 2013; Iskin 2014).

Experts within group could have somewhat different viewpoints or interpretation on the same decision criteria, which are very likely to result in certain degree of

Table 1.97 The potential experts identified from Web of Science database

Author	Affiliation	E-mail	Degree	Betweenness
Shah, N	Imperial Coll London, Ctr Proc Syst Engn, Dept Chem Engn, South Kensington Campus, London SW7 2AZ, Englandl	axelle.delangle15@alumni.imperial.ac.uk	10	24
Acha, S	Imperial Coll London, Ctr Proc Syst Engn, Dept Chem Engn, South Kensington Campus, London SW7 2AZ, England	salvador.acha@ic.ac.uk	4	0
Aryanpur, V	Sharif Univ Technol, Tehran, Iran	aryanpur@alum.sharif.edu	1	0
Bianchi, G	GSI Environm Inc, 2211 Norfolk, Suite 1000, Houston, TX 77098 USA	jaconnor@gsienv.com	5	0
Brannstrom, C	Texas A&M Univ, College Stn, TX 77843 USA	claudiofrate@unb.br; cbrannst@geos.tamu.edu	1	0
Carpignano, A	Politecn Torino, DENERG Energy Dept, Corso Duca Abruzzi 24, I-10129 Turin, Italy	raffaella.gerboni@polito.it	3	0
Castro-Santos, L	Univ A Coruna, Dept Enxenaria Naval & Ind, Escola Politecn Super, Esteiro 15471, Ferrol, Spain;	c.guedes.soares@centec.tecnico.ulisboa.pt	2	0

(continued)

Table 1.97 (continued)

Author	Affiliation	E-mail	Degree	Betweenness
Chiara, BD	Politecn Torino, Dept DIATI Transport Syst, Corso Duca Abruzzi 24, I-10129 Turin, Italy	raffaella.gerboni@polito.it	3	0
Connor, JA	GSI Environm Inc, 2211 Norfolk, Suite 1000, Houston, TX 77098 USA;	jaconnor@gsienv.com	5	0
Daim, TU	Portland State Univ, Portland, OR 97207 USA	tugrul.u.daim@pdx.edu	1	0
Delangle, A	Imperial Coll London, Ctr Proc Syst Engn, Dept Chem Engn, South Kensington Campus, London SW7 2AZ, England	axelle.delangle15@alumni.imperial.ac.uk; salvador.acha@ic.ac.uk	4	0

group disagreement. Kocaoglu (2014) explained that a modified variance index can be used for measuring the disagreement among experts. More specially, "the variance among (n) expert judgments is calculated for each of the (m) decision elements being measured." Then the disagreement can be calculated as the square root of the arithmetic mean of the (m) variances (Kocaoglu 2014).

Iskin (2014) discussed the following equations to calculate the disagreement index (Iskin 2014) (p. 101).

Let

m be the number of experts
$\overline{r_i}$ be mean relative value of the ith element for kth expert

Group relative value of the ith element for k expert can be calculated below:

$$R_i = \sum_{k=1}^{K} \overline{r_{ik}} \frac{1}{m} \text{for} \quad i = 1,2,3,\ldots,n$$

Then the disagreement index for k experts for n decision variables can be calculated below.

$$d = \sqrt{\frac{1}{n} \sum_{m}^{i=1} \left(R_i - \overline{r_{ik}} \right)^2}$$

Like the inconsistency, the threshold value for the disagreement index has been suggested as 10% (Gerdsri and Kocaoglu 2004; Estep 2016; Chan 2013; Iskin 2014). If there is high disagreement occurrence, more efforts devoted to explanation, and clarification are deemed necessary. Possible resolutions may include redoing the evaluation or omit the outliers for extreme cases.

With the advent of technology advancements and dramatic changing marketplace, the priority of the goals or objectives may change as the environment alters. The impacts of potential changes need to be responded, so that the resource allocation can be aligned with the strategic transformation. In HDM analytical setting, the impacts of potential changes are measured by conducting sensitivity analysis. Chen (2007) developed a comprehensive algorithm to analyze the sensitivity of HDM, which covered the issues of direct impact of changes to a local contribution value on decision alternatives' overall contribution, allowable range/region of perturbations, contribution tolerance, etc. (Chen 2007). This method was illustrated from the application to technology evaluation and energy portfolio forecast cases. Since then, the method has been applied in many cases.

Gerdsri et al. (2009) used this method to calculate and analyze the allowable range of perturbations, tolerance, and sensitivity coefficient of five objectives for showing the degree of sensitivity (Gerdsri 2009). Chan (2013) used this method to conduct sensitivity analysis on pharmaceutical technologies to the Ranks of Input Resources (Chan 2013). Iskin (2014) applied this method on assessment of emerging energy efficiency program, especially on the rank order of the best alternative and the rank order of all alternatives (Iskin 2014).

Despite of the comprehensive sensitivity algorithm proposed by Chen (2007), the impacts of changes in the value of HDM need to be periodically evaluated and monitored for assuring the alignment between priority and actions.

The government agencies, laboratories, non-profit organizations, or private companies, which have published Strategic Energy Technology Plan or other equivalent planning documents, are considered potential organizations to find experts from. The potential organizations are listed alphabetically (Table 1.95).

For this bibliometrics analysis, there are few steps needed to be conducted for identifying the potential experts.

1. Identify the key words such as Strategy, Technology, Plan, Energy, Maturity, or Roadmap
2. Search documents from the Compendex and Web of Science database and limit the search for selecting less than 500 documents for analysis
3. Save the documents files and upload to the RStudio software (Garces et al. 2016)
4. Perform analytical function in RStudio

5. Download the result files for data analysis
6. Check the results and select the experts with better Betweenness values

The Betweenness is defined as "Number of shortest paths between two actors that actor i resides on" (Daim et al. 2014). The high value of betweenness implies that "the node has significant influence on the network flow." The other metric used for measuring relationships between nodes is called "Degree," which refers to "the number of direct connections in a network for node i" (Garces et al. 2016).

The results of conducting the Bibliometrics analysis on the data collected from the Compendex and Web of Science databases are listed in the following tables (Tables 1.96 and 1.97).

Appendix 1: Meeting Notes with Expert A, BPA

1. Major tools for strategic technology planning or energy planning
   - The "Resource Planning," "Transmission Planning," "Non-Wire Transmission," and "Integrated Resource Planning (IRP)" studies all use "Load Forecasting" data. IRP will utilize the analysis from both the power side and transmission side of business. These planning efforts are often generated by policy requirements or direction. In other words, agency such as BPA needs to follow commission's requirement to submit the integrated resource plan.
   - For IRP, the commission will look at the adequacy of "Load Forecast" and the arrangement of the resource supply. The issues of whether the supply meeting the demand needs to be articulated. This includes the analysis on whether adopting adequate energy efficiency or demand response technology to meet our energy target.
   - In terms of the electricity consumption, there are generally low forecast, regional forecast, seasonal forecast, peak forecast, etc. Therefore, "Transmission Planning" is utilized to make sure that the transmission teams are capable to deal with the changes of the dynamic electricity load.
   - The utility company needs to use these planning efforts to meet the future customer need while considering if the approach is the least cost, sustainability beneficial, and meeting supply reliability requirements.
   - "Non-Wire Transmission" is an alternative method for tradition transmission and looking at a more cost-effective transmission approach in the context of the whole electricity infrastructure.
2. Technology Forecasting, demand forecasting, or load forecasting tools
   - BPA uses "Production cost model" for transmission planning.
   - The "Probabilistic model" is also used for contingency planning.
3. Technology Assessment tools or decision-making tools

- The HDM (Hierarchical Decision Model) has been presented to BPA Technology Innovation team. To my understanding, it seems that the tools have not been embraced by management team.
- BPA is now Technology Roadmap oriented. The second layer is the Portfolio Management, which is kind of building into it. The third layer is some organizational tools for the business line and facilitating the implementation of 3–5 years of commercialization programs.
- BPA is getting the "P3M3 (Portfolio, Program, and Project Maturity Management Model)" reevaluated again. Due to some budget limitation, the focus has been switched from long-term to short-term demonstration projects.

4. Strategic Planning tools

   - Strategic Planning tools are for further upstream analysis. "SWOT," "Scenario Analysis," or "Long-term Trend Tracking" have been adopted by BPA to identify who we are and where we are going to be.

5. Strategic Planning documents

   - BPA is now putting all these planning efforts into the "Integrated Resource and Transmission Plan." BPA used to do it long time ago and then stopped doing it. But, now BPA need to follow the requirement from Northwest Power and Conservation Council (NW Council) to prepare these plans and update the plans every 5 years. Those plans are on line and accessible. In addition, there are other IRP examples published by public or private utilities.
   - To my understanding, the BPA "Strategic Plan" did not show the transparency of strategic planning process or indicate the methodologies used for developing the strategic plan. However, it is used for communicating to customers/stakeholders about the prioritization of the future implementation plan.

Appendix 2: Meeting Notes with Expert B

1. Major tools for strategic technology planning or energy planning

   - PGE uses "Monte Carlo simulation" method to conduct a flexible capacity analysis. The variables needed include population growth, fuel price, carbon tax, etc. After many times of simulation with all the random variables, the results can be some distribution diagrams showing the pattern or status of energy usage in different scenario or assumption.
   - PGE also use "Economic Model" to evaluate options of energy resources.

2. Technology Forecasting, demand forecasting, or load forecasting tools

   - PGE works with manufacturing companies to test and analyze some emerging energy efficiency technology. Generally, a "Pilot" program will be initiated to see if it is cost effective and meet the future customer value expectation.

For example, the battery technology has been growing. With some R&D efforts and collaboration with manufacturing companies (via Request for Proposal), more efficient energy storage technology will be evaluated or adopted.
- Sun or solar power is easier to predict, whereas the wind power is difficult to predict. With "Probabilistic Modelling," the uncertainty or risk is expected to be minimized.
- For technology development in PGE, we rely on many "Data Analysis" to see if the technology will generate more value to the existing energy mix.

3. Technology Assessment tools or decision-making tools
   - The decision of energy portfolio mix sometimes is attributed to the state regulation. For example, the future renewable energy target is 70%. PGE will strive for meeting the Renewable Portfolio Standards.
   - I am not sure what you mean by multi criteria decision tools such as HDM. However, PGE does review the energy portfolio and conduct some assessments on the power plants to see if the future demand can be met.
4. Strategic Planning tools
   - PGE use "SWOT" for business or customer analysis.
   - PGE use "Scenario Planning" together with "Simulation Method" for analyzing future demand and energy resources.
5. Strategic Planning documents
   - "Integrated Resource Plan (IRP)" is the main document addressing the future planning about the energy portfolio mix.
   - PGE does have a "Strategic Plan." Generally, PGE will invite subject matter experts (SME) to provide their inputs and opinions. The focus of the strategic plan is on business matter or customer related issues. The technology planning is not emphasized in the strategic plan.

Appendix 3: Meeting Notes with Experts C and D

- I would like to begin with introducing some of the background information on our council. The website of the council (https://www.nwcouncil.org/energy/powerplan/7/home/) posts some materials related to understanding power planning. There is a video presentation on overview of the council's power plan development process. There are some review presentation materials in pdf format for introducing the power plan and how the plan has been developed.
- We developed and used lots of *models and simulation* to help develop a better resource strategy.
- We post most of the *tools and models* including excel files for the public to get access and understand how these tools can be utilized.

- The results of tools will go to *advisory group* for feedback and approval.
- We engage or educate the *stakeholders* to understand our tools and the results generated from using these tools.
- The people working in the council often have working experiences in the industry such as BPA, PGE, etc.
- During monthly council meetings, the *executive's committees* review and approve the planning documents in a way to show their *commitments*.
- It takes 3 years to develop the power plan and take about 2 years to implement. Power plan is a living document.
- We put specific actions into the power plan for guiding the future efforts required to meet the common goals. This is documented in Chapter 4 and served as a strategic roadmap for other agencies/utilities to follow.
- The *action plan* includes not only the recommendation for other utilities or agencies and but also the tasks for ourselves.
- For the tasks of tools development or analysis of energy data, we have project team and project managers who conduct the *monitoring and controlling* the progress of all the required tasks.
- *Action plan* will be *reviewed* periodically and constantly. Specifically, we also have mid-term assessment to monitor the progress.
- We *communicate and collaborate* among departments, teams, advisory committees, and council members to make sure the action plans or analytical works have been done correctly and satisfactorily.
- The *stakeholders* may include whoever is impacted by the power plan. For example, the utilities, the agency, the wind power companies, and some advocacy groups.
- You can check on https://www.nwcouncil.org/energy/ and find the information about the advisory committee in the areas of RTF (Regional Technical Forum) policy, conservation resources, demand forecasting, demand response, generating resources, natural gas, resource adequacy, resource strategies, and system analysis.
- For *resource allocation*, we have our virtual servers and use some cloud-based tools such as Amazon Web Services, and Box.com to develop our own tools or models.
- In terms of experiences in using *technology roadmapping*, the council used to ask our organization to develop a technology roadmap for internal IT (Information Technology) about 3 years ago. This roadmap showed our needs and how we would update to meet future IT requirements.
- We also use some *stochastic models* to predict the future prices of the natural gas under some assumptions or scenarios, for example, 800 future scenarios for evaluation.
- Understanding the uncertainty is also shown in our website at https://www.nwcouncil.org/energy/powerplan/7/planninguncertainty.
- We put all the energy resources into our *portfolio analysis* and run lots of simulation and come up with many different results. The result is documented in

Chapter 3, Resource Strategy. The way we did the analysis is included in Chapter 15 Analysis of Alternative Resource Strategies.

- We also employ *cost-effective analysis* to identify the least cost and the least risk options, and frequently we need to trade off or balance the both.
- For the *data collection and management*, we not only store the data inside our council but also we release some of the data to the public, so that the stakeholders can have some degree of understanding on how we come up with these results or suggestions.
- For the *review, update, and change management*, I think we have some regular review and update activities and tend to do change management on an ad hoc basis.
- Sometimes, the *stakeholder engagements* have been done more than what we want, because we have put so many information on the web and have arranged periodical and constant meetings with stakeholders.
- For the power planning, we do not use *SWOT analysis*, but we do use a lot of *scenario analysis* for resource planning purposes.
- All elements in the power plan appear to be *aligned* with each other.
- Our council is a government agency and is very transparent on many things such as the power plan, how it is developed, the tools and models, and others. Our council is funded by law as an independent agency and now is funded by BPA.
- The video presentation done by Tom Eckman is highly recommended, because he is the director of the power division and has extensive experiences on the development of all the power plans.
- There are only three kinds of agencies like us. I am initially not sure if your model is suitable to be applied to our council. However, you said that power plan is one type of strategic technology plan, because it contains some suggestions about what energy technologies are needed to be developed. Then it makes sense to me how your maturity assessment model links to our power plan.
- The *central staffs* are the *core team members* responsible for developing the power plan.
- We developed some tools and models specifically for the power planning. For example, the RPM (Regional Portfolio Model) tool is on line and welcome to use and give it a trial.
- These tools mentioned in the power plan have been developed by the council and may be migrated to proprietary tools.
- The chapter 4 "Action Plan" might be a high-level roadmap for utilities or agencies to implement resource strategy, while the whole power plan is more like a strategic power roadmap for the NW district.
- You can check both 6th and 7th power plans to see if the previous tasks have been monitored and completed.
- We report to our *council members* (https://www.nwcouncil.org/contact/members/) constantly and periodically (basically a month).
- We give presentation to council members every single month to show our progress.

- The *council members* are the eight appointees from Oregon, Washington, Idaho, and Montana states.
- We also *communicate and collaborate* with the staffs from each of the council members' offices to make sure that the work has been done with consensus.
- Whoever participate in the power plan are the *stakeholders*.
- The *stakeholders* may also include *advocacy groups* such as environmentalist group, trade group, regulatory group, and utilities commission..
- Some tools used for *forecasting* purposes may be just based on Excel spreadsheet, Excel Macro, or based on the R programming tool. For data repository, we use SQL.
- When we presented our results of simulation or forecasts to the council members, we will also provide the context, scenario, or assumptions for them to conduct a rational decision-making.
- In terms of *technology assessment* or decision-making tools, we have a RPM (Regional Portfolio Model) tool, which is used for identifying adaptive, least-cost resource strategies for the region by using a sophisticated risk analysis methodology.
- I would say the plan itself is the *guidebook* for our council and the utilities or agency in NW district.
- We are small in a sense, so we normally have *team training* and rely on some personal developments. Unlike the other big organizations which have some formal trainings, we tend to focus on personal needs and development and make sure that the required training has been completed.
- *Lesson learned* have been shown from the comparison between 6th and 7th Power Plan.
- The Fish and wildlife division deals with *environment analysis*.
- The *Northwest Power Act* gives the council the *mission and vision*.
- The power plan did include *resource strategy* and relevant *goals or objectives* with some *quantitative metrics*.
- The *action plan* is clearly articulated in the power plan.

Appendix 4: RI 1-Validation of Perspective

Q11

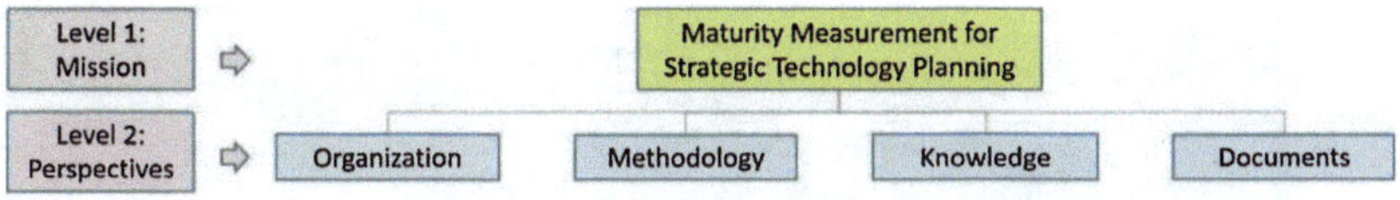

Q5

Description of Perspectives:

- Organization
  - The category of factors which are related to organizational inputs and support for enabling the implementation of strategic technology planning,
- Methodology
  - The category of factors pertaining to technical, analytical or managerial methods, approaches, tools and techniques that are utilized during strategic technology planning process.
- Knowledge
  - The category of factors related to knowledge assets and management for facilitating strategic technology planning process.
- Documents
  - The category of key factors or elements that are included in the content of the strategic technology plan for guiding future technology direction.

Q6 Please identify the perspectives that contribute to the maturity model for strategic technology planning

	Click to write Column 1	
	Yes	No
Organization	○	○
Methodology	○	○
Knowledge	○	○
Documents	○	○

Appendix 5: RI 2-Validation of Criteria

Q14

Criteria for organization perspective	Description
Executives' Commitments	The promises or assurance from high executives or senior managers, which support the implementation of strategic technology planning.
Teaming & Staffing	The team forming and responsibility assignment within or beyond the organization structure, which are structured specifically for strategic technology planning.
Communication & Collaboration	The communication & collaboration efforts across different departments within or beyond the whole organization, aiming to facilitate strategic technology planning.
Stakeholder Involvement	The proper arrangement, interaction, and involvement with various stakeholders for the sake of successfully managing strategic technology planning.
Resource Allocation	The financial or other organizational resources that are allocated to the usage for strategic technology planning.

Q15

Please validate the **criteria** included in **organization** perspective.

	Click to write Column 1	
	Yes	No
Executive's Commitments	○	○
Teaming & Staffing	○	○
Communication & Collaboration	○	○
Stakeholder Involvement	○	○
Resource Allocation	○	○

Q16

If there are other criteria needed to be added, please add them below.

Appendix 6: RI 3-Validation of Desirability Metrics

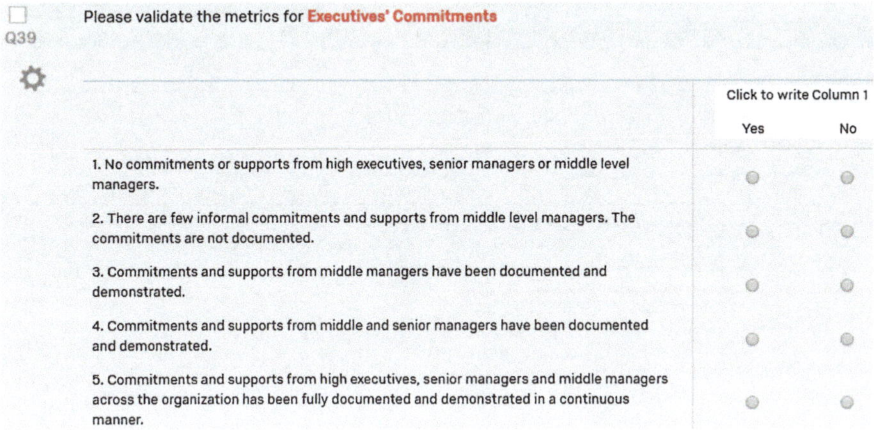

Appendix 7: RI 4-Quantification of Perspective

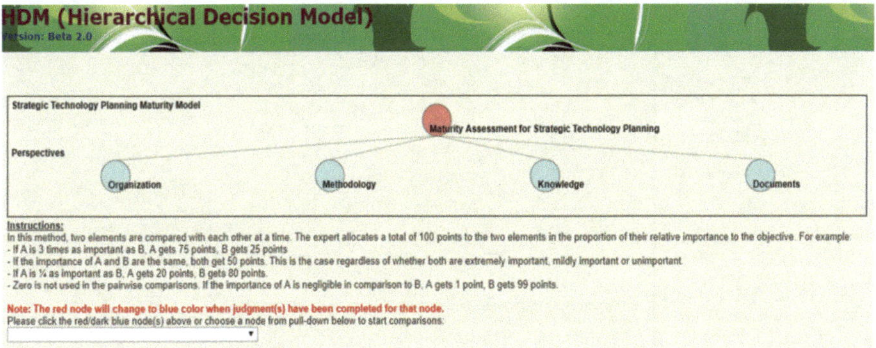

Appendix 8: RI 5-Quantification of Criteria

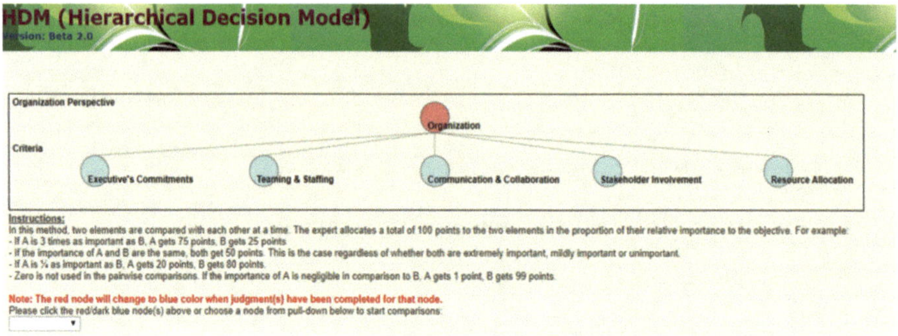

Appendix 9: RI 6-Quantification of Desirability Metrics

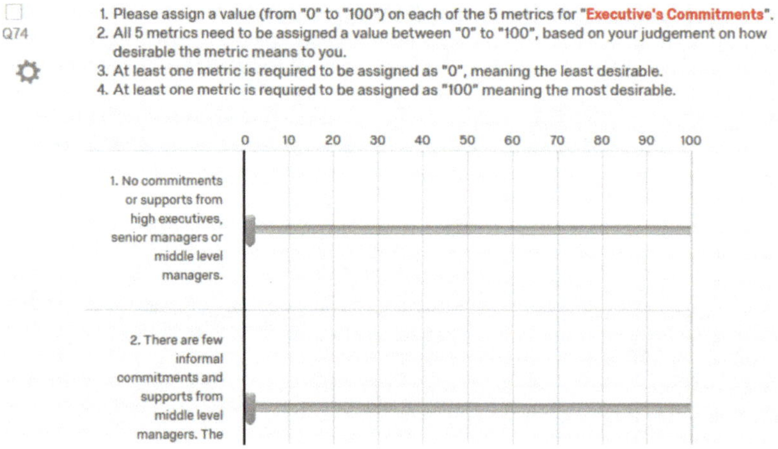

References

Antunes, P. M. B. (2014). A maturity model for energy management. Instituto Superior Técnico.

Aron, D. (2010). *Strategic planning: Key initiative overview*. Gartner, Inc, p. 2.

Ates, S. A., & Durakbasa, N. M. (2012). Evaluation of corporate energy management practices of energy intensive industries in Turkey. *Energy, 45*(1), 81–91.

Bai, R., & Lee, G. (2003). Organizational factors influencing the quality of the IS/IT strategic planning process. *Industrial Management & Data Systems, 103*(8), 622–632.

Bigwood, M. P. (2000). Applying cost of innovation to technology planning. *Research Technology Management, 43*, 39–46.

Bonneville Power Administration. (2017, March). *Bonneville power administration 2016–2021 energy efficiency action plan*. Bonneville Power Administration, pp. 1–109.

Bonneville Power Administration. (2018). *BPA 2018 – 2023 strategic plan*. Bonneville Power Administration, pp. 5–61.

Böttcher, C., & Müller, M. (2016). Insights on the impact of energy management systems on carbon and corporate performance. An empirical analysis with data from German automotive suppliers. *Journal of Cleaner Production, 137*, 1449–1457.

Bray, O. H., & Garcia, M. L. (1997). Technology roadmapping: The integration of strategic and technology planning for competitiveness. In *Innov. Technol. Manag. Key to Glob. Leadership. PICMET'97*, pp. 25–28.

Bryson, J. A. (1988). A strategic planning process for public and non-profit organizations. *Long Range Planning, 21*(1), 73–81.

Bull, J. W., et al. (2016). Strengths, weaknesses, opportunities and threats: A SWOT analysis of the ecosystem services framework. *Ecosystem Services, 17*, 99–111.

California Public Utilities Commision. (2008). California long term energy efficiency strategic plan. *California Public Utilities Commission*.

Caviggioli, F., De Marco, A., Scellato, G., & Ughetto, E. (2017). Corporate strategies for technology acquisition: Evidence from patent transactions. *Management Decision, 55*(6), 1163–1181.

Cayir Ervural, B., Zaim, S., Demirel, O. F., Aydin, Z., & Delen, D. (2018). An ANP and fuzzy TOPSIS-based SWOT analysis for Turkey's energy planning. *Renewable and Sustainable Energy Reviews, 82*, 1538–1550.

Chan, L. (2013). Developing a strategic policy choice framework for technological innovation: Case of Chinese pharmaceuticals (PhD Dissertation). Portland State University.

Checkland, P. (2000). Soft systems methodology: A thirty year retrospective. *Systems Research and Behavioral Science, 17*, 11–58.

Chen, T.-Y. (1999). Critical success factors for various strategies in the banking industry. *The International Journal of Bank Marketing, 17*(2), 83–91.

Chen, H. (2007). *Sensitivity analysis for hierarchical decision models* (PhD Dissertation). Portland State University.

Christensen, J. (1996). The dynamics of the diversified corporation and the role of central management of technology. In *Danish Research Unit for Industrial Dynamics Working Paper No. 98–4*, no. December.

Christensen, J. (1998). The dynamics of the diversified corporation and the role of central management of technology. In *Danish Res. Unit Ind. Dyn. Work. Pap. No. 98–4*, no. December, pp. 1–15.

Christensen, J. F. (2002). Corporate strategy and the management of innovation and technology. *Industrial and Corporate Change, 11*(2), 263–288.

Clarke, C. J. (1987). Acquisitions - techniques measuring strategic fit. *Long Range Planning, 20*(3), 12–18.

Crow, S. (2016). Seventh northwest conservation and electric power plan summary brochure. NWCOUNCIL, pp. 1–4.

Daim, T. U., Rueda, G. R., & Martin, H. T. (2005). Technology forecasting using bibliometric analysis and system dynamics. In *Technology Management: A Unifying Discipline for Melting the Boundaries*, pp. 112–122.

Daim, T. U., Islam, N., Ozcan, S., Van Blommestein, K., Hillegas, J., & Estep, J. (2014) Integrating data mining into technology Roadmapping. In *Portland International Center for Management of Engineering and Technology (PICMET) proceedings*, pp. 2972–2985.

Day, G. S. (1977). Diagnosing the product portfolio. *Journal of Marketing, 41*(2), 29.

Department of Sustainability and Environment. (2005). *Book 3 The engagement toolkit* (pp. 1–106). State of Victoria.

Drew, S. A. W. (2006). Building technology foresight: Using scenarios to embrace innovation. *European Journal of Innovation Management, 9*(3), 241–257.

Eastham, J., Tucker, D., Varma, S., & Sutton, S. (2014). PLM software selection model for project management using hierarchical decision modeling with criteria from PMBOK knowledge areas. *Engineering Management Journal, 3*, 13–24.

Edler, J., Meyer-Krahmer, F., & Reger, G. (2002). Changes in the strategic management of technology: Results of a global benchmarking study. *R and D Management, 32*(2), 149–164.

Ernst, H. (2003). Patent information for strategic technology management. *World Patent Information, 25*(3), 233–242.

Ernst, H., & Soll, J. H. (2003). Integrating market and patent portfolios for market-oriented R&D planning. In *Portland International Conference on Management of Engineering and Technology*, pp. 121–132.

Ernst, H., Fabry, B., & Soll, J. H. (2004). Enhancing market-oriented R & D planning by integrated market and patent portfolios. *Journal of Business Chemistry, 1*(1), 2–11.

Estep, J. (2016). Development of a technology transfer score for evaluating research proposals: Case study of demand response technologies in the Pacific Northwest (PhD Dissertation). Portland State University, Oregon, US.

European Commission. (2010). Energy 2020: A strategy for competitive, sustainable and secure energy. *European Commission*. European Union, pp. 4–23.

European Commission. (2017). The strategic energy technology plan. European Union, pp. 6–86.

European Commission. (2018, February). Renewable energy prospects for the European Union.

European Commission. n.d.). European Industrial Initiative on wind energy. *SETIS - European Commission*. [Online]. Available: https://setis.ec.europa.eu/european-industrial-initiative-wind-energy. Accessed 2 Apr 2018.

Federal Highway Administration U. S. Department of Transportation. (2002). Use of expert panels in developing land use forecasts. In *Proc. a Peer Exch. Spons. by Travel Model Improv. Progr.*, vol. 1–17.

Fenn, J., Linden, A., & Fairchok, S. (2003). Strategic technology planning: Picking the winners. *Gartner, R-20-3354*(July), 9–45.

Finnerty, N., Sterling, R., Coakley, D., & Keane, M. M. (2017). An energy management maturity model for multi-site industrial organisations with a global presence. *Journal of Cleaner Production, 167*, 1232–1250.

Garces, E., van Blommestein, K., Anthony, J., & Hillegas-Elting, J. (2016). ______Identification of experts using social network analysis (SNA). In *PICMET '16: Technology Management for Social Innovation*, pp. 1882–1896.

Garcia, M. L., & Bray, O. H. (1997). Fundamentals of technology Roadmapping. *Sandia National Laboratories*. [Online]. Available: http://prod.sandia.gov/techlib/access-control.cgi/1997/970665.pdf.

Gates, L. P. (2010). Strategic planning with critical success factors and future scenarios: An integrated strategic planning framework. Softw. Eng. Inst., no. November, p. 67.

Gerdsri, N. (2007). An analytical approach to building a technology development envelope (TDE) for roadmapping of emerging technologies. *International Journal of Innovation and Technology Management, 4*(2), 121–135.

Gerdsri, P. (2009). *A systematic approach to developing National Technology Policy and strategy for emerging technologies* (PhD Dissertation). Portland State University.

Gerdsri, N., & Kocaoglu, D. F. (2004). A quantitative model for the strategic evaluation of emerging technologies. In *PICMET*.

Gerdsri, N., & Kocaoglu, D. F. (2007). Applying the Analytic Hierarchy Process (AHP) to build a strategic framework for technology roadmapping. *Mathematical and Computer Modelling, 46*(7–8), 1071–1080.

Gerdsri, N., Assakul, P., & Vatananan, R. S. (2008). Applying change management approach to guide the implementation of technology roadmapping (TRM). In *2008 Portland International*

Center for Management of Engineering and Technology, Technology Management for a Sustainable Economy, PICMET '08, July 27, 20, no. c, pp. 2133–2140.

Gerdsri, N., Vatananan, R. S., & Dansamasatid, S. (2009). Dealing with the dynamics of technology roadmapping implementation: A case study. *Technological Forecasting and Social Change, 76*(1), 50–60.

Ghemawat, P. (1985). Building strategy on the experience curve. *Harvard Business Review, 63*(2), 143–149.

Gibson, E. C. (2016). A measurement system for science and engineering research center performance evaluation (PhD Dissertation). Portland State University.

Gopalakrishnan, B., Ramamoorthy, K., Crowe, E., Chaudhari, S., & Latif, H. (2014). A structured approach for facilitating the implementation of ISO 50001 standard in the manufacturing sector. *Sustainable Energy Technologies and Assessments, 7*, 154–165.

Grienitz, V., & Ley, S. (2007). Scenarios for the strategic planning of technology. *Journal of Technology Management & Innovation, 2*(3), 21–37.

Groenveld, P. (2007). Roadmapping integrates business and technology. *Research Technology Management, 50*, 49–58.

Harrison, J. P. (2010). Strategic planning and SWOT analysis. In *Essentials of strategic planning in healthcare* (pp. 91–97). Health Administration Press.

Haselkorn, A., Rosenstein, A. H., Rao, A. K., Van Zuiden, M., & Coye, M. J. (2007). New technology planning and approval: Critical factors for success. *American Journal of Medical Quality*, 164–169.

Hedley, B. (1977). Strategy and the 'business portfolio' (1). *Long Range Planning, 10*(1), 9–15.

Hervás Soriano, F., & Mulatero, F. (2011). EU Research and Innovation (R&I) in renewable energies: The role of the Strategic Energy Technology Plan (SET-Plan). *Energy Policy, 39*(6), 3582–3590.

Illinois Power Agency. (2017). Long-term renewable resources procurement plan.

International Renewable Energy Agency. (2017). *Renewable Energy Outlook: Thailand.*

Introna, V., Cesarotti, V., Benedetti, M., Biagiotti, S., & Rotunno, R. (2014). Energy Management Maturity Model: An organizational tool to foster the continuous reduction of energy consumption in companies. *Journal of Cleaner Production, 83*, 108–117.

Irfan, M., Putra, S. J., Alam, C. N., Subiyakto, A., & Wahana, A. (2017). Readiness factors for information system strategic planning among universities in developing countries: A systematic review. In *2nd international conference on computing and applied informatics* (pp. 1–6).

Iskin, I. (2014). An assessment model for energy efficiency program planning in electric utilities: Case of Northwest U. S. (PhD Dissertation). Portalnd State University.

Iskin, I., & Daim, T. U. (2014). Technology assessment for energy efficiency programs in Pacific Northwest. in *2014 Proc. PICMET'14 Infrustructure Serv. Integr.*, pp. 498–506.

Iskin, I., & Daim, T. U. (2016). An assessment model for energy efficiency program planning in electric utilities: Case of Northwest U. S. *Sustainable Energy Technologies and Assessments, 15*, 42–59.

Jeffrey, H., Sedgwick, J., & Robinson, C. (2013). Technology roadmaps: An evaluation of their success in the renewable energy sector. *Technological Forecasting and Social Change, 80*(5), 1015–1027.

Kappel, T. A. (2001). Perspectives on roadmaps: How organizations talk about the future. *Journal of Product Innovation Management, 18*(1), 39–50.

Koc, T., & Mutu, Y.. (2006). A technology planning methodology based on axiomatic design approach. In *PICMET 2006 Proceedings, 9–13 July, Istanbul, Turkey*, no. c, pp. 1450–1456.

Kocaoglu, D. F. (1983). A participative approach to program evaluation. *IEEE Transactions on Engineering Management, EM-30*(3).

Kocaoglu, D. F. (1987). Hierarchical decision modeling (HDM). *ETM, 530*(63), 1–31.

Kocaoglu, D. F. (2014). Inconsistency, disagreement and discrepancy measures in HDM. ETM 530/630 Lecture Presentation, Portland State University, pp. 1–8.

Kockan, I., Yildirim, A. M., Daim, T., Ergeneman, M., & Gerdsri, N. (2009). Application of technology development envelope (TDE) approach for future powertrain technologies: A case study of ford otosan. In *PICMET Portl. Int. Cent. Manag. Eng. Technol. Proc.*, pp. 3349–3363.

Kostoff, R. N., & Schaller, R. R. (2001). Science and technology roadmaps. *IEEE Transactions on Engineering Management, 48*(2), 132–143.

Kostoff, R. N., Eberhart, H. J., & Toothman, D. R. (1998). Database tomography for technical intelligence: A roadmap of the near-earth space science and technology literature. *Information Processing and Management, 34*(1), 69–85.

Kostoff, R. N., Boylan, R., & Simons, G. R. (2004). Disruptive technology roadmaps. *Technological Forecasting and Social Change, 71*(1–2), 141–159.

Kostoff, R. N., Tshiteya, R., Pfeil, K. M., Humenik, J. A., & Karypis, G. (2005). Power source roadmaps using bibliometrics and database tomography. *Energy, 30*(5), 709–730.

Krawiec, F., Thornton, J., & Edesess, M. (1980, April). An Investigatin of Learning and Experience Curves. In *Solar Energy Research Institute (SERI)* (pp. 19–20).

Lamb, A., Anderson, T. R. (Portland S. U.), & Daim, T.. (2010). Difficulties in R & D target-setting addressed through technology forecasting using data envelopment analysis. In *PICMET '10 Portl. Int. Conf. Manag. Eng. Technol.*, pp. 1–9.

Langfeldt, L. (2004). Expert panels evaluating research: Decision-making and sources of bias. *Research Evaluation, 13*(1), 51–62.

Larsson, A. (2005). Technology strategy formation from a resource-based view: Booz-Allen & Hamilton methodology revisited. Master program thesis, Luleå University of Technology.

Latham, J. R. (1995). Visioning: The concept, trilogy, and process. *Quality Progress, 28*, 65–68.

Lee, G., & Bai, R. (2003). Organizational mechanisms for successful IS/IT strategic planning in the digital era. *Management Decision, 41*(1), 32–42.

Lee, S., Lee, S., Seol, H., & Park, Y. (2008). Using patent information for designing new product and technology: Keyword based technology roadmapping. *R&D Management, 38*(2), 169–188.

Leritz, N., Energy, N., & Alliance, E. (2014). Strategic Energy Management – It's time to grow up! A maturity model for SEM implementation. In *ACEEE Summer Study Energy Efficiency in Buildings*, pp. 187–199.

Lichtenthaler, U. (2008). Opening up strategic technology planning: Extended roadmaps and functional markets. *Management Decision, 46*(1), 77–91.

Lingga, M. M. (2016). Developing a hierarchical decision model to evaluate nuclear power plant alternative siting technologies (PhD Dissertation). Portland State University.

Lingga, M. M. & Kocaoglu, D. F. (2013). Nuclear power plants alternative sitings: A literature review and research gaps. In *2013 Proc. PICMET '13 Technol. Manag. Emerg. Technol.*, pp. 382–392.

Linstone, H. A. (1974). Technology forecasting-planning: Toy or tool? *IEEE Spectr.*, no. April.

Lizaso, F., & Reger, G. (2004). Linking roadmapping and scenarios as an approach for strategic technology planning. *International Journal of Technology Intelligence and Planning, 1*(1), 68–86.

Martino, J. P. (1994). A technology audit: Key to technology planning. In *Proc. Natl. Aerosp. Electron. Conf.*, pp. 1241–1247.

McCarthy, J. J., Haley, D. J., & Dixon, B. W. (2001). Science and technology roadmapping to support project planning. In *PICMET '01. Portland international conference on Management of Engineering and Technology*, Vol. Supplement, pp. 193–194

McDowall, W. (2012). Technology roadmaps for transition management: The case of hydrogen energy. *Technological Forecasting and Social Change, 79*(3), 530–542.

Menon, A., Bharadwaj, S. G., Adidam, P. T., & Edison, S. W. (1999). Antecedents and consequences of marketing strategy making: A model and a test. *Journal of Marketing, 63*(2), 18–40.

Mittenthal, R. (2002). Ten keys to successful strategic planning for nonprofit and foundation leaders. TCC Gr., pp. 2–12.

MOD. (2006). MOD Roadmapping guidance. no. 1.1. Ministry of Defense, UK, pp. 1–12.

Mohaghegh, M., & Shirazi, B. (2017). Strategic assessment of power smart grid technology capabilities and attractiveness: A case study on Iran Power Distribution Company. *International Journal of Innovation and Technology Management, 14*(3), 1–13.

Natalense, J., & Zouain, D. (2013). Technology roadmapping for renewable fuels: Case of biobutanol in Brazil. *Journal of Technology Management & Innovation, 8*(4), 143–152.

Nauda, A., & Hall, D. L. (1991). Strategic technology planning - developing roadmaps for competitive advantage. In *Technology Management : the New International Language*, pp. 745–748.

Ngai, E. W. T., Chau, D. C. K., Poon, J. K. L., & C. K. M. To. (2013). Energy and utility management maturity model for sustainable manufacturing process. *International Journal of Production Economics, 146*(2), 453–464.

NICT. n.d.). Quantum ICT Road Map. *NICT-National Institute of Information and Communications Technology*. [Online]. Available: http://www.nict.go.jp/en/quantum/roadmap.html. Accessed 2 Apr 2018.

Northwest Energy Efficiency Alliance. (2014). Strategic plan: 2015–2019.

Northwest Energy Efficiency Alliance (NEEA). (2013). Strategic energy management maturity model. Northwest Energy Efficiency Alliance (NEEA).

NWCOUNCIL. n.d.-a). Council members. [Online]. Available: https://www.nwcouncil.org/contact/members/. Accessed 7 Apr 2018.

NWCOUNCIL. (n.d.-b). About the seventh power plan. [Online]. Available: https://www.nwcouncil.org/energy/powerplan/7/home/. Accessed 7 Apr 2018.

NWPCC. (2014). Overview of the Northwest Power and Conservation Council's Power Plan development process. Webinar Presentation, Northwest Power and Conservation Council, pp. 1–21.

NWPCC. (2016). Northwest power and conservation council statement of basis and purpose for the seventh power plan and response to comments on the draft seventh power plan. no. May. Northwest Power And Conservation Council, pp. 1–32.

NWPCC. (2018). Draft seventh power plan mid-term assessment report outline. Memorandum, Northwest Power and Conservation Council, pp. 1–5.

NWPCC. (n.d.). Fiscal year 2019 budget and 2018 revision. [Online]. Available: https://www.nwcouncil.org/reports/financial-reports/2017-3/history

Oliveira, M. G., & Fleury, A. L. (2015). A framework for improving the roadmapping performance. In *Portland International Conference on Management of Engineering and Technology*, vol. 2015–Sept, pp. 2255–2263.

Pagani, M. (2009). Roadmapping 3G mobile TV: Strategic thinking and scenario planning through repeated cross-impact handling. *Technological Forecasting and Social Change, 76*(3), 382–395.

Passey, S. J., Goh, N., & Kil, P. (2006). Targeting the innovation roadmap event horizon: Product concept visioning & scenario building. In *ICMIT 2006 Proc. - 2006 IEEE Int. Conf. Manag. Innov. Technol.* (Vol. 2, pp. 604–607).

Paulk, M. C. (2009). A history of the capability maturity model for software. *ASQ Software Quality Professional, 1*(1), 5–19.

Paulk, M. C., Curtis, B., Chrissis, M. B., & Weber, C. V. (1993). Capability maturity model for software – version 1.1. Software Engineering Institute, Carnegie Mellon University, Pittsburgh, Pennsylvania, pp. 1–47.

Paulk, M. C., Curtis, B., Chrissis, M. B., & Weber, C. V. (2006). The capability maturity model for software. *Softw. Eng. Proj. Manag.*, pp. 16–20.

Petrick, I. J. (2008). Developing and implementing roadmaps – A reference guide. [Online]. Available: https://www.sopheon.com/wp-content/uploads/WhitePaper-Petrick-RoadmappingReferenceGuide.pdf

PGE. (2016, November). Integrated resource plan.

Phaal, R., & Muller, G. (2009). An architectural framework for roadmapping: Towards visual strategy. *Technological Forecasting and Social Change, 76*(1), 39–49.

Phaal, R., Farrukh, C. J. P., & Probert, D. R. (2004). Technology roadmapping - a planning framework for evolution and revolution. *Technological Forecasting and Social Change, 71*(1–2), 5–26.

R. Phaal, C. J. P. Farrukh, and D. R. Probert, "Developing a technology roadmapping system," in Technology management: A unifying discipline for melting the boundaries, 2005, 1, pp. 99–111.

Phaal, R., Farrukh, C., & Probert, D. (2006a). Technology management tools: Generalization, integration and configuration. *International Journal of Innovation and Technology Management, 3*(3), 321–339.

Phaal, R., Farrukh, C. J. P., & Probert, D. R. (2006b). Technology management tools: Concept, development and application. *Technovation, 26*(3), 336–344.

Phan, K. (2013). Innovation measurement: A decision framework to determine innovativeness of a company (PhD Dissertation). Portland State University.

Porter, M. E. (2008). Strategy strategy the five competitive. *Harvard Business Review, 86*, 78–94.

Rahimi, N., & Koosawangsri, R. (2013). Selection of smart phone plans. In *PICMET '13 Technol. Manag. Emerg. Technol.*, pp. 426–448 .

Regional Technical Forum. (2015). Roadmap for the assessment of energy efficiency measures.

Saatchi, B., Pham, L., Pham, H., Pai, C. F., & Tran, Y. (2013). Decision model for selecting a sedan car. In *Technol. Manag. IT-Driven Serv. (PICMET),* 2013 *Proc. PICMET'13*, pp. 393–400.

Software Engineering Institute. (2010a). CMMI® for Acquisition, Version 1.3. *Carnegie Mellon University*. pp. 1–8.

Software Engineering Institute. (2010b). CMMI® for Services, Version 1.3. *Carnegie Mellon University*. pp. 1–7.

Software Engineering Institute. (2010c). CMMI for Development, Version 1.3. *Carnegie Mellon University*, pp. 1–60.

Suharto, Y., & Daim, T. (2013). Methods and tools applied in strategic technology planning. In T. Daim, T. Oliver, & J. Kim (Eds.), *Research and technology management in the electricity industry: Methods, tools and case studies* (pp. 1–12). London: Springer.

Tekin, I. M. J. N. (2014). Green index: Integration of environmental performance, green innovativeness and financial performance (PhD Dissertation). Portland State University.

Trainer, J. F. (2004). Models and tools for strategic planning. In *New directions for institutional research* (Vol. 123, pp. 129–138). Wiley Periodicals, Inc..

Tran, T. A., & Daim, T. (2008). A taxonomic review of methods and tools applied in technology assessment. *Technological Forecasting and Social Change, 75*(9), 1396–1405.

UNFCCC. (2013). Background paper on Technology Roadmaps. Technology Executive Committee of the United Nations Framework Convention on Climate Change, pp. 11–45.

United Nations. (n.d.). Strategic planning guide for managers. *United Nations*. [Online]. Available: https://hr.un.org/sites/hr.un.org/files/4.5.1.6_StrategicPlanningGuide_0.pdf.

United States Congress. (1980). *Pacific Northwest Electric Power Planning and Conservation Act*. pp. 1–40.

US Department of Energy. (2013). Guide to community strategic energy planning. no. March. US Department of Energy, pp. 1–95.

US DOE. (2012). Smart grid research & development - Multi-Year Program Plan (MYPP). In *U. S. Dep. Energy - Off. Electr. Deliviry Energy Reliab.*, no. March, pp. 1–63.

Vishnevskiy, K., Karasev, O., & Meissner, D. (2016). Integrated roadmaps for strategic management and planning. *Technological Forecasting and Social Change, 110*, 153–166.

Whalen, P. J. (2007). Strategic and technology planning on a Roadmapping foundation. *Research Technology Management, 50*, 40–51.

Yoon, B., & Lee, S. (2008). Patent analysis for technology forecasting: Sector-specific applications. In *Eng. Manag. Conf.* 2008. *IEMC Eur.* 2008. *IEEE Int. Estoril*, pp. 1–5.

Yuan, H. (2013). A SWOT analysis of successful construction waste management. *Journal of Cleaner Production, 39*, 1–8.

Chapter 2
Technology Policy Roadmap: Big Data Privacy

Maoloud Dabab, Husam Barham, Rebecca Craven, Elizabeth Gibson, and Tuğrul U. Daim

Big Data is increasingly becoming more mature and part of how to do business in almost all sectors. This is due to the valuable insights and analytics Big Data can offer to businesses. However, the expanded use of Big Data is not without risks. One major such risk is the breach of privacy and security of individuals and their data. This chapter examines how to address the privacy data concerns that the pervasiveness of data collection, analysis, and storage creates regarding individuals' ability to control their data. To conduct this analysis, the authors leveraged technology roadmapping (TRM) analysis in combination with quality function deployment (QFD) to assess the data privacy relevant factors. The focus was on the social problems, technologies, resources, and industries that are most relevant to this issue. The findings showed that when it comes to data privacy, the healthcare industry is one of the most critical industries to be considered due to the nature of the data generated through medical processes and technologies. Additionally, the paper found that enforcement mechanisms, specifically in the form of federal enforcement agencies, are the most effective approaches to ensure compliance by actors. It is also realized that there are extenuating political circumstances and increased costs that make the implementation of those policies challenging in the United States.

2.1 Introduction

The advances in information and communication technologies in the previous decade have spurred the generation and use of an unprecedented amount of data about almost everything surrounding our modern life (Acquisti et al. 2015; Kinder

M. Dabab · H. Barham · R. Craven · E. Gibson · T. U. Daim (✉)
Department of Engineering and Technology Management, Portland State University, Portland, OR, USA
e-mail: tugrul.u.daim@pdx.edu

T. U. Daim (ed.), *Roadmapping Future*, Applied Innovation and Technology Management, https://doi.org/10.1007/978-3-030-50502-8_2

2003). In response to this phenomenon, Big Data, a field in information technology, emerged as a viable way to handle and make use of this influx of data. Big Data does not have a single agreed-upon definition in the literature but is most often characterized as an entity by its volume, velocity, variety, veracity, and value. Other definitions addressed Big Data from a process perspective, referring to Big Data as being a holistic information management approach to acquire, clean, integrate, store, and analyze data that comes from multiple internal and external sources, that can be structured or unstructured, to generate insights and analytics to support decision-making (Kshetri 2014; Laney 2001; Bean 2016; Tsai et al. 2016; Barham 2017; Shaygan et al. 2017).

There are four main steps of data processing as part of Big Data: collection, storage, analysis, and usage. The technologies used in these steps are no longer novel and emergent; Gartner has not included Big Data from its annual hype cycle report since 2015 because Big Data has "gone mainstream" (Kho 2018). Big Data technologies depend on continuous streams of data, generated by individuals. Big Data aim is to process the data and find insights to help the business (Dabab et al. 2018a). However, the individuals may or may not be aware that their data is being captured and stored in data sets and is being used to generate analytics. This leads to rising concerns over the privacy and security of individuals' data. Such concerns echo those of previous eras of non-digital data, which are, at this point, well legislated and litigated in the American system. As such, the PCAST report on Big Data and privacy note that "…it is the use of data (including born-digital or born-analog data and the products of data fusion and analysis) that is the locus where consequences are produced" (Report to The President Big Data and privacy: A technological perspective 2014).

This chapter aims to address the privacy concerns that arise from the increasing ubiquity of Big Data. The findings of this study were presented at a conference in 2018 (Dabab et al. 2018b). First, the chapter provides an overview of literature related to the policy background and social problems related to privacy and Big Data. It then proceeds by identifying the most critical industries where Big Data privacy concerns have a severe impact. Then, potential solutions to these concerns are analyzed by using quality function deployment (QFD) and Technology Roadmapping analysis methods. Finally, recommendations are offered.

2.2 Literature Review

Big Data is now being used in most private and public sectors, as it can help solve a range of complex problems (Patterson 2017). However, there are challenges associated with using Big Data. According to literature review, the challenges imposed by Big Data includes issues like, privacy and security, data access and sharing, storage and processing, management, and technical issues (Report to The President Big Data and privacy: A technological perspective 2014; Kaisler et al. 2013; Katal et al. 2013; Sagiroglu and Sinanc 2013).

2.2.1 Big Data and Privacy

Big Data increasingly becomes part of every industry and every sector, as it introduces insights and analytics that can be used to solve a plethora of challenges. Examples include better addressing customers' needs, more accurate human behavior analytics, more effective medical treatment, improving food security, and preventing human trafficking, to name a few (Desouza and Smith 2014; LaValle et al. 2011; Chen et al. 2012). However, it also brings new social challenges with it. These social problems can be grouped into five general categories: privacy and security, data reuse, data accuracy, data access, and archiving and preservation (Report to The President Big Data and privacy: A technological perspective 2014). Each of these social problem areas have challenges related to each Big Data process, at which social externalities can occur and are discussed in both the PCAST report and in the broader Big Data literature (Report to The President Big Data and privacy: A technological perspective 2014).

Privacy and security are concerned primarily with the data being generated by, or about, individuals, including subsequent generation through association with other existing data sets. Data reuse, by contrast, is concerned with the repurposing of data from its intended recipients and processes for other uses. Data accuracy issues can arise when multiple data sources with different controls and verification processes influence the overall quality of the data and the degree to which the data is correct. Data access concerns the individuals and organizations that have access to any data that is part of the Big Data process, including the archiving and preservation of data, which refers to the historical cataloging of data once its initial use has passed. In all these cases, individuals that generate data have little to no control over that data once it comes into digital existence unless technology processes are constrained by social and political processes (Report to The President Big Data and privacy: A technological perspective 2014).

In the literature, concerns about Big Data privacy have explicitly been singled out for additional consideration by industry, government, and research consortia (The Federal Big Data Research and Development Strategic Plan 2016; Bertino and Ferrari 2018; Nunan and Di Domenico 2017). Many aspects of technology have been a party to the challenges to the bounds of the American right to privacy (Report to The President Big Data and privacy: A technological perspective 2014). However, others have noted that information technology poses privacy issues that are unique to digital-native data (Nissenbaum 1997), and that Big Data is particularly pertinent to technology privacy discussions (Van Dijck 2014). Because Big Data has implications for both public and private uses, there are exceptionally substantial implications for the government if policies are not enough to address privacy issues with Big Data (Williamson 2014), especially considering the historical precedent favoring the individual's right to privacy in the United States. Thus, while many aspects of Big Data pose potential risks and social problems, the issues concerning data privacy and the protection of individual data generators are the most important, and hence, will be the center of focus for the remainder of this report.

2.2.1.1 Data Privacy

According to Westin, information privacy is the ability of the individual to control the terms under which personal information is acquired and used. Information here refers to information identifiable to a person. In this context, data and information refer to the same thing, and this definition can be used for Big Data privacy as well (Westin 1968; Culnan and Armstrong 1999). Another popular definition by the Generally Accepted Privacy Principles (GAPP) standard is "the rights and obligations of individuals and organizations concerning the collection, use, retention, disclosure, and disposal of personal information" (Chen and Zhao 2012).

Data privacy concerns revolve around individuals losing the ability to control information related to them, by either unauthorized access to the data due to security breaches or using the data for purposes, the individual is not aware of, when initially providing consent. Regulations protecting privacy govern what can be revealed under different circumstances. However, such regulations vary by country and region. For example, in the United States, health records have strict regulations about how they can be used and shared (Nunan and Di Domenico 2017; Jagadish et al. 2014; Culnan and Armstrong 1999). Implementing privacy within Big Data is not a straightforward task; Big Data depends on sharing and consuming information from multiple internal and external sources to generate real value (Tsai et al. 2016; Sivarajah et al. 2017). This sharing of information among entities exposes data to increased risks of unauthorized access and unauthorized use. Furthermore, there is the possibility of identifying individuals, even if their identity is removed. For example, an attacker, with access to data representing traffic routes by individuals can analyze patterns of traffic routes to identify home and workplace of individuals and then cross-check that with targeted individuals, and then use that information in harmful or illegal ways, even if the source data is anonymized (Culnan and Armstrong 1999).

2.2.1.2 Data Privacy in Different Industries

Each industry has unique Big Data privacy concerns due to how Big Data is utilized in that sector. Following is a review of the leading Big Data privacy concerns in several industries where Big Data is widely adopted, including healthcare, telecommunication, retail, education, and utility.

Healthcare Patient medical records can be used by Big Data to offer tremendous insights, allowing for better diagnosis and treatment decisions. However, failing to address patients' privacy adequately could have a severe impact on patients related to their jobs, health insurance, and social life (Shaygan 2018; Murdoch and Detsky 2013).

Telecommunication Big Data is used to capture and analyze customers' communication and mobility data. Such Big Data systems help telecommunication companies to offer better customer service and to better target customers from marketing and sales points of view. However, failing to address customers' privacy adequately could result in breaches that expose business and personal secrets and plans, leading to economic losses and personal embarrassments (Bughin 2016; Kshetri 2014).

Retail There is tremendous growth in the amount of retail data being generated. Big Data offers valuable insights to retailers, like the ability to better target customer needs, as well as engage in the more efficient supply chain, operations, and inventory management. However, failing to address customers' privacy adequately could result in exposing customers' information and habits, which might have an impact on their jobs and personal lives (Nelson and Olovsson 2016).

Education Big Data has several educational applications, including offering better insights that can help in enhancing students' performance and offering students with educational methods that are customized to their individual skills. Furthermore, Big Data can play an essential role in addressing the higher education retention phenomenon. However, failing to address students' educational records adequately privacy could have a severe impact on their future jobs and future graduate studies (West 2012; Savin-Baden 2015).

Utility Big Data is used to enable smart grid initiatives by collecting and analyzing tremendous amounts of data about how power is being used by customers. Smart grid initiatives have many applications. For example, a smart grid can enable better forecasting of power demands and how to efficiently respond to it. Also, a smart grid allows for better use of intermittent renewable resources. However, like with other industries, failing to adequately address customers' records privacy could result in exposing personal information and make customers vulnerable for identity theft and other kinds of data attacks (Savin-Baden 2015; Chaichi et al. 2015; Song et al. 2013).

2.2.1.3 Examples of Big Data Privacy Cases

There are many known cases where Big Data resulted in legal or ethical violations of privacy that had severe impacts. Two well-known cases illustrate how Big-Data-related privacy breaches can have serious impact. Considered first is the Equifax case, which represents a data cybersecurity breach; then, the Target case that represents unauthorized use of data (Reddy 2014; Armerding 2017): Equifax is one of the most significant credit bureaus in the United States. One of Equifax's websites was hacked in 2017, resulting in the leak of more than a hundred million consumers' data that include their social security numbers, birth dates, and addresses, among other kinds of sensitive data, which can be used for identity theft. Another case is related to Target, the giant retailer. In 2014, Target implemented a Big Data system that can analyze purchasing patterns by customers to make predictions about them. Because of this system, a man received pregnancy-related promotions addressed to his teenage daughter, who had not yet disclosed her pregnancy. The system invaded the teenager's privacy as it used her purchase patterns without her consent to reach conclusions about her health and exposed this information to her father without her approval.

2.2.2 Policy Background

The full policy background on Big Data in the United States includes aspects of cybersecurity and privacy policies that are not necessarily specific to Big Data. Cybersecurity-specific policies are particularly well documented in the literature (Hart and Feenberg 2014). Additionally, much of the analysis of Big Data is from a legal and not technical perspective (Jahanian 2014). As such, the policy infrastructure primarily refers to existing privacy protection policies that are in place, which may not deal with digital privacy at all; these policies also operate under outdated assumptions of data isolation to specific contexts and fields (Horvitz and Mulligan 2015). In general, ingrained policies and protections are neither Big Data nor even technology-specific and do not reflect the evolving policy concerns that accompany each advance in information technology's abilities to collect and analyze data (Jeff Smith et al. 2011).

In the United States, policies regarding privacy and Big Data can be implemented at two levels: federal and state. Treatments of US Big Data policy generally show a lack of coherent federal policy that crosses sectors (Jahanian 2014; Stough and McBride 2014), with most policies specifically focusing on particular sectors like healthcare, education, and financial institutions. Data policies are contained within broader privacy legislation like HIPAA, FERPA, and GLBA/FCRA for those sectors, respectively. There are some precedents for state-level policies regarding data protection and privacy issues in the United States, though the scope of these laws and the strength of enforcement mechanisms vary considerably (Spidalieri 2015). California is mainly active in this regard, with statewide, cross-sector legislation like the California Online Privacy Protection Act (CalOPPA) addressing digital data privacy (California Online Privacy Protection Act (CalOPPA) 2015). However, this type of legislation remains uncommon. This process of distinguishing between states has the effect of further fragmenting data privacy policies and their reach.

Just as there are no comprehensive Big Data regulations at the federal level, there are also nobodies tasked specifically with Big Data compliance, monitoring, or enforcement. Legislation concerning data privacy is full of recommendations and self-enforcement requirements, but little in the way of coercive inducements to follow the guidelines or coordination to ensure equal adherence to those recommendations and requirements (Newman and Bach 2004). Given the acknowledged tension between corporate and consumer interests regarding data collection and usage (Culnan and Bies 2003), self-enforcement seems to protect corporate big data use over consumer privacy by default. Therefore, this policy regime also fails to explicitly consider the privacy concerns of individuals. This is not merely a corporate problem; there are also federal agencies that are engaged in Big Data research in cooperation with private industry (OSTP and NITRD are particularly crucial in this respect), though the research is actually carried out by legislatively created agencies that have much broader missions (such as NASA, EPA, NOAA all fit this characterization). Thus, there is a multitude of agencies and laws that impact data privacy, though the effects are inconsistent across jurisdictions and industries.

Some countries and regions are ahead of the United States in terms of the data privacy policies that they have either implemented or are in the process of implementing. Greenleaf (Greenleaf 2014) offers a comparative look at privacy laws, though most are not Big Data specific. Perhaps the best example of data privacy policies from which the US might learn is the European Union, which has had a binding, cross-sectoral Directive on the Protection of Personal Data since 1995 (Bowie and Jamal 2006). The new General Data Protection Regulation (GDPR), which takes effect in May 2018, represents further industry-spanning comprehensive Big Data policies that affect all data within the geographical territory and clarify existing conceptual understandings around matters like "consent" that would affect anyone generating, transmitting, or storing data in the EU (Elias 2014). Other countries are also considering implementing similar laws, which indicates that these types of policies might be possible to consider in the US context.

The gaps in the current US policies and mechanisms regarding Big Data privacy are identified in the PCAST report and what is implemented elsewhere. The United States has only focused on industry-specific privacy and data protection legislation and regulation (HIPAA, FERPA for instance) and not the comprehensive approaches like those in the EU. Given the recommendations of the PCAST report (Report to The President Big Data and privacy: A technological perspective 2014), especially recommendations 2 and 5, it is apparent that in order to meet increasing social needs regarding technology in the United States, a comprehensive policy is needed to address privacy issues that exist across jurisdictions and sectors.

2.3 Methodology

To analyze the policy solutions that could be applied to the social problems created by Big Data, this study utilized the technology roadmapping methodology (TRM), taking the multi-organizational approach as outlined in Phaal et al. (2004). Technology roadmapping is a comprehensive approach for strategic planning to integrate science/technological considerations into business planning and provide a way to identify new opportunities to achieve the desired objective from the development of new technologies (Phaal et al. 2004; Amer and Daim 2010; Vishnevskiy et al. 2016).

Moreover, Technology roadmapping plans have multiple related layers. Roadmaps are used to identify what needs to be achieved, the barriers and shortcomings preventing its achievement (the gap), and what needs to be done to overcome those barriers and shortcomings. A technology roadmap of social problems and technologies can provide insights for policymakers, industry leaders, and private citizens regarding the social challenges related to a precise technology, what type of actions needed to address those challenges and what is missing, or need further development, to address in act those actions.

This study used the technology roadmapping method to properly analyze and find out: What are the gaps in proper addressing of the privacy issues related to Big Data. Besides, what is needed to address them? Furthermore, to provide a prioritized list that is most related to help build a relevant technology roadmap, the study utilized the quality function deployment (QFD) method. QFD is defined as "a systematic way of ensuring that the development of product features, characteristics, and specifications, as well as the selection and development of process equipment, methods, and controls, are driven by the demands of the customer or marketplace" (Eureka and Ryan 1994, p. 2). Akao originally introduced the method at the Tokyo Institute of Technology in the late 1960s. Under this method, a house of quality (HoQ) is built, which consists of a matrix with alternatives presented in one dimension and the desired characteristics in another dimension. The cells of the matrix in the QFD method represent the weight of each alternative in achieving each desired characteristic by experts. QFD is useful in transferring the qualitative judgment of experts into quantitative parameters. In this study, using HoQ, experts determined how well each alternative meets the requirements on each of four levels. QFD is used at each level of the roadmap, starting from policies against social problems, then industry fields against each policy, then resources against each industry, and finally, technologies against each resource (Adiano and Roth 1994; Chan and Wu 2002; Mizuno et al. 1994; Sheikh 2017). The technology roadmap was divided into four phases.

As Big Data has vast effects on privacy across all sectors, and the literature varies considerably regarding projections and levels of certainty, it would be proper to have a technology roadmap with several relevant phases. Additionally, as outlined above, the groundwork is already laid for sector-specific approaches to data and privacy issues. Pavolotsky indicated that "… Because no two businesses are the same, if privacy policies are the same or substantially similar, at least one of the privacy policies is not on point" (Pavolotsky 2013). Therefore, we divide our roadmapping into four phases, which include social problem identification, policy analysis, industry assessment, technology assessment, and resource allocation as shown in Figs. 2.1, 2.2, 2.3, and 2.4, which leads us to identify a primary industry in which the need for policy change is most pertinent, and a prioritized list of technologies and resources.

The cells of the matrix in the QFD method were filled based on the experience of the authors since some of them have long experience with technology policy, and others have strong technical background and experience related to Big Data. Additionally, the information that they gained from the literature review helped them to give more accurate ratings. The weights represent how well each alternative addresses the requirements, and the index of weights is illustrated in Figs. 2.1, 2.2, 2.3, and 2.4. Then, the total scores for each alternative are calculated to identify the

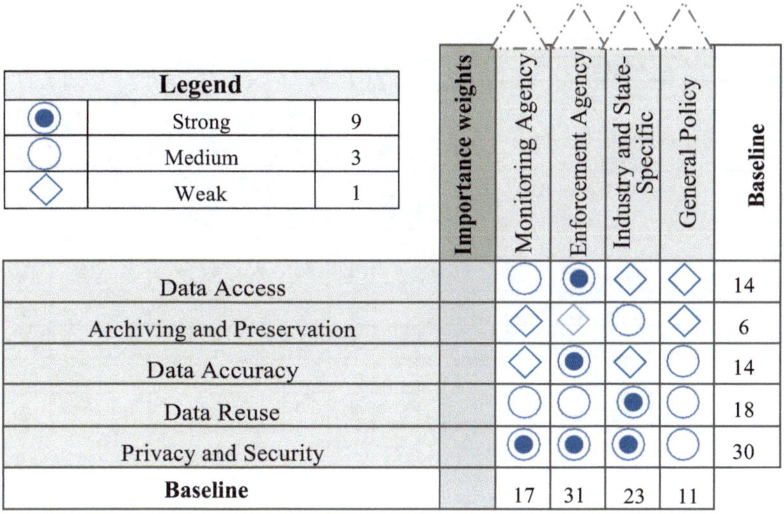

Fig. 2.1 First stage

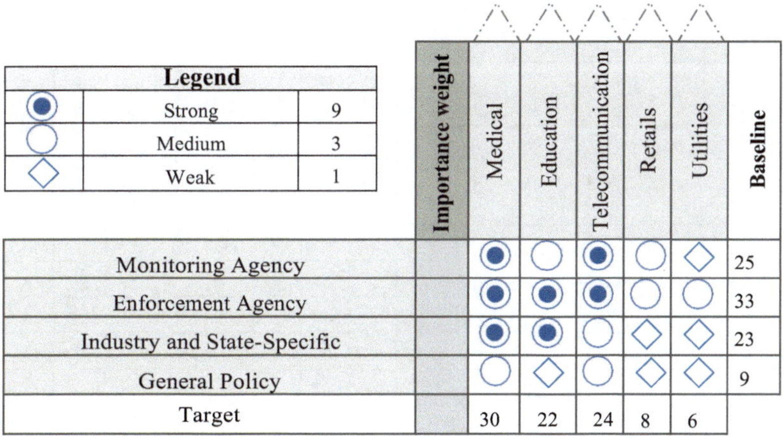

Fig. 2.2 Second stage

best alternatives. This analysis method was used throughout the four TRM phases to narrow down the aspects of Big Data and the policy environment that are most pertinent to a roadmap. Therefore, QFD helped in building a more accurate technology roadmap, by identifying the most important alternatives for each phase in the map and hence shedding light on the industries that are most affected by Big Data privacy challenges, and what actions should be done to address those challenges.

Legend		
◉	Strong	9
○	Medium	3
◇	Weak	1

	Importance weight	The NITRD Agencies	Health Care Provider Organization	Health Insurance Organization	Industry Organization	Legal Organization	Civil Sociality Organization	ONCHIT	Baseline
Medical		◉	◉	◉	◉	◇	◉	◉	55
Education		◉	◉	○	◉	◇	○	◇	35
Telecommunication		◉	◉	◇	◉	○	◉	◇	45
Retails		○	○	◇	○	◇	◇	◇	13
Utilities		◉	◇	◇	○	◇	◇	◇	17
Target		39	31	15	33	7	23	13	

Fig. 2.3 Third stage

Legend		
◉	Strong	9
○	Medium	3
◇	Weak	1

	Importance weight	All-Media Metrics	IoT Real-Time Analytics	IP	Predictive Analytics	Identity-Based Encryption	Network Security System	Baseline
The NITRD Agencies		◉	◉	◇	◉	◉	◉	46
Health Care Provider Organization		○	○	◇	○	◉	◉	28
Health Insurance Organization		◇	◇	○	◉	○	◉	23
Industry Organization		◉	○	○	◉	◉	◉	42
Legal Organization		◇	○	◇	◉	◇	◇	16
Civil Sociality Organization		◇	○	◇	○	◉	○	18
ONCHIT		◉	◉	○	◉	◇	◉	40
Target		33	31	13	51	41	49	

Fig. 2.4 Fourth stage

2.4 Analysis

Four phases of QFD analysis were used to generate the findings. The purpose of these phases of analysis is to identify in which industry Big Data privacy issues are more pertinent and what do policies need to address it and then to identify resources necessary for policy change to occur. The categories included in each analysis are based on the literature review and authors' experience (see above), as well as the PCAST report (Report to The President Big Data and privacy: A technological perspective 2014). In each stage, between four and seven characteristics were ranked on a scale of strong, medium, and weak, with strong indicating the most pronounced effects based on each pairing. The analysis tables for all four phases are shown above. In the first phase, policies that include enforcement mechanisms or enforcement agencies were selected as the most effective in addressing privacy and security issues. In the second phase, the medical industry was selected as the industry in which the effects of data policies are most felt. In the third phase, in terms of policy resources, industry organizations and NITRD agencies were selected as having the strongest role to play in policymaking that will affect data privacy and security in the medical field. Finally, in the fourth phase, the specific applications of Big Data technology that require the most consideration regarding privacy were identified as Internet of Things, Real-Time Analytics, and Predictive Analytics. Literature bolsters these findings; for example, Roski et al. (2014) identified these technologies as providing the most significant opportunities for advancement but also the greatest threats to privacy and security for individuals, in the healthcare context. Figures 2.1, 2.2, 2.3, and 2.4 show the first to fourth phases, respectively.

The information from the QFD analysis was then compiled into the technology roadmap framework to provide a pictorial representation of the analysis. Using the typology of roadmaps articulated by Phaal et al. (2004), these are best described as strategic planning roadmaps that use four main elements (social problem, policy, technology, and resources) that were then extended with elements specific to this analysis: addressing Big Data-related privacy issues. Because there is a variety of roadmap types, several depictions of technology roadmaps are included here. The QFD analysis informs each level and organizes the resulting information slightly differently. These roadmaps provide insight into the feedback loops that exist between technology, industry, and policy as Big Data becomes more ingrained in the processes of the healthcare industry. The steps in the development and adoption of any technology necessarily inform each other, and our analysis of relevant literature revealed the linkages shown in these roadmapping graphics (see Fig. 2.5).

2.5 Results and Discussion

Figure 2.5 shows that an enforcement agency with the ability to impose penalties on corporations, organizations, and industries that fail to adequately protect information is the best approach for effectively protecting individual privacy and providing data security. A good example of this approach is the enforcement mechanisms that are being implemented in the European Union (Raymond 2013). Indeed, enforcement agencies play an influential role in Europe to ensure the movement of goods, persons, services, and capital. Beyond EU-level enforcement, some countries also empower their state agencies to enact and carry out control of domestic

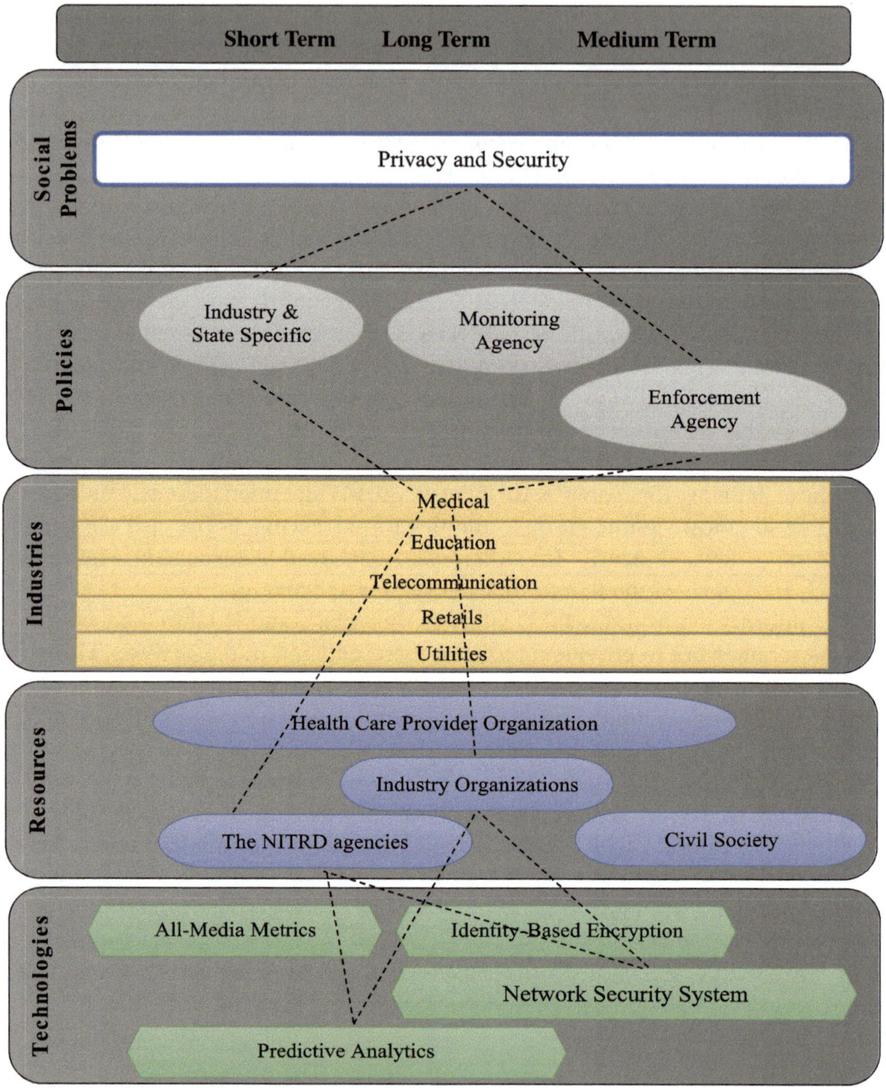

Fig. 2.5 Technology policy roadmapping frame

enforcement; one example of such an agency is the Sweden National Tax Board, which controls both civil and public matters (Kennett 2005). At the same time, the European Union has taken several steps to create data protection laws and enforcement agencies for protecting security and achieving justice. An excellent example of this is the European Union Agency for Law Enforcement Cooperation (Europol), which was created in 1998 to coordinate criminal intelligence and combat terrorism. The main goal behind Europol is to obtain a safer environment for the benefit of all the EU citizens through the coordination of information from numerous organizations (Kaunert et al. 2015).

While the policy tool emerged from the QFD analysis as the policy with the most effective approach to protect privacy, it also may face political feasibility issues that could derail it in its entirety. Nevertheless, there are examples of enforcement agencies in many sectors in the United States. One such example is the Consumer Financial Protection Bureau (CFPB). In the context of the 2008 economic crisis, the CFPB was established under the Dodd-Frank Act with the objective of protecting consumers' financial interests from "unfair, deceptive, or abusive acts or practices" of financial entities. The CFPB supervises financial entities' practices and has the authority to enforce policies by taking actions against violating financial entities. Over the years, CFPB issued fines totaling hundreds of millions of dollars and returned billions of dollars to harmed consumers; its enforcement prerogative allowed for better consumer protection (Townsend 2013; CFPB.gov 2019). However, even this agency remains susceptible to political feasibility issues, as legislation to repeal the Dodd-Frank Act and permanently remove all enforcement mechanisms from the CFPB gains momentum in Congress under the tutelage of the Trump administration.

The analysis raises several interesting points:

- While many fields are affected by Big Data, the healthcare industry poses both one of the most significant opportunities for Big Data to add value, as well as one of the biggest threats to individual privacy. Personal health data is considered among the most private and protected by strict laws, especially in the USA.
- The compilation of diverse data via typical processes and advancements like IoT-connected devices through processes that may or may not be evident to individuals generating data are new challenges to data that do not have an analog counterpart.
- The use of these data streams for predictive analytics purposes also is a more significant concern given the sensitivity of this data. As the history of privacy in the United States shows, this is an evolving legal process that will continually be challenged as technologies develop, and new uses for data are found.
- Data governance is a relatively new concept in the public sector in particular, which will undoubtedly result in the emergence of additional social problems around the related values, risks, and costs of Big Data that need to be considered while developing and implementing more Big Data governance policies (Tallon 2013).
- European Union Agency for Law Enforcement Cooperation is a significant departure from current US policies at all levels, which provide guidelines to industries but ultimately rely on self-reporting and the voluntary provision of compensation or civil litigation in the absence of this compliance. Criminalization of negligent data handling, like that proposed in the European Union, provides more significant incentives to comply but also requires political capital to facilitate such a significant policy shift.

2.6 Recommendations

Big Data policies that provide consistency across industries are desirable, as they create consistent expectations for individuals that are generating the data that is used in data sets and predictive analytics. For this reason, progress in the healthcare sector can be beneficial to other industries as well, as healthcare often provides a blueprint for other sectors seeking to protect individuals' data.

Some primary recommendations arise from the analysis. The United States government should keep funding of the agencies that participate in Big Data research and policy formation, including the NITRD, which is an organization that is created through executive order, so it is subject to presidential renewal with each new administration. NITRD also created a 5-year research plan for Big Data, much of which focuses explicitly on the privacy and security of data. Continuing to research the impacts and possible policies associated with Big Data in a collaborative way is imperative to protecting current and future data streams, and ensuring the continued existence and funding of agencies that are committed to this mission seems the best way to do this.

The complexity of the policy environment surrounding Big Data governance is a factor that makes a single analysis method unlikely to capture the full scope of the opportunities and threats that are facing all actors. We, therefore, recommend that additional research continue combining methods of analysis to attempt to better understand the extent of the interacting aspects of the ever-changing technology environment. While the opportunities that come from Big Data analytics are well documented, the particular technology and policy methods that can be used to best ensure data security and privacy remain unclear.

2.7 Conclusions and Future Work

Big Data poses new security and privacy risks as more enormous quantities of data are being harvested, processed, and used, often without the knowledge of the individuals creating the data streams. The United States has a complicated history with privacy that is only exasperated by the speed at which Big Data technology is adopted in industries that already collect and compile sensitive data. The use of Big Data is a primary concern for consumers, and those managing data privacy in general and modernization of privacy policies and data protection measures are needed to ensure individuals that their data are safe. The summary of the background information, QFD analysis, and critical aspects of the technology roadmapping are shown in Figs. 2.5 and 2.6.

This analysis shows that enforcement mechanisms in the form of agencies that can leverage civil and criminal penalties against those that fail to adequately guard the data generated by individuals provide the most robust protection. However, the social, political, and technical complexity of the policy environment continues to

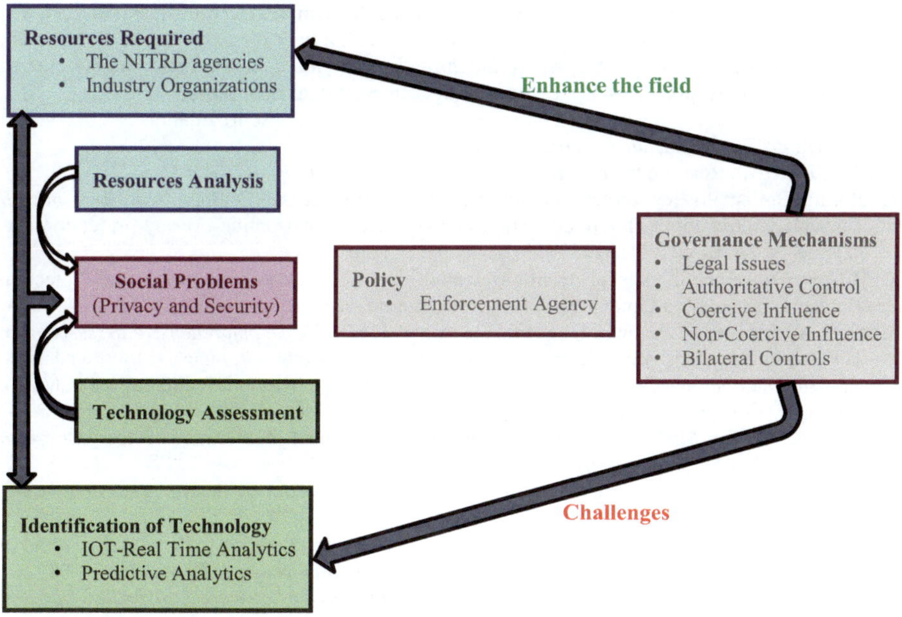

Fig. 2.6 Technology policy roadmapping frame

increase, and only time will tell if this analysis provides insight that is either possible or feasible to implement in the United States.

Additional research on Big Data's relationship to data privacy and security should also continue. Future research should incorporate the costs of technologies and policies that can ensure data security, which is acknowledged but not a part of this analysis. The legal and political costs of policy change in this area also warrant further analysis, as even the PCAST report acknowledges that "privacy protection cannot be achieved by technical measures alone" (Report to The President Big Data and privacy: A technological perspective 2014).

References

Acquisti, A., Brandimarte, L., & Loewenstein, G. (Jan. 2015). Privacy and human behavior in the age of information. *Science, 347*(6221), 509–514.

Adiano, C., & Roth, A. V. (1994). Beyond the house of quality: Dynamic QFD. *Benchmarking for Quality Management & Technology, 1*(1), 25–37.

Amer, M., & Daim, T. U. (Oct. 2010). Application of technology roadmaps for renewable energy sector. *Technological Forecasting and Social Change, 77*, 1355–1370.

Armerding, T. (2017, October 11). The 16 biggest data breaches of the 21st century. CSO Online. [Online]. Available: https://www.csoonline.com/article/2130877/data-breach/the-16-biggest-data-breaches-of-the-21st-century.html. Accessed: 03 Dec 2019.

Barham, H. (2017). Achieving competitive advantage through big data: A literature review.

Bean, R. (2016, February). Just using big data isn't enough anymore. Harvard Business Review.
Bertino, E., & Ferrari, E. (2018). Big data security and privacy. In S. Flesca, S. Greco, E. Masciari, & D. Saccà (Eds.), *A comprehensive guide through the Italian database research over the last 25 years* (Vol. 31, pp. 425–439). Cham: Springer International Publishing.
Bowie, N. E., & Jamal, K. (2006). Privacy rights on the internet: Selfregulation or government regulation? *Business Ethics Quarterly, 16*(3), 323–342.
Bughin, J. (2016). Reaping the benefits of big data in telecom. *Journal of Big Data, 3*(1), 14.
California Online Privacy Protection Act (CalOPPA). July 2015. [Online]. Available: https://consumercal.org/about-cfc/cfc-educationfoundation/california-online-privacy-protection-act-caloppa-3/. Accessed 3 Dec 2019.
CFPB.gov. Consumer financial protection Bureau official website. Consumer Financial Protection Bureau. [Online]. Available: https://www.consumerfinance.gov/. Accessed 3 Dec 2019.
Chaichi, N., Lavoie, J., Zarrin, S., Khalifa, R., & Sie, F. (2015). *A comprehensive assessment of cloud computing for smart grid applications: A multi-perspectives framework*. Presented at the Portland International Conference on Management of Engineering and Technology (PICMET), Portland, USA, pp. 2541–2547.
Chan, L.-K., & Wu, M.-L. (2002). Quality function deployment: A literature review. *European Journal of Operational Research, 143*(3), 463–497.
Chen, D., & Zhao, H. (2012). Data security and privacy protection issues in cloud computing, pp. 647–651.
Chen, H., Chiang, R. H. L., & Storey, V. C. (2012). Business intelligence and analytics: From big data to big impact. *MIS Quarterly: Management Information Systems, 36*(4), 1165–1188.
Culnan, M. J., & Armstrong, P. K. (Feb. 1999). Information privacy concerns, procedural fairness, and impersonal trust: An empirical investigation. *Organization Science, 10*(1), 104–115.
Culnan, M. J., & Bies, R. J. (2003). Consumer privacy: Balancing economic and justice considerations. *Journal of Social Issues, 59*(2), 323–342.
Dabab, M., Freiling, M., Rahman, N., & Sagalowicz, D. (2018a). A decision model for data mining techniques. In *2018 Portland International Conference on Management of Engineering and Technology (PICMET)*.
Dabab, M., Craven, R., Barham, H., & Gibson, E. (2018b). Exploratory strategic roadmapping framework for Big Data privacy issues. In *2018 Portland International Conference on Management of Engineering and Technology (PICMET)*.
Desouza, K. C., & Smith, K. L. (2014). Big Data for social innovation. *Stanford Social Innovation Review, 2014*, 39–43.
Elias, P. (2014). A European perspective on research and big data analysis. *Privacy, Big Data, and the Public Good: Frameworks for Engagement, 1*, 173–191.
Eureka, W. E., & Ryan, N. E. (1994). *The customer-driven company: Managerial perspective on quality function deployment* (2nd ed.). Dearborn, Mich.: Burr Ridge, Ill: ASI Press; Irwin.
Greenleaf, G. (2014). Sheherezade and the 101 data privacy laws: Origins, significance and global trajectories. *Journal of Law Information and Science, 23*, 4.
Hart, C., & Feenberg, A. (2014). The insecurity of innovation: A critical analysis of cybersecurity in the United States. *International Journal of Communication, 8*, 19.
Horvitz, E., & Mulligan, D. (2015). Data, privacy, and the greater good. *Science, 349*(6245), 253–255.
Jagadish, H. V., et al. (Jul. 2014). Big data and its technical challenges. *Communications of the ACM, 57*(7), 86–94.
Jahanian, F. (2014). The policy infrastructure for Big Data: From data to knowledge to action. *ISJLP, 10*, 865.
Jeff Smith, H., Dinev, T., & Heng, X. (2011). Information privacy research: An interdisciplinary review. *MIS Quarterly, 35*(4), 989–1015.
Kaisler, S., Armour, F., Espinosa, J. A., & Money, W. (2013). Big Data: Issues and challenges moving forward. In *2013 46th Hawaii International Conference on System Sciences*.

Katal, A., Wazid, M., & Goudar, R. H. (2013). Big data: Issues, challenges, tools and good practices. *2013 Sixth International Conference on Contemporary Computing (IC3).*

Kaunert, C., Sarah, L., & Occhipinti, J. D. (2015). *Justice and home affairs agencies in the European Union*. London: Routledge.

Kennett, W. A. (2005). *Enforcement of judgments in Europe*. Oxford: Oxford University Press.

Kho, N. D. (2018). The state of Big Data 2018. *EContent Magazine*. [Online]. Available: http://www.econtentmag.com/Articles/Editorial/Feature/The-State-of-Big-Data-2018-122572.htm. Accessed 03 Dec 2019.

Kinder, D. R. (2003). Communication and politics in the age of information. In *Oxford handbook of political psychology* (pp. 357–393). New York: Oxford University Press.

Kshetri, N. (2014). Big data' s impact on privacy, security and consumer welfare. *Telecommunications Policy, 38*(11), 1134–1145.

Laney, D. (2001). 3D data management: Controlling data volume, velocity and variety. *META Group Research Note, 6*, 70.

LaValle, S., Lesser, E., Shockley, R., Hopkins, M. S., & Kruschwitz, N. (2011). Big data, analytics and the path from insights to value. *MIT Sloan Management Review, 52*(2), 21.

Mizuno, S., Akao, Y., & Ishihara, K. (Eds.). (1994). *QFD, the customer-driven approach to quality planning and deployment*. Tokyo, Japan: Asian Productivity Organization.

Murdoch, T. B., & Detsky, A. S. (2013). The inevitable application of Big Data to health care. *JAMA, 309*(13), 1351.

Nelson, B., & Olovsson, T. (2016). Security and privacy for big data: A systematic literature review. In *2016 IEEE International Conference on Big Data (Big Data).*

Newman, A. L., & Bach, D. (2004). Self-regulatory trajectories in the shadow of public power: Resolving digital dilemmas in Europe and the United States. *Governance, 17*(3), 387–413.

Nissenbaum, H. (1997). Toward an approach to privacy in public: Challenges of information technology. *Ethics & Behavior, 7*(3), 207–219. Fig. 7. The Roadmapping feedback loop.

Nunan, D., & Di Domenico, M. (2017). Big data: A normal accident waiting to happen? *Journal of Business Ethics, 145*(3), 481–491.

Patterson, L. (2017, July 4). What social issues will Big Data solve in 2017?. https://www.technology.org/2017/07/04/what-social-issues-will-bigdata-solve-in-2017/

Pavolotsky, J. (2013). Privacy in the age of big data. *The Business Lawyer, 69*(1), 217–225.

Phaal, R., Farrukh, C. J., & Probert, D. R. (2004). Technology roadmapping—A planning framework for evolution and revolution. *Technological Forecasting and Social Change, 71*(1–2), 5–26.

Raymond, A. H. (2013). Data management regulation: Your company needs an up-to-date data/information management policy. *Business Horizons, 56*(4), 513–520.

Reddy, T. (2014, October 22). 7 Big Data blunders you're thankful your company didn't make. Umbel. [Online]. Available: https://www.umbel.com/blog/big-data/7-big-data-blunders/. Accessed 3 Dec 2019.

Report to The President Big Data and privacy: A technological perspective. May-2014. [Online]. Available: https://bigdatawg.nist.gov/pdf/pcast_big_data_and_privacy_-_may_2014.pdf. Accessed 3 Dec 2019.

Roski, J., Bo-Linn, G. W., & Andrews, T. A. (2014). Creating value in health care through big data: Opportunities and policy implications. *Health Affairs, 33*(7), 1115–1122.

Sagiroglu, S., & Sinanc, D. (2013). Big data: A review. In *2013 International Conference on Collaboration Technologies and Systems (CTS).*

Savin-Baden, M. (2015). Education and Big Data. In M. Peters (Ed.), *Encyclopedia of educational philosophy and theory* (pp. 1–7). Singapore: Springer Singapore.

Shaygan, A. (2018). Landscape analysis: What are the forefronts of change in the US hospitals? In T. U. Daim, L. Chan, & J. Estep (Eds.), *Infrastructure and technology management* (pp. 213–243). Cham: Springer International Publishing.

Shaygan, A., Gungor, D. O., Kutgun, H., & Daneshi, A. (2017). Adoption criteria evaluation of activity tracking wristbands for university students, pp. 1–7.

Sheikh, N. J. (2017). Developing a strategic roadmap for policy and decision making: Case study of ICT and disaster risk reduction in public safety networks, pp. 1–7.

Sivarajah, U., Kamal, M. M., Irani, Z., & Weerakkody, V. (2017). Critical analysis of Big Data challenges and analytical methods. *Journal of Business Research, 70*, 263–286.

Song, Y., Zhou, G., & Zhu, Y. (2013). Present status and challenges of big data processing in smart grid. *Power System Technology, 37*(4), 927–935.

Spidalieri, F. (2015). State of the states on cybersecurity. Pell Center for International Relations. Google Scholar.

Stough, R., & McBride, D. (2014). Big Data and US public policy. *Review of Policy Research, 31*(4), 339–342.

Tallon, P. P. (2013). Corporate governance of big data: Perspectives on value, risk, and cost. *Computer, 46*(6), 32–38.

The Federal Big Data Research and Development Strategic Plan. May 2016. [Online]. Available: https://www.nitrd.gov/pubs/bigdatardstrategicplan.pdf. Accessed 3 Dec 2019.

Townsend, A. M. (2013). *Smart cities: Big data, civic hackers, and the quest for a new utopia*. New York: WW Norton & Company.

Tsai, C.-W., Lai, C.-F., Chao, H.-C., & Vasilakos, A. V. (2016). Big Data analytics. In *Big data technologies and applications* (pp. 13–52). Cham: Springer International Publishing.

Van Dijck, J. (2014). Datafication, dataism and dataveillance: Big Data between scientific paradigm and ideology. *Surveillance & Society, 12*(2), 197.

Vishnevskiy, K., Karasev, O., & Meissner, D. (Sep. 2016). Integrated roadmaps for strategic management and planning. *Technological Forecasting and Social Change, 110*, 153–166.

West, D. M. (2012). Big data for education: Data mining, data analytics, and web dashboards. *Governance Studies at Brookings, 4*, 1.

Westin, A. F. (1968). Privacy and freedom. *Washington and Lee Law Review, 25*(1), 166.

Williamson, A. (2014). Big data and the implications for government. *Legal Information Management, 14*(4), 253–257.

Chapter 3
Technology Roadmap: Electric Shavers

Prakharvel Kulshrestha and Tuğrul U. Daim

The male grooming industry is expected to grow from a 5.36 billion dollar industry in 2016 to a 6.45 billion dollar industry by 2021 (Technavio 2017). Male grooming appliances also form a core component for profit generation due to the high margins per unit sale. This is evident from the fact that the 'grooming' category, which includes electric shavers contributes to 16% of the earning for 'Procter & Gamble' while it constitutes only 10% of the net sales. This trend is also visible in the strong performance of the 'Braun' brand for 'Procter & Gamble' that holds 25% of the global share for male dry/wet shaving and grooming appliances. The growth of the division in a sluggish global commodity market has been consistent with single digit volume growth in 2015 and 2016 followed by a double digit growth in 2017 (Procter and gamble 2016, 2015, 2017). Since the category has delivered considerable growth in the past years accompanied by constant profitability and market share increase, there has been a considerable push in terms of research and development to identify the technologies that can further drive growth and market penetration. Currently, 40% of men considered for the survey sport a beard, which shows a dicer correlation to the considerable market growth seen in the shaver industry in 2016 of 45% (Technavio 2017). What gives further hope of consistent growth to the industry in the coming years is the fact that the developing 'APAX' countries contribute to nearly a quarter of the revenue from grooming devices worldwide (Technavio 2017).

There are primarily two options for the design of the primary cutting element of a shaver that have gained traction in the market. The rotary shaving element that houses rotating blade components in a housing that can move as well and depending

P. Kulshrestha
Northern Institute of Technology, Hamburg, Germany

T. U. Daim (✉)
Department of Engineering and Technology Management, Portland State University, Portland, OR, USA
e-mail: tugrul.u.daim@pdx.edu

T. U. Daim (ed.), *Roadmapping Future*, Applied Innovation and Technology Management, https://doi.org/10.1007/978-3-030-50502-8_3

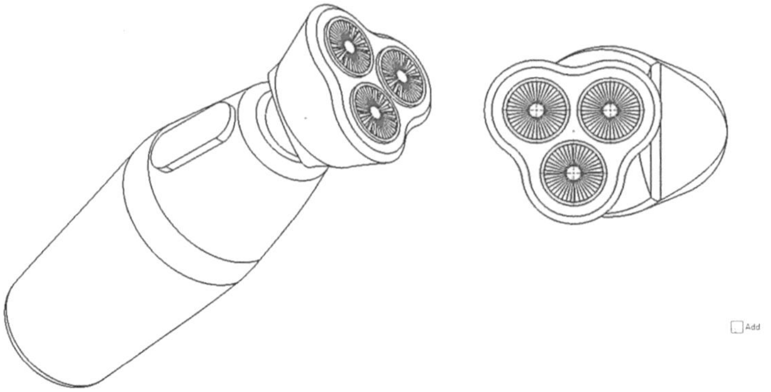

Fig. 3.1 Conventional rotary shaver design – self-drawn – based on the design adapted from Sch et al. (2015)

on the design the shaver can incorporate two, three or even four of these semi-independent cutter housings attached to a main body (Messinger et al. 1991). This concept has been adopted widely by companies like Philips and Remington, and many design improvements such as improved cutting blades and sliding elements have led to widespread acceptance by the market and global volume growth. The linear shaving element that includes a cutting mechanism housed in elongated cutter heads extending in longitudinal directions, which are basically parallel to each other and also incorporate a third parallel element in between which enables the other two cutting heads to pivot along the axis of the third cutting element (Franke and Klauer 1998). This apparatus is also available in similar formats that include no pivoting mechanisms and house a rolling cutter mounted on a spindle in a longitudinal housing that is separated from the skin by a perforated thin sheet and cuts hair that protrude through the holes in the sheet (Tietjens 1979). This concept has been widely adapted and developed by brands like Braun and Panasonic that have cultivated the design over the past four decades. This thesis aims to dig deeper into the customer preferences, which drive product features and the associated technologies (Figs. 3.1 and 3.2).

3.1 Literature Review

The boom in global sales of electric shavers driven by the improving standard of living and trend of facial hair styling has also driven the industry to fast track innovation and research to fend of competition and further grow market shares. It has also been observed that if a firm is unable to either develop innovation from within or work with alliances to develop their products they are bound to see severe setbacks and market share loss (Garcia and Bray 1997). Hence, the process of roadmapping (Hillegas-Elting 2016), which is the technique or process that can be used

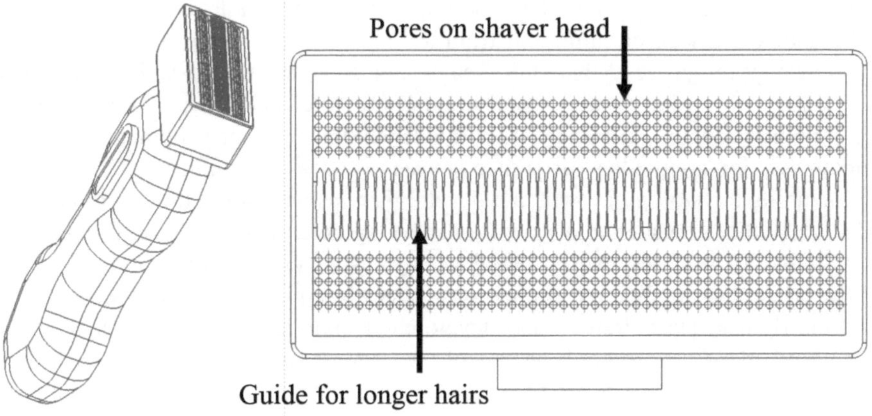

Fig. 3.2 Conventional linear shaver design isometric view (left) and top view (right) – self-drawn – based on a design adapted from Nishizawa (2006)

to get to a target can be utilised here to achieve the objective of developing the blueprint of the next generation male shaving apparatus. There are in addition four essential components: technology roadmapping, portfolio management, project management, and technology transfer, which a firm must manage to effectively balance the technology push and the pull from the market while ensuring profitability and sustainable growth (Barham et al.). The technology roadmap artefact at the end should be a visual representation of the technologies that can be utilised by a firm to achieve its targets in the form of a map, and the process of generating this map is the roadmapping (Barker and Smith 1995).

This has led to the development of many new technologies based on three fundamental choices which can be, firstly, to develop an innovative and novel component or set of components to be used in existing systems to improve customer value creation or it could be the development of a novel system using already existing internally or externally sourced components, and lastly a firm could chose to start from scratch and develop an entirely new system from new components (Petrick and Echols 2004). The first choice of simply switching components is generally used for small-scale innovation and product improvement projects, whereas the other two choices are utilised primarily for large-scale development and improvement projects driven by pull forces in the market and push forces of the firms to establish dominance using breakthrough and innovative products (Petrick and Echols 2004).

To enable successful roadmapping, proper organisational structures, processes, stakeholder engagement and change management techniques need to be used and linked to effective portfolio management in order to deliver a successful roadmap and the associated business deliverables (Cosner et al. 2007). The success factors include having the right expertise at the right time, including the appropriate target audience, clearly defines goals and proper documentation amongst other essentials

(Cosner et al. 2007). On the other side key factors that can deter the roadmapping effort include lack of commitment from top management to the roadmap itself or the roadmapping process, distraction caused by short-term tasks and goals and the resistance caused due to the organisational structure and politics (Lupini 2002). Keeping these factors in mind during the development of the roadmap and during implementation are the key difference between success and failure.

Including technology growth and development as part of the long-term business plan is essential to sustainable growth and development of any organisation (Bitondo and Frohman 1981). It is also essential for any business involving technology as part of the product to amalgamate technology strategy and business strategy in order to find the right growth strategy and development plan (Matthews 1992). It is also imperative that the different branches of the firm are in coherence with each other and aligned in all activities towards the technological and business strategy in order to sustainably deliver growth in the market (Porter 2008).

Several techniques and maps have been used to generate a link between technology development in a firm and the strategy of the firm, and amongst these the most widely used is perhaps the S-curve as shown in Fig. 3.1 (Nieto et al. 1998). The curve reflects the fact that the growth of technology over a period of time is not linear and is influenced by lack of data, research, intention, market pull and funding in the initial phases causing it to slow down and is limited by physical and technological barriers after it reaches maturity (Nieto et al. 1998) (Fig. 3.3).

The S-curve represents the time and effort required versus the technical performance of the product group (Foster 1985). The shape of the curve, however, can be unique to a product or industry depending on the market forces and the scientific knowledge and innovation in the field (Bower and Christensen 1995). The stages of the S-curve can be split into three phases: the introduction stage followed by the growth stage and finally the maturity stage (Sood 2010). The initial stages of introducing any technology to the market are marked by slow technological growth driven by lack of knowledge, market pull forces, and funding in the area (Sood 2010). This stage is then followed by sustained market pull and investments in the field leading to sporadic technology growth and development characterizing the growth stage (Sood 2010). After sporadic growth that can be short lived or drawn

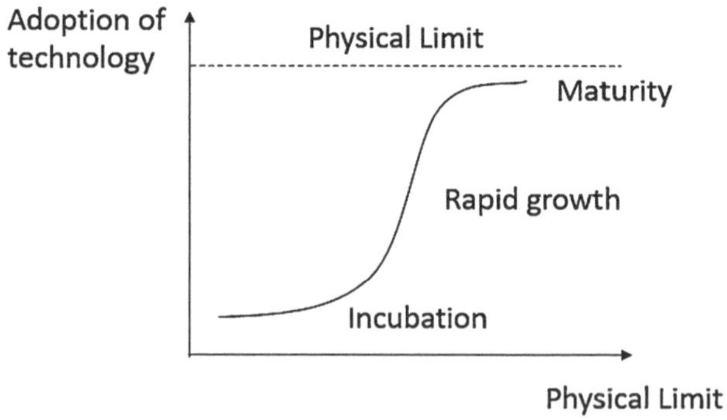

Fig. 3.3 S – curve of technology with respect to time – self drawn

out over decades as seen in the automobile industry, inevitably the field reached a maturity phase where technological growth slows down yet again due to a combination of factors including technological limits, material property boundaries and lack of need of additional features by consumers (Sood 2010).

The industry being analysed in this review is also nearing maturity with over five decades of research that has already developed a strong technological base for the shavers. The key challenges now include developing advanced technologies into a shaver while maintaining cost for customers. Step changes can still be expected in the coming years, which may include laser hair removal and nanotechnology, but these will not be considered within the scope of this thesis as they essentially regard a drastically different product group and share no technological similarities to the shavers currently available in the market. It has also been observed that as the market reached maturity and technological advance is slow, the available technologies compete till a dominant design appears and maintains a technological monopoly known as the 'technological discontinuity' (Anderson and Tushman 1990). This can also be observed in the shaver industry, which has narrowed down to two primary competing designs, namely the linear blade arrangement and the rotary design discussed before. In this phase of market maturity, competing firms strain to prove to consumers the technological superiority of their product, and the marketing method and peripheral aspects such as 'brand value' and 'prestige symbol' play an essential role in the choices made by customers.

The inception of technology roadmapping lies in the early American automotive industry where the 'Motorola' company (Willyard and McClees 1987) and 'Corning' developed methodical approaches to direct the research and development effort in the 1970s and 1980s to attempt to develop a bridge between the rapidly developing technologies and innovative application landscape (Probert and Radnor 2003). The techniques were quickly adapted by various major global firms in the rapidly progressing electronic technologies sector such as 'Philips' (Groenveld 1997), 'Lucent Technologies' (Albright and Kappel 2003) and the 'Semiconductor Industry Association' (Kostoff and Schaller 2001). Technology roadmapping can also be perceived as a planning methodology, which takes into account the unlimited goals, targets and mission of a firm and catalogues the barriers that can be faced along with the corresponding strategies and processes that can be used to overcome the barriers (Winebrake and Creswick 2003).

Technological and scientific roadmaps are used across the academic scientific and administrative disciplines to establish a connection between the technological aspects and applications of a product or service so that limited resources and time available can be allocated in order to achieve the maximum yield by synergizing multiple activities in an organisation in a complex market landscape (Kostoff and Schaller 2001). Figure 3.2 represents a generic roadmap made up of nodes that have both qualitative and quantitative assets from a perspective of an organisation or firm connected via links (Kostoff and Schaller 2001). The figure also shows that the roadmap has a temporal aspect associated with it besides the spatial aspect made up of the various research and development projects that lead to technologies which in turn impact the various products of the firm and the competition in the market

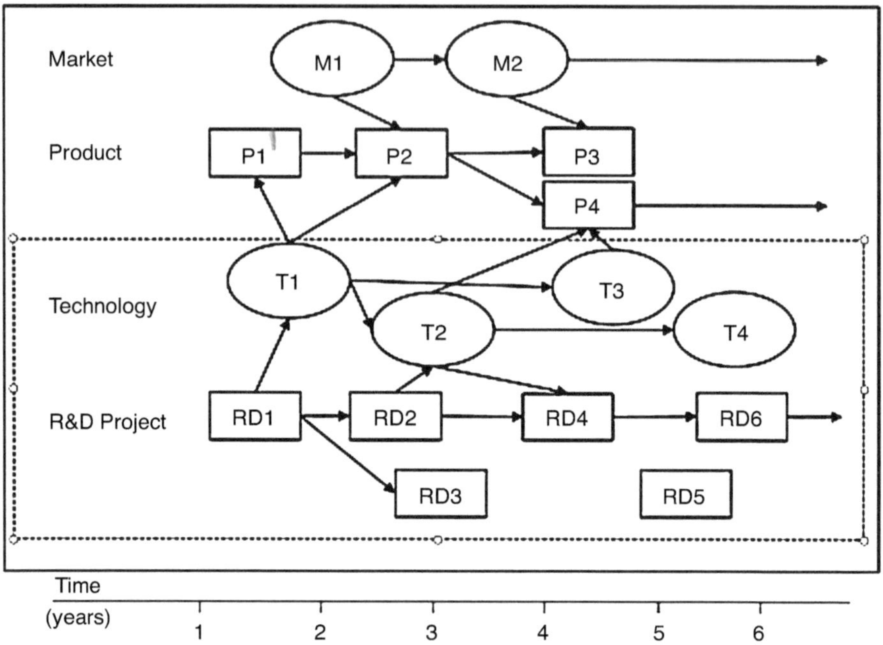

Fig. 3.4 Generic science and technology roadmap with nodes and links adapted from Kostoff and Schaller (2001)

(Kostoff and Schaller 2001). A robust technology roadmap can allow for analysis of any of the spatial factors at a moment in time via the links that connect the different nodes in that time frame (Kostoff and Schaller 2001). Hence, the links themselves need to be defined as vectors with both magnitude and direction in order to establish a full description of the link and the association it generated between two or more spatial nodes (Kostoff and Schaller 2001). In addition, since a roadmap can be used for prospective and retrospective analysis, a robust roadmap should reveal the development history of a product in the market and help analyse future growth (Kostoff and Schaller 2001) (Fig. 3.4).

A compilation of roadmaps that can be used for different industries and scenarios was released as a digest to aid industries undergoing dramatic change bought on by the onset of technological growth and innovation leading to rapid product development and phasing out (Force 1997). Different industries incorporate different types of technology roadmaps varying in degree of detail, timeframe and intended application based on the needs and resources. A naming convention based on the application and objective space was developed classifying the technology roadmaps into four distinct groups shown in Fig. 3.3. These roadmaps fall into four major groups as shown in the figure: science technology roadmaps, industry technology roadmaps, portfolio management roadmaps and corporate technology roadmaps (Fig. 3.5).

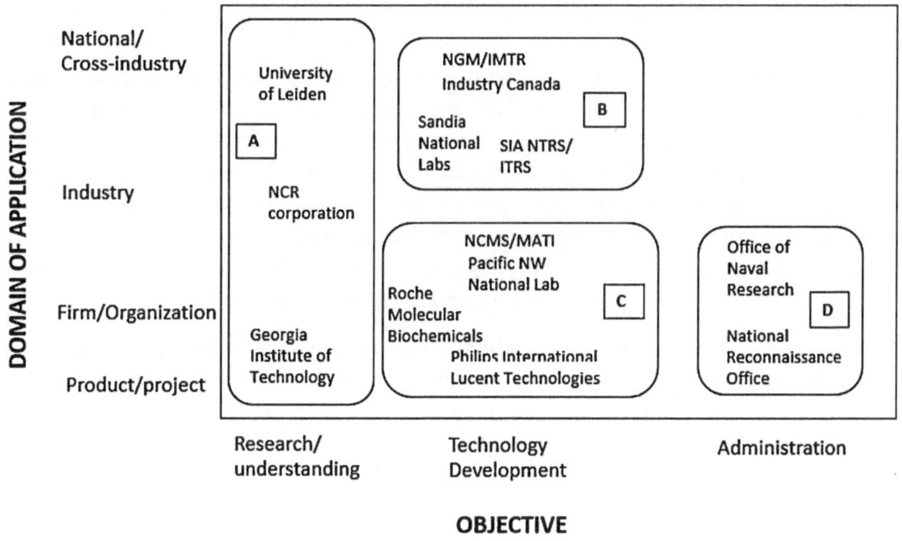

Fig. 3.5 Taxonomy of technology roadmaps adapted from Roadmaps and roadmapping for commercial applications (1998)

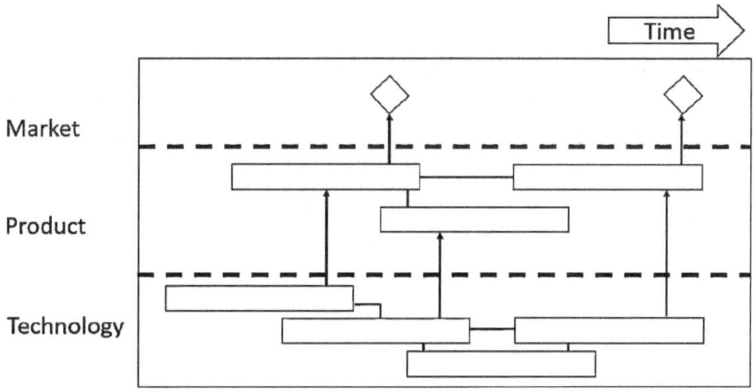

Fig. 3.6 Technology roadmap linking the technologies product and the market adapted from Kostoff and Schaller (2001)

The most commonly used approach to technology roadmapping is depicted is Fig. 3.4, wherein the roadmap consists of a multilayer strategy that connects the technologies that form the basis of the products to the market where the product is to be consumed (EIRMA 1997). The map has a temporal aspect to it to depict the progression of the technologies over time to respond to the maturing market needs (EIRMA 1997). This map can be used to analyse the technological development is the past, as well as the strategies needed for sustainable growth in the future (Fig. 3.6).

Though research data are abundant and technology roadmaps have been proven to be effective in establishing links between the technology progression and business needs and drivers, most firms struggle with practical usage and application of the developed roadmaps. This is perhaps due to the fact that most specific problems require the development of roadmaps tailored to the problem and the needs of the firm using the roadmap. The lack of practical support and help during implementation results in reinvention of the process by firms during development and application. Addressing this flexibility, technology roadmapping can be used to solve a variety of business cases and eight types have been listed depending on the purpose for which the technology roadmap is intended (Phaal et al. 2004).

1. Product planning technology roadmap: This roadmap is widely used and commonly developed to analyse the utilization of different technologies for manufactured products and the development of a variety of products using a common technology base (Fig. 3.7).
2. Service or capacity planning technology roadmap: This type of roadmap as the name suggests is most appropriate for use by service-based industry. It analyses the interaction between the capability and technology existing within firm and the ability to meet business drivers over a period of time. It can be used to identify capability or technology gaps that need to be filled to identify specific business needs. The Fig. 3.6 below shows the roadmap used by the 'Royal Mail' of the United Kingdom to analyse the impact of the developing technologies on the growth of the firm (Fig. 3.8).
3. Strategic planning roadmap: This roadmap is most commonly used to analyse the robustness of a strategy used for a specific business case by analysing the future ideal situation with the current business condition. This type of roadmap helps comparing the opportunities and risks in the market and identifying gaps in the strategy currently defined. This roadmap can be used with a very wide scope

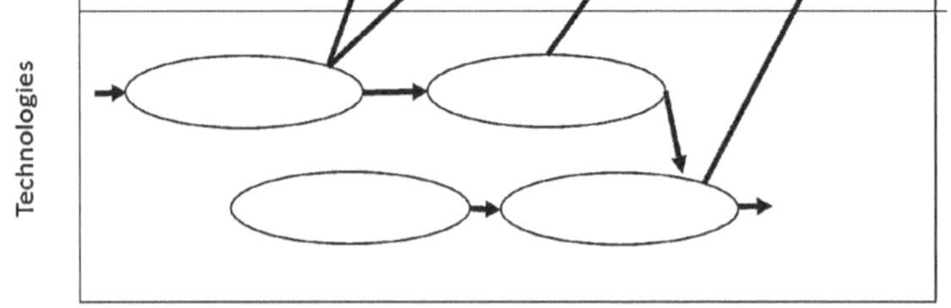

Fig. 3.7 Product planning technology roadmap adapted from Johnson et al. (2008)

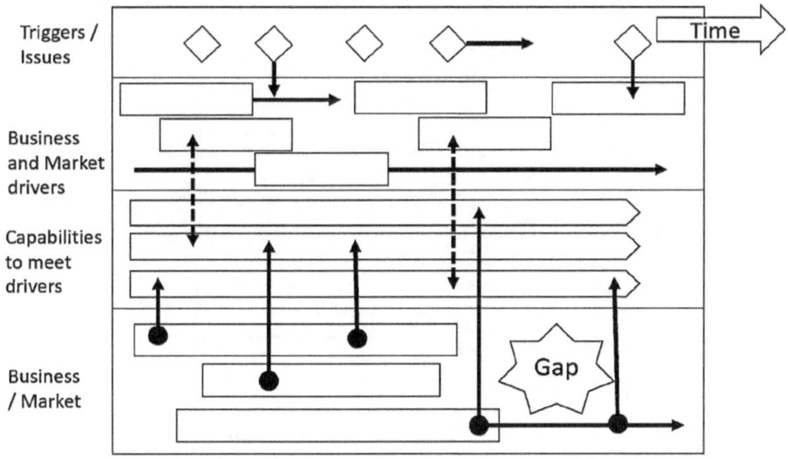

Fig. 3.8 Service or Capacity planning technology roadmap adapted from de Wet (1992)

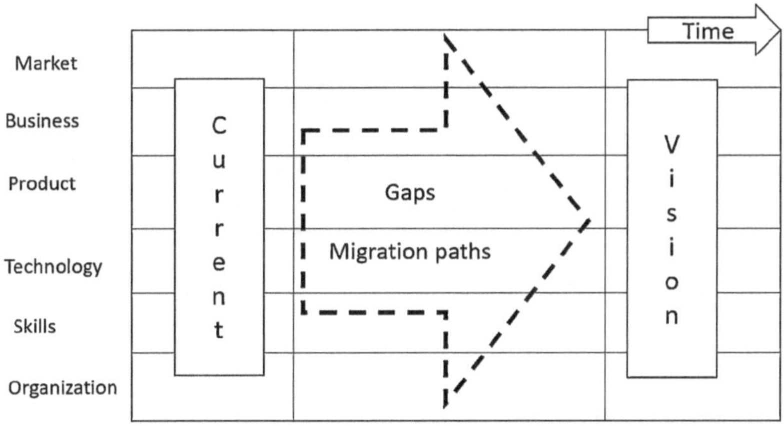

Fig. 3.9 Strategic planning roadmap adapted from Phaal et al. (2004)

and can help in defining a proper long-term strategy for the market spaces of interest, the capabilities of the Fig. 2.5: Strategic planning (Phaal et al. 2004) (Fig. 3.9)

4. Long range planning technology roadmap: This type of technology roadmap is developed for large-scale operations such as national or regional or sector levels to identify market opportunities in the long term and potential technologies that can act as market disruptors. Fig. 3.8 shows an example used by Integrated Manufacturing U.S (Fig. 3.10).
5. Knowledge asset planning technology roadmap: This type of technology roadmap is used to manage the technical and business knowledge within a firm to align with the business growth and objectives in the long term. The Artificial

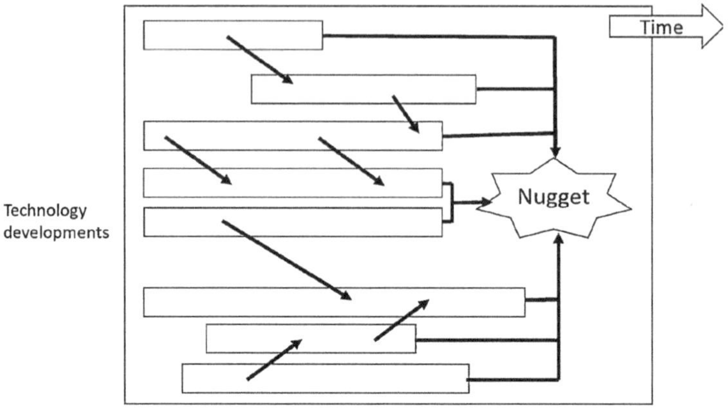

Fig. 3.10 Long-range technology roadmap adapted from Mitchell (1985)

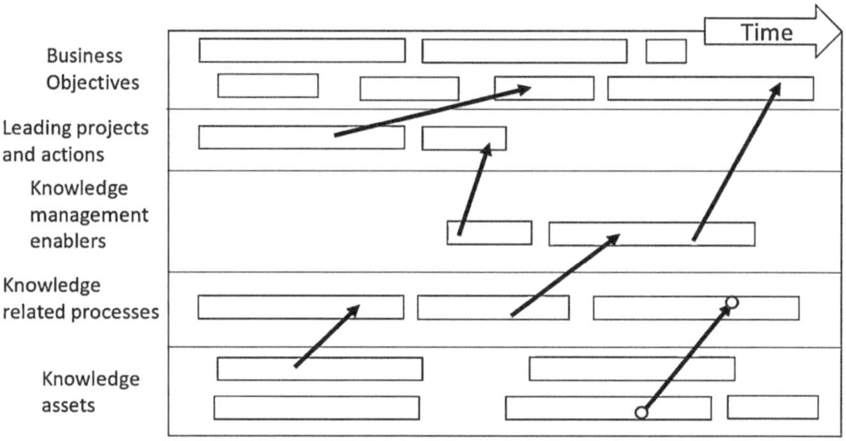

Fig. 3.11 Knowledge asset planning technology roadmap adapted from Foster (1985)

Intelligence Applications Unit of the University of Edinburgh used this type of technology roadmap to align the internal technical knowledge assets to better align with the market needs (Macintosh et al. 1998) (Fig. 3.11).

6. Program planning technology roadmap: This type of roadmap focuses on implementation of developed strategies across the different organizational sub-sectors, projects and programs. It takes into account the different projects implemented as per the strategy and the technological developments and decision-making that needs to take place at critical points in the timeline to ensure successful implementation (Fig. 3.12).
7. Process planning technology roadmap: Process planning roadmaps can be used by organisations to reassess and further develop several processes that are business critical. It has a wide range of applications and can be used to develop

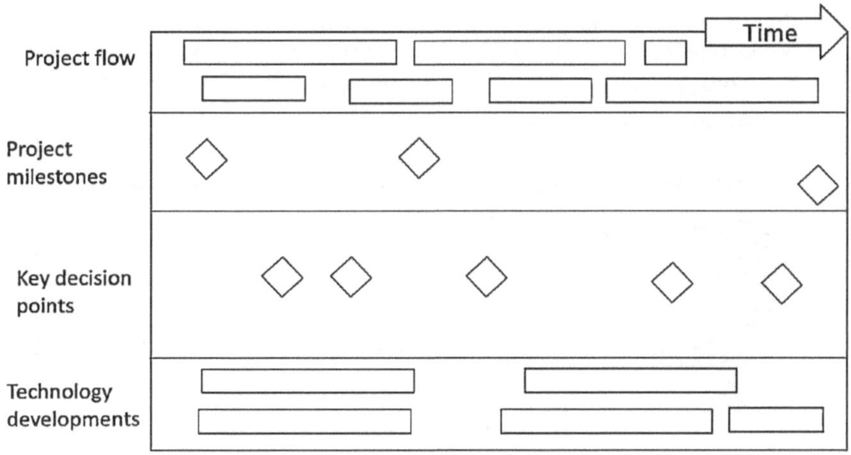

Fig. 3.12 Program planning technology roadmap adapted from Bower and Christensen (1995)

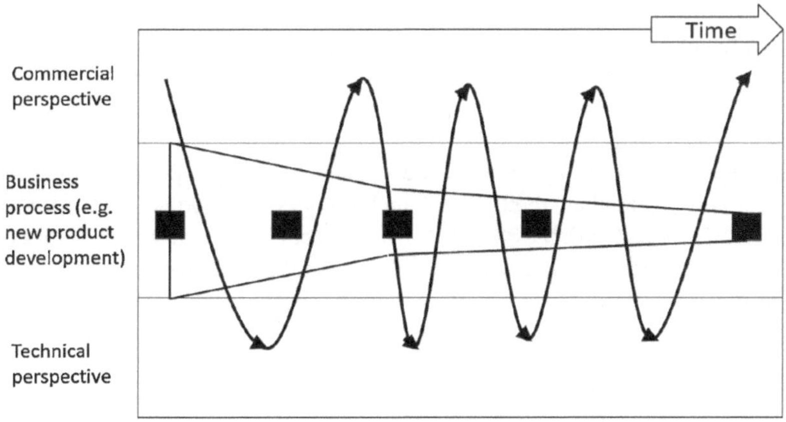

Fig. 3.13 Process planning technology roadmap adapted from Phaal et al. (2004)

specific technology-based processes or be focussed on a broader process involving several functions and individuals like products development (Fig. 3.13).

8. Integration planning technology roadmap: This type of technology roadmap can be used to amalgamate several technologies or subsystems in order to develop a larger more complex product or process. This type of technology roadmap is often used when several highly complicated technologies need to interface in order to develop a complete product. The following roadmap was used by NASA for creating a program for developing and directing technological developments and innovation towards technological and scientific missions managed by the organization (Fig. 3.14).
9. Technology development envelope (TDE): Since firms invest considerable capital in competitive economies to survive by discovering the next successful technological breakthrough (Radhakrishna and Vardarajan 1991) and use it to

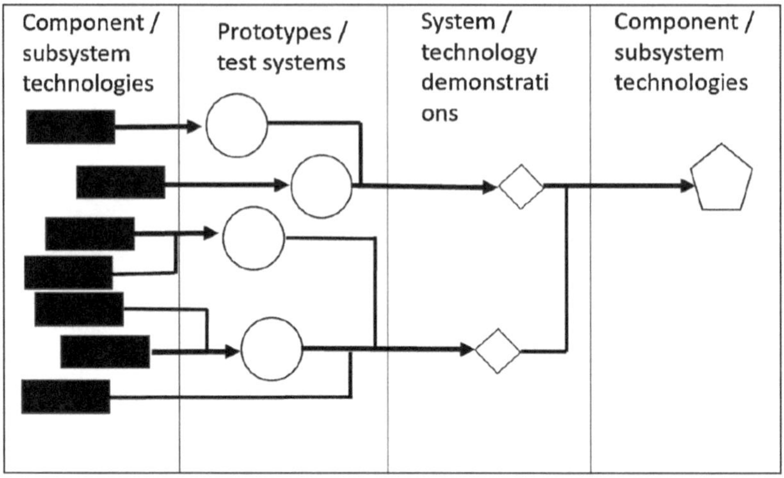

Fig. 3.14 Integration planning technology roadmap adapted from Bower and Christensen (1995)

generate market value, the choice of investment area is key to the success or failure of the firm (Gerdsri 2007). TDE utilises several steps to identify the emerging technologies in the academic and industrial spheres of the intended focus area by using the 'Delphi Process' to accumulate the technologies being developed and the time it would take for them to be complete with the technological and physical attributes of the technology or components from a group or team of experts (Gerdsri 2007). Criteria are also identified which would enable the firm to succeed in the market or which would be able to provide the best user experience in relation to a product which would in turn enable a firm to gain a dominant position in the market (Gerdsri 2007). These criteria are then associated with factors at the different technological and market hierarchies of the product, which need to be developed or enhanced to achieve the optimal position with regards to each of the criteria (Gerdsri 2007). These criteria are then associated with the different technologies that would enable the development of the factors that lead up to the criteria and the final product goal in the end.

The technologies form the very base of the hierarchical model and by finding the right correlation of the technologies to the end goal the firm can focus on the right focus are of the R&D teams or invest capital in the correct technology acquisition. The degree of correlation between the different available technologies and the factors that are influenced is also critical to identify the focus areas and more importantly find out which factors can be entirely ignored or paid less attention while still maintaining a relatively high customer use experience. The correlation of the factors to the end success criteria can also be measured, and a comprehensive evaluation with expert panellists can help rank the various criteria to identify which criteria contribute the most to the organizational success and in turn which technologies need to be prioritised (Fig. 3.15).

The following diagram is a visual representation of the process followed in a TDE development (Fig. 3.16)

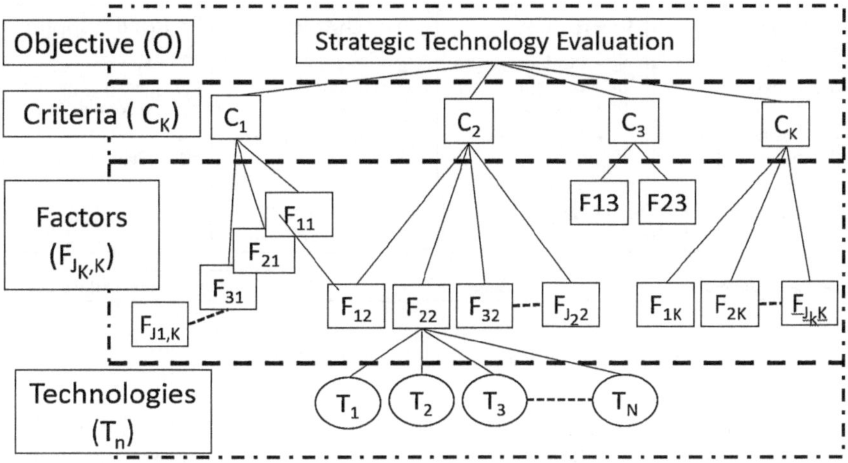

Fig. 3.15 Hierarchical model to analyse and evaluate upcoming technologies adapted from Gerdsri (2007)

Fig. 3.16 TDE development steps adapted from Gerdsri (2007)

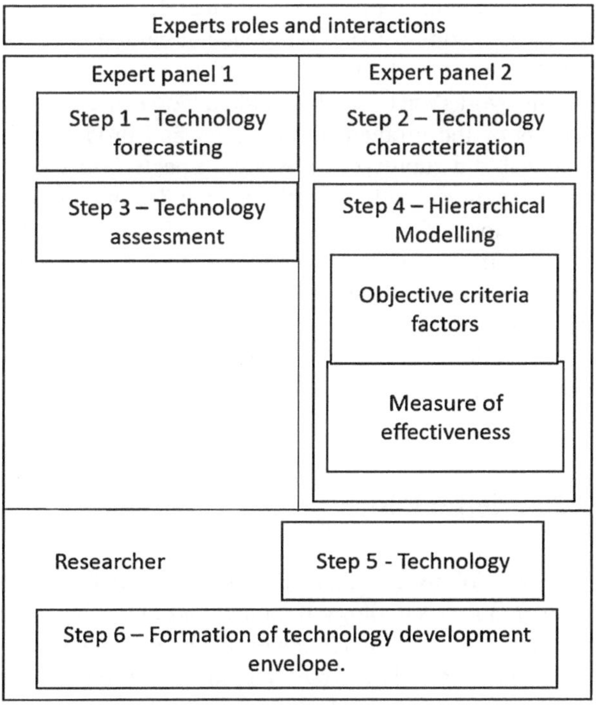

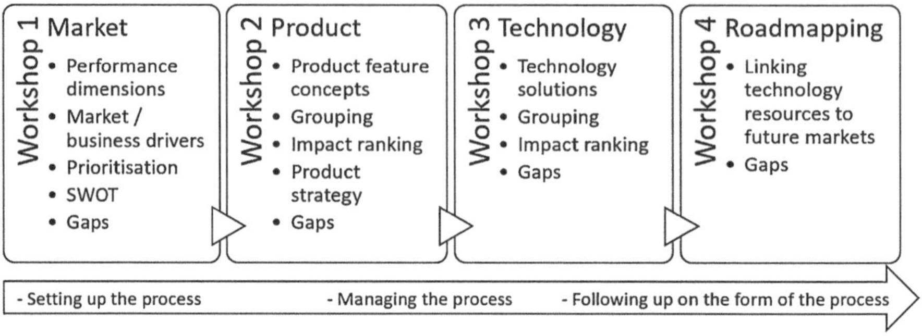

Fig. 3.17 The T-plan work process adapted from Phaal et al. (2003)

The involvement of the right experts during inception points the TDE in the correct direction and ensures the project focusses on the correct upcoming technologies and has the right information about the attributes and potential uses of the technologies.

Since considerable resources and time are devoted towards the development of the product in a firm that can help them capture the desired market space, it helps to have a directional approach to product development. The following diagram shows the basic layout of the T-plan that can be used to guide and accelerate the product development efforts (Phaal et al. 2003) (Fig. 3.17).

The workshops ideally would involve a cross functional group covering all aspects of the product development and product delivery to customer including financial and supply-chain-related aspects (Phaal et al. 2003). The core group of people involves should ideally be included in all workshops to ensure continuity of information and product development (Phaal et al. 2003). The workshops start from the customer centric view and move towards the linking of customer needs to the technology needed to meet the customer needs (Phaal et al. 2003). To facilitate the decision-making and information processing, linked analysis grids like the one shown in the figure below can be used in order to draw links between market needs and priorities and the technology that can enable this (Fig. 3.18).

3.2 Methodology

The steps followed in this thesis start with analysing the different tools available for developing a technology roadmap and analysing their strengths and weaknesses. The next step was to pick the optimal technique based on project requirements and time constraints. The T-plan method was used to develop basic knowledge of the customer needs and the technological requirements along with the cost and technology constraints currently faced by the firm. This enabled the identification of factors that were necessary for customer satisfaction and the key business drivers. This step in turn led to definition of product features that were key to achieving these market

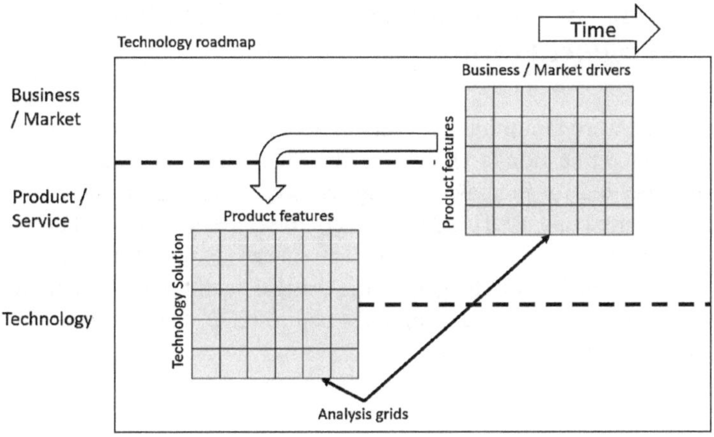

Fig. 3.18 Linked analysis grid adapted from Phaal et al. (2003)

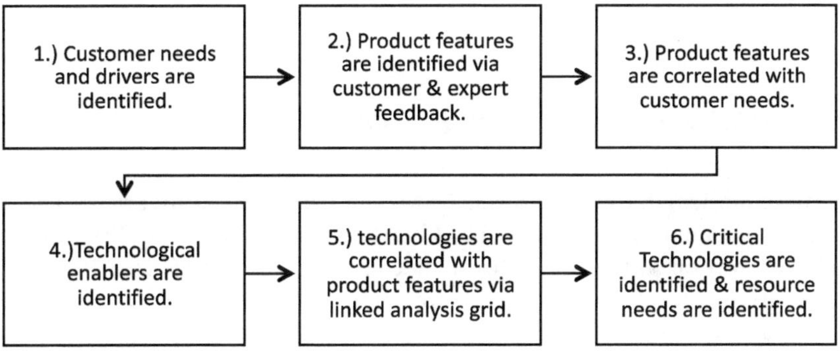

Fig. 3.19 Thesis methodology

requirements. This was then followed by technology identification that could support the required featured. This information was organised using a technology development envelope for consolidation and diagrammatic depiction. The next step involved using expert opinions to plot out a linked analysis grid in order to identify the technologies we should focus on in order to achieve maximum benefits.

An analysis of critical customer needs was also performed in order to identify the market preferences and standing of male dry shaving equipment in the current market place and the effect of competition, trends and competing technologies on the male dry shaving products. This also helped better understand the business needs and business drivers that are at play. The analysis also helped identify the key customer needs besides the expert interviews, interactions and questionnaires (Fig. 3.19).

3.3 Customer Needs Identification and Correlation with Product Features

Customer needs were identified by referring to test groups and filling out forms that contained detailed questions to help identify what are the critical features that customers were looking for in shavers along with the weightage if these features. The customers identified had used electric shavers for at least 1 year, and customers who had used multiple bands were preferred. The questions posed were independent of brand and the intention was to gather the critical features necessary to influence loyalty and buying decisions, as well as 'good to have' features that could set one shaver apart from the rest.

The test sheet also included a weight factor that the individuals could utilise to emphasise the importance of a feature or mark it as unnecessary. Both technical and aesthetic features were assessed in order to develop a more holistic understanding of the market and identify what influences customers to pay premium pricing for the products. The individuals were also encouraged to communicate desired features verbally if the test form did not encompass their needs and requirements.

3.3.1 Minimum Hair Length Possible

The minimum length to which a shaver can cut hair is primarily driven how close the cutting apparatus can come to the skin (Buras Jr and Harper 1979). Since electric shavers rely on cutting apparatus shielded behind a thin foil (Jacob 1930), which is generally thinner than 150 micrometres, it is harder to match the same shaving results that are achieved with a blade directly applied to the skin that even cuts below the level of the skin to deliver an extra clean shave. Though modern shavers are capable of delivering exceptional results using a variety of technologies, such a vibration to stimulate the hair to stand straight, high revolutions per minute cutting apparatus and foils as thin as 45 micrometres, the clean shave desired by some customers is achievable only after multiple repetitions and a proper pre-shave process.

3.3.2 Skin Irritation

Relying on information from experts in the field with more than 5 years of experience, most leading brands such as Braun, Remington, Phillips, Panasonic and so on have launched premium shavers with a motor delivering more than 10,000 revolutions per minute with the best-in-class shavers from Panasonic boasting up to 14,000 revolutions per minute resulting in around 70,000 cross-cutting actions per minute. This rapid movement of cutting elements along with choice of thinner foils to result in closer shaves unfortunately comes with the negative impact of more uncomfortable shaves that can result in skin burns, rashes, and irritations depending on the method and duration of use. The shaving elements tend to heat up during use, and based on the design of the shaver can end up dissipating heat directly to the skin of the user. This is an undesirable effect and can cause irritation in two ways.

The first reason for skin irritation is the direct heat applied by the metallic foil or rotary protection caps in case of rotary shavers directly on the skin, and the second reason for skin irritation arises from the fact that the skin in contact with the heated element dries out faster causing it to harden and lose its flexibility and gel like behaviour when wet. Once the skin becomes rigid, it does not interact well with the pores in the foil or protection cap of the shaver in order to push the hair through the pores to enable a closer shave. This effect is more prominent in low lying hair that grow at an angle horizontal to the skin. The lack of gelatinous behaviour of the dried skin results in multiple runs over the same area to get an even and close shave, and this can result in inflamed patches and irritated skin. Repeated cutting actions over dried skin can also cause the shaving element to peel off the top layers of skin, which may result in hypersensitivity in the damaged skin patches along with ingrowth of the hair cut below skin level, especially if the hair in the area tends to grow in a more horizontal direction with respect to the skin.

Hence, the holy grail of electric shavers is to be able to deliver an extremely close shave similar to that achieved by using blades directly on the skin under wet conditions while delivering the least possible amount of skin irritation delivered due to the mechanical vibrations of the cutting elements and the heat dissipated.

3.3.3 Battery Operation Period

Though the average shave takes only 5–7 minutes, industry leading brands deliver shavers with up to an hour of operational life due to high consumer demand. The primary reason for this is users often need to use the shaving apparatus multiple times before recharging it, for example, during travel. Since the device is intended for daily use and often becomes an integral part of the morning routine for many users, it is essential for the shaver to deliver multiple operational cycles before discharging. Though this use case is very exceptional, most users prefer to plan for this event as well considering they are purchasing a premium product that often costs a significant sum (ranging between 150 and 600 Euros). Another concern from users stems from the fact that the shaver offers versatile use over the entire body, and some users intent to use the full functionality of the shaver. This would mean a significant increase in the use time of the shaver, and hence, it is essential to have a long operational period of the battery. A common workaround for this is to provide charging cables and use while charging capabilities in the shaver. This, however, has the obvious drawback or reduced convenience caused by dangling wires during shaving. In addition, most premium shavers boast cordless charging and do not have a 'charge while shaving' feature to begin with. The first hurdle is the increased weight of the shaver that comes with batteries having increased capacities. This makes the user experience cumbersome and tiring, and a shaver weight of above 500 g is highly undesirable. The second hurdle is that higher energy batteries require higher recharge times. With current market leading brands delivering three- to four-hour charging cycles, it is already causing consumer complaints regarding long recharge times and lack of versatility in use during travel or circumstances where the user forgot to charge the device. The third barrier regarding high-capacity batteries is the high cost that comes with using modern innovations in battery technology. The current industry standard is to use Lithium ion batteries having different bases such as manganese, phosphate and cobalt, with cobalt delivering the highest energy densities higher than 160 Wh/kg (Panasonic 01/2007). However, the most commonly used Li-ion batteries are manganese based due to high stability and capability to deliver high ampere current.

3.3.4 Battery Recharge Time

As discussed briefly in the previous section, the battery life of the device is also correlated to the recharge time of the device with users demanding shorter recharge cycles to enable more versatile usage in the daily routine and especially during travel. The common complaint is that devices do not deliver even a single shave after short charging cycles of less than 30 minutes. Though most shavers are designed to deliver recharge times in accordance with morning and evening grooming routines than users most often follow, the user group that requires fast charging options due to needs while travelling or being in regions with limited charging options for work or vacation is growing steadily.

3.3.5 Product Versatility

Though shavers were primarily invented in a commercially viable form by Jacob Schick in 1930 for facial grooming only with the intention of delivering a clean shave like a razor, the onset of facial hair for style and fashion purposes in the last four decades has led to the fact that the latest shavers must incorporate additional features to allow for precision trimming and hair clipping. Shavers usually offer hair clipping and limited grooming options such as clippers to cut sideburns in a straight line, but the user demand for a more versatile tool is increasing rapidly. The end goal is to deliver a single tool to serve all the body grooming needs of the customer in a single device by using either accessories or new technologies to groom multiple hair lengths with a single shaving head.

3.3.6 Design Aesthetics and Anthropometric Features

The user group assessment also revealed that a large proportion of users make final purchasing decisions based on the look of the shaver. Since most top line products promise similar results and reviews online present a mixed result, the personal aesthetic preferences of the individual play a big role in final product selection. Over the last two decades, there has been a heavy focus on more curved designs with softer edges. This is a deviation from the original shavers from the 1950s to the 1980s. The softer edges also come into being with anthropometric studies revealing ideal gripping curvatures and comfortable handling grip designs. Users also prefer shavers that sit better in the hand and can be manoeuvred across the face with relative ease. All top brands offer shaver body designs that are engineered to fit the hand with ease and enable maximum manoeuvrability.

3.3.7 Accessories Available

Another common customer demand are the accessories that accompany the shaver. Some accessories such as trimmers and cleaning canters serve to enhance the versatility of the shaver or improve performance by allowing the user to groom and shape the beard with the trimmer or to enable better smoother shaves by enabling cleaning with special cleaning fluids in the cleaning canter. Travel pouches are another accessory demanded commonly by users to allow the user to take the shaver along on journeys. The travel pouch, however, adds no technological improvement to the shaver itself and only contributes to the ease of use. There is, however, a direct correlation between sales uptake and accessories offered, which should not be ignored.

3.3.8 Product Lifecycle

The product life is currently limited by two factors: the battery life and the life of the cutting elements. Most manufacturers advise to change the cutting element every 6 months. The cutting element is limited by the materials used to forge the blade, and as the blade becomes blunter with every cutting action, it tends to pull on the hair instead of cutting through it. The most prevalently used materials are stainless steels with chromium as an alloying element but more premium blades offer carbide steels with chromium, molybdenum, manganese, and silicon-based alloys coated with tungsten or titanium to offer a better cutting edge and corrosion resistance. The second limitation comes with the battery design that enables only about 2,000 recharges, which get consumed in about 2–3 years (Table 3.1).

Table 3.1 Customer needs ranking

Customer needs ranking	Expert 1	Expert 2	Expert 3	Expert 4	Expert 5	User 1	User 2	User 3	User 4	User 5	SUM	Rank	Points
1. Hair length after cut	8	3	8	7	8	5	4	6	7	7	63	1	8
2. Skin irritation	7	6	5	8	2	7	5	7	3	8	58	3	6
3. Battery operational time	1	5	3	5		4	3	5	4	2	32	7	2
4. Battery recharge time	2	4	6	3	1	3	1	4	5	6	35	6	3
5. Product versatility	5	7	4	4	4	2	2	3	6	4	41	4	5
6. Aesthetics	6	8	7	6	5	6	8	2	8	3	59	2	7
7. Accessories	3	2	1	1	6	8	6	8	1	5	41	4	5
8. Product lifecycle	4	1	2	2	3	1	7	1	2	1	24	8	1

Rank assigned on basis of score sum from experts and users

Points = 9- Rank

Points are used to determine the most important product features by multiplying them to the correlation table of customer needs and product features

Nine individual correlation tables between customer needs and product features was made with the help of nine experts by assigning correlation scores between customer needs and product features where

9 = Strong correlation; 3 = Moderate correlation; 1 = Weak correlation; 0 = No correlation

Then all tables were added up and each cell multiplied by the customer needs points from Table 3.1 to get Table 3.2

Table 3.2 Correlation matrix of customer needs with product features – summation of individual tables and multiplication with points of each customer need from Table 3.1

Customer needs	Product features: Thinner foil + foil design	Motor speed	Battery capacity	Motor power	Travel accessories	Long hair trimmer attachment	Battery life – recharge cycles	Blade sharpness/ cutting actions + durability	Anthropometric design + grip comfort	Curved edges/ metallic body design	Quick charge features	Shaver head durability	Use during charging	Cleaning centre for shaver head
1. Hair length after cut	504	384	312	416	0	48	24	456	96	0	0	320	48	264
2. Skin irritation	294	342	102	210	0	42	0	378	306	0	0	306	54	270
3. Battery operational time	6	114	150	138	0	66	34	0	0	0	126	0	162	0
4. Battery recharge time	0	0	189	0	0	0	93	0	0	0	243	0	0	0
5. Product versatility	165	65	315	195	315	405	20	255	105	0	105	315	345	215
6. Aesthetics	119	0	0	0	259	161	0	0	567	567	0	0	0	119
7. Accessories	0	0	0	0	405	255	0	0	0	0	0	0	0	0
8. Product lifecycle	0	0	28	13	0	0	81	57	0	0	63	51	63	35
SUM	1088	905	1096	972	979	977	252	1146	1074	567	537	992	672	903
Rank	3	9	2	8	6	7	14	1	4	12	13	5	11	10
Points	12	6	13	7	9	8	1	14	11	3	2	10	4	5

3.4 Technological Enablers

3.4.1 Protective Foil Thickness and Material Used

The protective foils currently used in the industry serve the purpose of acting as a shield between the user's skin and the cutting elements within the shaver. The essential features of the foil include corrosion resistance, durability, and shear resistance. The pores of the foil allow for the hair to come in contact with the cutting element in the shaver head and in the ideal condition no foil at all would be needed to enable all hair to be cut directly by the cutting elements in one shot. This is currently not possible as the cutting elements cannot distinguish between hair and skin and without the protective foil would severely damage the user's skin.

Current foil thickness can be as low as 45 micrometres, and thinner foils are still not possible due to shearing damage and brittle fracture causing the foil to break down entirely and lower thickness. The lower thickness also prevents corrosion resistance due to development of corrosion in damaged areas that occur more often in thinner foils. The foils used often consist of a base of stainless steel having any desired design of the apertures coated with a harder material such as nickel, chromium or rhodium. The base layer provides the desired stability and shear resistance, while the hardened coating interacts with the skin and hair to create proper uptake for the shaver. The design of the coating and how it protrudes above the base sheet is essential to enable it to interact with the hair and lead it into the perforations for cutting. The foil should be as thin as possible to enable a closer shave (Blume et al. 1977). This is coupled with the challenge of depositing the coating on the base plate with appropriate design parameters to allow for hair uptake in both shaving directions (Bodo 1968). Negative cutting angle at the edges of the holes also need to be avoided as they can cause skin irritations and undercuts (Blume and Voigtmann 1985).

The most modern shavers use large holes of various profiles such as elongated ovals in the foil surface with protuberances on the foil surface and along the edge of the hole to allow multiple sections to engage the hair and increase uptake. Further research is currently aimed towards reducing foil thickness below 30 micrometres and developing orifice designs that direct the hair towards the blades. The use of vibrations has also been incorporated that causes high-frequency foil oscillations that are transferred to the skin causing hair to stand on end. This reduces the need for the user to apply force on the devise to force skin through the orifice and enable a closer shave in that way. This in turn will enable thinner shaving foils as risk of fractures caused by excessive force are reduced.

3.4.2 Cutting Element Sharpness and Materials Used

Conventional cutting elements in both rotary designs (Uchiyama and Okabe 1995) and linear shavers utilise components made of stainless steel with hardened sections that are in contact with the hair (Blaauw 2010). These sections are often hardened by nitriding or surface coatings (Blaauw 2010). Blades that are developed from carbon or stainless steels coated with resins often face the problem that they do not have sufficient hardness due to softening during heat treatment, which reduces the sharpness of the blade (Nakatsu and Tamura 2004). The primary reason for the lack of hardness is the formation of coarse carbides in the crystal structure during heat treatment during the resin addition process that takes place at 300–400 °C, which results in defects in the lattice structure (Nakatsu and Tamura 2004). This effect can lead to deteriorated cutting edges having reduced durability, reduced corrosion resistance and even result in breakage during use (Nakatsu and Tamura 2004). Hence, blades were modified in terms of composition to have reduced carbon percentage as low at 0.5% with chromium in the range of 15% and molybdenum in the range of 8% with boron and silicon added up to 8% in combination depending on the net crystal structure desired with carbon precipitates having diameter less than 0.1 micrometre, hence increasing the durability and corrosion resistance of the blades (Nakatsu and Tamura 2004; Blaauw 2010). However, this hardening of the cutting elements then results in wear of the cutting element cap depending on the relative hardness of the cap materials in addition to corrosion that may occur if the shaver is exposed to wet conditions during use (Blaauw 2010). To tackle this problem, the cutting blade extremities are designed to come in contact with the cap in such a manner that a section of reduced hardness is in contact with the cap surface while the section of increased hardness interacts with the hair (Blaauw 2010).

Blades are also being developed using ceramic-coated edges that offer increased sharpness and durability, which comprise of blade edges coated with boron, titanium, carbon, chromium, titanium and zirconium carbides (King and Crawford 2009). This production method, however, is very expensive and slow and has gained no traction in the industry due to the price impact.

Obsidian has also been considered for potential use as a cutting material as its blade can be ground down to a cutting edge thickness of around 3 nanometres (Buck 1982), whereas the best shaving blades range in 20–40 micrometres (20,000–40,000 nanometres). However, the cost and durability of the blade under wet and dry conditions over a longer period prevent its use. The blade does not bond well with metals, and it is not malleable enough to be used for manufacturing the complete cutting element and attach to the motor for movement at high frequencies. The common problems are brittle fracture and edge cracking at high frequencies.

3.4.3 Battery Design and Material

As previously discussed, the battery life and recharge time are key features that consumers look for when purchasing shavers. The average shaving time offered by most leading brands ranges around an hour with a full recharge duration of 2–4 hours. Manganese-based lithium ion batteries are currently the most commonly used batteries in the market; however, experimentation with other bases such as cobalt and molybdenum is also in place to increase the energy density of the battery.

A recent development in shaving devices is the option of having wireless charging in the device, which leads to the capacity to have zero open ports to the electronic components of the shaver and enables use in wet conditions as well. Though both inductive and capacitive charging technologies are available in the market, exclusively inductive wireless charging is used for shavers and other small portable electronic devices (Hui 2013), this is due to the fact that capacitive wireless charging requires a comparatively larger area of coupling (A contactless power transfer system with capacitively coupled matrix pad 2011). Though inductive charging does have the disadvantage of generating eddy currents in metallic components (Hui 2013), this can be overcome by using thin electromagnetic shields on top of the receiving coil and beneath the charging coil (An analysis of a double-layer electromagnetic shield for a universal contactless battery charging platform 2005). This along with the higher energy density offered at lower coupling areas (Hui 2013) has made inductive charging the leading choice. Energy transfer via magnetic flux between a winding that generates the magnetic flux and a coil circuitry tunes to this winding is essentially the concept used for wireless charging (Hui and Ho 2005). Within this as well, vertical flux transfer devices are preferred as they offer a more versatile capability to charge devices when vertically aligned with respect to the charging pad due to the vertical flow of magnetic flux (Hui and Ho 2005).

Molybdenum disulphide two-dimensional 'rectennas' are also being researched for use with shavers to enable them to be passively charged using the commonly available Wi-Fi frequencies. The thin two-dimensional structure of the MoS_2 makes it ideal for use in portable and slim design electronic devices and enables them to harvest energy passively (Zhang et al. 2019). This technology would eliminate the need for a charger altogether as most household have several Wi-Fi signals permeating the area. The cost of incorporation of the technology to current shavers and low charging achieved, however, holds back the full application.

To address the quick charging needs of batteries, several approaches that revolve around replacing the graphite electrodes in the conventional lithium ion battery are being researched. Silicon-based electrodes already provide a much higher charging rate but come with the danger of battery overheating and explosions. Companies such as 'StoreDot' are using proprietary materials such as silica combined with organic peptides that protect the active materials while being fast charged. The Lithium ion structures used in the battery have also been modified along with the use of nanotechnology to enable full charging in less than 5 minutes (StoreDot). Several similar technologies are being researched across the globe, but the non-repetitive test results and high costs are holding back application.

3.4.4 Integrated Accessories

The accessories provided with a shaver are not necessarily technologically more advanced than the shaver and rarely utilise novel concepts, deigns, proprietary materials or patented technologies. However, the accessories offered often tend to determine the final buying choice of the consumer and has been known to even dissuade consumers from buying a particular product due to lack of key accessories. The primary challenge revolves around offering a full grooming kit integrated with a conventional electric shaver.

Though some devices offer the possibility to replace the head of the shaver containing the cutting mechanism with other accessories that can enable grooming longer hair or body grooming, the attachments come with an additional cost and usually do not offer the same comfort as a conventional grooming kit that is much cheaper than the shaver. Most premium shavers do not offer this option at all and at best have an integrated clipper attached to the back side of the shaver to enable basic grooming. Customers, however, are demanding features such as variable head attachments of different length shaving components to enable access to hard-to-reach crevasses or shaving smaller narrower patches like behind the ear or between the eyebrows. Another request is for having an ear and nose hair trimmer integrated into the shaver or modification of the shaver made possible by a replaceable head. Though almost all leading brands offer a multi-grooming kit, the key challenge is to incorporate the features into the premium flagship shavers. This is challenging due to the higher performance requirements of the components of the replaceable heads when coupled with the high-frequency motors used in the flagship shavers. Developing cost-effective components to be used with motors running above 10,000 revolutions per minute means using metal or ceramic components in the replaceable heads instead of the current industry standard plastic internal components. This comes with additional costs and requires research and testing before launch.

3.4.5 Anthropometric Handle Design/Customised Handles

The effort put into the shaver handle seems disproportionate considering it as a part of the shaver that does not even make contact with the skin but in fact it serves two essential purposes that enable a perfect shave. The first and most important feature of a handle or shaver body must be that it fits easily into the palm of a user enabling easy handling and movement to ensure a comfortable user experience. The advances in this area primarily stem from the research done on traditional blade handles with many of the aspects being reused and reapplied. This is because there is a long history and vast data set to draw from pertaining to a very similar problem with regards to electric shavers and traditional blades.

The first challenge is to enable the user to have control over the shaver such that the shaver can be pressed upon the skin with light force, yet can be easily pivoted to change the angle depending on the contour of the face or comfort. To enable this shaver, commonly used rubberised sections moulded into the shaver body intended to provide points where the users hand can firmly stick to the rubber of the shaver without slippage. The elastomeric rubber sections are often placed to allow the finger tips to attach onto them with a separate section often provided behind the shaved near the head so that the index finger can be used to apply additional pressure on the shaver. This stems from progress made in the blade shaver handle design, which included adding gripping points and pressure application areas as described in the following US patents 'Shaving system' (Miller 1991) and 'Razor handle having ergonomic gripping areas (Bozikis et al. 2008). Since modern shavers also feature wet usage capabilities, it is also preferred that the material used is water repelling, and the groves in the rubber are designed to draw water away from the contact area of the user's skin and the shaver.

The handle can also incorporate additional accessories such as trimmers used to shape denser hair. Since the trimmer must stay out of the way while it is used in shaving, it is essential that it be designed into the handle in such a manner that it does not obstruct the primary shaving experience. To enable this, shavers often feature a slidable trimmer that can be drawn out to be used or be tucked away while the primary shaving element is being used (Yasunaka et al. 1989).

The other important role played by the shaver body is that it must block the motor vibrations from being transmitted to the user's hand and also insulate the heat and noise within the body. The primary tactic used here is diversion of the motor head and vibration the shaving foil that is in contact with the user's skin to offer a more comfortable shave. The shaver body also needs to be water proof and protect the internal circuitry from water contact.

3.4.6 *Shaver Head Design*

The external design and appearance of the shaver head plays an essential role in delivering a smoother and cleaner shave as it enables sufficient contact between the skin and the shaving mechanism. A shaver head that is too large would be able to cover flat areas faster but not be able to access the crevasses or upper lip without coming in contact with the lip itself. On the other hand, a shaver head that is too small would result in a longer shave period. Though the rotary and linear shaver designs approach this problem in different ways, the target is to achieve a shaver head design that enables speed over flat areas and finesse in narrow patches based on how the user holds the device.

The shaver head also plays an aesthetic role in the shaver design and since it is the functional part of the razor, user reviews show that a shaver head that appears more technologically advanced is preferred by most users. Firms try to achieve this by using chrome-plated surfaces and smoother edges with a combination of metal and plastic/polymer fusion designs. Several patents for designs of shaver heads are available and implemented by different companies, but the two prevailing

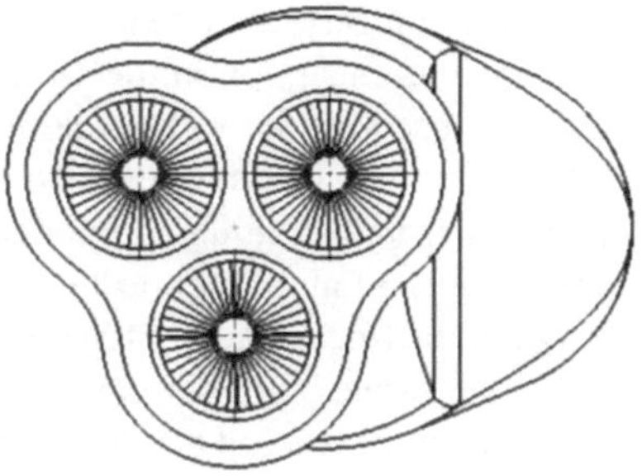

Fig. 3.20 Multiple cutter rotary shaver head – self drawn based on design adapted from Ottone (1957)

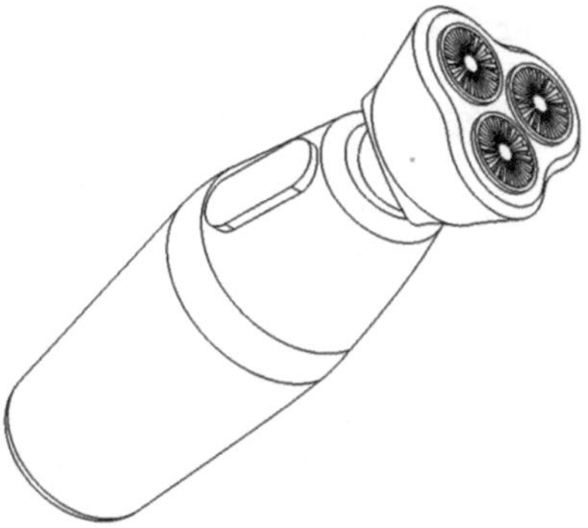

Fig. 3.21 Three head dry shaver head design – self drawn based on design adapted from Prat-Pfister (2012)

technologies still use the rotary or linear concepts. The rotary shaver with multiple individual circular shaving modules incorporated on the head was conceptualised and developed as early as 1957, and the following diagram shows the basic concept. These types of shavers take advantage of the smaller rotary units and the accessibility they offer in difficult to use areas (Trusco) (Fig. 3.20).

There have been several advancements regarding the rotary shaver heads especially regarding the arrangements of the cutter modules and the flexibility offered by using pivots to mount the shaver head on the shaver body allowing for a greater flexibility and shaving comfort. The patent of a triple cutter rotary shaver developed in 2007 has become as the base for several premium Philips shavers and have received positive customer feedback due to the pivotable shaver head and the curved upper surface offering access to difficult crevasses and a comfortable shave as the user does not have to pivot the device excessively with the hand to achieve the prefect shaving angle (Fig. 3.21).

The downside to the multiple head rotary shavers is still the gap that exists between each of the cutter blocks and also the gap existing between the cutter and the outer frame of the shaver. This makes it hard to use the shaver along the cheek bone, especially under the ear and the area is hard to access, and the fact that the full area of the shaver does not offer cutting (only the rotary cutters on the surface – not the area in between) makes it hard for the user to make contact between the cutting area of the shaver and the skin.

On the contrary, the linear shaver head has maintained a fairly stable design concept since its inception in the 1950s. There have been several advancements in the materials used on the foil to add to the corrosion avoidance properties, micro-foil technology that allows for the best possible contact between the cutting components and the hair and aligning components between the linear cutting components to cut longer hair and align then to be fed into the primary cutting elements, but the key design of the linear shaver of having one or more parallel linear cutting elements has stayed the same. The user can use the full surface of the linear cutting element as each portion of the foil will shave if contact is made with the skin and with recent premium linear shavers being mounted on pivoted heads, the shavers are very. This means that the user can achieve shaving in any skin surface with the least hand pivoting of the device. A drawback of the linear shave.

The key progress in the area of shave heads revolves around using lightweight materials to make the shaver lighter and to design the shaver head to expose the maximum possible area of the shaving cutting elements to the user's skin without causing and skin damage. The shaver heads currently use plastic components coated with chrome paint to give a metallic appearance. This enables lightweight designs and eliminates any corrosion-related problems. Another area being researched is the possible use of the drainage devices that work on suction to remove the foam, water or hair from the user's skin as the shaver is used. There have been no practical breakthroughs in this area, but two approaches are being researched: one involves using a drain pipe to continuously drain the waste that comes with the obvious drawback of a dangling pipe during shaving, which would be inconvenient for the user. The other approach revolves around using a storage chamber inside the shaver, but this adds to the size of the shaver and also makes the shaver progressively heavier as waste is accumulated. Both approaches would also need a more powerful battery to support the suction pump and a bigger shaver body.

3.4.7 Motor Power and Frequency

The electric motor driving the shaving mechanism is perhaps one of the two most critical components of the shaver, the other being the cutting apparatus itself. With motor speeds already having hit 14,000 rpm, the main industry focus is not on delivering higher frequencies. The speed delivered by current motors is sufficient to deliver the necessary cross-cutting actions to deliver a smooth shave. This effect is further amplified by the fact that a single motor can be used to derive four or five separate cutting units delivering over 70,000 cutting actions per minute. This is sufficient to deliver a perfect shave in a single movement across the skin if the efficiency of the cutting is high. The primary work being done is to reduce motor costs and increase efficiency to deliver a cheaper motor that draws lower battery power, hence leading to longer shaving times or multiple shaves before recharging.

To achieve higher motor efficiencies, the primary approach used is to have higher conductivity copper in the windings and using a higher cross-sectional area for decreasing losses (Yoon et al. 2002). This approach, however, results in net higher construction costs and the offset offered by increased performance needs to be recovered from the customer via increased prices. On the other hand, improved performance of the rotor bearing and reducing the gaps between the stator and rotor combined with lower tolerances for the machinery (Yoon et al. 2002) have proven to be more beneficial from a cost point of view as they rely on one-time design investments and production accuracy improvements instead of ongoing material costs. The other area of investigation is regarding the cooling mechanism of the motor. Since the motor efficacy is higher at lower running temperatures (Yoon et al. 2002), several newer shaver designs focus on diverting the heat generated by the motor to the shaving foil in contact with the user's skin to help achieve a more comfortable shave in order to generate a benefit from the otherwise useless motor heat.

Another technology feature uses a feedback loop to modulate the motor frequency to deliver the closest shave possible depending on the different beard thickness. This is a critical technology developed in the 1990s when the electric shaver market spread from Europe and North America to the east and manufacturers faced a wide variety of beard thickness to deal with. Since no one setting can serve the full spectrum to beard thickness and density, modulating the motor is critical to deliver an optimal shave. The development is based on sensing the motor load during shaving (Okada 1994). When the motor load is higher due to beard thickness, the motor speed is increased resulting in more cutting actions per second, hence delivering a cleaner shave over rougher hair (Okada 1994). On the other hand, the motor speed is reduced when motor load is low so that the battery lasts longer during use (Okada 1994) (Tables 3.3, 3.4, 3.5, 3.6, 3.7, and 3.8; Figs. 3.22 and 3.23).

3.5 Results and Discussion (Fig. 3.24, Table 3.9)

After consolidating the results of the ratings made by experts, the results obtained first allow us to correlate the product features with the customer needs in the first step and further correlate the product features to the technologies that can enable the development of better shavers that suit customer needs. The technology that needs the most emphasis based on a current market situation is surprisingly the materials of the battery and the battery capacity. This result is contrary to common perception that the cutting elements, being the core of the shaver, should receive prime focus. The reason the battery technology used is so essential is due to the impact it has on every single aspect of the shaver. The longer battery life can enable a slower, smoother shave, use of the shaver for body grooming, multiple uses before charging and even provide for quick charge options. Though the technology is not expected to diverge from the traditional Li-ion technology, use of different bases such as molybdenum and cobalt can be considered to increase capacity along with larger batteries incorporated in shavers as the extra weight may be an acceptable trade-off for a longer operational period. At the same time it is surprising to see that associated battery technologies such as wireless inductive charging (rank 11) and

Table 3.3 Product features and technology enablers correlation matrix (part 1 of 3)

	Technology enablers									
Product features	Stainless steel/Titanium alloys coated with Ni/Cr/Rh	Foil orifice design	Vibrations transferred via foil	Heated foils	Stainless steel alloy/ titanium alloy blades	Surface coating/ nitriding of blades	B/Ti/C/Cr/Zr & Ti carbide-coated ceramic blades	Blade edge heat treatment	Edge thickness reduction <10 μm	Co/Mo Li ion batteries – increased capacity
Thinner foil + foil design	48	44	22	30	0	0	0	0	0	0
Motor speed	0	0	13	8	0	0	0	0	0	46
Battery capacity	0	0	0	12	0	0	0	0	0	60
Motor power	0	0	10	5	0	0	0	0	0	0
Travel accessories	0	0	0	0	0	0	0	0	0	0
Long hair trimmer attachment	0	0	0	0	0	0	0	0	0	42
Battery life – recharge cycles	0	0	0	0	0	0	0	0	0	16
Blade sharpness/ cutting actions + durability	0	12	5	0	42	66	48	54	40	0
Anthropometric design + grip comfort	0	0	0	0	0	0	0	0	0	0
Curved edges/ metallic body design	0	0	0	0	0	0	0	0	0	0
Quick charge features	0	0	0	0	0	0	0	0	0	12

	Technology enablers									
Product features	Stainless steel/Titanium alloys coated with Ni/Cr/Rh	Foil orifice design	Vibrations transferred via foil	Heated foils	Stainless steel alloy/ titanium alloy blades	Surface coating/ nitriding of blades	B/Ti/C/Cr/Zr & Ti carbide-coated ceramic blades	Blade edge heat treatment	Edge thickness reduction <10 μm	Co/Mo Li ion batteries – increased capacity
Shaver head durability	138	46	22	12	28	66	48	46	4	0
Use during charging	0	0	0	0	0	0	0	0	0	72
Cleaning centre for shaver head	0	0	0	0	0	0	0	0	0	0
Sum	186	102	72	67	70	132	96	100	44	248
Share	6%	3%	2%	2%	2%	4%	3%	3%	1%	8%
Rank	2	13	22	24	23	5	16	14	27	1

Table 3.4 Product features and technology enablers correlation matrix (part 2 of 3)

	Technology enablers									
Product features	Wireless inductive charging	Passive Wi-Fi charging	Silicon based battery electrodes	Swappable shaver heads	Long hair/ ear nose trimmer integration to shaver body	Anthropometrically designed handles	Pivotable shaver heads	Elastomeric rubber on handles	Water repelling grooves in handles	Vibration dampers/ redirectors
Thinner foil + foil design	0	0	0	0	0	0	0	0	0	0
Motor speed	0	0	8	0	0	0	0	0	0	0
Battery capacity	0	5	54	0	0	0	0	0	0	0
Motor power	0	0	0	0	0	0	0	0	0	0
Travel accessories	0	0	0	66	54	0	0	0	0	0
Long hair trimmer attachment	0	0	0	0	72	2	0	0	0	0
Battery life – recharge cycles	52	10	11	0	0	0	0	0	0	0
Blade sharpness/ cutting actions + durability	0	0	0	0	0	0	0	0	0	0
Anthropometric design +grip comfort	0	0	0	0	0	66	16	38	24	28
Curved edges/ metallic body design	0	0	0	2	0	60	0	60	54	0

	Technology enablers									
Product features	Wireless inductive charging	Passive Wi-Fi charging	Silicon based battery electrodes	Swappable shaver heads	Long hair/ ear nose trimmer integration to shaver body	Anthropometrically designed handles	Pivotable shaver heads	Elastomeric rubber on handles	Water repelling grooves in handles	Vibration dampers/ redirectors
Quick charge features	66	2	54	0	0	0	0	0	0	0
Shaver head durability	0	0	0	11	0	0	72	0	0	0
Use during charging	0	2	0	0	0	0	0	0	0	0
Cleaning centre for shaver head	0	0	0	4	0	0	0	0	0	0
Sum	118	19	127	83	126	128	88	98	78	28
Share	4%	1%	4%	3%	4%	4%	3%	3%	3%	1%
Rank	10	29	7	19	8	6	18	15	20	28

Table 3.5 Product features and technology enablers correlation matrix (part 3 of 3)

	Technology enablers								
Product features	Shaver head chrome coating	Suction properties on shaver head	Increased cutting elements	Hair guiding elements on shaver head	Reduced stator rotor gaps in motor	Motor heat redirection	Hair thickness modulated motor frequency	Motor winding material – Ag/ copper alloys	Cleaning fluid + cleaning jet technology
Thinner foil + foil design	0	16	0	0	0	0	0	0	0
Motor speed	0	0	0	0	54	32	52	36	0
Battery capacity	0	0	0	0	30	14	12	9	0
Motor power	0	0	0	0	0	5	0	52	0
Travel accessories	0	0	0	0	0	0	0	0	0
Long hair trimmer attachment	0	0	0	0	40	0	0	20	0
Battery life – recharge cycles	0	0	0	0	0	19	0	0	0
Blade sharpness/ cutting actions + durability	0	18	48	40	0	0	0	0	48
Anthropometric design +grip comfort	0	0	0	0	0	0	0	0	0
Curved edges/ metallic body design	66	0	0	0	0	0	0	0	0
Quick charge features	0	0	0	0	0	0	0	0	0
Shaver head durability	0	0	46	66	0	4	72	0	42

	Technology enablers								
Product features	Shaver head chrome coating	Suction properties on shaver head	Increased cutting elements	Hair guiding elements on shaver head	Reduced stator rotor gaps in motor	Motor heat redirection	Hair thickness modulated motor frequency	Motor winding material – Ag/ copper alloys	Cleaning fluid + cleaning jet technology
Use during charging	0	0	0	0	0	0	0	0	0
Cleaning centre for shaver head	0	18	0	0	0	0	0	0	72
Sum	66	52	94	106	124	74	136	117	162
Share	2%	2%	3%	4%	4%	3%	5%	4%	6%
Rank	25	26	17	12	9	21	4	11	3

Table 3.6 Technology roadmap: drivers 1–4

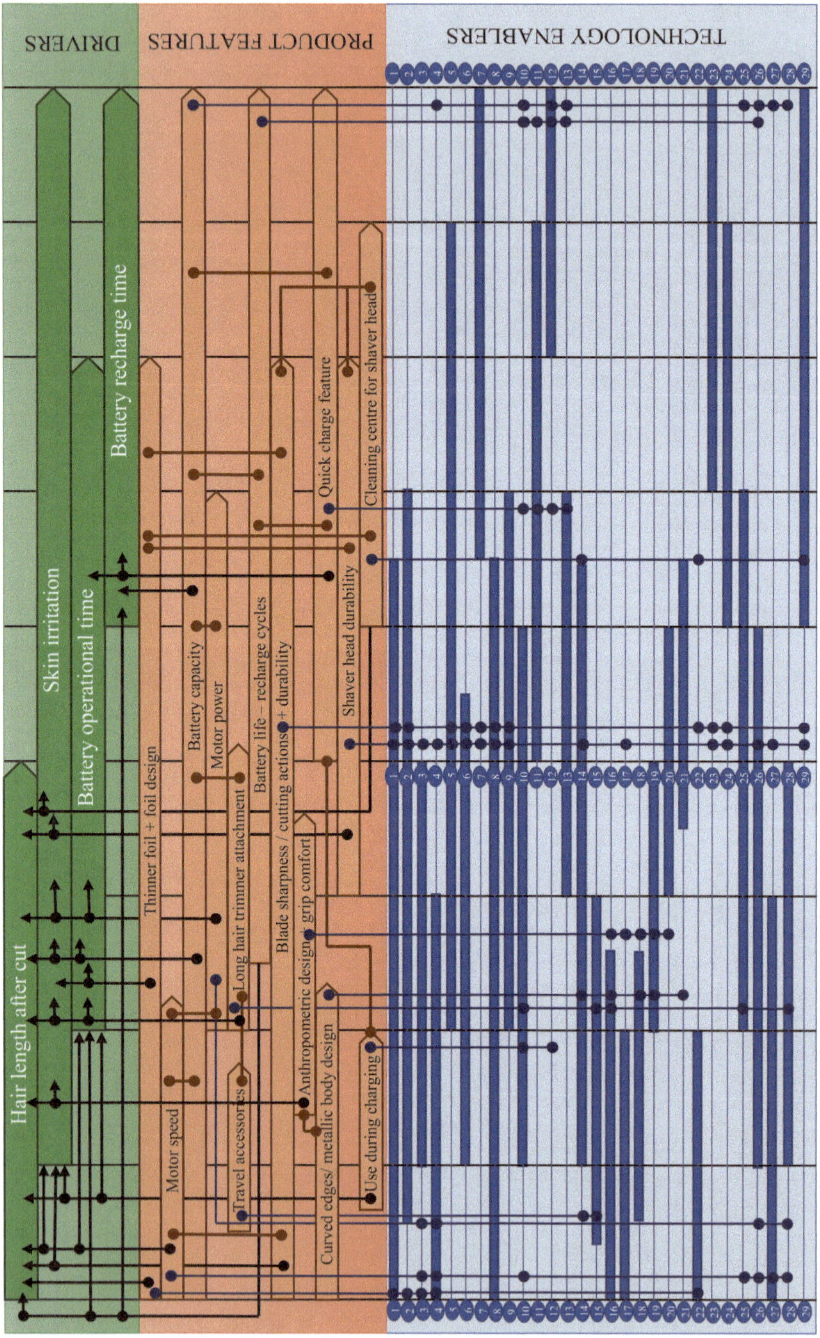

Table 3.7 Technology roadmap: drivers 4–8

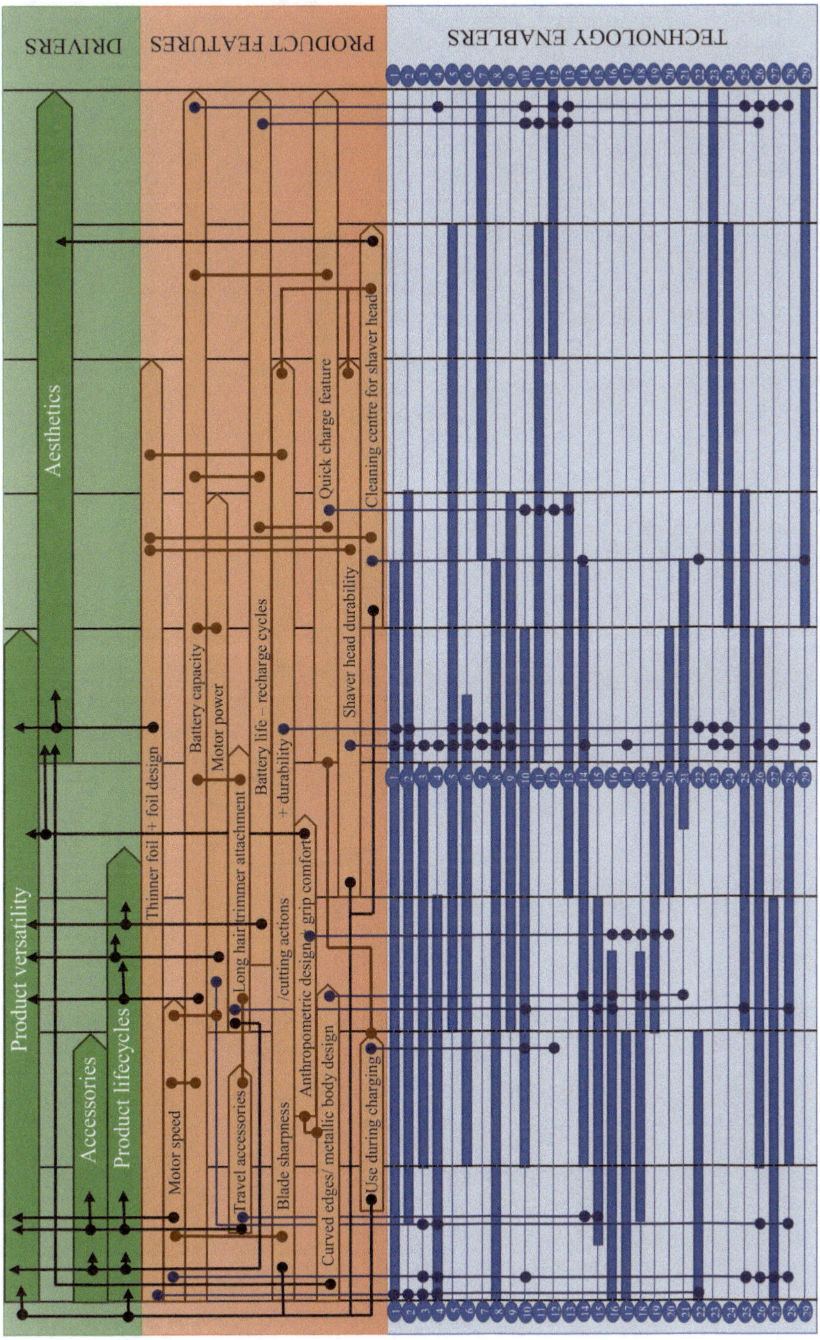

Table 3.8 Technology roadmap: technology enablers and resources

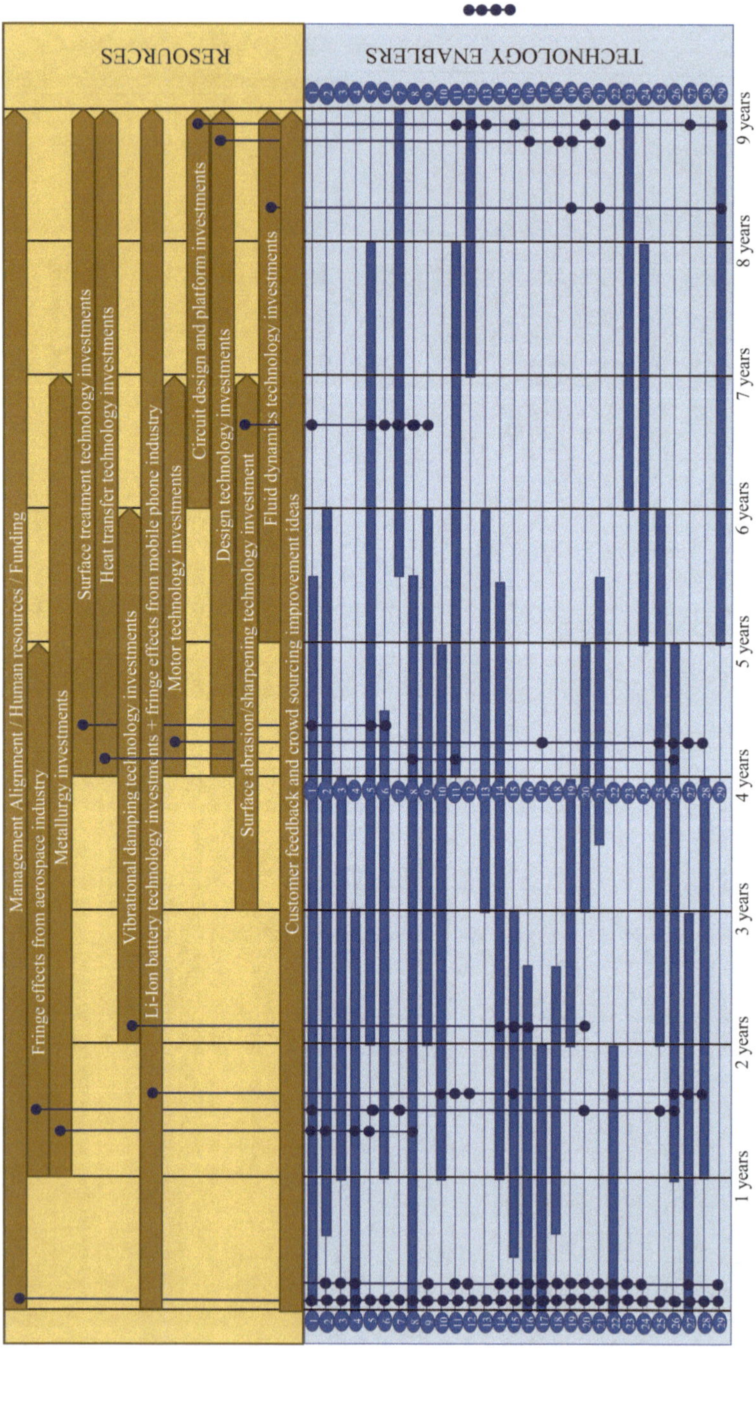

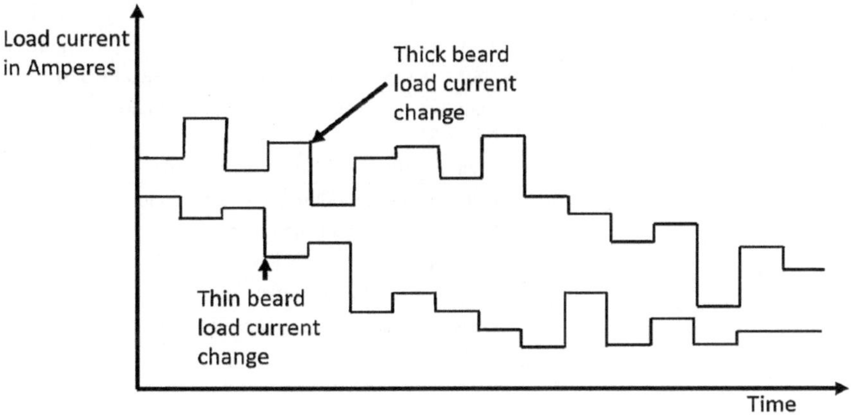

Fig. 3.22 Changing load current pattern while shaving – self-drawn – based on concept adapted from Okada (1994)

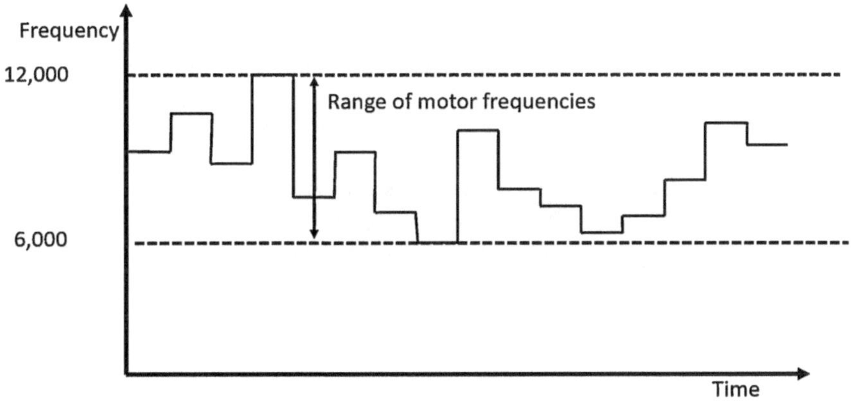

Fig. 3.23 Modulating motor speed based on motor load – self-drawn – based on concept adapted from Okada (1994)

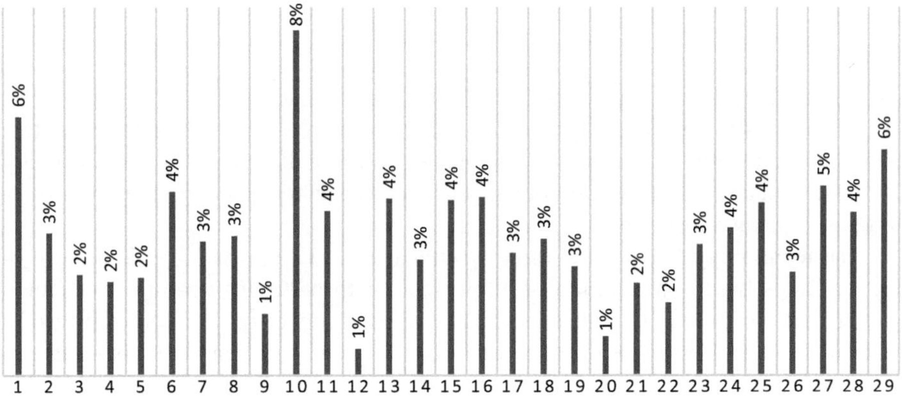

Fig. 3.24 Technology enablers relative influence

Table 3.9 Ranking and key of technology enabler for Fig. 3.23

Rank	Technology enabler	Key for figure
1	Co/Mo Li ion batteries – increased capacity	10
2	Stainless steel/titanium alloys coated with Ni/Cr/Rh	1
3	Cleaning fluid + cleaning jet technology	29
4	Hair thickness modulated motor frequency	27
5	Surface coating/nitriding of blades	6
6	Anthropometrically designed handles	16
7	Silicon-based battery electrodes	13
8	Long hair/ear nose trimmer integration to shaver body	15
9	Reduced stator rotor gaps in motor	25
10	Wireless inductive charging	11
11	Motor winding material – Ag/copper alloys	28
12	Hair guiding elements on shaver head	24
13	Foil orifice design	2
14	Blade edge heat treatment	8
15	Elastomeric rubber on handles	18
16	B/Ti/C/Cr/Zr & Ti carbide-coated ceramic blades	7
17	Increased cutting elements	23
18	Pivotable shaver heads	17
19	Swappable shaver heads	14
20	Water repelling grooves in handles	19
21	Motor heat redirection	26
22	Vibrations transferred via foil	3
23	Stainless steel alloy/titanium alloy blades	5
24	Heated foils	4
25	Shaver head chrome coating	21
26	Suction properties on shaver head	22
27	Edge thickness reduction <10 μm	9
28	Vibration dampers/redirectors	20
29	Passive Wi-Fi charging	12

silica-based electrodes (rank 13) did not reach the first one third of the ranking table. This indicates that even though fast charging and wireless charging are important features for the shaver since they ranked in the top half of the table, the primary focus should be on battery operational period and capacity, since these most directly influence the user experience.

The technology that ranked directly after Co/Mo Li ion batteries is the use of better alloys based on titanium or alloy steels with Ni/Cr/Rh coatings to develop better cutting blades. This of course has a direct impact on the core utility of a shaver to shave. This aspect of the shaver is also gaining increased attention due to the increase of purchasing power in countries of Africa and the Indian subcontinent along with the Middle-east. The increased buying power in these countries has put them on the map as potential growth markets for electric shavers, but the higher beard density and thickness of the people of these regions is a challenge for shaver manufacturers who have devoted the last five decades focussing on European and North American beard profiles and thinner facial hair. An improvement in the cutting element would not have a drastic impact on the comfort experienced during the first shave as most shavers have cutting elements able to cut hair with minimal discomfort for the first few shaves, but the improved cutting element material enables a comfortable shave even after several months of use, and this is what truly sets the best shavers apart from the competition and develops customer loyalty. A more durable blade also results in a quicker shave even after a few months of use and no need to repeatedly shave rough patches causing skin burn and irritation. The increased durability would also result in a longer lifecycle of the shaver head and lower expense for the consumer, considering they have to replace the shaver head less often. The fact that prolonged comfort of shaving is essential to users is proven further by the fact that the cleaning centre and cleaning jet technology rank third in the list while swappable shaver heads rank only 19th on the list. The cleaning centre available with some shavers is used to clean the shaver head with ethanol-based fluids to remove all remnants of hair from the cutting elements to enable smoother internal operations and a clean contact of the cutting element of the blade with the hair upon use the next time. If there are any remnants of hair or skin on the edge of the blade, it may result in pulling of the hair and undercuts, which may leave rough patches and injured skin. Further improvements to the blade can also be delivered by surface coating and nitriding of blades, and this again ranks at number 5 in the list along with heat treatment of the blade edge to deliver increased localised hardness and better hair cutting properties, which hanks in the first half of the list at number 14. Along with conventional metal-based blades, testing on coated ceramic blades is in progress and may lead to the next breakthrough in shaving technology, as it would deliver a much higher lifecycle (2–3 times more), hence earning it the seventeenth position on the list. The fact that four of the top five focus technologies focus on a cleaner, smoother, more comfortable shave over a longer time frame shows the true focus of shaving technology has not yet deviated from the core aspect unlike some other fields such as mobile phones where the peripherals such as cameras and processor speed have overtaken the key utility of the mobile phone, which is that of communication.

The anthropometric design of the shaver ranks sixth on the list. The design of the shaver is not only essential during use so that the shaver fits perfectly into the user's hands but also improves the aesthetics of the shaver making it appear more technologically advanced, resulting in better sales. This is a fringe effect of the anthropometric design of these shaver body and has contributed to success of the product over other similar products in several cases. This is also the reason that rubber grips and water repelling grooved on the handle design come up at rank 15 and 20, respectively, in the list putting them in the first two-thirds of the focus technologies. The chrome coating of shaver heads also has more of a cosmetic impact on the shaver than functional and is already a prevalent design feature in most premium shavers since a decade hence landing it the 25th position on the list. Another surprise entry into the top ten is the long hair trimmer attachment, which has not traditionally been considered a core shaving component and is not present on several premium shavers. The long hair trimmer, however, is key in transforming the shaver from a device for use only on the beard to an overall grooming apparatus that enables the user to modify sideburns, do basic beard styling and even groom other parts of the body. The versatility added to the shaver is key in ensuring success over the coming years considering more and more users prefer a light trim over a clean shave.

The technologies relating to motor efficiency such as reduced stator motor gaps and more efficient silver and copper-alloy-based winding materials rank number 9 and 11 on the list signifying that motor efficiency and power is a key contributor to the efficacy of the shaver but not as critical as the blade design, battery or the body design. This also stems from the fact that motor deign has received critical attention in almost all field ranging from shavers to satellites, and the currently available technologies are already sufficient to meet the needs of the shaving industry with further research proving to be superfluous and expensive. Instead of focussing on higher motor frequencies and efficiency, the increase in cutting actions can be achieved by increasing the number of cutting element arrays on the shaver head driven by the same motor essentially multiplying the cutting actions by the number of arrays included in the shaver; this technology also ranks in the top two-thirds of the list at number 17.

The last essential group of technologies on the list pertains to guiding the hair to the cutting elements to achieve a comfortable shave. This technology founds the very basis of shaver research since the 1950s and is essential because if the blades make contact with the hair at a wrong angle, they will result in painful undercuts and tugging despite exceptional blade design. This is the reason positions 12 and 13 on the list are taken up by hair guiding elements on shaver head and foil orifice design, respectively. The guiding elements are more essential in today's market than even before considering the number of users who like to maintain a light beard. The longer hair does not flow into the minute orifices of the shave head as easily and short hair that essentially stands more perpendicular to the skin in comparison to the long hair, going into the shaver foil orifice more easily and interacting with the cutting element at perpendicular angles. The long hair on the other hand tends to lay flat on the skin, and the guiding elements on the shaver head play an essential role of feeding the hair tips into the shaver orifices to be cut by the cutting elements. The further implementation of pivotable heads adds to the comfort of shaving and enabling the shaver to be adjusted to the best contact angle to deliver the best shave for each section of the skin, hence putting this technology at number 18, in the first two thirds of the list.

Moving to the less critical technologies, we can see motor heat redirection, vibration transfer to the foil and heated foils at number 21, 22 and 24, respectively, on the list. All three technologies and fairly new and research has grown in the past years due to the fact that increasingly powerful motors are used in shavers over the years with more undesirable heat and vibration by-products that have not been efficiently utilised to date. Unfortunately, heated shaver foils and even conventional shaving blades have not gained much market traction as users look for a cool soothing experience during a shave. The heated blade may have some use in regions where bathrooms are traditionally not heated even in winters such as Japan or in regions where indoor heating is not yet prevalent even in winter, such as the Indian subcontinent, but besides that user feedback does not point to a more comfortable shave and hence the added investment in developing this technology does not seem prudent. Another counterintuitive fact is that most users consider the shaver to be doing a better job if they can feel a certain degree of vibration in the hands while using the shaver. This essentially means that little or no further investment is needed to redirect motor vibration till user perception shifts over the decades.

In the end, technologies such as suction properties on shave heads during use, vibration dampers and Wi-Fi charging and even super sharp blades with less than 10 μm blade thickness rank at the bottom of the list, as they are not sufficiently developed to be marketed as such with current shavers, and the technology has not reached a point where it can be launched over the next 5 years. These technologies may be critical over the next decade as other technologies plateau out, but in their current formats, launching them may lead to product failure and reputation damage for any firm. In case of techno-products, being too early has an effect as adverse as being too late and hence it would be worthwhile to perfect these technologies slowly, learning from competition and monitoring entrepreneurs in the field to deliver the best product for the user.

Publication Bibliography

A contactless power transfer system with capacitively coupled matrix pad (2011) With assistance of C Liu, AP Hu, X Dai. 2011 IEEE Energy Conversion Congress and Exposition: IEEE.

Albright, R. E., & Kappel, T. A. (2003). Roadmapping in the corporation. *Research-Technology Management, 46*(2), 31–40.

Anderson, P., & Tushman, M. L. (1990). Technological discontinuities and dominant designs: A cyclical model of technological change. *Administrative Science Quarterly*, 604–633.

Barham, H., Estep, J., Daim, T. U., & Oliver, T. (2018). Research and Development (R&D) Management in the Utility Industry. In (*World Scientific series in R&D management*), Technology Roadmapping, World Scientific, Singapore. (p. 113).

Barker, D., & Smith, D. J. H. (1995). Technology foresight using roadmaps. *Long Range Planning, 28*(2), 21–28.

Bitondo, D., & Frohman, A. (1981). Planning tools for effective research management in the "80s: Linking technological and business planning. *Research Management, 24*(6), 19–23.

Blaauw, H. S. (2010). Cutting element, electric shaver provided with a cutting element and method for producing such element: Google Patents.

Blume, F., Voigtmann, L. (1985). Shear foil having protrusions on its skin-contacting surface thereof: Google Patents.

Blume, F., heir Monika, I. B., heir Friedrich, B., Messinger, W., Schiffer, J. (1977). Shear foil for a dry shaver: Google Patents.

Bodo, F. (1968). Shaving head for dry shavers having a coated outer surface: Google Patents.

Bower, J. L., & Christensen, C. M. (1995). Disruptive technologies: Catching the wave., Harvard Business Review.

Bozikis, I., Psimadas, Y., & Angelides, A. (2008). Razor handle having ergonomic gripping areas: Google Patents.

Buck, B. A. (1982). Ancient technology in contemporary surgery. *Western Journal of Medicine, 136*(3), 265.

Buras Jr, E. M., & Harper, A. C. (1979). Electric shaver: Google Patents.

Cosner, R. R., Hynds, E. J., Fusfeld, A. R., Loweth, C. V., Scouten, C., & Albright, R. (2007). Integrating roadmapping into technical planning. *Research-Technology Management, 50*(6), 31–48.

Wet, G. de (1992). Corporate strategy and technology management: creating the interface. In *CSIR, Pretoria*.

EIRMA, W. G (1997). *Technology roadmapping: Delivering business vision*. In *Management summary*. European Industry Research Management Association, Working Group Report No. 52, Paris.

Force, N. Roadmapping Task G.M. (1997). Volume III: Digest of US industry roadmaps. In *Next-generation manufacturing: A framework for action*.

Foster, R. N. (1985). Timing technological transitions. *Technology in Society, 7*(2–3), 127–141.

Franke, W., & Klauer, H.-D. (1998). Dry shaving apparatus with a pivotally mounted long-hair trimmer: Google Patents.

Garcia, M. L., & Bray, O. H. (1997). *Fundamentals of technology roadmapping*: Citeseer.

Gerdsri, N. (2007). An analytical approach to building a technology development envelope (TDE) for roadmapping of emerging technologies. *International Journal of Innovation and Technology Management, 4*(02), 121–135.

Grand View Research. (2018). U.S. electric shavers market size, share & trends analysis report by product type (Trimmers/Clippers, Rotary Shavers, Foil Shavers), competitive landscape, and segment forecasts 2018–2025.

Groenveld, P. (1997). Roadmapping integrates business and technology. *Research-Technology Management, 40*(5), 48–55.

Hillegas-Elting, J. V. (2016). Roadmapping & maturity models: Coming to a view of the forest.

Hui, S. Y. (2013). Planar wireless charging technology for portable electronic products and Qi. *Proceedings of the IEEE, 101*(6), 1290–1301.

Hui, S. R. Y., & Ho, W. W. C. (2005). A new generation of universal contactless battery charging platform for portable consumer electronic equipment. *IEEE Transactions on Power Electronics, 20*(3), 620–627.

Jacob, S. (1930). Shaving machine: Google Patents.

Johnson, G., Scholes, K., & Whittington, R. (2008). *Exploring corporate strategy: text & cases*. Pearson Education.

King, R. L., Crawford, T. C. (2009). Ceramic blade and production method therefor: Google Patents.

Kostoff, R. N., & Schaller, R. R. (2001). Science and technology roadmaps. *IEEE Transactions on Engineering Management, 48*(2), 132–143.

Liu, X., & Hui, S. Y. R. (2005). An analysis of a double-layer electromagnetic shield for a universal contactless battery charging platform. In *2005 IEEE 36th Power Electronics Specialists Conference*. IEEE.

Lupini, S. (2002, June 21). Roadmapping software survey report: White Paper, Cambridge University. Available http://www.strateva.com/resources/white papers.htm

Macintosh, A., Filby, I., & Tate, A. (1998). *Knowledge asset road maps*. University of Edinburgh, Artificial Intelligence Applications Institute.

Matthews, W. H. (1992). Conceptual framework for integrating technology into business strategy. *International Journal of Vehicle Design, 13*(5–6), 524–532.

Messinger, W., Trolltsch, K., & Tewes, B. (1991). Electric dry shaving apparatus with a movable shaving head arrangement: Google Patents.

Miller, G. R. (1991). Shaving system: Google Patents.

Mitchell, G. R. (1985). New approaches for the strategic management of technology. *Technology in Society, 7*(2–3), 227–239.

Nakatsu, H., & Tamura, Y. (2004). Razor blade material and a razor blade: Google Patents.

Nieto, M., Lopéz, F., & Cruz, F. (1998). Performance analysis of technology using the S curve model: the case of digital signal processing (DSP) technologies. *Technovation, 18*(6–7), 439–457. https://doi.org/10.1016/S0166-4972(98)00021-2.

Nishizawa, T. (2006). Electric shaver: Google Patents.

Okada, T. (1994). Electric shaver in which motor rotational speed is controlled according to beard thickness: Google Patents.

Ottone, B. (1957). Multiple cutter rotary shaver having uninterrupted shaving surface: Google Patents.

Panasonic. (01/2007): Overview of lithium ion batteries: Meeting the needs of portable electronic devices. Available online at https://web.archive.org/web/20111107060525/ http://www.panasonic.com/industrial/includes/pdf/Panasonic_LiIon_Overview.pdf

Petrick, I. J., & Echols, A. E. (2004). Technology roadmapping in review: A tool for making sustainable new product development decisions. *Technological Forecasting and Social Change, 71*(1–2), 81–100.

Phaal, R., Farrukh, C., Mitchell, R., & Probert, D. (2003). Starting-up roadmapping fast. *Research-Technology Management, 46*(2), 52–59. https://doi.org/10.1080/08956308.2003.11671555.

Phaal, R., Farrukh, C. J. P., & Probert, D. R. (2004). Technology roadmapping—a planning framework for evolution and revolution. *Technological Forecasting and Social Change, 71*(1–2), 5–26.

Porter, M. E. (2008). *On competition*. Boston: Harvard Business Press.

Prat-Pfister, G. (2012). Three head dry shaver: Google Patents.

Prescient and Strategic Intelligence. (2015). Global shavers market size, share, development, growth and demand forecast to 2020.

Probert, D., & Radnor, M. (2003). Frontier experiences from industry-academia consortia. *Research-Technology Management, 46*(2), 27–30.

Procter & gamble. (2015). Annual Report 2015: Procter & Gamble.

Procter & gamble. (2016). Annual Report 2016: Procter & Gamble.

Procter & gamble. (2017). Annual Report 2017: Procter & Gamble.

Radhakrishna, A. V., & Vardarajan, A. (1991). Maximizing innovation in industry and adopting to change. *Industrial Management, 33*(6), 19–21.

Roadmaps and roadmapping for commercial applications (1998). Proc. Technol. Roadmap Workshop.

Schmitt, P., Nitz, W., Nickel, M. J., & Boudreau, C. R. (2015). Rotary electric shaver: Google Patents.

Sood, A. (2010). Technology S-Curve. In *Wiley International Encyclopedia of Marketing*.

StoreDot: Business Units. Mobile flash batteries for smartphones. Available online at https://www.store-dot.com/business-units, checked on 3/10/2019.

Technavio. (2017). Global electric shaver market: Key rivers and figures.

Tietjens, E. W. (1979). Dry-shaving apparatus with hair-pulling lead cutters: Google Patents.

Trusco, K. Li-Ion prismatics. Available online at https://web.archive.org/web/20111107060525if_/ http://www.panasonic.com/industrial/includes/pdf/Panasonic_LiIon_Overview.pdf, checked on 1/27/2019.

Uchiyama, H., & Okabe, M. (1995). Cutter for electric shaver: Google Patents.

Willyard, C. H., & McClees, C. W. (1987). Motorola's technology roadmap process. *Research Management, 30*(5), 13–19.

Winebrake, J. J., & Creswick, B. P. (2003). The future of hydrogen fueling systems for transportation: an application of perspective-based scenario analysis using the analytic hierarchy process. *Technological Forecasting and Social Change, 70*(4), 359–384.

Yasunaka, S., Ganse, K., & Nakano, H. (1989). Dry shaver with a slidable trimmer handle: Google Patents.

Yoon, M. K., Jeon, C. S., & Kauh, S. K. (2002). Efficiency increase of an induction motor by improving cooling performance. *IEEE Transactions on Energy Conversion, 17*(1), 1–6. https://doi.org/10.1109/60.986430.

Zhang, X., Grajal, J., Vazquez-Roy, J. L., Radhakrishna, U., Wang, X., Chern, W., et al. (2019). Two-dimensional MoS 2-enabled flexible rectenna for Wi-Fi-band wireless energy harvesting. *Nature, 566*(7744), 1.

Chapter 4
A Strategy Roadmap for Post-quantum Cryptography

Grishma R. Pandeya, Tuğrul U. Daim, and Adrian Marotzke

Abbreviations

AES	Advanced encryption standard
BSI	Bundesamt für Sicherheit in der Informationstechnik
CPU	Central processing unit
DARPA	Defense Advanced Research Agency
DLP	Discrete logarithm problem
ECC	Elliptic curve cryptography
ECU	Electronic control unit
ETSI	European Telecommunications Standards Institute
FCM	Fuzzy cognitive map
IoT	Internet of Things
ISGQSC	Industry Specification Group Quantum Safe Cryptography
LiDAR	Light detection and ranging
NIST	National Institute of Standards and Technology
NPD	New product development
NSF	National Science Foundation
OTA	Over-the-Air
PEST	Political Economic Social Technological
PQC	Post-quantum cryptography
RAM	Random access memory
RSA	Rivest–Shamir–Adleman
SSL	Secure socket layer
SWOT	Strength, weakness, opportunities, threat

G. R. Pandeya
Northern Institute of Technology, Hamburg, Germany

T. U. Daim (✉)
Department of Engineering and Technology Management, Portland State University, Portland, OR, USA
e-mail: tugrul.u.daim@pdx.edu

A. Marotzke
NXP, Hamburg, Germany

T. U. Daim (ed.), *Roadmapping Future*, Applied Innovation and Technology Management, https://doi.org/10.1007/978-3-030-50502-8_4

TEC	Technology Executive Committee
TLS	Transport layer security
TRM	Technology roadmapping
UNFCC	United Nations Convention on Climate Change
V2X	Vehicle to everything

This work outlines a three-tier strategy for a European semiconductor vendor to align its internal R&D endeavors with National Institute of Standards and Technology (NIST)'s post-quantum cryptography (PQC) competition. We explore how a European semiconductor vendor is affected by a technological breakthrough in an adjacent technological field and policies from regulatory and nonregulatory organizations like NIST, European Technical Standards Institute (ETSI), and Bundesamt für Sicherheit in der Informationstechnik (BSI). Strength, weakness, opportunities, threat (SWOT) analysis for the semiconductor vendor is conducted to showcase why there is a need to align the vendor's internal R&D efforts with the vendor's innovation portfolio and NIST's PQC standardization effort. The technology roadmapping framework is adopted for this case study where the drivers and their intrinsic motivation to drive PQC standardization effort are identified. The features that the drivers identified earlier want in their products and services are identified. We look into the resources at the disposal of the semiconductor vendor to achieve the features identified earlier. We outline a visual technology roadmap at the end of the analysis and a short-term, mid-term, and long-term strategies for the semiconductor vendor to achieve its goals.

4.1 Introduction

Some inventions have a ripple effect across several disciplines. Such inventions create a paradigm shift in the intended domain and have far-reaching implications in several adjacent fields. A quantum computer is such an example. A large-scale quantum computer, in theory, can render the confidentiality, privacy, and authenticity offered by classical cryptographic algorithms obsolete (Grover 1996; Shor 1997). These features are critical to keeping the idea of connectivity and digitalization alive. A large-scale quantum computer might still be decades away, but National Institute of Standards and Technology (NIST), academia, and industry have already begun to look into a new class of algorithms that resist attacks from quantum computers (NIST 2017b). These algorithms are classified as post-quantum cryptography (PQC) (Niederhagen and Waidner 2017).

For this work, we cooperate with a European semiconductor vendor (ESV hereafter), who wishes to rename unnamed. ESV is a global semiconductor company that designs and manufactures embedded processors among other things. These processors run the cryptographic algorithms and provid the features to enable the idea of connectivity and digitalization. ESV's products are used in a wide array of applications ranging from automotive and personal identification to IoT. The replacement of the classical cryptographic algorithms with the novel cryptographic algorithms will affect the ESV's product portfolio primarily because the current products offered by ESV were designed to support current cryptographic standards. The new (proposed) cryptographic algorithms require more resources (computation, memory, and communication bandwidth) than the classical ones (Niederhagen and Waidner 2017; Valyukh 2017). Standardizations, in general, are lengthy processes. There have been a number of standardizations processes in the past. The usage of elliptic curves in cryptography was first proposed in the mid-1980s and took about two decades before it was widely adopted by industry (Koblitz 1987; Miller 1985). Similarly, advanced encryption standard (AES) was analyzed for 4 years and took a decade to gain a wider adoption (NIST 1997). NIST started the selection procedure for PQC algorithms in the year 2017 and is currently in Round 2 of the selection process. It is expected to take a couple of more years for NIST to announce PQC standards (Moody 2017). Currently, 17 submissions are being considered and ESV is a co-author in several of the submissions. Once NIST announces the standards and quantum computers are perceived as an imminent threat by the customers, it is expected that there will be a demand for the change in (digital) infrastructure to support the new standards. This poses a unique challenge for ESV. Businesses usually have a sharp focus for only a few quarters ahead. Reacting to PQC standards requires alignment with ESV's long-term strategy.

4.1.1 An Introduction to Technology Roadmapping

High-tech companies must constantly scout for next trends in the industry to stay ahead of the curve. Anticipating a trend and funneling resources to innovate in a new direction are usually met with resistance and uncertainties within and outside of the company. A company must make a series of complicated decisions before bringing out a new product or service to the market. In a semiconductor company, a new business creation (or a new product development) can be a complex process which involves a cross-functional collaboration (Nauta 2011). It is an absolute necessity that the key stakeholders within the company agree to the vision of the company (both short-term and long-term). A planning technique that involves the key stakeholders across different teams that is consistent with the company's strategic goals is a technology roadmapping (Olin et al. 1997). The technology roadmap is the outcome of the planning process (Olin et al. 1997). Such a formal approach was first adopted by Motorola and Corning to link product and technology plans (Willyard and McClees 1987). This procedure is now widely used in small enterprises and large companies alike. Several projects of national and international importance have adopted the technology roadmap (TRM) approach (Olin et al. 1997). The United Nations Framework Convention on Climate Change (UNFCCC)

Technology Executive Committee (TEC) defines technology roadmap (TRM) as follows (UNFCC 2013):

> A Technology Roadmap (TRM) serves as a coherent basis for specific technology development and transfer activities, providing a common (preferably quantifiable) objective, time-specific milestones and a consistent set of concrete actions; developed jointly with relevant stakeholders, who commit to their roles in the TRM implementation.

At the very core, TRM process aims to answer three basic questions (Olin et al. 1997).

- Where do we want to go?

This question addresses the strategy and vision aspect of a company. We explore what kind of product features and services we would want at some point in the future. The forward thinking must align with the organizational goals and ambitions. To address this question, one must have a clear understanding of the trends in the industry and why the organization must respond to such trends. Such an endeavor requires a sponsorship from the higher management (Olin et al. 1997).

- Where are we now?

In this part, we explore the current capabilities of an organization. The findings of this question give an insight on how much of an effort it would be to achieve the vision set in the first question. Furthermore, the findings of this question tell us about the dynamic capabilities of the organization in question (Barney et al. 2001).

- How do we get there?

In this question, we explore the available resources, possible collaborations, etc. to help the organization to achieve the goals set in the first question.

"Roadmap" is a generic umbrella terminology used by organizations to map the relationship between science, technology, and its applications (Kostoff and Schaller 2001). Such a formal approach was first adopted by Motorola to link product and technology plans (Galvin 1998). In their paper, Willyard and McClees (Willyard and McClees 1987) argue that Motorola's products were getting much more complex which led Motorola to develop and maintain a roadmap so that Motorola did not miss out on emerging technologies. These roadmaps ensured an efficient communication between technology experts and market experts within Motorola. To achieve the aforementioned goals, Motorola worked on two separate roadmaps: Emerging Technology Roadmap and Product Technology Roadmap (Willyard and McClees 1987). The former is aimed at addressing general strategic challenges (also known as S-plan), while the latter (also known as T-plan) is more granular and more focused on a particular product (Phaal et al. 2013). The contrast between these two roadmaps can be shown visually. An innovation funnel which is widely being used by corporations to foster innovation (Koen et al. 2001). The innovation funnel has three distinct phases, that is, fuzzy front end, new product development, and commercialization (Turner 1989). The S-plan corresponds to the left part of the innovation funnel (Phaal et al. 2013). The situation in this phase usually corresponds to a lot of fuzzy ideas and no concrete strategy. The roadmapping process (S-plan) carried out in this stage deals with the challenges at business, corporate sector, and policy levels (Phaal et al. 2013). Once a concrete strategic outline is defined in this stage, the next stage we move on to is the new product development (NPD). The right-hand side of the funnel represents the T-plan or product level roadmap where the requirements are more granular and well defined.

The usage of TRM by the firm Motorola illustrates the importance of a roadmap in the overall strategic planning process. A roadmap creating process acts as a crucial link between the initial idea generation phase and the concrete implementation phase. The roadmap generation process brings in the people across several disciplines within a company together to brainstorm ideas and to define future products and services. Such a collaborative approach can also be seen as a knowledge management tool, which in turn can be used as a decision-making tool (Guo 2010). The technology roadmap that we create at the end of the roadmapping process is a visual representation with multiple layers. Such a roadmap shows a visual relationship between technology, products, and markets. The top layer of the roadmap showcases market and business drivers. In the middle layer, we have the product features or services that will help the organization achieve its vision or strategy. On the bottom layer, we have the R&D programs and the resources available to the organization to achieve the goals. TRM is not the only strategic planning tools used by firms to identify opportunities (Schneier 1999). There exist several strategic planning tools like SWOT (Helms and Nixon 2010), Political Economic Social Technological (PEST) (Gupta 2013), and Potter's five forces. Because of the collaborative nature of the TRM framework and the ease of interpretation, TRM has found its applications in diverse industrial use cases (Schimpf and Abele 2019). TRM has been adopted in use cases like Russian Media Industry (Milshina and Vishnevskiy 2019), cattle farming (Rivero and Daim 2017), and robotics technology for the power industry (Daim et al. 2018). The wider adoption of TRM is further evident from the bibliometric analysis carried out by Carvalho et al. (Huang et al. 2014). The trend is even more pronounced in tech sectors, such as mobile communications, automotive and energy, chemical products, software, nanotechnology, and mining (Schneier 1999). In some use cases, TRM has been even extended to be used with fuzzy cognitive map (FCM)-based scenarios (Amer et al. 2016) and also in tandem with marketing and decision methodologies (Fenwick et al. 2009), etc. among others. In Amer et al. (2016)), the authors point several possible future scenarios in the wind sector of a country, which is then used to develop a national-level wind energy roadmap of the country. This wider adoption of TRM framework and its suitability to be used along with other strategy frameworks show its effectiveness as a strategic planning tool.

4.2 Methodology

This work makes use of the technology roadmapping process outlined in Fundamentals of Technology Roadmapping (Albrecht et al. 2019). As outlined in Albrecht et al. (2019)), it was essential to lay the foundation for this work, that is, why ESV should have a strategy roadmap aligned with NIST's PQC standardization effort? We show that there is a need for ESV to react to NIST's standardization effort. During the time of registration of the study, it was clear that ESV acknowledges NIST's standardization effort. However, ESV does not yet have a decisive answer about the resources ESV is willing to invest for this initiative. That being said, seeking sponsorship from higher management for ESV's official PQC endeavors is beyond the scope of this study. It is encouraging, nevertheless, to know that ESV has an interest to invest in this direction. For this study, we identify experts in the field of cryptography and hardware architecture that are employed by ESV. We consult these experts throughout the study to gather the relevant information for the study. In the second phase of the study, we identify the critical features that we want our product to support in a post-quantum world. We also identify the current capabilities of the products offered by ESV. We identify the resources that ESV should mobilize to achieve its PQC endeavors. We recommend a three-tier strategy for ESV to align its internal PQC endeavors with NIST's PQC standardization effort (Fig. 4.1).

4.2.1 *Risk Analysis*

The society today is increasingly relying on connectivity, that is, data generation, data sharing, and data storage. As the IoT trend accelerates, we find ourselves in the web of connectivity in every aspect of our modern society. As an individual, we are constantly transmitting our messages, our banking information, health data, etc. through the Internet. Through Industry 4.0 (convergence of industrial production

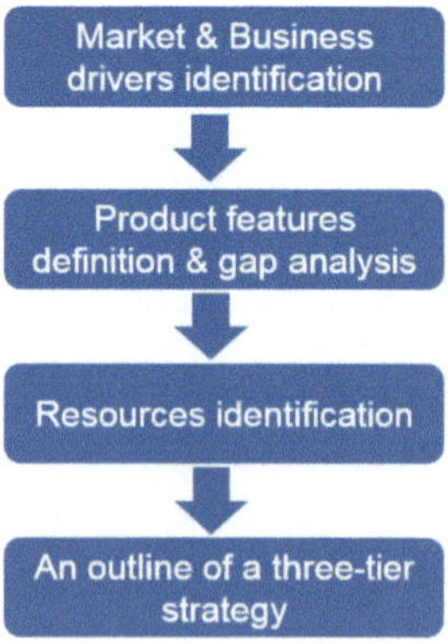

Fig. 4.1 Methodology used in this study

and information and communication technologies (Popp 2009)), our industries, automotive, and digital infrastructures, in general, are also connected to the Internet and are continuously transmitting and receiving information. The concept of connectedness and digitalization finds its confidence because of the cryptographic measures that we deploy. The cryptographic schemes that we use today ensure the confidentiality, authenticity, and integrity of the data that we transmit in the Internet. Without such schemes in place, we would not see such a wider adoption of connectivity.

At the core, public-key cryptographic schemes are based on mathematical (hard) problems that are conjectured to be computationally infeasible for an attacker to solve on a classical (silicon-based) processor. Such schemes ensure that our data are safe on the Internet in the presence of eavesdropping adversaries. Quantum computers can, in theory, solve many such hard problems quickly. Therefore, an adversary with a large-scale quantum computer could (in theory) render the confidentiality, authenticity, and integrity offered by public-key cryptosystems obsolete. A large-scale quantum computer capable of breaking today's cryptosystem might still be years if not decades away. But some forward-looking standardization institutes, academia, and industries have already begun looking into the cryptographic schemes to safeguard our communication against adversaries with a quantum computer. The National Institute of Standards and Technology (NIST) started a competition on post-quantum cryptography in the year 2017. The competition is currently in the second round, where 17 key-exchange algorithms and nine digital signature algorithms have been shortlisted for further security evaluation. Given the daunting task of the security evaluation of the selected algorithms, it will take NIST several years to prepare the draft standards. The ESV acknowledges the fact that majority of the market will not migrate their infrastructure to the PQC standards by their own violation. Significant amount of work lies ahead in terms of educating the market of the risk from an adversary with a quantum computer (see Mosca Theorem 4.1), educating the current workforce about the new class of algorithms, developing an ecosystem that is compatible with PQC standards, and so forth.

The cryptographic algorithms are executed in the processors that are embedded on ubiquitous electronic devices. Such processors are the infrastructures of the digital world. ESV manufactures such processors that are used in various applications. The product portfolio of ESV can be broadly categorized into the following verticals:

1. Automotive
2. Industrial
3. Smart city
4. Communication infrastructure
5. Mobile
6. Smart home

Quantum computers are a not only a threat in the post-quantum era but they also pose threat to sensitive data even in the prequantum era. Some data need to be secure for a long period of time. Examples for such cases are personal health records, data on national identity cards, military communication, and a company's intellectual property. An adversary with enough resources could record all the encrypted traffic and wait until quantum computers are available to de-crypt the sensitive data. This attack model is called "store now, decrypt later" (Kabanov et al. 2018). Similarly, some digital infrastructures and devices will remain in the field for a duration of 5–10 years. Automotive in general has a field life of 10 years (NIADA 2019). Similarly, e-passports also have a field-life of 10 years. If we assume that a large-scale quantum computer can be realized in 10 years from now, we must have cryptographic measures in place well in advance (Niederhagen and Waidner 2017).

4.2.2 *SWOT Analysis*

4.2.2.1 Strengths

ESV has arguably been one of the best competencies in terms of cryptography, secure implementations, and hardware design. It is evident from the list of patents assigned to ESV and ESV's inclusion in the Forbe's 100 most innovative companies. Out of the 17 key-exchange submissions made to the NIST's PQC competition, ESV employees have contributed to several submissions. ESV has developed in-house competency over the years, which it can leverage to position itself strongly in terms of post-quantum cryptography.

4.2.2.2 Weaknesses

ESV acknowledges NIST's PQC standardization effort. However, resource allocation for ESV's PQC endeavors is not sufficient. Only one head count has been approved to work on ESV's PQC initiative. Most of the researchers, engineers, and architects are working to support the current generation products or the products that are already in the pipeline. Furthermore, talent pool working on the PQC is very small and most of them are working at a university or a research institute. Attracting this talent to ESV might be a challenge because cyber-security skills are one of the most sought after skills (Leaser 2019).

4.2.2.3 Opportunities

ESV researchers argue that Bundesamt für Sicherheit in der Informationstechnik (BSI) might publish a recommendation for PQC algorithms for products that handle sensitive data and for products that go to safety critical infrastructures. ESV researchers argue that BSI will adopt very conservative approach, that is, they will recommend algorithms that have been studied for a long time but are comparatively inefficient on current hardware. This will affect all semiconductor companies because none of the products from these companies are optimized for such algorithms. However, once the recommendation from BSI is published, semiconductor vendors have to comply. Thus, if such a scenario were to play out, ESV could capture greater market share by being the first one to bring out products optimized for the recommended algorithms.

4.2.2.4 Threats

Infineon Technologies AG (Infineon hereafter) is a direct competitor to ESV in most of the product verticals. Infineon is actively marketing its PQC readiness on its products. Infineon was even awarded two SESAMES award for implementing post-quantum cryptographic algorithm on contactless security chips (Braeckle 2017). Infineon has been actively marketing its PQC readiness, while ESV is yet to make any official statement on PQC on its products. Infineon has not announced any product optimized for PQC algorithms, but its marketing activities could be seen favorably by the customers. The customers who are planning to prepare their infrastructure for PQC algorithms might go to Infineon for consultation rather than come to ESV.

The high cost of IP licensing fee from ARM and the high R&D cost to develop brand new IP can be seen as a barrier to market entry for new companies. With its open BSD license, the RISC-V architecture offers an alternative to the ARM architecture (Asanović and Patterson 2014). Historically, Europe, the United States, South Korea, Japan, and Taiwan have been the leader of the semiconductor industry. RISC-V architecture might potentially change that and allow smaller entities to develop competitive semiconductors. This might also support the semiconductor ambitions that China has. China has a goal to increase domestic semiconductor manufacturing as a share of domestic consumption to 80% by 2030 (Allen 2019). Further, China aims to reduce its external dependencies for its semiconductor industry and is heavily investing in this area (Allen 2019). China will enormously benefit from the open-source hardware architecture initiative like RISC-V. Further, there are ongoing research initiatives to develop optimized processors for PQC upon RISC-V architecture (Niederhagen 2019). Unless there is a trade war between Europe and China or sanctions on exporting IP related to cryptography, Chinese firms should be able to buy IP from the research initiative in (Niederhagen 2019). Chinese firms or any other firms can also buy IP from companies like PQSecure and PQShield (Azarderakhsh 2020; Kaafarani 2020). These Chinese firms will then be directly competing with ESV products in certain verticals. It would be too ambitious for such Chinese firms to immediately compete with ESV in European or North American market. However, such firms could start small like Xiaomi and Huwawei and slowly compete with ESV's products in the Chinese market. Over the past years, we have seen that China has significant soft power in Asia through its one-road-one-belt initiative and in Africa through its economic engagements (Ferdinand 2016; King 2013). Such Chinese firms could directly compete with ESV's products in Asian and African market as well.

4.2.2.5 Automotive

The three major trends in the automotive sector are Vehicle Autonomy, Vehicle Over-The-Air (OTA) Updates, and Vehicle-to-Everything (V2X) connectivity (Patel and Hodgson 2019). Powerful processors (electronic control units, ECUs) are fueling these trends. Today, there are about 100 ECUs in a car-enabling feature such as safety airbag, media, and LiDAR sensor data processing. These trends bring added value to customers while opening new potential attack vectors. ESV offers several products to support these trends in automotive. Please refer to Annex 1 for the list of the product offerings. The product line of primary interest in terms of the security of the automotive would be the ECUs themselves. In-vehicle network products are also relevant in terms of security of the automotive. However, an attacker would require a physical access of the automotive to mount an attack on the in-vehicle network. For the risk analysis in this chapter, we will primarily focus on the remote attacks.

The software components in the automotive require regular updates. The main electronic processor of the automotive acts as an interface between these components and the outer world. These updates leave the automotive susceptible to remote attacks. Thus, the main automotive processors are used in a tandem with a security controller. Security controllers provide cryptographic functions (key storage, encryption, authentication, etc.) that thwart any intrusion attempts from an (remote) attacker.

4.2.2.6 Security and Authentication Products

ESV has several product offerings in security and authentication domain. The products are used as national identity cards, health cards, payment cards, industrial security controllers, etc. The products in this domain can be of high interest to attackers because of the financial motive and the sensitive data these products process.

As mentioned earlier, e-passports are expected to remain in the field for a period of 10 years. The e-passports issued in the prequantum era will overlap with the post-quantum era if we assume a large-scale quantum computer will be built within 8–10 years. Therefore, cryptographic schemes used in e-passports must withstand threats from quantum computers.

4.2.2.7 Threat Factors

ESV's security products are used in a wide range of applications. They process data of varying sensitivity. Thus, "store now, decrypt later" model does not apply to all the products. Similarly, different products and applications will have different level of impacts if compromised. We have identified features that would determine the level of impacts on applications that make use of ESV's security products. These threat factors help ESV's customers to assess the threat level from an adversary with a quantum computer. It is advisable to consider Mosca's theorem (see sect. 4.1) while accessing the risk from an adversary with a quantum computer.

- Ease of recall

Vulnerabilities in consumer products are discovered regularly, and they are patched if possible over the air. In some instances, the products have to be recalled to fix the detected vulnerabilities (Eikenberg 2019). Products like Logitech presenters are easier to recall; however, recalling products like automotive or processors used in smart grids in far off locations are a logistic nightmare.

- Expected duration in the field

Devices remain in service for varying durations. Smart cards have a field life of 1 year, while e-passport remains in the field for a period of 10 years. The longer the field life, greater are the risks of attack from quantum computers.

- Sensitivity of the data

As discussed in the earlier section of the chapter, different security products handle different data. Health data, personal data on e-passports, etc., are more sensitive compared to, for example, the data that secure car key handles. Thus, depending on the type of data handled by the products, the impact of quantum computers can be moderate to high.

- Ease of update

The device's Over-The-Air (OTA) update feature plays a crucial role in extending the device usage in the field. A secure update of a device in the field helps patch the vulnerabilities identified in the device or any of its software components.

- Financial impact

ESV offers products in the secure payment sector. Vulnerabilities in such system could cause a financial havoc to ESV's customers and the end users. Similarly, vulnerabilities in the IT infrastructures that cause elevation of privileges and disclosure of sensitive information (such as company secrets and government's trade strategy) could cause a colossal financial damage.

- Scalability

The vulnerabilities in the IoT architecture can be scaled up because of the distributed and the connected nature of the IoT systems. Thus, any vulnerability can be exploited to affect a large number of devices (or end users).

4.3 Market and Business Drivers

In this chapter, we explore the key element that created the need to develop the new cryptographic primitives and the agents that are driving the post-quantum cryptography standardization process. Furthermore, we classify private business sectors into two major groups and discuss their motivation on why they should drive post-quantum cryptography standardization process.

4.3.1 Perceived Threat of Quantum Computers

The idea of a quantum computer was first proposed by Paul Benioff (Benioff 1965). Quantum computers got a wider traction in the cryptographic community when Peter Shor showed in his paper that it was possible to break two widespread hard problems using a quantum computer in polynomial time (Shor 1997), namely integer factorization and the discrete logarithm problem (DLP). Integer factorization (finding the prime factors of a large integer) is the underlying hard problem used in Rivest–Shamir–Adleman (RSA) (Rivest et al. 1983). Similarly, the DLP is the underlying hard problem used in elliptic curve cryptography (ECC) (Koblitz 1987; Miller 1985). It is computationally infeasible to factorize large integers or to solve the DLP problem used in ECC using a silicon-based processor. It is, however, possible to factorize such integers and solve the DLP problem using a large-scale quantum computer in polynomial time (Grover 1996; Shor 1997). RSA and ECC are widely used to encrypt the modern-day communication. Thus, a large-scale quantum computer would render the modern-day asymmetric cryptography useless.

Anybody with a large-scale quantum computer can, in theory, decrypt most of the encrypted data in the web today. Governments and military organizations are naturally interested in developing such a device. Similarly, private organizations like Google, IBM, and research universities are also taking initiatives in developing their own quantum computers. Google recently published a paper where they claim to have a functioning quantum computer with 53 qubits (Arute et al. 2019). They claim that this quantum computer showed quantum supremacy, aka, that it was able to solve a problem that is infeasible for a classic computer (Arute et al. 2019). A quantum computer capable of breaking RSA and ECC would require a quantum computer with at least 4096 logical qubits and 2330 logical qubits, respectively (Roetteler et al. 2017). We might still be far away from building such a powerful quantum computer. Quantum computer research, nevertheless, is moving in the right direction in an accelerated pace. Some experts argue that there might be such a powerful a quantum computer within the next decades (Niederhagen and Waidner 2017).

Due to the inherent uncertainty in terms of the prediction of the arrival of a large-scale quantum computer and the "Store now, decrypt later" attack model, it is imperative that organizations have a robust and agile migration plan in place.

Assessing the risk from quantum computing and outlining a migration plan is a rather complicated process. The risk assessment and the migration plan depend on the assets that need to be protected, current digital infrastructure in place, and resources that are at company's disposal to facilitate the migration process. Dr. Michele Mosca simplifies the risk assessment process in a seemingly easy theorem (Mosca 2015):

If X + Y > Z, then worry!

What is X?

Mosca defines X as the duration for which the information needs to be secure or the shelf life of the cryptographic primitives used to protect the information. Different organizations will have different assets to protect. Similarly, different organizations use different cryptographic schemes to protect those assets. The assets and the state of the current deployed cryptographic system will determine the X.

What is Y?

Mosca defines Y as the time that would be required to migrate the current digital infrastructure to a fully quantum-secure infrastructure. The path to achieve this varies from organization to organization. The Y depends on several factors such as the current state of the digital infrastructure and the financial resource of an organization.

What is Z?

Mosca defines Z as the number of years until a large-scale quantum computer can be realized. There is no consensus among researchers for the about the expected timeline in which a large-scale quantum computer can be built. However, experts argue that 20+ years is a conservative assumption to determine Z (Niederhagen and Waidner 2017) (Fig. 4.2).

If X > Z, then the organization does not have to worry about the privacy of the information in the prequantum era. If your X + Y > Z, then it means that the organization has a good migration plan in place. Thus, the organization need not worry about the threat of quantum computers. The threat becomes imminent if X > Z or X + Y > Z. In these two scenarios, the organization might need to panic. A remedy for this situation might be pouring additional resources to escalate the migration process which will be extremely expensive and risk prone. If an organization has any asset to protect for a longer period, the organization must conduct a thorough assessment of the assets they want to protect, the current state of their infrastructure, and the resources they have for the migration process.

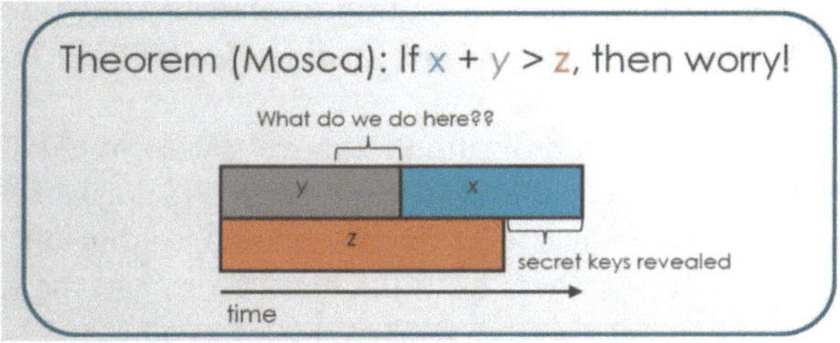

Fig. 4.2 Illustration of Mosca theorem (Mosca 2015)

4.3.1.1 Government

BSI, in their annual report, has listed three levels of cyber threats to a country (Germany in this report) (BSI 2019).

- The level of threat to the Federal government

Government bodies such as military and foreign policy office must constantly deal with the sensitive data that must remain secret for a considerably long period of time. Any disclosure of such data (e.g., personal data of the residents of a country and military secrets) could undermine the national security of the country.

- The level of threat to critical infrastructures and the business world

In the recent years, digitalization and connectivity have changed the way the economy performs. Industry 4.0, connected infrastructures, V2X, etc. have opened new avenues for the economy of a state to flourish while leaving them very vulnerable to plethora of attack vectors. A successful attack in any of the critical infrastructures could cripple the economy, create a distrust among citizens in the adoption of new digital technology, and force the government to select conservative approaches to digitalization which in turn will subvert any further digitalization initiatives.

- The level of threat to the society

Several facets of digitalization are so pervasive in our society that it is impossible to imagine a society without the amenities that we enjoy today. Citizens are constantly connected to the Internet through their smart phones, PCs, smart wearables, etc. and are sharing confidential data. It can be assumed that the state has a certain responsibility to protect citizens' confidential data from the prying eyes of an adversarial state. The recent episodes of meddling in the US election, German Bundestag election, and the targeted widespread of fake news on the social media show how far-reaching the implications of data at wrong hands can be (Van De Velde 2017).

Historically, the US nonregulatory organization NIST has taken initiatives to standardize cryptographic primitives and protocols (Burr et al. 2014; NIST 1997; NIST 2017b). Such standards have been widely adopted and deployed in the systems all around the world. In the year 2015, NSA acknowledged the risk of quantum computers on existing cryptographic schemes. NSA wanted a transition of the so-called Suite B cryptographic algorithms toward quantum-resistant algorithms (Schneier 2015). Suite B consists of a range of algorithms approved by the NSA for top secret and military operations (Schneier 2015). Toward the end of 2016, NIST called for a proposal for PQC candidate algorithms. The official call from NIST was a huge impetus (Table 4.1).

Thus, the standardization initiative started by NIST in November 2017 was the first step toward the transition process. NIST asked the cryptographic community (researchers, industries, etc.) to submit algorithms that are conjectured to be quantum safe. Researchers from the universities and the industry alike submitted their

Table 4.1 NIST PQC standardization time line

Fall 2016	Formal call for proposals
November 2017	Deadline for submissions
Early 2018	Workshop submitters' presentations
Early 2019	Second round candidates announced
3–5 years	Analysis phase
2 years later	Draft standards ready

algorithms for the evaluation. We are currently in the second round of evaluation. NIST is expected to announce the standards in 2023 (tentative) (Moody 2017).

Similarly, European Technical Standards Institute (ETSI) created Industry Specification Group on Quantum-Safe Cryptography (ISG QSC) (ETSI 2015). The primary aim of the group is to analyze the performance, implementation capabilities, benchmarking, and practical architectural consideration for specific applications (ETSI 2015).

The institutes like NIST and ETSI are taking leadership roles in the standardization of the post-quantum cryptography and are one of the major drivers of PQC standardization process. The US government is promoting the research into PQC through funding (DARPA, NSF, etc.) (see Table 4.4). Similarly, European Union is promoting the research into PQC through funding program like HORIZON2020 (see Table 4.4).

4.3.2 Industry

In this section, we make a distinction between two types of industries. Type I: the businesses whose business model involves around gathering, storing, and processing user data. Companies like Facebook, Amazon, and Google fall under this category. Type II: the businesses whose business model is about providing digital infrastructure (processors, sensors, etc.) to enable the idea of connectivity and digitalization. Companies like ARM, ESV, Infineon, Intel, and NXP fall under this category.

4.3.2.1 Type I Industries

As mentioned earlier, the core business model of companies like Facebook, Google, Amazon, and banking sectors relies around user data. These data are usually generated by users using the end devices such as PCs, mobiles, and smart cards. The user data are transferred from the user device to the central server of the company. Encryption is used to protect it from eavesdropping and manipulation. An eavesdropping adversary can in theory record all the data in transit. However, due to the encryption schemes used, the adversary would not be able to get any meaningful information out of the encrypted data.

If an adversary with a powerful quantum computer were able to decrypt the intercepted communication, the business model of these companies would crumble, and the end users would lose confidence in using such services. For example, the IT Governance UK lists the following consequence of such data breaches for a business (Graham 2019):

- Compensating affected customers
- Setting up breach response efforts, like helpdesks for affected customers and complementary credit check
- Investigating the incident, which might include hiring a third party or paying your own security staff overtime
- Falling share prices
- GDPR fine of up to 20 million Euros or 4% of annual global turnover

These are grave consequences for any business establishment. To avoid such consequences, Google has already begun looking into feasibility of deploying quantum secure cryptographic schemes in their services (Braithwaite 2016). Google researchers are collaborating with researchers from NXP, Microsoft, Centrum Wiskunde & Informatica, and McMaster University to promote the research in post-quantum cryptography (Alkim et al. 2015; Bos et al. 2016). Big companies like Google, Microsoft, IBM, and Facebook have the resources and the incentives to drive the PQC research.

4.3.2.2 Type II Industries

As mentioned earlier, the companies that provide services, products, or intellectual property that enable the digitalization and connectivity fall under this category. These industries provide microprocessors, sensors, processors, cryptographic libraries, etc. For our analysis, we will consider the ESV as an example.

ESV is a semiconductor manufacturer. ESV manufactures products that are used in automotive, identification, wired and wireless infrastructure, lighting, industrial, consumer, mobile, and computer applications. The risk analysis suggests that the products that go into critical infrastructure and process sensitive data will be at high risk (see Chap. 3). The devices that process sensitive data (e.g., eID cards) need to comply to the guidelines outlined by BSI (or other regulatory organizations). Thus, ESV needs to closely follow the latest developments in the PQC research area and try to influence the prestandard processes.

A researcher from the ESV expects BSI to publish recommendations on PQC as a short-term strategy. The researcher argues that BSI will take the conservative approach and recommend the algorithms that are well researched in the academia for decades. Such algorithms might be (conjectured) quantum secure but are

Table 4.2 Drivers for PQC standardization process

Category	Drivers	Remark	Code
Technological advancement	Quantum threat	The perceived threat that quantum computers can break foundations of modern-day cryptography	D1
	University and research institutes	The institutes that drive research forward	D2
Business need	Government	Government organizations that have secrets to protect for a longer period	D3
	Type I industry	The companies whose business model involves around gathering, storing, and processing user data	D4
	Type II industry	The companies whose business model is about providing digital infrastructure (processors, sensors, etc.) to enable the idea of connectivity and digitalization connectivity and digitalization	D5

computationally expensive and require more memory resources and bandwidth compared to the classical cryptographic algorithms (Niederhagen and Waidner 2017; Valyukh 2017), as well as more modern PQC schemes. Since ESV has significant product offerings that are resource constrained, running those BSI recommended algorithms on such devices will not be an attractive solution to customers if ESV's competition offers a better solution.

ESV is an industry leader in its domain. Some of the leading automakers and smartphone companies are among its customers and the early adopters of new technology. These early adopters constantly turn to ESV for consultation for latest trends in technology. The researcher mentioned that Daimler is concerned about the latest developments in the PQC area and is interested in making its products future proof. When such major customers turn to ESV with questions and if ESV does not have proper answers for that, ESV might lose its loyal customer base. ESV must have recent developments in PQC sphere aligned with its own strategy and vision. Customers like Daimler or Apple have their own strategy and product roadmaps outlined for few years ahead. There must be an alignment between ESV's roadmap and customers' roadmap to create a synergy (Table 4.2).

4.4 Technology Features and Gap Analysis

In this chapter, we explore the technology features that address the needs of the drivers that we identified in the earlier chapter.

4.4.1 Technology Features

4.4.1.1 The Role of Cryptography

Cryptography, the science of hiding a message, has been around for a while in forms such as Caesar's Cipher and ROT13 (Kahn 1996). The world wars saw a heavy use of cryptographic schemes and cryptanalysis in encrypting and breaking messages. As the World Wide Web and personal computing devices were widely adopted by the general public, the field of cryptography garnered its attention since security and privacy concerns are of utmost importance while communicating and making financial transactions over networks. We use cryptography in our communication to achieve the following five primary goals (Paar and Pelzl 2010).

4.4.1.2 Key Establishment

Asymmetric key-exchange protocols are used to establish a shared secret between two parties over an insecure communication channel. A shared secret, thus, established is then used to encrypt further communications. In an asymmetric key-exchange protocol, an individual (Alice) has a pair of keys, a public key, and a private key. The public key is publicly known to everyone who wants to initiate a communication with Alice. The private key, however, should remain private and only Alice should have it. Anyone (Bob) who wants to send a message to Alice would use Alice's public key to encrypt the message. Bob can now send the encrypted message to Alice via an insecure channel. This encrypted message can only be decrypted using Alice's private key. There are three major classes of asymmetric key-exchange schemes based on the underlying mathematical problems.

- Integer-factorization problem

  RSA algorithm is based on this mathematical problem.

- Discrete logarithm problem (DLP)

  Diffie–Hellman key exchange, Elgamal encryption, etc., are based on this class of mathematical problem.

- Elliptic curve DLP

  Elliptic curve Diffie–Hellman key exchange (ECDH) is based on this class of mathematical problem.

4.4.1.3 Confidentiality

Using cryptographic schemes, we can encrypt sensitive data before sending it out. Even if the encrypted data are intercepted during the transit by an eavesdropper, the eavesdropper would not be able to extract any information from the encrypted data. The encrypted data can only be read by the owner of proper decryption key.

4.4.1.4 Integrity

Cryptographic schemes are used to make sure that the data have not been modified in the transit.

4.4.1.5 Authentication

If Alice wants to communicate with Bob over an insecure channel, she must convince Bob that she is indeed Alice and not some impersonator. This can be achieved via challenge-response protocols. Bob asks questions (challenges) and Alice must send responses computed by her private key which Bob can verify using Alice's public key.

4.4.1.6 Nonrepudiation

Nonrepudiation is a cryptographic property to ensure that the sender of a message cannot falsely deny sending the message. This is done via digital signature schemes. A sender generates a signature on a message using their private key, and then sends the message along with the original message. Since the signature could have only been generated by the private key that the sender has, the sender cannot deny sending the message. In an e-commerce example, Bob places an order of 1000 apples and later changes his mind and denies placing such an order and blames Alice for forging such an order.

We have classical cryptographic schemes and protocols that help us achieve the above-mentioned goals. The trust on these schemes and protocols comes from the underlying (conjectured) hard problems and the standardization effort from organizations like NIST, BSI, and ETSI. In a post-quantum world, we would not be able to use the classical hard problems and the protocols built upon those classical hard problems. We would nevertheless need the similar set of cryptographic features built upon new classes of hard problems and new standards. There are several obstacles (mathematical and engineering challenges) to overcome before we can achieve the said cryptographic features suitable to be used in a post-quantum world.

There is a wide array of features that the academia and the industry need to engineer together. This is a massive undertaking for any single entity. The drivers that we identified earlier are working on different sets of problems that are more critical to their business rather than trying to solve the whole puzzle. Below, we discuss the features that different entity is working on.

4.5 Universities and the Affiliated Research Institutes

Researchers at the universities and the affiliated research institutes are primarily interested in the fundamental research. Professors of mathematics, cryptography, theoretical computer science etc., along with their PhD students and postdocs are interested in studying new research areas. If we survey the submissions made to NIST in Round 2 of the PQC competition, it is evident (Fig. 4.3) that the majority of the authors of the submissions are affiliated to universities in some capacity. These groups of people are interested in the following features.

4.5.1 Hard Problems

A large-scale quantum computer could (in theory) break the currently used hard problems, integer factorization, and DLP. This means researchers must explore new classes of hard problems. The talent pool to tackle these problems is concentrated in the universities and the research institutes.

4.5.2 Cryptographic Schemes

Once we have conjectured hard problems, we build cryptographic schemes upon those hard problems. We need to engineer the cryptographic features mentioned earlier in the chapter using the new classes of hard problems.

Scheme	Type	Time [*ms*]			Communication [*bytes*]		Security bits	
		Alice0	Bob	Alice1	$A \to B$	$B \to A$	classical	quantum
RSA 3072 bit	RSA	-	0.08	3.76	384	384	128	-
ECDH nistp256	ECDH	0.225	0.586	0.268	32	32	128	-
BCNS15	R-LWE	0.881	1.41	0.179	4 096	4 224	86	78
NewHope	R-LWE	0.055	0.084	0.015	1 824	2 048	229	206
NewHope MSR LN16	R-LWE	0.046	0.079	0.014	1 824	2 048	128	128
NTRU EES743EP1	LWE	1.08	0.116	0.068	1 027	1 022	256	128
Frodo	LWE	1.95	2.30	0.091	11 280	11 282	144	130
SIDH	SIDH	113	251	106	576	576	192	128
SIDH (compressed)	SIDH	387	586	158	336	336	192	128
McBits	Error-correcting codes	129	0.038	0.106	311 736	141	157	157

Fig. 4.3 Comparison of classical cryptographic vs. PQC algorithms

4.5.3 Mathematical Attacks Against Cryptographic Protocols

The trust on any new cryptographic protocols stems from the years of research and failed attempt at attacking those protocols. There have been numerous attempts to break RSA and the hard problem of integer factorization (Boneh 1999). We trust RSA because it has withstood those attacks. The new PQC algorithms will be under scrutiny from academia and the industry. The research into mathematical attacks is needed to weed out the weak crypto systems and to develop trust in the new algorithms.

These three features are the integral parts of defining a cryptographic standard. NIST alone cannot investigate all the proposed algorithms. That is a massive assignment for any institute.

The government promotes partnership for research by providing research grants to several universities and research institutes to foster collaboration and innovation. This is evident from the fact that almost all the submissions made to NIST were supported by government funds in one way or another (Fig. 4.4).

4.6 European Semiconductor Vendor

As a Type II industry, ESV's products enable digitalization and connectivity through its products. As mentioned in the earlier chapter, ESV's products must comply with the standards outlined by the regulatory bodies. In a post-quantum world, to have the competitive edge, ESV's products must not only comply with the outlined standards but also offer a better solution than the competition (efficiency, security, and price). In this section, we will discuss the features ESV should aspire to bring into its products and services to be competitive in the post-quantum world.

4.6.1 PQC Standards

One of the primary objectives of the standards is to promote interoperability of processes (Blind 2013). There are multiple stakeholders in the semiconductor industry, and it is not desirable to have a scenario where each player promotes its own proprietary solutions which are most likely to be covered by patents. Such a scenario hinders the wider adoption of digitalization and makes it expensive for the end customers. Furthermore, Schneier's law states that "any person can invent a security system so clever that she or he can't think of how to break it" (Schneier 2011). Thus, it is not advised to develop and deploy crypto systems that have not been under the scrutiny from wider experts. ESV created a proprietary encryption algorithm in the past and used it in RFID products. Researchers later showed that the security of the cipher used was close to zero. This incident exemplifies the risk of proprietary crypto algorithms.

Scheme		Public Key size (bytes)	Data size (bytes)
Public-key signatures:			
• Hash based:			
– XMSS (stateful)	[17]	64	2,500 - 2,820
– SPHINCS (state free)	[9]	1,056	41,000
• Multivariate based:			
– HFEv-*	[51]	500,000 - 1,000,000	25 - 32
Public - key encryption:			
• Code based:			
– McEliece	[10]	958,482 - 1,046,739	187 - 194
• Lattice based:			
– NTRUEncrypt	[35, 37]	1,495 - 2,062	1,495 - 2,062
Key exchange:			
• Lattice based:			
– NewHope	[3]	–	1,824 - 2,048
• Supersingular isogenies:			
– SIDH	[21]	–	564
Classical schemes:			
• RSA:			
– RSA-2048		256	256
– RSA-4096		512	512
• ECC:			
– 256-bit		32	32
– 512-bit		64	64
• Key exchange:			
– DH		–	256 - 512
– ECDH		–	32 - 64

* Values using field F_2 and parameter n (number of variables) between 200 and 256.

Fig. 4.4 Comparison of classical cryptographic vs. PQC algorithms

The standardization process is like a rudder of a ship, providing direction to the industry. Further, the standards shift the risk associated with the failure of algorithms (from mathematical attacks) from the companies to the standardization bodies (NIST, ETSI, BSI, etc.). In the absence of standards, ESV would not know what to implement on its products. There would also be no clear strategy on where to drive the next-generation products. Thus, ESV (along with other Type II industries) needs standards that are outlined by regulatory organizations.

4.6.2 PQC Standards for Resource-Constrained Devices

As discussed in the earlier chapter, ESV's product offering varies from high-end processors to resource-constrained devices (see Chap. 3). The current generation of crypto coprocessors is highly optimized to support classical cryptographic operations. The PQC research shows that the PQC algorithms are computationally expensive and require more resources (computation, memory, and communication

Table 4.3 Overview of submissions (key-exchange algorithms) made to NIST

Algorithms	Funding agency	No. of authors	Authors affiliated to universities
BIKE	Horizon 2020, Israel Science	16	13
	Foundation, Intel, FAPESP		
Classic McEliece	The CISCO University Research Program, NSF, NWO, EU(1)	12	8
CRYSTAL-	EU, Swiss National Science	10	6
KYBER	Foundation, NWO, DFG		
FrodoKEM	Microsoft, Google	12	3
HQC	NA	9	7
LAC	NA	10	NA
LEDAcrypt	NA	5	5
NewHope	EC (ICT), NWO, TÜBITAK, CWI, NWO, FCR	14	7
NTRU	EC (ICT), EC (ERC)	9	6
NTRU Prime	Cisco, NSF, EC (Horizon 2020), NWO	4	4
NTS-KEM	NA	NA	NA
ROLLO	NA	13	11
Round 5	NA	13	4
RQC	NA	10	8
SABER	NA	4	4
SIKE	NA	15	6
Three Bears	NA	NA	NA

Table 4.4 Various sources of funding for the submissions made to the NIST

Funding source	Remark
CWI	Centrum Wiskunde & Informatica, the Netherlands
NSF	National Science Foundation
NWO	Netherlands Organization for Scientific Research
EU(1)	PQC for long-term security, European Commission
DFG	Deutsche Forschungsgemeinschaft
FCR	Free Competition Grant

bandwidth) compared to the classical algorithms. The high-end processors might be flexible enough to support the PQC algorithms, but the resource-constrained devices might not be the best choice to run these algorithms.

If we look at Tables 4.3 and 4.4, we can note a dramatic shift in terms of resource requirements. To exemplify two examples, SIDH computation is just over a 1000 times slower than current ECC algorithm for comparable security. Similarly, the Frodo scheme requires 352 times more bandwidth to do the key exchange compared to current ECC algorithms. It is important to note that the timing benchmarks were done on processors running at 4 GHz (Valyukh 2017), while the resource-constrained devices such as smart cards and e-ID have processors running at around 100 MHz (ESV 2020a). Thus, the timing performance is expected to be even worse on such devices. Thus, it is desirable to have a range of algorithms to choose from so that an optimal one can be chosen for different target devices.

4.6.3 Optimized Crypto Coprocessor

Crypto coprocessors are special purpose processors highly optimized for performing complex cryptographic operations (Smith 2011). Current generation of crypto coprocessors is optimized to perform classical cryptographic schemes (RSA, ECC). There has been research to test the feasibility of implementing PQC algorithms on current crypto coprocessors (Albrecht et al. 2019). The findings suggest that there might be limitations on current crypto coprocessors to speed up the computations involved in the PQC algorithms.

ESV's current generation crypto coprocessors have a single arithmetic logic unit (ALU). This does not allow us to execute multiple instructions in parallel (Albrecht et al. 2019; Strenzke 2010; Laan et al. 2018). Research on efficient implementation of PQC algorithms shows that we could achieve significant speed-up in the implementation of PQC with data level parallelization (for isogeny and lattice-based cryptosystems) (Jalali et al. 2019). Similarly, memory allocation to these devices has been made to support the security level suggested by regulatory bodies with a margin to support extended length keys for classical cryptographic schemes. However, the key size of the PQC algorithms is massive compared to the classical algorithms (see Fig. 4.4), requiring more RAM to process. This limitation in RAM memory would force us to use the FLASH memory in devices which is significantly slower than RAM memory, thus further slowing down the overall computation (Rankl 2010). In addition, contactless radio technology such as NFC does not have high data transfer rates. This means that even the transfer of the larger keys can take significantly longer.

The current generation of smart cards takes 300 ms (approximately) to process a transaction (NXP 2020). The time taken to process a transaction is a crucial part of the user experience in, for example, contactless payment in a supermarket. Given the fact that the PQC algorithms require more resources compared to the classical algorithms, the transaction time using PQC (on the current generation hardware) is expected to be higher. Such limitations would make the current devices unattractive in the long run. Furthermore, a marginally better product offering from the competition would affect ESV's market share.

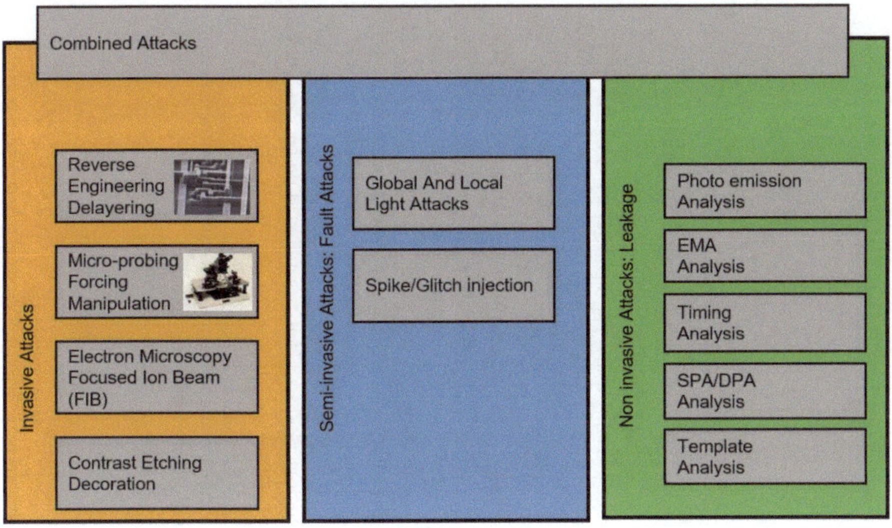

Fig. 4.5 Different implementation attacks (ESV 2020b)

4.6.4 Secure Implementation

From e-ID to payment cards and from ECUs to edge processors, ESV processors are constantly handling security critical data. Such processors are vulnerable to a wide range of implementation attacks (see Fig. 4.5). These are attacks that do not focus on the maths or hard problems behind the schemes, but rather on the concrete implementation. These kinds of attacks are also called side channel attacks. An example for such an attack is where an attacker tries to gain knowledge of a secret key by observing the power consumption of the device during cryptographic operations (this is called a simple power analysis (SPA) attack). Preventing such attacks requires deep knowledge of the cryptographic scheme, physics, and the underlying hardware, together with a carefully designed implementation (Fan et al. 2010).

NIST's PQC standards can only provide guidelines for the functional implementation of the PQC algorithms. Now, building those cryptographic algorithms to devices that will be deployed in the field is not a trivial assignment. Many things can go wrong during the implementation from trust model, the system design, the algorithms and protocols, the implementations, the source code, the human–computer interface, the procedures to the underlying computer system, etc. (Schneier 1999).

PQC research is a new research area. PQC research got a wider attention because of NIST's initiative to standardize PQC algorithms. There has not been significant research on implementing PQC algorithms on ESV's processors and securing the implementation from the known attacks. This requires a significant amount of in-house expertise and intellectual property generation. Furthermore, these novel PQC algorithms have fundamentally different design attributes and mathematical foundations, and there might be novel ways of attacking the PQC schemes and the implementation (NIST 2017a). Finding such weaknesses and mitigating them requires significant effort (Table 4.5).

Table 4.5. Product features and gaps

Code	Features we want	Current level	Where we want to be	Gap
PF1	Hard problems	Integer factorization, discrete algorithm problem	Quantum secure hard problems	NIST is reviewing candidate algorithms
PF2	Key-exchange algorithms	RSA, ECDH	Selection of key-exchange algorithms from NIST's Round 2 PQC competition	NIST is investigating the candidate algorithms submitted to NIST's PQC competition
PF3	Signature algorithms	RSA Signature, Elliptic curve, Digital signature scheme	Selection of signature algorithms from NIST's round 2 PQC competition	NIST is reviewing candidate algorithms submitted to NIST's PQC competition
PF4	Mathematical attacks	Please refer to following papers	Research into candidate algorithms submitted into NIST's PQC competition	Extensive research required
PF5	PQC standards/ recommendations	ANSI X9.62, ANSI X9.63, RFC 8017, etc.	Standards and recommendations from NIST at the end of NIST's PQC competition	NIST is expected to announce standards/recommendations in year 2022
PF6	Optimized crypto coprocessors	Secure elements optimized for current industry standards and practices	New generation of optimized processors to accelerate the computations involved in PQC algorithms	New generation of processors required
PF7	Secure implementation	Secure implementation of classical cryptography on the current generation of processors	Know-how of secure implementation of the new PQC algorithms across different target devices	In-house research and competence required

4.7 ESV's PQC Strategy

In this chapter, we present ESV's strategy to align its internal R&D efforts with PQC standardization process. We divide the strategy into three stages, that is, short-term strategy, mid-term strategy, and long-term strategy. We further present how ESV could utilize its resources to achieve its goals.

4.7.1 ESV's Strategy

As mentioned earlier, ESV recognizes NIST's initiative to standardize PQC algorithms (ESV 2019). ESV employees have submitted several candidate algorithms to NIST's PQC competition.

ESV has recently approved of a task force to work on PQC algorithms to generate in-house expertise and intellectual property on PQC algorithms. The task force plans to work on following three strategies.

4.7.1.1 Short-Term Strategy

NIST is currently on Round 2 of the PQC competition where it is evaluating 17 key-exchange algorithms and 9 signature algorithms. As seen in Table 4.3, most of the authors of the submissions are professors or researchers associated with universities and the affiliated institutes. Since it is a relatively new research area compared to the classical cryptography, one of the primary goals of ESV should be to develop talents in-house. Since most of the senior engineers and architects at ESV are most likely working on products and services that are already in the pipeline, they might not have hours to contribute to ESV's PQC efforts.

The lack of talent pool to work on ESV's PQC efforts can be addressed by establishing research tie-ins with universities. ESV working students along with master study students working on problems relevant to PQC would be an excellent way to attract new talents. In the early stage, ESV should focus on educating the PhD students on PQC algorithms. ESV should also study the suitability of running the PQC algorithms on ESV's hardware. This should help to find the limitations of current hardware architecture to run the PQC algorithms.

4.7.1.2 Mid-Term Strategy

There are around 20 billion connected devices in the world. The number is expected to grow up to 30 billion connected devices by 2022 (Ericsson 2019). Once NIST announces PQC standards, there will be a demand from customer (Early adopters) who would like to have ESV products running those algorithms. Recalling the devices in the field and replacing them with the new ones will not be a pragmatic approach. The better approach would be to update the devices in the field through a software update. ESV's current crypto processors are flexible to support the update feature. However, performance, memory consequences, attack resistance, etc. need to be considered (ESV 2019). As a mid-term strategy, ESV should look into upgrading devices in the field through a firmware update.

4.7.1.3 Long-Term Strategy

As mentioned in the earlier chapter, the current generation of hardware is only optimized to support current cryptographic standards. The early research into PQC algorithms shows that PQC algorithms require more resources (computational, memory, energy, bandwidth, etc.) (Valyukh 2017). Since there are 17 key-exchange algorithms and 9 signature algorithms, it would not be practical to research optimized crypto coprocessors for each of the algorithms. Once there is a better outlook about the winning algorithms in the later ESV of the PQC competition, ESV should look into developing the processors optimized for the PQC algorithms. As mentioned in Chap. 5.3.4, ESV should research secure and efficient implementation of the different PQC algorithms for different target platforms. This includes safeguarding the implementations from the known and unknown hardware attacks (see Fig. 4.5).

4.7.2 Resources

Execution of a strategy requires resources that might or might not be available within an organization. The execution of this strategy primarily requires a talent pool that would generate intellectual property for ESV. As seen in Table 4.3, the majority of people working on PQC research have very specialized skill sets and majority of the PQC researchers are associated with universities. So, ESV will have to develop in-house competence and collaborate with universities to develop PQC intellectual property. In this section, we discuss the resources that are at ESV's disposal to achieve its goals.

4.7.2.1 Collaboration with Universities

University–industry collaboration is a standard practice to transfer technology from academia to the industry and provide direction and resources to the university to work on the problems that are relevant to the industry (Anderson et al. 2007). For an industry, the internal R&D team is not always enough to work on innovation projects. At times, there is also a need for a company to innovate into a new direction. For such scenarios, universities provide a large pool of talented graduates and researchers. Furthermore, it is also possible for a company to license intellectual property generated by research institutes to accelerate the growth into areas.

4.7.2.2 NIST's PQC Standardization Effort

There is a close-knit relationship between the standardization efforts and innovative research. Often, standardization efforts are considered as a channel for technology transfer thus facilitating research and innovation (Blind 2013). Most of the submissions made to NIST PQC competition were funded by some public grants (see Table 4.4). Such submissions are not covered by patents and, thus, become public goods via standards (Blind 2013). Figure 4.6 shows a general standardization process. In the beginning, research is done by heterogenous group of stakeholders which results in the generation of a pool of publications and patents. Organizations like ETSI, BSI, and NIST take the leadership and evaluate the submissions made to the standardization competition. The standards that come out of this process address the needs of the heterogenous group of stakeholders involved in the process.

The intellectual property generated during the NIST's PQC standardization effort is a huge knowledge pool for ESV. ESV, of course, needs to filter through the huge amount of papers published and find what is relevant for its applications. Generating that amount of knowledge would not have been possible to any single entity. Furthermore, it is an excellent platform to headhunt future employees based on their publication lists.

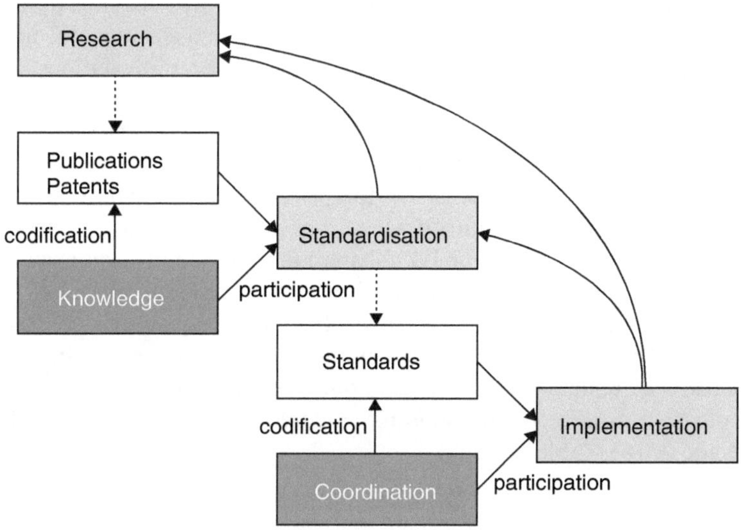

Fig. 4.6 Standardization process

4.7.3 Internal R&D

One of the core competencies of ESV is designing crypto coprocessor and secure and efficient implementation of the crypto algorithms. Once PQC standards are announced and the long-term strategy (see Sect. 6.1.3) is in its execution phase, the internal R&D team would play a crucial role in bringing the products in the market.

4.7.3.1 Collaboration with Startups

We showed in Table 4.3 that working in PQC algorithms requires a very specialized skillset, and these talent pools now are concentrated in the universities and research institutes. A few startups have emerged from the research institutes that are aiming to commercialize quantum secure security solutions. PQShield is one of such startups that spun off from Mathematical Institute of Oxford University (Kaafarani 2020). PQSecure is another such example. PQSecure is a spin-off from computer engineering lab at Florida Atlantic University. PQSecure aims to provide quantum secure security solutions from servers and cloud computing all the way down to small IoT devices (Azarderakhsh 2020). It could be possible to license intellectual property from such startup companies to supplement ESV's internal R&D efforts.

4.7.3.2 Collaboration with Customers

ESV collaborates with its customers to help them bring new features to their products. One such example is when ESV collaborated with a car manufacture to help them bring a secure car access feature using ESV's new radio solutions. Such collaboration effort between ESV and the car manufacture creates a synergy and helps foster innovation between these two companies. It would be safe to assume that big customers like Volkswagen, Siemens, or Daimler are equally concerned about securing their products and services in the post-quantum world. Such customers most likely have a roadmap outlined where they want to bring quantum safe solutions to their products. If ESV could collaborate with one of such customers and prepare them for the post-quantum world, it would create an excellent story and ESV could signal the market that ESV is ready to deliver quantum safe solutions to its customers. This can be achieved by helping them to provide quantum safe solution as a software update over the air as outlined in mid-term strategy. Additionally, ESV could also provide hybrid solution as deployed by Google on its browsers (Braithwaite 2016) (Table 4.6).

Table 4.6. Resource analysis

Code	Resources	Remark	Technology features
R1	Universities	Talent pool from universities and technology transfer from universities	PF1, PF4
R2	NIST's PQC standardization effort		PF1, PF2, PF3, PF5
R3	Internal R&D	ESV's internal R&D resource for secure implementation and design of next-generation processors	PF6, PF7
R4	Collaboration with customers	Demonstration of ESV's capabilities	PF7
R5	Collaboration with startups	Licensing of intellectual property from startups	PF6, PF7

4.8 Conclusion and Future Work

4.8.1 Conclusion

In this study, we explored the application of TRM as a decision-making tool to help a semiconductor company align its R&D activities with NIST's PQC standardization effort. The products from semiconductor industry are used to process sensitive data and are deployed in critical infrastructures. Thus, in addition to the disruption from technological breakthroughs, the semiconductor industry market is also susceptible to policies from regulatory organizations. In this study, we showed the impact on semiconductor industry not only from quantum computing but also from the regulatory organizations. We adopted the standard TRM framework and identified the key drivers and their motivation for driving the PQC standardization effort forward. We also analyzed the desirable features on products and services from semiconductor industry in a post-quantum world. We linked the features to the drivers along with the explanation for the relevance of the features for the drivers. We also identified the resources available to the semiconductor company and the possible collaborations needed to bring the features (see Chap. 5) into the products. As a summary of the TRM framework applied to this specific use case, we present a visual representation of the TRM (Fig. 4.7) with the important milestones aligned with NIST's PQC timeline.

4.8.2 Future Work

As mentioned in the earlier chapter, TRM is an iterative process. There are several variables to account for in the NIST's PQC standardization process itself. A discovery of a novel mathematical attack on any of the proposed schemes can put that scheme out of the NIST's standardization race (Albrecht et al. 2019). If such a

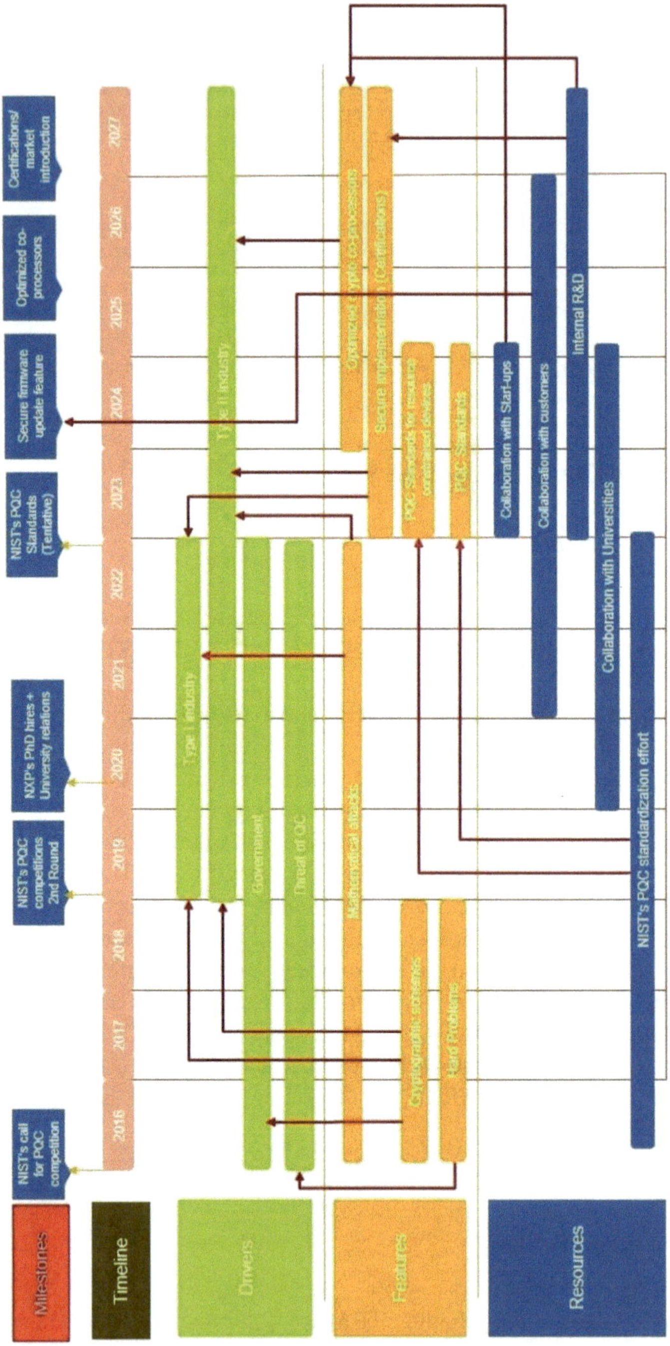

Fig. 4.7 A proposed roadmap for a semiconductor company to align its R&D efforts with NIST's PQC standardization effort

scenario were to develop, it would need to be reflected on the most recent roadmap maintained by the semiconductor company. Similarly, there might be further allocation or cutback of resources which might affect the milestones presented in the final roadmap. This also has to be reflected in the roadmap. Thus, maintenance of the roadmap and communicating it with several teams within the semiconductor company would be a natural extension of this work.

In this study, we have outlined a general strategy roadmap for a semiconductor company. It lacks the granularity and fine details on the product level. This is because the final product level roadmap relies on the algorithms that are selected by NIST. Since there are several candidate algorithms, outlining product-level roadmap with this level of uncertainty is a daunting task. Once we have a clear outlook on the winning algorithm, it would be a natural extension of this work to develop a product level roadmap to bring the winning algorithms to the products.

References

Albrecht, M. R., Hanser, C., Höller, A., Pöppelmann, T., Virdia, F., & Wallner, A. (2019). Implementing RLWE-based schemes using an RSA co-processor. *IACR Trans. Cryptogr. Hardw. Embed. Syst., 2019*(1), 169–208.

Alkim, E., Ducas, L., Pöppelmann, T., & Schwabe, P. (2015). Post-quantum key exchange – A new hope. *Cryptology ePrint Archive*, Report 2015/1092. https://eprint.iacr.org/2015/1092

Allen, G. (2019). *Understanding China's AI Strategy: Clues to Chinese strategic thinking on artificial intelligence and national security*. https://s3.amazonaws.com/files.cnas.org/documents/CNAS-Understanding-Chinas-AI-Strategy-Gregory-C.-Allen-FINAL-2.15.19.pdf?mtime=20190215104041. Last accessed on 18/02/2020.

Amer, M., Daim, T. U., & Jetter, A. (2016). Technology roadmap through fuzzy cognitive map-based scenarios: The case of wind energy sector of a developing country. *Technology Analysis & Strategic Management, 28*(2), 131–155.

Anderson, T. R., Daim, T. U., & Lavoie, F. F. (2007). Measuring the efficiency of university technology transfer. *Technovation, 27*(5), 306–318.

Arute, F., Arya, K., Babbush, R., Bacon, D., Bardin, J. C., Barends, R., Biswas, R., Boixo, S., Brandao, F. G. S. L., Buell, D. A., Burkett, B., Chen, Y., Chen, Z., Chiaro, B., Collins, R., Courtney, W., Dunsworth, A., Farhi, E., Foxen, B., Fowler, A., Gidney, C., Giustina, M., Graff, R., Guerin, K., Habegger, S., Harrigan, M. P., Hartmann, M. J., Ho, A., Hoffmann, M., Huang, T., Humble, T. S., Isakov, S. V., Jeffrey, E., Jiang, Z., Kafri, D., Kechedzhi, K., Kelly, J., Klimov, P. V., Knysh, S., Korotkov, A., Kostritsa, F., Landhuis, D., Lindmark, M., Lucero, E., Lyakh, D., Mandrà, S., McClean, J. R., McEwen, M., Megrant, A., Mi, X., Michielsen, K., Mohseni, M., Mutus, J., Naaman, O., Neeley, M., Neill, C., Niu, M. Y., Ostby, E., Petukhov, A., Platt, J. C., Quintana, C., Rieffel, E. G., Roushan, P., Rubin, N. C., Sank, D., Satzinger, K. J., Smelyanskiy, V., Sung, K. J., Trevithick, M. D., Vainsencher, A., Villalonga, B., White, T., Yao, Z. J., Yeh, P., Zalcman, A., Neven, H., & Martinis, J. M. (2019). Quantum supremacy using a programmable superconducting processor. *Nature, 574*(7779), 505–510.

Asanović, K., & Patterson, D. A. (2014). *Instruction sets should be free: The case for risc-v* (Tech. Rep. UCB/EECS-2014-146). Berkeley: EECS Department, University of California.

Azarderakhsh, R. (2020). *Post-Quantum, classical, and hybrid cryptography IPs*. [online] http://www.pqsecurity.com/products/

Barney, J., Wright, M., David, J., & Ketchen, J. (2001). The resource-based view of the firm: Ten years after 1991. *Journal of Management, 27*(6), 625–641.

Benioff, P. (1965). Information theory in quantum statistical mechanics. *Physics Letters, 14*(3), 196–197.
Blind, K. (2013). *The impact of standardization of standards on innovation*. Nesta Working Paper 13/15. www.nesta.org.uk/wp13-15
Boneh, D. (1999). Twenty years of attacks on the RSA cryptosystem. *Notices of the AMS, 46*, 203–213.
Bos, J., Costello, C., Ducas, L., Mironov, I., Naehrig, M., Nikolaenko, V., Raghunathan, A., & Stebila, D. (2016). Frodo: Take off the ring! Practical, quantum-secure key exchange from LWE. In *Proceedings of the 2016 ACM SIGSAC conference on computer and communications security* (pp. 1006–1018).
Braeckle, K. (2017). Two "SESAMES awards" for post-quantum cryptography on contactless security chip. https://www.infineon.com/cms/en/about-infineon/press/market-news/2017/INFCCS201711-013.html. Last accessed on 18/02/2020.
Braithwaite, M. (2016). *Experimenting with post-quantum cryptography*. [online] https://security.googleblog.com/2016/07/experimenting-with-post-quantum.html. Last accessed on 10/01/2020.
BSI. (2019). *The State of IT Security in Germany 2018*. [online] https://www.bsi.bund.de/EN/Publications/SecuritySituation/SecuritySituation_node.html. Last accessed on 15/0/2020.
Burr, W., Ferraiolo, H., & Waltermire, D. (2014). NIST and computer security. *IT Professional, 16*(2), 31–37.
Daim, T. U., Yoon, B.-S., Lindenberg, J., Grizzi, R., Estep, J., & Oliver, T. (2018). Strategic roadmapping of robotics technologies for the power industry: A multicriteria technology assessment. *Technological Forecasting and Social Change, 131*, 49–66.
Eikenberg, R. (2019). *Logitech keyboards and mice vulnerable to extensive cyber attacks*. heise online. Last accessed on 15/02/2020. https://www.heise.de/ct/artikel/Logitech-keyboards-and-mice-vulnerable-to-extensive-cyber-attacks-4464533.html
Ericsson. (2019). *Ericsson mobility report*. [online] https://www.ericsson.com/en/mobility-report/reports/november-2019. Last accessed on 18/12/2019.
ESV. (2019). *An internal memo of the European semiconductor vendor to the basics of PQC and outlook*. Available under NDA. https://www.heise.de/ct/artikel/Logitech-keyboards-and-mice-vulnerable-to-extensive-cyber-attacks-4464533.html
ESV. (2020a). *An internal technical document on the architecture of smartcard processors*. Available under NDA from ESV.
ESV. (2020b). *Security threats landscape*. Available under NDA from ESV.
ETSI. (2015). *Quantum-Safe Cryptography (QSC)*. [online] https://www.etsi.org/technologies/quantum-safe-cryptography. Last accessed on 12/10/2019.
Fan, J., Guo, X., De Mulder, E., Schaumont, P., Preneel, B., & Verbauwhede, I. (2010). State-of-the-art of secure ECC implementations: A survey on known side-channel attacks and countermeasures. In *2010 IEEE international symposium on hardware-oriented security and trust (HOST)* (pp. 76–87). IEEE. New Jersey, USA
Fenwick, D., Daim, T. U., & Gerdsri, N. (2009). Value driven technology road mapping (VTRM) process integrating decision making and marketing tools: Case of internet security technologies. *Technological Forecasting and Social Change, 76*(8), 1055–1077.
Ferdinand, P. (2016). Westward ho—The China dream and 'one belt, one road': Chinese foreign policy under xi Jinping. *International Affairs, 92*(4), 941–957.
Galvin, R. (1998). Science roadmaps. *Science, 280*, 803.
Graham, A. (2019). *The damaging after-effects of a data breach*. [online] https://www.itgovernance.co.uk/blog/the-damaging-after-effects-of-a-data-breach. Last accessed on 02/02/2020.
Grover, L. K. (1996). A fast quantum mechanical algorithm for database search. In *Proceedings of the twenty-eighth annual ACM symposium on the theory of computing, 22–24 may 1996* (pp. 212–219). Philadelphia.
Gupta, A. (2013). Environment & PEST analysis: An approach to the external business environment. *International Journal of Modern Social Sciences, 2*(1), 34–43.

Helms, M. M., & Nixon, J. (2010). Exploring SWOT analysis–where are we now? A review of academic research from the last decade. *Journal of Strategy and Management, 3*(3), 215–251.

Huang, L., Zhang, Y., Guo, Y., Zhu, D., & Porter, A. L. (2014). Four dimensional science and technology planning: A new approach based on bibliometrics and technology roadmapping. *Technological Forecasting and Social Change, 81*, 39–48.

Jalali, A., Azarderakhsh, R., Kermani, M. M., Campagna, M., & Jao, D. (2019). ARMv8 SIKE: Optimized supersingular isogeny key encapsulation on ARMv8 processors. *IEEE Trans. on Circuits and Systems, 66-I*(11), 4209–4218.

Kaafarani, A. E. (2020). *Quantum is a reality*. [online] https://pqshield.com/

Kabanov, I. S., Yunusov, R. R., Kurochkin, Y. V., & Fedorov, A. K. (2018). *Practical cryptographic strategies in the post-quantum era*. Author(s).

Kahn, D. (1996). THE CODEBREAKERS; THE STORY OF SECRET WRITING [The Comprehensive History Of Secr. Scribner.

King, K. (2013). *China's aid and soft power in Africa. The case of education and training*. Woodbridge: James Currey.

Koblitz, N. (1987). Elliptic curve cryptosystems. *Mathematics of Computation, 48*(177), 203–203.

Koen, P., Ajamian, G., Burkart, R., Clamen, A., Davidson, J., D'Amore, R., Elkins, C., Herald, K., Incorvia, M., Johnson, A., Karol, R., Seibert, R., Slavejkov, A., & Wagner, K. (2001). Providing clarity and a common language to the "fuzzy front end". *Research-Technology Management, 44*(2), 46–55.

Kostoff, R. N., & Schaller, R. R. (2001). Science and technology roadmaps. *IEEE Transactions on Engineering Management, 48*(2), 132–143.

Leaser, D. (2019). *The demand for cybersecurity professionals is outstrip-ping the supply of skilled workers*. https://www.ibm.com/blogs/ibm-training/new-cybersecurity-threat-not-enough-talent-to-fill-open-security-jobs/. Last accessed on 18/02/2020.

Miller, V. S. (1985). Use of Elliptic Curves in Cryptography. In *Advances in Cryptology - CRYPTO '85, Santa Barbara, California, USA, 18–22 August 1985, Proceedings* (pp. 417–426).

Milshina, Y., & Vishnevskiy, K. (2019). Roadmapping in fast changing environments – The case of the Russian media industry. *Journal of Engineering and Technology Management, 52*, 32–47. SI: Smart Roadmapping.

Moody, D. (2017). *The ship has sailed. The NIST Post-Quantum Crypto "Competition"*. AsiaCrypt. https://csrc.nist.gov/CSRC/media/Projects/Post-Quantum-Cryptography/documents/asiacrypt-2017-moody-pqc.pdf. Last accessed on 15/02/2020.

Mosca, M. (2015). *Announcing development of a federal information processing standard for advanced encryption standard*. [online] https://csrc.nist.gov/csrc/media/events/workshop-on-cybersecurity-in-a-post-quantum-world/documents/presentations/session8-mosca-michele.pdf. Last accessed on 15/02/2020.

Nauta, S. (2011). *Investigation project complexity at NXP semiconductors B.V. Master's study*. Delft University of Technology. New Jersey, USA.

NIADA (2019). *Used car industry report*. https://www.niada.com/uploads/dynamic_areas/5Nd7sQuZuYjQ5FCDUBXL/34/UCIR_2019_web.pdf? Last accessed on 16/02/2020.

Niederhagen, R. (2019). *Next-generation cryptography for embedded systems*. https://www.sit.fraunhofer.de/en/quantumrisc/. Last accessed on 18/02/2020.

Niederhagen, R., & Waidner, M. (2017). *White paper: Practical post-quantum cryptography*. [online] https://www.sit.fraunhofer.de/fileadmin/dokumente/studien_und_technical_reports/Practical.PostQuantum.Cryptography_WP_FraunhoferSIT.pdf?_=1503992279. Last accessed on 18/02/2020.

NIST. (1997). *Announcing development of a federal information processing standard for advanced encryption standard*. [online] https://csrc.nist.gov/news/1997/announcing-development-of-fips-for-advanced-encryp. Last accessed on 18/01/2020.

NIST. (2017a). *Post-quantum cryptography*. https://csrc.nist.gov/Projects/Post-Quantum-Cryptography/Post-Quantum-Cryptography-Standardization/Call-for-Proposals. [Online; accessed 08.06.2019].

NIST. (2017b). *Post-quantum cryptography: Round 1 submissions*. [online] https://csrc.nist.gov/Projects/Post-Quantum-Cryptography/Post-Quantum-Cryptography-Standardization/Call-for-Proposals

NXP. (2020). *Smart and secure identity services*. Available at https://www.nxp.com/docs/en/brochure/NXP-Smart-Governance-Brochure.pdf

Olin, M. G., Garcia, M. L., & Bray, O. H. (1997). *Fundamentals of technology roadmapping*. Albuquerque: Sandia National Laboratories.

Paar, C., & Pelzl, J. (2010). *Understanding cryptography*. Berlin Heidelberg: Springer.

Patel, S., & Hodgson, J. (2019). *Used car industry report*. Available to ESV employees.

Phaal, R., Farrukh, C., & Probert, D. R. (2013). Fast-start roadmapping workshop approaches. In M. G. Moehrle, R. Isenmann, & R. Phaal (Eds.), *Technology roadmapping for strategy and innovation* (pp. 91–106). Berlin, Heidelberg: Springer.

Popp, T. (2009). An introduction to implementation attacks and countermeasures. In *2009 7th IEEE/ACM international conference on formal methods and models for co-design* (pp. 108–115).

Rankl, W. (2010). *Smart card handbook*. Wiley. New Jersey, USA.

Rivero, A. R. G., & Daim, T. (2017). Technology roadmap: Cattle farming sustainability in Germany. *Journal of Cleaner Production, 142*, 4310–4326.

Rivest, R. L., Shamir, A., & Adleman, L. M. (1983). A method for obtaining digital signatures and public-key cryptosystems (reprint). *Communications of the ACM, 26*(1), 96–99.

Roetteler, M., Naehrig, M., Svore, K. M., & Lauter, K. E. (2017). Quantum resource estimates for computing elliptic curve discrete logarithms. In *Advances in cryptology – ASIACRYPT 2017–23rd international conference on the theory and applications of cryptology and information security, Hong Kong, 3–7 December 2017, proceedings, part II* (pp. 241–270).

Schimpf, S., & Abele, T. (2019). How German companies apply roadmapping: Evidence from an empirical study. *Journal of Engineering and Technology Management, 52*, 74–88. SI: Smart Roadmapping.

Schneier, B. (1999). Cryptography: The importance of not being different. *Computer, 32*(3), 108–109.

Schneier, B. (2011). *Schneier's law*. [online] https://www.schneier.com/blog/archives/2011/04/schneiers_law.html. Last accessed on 18/02/2020.

Schneier, B. (2015). *NSA plans for a post-quantum world*. [online] https://www.schneier.com/blog/archives/2015/08/nsa_plans_for_a.html. Last accessed on 18/02/2020.

Shor, P. W. (1997). Polynomial-time algorithms for prime factorization and discrete logarithms on a quantum computer. *Society for Industrial and Applied Mathematics Journal on Computing, 26*(5), 1484–1509.

Smith, S. W. (2011). *Secure coprocessor* (pp. 1102–1103). Boston: Springer US.

Strenzke, F. (2010). A smart card implementation of the McEliece PKC. In P. Samarati, M. Tunstall, J. Posegga, K. Markantonakis, & D. Sauveron (Eds.), *Information security theory and practices. Security and privacy of pervasive systems and smart devices* (pp. 47–59). Berlin, Heidelberg: Springer.

Turner, A. N. (1989). Dynamic manufacturing: Creating the learning organization, by Robert H. Hayes, Steven C. Wheelright, and Kim B. Clark. New York: The free press, 1988, 429 pp. $24.95. *Human Resource Management, 28*(2), 297–299.

UNFCC. (2013). *Using roadmapping to facilitate the planning and implementation of technologies for mitigation and adaptation*. [online] https://www.ctc-n.org/file/20614/download?token=HTkpFOr4. Last accessed on 16/01/2020.

Valyukh, V. (2017). *Performance and comparison of post-quantum cryptographic algorithms. Master's study*. Linköping University.

Van De Velde, J. (2017). The law of cyber interference in elections. Available at SSRN 3043828.

van der Laan, E., Poll, E., Rijneveld, J., de Ruiter, J., Schwabe, P., & Verschuren, J. (2018). Is java card ready for hash-based signatures? In A. Inomata & K. Yasuda (Eds.), *Advances in information and computer security – 13th international workshop on security, IWSEC 2018,*

Sendai, Japan, September 3–5, 2018, proceedings, volume 11049 of lecture notes in computer science (pp. 127–142). Springer.
Weidong, G. (2010). Technology roadmapping as a new tool of knowledge management. In *2010 Chinese control and decision conference* (pp. 1658–1661).
Willyard, C. H., & McClees, C. W. (1987). Motorola's technology roadmap process. *Research Management, 30*(5), 13–19.

Chapter 5
Technology Roadmap: Smartwatches

Shivani Purwar and Tuğrul U. Daim

This chapter examines development of technology roadmapping for smartwatches such as the Fitbit, Apple Watch, LG Watch Sport, and Samsung Gear S3 Frontier.

Wristwatches have different purposes; therefore, they are built in a variety of different ways according to the requirements. On the basis of their functionalities, they are classified into two categories:

1. *Smartbands:* These do not have all of the same features as smartwatches. In other words, they have only limited features in comparison with smartwatches. The main purpose of a smartband is to monitor the sleep pattern or the physical activity of the user.
2. *Smartwatches:* These are more advanced versions of smartbands. They have an embedded operating system (OS), which allows installation of third-party applications.

This classification is most relevant to research in specific areas. The ways in which data are collected in both categories are very different. Smartbands do not have the same facility to transfer data to third-party systems that smartwatches do.

Smartwatches motivate people a lot by counting their steps or by measuring the number of calories they burn over the course of a day. As a result, people are getting more attracted to this type of device.

We have conducted research by interacting with people, interviewing them, and trying to understand how these technologies (smartwatches) impact their daily life. In our survey, we observed that these smartwatches have a direct effect on people's lives, such as in physical activity and sports participation.

S. Purwar · T. U. Daim (✉)
Department of Engineering and Technology Management, Portland State University, Portland, OR, USA
e-mail: tugrul.u.daim@pdx.edu

T. U. Daim (ed.), *Roadmapping Future*, Applied Innovation and Technology Management, https://doi.org/10.1007/978-3-030-50502-8_5

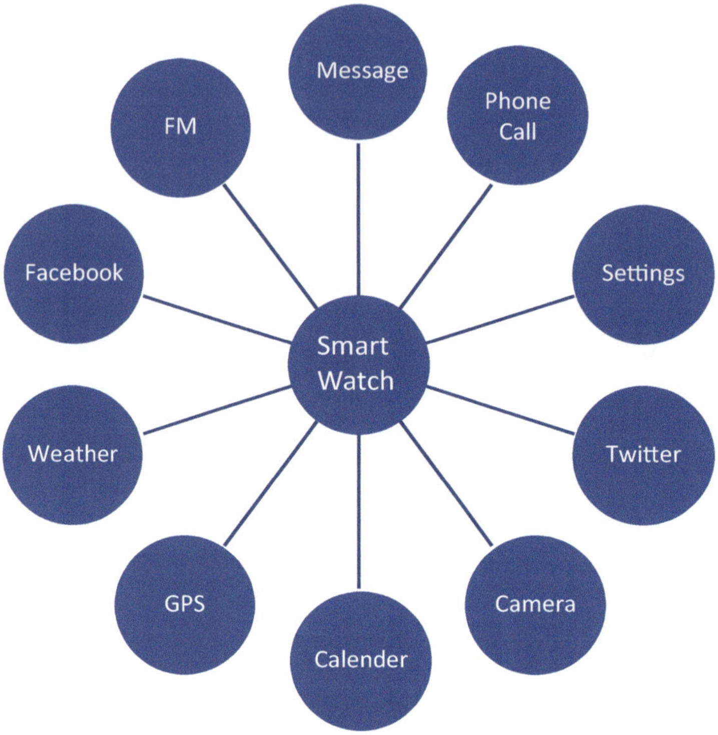

A smartwatch is a wearable touchscreen computer device in the form of a wristwatch, which is designed to support physical activities. A smartwatch offers not just the functions of a regular watch but also more advanced smart features. Like smartphones, smartwatches have touchscreens and support apps, and can monitor the user's heart rate and other vital signs (Darmwal 2015). Smartwatches have been touted as the next generation of devices set to transform consumers' lives. In the coming years, it is intended to increase the capabilities of smartwatches through the addition of features and functionality. It was previously estimated that by 2021, the shipment of smartwatches would increase to about 70 million units. The day is not far off when smartwatches will replace smartphones.

One of the greatest advantages of smartwatches is that by use of a number of different behavioral and physiological sensors, they help the user to develop good habits and to be on time and up to date. Smartwatch brands include Apple, Samsung, Sony, Fitbit, and LG (IIPRD 2014).

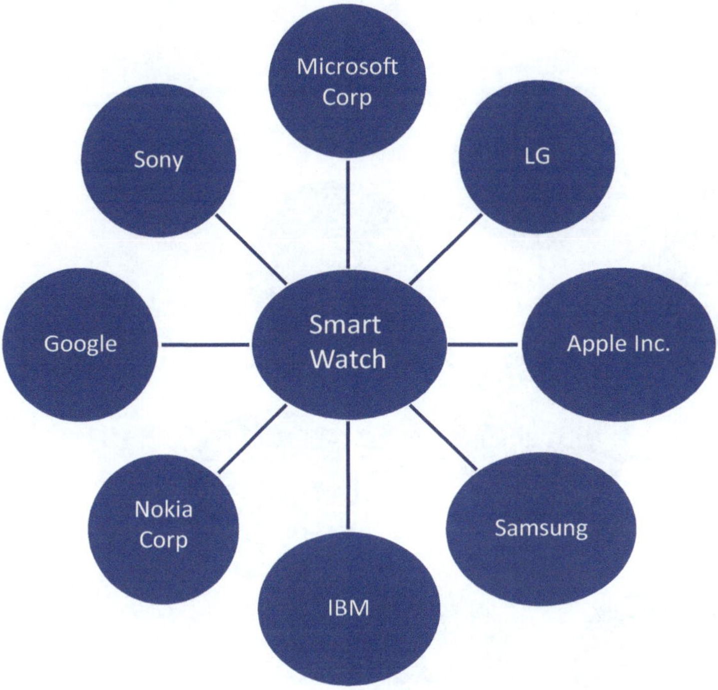

Smartwatches are not only involved in the health care sector; they are also integrated with other services such as checking e-mails or setting up appointments and meetings. A smartwatch can be connected to your smartphone, and you can easily control your phone without having the phone out all the time. For example, you can make a call while driving and you can easily listen to music while exercising or driving. Other advantages of smartwatches such as the Fitbit or Apple Watch are that they can be used as accelerometers and altimeters to calculate the distance that you cover and your caloric expenditure, and they can also chart your heart rate and the quality of your sleep. A smartwatch also makes it easier for you to connect with other activities while doing work.

A smartwatch can become one of your best friends when your phone battery goes flat, as a smartwatch can keep you connected for up to 10 days on a single charge. The popularity of smartwatches is growing not only in the fitness area but also in the fashion market. Today, there are lots of fashionable smartwatch brands on the market, such as Michael Kors and Fossil.

5.1 Benefits of a Smartwatch

Some of the benefits of a smartwatch are described briefly here.

Ease of Communication A smartwatch is compatible with other devices such as smartphones and computers, and provides smart communication with those devices. A smartwatch also supports various types of communication such as internet communication, Bluetooth, and infrared.

Health Monitoring A smartwatch is equipped with various sensors and applications that help you to monitor your health and day-to-day activities; for example, it can monitor the number of steps you have taken or how many calories you have burned over the course of a day.

Multiple Purposes A smartwatch can be used for various purposes such as listening to music, checking e-mails, internet browsing, online shopping, reading books, and interacting with social media (such as Facebook), thus making it a multipurpose gadget.

Wearable Convenience A smartwatch is worn on your wrist, making it easy to take a call, see a notification, or use a smartwatch application.

Portability A smartwatch is convenient and comfortable for the user because of its compact, lightweight, and portable size (Carter n.d.).

Security Because a smartwatch is worn on your wrist, you do not need to always keep an eye on it. In contrast, smartphone users always need to keep an eye on their costly phones to avoid losing them.

Stylishness A smartwatch is not only a multipurpose gadget but also a fashion-meets-tech accessory.

5.2 Drawbacks of a Smartwatch

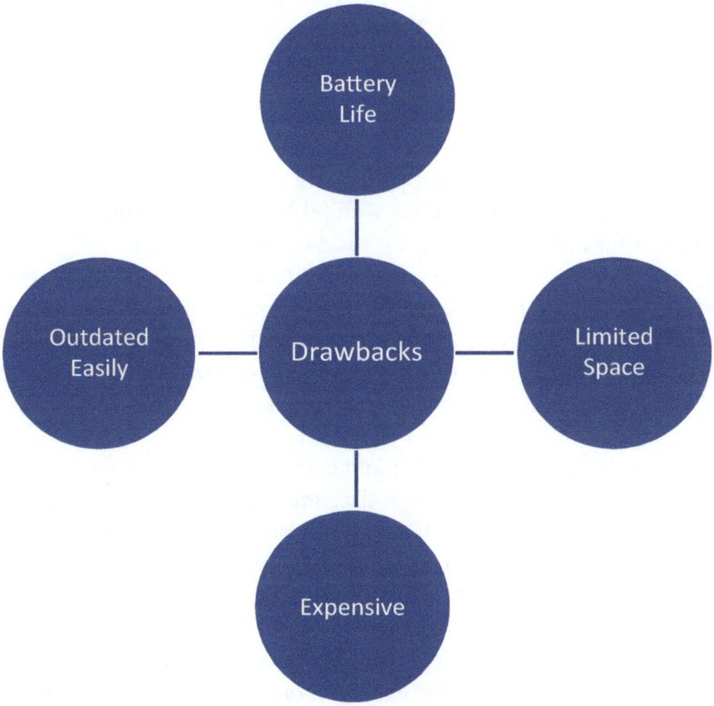

Just as every coin has two sides, a smartwatch does have some drawbacks.

Limited Space Sometimes it is difficult for the user to read a whole message without having to scroll down because of the limited display size. For example, watching movies and navigation require extra effort for your eyes to look at such a small screen.

Limited Battery Life One of the biggest disadvantages of smartwatches is the battery life. Typically, smartwatches can be used for only 1–2 days after a full charge, and some watch batteries do not even last for a full day. This may prevent the smartwatch from sleep tracking when the battery is low, and it is necessary for the user to charge it fully before using it.

Rapid Obsolescence The operating systems used in smartwatches are upgraded frequently. If you buy a new smartwatch, it will probably be out of date within a year or two.

Expense Smartwatches such as the Sony Smartwatch, Apple Watch, Neptune Pine, and Garmin Fēnix 3 are luxury smartwatches, and their prices vary between $300 and $500.

5.3 Literature Review

The main focus of this section is a review of the most recent research on smartwatches and their features such as tracking functions for daily activity and fitness or exercise routines. Fitness trackers are popular with athletes because smartwatches are very helpful to them for monitoring their exercise routines.

A survey was done in which it was observed that one in ten adults in the USA had a smartwatch. Today, people across the country are buying fitness trackers to monitor their activities, such as how many calories they burn over the course of a day and how many steps they take, as well monitoring their sleep patterns. It has also been reported that 36% of people who own a fitness tracker are between 35 and 54 years old.

The digital technology industry provides the foundation for information exchange in all sectors, including health care and retail. Today, smartwatches are successful in the market because of the advanced smart features that are built into them. Several companies such as Fitbit, Apple, and LG have gained popularity in the market by launching various high-tech smartwatches. It was previously predicted that there would be 35 million smartwatches in use by the end of 2018.

Smartwatches are electronic devices that are designed to encourage physical activity or other health-related activities. For example, on a smartwatch, you can set up daily goals such as how many calories you want to burn today and, on the basis of your goals, the watch will notify you that it is time to stand up, walk, etc. Users perceive that their smartwatches have an awareness regarding their physical activity, and they can use their daily data to make plans for the next day, such as how many extra calories they need to burn or how many calories they need to ingest.

5.3.1 Findings from Smartwatch User Interviews

Smartwatch users were interviewed to ascertain their opinions on their devices. Some of their comments are presented here.

Interviewee 1 stated:

> My smartwatch has changed my life. Now I don't need to take out my phone every time or need to search for my phone if I get a notification or a call. Now my whole world is in my hand. Another great advantage I've observed is that it helps me a lot in analyzing my sleep pattern.

Interviewee 2 stated:

> My smartwatch was a special gift that my husband gave me on my birthday. Before the smartwatch, I was not able to keep track of my calories and my daily exercise routine. Since I started using the smartwatch, I have been able to keep track of my daily caloric intake, as well as being able to track my daily exercise. It has helped me a lot in reducing my weight as well as making my life easier when I am doing other activities, especially swimming.

Interviewee 3 stated:

> When I found out about the benefits of using a smartwatch a year ago, I went straight to the store and bought an Apple Watch. It has almost all of the same features as my iPhone. This product is awesome. It's very easy to use and is user friendly.

5.4 Roadmapping

5.4.1 What Is Roadmapping?

A technology roadmap is a strategic planning technique in which planning is done through understanding of the short- and long-range goals of a project with a specific technology solution.

Roadmapping enables a team to design and execute a way to achieve its objectives. It acts as a guide during the team's journey and helps them to act on the events that require a change in direction.

Technology roadmapping helps an organization to identify what technology it requires to adapt in order to achieve specific business goals. It also states when and how the technology process will need to be implemented. It helps the organization to answer questions such as what the current state of all of the technology is, and whether the organization will need different technology for leverage.

5.4.2 Advantages of Roadmapping

1. It is helpful for prioritization of goals.

2. It is helpful for development of forecasts.
3. It is helpful for development of a project in a more structured way.
4. It engages people to collaborate and think creatively and differently.
5. It supports budget planning and forecasting.
6. It is helpful for strategic planning.

Roadmapping helps to answer these three most important questions:

1. What are we doing?
2. Why are we doing it?
3. How does it tie back to our objectives?

5.5 Market Drivers

In the area of heath, smartwatches play a vital role in supporting healthy everyday living. A smartwatch is helpful for monitoring personal activity and obtaining feedback based on day-to-day activities. For example, it helps to keep track of your resting heart rate (the average rate is 80–100 beats per minute) and, on that basis, you can improve your exercise as required. Another important feature of a smartwatch is the reminder function. On a smartphone, it is very easy to miss a reminder notification, but on a smartwatch, it is very difficult to miss one, because the reminder causes the smartwatch to vibrate on your wrist, which is very difficult to ignore.

Another advantage of a smartwatch is that it monitors your movements and, on that basis, it helps to determine your sleep pattern. With the help of Bluetooth, this information is stored in your smartwatch, and that helps you to determine your sleep pattern for a whole week. Thus, it makes it easier for you to improve your sleep pattern. However, smartwatch technology does not only benefit the common man; this technology also plays a vital role in an athlete's life. Parameters such as the athlete's heart rate and stress level are very important for the trainer to determine. A smartwatch helps the trainer to determine the athlete's heart rate, stress level, diet, and sleep pattern, assisting them to make adjustments in their training pattern (Smartsheet 2017).

Apart from fitness applications, a smartwatch is helpful for instant messaging and e-mail. Instead of looking at a phone for a message or e-mail, you can just look at your arm while running or doing some other task. For example, if you are exercising on a treadmill and get an e-mail, you can easily check it on your smartwatch instead of having to wait and check it on your phone later.

Another benefit of a smartwatch is that you no longer have to haphazardly pull your phone out to see who is calling or texting you. For example, when you are driving, it is difficult to take out your phone from your pocket or bag to check who is calling you. A smartwatch overcomes this issue. Instead of taking your phone out of your bag, you can take a call from a smartwatch on your wrist directly.

Smartwatches have the potential to support health in everyday life by analyzing behavior patterns and, on that basis, the user can choose his or her physical activities or exercise. Another key feature of the smartwatch is the log history, which makes it easier for the user to understand his or her dietary patterns and make healthier dietary choices in the future. Smartwatches complement smartphones and, as a result, companies are taking interest in selling and promoting smartwatches. As more and more companies do so, the functionality of smartwatches will improve more and prices will also drop significantly. Hence, smartwatches are an emerging technology in today's market. (Rebelo 2017). In 2017, it was observed that the smartwatch market was experiencing good growth and was expected to reach around $1.8 billion the following year (Smartsheet 2017).

5.6 Market Drives Ranking

Category	Code	Market driver	Weighting[a]
Social/consumer	D1	Fitness/wellness	4
Interest	D2	Lifestyle	3
	D3	Adaptation of smart technologies	4
	D4	Brand loyalty	2
	D5	Social media participation	3
Health	D6	Stress level monitoring	2
	D7	Rise in diabetes prevalence	3
	D8	Aging population	2
	D9	Obesity	4
	D10	Desire for data	3
Technical	D11	Connectivity with multiple devices	4
	D12	Interest for athletes	4
Economical	D13	Inexpensive components	3
	D14	Work productivity	3

[a]The weighting is ranked from 1 (the lowest) to 4 (the highest)

5.7 Product Security and Future Features

Today, smartwatches such as the Fitbit Ionic offer a variety of attractive features and advanced technology to attract many users in the market. They are excellent gadgets that make life easier and smoother. However, as with other gadgets, there are some security concerns associated with them.

The Stevens Institute of Technology did some security research on smartwatches and observed that it was entirely possible for programmers and hackers to get unauthorized access and steal users' ATM [automated teller machine] PINs [personal identification numbers] from their smartwatches (Reeder and David 2016). Another important security concern is users' personal data, such as their day-to-day monitoring data, which could be of great interest to advertising sponsors and insurance providers. As an example, if a user's health insurance provider had access to his or her smartwatch data showing heart rate irregularities or high daily consumption of fat, the provider could classify that user as high risk and raise their health insurance premium.

According to Diane Stapley of Advanced Micro Devices (AMD), these devices contain many vulnerabilities, and that makes it risky to rely on them in two-factor authentication for access purposes. It is therefore important for vendors of such devices to be careful with this type of functionality and make sure it cannot be stolen or blocked (Best Doctors n.d.).

Connections between a smartphone and a smartwatch via Bluetooth are also very important. It has been estimated that the range of connectivity between a smartphone and a smartwatch should be at least 10 meters. Various manufacturers are working on overcoming this limitation and trying to increase the range between them.

The biggest problem with smartwatches is their battery life, which is not long. Manufacturers are focusing on overcoming this limitation and providing batteries that will work for a longer period of time—for example, by using solar energy to charge the battery, as well as using wireless charging technology to implement this.

Product feature	Current level	In 2 years	In 5 years	Where we want to be (in 15 years)
Data encryption	Most smartwatches send unencrypted data	Encryption of 80% of data	Encryption	Full encryption of confidential data
Battery	Lithium ion batteries have a short battery life and frequent need for recharging	Longer-lasting batteries that can withstand plenty of recharging	Less reliance on lithium	High-capacity, small, and light batteries
Image projection	Only a coin-sized display, which is difficult to read	Projection of the user interface screen onto the user's hand	Improvements in projection	Higher-quality image projection; integration of devices with a camera to capture the user's selections on the projected image
5G	Limited coverage	Cellular network coverage expansion	Introduction of 5G	5G compatibility and accessibility to health care anytime and anywhere
Apps	A large number of apps that provide specific medical references	Access to massive data sets through the cloud		Cost-effective solutions and ability to interact with other apps

Product feature	Current level	In 2 years	In 5 years	Where we want to be (in 15 years)
Seamless syncing	Connection to limited devices only		Ability to connect with health professionals for monitoring	Ability to communicate with health professionals
Notifications	Not all notifications are shown	Increased reliability of notifications	Customized notifications for health care	Integrated notification system with calendar and other health apps

5G fifth-generation technology standard for cellular networks

Data Encryption A smartwatch is a multipurpose device, which collects data with the help of sensors and communicates with a paired smartphone. On the other hand, smartwatches are being increasingly used to store more sensitive information, creating a potential security breach for hackers to obtain sensitive data such as text messages or other private data. As smartwatch capacities for storage and communications increase, these devices become more attractive to hackers. At present, most smartwatches transmit unencrypted data. People are working on improving this technology; it is believed that within 2 years it will be possible to encrypt 80% of the data, and within 15 years it will be possible to encrypt the data fully and securely.

Fingerprint Unlocking Today's smartwatches use a passcode method for locking the device. People are currently working on face recognition features, and it is expected that within 15 years, this technology will be implemented in smartwatches. Face recognition technology will also help to overcome the security concerns associated with smartwatches.

Longer Battery Life Most smartwatches use lithium ion batteries. A short battery life is one of the drawbacks of smartwatch technology. Various companies are working on this technology and trying to overcome its limitations by making small, lightweight batteries that can go without charging for at least 30 days.

Image Projection Because smartwatch screens are only the size of a coin, it can be very difficult to read an entire message on the screen without needing to scroll down. Various companies are working on this technology and trying to overcome this limitation through use of high-quality image projection features.

Body Monitoring Today, smartwatches can monitor only certain parts of the user's body, such as the heart rate and blood pressure. Manufacturers are trying to enhance the technology to enable smartwatches to also measure other parameters such as cholesterol levels and thyroid function.

5G Enhancements in 5G [the fifth-generation technology standard for cellular networks] will allow wearable devices to communicate within less than 1 millisecond. If wearable devices become able to communicate 50 times faster, they will be able to perform functions such as steering a car, piloting a drone, and other functions that demand real-time data links for remote control.

Communication with Health Professionals (Notifications) At present, smartwatches can detect only some vital signs. Manufacturers are therefore trying to devise smartwatch technologies that can transmit patients' health data directly to their clinicians so they can take immediate steps to resolve any health issues.

Apps Current smartwatches offer a number of good apps that are beneficial for monitoring fitness and wellness programs. Software developers are working on trying to provide more advanced and cost-effective apps that can interact with other apps, devices, and wearables, and leverage the opportunity to personalize wellness programs by learning and adapting to user needs/habits.

5.8 Technologies

A smartwatch is a wristwatch with a touchscreen, which works as a wearable computer. Early models had basic functionality such as calculation, a digital clock, and games. Then, from 2010 onward, additional functionality was added to smartwatches, such as Bluetooth connectivity, mobile apps, and FM radio. Today, smartwatches have almost all of the same functionality as smartphones. Internet access connects a smartwatch with a whole world of potential capabilities, such as message notifications, GPS [Global Positioning System] navigation, and Bluetooth connectivity to help make phone calls and to send and receive messages.

The smartwatch market is growing. According to the International Data Corporation (Mahadevan 2017), about 50 million smartwatches were sold worldwide in 2016, and this number was expected to grow to 150 million by 2021.

Today, wearable devices are changing human–machine interactions. The current technology provides various features such as voice activation and gesture control. Operating systems provide the silent feature of all smartphones. Currently, there are four primary operating systems that are dominating the smartwatch market: Android, Android wear and Linux. Two more upcoming operating system that will soon hit the market such as iOS for wearable devices (Xu et al. 2015).

Furthermore, if smartwatch technology joins hands with nanotechnology and other applied developments in the field, we can expect tremendous growth with improved products such as smart devices that are bendable or can be charged using body heat, with a greater range of functions. Because of the multipurpose nature of smart devices, different users can use the same device for different needs. Smartwatches also have the potential to integrate forthcoming technologies. In the future, nanotechnology may not only remove the limited processing capability barrier but also provide new dimensions for portable smartwatch technology in all aspects.

Even though there are still a few concerns and limitations (including battery issues), in comparison with the benefits of smartwatches, the risks can be relatively quantified and thereby controlled and managed. Once the smartwatch battery capacity barrier is solved by nanotechnology (or its underlying technologies) or solar power, there is bound to be rapid growth in the smartwatch field.

5.9 Future Scope

Heart disease is deemed to be the leading cause of death in the USA, and there is statistical evidence that close to 600,000 Americans die from it each year. There is a focus on reducing this number considerably by alerting first responders and providing timely emergency care in the case of any cardiac anomaly (Lemos 2018). Current smartwatches have the capability to provide heart rate information as wearable devices, along with the physical location of the wearer. However, heart rate information is not sufficient to determine whether a person is experiencing a heart attack. Smartwatch services are extensible; other critical data such as the respiratory rate, blood pressure, and fall detection could be added later. This would improve the fidelity of the system. For this, additional sensors could be linked to a wearable smartwatch, and collection of these extra data points would be helpful for more accurate decision-making regarding the sending of alerts to first responders in the event of an emergency. The care of elderly people is another area with great potential for future development of smartwatch technology. Elderly people are more prone to diseases and, in the future, smartwatches will be able to collect various health-related data (e.g., the heart rate and blood pressure) for transmission to clinicians for analysis of health patterns.

5.10 Conclusion

Smartwatches offer numerous benefits, such as the following:

Travel buddy A smartwatch is one of your best friends if you are traveling alone or in a taxi cab. It helps by giving you directions, using GPS, and it helps to save you time by telling you the shortest way to reach your destination.

Fitness tracker A smartwatch helps you to reach your fitness goals. If you are ever thinking of using a fitness tracker or a pedometer, you can replace it with a good smartwatch. It helps by counting your steps, caloric consumption, heart rate, sleep pattern, etc.

SMS messaging and phone calls If a smartwatch is on your wrist, you do not need to think about taking your phone out of your pocket or your bag. You can reply to SMS messages immediately via your smartwatch, as well as being able to receive phone calls with the help of your smartwatch.

Help for seniors A smartwatch is the best device for seniors. In the event of an emergency, it enables them to have a conversation with their caregiver. Moreover, the GPS functionality built into the smartwatch can inform the caregiver of the user's location.

Finding your phone It is not uncommon to forget where you have put your phone. You can solve this problem by using your smartwatch. Most smartwatches have phone-finding features. You can connect your phone or any other compatible device with the smartwatch and use it via your watch whenever you wish.

Connectivity during physical activities It is generally not possible to keep hold of your phone while doing many types of exercise (swimming, cycling, etc.). As a result, a smartwatch is your best friend while you are doing these activities, particularly if it is waterproof. For example, the Apple Watch (from Series 2 onward) is waterproof, which means it can be worn even during swimming (Carron 2017; Yamada 2014; HITC Staff 2015).

References

Best Doctors. (n.d.). *Smartwatch: the benefits*. https://www.bestdoctorsblog.com/the-benefits-of-a-smart-watch/. Accessed.

Carron, C. (2017). A new sensor increases smartwatch battery life five times. *EPFL*. https://actu.epfl.ch/news/a-new-sensor-increases-smartwatch-battery-life-fiv/. Accessed.

Carter, J. (n.d.) *How will wearables like smartwatches and trackers use 5G?* 5G.co.uk. https://5g.co.uk/guides/how-will-wearables-use-5g/. Accessed.

Darmwal, R. (2015). Wrist wars: smart watches vs traditional watches. Telecom Business Review, 8(1).

HITC Staff. (2015). Future role of smartwatches in personal health (infographic). *Healthcare IT News*. https://hitconsultant.net/2015/09/04/future-role-of-smartwatches-in-personal-health/. Accessed.

IIPRD. (2014). *Patent landscape analysis—smart watch*. https://www.iiprd.com/wp-content/uploads/2015/11/Sample-Patent-Landscape-Analysis_Smart-Watch.pdf. Accessed.

Lemos, R. (2018). Watch out: smartwatches need smarter security. *TechBeacon*. https://techbeacon.com/watch-out-5-reasons-smartwatches-need-smartersecurity. Accessed.

Mahadevan, S. (2017). Your smartwatch actually makes you fitter, athletes are swearing by it too. *The News Minute*. https://www.thenewsminute.com/article/your-smartwatch-actually-makes-you-fitter-athletes-are-swearing-it-too-60964. Accessed.

Rebelo, P. (2017). The product roadmap as a driver for innovation and engagement. *Front Row Agile*. https://www.frontrowagile.com/blog/posts/141-the-product-roadmap-as-adriver-for-innovation-and-engagement. Accessed.

Reeder, B., & David, A. (2016). Health at hand: a systematic review of smart watch uses for health and wellness. *Journal of Biomedical Informatics, 63*, 269–276. https://doi.org/10.1016/j.jbi.2016.09.001.

Smartsheet. (2017). *Technology roadmapping: getting started and staying focused.* https://www.smartsheet.com/technology-roadmapping-getting-started-and-staying-focused. Accessed.

Xu, C., Pathak, P.H., & Mohapatra, P. (2015). Finger-writing with smartwatch: a case for finger and hand gesture recognition using smartwatch. *HotMobile '15: proceedings of the 16th International Workshop on Mobile Computing Systems and Applications.* https://dl.acm.org/citation.cfm?id=2699350. Accessed.

Yamada, K. (2014). What operating systems do wearable devices run on? *MakeUseOf.* https://www.makeuseof.com/tag/what-operating-systems-dowearable-devices-run-on/. Accessed.

Chapter 6
Technology Roadmap: Hyperloop One

Ali Alkhafaji and Tuğrul U. Daim

There are many reasons for humans' search for alternative modes of transportation across the globe and in the USA specifically, including (but not limited to) the following: as population densities continue to increase, traffic conditions continue to deteriorate with the existing infrastructure; a societal shift toward adoption of new methods utilizing green energy demands development of efficient modes of transportation utilizing renewable energy; we need to reduce our dependency on fossil fuels; and the logistical obstacles of airlines' long security checking lines create enormous inconvenience for air travelers. All of these issues point toward the need for an intelligent, comfortable, and autonomous means of transport; that is people's impression of the transport of tomorrow. A hyperloop is a train capable of reaching the speed of sound. While many people are skeptical about this concept, researchers and entrepreneurs from all over the world are working on making the hyperloop a reality. This chapter describes a study with the goal of developing a technology roadmap for a hyperloop company, in order to prioritize items on the basis of an expert panel's judgment. With use of multielement tables, the team was able to shape a final hyperloop roadmap for the next 20 years. This roadmap is divided into 5-year intervals on the basis of the importance, feasibility, and availability of the necessary technologies.

Over time, human nature and our obsession with discovering new methods and useful tools have increasingly enabled us to travel and reach new places in new and better ways. This dream stimulates our desire to move faster and further and to use innovation and discovery to conquer nature's rules and overcome the boundaries (imposed by nature's power) of gravity and distance. Transportation and mobility have allowed humans to fulfill our dreams and find new habitats on earth, allowing us huge freedom and providing ease of multidirectional freedom of movement for

A. Alkhafaji · T. U. Daim (✉)
Department of Engineering and Technology Management, Portland State University, Portland, OR, USA
e-mail: tugrul.u.daim@pdx.edu

T. U. Daim (ed.), *Roadmapping Future*, Applied Innovation and Technology Management, https://doi.org/10.1007/978-3-030-50502-8_6

global exchange of knowledge and goods. Over time, the evolution of human transportation systems has progressed from travel on foot to transportation using carts, floated wood trunks, hot air balloons, coal and steam railways, vehicles powered by internal combustion engines, primitive airplanes (made of wood, fabric, and wire connections and joints), jet aircraft, supersonic jet aircraft, and, more recently, fuel-powered rockets to convey astronauts into the heavens.

Important factors that demonstrate the urgent need for alternative modes of transportation are increasing population densities; continuing deterioration in traffic conditions with the existing infrastructure; extreme congestion in metropolitan areas and on interstate highways; societal shifts toward adoption of new methods utilizing green energy, with growing demand for development of more efficient modes of transportation utilizing renewable energy to reduce dependency on fossil fuels and reduce per capita energy consumption for travel; consumers' "on demand" requirement for faster, more convenient modes of travel; and the logistical obstacles of airlines' long security checking lines, overbooking, and minimal control for consumers. These factors have stimulated the desire to find new and improved forms of mass transport, pushing engineers, researchers, and entrepreneurs to embrace change and innovation. This brings us to the concept of the hyperloop.

A hyperloop is a train capable of reaching the speed of sound. While many others reject this idea, Hyperloop One researchers and entrepreneurs from all over the world are working very hard on turning the hyperloop into a reality. A new, intelligent, comfortable, and autonomous hyperloop is our conception of future transport. Transportation has a big share of the global economy; it employs a significant part of the labor force and has been estimated to account for 13% of the global economy in financial terms. The transportation sector has developed and expanded greatly over the years, and it will continue to do so because of the increase in the human population, accompanied by high demand for travel (Bureau of Transportation Statistics 2018).

The hyperloop—a concept that was first proposed in 1909 or even earlier (as science fiction) as a means of high-speed travel through low-pressure tubes—is among the most dramatic and astounding inventions, in terms of mass transit, that the world has ever seen. Indeed, to even describe its workings is to venture deep into the realm of science fiction. The idea of the hyperloop came from the ambitious and prolific inventor and entrepreneur Elon Musk. His impression led him to imagine a pod, or capsule, magnetically suspended in and traveling through a vacuum tube. He also drafted a white paper describing the future of this revolutionary dream and the difficult obstacles that would be faced in making this invention a reality. His paper focused on the Hyperloop One project and how it could be developed and brought to life as a commercial project. One of the biggest questions about constructing and commercializing the first hyperloop was "How can we compete with other hyperloop companies and seize this fleeting opportunity?"; the solution was to use a roadmap to plot the future of this venture for the next 20 years and start working on it. The purposes of this project were to develop a roadmap for the hyperloop, to drive it on to the next level, and to make sure that this big new transportation concept is on the right track.

6.1 Background

The majority of well-regarded enterprises and agencies do long-range planning to develop their own roadmaps in order to clarify and determine the course they should follow in the future. A roadmap is the greatest tool for tracking and communicating both a long-term strategy and short-term objectives for its users without expending unnecessary time or effort. A strategic roadmap for a target firm is heavily involved in and contributes to all of the firm's activities on a time scheme from now to the future, via a predetermined course (Bitondo and Frohman 1981; Matthews 1992; Peck and Goto 1981). It is a visual representation that organizes and presents important information related to future plans. Roadmapping acts as a focusing tool that directs and focuses efforts toward achieving important goals. With too many companies vying for product perfection and a continuous cycle of product launches and discontinuations, roadmaps have become an essential part of figuring out an organization's future direction, along with both future market needs and future customer needs (Willyard and McClees 1987; Groenveld 1997; Kostoff and Schaller 2001; Albright and Kappel 2003; de Laat and McKibbin 2003; EIRMA 1997).

A strategic roadmap is an achievable fantasy that dictates a company's plans on what is believed to be the optimal path to the company's future destination. Hence, it requires all of the participating company's members' daily processes to work in harmony with the company's future vision. These techniques are considered very expressive and meaningful in a range of sectors (including companies, governments, defense, aerospace, electronics, manufacturing, materials, technologies, health care, and many other big agencies) to meet their needs to plan for modernization in an all-inclusive strategic manner that points precisely toward critical business drivers and market trends (Phaal 2011; Phaal 2004; Phaal et al. 2004; Willyard and McClees 1987; Winebrake 2004). Utilities and other organizations in the energy sector have also increasingly been developing technology roadmaps (Amer and Daim 2010; Daim et al. 2012; Daim and Oliver 2008; Daim et al. 2013; Phaal 2011).

Robert Galvin (Galvin 1998), a former chairman and chief executive officer of Motorola, invented this new terminology in 1970 and defined it as follows: "A 'roadmap' is an extended look at the future of a chosen field of inquiry composed from the collective knowledge and imagination of the brightest drivers of change in that field." Uncertainty is frustrating to deal with but cannot be avoided; hence, we distinguish between two types of uncertainty. The major one—which is unavoidable—is the future itself, since strategy is always a time function and we are trying to make decisions now for long-term complications, which are serious, especially with regard to the future. The second type of uncertainty concerns the quality of information and knowledge on the basis of which a strategy should be built. Clearly, a high-quality strategy demands high-quality information. In the 1970s, Motorola initiated and adapted concepts such as integrated product technology through the use of roadmaps. Motorola was recognized as the first major corporation to use roadmaps to support improved alignment between technology and product development (Kostoff and Schaller 2001; Phaal et al. 2004; Bray and Garcia 1997; Nauda and Hall 1991). Since the inception of roadmapping, research has shown that roadmapping processes have been adopted and modified by governments, businesses, and industries at multiple levels (Phaal 2011). Phaal noted that roadmapping techniques spread from the consumer electronics sector to other technology-driven sectors such as aerospace and the defense industry.

The strategic management process is about making a transition between two points—from X to Y—in a more effective and efficient manner, and avoiding barriers, as far as possible, in a safe manner to ensure a comfortable transition, while keeping a record of what has been learned along the way. It is a two-part cycle model: the devised strategy and the means of bringing it to life. Being armed with an excellent and valuable strategy will definitely help one to reach one's destination safely with minimal loss and enormous gain in a predetermined interval. However, there is no law governing the whole process and there are no rules that control the pace of it; it is changeable and depends on the organization's requirements and resources. It also differs between organizations and follows no fixed course.

The main breakthrough for the roadmapping tool was the adoption of this technique by the semiconductor industry (Schaller 2004). Semiconductors have made electronics smaller, lighter, and more portable, allowing consumers to enjoy the benefits of cell phones, laptops, drones, and other popular modern electronic devices. The first industry-wide semiconductor roadmap was published in 1992. This demonstration of the procedure of roadmapping played a significant role in its growing acceptance and ignited the evolution of the semiconductor industry because its industry-wide roadmap provided harmony and ideal conditions for development and production. Hence, Phaal believes that roadmaps helped serve as a stimulant for rapid innovation in the semiconductor industry.

In most cases, a roadmap is kept private and not released to the public. However, some roadmaps are published and made publicly available—for example, the International Technology Roadmap for Semiconductors (ITRS)—when there is agreement for that to be done in order to draw more widespread attention to the roadmap.

In practical terms, a technology roadmap is an influential, concentrated, multiyear business planning methodology. For the future of any organization, a roadmap's usability is as important as its strategic value. Hence, it is helpful to find appropriate solutions to the organization's needs by adapting a multiday workshop model into two workshops. Evolving a detailed, explanatory, and diagnostic workshop is necessary for getting adequate, productive, and powerful results. A successful technology roadmap should be based on numerous workshops. By following the aforementioned procedure, a team was able to develop a technology roadmap for a new and revolutionary concept of transportation in a very fast and aggressive medium (where there is no place for a lagging organization), in connection with financial robustness (Groenveld 1997).

A science and technology (S&T) roadmap provides a vision of the future. The roadmap methodology uses different tactics to recognize, assess, and direct participants toward selected optimal substitutes to attain appropriate S&T core goals. The S&T roadmap can be seen as existing conceptually of connected joints and their extensions. These roadmap nodes and links can, in the most general case, have quantitative and qualitative attributes.

An S&T roadmap can show qualitative effects that will influence a science program, which could have effects in one way or another. This influences the technology agenda in a restricted time interval, allowing shifts between science and technology.

Since technology development behaviors are nonlinear and unpredictable, and since roadmaps are used for both retrospective and prospective studies in time, dynamic-featured link vectors can deal with both directions in time. Hence, work on a roadmap requires use of skills and experiences to perfectly identify the nodes and joining vectors.

According to Radnor (1998a, b) roadmaps have a huge variety of types and have been increasingly used specifically in technical organizations. However, most of the roadmap literature has been gathered from practical-side information. Some of the most common roadmaps have been presented in diverse areas such as science, industry, and technology.

6.2 Methodology

After several discussions and workshops, our team reviewed many roadmap variants and decided to go with S&T roadmaps. The team also chose to go with an expert-based approach. Then we used quality function deployment (QFD) methodology, which, through its sequences of tables, allowed us to archive all of the gained observations through which the team produced the technology roadmap. The above technique followed a schematic engineering model. Hence, our team of experts held many meetings to identify and develop attributes for the nodes and links of the roadmap. Starting to build the roadmap led the team to participate through all divisions of technology, including design, infrastructure, new emerging technology (maglev), and green energy, as well as cross-cutting technology fields such as environment, safety. Furthermore, the team was composed of a mixture of multinational personnel from different backgrounds, mixed with skilled senior academic colleagues to ensure a balance of expertise and views. We tried to develop the Hyperloop One roadmap considering technology, strategy, timeframes, and structure, to publish a roadmap that should be seen as one coherent block system. We joined the nodes of drivers and product features together and then created a QFD table to find the scores and ranks of all of the suggested product features from the most important to least important drivers. After selecting the top-level product requirements from customer needs that fitted product concepts, we went through an evaluation of product concepts to select the most optimum ones for the hyperloop project. In the resultant QFD diagram, the market drivers and product features were itemized on the axes. The x-axis represented market drivers, while the y-axis represented product features. Measurements of the market drivers were calculated on a scale of 1 to 10 by the panel members, using the QFD chart to figure out the correlation between the market drivers and the product features, which had a value of 0–4, with 4 representing a high correlation, 2 a medium correlation, 1 a low correlation, and 0 no correlation. Then the team determined the technologies that were needed to cover all features, combined with resources and materials, either internally within the Hyperloop One project's organizational capability or externally through joining with new organizations to create new partnerships to cover all of those technologies. The work also involved creation of another QFD to rank the technologies chosen to cover all product features and then link all nodes together in the timeline. Finally, the team used process gap analysis to identify any significant shortcomings between the current and desired end states. On the basis of these findings, the team started developing alternative strategies and related initiatives to address aspects that had not been covered. The complete roadmap for the Hyperloop One organization was then ready to be published.

6.3 Technology Roadmap

6.3.1 Market Drivers

The first business driver is renewable energy, since the world in general is moving toward green technology while, generally speaking, this kind of energy is considered to be the most cutting-edge technology. It intelligently exploits nature's potential and converts it to meet different human needs. Renewable energy is increasingly gaining popularity because of the need to replace hydrocarbon deposit–based fuels with clean energy sources that do not generate highly toxic emissions. Since the technology is evolving fast, with great steps, and we are continuously on the hunt for more ingenious and inexpensive methods to commercialize solar energy, renewable energy sources are seen as more attractive options. Other factors that are helping to make renewable energy more popular are innovations that makes it less expensive and the potential it offers for improved air quality through reduction of toxic emissions.

The second driver in the list is a reduced environmental impact, as it is clear that most energy consumption in the world involves burning of fossil fuels. For more than a century, natural gas, petroleum, and coal have accounted for a very big share of this (e.g., 81% in the USA, according to the US Energy Information Administration (EIA)). However, humans' use of hydrocarbon deposit–based fuels has contributed greatly to the environmental damage we have caused on our planet and has also caused serious health problems for humans. The only solution—for both our own health and that of the planet—is for humans to discontinue our reliance on hydrocarbon deposit–based fuels.

The third market driver is transportation efficiency, as we know that transportation accounts for one third of the total yearly energy consumption in the USA, and road vehicles consume more than 80% of all energy used for transportation. In fact, vehicle combustion engines are very inefficient. Diesel engines are notorious for their low thermal efficiency, which is estimated to be less than 50%. Thus, for every gallon of diesel that is burned in a combustion engine, less than half of the energy generated is converted into mechanical energy. All internal combustion engines in transportation vehicles have low efficiency, which means that more than half of the energy that is consumed is dissipated as heat and carbon emissions. All of this leads us to think about new ways to enhance transportation efficiency. Energy implementation may alter completely, as people's needs and attitudes, in combination with the economic situation, will definitely lead to adoption of new transportation methods. The aim here is to look for new energy-efficient mobility system programs by investigating opportunities to increase mobility energy productivity. While conventional vehicles may make use of only 50% of the energy released by burning fuel in terms of physical displacement, lots of other vehicles in use by many people waste even greater proportions of their fuel—more than two thirds, according to some experts.

The fourth driver is fossil fuel costs. Industry is extremely dependent on hydrocarbon deposit–based fuels, leading the world to pay a high price for that flawed approach, which continues to exhaust national economies by sending valuable dollars overseas and out of domestic economies. Moreover, our dependence on fossil fuels is threatened by their dwindling supply at a time of growing demand from industrialized societies. Meanwhile, fossil fuel use continues to impose massive environmental and economic costs. Now it is vital for us to abandon the current status quo and invest in a new energy future. Use of hydrocarbon deposit–based fuels costs humanity twice, first for its extraction and secondly for its refinement cycle. In both situations, it contributes to pollution of the surrounding environment, which damages ecosystems and the species that live in them. All of the aforementioned impacts must be taken into account during assessment of the unseen impacts of hydrocarbon deposit–based fuel use when we are choosing a path for future energy implementation.

The fifth driver is aging infrastructure. The history of railroads in the USA dates back about two centuries to an early stage of the nation's history. Railroads played a major role in the growth and prosperity of the country. The period between the 1880s and the 1920s was prosperous for railroads, until the appearance of road vehicles and subsequently airplanes. Problems with America's public transportation systems translate into less efficiency and productivity in the economy. Hyperloop Transportation Technologies (HTT) now offers new hope for reviving this sector and could enable it to contribute effectively to the economy again. In 2018, HTT unveiled its first full-sized passenger capsule in Cadiz, Spain. The vessel was built to scale to transport passengers at superfast speeds through magnetic tubes.

The sixth driver is transportation innovation. The hyperloop is a new form of hybrid transportation, combining aspects of flying and land gliding, adapted by a few companies to commercialize this new concept, which could move people at approximately the speed of sound in a hovering magnetic glide within low-pressure tubes. The pace was set by Elon Musk, who suggested that hyperloop infrastructure could be installed on the ground or underground. The difference between a traditional railroad and the hyperloop is the fact that the transportation pod in the hyperloop moves through an air-evacuated tunnel to reduce friction, allowing the pod to reach supersonic speeds. Unlike standard ground vehicles, which are supported by moving wheels, the hyperloop pod uses frictionless magnetic levitation to hover within the evacuated tunnel.

Increased automation: Integrating automation with this new technology into operations will produce promising results. Rapid achievements promote successful firms even more by enhancing and increasing worker productiveness. How does this happen? By minimizing tension and the need to concentrate on boring, repetitive, and basic daily tasks and the consequences they have for the people involved. Automation is a great environment for exploring and gaining new perspectives in completely new areas and thereby creating new jobs. It also builds profitability by reducing the time taken to complete tasks while, at the same time, enhancing both workers' jobs in general and product quality. In the context of the hyperloop, the whole process could be fully automated, from ticket booking to baggage handling. A fully smart system would have the ability to adapt to many human decisions.

Security: Any sudden and unexpected destruction of transportation infrastructure in any nation rapidly causes public chaos, with unpredictable ramifications for society. The rapid increase in worldwide travel and the immense expansion in transportation infrastructure have generated a sequence of pressures in terms of security issues. One type of these huge and important facilities is airports. Although the process of checking travelers differs from one airport to another, they share one common fact: these procedures place huge pressures on travelers and continue to increase, leading to enormous pressures on the screening process. A hyperloop could transport passengers between cities at the same speed as a commercial jet, if not faster, in capsules that float in partial-vacuum tubes. Hyperloops could possibly be one of the best alternatives for governments and societies that wish to expand their network infrastructure. People will gain ease of implementation, less exposure to security screening, and freedom from fluctuating and high fossil fuel prices, which lead to frequent price rises.

New methods of transportation: Given humans' needs for a new transportation method to travel great distances within a short period of time at less expense, future transportation will either be a development of current technology or a completely new innovation, obviously. Future forms of travel will come in different varieties, depending on people's requirements: some will be magnetically levitated, some will use air cushions, some will use jet propulsion, and someday we may even use teleportation. The public transport system suggested by Elon Musk involves a very sophisticated form of supersonic travel between places, and potentially even between continents, through evacuated tunnels or pipes either on the ground or underground, in pods traveling at a speed of 760 miles per hour (1220 km per hour) or faster. This means of transportation will be available for both cargo and people, using green energy. Hyperloop One has been planned to launch in a very short time interval, given people's need for a better transport system, and is based on the pod principle, which means that each pod could transport a number of people and hence would greatly improve public time management.

Table 6.1 lists the market and business drivers, with their codes.

Table 6.1 Market and business drivers

Code	Driver
D1	Renewable energy
D2	Reduced environmental impact
D3	Transportation efficiency
D4	Fossil fuel costs
D5	Aging infrastructure
D6	Transportation innovation
D7	Increased automation
D8	Security
D9	Transportation time
D10	New methods of transportation

6.3.2 Technologies and Product Features

Most successful organizations have adopted a marketing concept based on the right principles. The team saw that the product features mentioned in this chapter could be used to find a balance to satisfy both market and business drivers. As is known, the marketing concept is the use of marketing data to focus on the needs of customers in order to develop marketing strategies that not only satisfy customers' needs but also accomplish the goals of the organization. Hence, combining the most suitable combination of product features with market drivers should produce optimal results while also ensuring that time and financial aspects are monitored throughout the whole process.

The first feature is self-power, and the goal at Hyperloop One is to run the hyperloop exclusively on renewable electricity by making it self-powered through installation of energy units in the form of solar panels on the top of the tunnel. The generated energy will be used directly to power the hyperloop, while the excess will either be stored in high-capacity batteries for use in cloudy or nighttime conditions, or could be converted back into usable electricity for use in many other new applications.

The second feature is a vacuum tube or low-pressure tube. Creating a perfect vacuum is dangerous, very costly, and very hard to manage, especially in a huge infrastructure, but the goal is to reduce the air pressure to a level that fits with the pod speed—in some ways, similar to the atmospheric layer that is used for commercial air flights—otherwise, supersonic shock waves would be caused.

The third feature is propulsion to operate both the magnetic drag and magnetic levitation. The pod will behave like a rotor in a traditional electric motor, while the tube itself will behave like a starter in a regular motor; that is how a linear motor works. This causes the pod to accelerate in a low-pressure unit very smoothly, up to the speed of a bullet. Technically, an advanced linear motor system is being developed to accelerate the capsule to above 760 miles per hour (1220 km per hour), at a maximum of 1 *g* for comfort, in order to propel the vehicle at the required travel speed.

The fourth feature is passenger transportation, and that will be achieved by use of pods to carry passengers. These pods will travel through the tubes at speeds topping out over 700 miles per hour. These pods will be propelled forward by magnetic accelerators and housed by the tubes in a low-pressure environment. The cradle container is carried by the pod from one city to another, comfortably and with very low vibration and no sensation of acceleration despite the very high speed. In fact, it has been said that its speed will be ten times as fast as that of other vehicles.

Station construction: Stations will be designed for optimal loading and unloading of both passenger versions and passenger-plus-vehicle versions of the hyperloop capsules. Two of the hyperloop's goals are to minimize the boarding time and provide more comfort with less stress for travelers by making it more fun and far better than any other means of transportation, supported by lower prices and freedom from market fluctuations without the types of delays that occur in air travel systems.

Tube construction: A fixed-thickness steel tube, reinforced with stringers, was selected as the material of choice for the inner-diameter tube, in order to keep costs to a minimum. The tunnel parts will be manufactured and tested at facilities, then moved to the site to be assembled there. Studies have shown that the optimal length of these sections will be 30 m, and they will be supported by special-purpose pillars.

Cargo transportation: With the Internet Of Things, the world changing into a global village, and the huge growth of trading between cities, goods transportation is an increasing priority for organizations. To provide an emergency evacuation system, safety emergency exits and pressurization ports will be added in key locations along the length of the tube. These openings will be designed to work in harmony with a very accurate cleaning vehicle to allow very smooth hovering and drag for the transporting capsule.

Onboard power: The passenger capsule power system will be equipped with an array of batteries to supply all of the onboard compressor and capsule systems in addition to the compressor motor and coolant. A transcontinental or transoceanic evacuated tunnel train system would be a future step, since it will face many engineering challenges. Its thermal expansion would be measured in tens of feet. Continental drift would need to be compensated for. Ocean and river traffic would be a significant concern.

Table 6.2 lists product features and the supporting technologies they require, and Table 6.3 summarizes the gap analysis of those technologies.

6.3.3 Quality Function Deployment

After a few workshops, the team members engaged with visual objects using visual methods on the basis that "if you can't see it, you can't manage it," and they found that with use of a visual management system, the market driver scoring system could be measured and scaled on the basis of the importance of these drivers to the hyperloop roadmap. Renewable energy, a reduced environmental impact, transportation efficiency, transportation innovation and leadership, and transportation time

Table 6.2 Product features and supporting technologies required

#	Product feature		Supporting technology	
	Code	Feature	Code	Technology
1	PF1	Self-powered	1a	Solar panel efficiency
2	PF2	Vacuum tube	2a	Pressure sensors
			2b	Depressurization
			2c	Solar panel efficiency
3	PF3	Propulsion	3a	Advanced linear motor system
			3b	Magnetic levitation
			3c	Solar panel efficiency
4	PF4	Passenger transportation	4a	Solar panel efficiency
			4b	Magnetic levitation
			4c	Propulsion
			4d	Vacuum system
			4e	Advanced linear motor system
			4f	Pressure sensors
			4g	Earthquake proofing
			4h	Depressurization
			4i	Magnetic braking
5	PF5	Station construction	5a	Vacuum system
			5b	Depressurization
			5c	Magnetic braking
			5d	Earthquake proofing
6	PF6	Tube construction	6a	Depressurization
			6b	Earthquake proofing
			6c	Propulsion
7	PF7	Cargo transportation	7a	Solar panel efficiency
			7b	Magnetic levitation
			7c	Propulsion
			7d	Vacuum system
			7e	Advanced linear motor system
			7f	Pressure sensors
			7g	Earthquake proofing
			7h	Depressurization
			7i	Magnetic braking
8	PF8	Onboard power	8a	Magnetic levitation
			8b	Propulsion
			8c	Vacuum system
			8d	Advanced linear motor system
			8e	Pressure sensors
			8f	Depressurization
			8g	Magnetic braking
9	PF9	Transcontinental hyperloop route	9a	Solar panel efficiency
			9b	Earthquake proofing
			9c	Depressurization

Table 6.3 Technologies and gap analysis

Code	Technology	Gap analysis
T1	Solar panel efficiency and batteries	Efforts should be made to scale up production of existing technologies; battery technology must be enhanced further
T2	Magnetic levitation	Technology needs to be scaled up for weight lifting; nowadays, technology requires 1 kW metric ton – production level usage
T3	Propulsion	Technology needs to be scaled up in terms of the speed limit; the current world speed record for maglev trains is 581 km per hour; the goal is to overcome the Kantrowitz limit
T4	Vacuum system	Undeveloped on a large scale; there is also the problem of thermal expansion, which threatens to buckle any large structure without proper thermal expansion capabilities; dealing with long metal surfaces generates a thermal expansion complication, contravening the evacuation principle, which requires no openings in the structure
T5	Advanced linear motor system	Enhanced efficiency for production-level usage
T6	Pressure sensors	Pressure sensors and control networks lack of trail need to be investigated thoroughly
T7	Earthquake proofing	More examination of damper technology, which could be adapted to resist earthquakes and their reflections at each hub end point
T8	Depressurization	Needs to be tested and researched further, especially over long distances
T9	Magnetic braking	Needs to be tested, especially in emergency scenarios, but has not been tested in a hyperloop

each got a high score (4 points) on the three-grade scale (4 for high, 2 for medium, and 1 for low). The other two drivers—aging infrastructure and increased automation—each got a medium score (2 points), while fossil fuel costs and security each got a low score (1 point).

Using QFD between market drivers and product features to identify product feature importance, the team found that the transcontinental hyperloop route had the highest score, followed by the passenger transportation feature in second place, and tube construction came third in the list. Including all of these features will help to make the hyperloop project a reality and will provide a new transportation method for humanity. The last three features in the list were the vacuum tube (number 8 in the list), onboard power (number 9), and the emergency evacuation system (number 10) (Table 6.4).

A second QFD comparison was done to choose the most important technology to help develop product features and also to help define the gap and choose the most suitable solution to cross it. Propulsion, cargo transportation, passenger transportation, tube construction, and the transcontinental hyperloop route each achieved the highest rank. Self-power, station construction, and the vacuum tube each achieved a medium rank, while onboard power and the emergency evacuation system each achieved the lowest rank among the product features. Completion of the QFD calculation led to the following results, ranked from highest to lowest: magnetic

Table 6.4 Market drivers versus product features: quality function deployment (QFD)

		Drivers											
		Renewable Energy (High 4)	Reduced Environmental Impact (High 4)	Transportation Efficiency (High 4)	Fossil Fuel Costs (Low 1)	Aging Infrastructure (Medium 2)	Transportation Innovation and Leadership (High 4)	Increased Automation (Medium 2)	Security (Low 1)	Transportation Times (High 4)	New Methods of transportation (Medium 2)	Score	Ranking
Product Features	Propulsion	(4 × 1)4	(4 × 4)16	(4 × 4)16	(1 × 2)2	0	(4 × 2)8	0	(1 × 2)2	(4 × 4)16	(2 × 4)8	72	5
	Cargo Transportation	(4 × 1)4	(4 × 4)16	(4 × 4)16	0	(2 × 2)4	(4 × 4)16	(2 × 2)4	(1 × 2)2	(4 × 2)8	(4 × 2)3	78	4
	Passenger Transportation	(4 × 1)4	(4 × 4)16	(4 × 4)16	0	(2 × 2)4	(4 × 4)16	(2 × 2)4	(4 × 1)4	(4 × 4)16	(2 × 2)4	84	2
	Self-Powered	(4 × 4)16	(4 × 4)16	(4 × 4)16	0	0	(4 × 4)16	0	0	0	(2 × 2)4	68	7
	Station Construction	(4 × 1)4	(4 × 1)4	0	(1 × 2)2	(4 × 2)8	(4 × 4)16	(4 × 2)8	(4 × 1)4	(4 × 4)16	(4 × 2)8	70	6
	Tube Construction	(4 × 1)4	(4 × 2)8	(4 × 4)16	(1 × 2)2	(2 × 2)4	(4 × 4)16	0	(4 × 1)4	(4 × 4)16	(4 × 2)8	82	3
	On Board Power	(4 × 1)4	(4 × 2)8	(4 × 2)8	0	0	(4 × 2)8	0	(1 × 1)1	0	0	29	9
	Vacuum Tube	0	(4 × 2)8	(4 × 4)16	0	0	(4 × 2)8	0	0	(4 × 4)16	(4 × 2)8	56	8
	Emergency Evacuation System	0	0	0	0	0	0	(1 × 2)2	(1 × 4)4	0	(1 × 2)	8	10
	Trans-continental Hyper Loop Route	(1 × 4)4	(4 × 4)16	(4 × 4)16	(2 × 1)2	(4 × 2)8	(4 × 4)16	(1 × 2)2	(1 × 2)2	(4 × 4)16	(4 × 2)8	90	1

braking, pressure sensors, the vacuum system, magnetic levitation, the advanced linear motor system, earthquake proofing, the emergency evacuation system, depressurization, and solar panel efficiency. It is obvious why magnetic braking achieved the highest score in the list, since such a supersonic speed would need a very powerful braking system to prevent emergencies or stop at a specific station. Also, pressure sensors and the vacuum system are critical for passenger safety in case any leakage should occur in pipes almost completely emptied of air (Table 6.5).

6.3.4 Resources

The team found that one of the most important resources is the government, since it will play an essential role in the Hyperloop One development path. Hyperloop One is looking forward to working with the government's Department of Transportation (DOT) to move hyperloop technology forward in the USA (via the newly announced Council to Support Hyperloop Commercialization) and also to identify and resolve jurisdictional and regulatory gaps, which will help bring the technology to commercialization. The other way in which the government could help is by offering its testing facilities (military and many other facilities) for evaluation of the new emerging technologies from Hyperloop One. Another important aspect that the government could help with is political decision making. The politics opposing such a proposal would be extremely difficult to overcome in the short term. Even if Hyperloop One demonstrates its prototype to other companies and government officials and shows that this idea actually works, there will still be a lot of political obstacles to overcome before a point is reached where there is a push on companies to do this.

The other important resource is pod manufacturers. A good example here is HTT—a great competitor and one of several companies racing to make Musk's vision a reality. HTT and its Spanish partner Airtificial have experimented with a pod model for passenger transportation at an aerospace testing facility in Spain.

This joint work will help a lot in expediting the Hyperloop One project.

Another resource is university partnerships, which will be very beneficial for the company's future by encouraging and motivating young people to file new patents and apply out-of-the-box thinking to close project gaps. Another motivating idea put forward by the Hyperloop One administration team is to hold a yearly worldwide competition for many participants—including the private sector, universities, and different companies—to develop new ideas that can subsequently be used by the company to grow and develop its concepts almost freely. That will be helpful for implementation strategies, key stakeholder involvement by both the public and private sectors, and innovative and creative applications of hyperloop systems.

Construction partnerships are another resource that will help and support the hyperloop project, through joining of forces with leading companies that share Hyperloop One's vision and its ambition to resolve today's transportation challenges. Work with cutting-edge technology partners will push forward technology

Table 6.5 Scores and ranking of all technologies

		Product Features											
		Propulsion (High 4)	Cargo Transportation (High 4)	Passenger Transportation (High 4)	Self-Powered (Medium 2)	Station Construction (Medium 2)	Tube Construction (High 4)	On-Board Power (Low 1)	Vacuum Tube (Medium 2)	Emergency Evacuation System (Low 1)	Trans-Continental Hyperloop Route (High 4)	Score	Ranking
Technical Features	Magnetic Levitation	(4 × 4)16	(4 × 4)16	(4 × 4)16	(1 × 4)4	0	(2 × 4)8	0	0	0	(4 × 4)16	74	4
	Advanced Linear Motor System	(4 × 4)16	(4 × 4)16	(4 × 4)16	(1 × 2)2	0	(1 × 4)	0	0	0	(4 × 4)16	70	5
	Vacuum System	(2 × 4)8	(4 × 4)16	(4 × 4)16	0	0	(4 × 4)16	0	(4 × 2)8	0	(4 × 4)16	80	3
	Emergency Evacuation system	(4 × 2)8	(4 × 2)8	(4 × 2)8	(4 × 2)8	0	(1 × 4)4	(1 × 4)4	0	(1 × 1)1	(1 × 4)4	45	7
	Solar Panel Efficiency	0	0	0	(4 × 2)8	(2 × 2)4	(4 × 4)16	(2 × 1)2	0	(1 × × 1)1	(1 × 4)4	35	9
	Pressure Sensors	(4 × 4)16	(4 × 4)16	(4 × 4)16	0	(4 × 2)8	(4 × 4)16	0	(2 × 2)4	(4 × 1)4	(1 × 4)4	84	2
	Magnetic Braking	(4 × 4)16	(4 × 4)16	(4 × 4)16	(2 × 2)4	(2 × 2)4	(4 × 4)16	(1 × 1)1	0	(1 × 4)	(4 × 4)	93	1
	Depressurization	0	0	(4 × 4)16	(2 × 2)4	0	(2 × 4)8	(2 × 1)2	(2 × 1)4	(4 × 1)4	(2 × 4)8	44	8
	Earthquake Proofing	0	(2 × 4)8	(2 × 4)8	0	(2 × 4)8	(4 × 4)16	0	0	(4 × 1)4	(4 × 4)	60	6

Table 6.6 Resources

Code	Resource	Examples
R1	Government	Federal, state, local
R2	Pod manufacture	Tesla, Voxeljet
R3	Private capital	US$150 million; Sherpa Capital, DP World Group
R4	University partnerships	Research funding and internships
R5	Construction partnerships	Tube and station construction firms
R6	Software	IBM
R7	Batteries and solar power	Solar City, Tesla
R8	Vacuum technology	NASA, AMAT
R9	Political	Lobbyists, political action committee funds, industry coalitions

acquisition and bring the hyperloop concept to life in a shorter time, allowing a more efficient mode of transportation to be adopted.

Batteries and solar power: This is another very useful and important resource for the Hyperloop One project. It plays a major role in feeding electricity to the hyperloop system when it is in action, using clean and green energy without any carbon oxide emissions, which will improve the surrounding environment by reducing air pollution.

Vacuum technology (the transporting medium): An essential part of the whole scheme is the vacuum within which the passenger capsules will travel. Use of a vacuum significantly reduces air resistance, along with low-friction propulsion and levitation technologies based on air cushion or magnetic levitation concepts. Partnerships with NASA and AMAT will further enrich this new concept of transportation.

Software: This is a very important resource, handling various items and objects, from the smallest to the biggest: coordinating the very smooth lifting of the pod, determining the exact time to start running the braking system when reaching the next station at such high (supersonic) speeds, monitoring the pressure within the pod, and creating a perfect balance between the speed of the pod and the vacuum within the surrounding medium, along with many other tasks. Partnerships with giant software companies such as IBM will be helpful for finding solutions to these gaps (Table 6.6).

6.4 Conclusion

A technology roadmap commonly embodies a master diagram showing detailed work in relation to time intervals, with respect to essential rules that determine what key technology will be implemented (and when and why) to support an organization's decision makers in finding optimal and low-risk solutions within the available budget and resources. It also highlights today's technological limits and the life cycle, while maintaining alertness to the possibility of relevant but unexpected technologies. In this project, the team built a technology roadmap model considering the most important market drivers and examined them against the product features, with rankings and scores to help achieve optimal results in terms of market drivers and also to enable definition of both current and future barriers. The team proposed technologies that could help in closing the gap in hyperloop product features by focusing on the available resources and budget to match the product features with the emerging technologies. That will further enhance our ability to overcome some significant hurdles, using the most suitable techniques in terms of product features, and also highlights the resources that will be necessary for linking product features to newer technologies. To point this out, a QFD technique was implemented, analyzing technological features with product features and then analyzing product features with market drivers. The next step will be analyzing each part of the roadmap. The final technology roadmap is shown in Fig. 6.1.

The final roadmap is divided into five main time periods:

1. *0–5 years:* Following publication of Elon Musk's white paper, work started on establishing a hyperloop company. This has involved prioritization of market drivers and product features to choose the best technology for the company to pursue for the project's success. The initial focus has been on several market

Fig. 6.1 Final roadmap

drivers, starting with renewable energy, a reduced environmental impact, and fossil fuel costs.

2. *5–10 years:* Hyperloop One needs to prove that the technology works, and has done so. The next step is to start work on establishing the first route, creating the first open sled, and performing a propulsion sled test, then the first full flight test to be conducted will occur after 2 years. Also, the company should work with the government—specifically, with regulators—to certify this technology by going through the process of making sure it is safe and properly tested, and thoroughly investigating the potential emergency scenarios that are important for the whole project. The company also should concentrate on government support and the financial aspect, which will require an economic environment that can raise the necessary capital—maybe through a public–private partnership—to put that technology in place.
3. *10–15 years:* This is the commercialization phase for Hyperloop One. Talent has been brought into the organization's leadership, which will allow the company to commercialize this technology by building a prototype. Work will start on cargo trips. The route will initially go from Los Angeles to San Francisco, followed by expansion of commercial trips to three networks.
4. *15–20 years:* Hyperloop One should continue developing product solution software experiences that will take full advantage of the technology and finding customers who want to be first because they want to take advantage of it to achieve a major transformation. The economic disruption that this technology could cause in various economies should not be overlooked. Hence, the company is focusing on this and looking for suitable spots around the world. At the same time, commercial passenger trips from Los Angeles to San Francisco should be achieved. After 2 years, international rail networks could be expanded to many cities.
5. *Beyond 20 years:* At this stage, development should be continued to maintain the company's status as a pioneer in this field. Research on superconductors and maglev should be directed toward a more efficient system. This work should be done in parallel with starting transcontinental rail from Los Angeles to New York, with passenger trips beginning after a few years and work being done on a US network to also connect other major cities.

References

Bureau of Transportation Statistics. (2018). TET 2017—chapter 2—transportation's contribution to the economy. https://www.bts.gov/browse-statistical-products-and-data/transportation-economic-trends/tet-2017-chapter-2. Accessed .

Bitondo, D., & Frohman, A. (1981). Linking technological and business planning. *Research Management, 24*(6), 19–23. http://www.jstor.org/stable/24118805. Accessed .

Matthews, W. H. (1992). Conceptual framework for integrating technology into business strategy International Journal of Vehicle Design, 13(5–6), 524–532.

Peck, M. & Goto, A. (1981). Technology and economic growth: the case of Japan. Research Policy, 10(3), 222–243. https://doi.org/10.1016/0048-7333(91)90039-S.

Research Management, 24(6), 40–41. (1981b). http://www.jstor.org/stable/24118821. Accessed.

Willyard, C. H., & McClees, C. W. (1987). Motorola's technology roadmapping process. *Research Management, 30*, 13–19.

Groenveld, P. (1997). Roadmapping integrates business and technology. Research-Technology Management, 40, 48–55.

Kostoff, R. N., & Schaller, R. R. (2001). Science and technology roadmaps. *IEEE Transactions on Engineering Management, 38*(2), 132–143.

Albright, R. E., & Kappel, T. A. (2003). Roadmapping in the corporation. *Research-Technology Management, 42*(2), 31–40.

de Laat, B. & McKibbin, S. (2003). The effectiveness of technology Roadmapping—building a strategic vision. The Hague: Ministry of Economic Affairs.

EIRMA. (1997). Technology roadmapping: delivering business vision. Working Group Report No. 52. Paris: European Industrial Research Management Association.

Amer, M., & Daim, T. (2010). Application of technology roadmaps for renewable energy sector. *Technological Forecasting and Social Change, 77*(8), 1355–1370.

Daim, T., Harell, G., & Hogaboam, L. (2012). Technology roadmapping: wind energy for Pacific NW. *Journal of Cleaner Production, 20*(1), 27–37.

Daim, T., & Oliver, T. (2008). Implementing technology roadmap process in the energy services sector: a case study of a government agency. *Technology Forecasting and Social Change, 75*(5), 687–720.

Daim, T., Oliver, T., & Kim, J. (2013). *Research and technology management in the electric industry*. London: Springer International Publishing.

Phaal, R. (2011) Public-domain roadmaps. University of Cambridge. http://www.ifm.eng.cam.ac.uk/uploads/Research/CTM/Roadmapping/public_domain_roadmaps.pdf. Accessed 30 Mar 2015.

Phaal, R. (2004). Technology roadmapping. In R. Seidl da Fonseca (Ed.), *Foresight methodologies: training module 2* (pp. 129–153). Vienna: United Nations Industrial Development Organization.

Phaal, R., Clare, J. P., Farrukh, C. J. P., & Probert, D. R. (2004). Technology roadmapping—a planning framework for evolution and revolution. *Technological Forecasting and Social Change, 71*(1–2), 5–26.

Winebrake, J. J. (2004). Technology roadmaps as a tool for energy planning and policy decisions,". Energy Engineering, 101(4), 20–36.

Galvin, R. (1998). Science roadmaps. *Science, 280*, 803.

Bray, O. H. & Garcia, M. L. (1997). Technology roadmapping: the integration of strategic and technology planning for competitiveness. Presented at the Portland International Conference on Management of Engineering and Technology, Portland, 27–31 Jul 1997.

Nauda, A. & Hall, D. L. 1991. Strategic technology planning—developing roadmaps for competitive advantage. Presented at the Portland International Conference on Management of Engineering and Technology, Portland, Oct 1991.

Schaller, R. R. (2004). Technological innovation in the semiconductor industry: a case study of the International Technology Roadmap for Semiconductors (ITRS). IEEE Xplore. https://ieeexplore.ieee.org/document/951917. Accessed .

Radnor, M. (1998a). Roadmapping: how does it work in practice? In: Proceedings of the National Center for Manufacturing Sciences Conference and Exposition, Orlando, 14 May 1998, p. 14.

Radnor, M. (1998b). Corporate technology and product roadmapping: comparing hopes and realities. In: Proceedings of the Technology Roadmap Workshop, Washington, DC, 29 Oct 1998, p. 12.

Chapter 7
Technology Roadmap: Smart Apartments

Mohammad Al Gafly and Tuğrul U. Daim

Companies all over the world are facing the pressures of dealing with competition and an endless need to improve on cost-efficiency and cost-effectiveness. Technology companies are particularly facing the challenge of maintaining their competitiveness given the speed of technological advancements. As such, tech companies need to create strategic plans for technology advancement to enable them to deal with the market challenges. A technology roadmap is a tool that companies can use in planning short-term and long-term goals and aligning them with technology adoption and advancement. This chapter is a technology roadmap for ABC Inc. ABC is a technology startup in Portland, Oregon, that provides smart apartment solutions in Oregon and in the US market. In developing this technology roadmap, industry experts were asked to identify the key market drivers in the smart apartment market, the product features, technologies, and resources. The study also identified the gaps in product features, technologies, and resources that need to be addressed. We then used Quality Function Development (QFD) method in mapping the four components using a 4-scale scoring model. Finally, a timeline plan for each of the drivers, product features, technology, and resources is done to indicate the appropriate market timing for deployment.

7.1 Overview

Advances in technology are permitting increased connection of electronic devices through digital communication. With increased connection of devices, it is now possible to operate and manage reprocess remotely using a smartphone or any other

M. Al Gafly · T. U. Daim (✉)
Department of Engineering and Technology Management, Portland State University, Portland, OR, USA
e-mail: tugrul.u.daim@pdx.edu

T. U. Daim (ed.), *Roadmapping Future*, Applied Innovation and Technology Management, https://doi.org/10.1007/978-3-030-50502-8_7

smart devices such as tablets and smartwatches. This interconnection between different devices is commonly referred to as Internet of Things (IoT). Smart apartments are built by considering the concepts of IoT in developing home equipment and devices that are connected through the Internet and can be controlled remotely. This can involve connecting different home products including smartphones, cameras, televisions and sound systems, lighting and energy systems, computers, security systems, cars, and home appliances among other things. The connection between different components is accomplished through the use of central hub that the owner can then access and control using a smartphone or through any other human machine interface.

Several decades ago, smart homes existed as mere concepts that envisioned future residential homes. However, this concept has been actualized in today's business world. Advancement in digital technologies is the major drive that is informing the growth in adoption of smart homes in recent times. The demand for smart homes is increasing at a fast speed. According to a white paper, the market share of smart home by 2018 was worth $70 billion ("ABC") and is expected to increase to $127 billion by 2022 (VeCap 2017).

The motivation to develop smart apartments stems from different things including creating enhanced security and safety, saving energy, ease in handling household tasks as well as making home more livable, and entertaining. Smart homes are also developed to make it easier for people with health problems to handle household tasks. Additionally, past studies have established that some smart homes are designed with the objective of reducing costs through optimization of energy management (Cempel & Mikulik 2013).

Generally, the primary aim of developing smart homes is to automate the building. Smart homes are becoming increasingly common across the world. The building secret is also undergoing a paradigm shift in which the design for smart homes is taking center stage. The objective of this paper is to do comprehensive analysis of technology roadmap for smart apartments. The study will specifically examine ABC Company which deals with design and installation of smart gadgets for automating apartments.

7.2 Objective

The objective of this study is to do technology roadmap for smart apartments. The specific objectives of the study include the following:

(i) To carry out past, current, and future market analysis of smart apartments
(ii) To examine market drivers of smart apartments
(iii) To perform product feature gap analysis of smart homes
(iv) To explore the resources needed in the development of smart homes

7.3 ABC: Company Background

ABC Inc. is a real-estate startup that deals with home automation. It was started in 2014 by Ms. Sce Pike who is the founder. The company headquarters is located in Portland, Oregon. Currently, the company has a workforce of 30 employees which grew from 12 in 2017. The company's management team is composed of people with different professional and business backgrounds, including Real estate experts, strategists, entrepreneurs, as well as technologists. Sce Pike the founder of the company is a strategist who has over 18 years' experience in different fields.

The company was started on the platform of using technology and leveraging on the capabilities of IoT in automating homes. The company focuses on multifamily homes/apartment developers as the target market for their services and not the end-user homeowners. The company produces both hardware products its range of products include smart locks, door sensors, smart switches, motion sensors, window coverings, ceiling fans, humidity and leak sensors, and garage door controls among other products.

ABC has set its Mission statement as to "Help the real estate industry digitize itself and create new revenue sources." The vision of the organization is to "Help residents generate passive income through brokering of their own data in the way they see fit." The overall objective of the company is to help property managers to maximize their profits and at the same time make a difference in the market by providing a faster way through which property managers can showcase vacant rooms ("ABC, Inc." 2016).

7.4 ABC Competitive Strategy

The company's competitive strategy is based on four key areas. The first one is market differentiation in which the company targets creating a new market segment in which the target clients are property managers and developers. The company sells its products to this category of customers who then offer their clients the service for free and charge at a markup to recover their costs. By adopting the smart apartment technology, developers, and property managers seek to provide differentiated products to their clients. This helps in attracting more clients and providing a rationale for premium pricing.

The company focuses specifically on multifamily homes and not single-family homes in this market segmentation is that smart apartment technologies are expensive and thus unattractive to single-family home market. Secondly, smart home automation provides little Return on Investment (ROI) with regard to energy saving. For this reason, the most likely clients to adopt the technology are apartment owners and managers.

Another aspect of the competitive strategy is creating a long-term goal of data management. In home automation, ABC connects the hardware components with a software and this enables users to collect and store data. With this, a user can decide on how to use that data for his/her own advantage. Lastly, the company employs an expansion strategy that focuses on creating partnerships with multifamily apartment entities. Currently, the company has more than 20 partners nationwide and is continuing to create more relationships with partners. Some of the partners in the real-estate segment include Kaktu Life, TRU Apartments. Both of these companies operate in Las Vegas, Nevada (Dillard 2018).

7.5 Competitors

Home automation is a relatively new investment platform and there are no well-established competitors, especially in the Multifamily homes segment. Majority of the companies are indirect competitors who either provide isolated products or are involved in the development of parts that can then be integrated into a system. For example, some competitors can specialize in security and access control systems, home appliances, energy management as well as in lighting and window control systems. Some of ABC's indirect competitors include.

OTAS also faces several direct competitors in this market. These competitors include AT&T Digital Life, WEMO, Apple Homeki, Samsung Smart Things, Control 4, and Philips Hue. Others include Remotely Inc. Dwelo, Entrata, Stratis, and Apple Homekit. Although these companies are ABC's direct competitors, they offer discrete solutions which are integrated into a hub but are not specifically under the smart apartment sector. Unlike its competitors, ABC has the capability of deploying smart apartment technology in a large scale but with minimal overheads ("ABC Inc." 2016).

7.6 Market Analysis

Smart home market is growing at a fast pace since 2010. The growth is largely attributed to the increased adoption of smart devices and Internet of Things. According to market watch, the smart homes market was estimated to be about $12.5 billion in 2017. The annual growth rate sector is expected to be about 15% on average for the next 5 years. The main drivers of growth are projected to be increased adoption of IoT, and growing demand for security in residential homes. However, it is also indicated that the major hindrance to the growth of the sector is the cyber security risks.

The compound annual growth rate (CAGR) in the smart home market is expected to be 16.9% in the period between 2019 and 2023 (Market Watch 2019). It is projected that Google and Amazon will remain dominant in the industry in the next 5 years. However, more companies are expected to enter into the market.

7.7 Technology Roadmapping Background

The term technology roadmap is used to mean an elaborate plan that is drafted as a guide for reaching a specific objective. Technology roadmapping is considered to be an important tool that can help in strategic decision-making. Use of technology roadmaps started in the private sector. However, it is now widely used in both private and public sectors. Technology roadmaps are used to support development of strategy and its implementation. A good technology roadmap is that which satisfies three requirements. First, is should outline the trends and drivers that influence development as well as use of a given technology. Second, a technology roadmap should also outline the applications, products, as well as services that are enabled by technology innovations. Lastly, the roadmap should outline the resources and other capabilities that are created as a result of the technology (Phaal et al. 2004).

According to Garcia and Bray (1997), the roadmapping process is as important as the roadmap itself. Roadmapping process is essential because it can promote communication between the key stakeholders who are involved in the development of the technology. It also aids consensus building and can also create interest for potential investors. Additionally, it can serve as a basis for future development.

It is important to note that there is a fundamental difference between roadmaps and forecasts or scenarios. Forecasts and scenarios are essentially descriptive in nature. They are used as tools for exploring the future but they do not evaluate whether the futures are good or bad. Unlike forecasts and scenarios, technology roadmaps are action oriented. In other words, they look at the actions that can be taken in order to address the identified issues. Roadmaps are also normative in nature, which means that they attempt to give and show what ought to be with regard to the development of a specific technology.

Technology roadmaps have certain things in common with visions and backcasting. Notably, a vision or a backcasting study can be used as a starting point when doing technology roadmapping. Essentially, a vision outlines a desirable future however, it does not show how that future will be reached. One the other hand, a backcasting study starts with drawing a desirable future and then works backwards in exploring the programs or policies that will help in linking that future to the present.

Technology roadmapping is essential for several reasons. Companies across the world are facing many challenges while products are increasingly becoming more customized and complicated. In addition, the time to market a product and product life are getting considerably short. The impact of this short-term focus is that funding is also taking a short-term approach. Competition is also becoming more intense. In order to address these challenges, companies are required to understand the market and the industry and become more focused. Technology roadmapping is a tool that can help companies to deal with these challenges given that it helps in reaching the most appropriate investment decisions. This is possible because technology roadmapping helps in identifying a technology that can help in reaching the most critical product needs. Secondly, it can help to identify technology alternatives that can be used in fulfilling the product needs (Garcia 1997).

There are different types of roadmaps. The most common types include industry roadmap, cross industry roadmap, project roadmap, product roadmap, and technology roadmap. Each of these types of roadmaps has its use on strategic planning. A technology roadmap entails a systematic analysis of the industry in terms of the needs and barriers. The outcome of technology roadmapping is a technology roadmap which is a summary that outlines opportunities and the critical drivers of the industry. Technology mapping usually takes a long-term view of the industry for a span of 10–15 years (Garcia 1997).

In strategic planning, there are a number of processes involved such as evaluation of the competitive environment, outlining objectives and target and defining a company's positioning in the value chain. Another important process is developing a business model. In strategic planning, a technology roadmapping helps in establishing the connection between business drivers and technology. A technology roadmap has three layers. The first one is market/business layer, while the second one outlines products and services. The third layer is technology. Roadmapping integrates the three layers by establishing a connection between technology and market opportunities and gaps (Phaal et al. 2004).

7.8 Methodology

The primary objective of this study is to identify the key drivers that ABC requires in order to enhance its competitiveness in the smart home industry in the United States. In total, 15 drivers were identified to influence adoption of Smart home technology. The 15 drivers were categorized into five major categories comprising of: (1) social, (2) safety & security, (3) application & convenience, (4) political and (5) economic categories. Each driver is linked with the products then with technologies and resources. Drivers are given weights on a scale of 1–10 with a weight of 1 representing the least important driver and 10 being the most important driver. Experts were asked to assign each driver the corresponding weight. A quality function deployment (QFD) table was used in listing the drivers as per the results from the experts from the most important to the least important. A QFD chart was also generated in which the x-axis represents the drivers while the y-axis represents the products. A correlation is also performed on the drivers and the products using 0–4 scale of correlation. A score of 4 is assigned where the correlation is high, 2 for medium, 1 for low, and 0 for zero correlation.

Two more QFDs were done for ranking key technologies and resources while following the same procedure as outlined for the drivers and project features. A timeline plan for each of the drivers, product features, technology, and resources was done to indicate the appropriate market timing for deployment. The entire

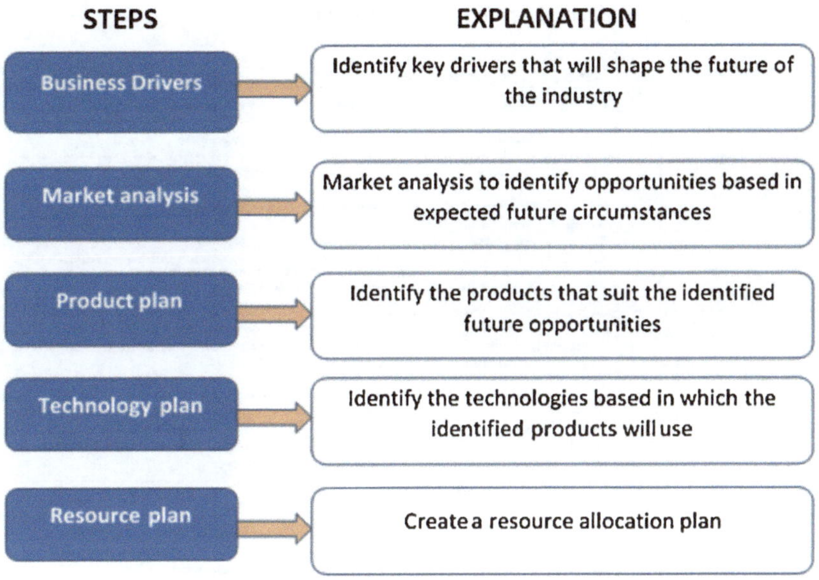

Fig. 7.1 Steps for technology roadmapping

process of technology roadmapping for ABC entailed five main steps. The first one is analyzing business drivers which entailed identifying the forces or the drivers that are key in influencing the future of the smart home industry and identify the impact of each of the drivers. The second step is market analysis to identify gaps which is determination of the market potential as well as the opportunities based on expected future circumstances in the industry. The third step is product analysis where the key products that can be offered based on the anticipated market opportunities are outlined. The fourth step is technology analysis where the key technologies that the proposed products are highlighted. The last step is developing a resource allocation plan. These steps are summarized in the framework shown in Fig. 7.1.

7.9 Roadmap Elements

7.9.1 Market Drivers

In our analysis, we identified 15 key market drivers for ABC smart home products. These drivers were further put into five broad categories. These categories include the following: (1) social, (2) safety and security, (3) application and convenience, (4) political, and (5) economic. There are three key drivers under each of the five categories. A summary of the drivers and their categorization is shown in Table 7.1.

Table 7.1 Summary of market drivers for smart apartments

Category	Label	Driver	Definition
Social	D1	Adoption	Customers adopts the smart technology.
	D2	Ease of use	User friendly, easy to use via application.
	D3	Networking	Network with IOTAS users/property managers.
Safety & Security	D4	Surveillance	Secured apartment via monitored products.
	D5	Data privacy	Secured customer information.
	D6	Access control	Control apartments with secured login/password. Access apartment through mobile application/web.
Application & Convenience	D7	Connectivity	All authorized devices connected to the Hub.
	D8	Customization	Custom profiles for each user.
	D9	Algorithms	Automatically learn user patterns.
Political	D10	Regulations	Local laws, license restrictions, and geographic restrictions.
	D11	Globalization	Export regulations, and culture.
	D12	Policies	Privacy and user policy.
Economic	D13	Lower management and utility cost	Reduce energy, monitor unoccupied units.
	D14	Affordable	Bundled products at an affordable prices.
	D15	Investment	Increase apartment value and demand.

7.9.2 Social Drivers

Under social category, the three key drivers include adoption, ease of use, and networking. Social drivers are factors that relate to the influence of our social groups in adopting smart home technologies and the aspects of technology that improve user's lives and experiences.

7.9.2.1 Adoption

"Adoption" as a driver for the increased uptake of smart home technology in the market relates to the desire to adopt the emerging smart technology. With increased adoption of smart devices, people are generally seeking to experience the new capabilities of living in a connected environment. Specifically, the desire to live in an environment where intelligent devices are interconnected and can be controlled remotely is playing a pivotal role in encouraging adoption of smart home technologies (Yang, Lee & Lee 2018). The most critical aspects that people are looking in smart home technologies include convenience, health, reduction of household labor, comfort as well as energy efficient systems.

7.9.2.2 Ease of Use

Based on research, we identified that having user-friendly and easy-to-use applications is playing a big part in informing a decision to adopt the technology. Generally, control and operation of connected devices is made possible through the use of user-friendly smartphone applications that is simplified to make it convenient for all residents without requiring users to have deep knowledge on how the systems work. Smart home technologies provide for a high level of personalization in order to make it convenient and easy to use. Generally, by simplifying their use, it is possible to include a high level of functionality, for instance, by creating alerts. For instance, a user can get alerts for expiring food items, automated watering of lawns and flowers, automatic air air-conditioning and more (Cooper 2019).

7.9.2.3 Networking

The last driver in the social category is "Networking." Users and property managers who have already purchased ABC products have a strong influence on the people in the social and business networks in encouraging them to adopt the smart home technology. Early adopters of smart home technologies have a strong influence on others to also adopt. For ABC, property managers are the primary target. Property managers who have strong networks with other managers and property companies have played a pivotal role in encouraging adoption in the multifamily segment. Property managers have the power to convince others in the same field to adopt smart home technologies based on its real benefits in creating a competitive edge. Networking as a driver also encompasses the aspects of ABC users being able to connect with their neighbors who also use ABC smart home solution.

7.9.3 Safety and Security Drivers

The second category is safety and security. There are three drivers under this category: surveillance, data privacy, and access control.

7.9.3.1 Surveillance

In surveillance, we identified that a bulk of ABC's clients are motivated by the promise of a secured apartment in which everything is closely monitored through a surveillance system. The application permits real-time surveillance of the connected systems which makes a user to have a complete view of their homes at the click of a button. Again, the user is able to monitor what is going on remotely. Automated surveillance and alerts make it even easier as a user that does not have to necessarily keep track of everything that is going on in their homes.

7.9.3.2 Data Privacy

We identified that customers are motivated to adopt ABC smart home products based on a guarantee that customer information will be protected. User's perception about their personal privacy and the privacy of their family members is a key factor that influences the demand for Smart homes. Data privacy is one of the major challenges that affect technology adoption, especially with IoT platforms. The devices monitor and collect data continuously. Therefore, there is a concern whether the systems have adequate security features to protect data from access by authorized persons and entities external to the owner. A past survey identified that a majority of owners usually trust smart home solutions from large companies such as Amazon and Google based on the perception that such companies have a technical capacity to protect their data. Uptake of ABC smart home solutions depends on the trust that the company cannot use the collected data to carry out malicious activities.

7.9.3.3 Access Control

Access control is closely related with privacy. Users and concerned about the measures that a smart home solution incorporates in order to ensure that only legitimate users can access the system. This includes having secured login and access passwords and accessibility of controls through mobile application or on the web.

7.9.4 Application and Convenience

The third category of drivers is "application and convenience." In this group, key drivers include connectivity, customization, and algorithms.

7.9.4.1 Connectivity

The primary concern with regard to connectivity is the extent of interoperability of the connected devices. Customers are interested in systems that provide seamless interaction of the devices in the IoT environment. ABC has a wide array of products and software platform that enables seamless interaction between devices. If customers decide to buy devices from different manufacturers, they can be forced to have the devices run on separate platforms. This is unlike having all devices from one supplier such as ABC where a single integrated application is used.

7.9.4.2 Customization

The scale of customization that a smart home solution provides is a key driver of smart home technology adoption. Users are interested in solutions that can be customized based on their needs and preferences. A solution that provides a good level of customization provides a good experience, as users are able to understand how the system works. Customization entails configuring the system in areas such as system functionalities, content, as well as layout in order to suit one's priorities. By permitting customization, smart home applications give the users a change to control how the devices in the smart home ecosystem interact.

7.9.4.3 Algorithm

Algorithm as a driver relates to the capability of the system to learn user patterns for the purpose of informing automated responses when similar patterns are detected in future. The ability of the system to record and analyze patterns is a critical requirement for automated systems. For example, the air conditioning functionality, the smart application system should be able to record details such as time and room temperature and create a log. The system can then use the logs for analysis to establishing a pattern of events, which becomes the basis for automated control of the air conditioner. This can also apply to other connected devices.

7.10 Political

The political factors that we identified to have a strong influence pattern of adoption of smart home technology include government regulation, globalization, and policies.

7.10.1 Regulations

Regulations that play a pivotal role in encouraging adoption of smart home technologies include local, state, and federal laws that favor adoption on energy effect technologies, especially in the area of energy consumption and security. In some controlled development areas such as in gated communities, the need to enhance the security of the residents is one of the forces that is driving adoption of smart home technologies. In this case, local laws and regulation can demand homeowners to invest in home surveillance systems as a security measure and energy effect systems as a sustainable approach in real-estate development. Other than this, fewer barriers in licensing of smart home technologies is also playing a part in encouraging smart home adoption.

7.10.2 Globalization

The world is swiftly moving toward full adoption of smart technologies in all sectors. Apart from the United States, other countries such as China, Japan, Australia, and a majority of European countries are at the forefront with regard to adoption of the "Smart globalization" where the main focus is connecting people, institutions, and communities with technologies that are geared toward building a better future (Rockefeller Foundation 2007).

7.10.3 Policies

Government policies on technology use and data and information management have a strong influence on adoption of smart technologies. Over the decades, the government in the United States has been implementing policies that are aimed at enhancing security in cyberspace. The most important policies in this regard include personal privacy and data protection policies. Companies such as ABC are striving to abide by such policies by adhering to regulations and policies for enhancing data security and user privacy.

7.11 Economic Drivers

Economic drivers are forces that relate to the prevailing market conditions that play a part in encouraging increased adoption of smart home solutions. The economic drivers we found to be most influential include lower management and utility costs, affordability, and investment.

7.11.1 Lower Management and Utility Costs

Smart home solutions are relatively cheap to manage. With technology advancements, newer smart home technologies are designs to meet not only the functional needs but also the economic benefits through cost reduction. In particular, developers of smart home applications focus on enhancing energy efficiency through the use of smart low-power devices and components in order to minimize the cost of energy that goes into powering the systems and the controlled devices. For example, last technologies enable monitoring of unoccupied units and suppress the functionality of devices such as air conditioners. Such devices are only powered based on the number of occupants in a given unit. It is estimated that installation of smart home solution can reduce utility costs by up to 27% (Higginbotham 2017).

7.11.2 Affordability

Installation of smart home technologies translates to extra costs to the new homeowner. For this reason, developers of smart home solutions are finding ways of reducing costs in order to accommodate the diverse needs of customers. It is possible to convert a regular into a smart one on budget. ABC technologies only incur apartment managers approximately $1 extra per square foot for the service (Higginbotham 2017). Another thing to note is that ABC offers their smart home solution as a service meaning that apartment managers pay a monthly fee for the service depending on the number of connected units. This makes the solution affordable for adoption by a wider market ranging from the high-end segment to the economy segment.

7.11.3 Investment

A good number of managers and developers are adopting smart home solutions based on benefits it presents as an investment opportunity. From a business perspective, investing in smart home solutions provides a competitive edge given that customers are more interested in products that enhance their experience on top of the normal functionality. The rising demand for smart apartments is pushing property developers and managers to invest in smart home technologies in order to serve the rising population of smart home enthusiasts. Investors are looking at how to address the needs of special groups such as the elderly, people with health conditions, and the physically challenged by incorporating smart technologies that make the lives of these categories of residents.

Table 7.1 gives a summary of the drivers and the codes assigned to each driver

7.12 Technology and Product Features

Technology and product features are attributes of a smart home solution that are key in driving demand. This chapter discusses the technologies and product features that customers consider to be most important when making decisions on whether to invest in the product or not. For ABC smart apartment solution, we came up with 20 product features and 10 key technologies. Each of these technologies and product features is discussed below. It will be seen that every type of technology is related closely with a specific product feature.

As mentioned in the foregoing, we identified 20 product features which are categorized into four main groups including (1) smart, (2) value and system efficiency, (3) user friendly and ease to use, and (4) security. We also identified 10 important technologies that are critical in driving the demand for Smart home products. These

10 technologies are further classified into four primary categories including IOT, Sensor Applications, Cloud computing, and energy.

The first category of product feature is the "Smart" category. This category encompasses all the features that relate to the aspect of connecting, integrating, and sharing of data and information between devices. Smart category also entails the aspects of self-monitoring, analysis, as well as reporting of real-time situations in a given environment. Under this category are six product features. The first product feature is smart outlets, which involves controlling of the connected devices through wireless connection and using the ABC smartphone app. The second product feature is smart thermostats which are devices that control heating as well as air conditioning systems. Smart switches allow control of the lighting systems remotely using a mobile app. Another feature is the smart user profile which the capability of the system to learn user routines and patterns. This is used for the purpose of executing controls automatically. Another product feature under smart category is the "Hub" which is a central control unit that communicates with all connected devices through remote control via Internet connection. The last product feature under smart category is automated supply where the interest is the ability for the devices to work automatically as required.

The second category of product features is the "value and system efficiency" category. This category encompasses five product features. The first one is data collection where we identified that customers seek for a product that can collect user data and use it to recover information from users via the ABC application. The second feature is the product management feature in the ABC app which permits users to connect directly with property managers for the purposes of solving issues that relate specifically to the smart home product. Motion sensors are another product feature that detects occupancy in a given unit. This is used for surveillance as a security measure and well as energy saving where devices such as writing and air-conditioners are switched on automatically when the unit becomes occupied. We also identified that hydrogen storage is an important feature to some customers. A small-scale hydrogen storage can be used as car fuel, for heating and for generating electricity through a fuel cell. The last product feature in this category is fuel switching which allows users to switch between different types of energy sources, for instance, from grid to solar power.

The third category of product features is "User-friendliness and ease of use." With regard to this category, we identified that consumers are interested in features as a phone application which is compatible with IOS and Android Operating systems. Another feature is remote access which is the ability to monitor, operate, and control connected devices and systems remotely using a mobile phone. Another product feature under this category consists of microgrids which are community-level networks that enable neighbors to share power. This feature enables connected apartments to isolate themselves in case of a power outage and also permits the apartments to share excess power with the main power grid. The last product feature in this category is wireless charging driveways which consist of a charging plate that is buried in the ground where electric cars can pass for a recharge through induction without having to plug a charger.

The fourth category of product features is the "Security" features. These features include door sensors which monitor access and giver alert in case of unauthorized access. Another security feature is sprinklers which consider of smoke detectors and sprinklers which are activated in case of fire or smoke. Video cameras also provide surveillance functionalities and consist of the first line of defense in case of a security issue. Another feature is the security system, which comprises a network of integrated security devices. Lastly, there are door locks that can be controlled remotely to allow access to authorized people only.

The key technologies that support the above-discussed product features are also categorized into four groups: (1) IoT, (2) sensor applications, (3) cloud computing, and (4) energy. IoT technologies encompass IoT platform which comprise mainly of software applications that permit building of the IoT ecosystem. Smart monitoring technologies allow real-time tracking and monitoring the environment, including number of occupants in a unit and other conditions in order to control how energy is consumed and for surveillance purposes. Remote access technologies permit operation of the connected devices from a remote location and include all relevant features such as encryption, and use of wireless and wired protocols in order to access devices remotely.

Sensor application is another technology that supports the products features that drive demand. Sensor applications include sensor networks and face recognition applications. These technologies are specifically meant for data collection, monitoring, and surveillance of the environment based on the system configurations. Sensor applications give alerts or trigger action in specific areas when the system detects conditions that warrant a response based on the configurations. For example, facial recognition applications will automatically detect a facial image that has already been configured into the system and permit access to authorized persons in certain areas within the apartment.

Cloud computing technologies are used specifically for data analytics and for other cloud-based solutions that use Artificial Intelligence (AI). Data analytics help in developing user patterns and executing machine learning where the system is able to automatically learn and make improvements based on the learned patterns and experiences without requiring programming. An automated replenish system can use past experiences to send orders automatically.

The last category of technologies is energy technologies which comprise technologies such as energy storage, fuel cells as well as wireless charging and electromagnetic resonant pad. Under normal conditions, the Smart Apartment ecosystem is powered using grid power. However, in a microgrid environment, alternative sources of power can be used such as solar power, fuel cells, or wind turbines. A fuel cell, for instance, combines hydrogen and oxygen to produce electricity. On the other hand, wireless charging allows charging of electric cars without plugging in the charge when the car is up to 9 inches away from the charging pad.

Tables 7.2 and 7.3 provide a summary of the product features and the key technologies that relate to the drivers of smart home solution.

Table 7.2 Product features

Category	Label	Feature	Definition
Smart	P1	Smart outlets	Control the connected devices wirelessly via an app
	P2	Smart thermostats	Devices used with home automation that controls heating / air conditioning system
	P3	Smart switches	Control the lights wirelessly via an app
	P4	Smart user profiles	Learns routines / patterns of customers
	P5	HUB	Device that communicates with and controls all the devices and sensors in each apartment. It's also connected to the internet, which is what allows remote control from anywhere in the world
	P6	Automated supply	Smart devices will automatically work when needed
Value and system efficiency	P7	Data collection	Recover information from users via app
	P8	Property Management solution	Residents can contact property managers for any concerns in regard to the smart apartment through the app
	P9	Motion sensors	Used for lightning (saving energy), and security
	P10	Hydrogen storage	Some homeowners choose to install small-scale hydrogen storage tanks which can be used as a car fuel, for heating or to generate electricity through a fuel cell
	P11	Fuel switching	Homes can switch between multiple sources of energy
User friendly and ease to use	P12	Phone application	Works with IOS and Android operating systems
	P13	Remote access	Manage smart devices wirelessly
	P14	Microgrids	Community-level networks known as "microgrids" allow neighbours to share power, isolate themselves from outages and provide energy to the broader power grid
	P15	Wireless charging driveway	A wireless driveway charger consists of a charging plate buried in the ground that charges an electric car through induction, without having to plug it in
Security	P16	Door sensors	Will alert unauthorized access
	P17	Sprinklers	Will activate in case of smoke / fire
	P18	Video cameras	First line of defense
	P19	Security systems	Networks of integrated electronic devices
	P20	Door locks	Manage doors remotely

Table 7.3 Technologies

Category	Label	Driver	Definition
IOT	T1	IOT platform	They are essentially software products, which offer comprehensive sets of application-independent functionalities that can be utilized to build IoT applications.
	T2	Smart Monitoring	• It's implemented to keep a track of the number of occupants and manage the utilities within the building. • Efficient energy consumption can be achieved by continuously monitoring every electricity point within a house and using this information to modify the way electricity is consumed. • Surveillance: Helps track targets, identify suspicious activities, detect left luggage and monitor unauthorized access.
	T3	Remote access	Get access to a computer / HUB / network from a remote distance. Use of protocols, encryption.
Sensor Applications	T4	Sensors	Will send an alert when there's motion around your house, and they can even tell the difference between pets and burglars.
	T5	Facial recognition	It allows to identify a person from a digital image / video source. It compares selected facial features with faces within a database.
Cloud computing	T6	Data analytics	It's the use of data collection and analysis to optimize decision making.
	T7	Automated replenishment	Monitor consumption of a device and buying patterns of of the end user. It automatically places orders when needed.
Energy	T8	Energy Storage	It's a localized grouping of electricity generations, energy storages, and loads. In the normal operation, it is connected to a traditional power grid . The users in a microgrid can generate low voltage electricity using distributed generation, such as solar panels, wind turbines, and fuel cells.
	T9	Fuel cells	A fuel cell combines hydrogen and oxygen to produce electricity, heat, and water. Fuel cells are often compared to batteries. Both convert the energy produced by a chemical reaction into usable electric power.
	T10	Wireless Charging - Electromagnetic resonant pad	Allows power to transfer at distances of up to about nine inches away from a charging pad. That allows electric cars to charge just by parking on top of a large charging pad

7.13 Technology and Product Features Gap Analysis

For the product features and technologies that we identified to be critical in driving the demand for smart home solutions, we carried out an investigation on the current status of ABC smart home product and made predictions on how these technologies and features need to be improved and advanced to attain a desired status. Product features gap analysis and technology gap analysis are summarized in Tables 7.4 and 7.5, respectively.

Table 7.4 Product features gap analysis

Category	Label	Feature	Current Level	Where we want to be	Gap
Smart	P1	Smart outlets	120 VAC, 60 Hz power input	Built in backup battery in case of power failure and rise - drop voltage	Globalization
	P2	Smart thermostats	Works for all the apartment's zones	Will work each zone separately	We can customize the whole apartment depending on the resident's needs
	P3	Smart switches	Only compatible with Android and IOS devices	Integrate with main operating systems	Globalization
	P4	Smart user profiles	Your profile goes with you to your next IOTAS Smart Home	Pop up window with tips on how to improve the profile settings	Users don't obtain the most optimal use and performance from the smart stories
	P5	HUB	The device connects only with other compatible hardware	Increase compatibility with other common smart technologies	IOTAS is working so the HUB can incorporate additional smart home technologies
	P6	Automated supply		Automatically detect missing groceries and order them online	Smart
Value and system efficiency	P7	Data collection	Only compatible with Android and IOS devices	Connecting all devices and implement Business Intelligence	Through data analytics, IOTAS can offer a better smart home experience
	P8	Property Management solution	Customers don't often use this option	Customers learn the importance of this option and use it	Customers can interact with the property managers and solve many issues with the apartments
	P9	Motion sensors	Indoor use only	Outdoor use	Property managers can implement it in the buildings
	P10	Hydrogen storage		Residents will be able to use and store hydrogen for multipurpose needs	System efficiency
	P11	Fuel switching		Customize energy usage based on residents needs	Ease of use and system efficiency
User friendly and ease to use	P12	Phone application	Works with IOS and Android operating systems	Compatible with all frequently used mobile operating systems	Meetings with the mobile phone brands to develop the smart home app for IOTAS
	P13	Remote access	Only compatible with Android and IOS devices	Integrate with main operating systems	The main purpose of smart technology is that remote access works efficiently with other OS
	P14	Microgrids		Share electricity with neighbors and avoid outages	Power outage
	P15	Wireless charging driveway		Charge electric vehicles upon arrival to house	Ease of use and efficiency
Security	P16	Door sensors		Only detect humans	Fake alarm alerts
	P17	Sprinklers		Indoor / Outdoor use and implement fire detector system	Property managers can implement it in the buildings
	P18	Video cameras		Have one 360 degree camera	Provide one camera for the whole apt
	P19	Security systems		Integrated dashboard - call 911 automatically	More safety for customers
	P20	Door locks		Facial / Iris / fingerprint recognition	Improve security

Table 7.5 Technology gap analysis

Category	Label	Feature	Current state	Future state
IOT	*T1	IOT platform	IOS/Android	Fully compatibility with other products
	T2	Smart Monitoring	Sensors that detect current environment	Optimized physical size
	*T3	Remote access	IOTAS user app	Community connection
Sensors Applications	T4	Sensors	Door & Motion	Monitor health, safety, and security
	T5	Facial recognition	N/A	Biometrics
Cloud computing	*T6	Data analytics	N/A	Providing 3rd party with important consumer data
	T7	Automated replenishment	User Profiles	Pattern Recognition
Energy	T8	Energy Storage	N/A	Microgrid
	T9	Fuel cells	N/A	Battery replacements
	T10	Wireless Charging – Electromagnetic resonant pad	N/A	Smart parking spots

7.14 Quality Function Deployment (QFD)

7.14.1 Drivers Versus Product Features

For each of the 15 drivers, a score was awarded depending on the importance of that driver to ABC Inc. In this ranking, a score of 4 is high in importance, 2 is medium, and 1 is low, while a 0 is assigned where it is not applicable. The first QFD was done for market drivers and products. The drivers that were ranked as high (Score of 4) include Ease of use, Data privacy, connectivity, regulation, globalization, and investment. All other drivers were ranked as medium meaning that they were assigned a score of 2. These scores were given by the authors who conducted a thorough study to identify the relative importance of the drivers among ABC's customers. Table 7.6 shows the ranking of each of the drivers and product features as well as the total score.

From Table 7.6, the combination of drivers and product features that had the highest total score were selected and tabulated to show the most important product features and how each contributes in driving adoption of ABC smart home products as shown in Table 7.7:

Table 7.7 shows that "smart user profiles" and "phone application" are the product features that home developers consider to lead in importance for OITAS Smart home products. We identified that the ability to create smart user profiles is most important in meeting the need of ease of use and for data privacy. Smart user profile is also important as it allows access to all the connected devices and a means of boosting the value of the investment. On the other hand, the "phone application" feature is most important in meeting the needs of ease of use and connectivity and to a lesser extent data privacy and increasing investment. "Hub" is another product

Table 7.6 QFD of drivers versus product features

		MARKET DRIVERS															
		D1	D2	D3	D4	D5	D6	D7	D8	D9	D10	D11	D12	D13	D14	D15	
WEIGHT		2	4	2	2	4	2	4	2	2	4	4	2	2	2	4	
PRODUCTS	Products	Adoption	Ease of use	Networking	Surveillance	Data privacy	Access control	Connectivity	Customization	Algorithms	Regulations	Globalization	Policies	Lower management and utility cost	Affordable	Investment	SCORE
P1	Smart outlets	0	0	0	0	0	1	0	0	0	0	0	0	2	2	1	14
P2	Smart thermostats	0	0	0	0	0	2	0	0	2	0	0	0	4	2	1	24
P3	Smart switches	0	0	0	0	0	2	0	0	1	0	0	0	4	2	1	22
P4	Smart user profiles	4	4	2	0	4	2	2	4	4	0	1	0	0	0	2	84
P5	HUB	0	0	2	0	0	0	4	0	0	1	0	0	0	4	4	48
P6	Automated supply	1	4	1	1	0	0	1	2	4	0	0	1	0	1	0	42
P7	Data collection	0	0	0	0	4	1	0	2	1	4	0	4	2	0	0	52
P8	Property Management solution	2	0	4	0	2	4	2	0	0	0	0	0	1	0	2	46
P9	Motion sensors	0	0	0	0	0	0	0	0	2	0	0	0	4	1	1	18
P10	Hydrogen storage	0	1	0	0	0	0	0	2	1	2	1	4	1	1	0	34
P11	Fuel switching	0	4	0	0	0	0	0	2	4	2	0	4	1	1	0	48
P12	Phone application	4	4	4	0	2	4	4	2	0	0	1	0	0	2	2	84
P13	Remote Access	4	4	0	2	0	4	2	0	0	0	0	0	0	0	1	48
P14	Microgrids	1	2	4	0	1	0	4	0	2	2	0	0	1	0	0	52
P15	Wireless charging driveway	1	4	0	1	1	0	2	0	0	4	0	0	1	1	0	52
P16	Door sensors	0	0	0	2	0	0	0	0	0	0	0	0	0	0	1	8
P17	Sprinklers	0	0	0	0	0	0	0	0	0	0	0	0	0	0	1	4
P18	Video Cameras	0	0	0	4	1	1	0	0	0	0	0	4	0	0	1	26
P19	Security Systems	2	0	0	4	0	4	0	0	0	0	0	0	0	0	4	36
P20	Door locks	0	0	0	2	0	2	0	0	1	0	0	0	0	0	1	14

Market Drivers vs Product Features	
High	4
Medium	2
Low	1
N/A	0

Table 7.7 Most important product features versus drivers

Market (customer) Driver Area

Product Feature Areas	4	4	4	4	4	4	
	Ease of use	Connectivity	Data privacy	Regulations	Globalization	Investment	Score
Smart user Profiles	4	2	4	0	1	2	52
HUB	0	4	0	1	0	4	36
Phone Application	4	4	2	0	0	2	52
Data Collection	0	0	4	4	0	0	32
Property management solution	0	2	2	0	0	2	24

feature that ranks second in importance in driving uptake of ABC smart home products. Based on the ranking, we identified that having a hub that links all the devices is important in providing connectivity. Owners also consider a hub to be important in increasing the value of the investment. "Data collection" and "Property management solution" are ranked to be the fourth and fifth most important product features, respectively. Both features are important in terms of ensuring data privacy and adhering to government regulation.

7.14.2 Drivers Versus Product Features

A second QFD was done in which, each of the ten (10) technologies was ranked against the twenty (20) product features and scores given. The ranking criteria was similar to the previous QFD in which score of 4 is "High score," 2 is "medium score," 1 is a "low score" and zero is "not applicable" as shown in Table 7.8. Depending on the total score, the four most important technologies were selected from Table 7.8 and tabulated as shown in Table 7.9.

Table 7.9 shows that the most important technology is remote access followed by IOT Platform and Data analytics. Smart monitoring technologies are the fourth most important technology based on our ranking. In relation to product features, a hub, phone applications, and security systems product features are high ranking for the "Remote Access" technologies. On the other hand, hub, data collection, and security systems product features are high ranking for the "IOT platform" technology and for data analytics, hub, phone applications, and data collection are high ranking. For the "Smart monitoring" technologies, security systems are rated as a high-ranking product feature, while hub, phone applications, and data collection are ranked as medium.

Table 7.8 Technologies versus product features

Technologies		P1 Smart outlets	P2 Smart thermostats	P3 Smart switches	P4 Smart user profiles	P5 HUB	P6 Automated supply	P7 Data collection	P8 Property Management solution	P9 Motion sensors	P10 Hydrogen storage	P11 Fuel switching	P12 Phone application	P13 Remote Access	P14 Microgrids	P15 Wireless charging driveway	P16 Door sensors	P17 Sprinklers	P18 Video Cameras	P19 Security Systems	P20 Door locks	
	Weight	2	2	2	4	4	2	4	2	2	2	2	4	4	2	2	2	1	2	4	2	
T1	IOT platform	2	2	2	2	4	2	4	1	2	1	1	2	2	1	2	2	1	2	4	1	115
T2	Smart Monitoring	2	2	2	1	2	2	2	1	2	1	1	2	1	1	1	2	1	2	4	1	89
T3	Remote access	2	4	2	2	4	1	2	2	1	1	1	4	4	2	0	1	1	2	4	2	123
T4	Sensors	1	1	2	0	2	1	2	1	2	1	1	1	0	0	2	2	0	2	2	2	64
T5	Facial recognition	0	0	0	1	1	1	2	1	1	1	1	4	2	0	0	1	0	1	4	2	74
T6	Data analytics	2	2	2	2	4	2	4	1	2	1	1	4	2	1	1	2	1	1	2	1	101
T7	Automated replenishment	0	0	0	0	2	4	4	1	2	0	0	2	1	0	0	0	0	1	0	0	52
T8	Energy Storage	0	0	0	1	1	0	2	2	0	4	4	1	0	4	4	0	0	1	2	1	68
T9	Fuel cells	0	0	0	0	1	0	2	2	0	4	4	1	0	2	4	0	0	1	2	0	58
T10	Wireless Charging - Electromagnetic resonant pad	1	0	1	0	1	0	1	1	0	2	2	0	0	4	4	0	1	1	2	0	49

High	4
Medium	2
Low	1
N/A	0

Table 7.9 Most important technologies versus product features

Technologies	Smart user profile	Hub	Phone application	Data collection	Property management solution	Security Systems	Score
	4	4	4	4	2	4	
IOT Platform	2	4	2	4	1	4	66
Remote Access	2	4	4	2	2	4	68
Data Analytics	2	4	4	4	1	2	66
Smart Monitoring	1	2	2	2	1	4	46

7.15 Resources

We conducted a thorough study to identify some of the resources that ABC require in order to develop a smart home solution that meets the market needs and that which satisfies the key drivers. We identified 12 key resources which are further classified into four categories: (1) research and development, (2) partnership, (3) privacy, and (4) capital. The first category is research and development includes software development, data mining, and hardware development. In software development, ABC can rely on engage software development partners including Intel,

Google, and Tesla. ABC must also ensure the systems are able to mine useful data and provide reliable access and in right type and format. The company can rely on cloud services offered by Google, Amazon, SAP, or Microsoft. In hardware development, the company can rely on companies such as Apple, Brazen, and Tek.

Partners are also critical resources for ABC. Under the category of partnerships, there are high-tech companies, tech-support, and solar companies. Being a technology-based company ABC will require to establish very close partnerships with reputable high-tech companies who provide smart home solutions such as Amazon, August Smart Lock, Nest, and Home Seer. These companies should be engaged on a long-term basis to supply high-quality smart home products and solutions. The company should also establish strong partnerships with tech-support companies such as Tech to Us and iYogi. These companies provide remote access support services in different areas including smart home devices, smartphones, and cameras among other products. Tech-support is important, as it will ensure that users can enjoy the smart home services without interruption. It is also important for ABC to engage Solar companies to install solar systems that will provide an alternative, renewable source of energy in all the connected apartments. As indicated in the foregoing, energy switching was identified as a key driver in this sector. Cella Energy and Solarcity were identified as companies with the highest potential for establishing long-term partnerships with ABC.

The third category of resources is the "privacy." Again, we identified data privacy is a key driver hence ABC must ensure that data privacy is guaranteed. The resources under this category include government resources, piracy resource, and business resources. Under the government, ABC should ensure that it meets all the regulatory requirements as per the applicable regulations and policies in the United States. Under privacy, ABC should also market the brand in order to create brand equity and also license its products in accordance with the applicable copyright and intellectual property laws. With regard to business resources, ABC must also identify quality smart home products that provide the highest level of security of customer data. Business resources such as wireless sensor connections must also have adequate security features to disallow illegal access.

The last category of resources is "capital" resources and includes private funding, grants and contracts. In order to fund the developments, ABC must identify a strategic source of capital, one of them being the owner's investment. This is critical because all the proposed technological advancements require significant funding in order to achieve the intended goals. ABC must also look for grants as an external source of capital. Grants are issued in the form of government grants or bank loans. These external sources of capital are important if owner's private investment will not be enough to fund all the proposed developments. A summary of the key resources based on their respective categories is shown in Table 7.10.

Table 7.10 Resources

Resources

Category	Label	Resource	Example
Research & Development	R1	Software development	Intel, Google, Tesla
	R2	Data	Google, Amazon, SAP, Microsoft
	R3	Hardware development	Apple, Brazen Tek
Partnership	R4	High-tech companies	Amazon, August Smart Lock, Nest, Home Seer
	R5	Tech-support	Tech to Us, iYogi,
	R6	Solar companies	Cella energy, SolarCity,
Privacy	R7	Government	US government regulation and policies, permits
	R8	Piracy	Market brand, licensing
	R9	Business	Customer privacy
Capital	R10	Private funding	Investments
	R11	Grants	Government grants, bank loans
	R12	Contracts	Sales contracts

7.16 Technology Roadmap Timelines

After outlining the key drivers, product feature areas, technologies and resources, we developed a timeline for each of these aspects of technology roadmapping that ABC can use in order to bridge the gaps and to achieve the ultimate goal. In each of the four timelines, a maximum period of 20 years is set. This period is then divided into 5 periods: (1) present to 5 years, (2) 5–10 years, (3) 10–15 years, (4) 15–20 years, and (5) 20 years and above. A summary of the timelines for the market drivers, product features, technologies, and resources is shown in Table 7.11.

As shown in the Table 7.11, each item which is identified as a driver, product features, technology, or resource has a timeline ABC requires to implement for the purpose of bridging the gaps and creating new capabilities. The timelines can be viewed in the perspective of short-term actions (present to a maximum of 5 years), medium term (for up to 10 years) and long-term actions (15 years and above).

7.16.1 Present to 5 Years

From Table 7.11, we can see that the actions that ABC need to focus more to accomplish in 5 years time with regard to addressing the drivers include encouraging adoption, improve on access controls, and customization of its products. This should be combined with improving product features and coming up with more advanced features such as smart outlets, smart thermostats and smart switches, door sensors, and door locks. It should also enhance the property management product feature. In the area of resources, ABC should prioritize regulatory compliance and measures to prevent privacy.

Table 7.11 Technology roadmap timelines

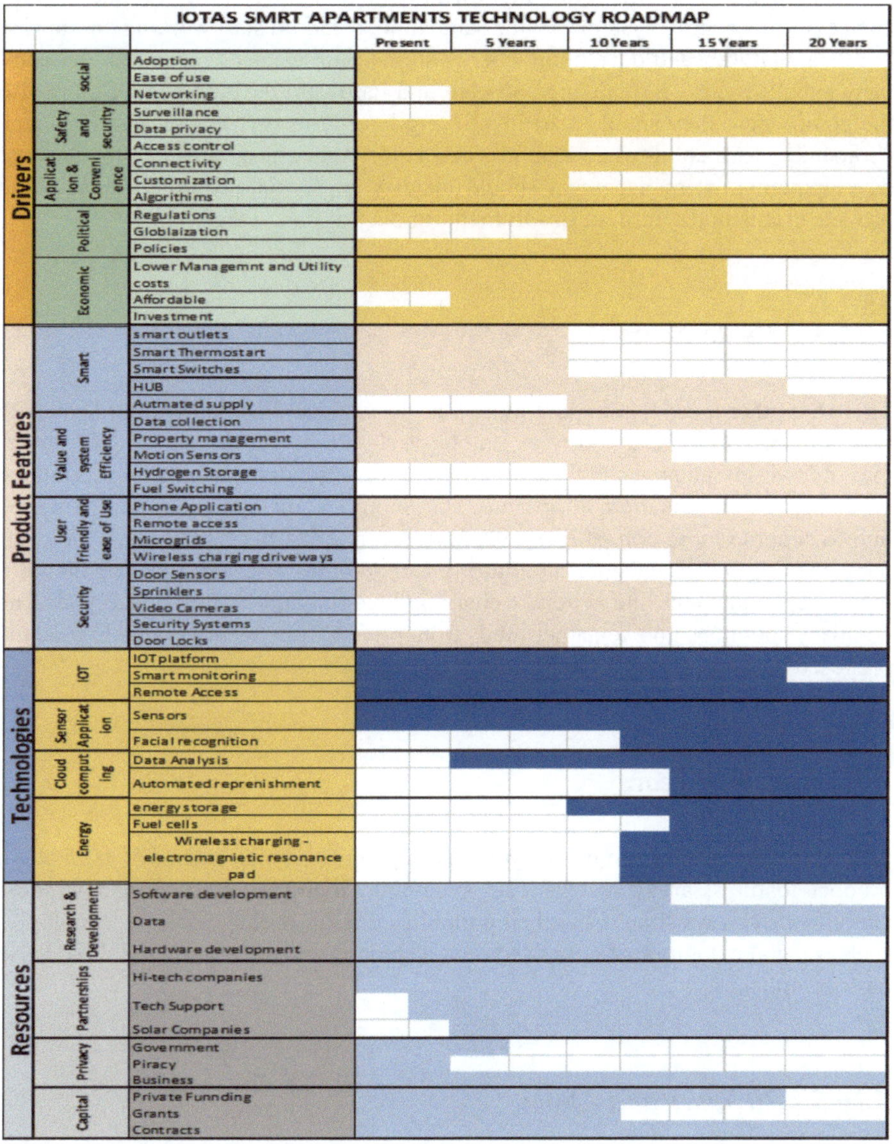

7.16.2 5–10 Years

Some actions in the implementation of the technology roadmap require more than 5 years and a maximum of 10 years. Again, in this period, ABC should focus more on enhancing connectivity as a market driver and also focus on improving some product features such as developing or improving motion sensors and start with developing automated supply, hydrogen storage, and microgrids. Other areas of product features that will require mobilization include developing sprinklers, video cameras, and security systems. At the same time, it should start working on improving data analytics and developing energy storage. In terms of resources, the company should focus on finalizing software and hardware development. This is also the point where they should start identifying long-term partners such as high-tech companies, tech-solution companies, and solar companies. In order to fund the developments, the management should identify appropriate external financiers to provide grants in the form of government funding or bank loans.

7.16.3 10–15 Years

There are other areas that require more than 10 years and a maximum of 5 years to accomplish. With regard to drivers, emphasis should be put on finalizing on strategies of lowering management and utility costs and also start working on making the smart home solutions more affordable. This should be matched with improving the hub to become more compatible with devices. In terms of technologies, the company should finalize on smart monitoring technologies and start working on automated replenishment and wireless charging-electromagnetic resonance pad. The company should be able to gather funds from private investment in order to fund the proposed developments and improvements.

7.16.4 15–20 Years

Some aspects will need to be delayed and implemented after later. For example in product features, areas such as fuel switching, hydrogen storage, and wireless charging driveways should start being mobilized at this point. Several technologies such as facial recognition, automated replenishment, and energy storage should also be accomplished.

7.16.5 20 Years and Above

As identified before, some actions will require a long-term approach. These are particularly aspects that were identified to be most important as drivers, product features, and technologies. In Table 7.11, such areas are represented by bars closing from present period across all durations. This means that ABC must prioritize in this area and work toward improving and enhancing them as part of the long-term strategic plan of ensuring that the firms remain competitive. For the drivers, ABC should focus on aspects such as ease of use, data privacy, improving alorithims, adhering to regulations, and aligning its operations as per government policies. The company should also focus on making smart home products an attractive investment for apartment managers and developers.

Product features that require a long-term focus include data collection and remote access. This should be matched with improvements on technologies such as IoT, remote access, and sensor networks. In terms of resources, ABC must continuously seek to create a data resource that is useful and reliable for more advanced developments. The company must also look for high-tech companies it can engage in long-term relationships to supply high quality smart home products. In addition, the company should continuously improve its business by developing security measures to prevent breaches of customer privacy or illegal access to the system.

References

Cempel, A., & Mikulik, J. (2013). Intelligent building reengineering: Adjusting life and work environment to occupant's optimal routine processes. *Intelligent Buildings International, 5*(1), 51–64.

Cooper, A. (2019). The emerging trend of smart home technology in real estate. *Entrepreneur.* Retrieved from https://www.entrepreneur.com/article

Dillard, C. (2018). *OTAS and TRU Development enter Portfolio Partnership.* ABC. https://www.ABChome.com/ABC-tru-development-enter-portfolio-partnership/

Garcia, M. L. (1997). *Fundamentals of technology roadmapping.* Albuquerque: Sandia National Laboratories.

Garcia, M. L., & Bray, O. H. (1997). Fundamentals of technology roadmapping. Albuquerque: Sandia National Laboratories.

Garcia, M. L., & Bray, O. H. (1997). *Fundamentals of technology roadmapping.* Albuquerque: Sandia National Laboratories.

Higginbotham, S. (2017). *Startup profile: ABC is using Z-wave for smart apartments.* StaceyonIot.com. https://staceyoniot.com/startup-profile-ABC-is-using-z-wave-for-smart-apartments/?

Market Watch. (2019). *Global Smart Homes market: Advanced technologies & growth opportunities by 2018–2023.* MarketWatch.com. Retrieved from https://docs.google.com/document/d/10NVmBFQmxah9PwbwX8uZZTss3_UHcPZqa4odd-AjdY0/edit

Phaal, R., Farrukh, C. J. P., & Probert, D. R. (2004). Technology roadmapping – a planning framework for evolution and revolution. *Technological Forecasting and Social Change, 71*, 5–26.

Rockefeller Foundation. (2007). *Smart globalization: Benefiting more people, more fully, in more places.* Rockefeller Foundation. https://assets.rockefellerfoundation.org/app/uploads/20150530121738/Annual-Report-2007.pdf

VeCap. (2017). *Next generation smart homes.* White Paper. https://vecap.io/pdf/whitepaper/VeCap_Whitepaper.pdf

Yang, H., Lee, W., & Lee, H. (2018). IoT smart home adoption: The importance of proper level of automation. *Journal of Sensors, 2018*, 1–11. https://doi.org/10.1155/2018/6464036.

Chapter 8
Technology Roadmap: Autonomous Bus Service

Bobby Romanski, Dave Sherman, Janet Rosenthal, Deemah Alassaf, Jacqueline Nayame, and Tuğrul U. Daim

Organizations use technology roadmapping (TRM) to discover a critical path of action to pursue in their research and development of future technologies and products. This provides a technology roadmap for creating autonomous bus service in the Tri-County Metropolitan Transportation District of Oregon (TriMet). A review of related literature was performed and the essential market drivers, products, and technologies were researched and integrated into this technology roadmap, inferring that autonomous bus service be made available in the year 2042 while continuously enhancing intelligent transportation systems and its related products from the year 2018. The QFD (Quality Function Deployment) method was used to discover and evaluate relationships among market drivers, products, and technologies.

8.1 Introduction

The future of transportation is in the balance. Transportation professionals face challenges with capacity, reliability, pollution, and energy efficiency. Travel demand is constantly increasing with economic development and population growth. The increasing complexity of transportation systems challenges professionals. Every major automaker and industry supplier are racing to put autonomous vehicles on the road. With the introduction of artificial intelligence (AI), safety, efficiency, and reliability will be increased. Environmental and social impact will be minimized. AI has the power to deliver forward-looking analysis of data with predictive analytics. Through the creation of a technology roadmap, this chapter attempts to understand the current and future technologies that will be crucial to the future of

B. Romanski · D. Sherman · J. Rosenthal · D. Alassaf · J. Nayame · T. Daim (✉)
Department of Engineering and Technology Management, Portland State University, Portland, OR, USA
e-mail: tugrul.u.daim@pdx.edu

T. U. Daim (ed.), *Roadmapping Future*, Applied Innovation and Technology Management, https://doi.org/10.1007/978-3-030-50502-8_8

transportation, specifically with autonomous vehicle technologies and TriMet, a mass transit agency that serves 1.5 million people in a 533 square mile area. Most of its services and assets are owned by the agency, while most of its on-demand and frequently changing services are contracted out via a public bidding process.

8.2 Literature Review

As the level of competition increases in today's economy, companies need to be able to identify and make the technologies required to successfully meet their mission. And, in order to do so, they need to develop an effective technology planning tool. Technology roadmapping is a technology forecasting tool that helps finding, choosing, and constructing technology options in order to fulfill product requirements (Garcia and Bray 1997; Hansen et al. 2016). It is an effective tool for technology planning and organization, where it covers a broader set of planning activities. In addition, the application of this tool can occur at both the individual corporate and industry levels. In technology roadmapping, a team of subject matter experts come together to set an outline for forming and presenting the important technology planning information which includes detecting critical technologies and the gaps associated with them, and finding means to leverage R&D investments as well. Using a technology roadmapping becomes necessary when the technology investment decision is not direct. This happens when it is not clear which alternative to choose, how rapidly the technology is needed, or when managing the development of several technologies simultaneously. The outline of the roadmap enables the company to make better and more informative technology investment decisions (Garcia and Bray 1997). It also provides a method to aid experts estimate technology developments in selected areas. The process offers a scheduled solution that incorporates critical system requirements and performance objectives that need to be accomplished by a definite time frame, as well as finding the necessary technologies to achieve these goals. With the information it provides, it helps in choosing the best technology alternative more efficiently. The technology roadmap process starts by identifying the critical dimensions and boundary conditions that must be satisfied by emerging technologies related to the developed system. Yet, before starting the roadmapping process, limiting the scope and identifying the stakeholders of the project are necessary. After forming the scope and choosing the stakeholders, the company has to decide which process of technology roadmapping to use. They can choose either the market pull process, which focuses on customer needs, or the technology push process that focuses on opportunities (Marinakis et al. 2017).

In the past, artificial intelligence was described as computer science discipline that illustrates the intelligence of human behavior. Today, AI has a wider meaning. It refers to computer systems that have a complex performance similar to living systems as neural systems and microbiology (National Research Council 2012). The main difference between normal computer systems and AI is the AI computer system's complexity level in programs, algorithm performance, and the subject of the AI system itself depending on the amount of data it analyzes, the part of the reality it demonstrates, or the performance it has to do. AI programs are able to identify and cope with shapes as part of its shape recognition and manipulation feature. Within an AI system, applications extract special features as facial recognition, plate number, and type of an automobile by processing the graphical data. The complexity of a system depends on data analysis. AI applications offer durability that avoids losing valuable knowledge as an individual or group of members retire or leave the organization (National Research Council 2012). Furthermore, multiple research and applications in transportation have confirmed the argued advantages of AI in general. It minimizes costs, as it reduces personnel need. An organization can save significant staff time by implementing the right AI applications to support their decision-making process, which leads to decreasing operational costs. In addition, by integrating existing knowledge with probabilities and probability inference computations, AI models identify the uncertainty between real-life cause and effect scenarios (Patterson 1990). In transportation, AI has been used in many applications such as transforming traffic sensors into intelligent agents allowing them to automatically find and report traffic accidents or even forecast traffic conditions (Schleiffer 2002). Also, AI tools can be beneficial when used to detect security violations, and in the improvement and management of automated response and control plans. The five general applications of AI in transportation were focused on traffic operations, modeling travel demand, safety and security of transportation, public transportation, and infrastructure design and construction. Additionally, it has been argued that the development of the technology of autonomous/connected vehicles is promising to deliver a large enhancement in safety, congestion reduction, energy efficiency, and raise mobility alternatives (McDermott 2017). Also, it is assumed that as the amount of autonomous vehicles increase, it reduces gaps or adds more capacity to the existing transportation system. According to the National Highway Traffic Safety Administration (NHTSA), the process of transforming vehicles to be fully automated will happen in five phases (Scherer 2017). The first phase is the no-automation, level 0, where the driver has full control over the vehicle functions at all times. Then comes the second phase the function-specific automation, Level 1, where one or more specific control functions can be automated as electronic stability control or pre-charged brakes. In the third phase of combined function automation, level 2, at least two main control functions become automated where they work in unison to release the driver of controlling those functions. Next, will be the fourth phase of limited self-driving automation, level 3; at this level, the driver will give full control of all safety-related functions under some traffic or environmental conditions. The Google car is an example of limited self-driving automation. Finally, at the fifth phase, full self-driving automation, level 4, the vehicle will do all safety-critical driving tasks and observe roadway conditions for the whole trip. In this phase, the driver's role will be only limited to providing destination or navigation feedback (Scherer 2017).

Activists believe that autonomous vehicles will have a significant impact on user convenience, security, traffic congestion reductions, energy efficiency, and decreasing pollution (Litman 2014; KPMG 2012; Fagnant and Kockelman 2013). For instance, they argue that since 90% of traffic accidents happen because of the driver's fault, autonomous cars will decrease accidents by 90%. In addition, the projected decrease in congestion and parking cost, energy savings, and emission are also uncertain due to cooperative effects. For instance, self-driving taxis and self-parking cars may increase empty vehicle travel. Though the extra vehicle travel provides user advantages, it can increase external costs such as congestion, roadway, and parking facility costs, accident risk forced on the other driver on the road. Therefore, some strategies such as platooning should be narrowed to grade-separated roadways; so non-automated vehicles could raise congestion on surface streets (Litman 2014). On the other hand, the increased costs of manufacturing autonomous vehicles are uncertain. They demand a range of special components such as sensors, computers and controls, which cost multiple thousands of dollars today, but could become cheaper with mass production in the future (KPMG 2012). Also, since system failures could be deadly to vehicle riders and other road users, all life-threatening components will need to meet high standards as in the aircraft components, which will possibly be fairly expensive (Litman 2014).

8.3 Technology Roadmap Application

The roadmapping process that was applied to this study was the QFD (Quality Function Deployment) method. This method establishes interconnections among market conditions, technologies, and products that would be used in an autonomous bus system. By using QFD, this paper gives the understanding of the future technologies' impact on transportation development. By using QFD method, the roadmap developed mid- and long-term planning for TriMet bus services.

QFD is a management innovation tool based on a matrix approach to mapping customer requirements and engineering attributes of products. It has been widely used as a communication tool for cross-functional teams (e.g., manufacturing and marketing) in order to establish relationships and trade-offs. QFD has been identified for roughly the last 15 years as providing a reliable approach for linking the different layers of roadmaps. Different roadmap layers can be coordinated with QFD through cross-functional collaboration to determine which product features should be given development priority on the basis of customer/market needs (Groenveld 1997; Phaal et al. 2004).

8.3.1 Market Drivers

8.3.1.1 Accessibility

Accessible public transportation provides a service for people to be transported safely and inexpensively to places that provide opportunity for enrichment of their life. This service is also available to people experiencing various bodily conditions and thus also includes the "LIFT Paratransit Service (which) Provides door-to-door service for people with disabilities who are unable to ride TriMet's regular buses or trains" (TriMet Business Plan for Fiscal Years 2018–2022 2014d, p. 9). This most specifically applies to those with, "Limited Mobility, Blind or low vision, Deaf or hard of hearing" (TriMet.org/access).

As for the general public, these "Consumers will be choosing from a range of options; convenience and cost will therefore be critical factors" (McKinsey and Company 2017, p. 12). Also, the diversity of destinations plays a factor in this, including the ability for people outside the service area of TriMet to be transported to one of its terminals by utilizing "community connectors" which are routes whose service areas overlap with areas served by other transportation providers (McKinsey and Company 2017, p. 12). This includes shared bus terminals, collaborative scheduling, and presenting customers with service offerings from these other organizations as an option for their commute.

The demand for more accessible transportation is increasing as vehicle ownership rates continue to decline (McKinsey and Company 2017, p. 16), and car-free zones become more common (McKinsey and Company 2017, p. 7). In fact, with dynamic routing technology, public transportation may begin to resemble shuttle service (McKinsey and Company 2017, p. 10) made easier by maximizing the benefits and potential of the Hop Fastpass (TriMet Business Plan for Fiscal Years 2018–2022 2014d, p. 6).

8.3.1.2 Air Quality Improvement

The quality of air is a part of environmental stewardship and is also a goal that is mentioned in the Metro 2040 plan (Metro 2017). Also, "TriMet is procuring five grant-funded, fully-electric battery-powered buses for a pilot, to inform a strategy about whether and when to begin switching to battery buses as performance and cost warrant" (TriMet Business Plan for Fiscal Years 2018–2022 2014d, p. 6).

8.3.1.3 Noise Improvement

Sound that distracts from the serenity of the environment is a form of pollution that can be controlled by selecting quieter engines and reducing the noise heard from louder engines by installing sound barriers. Public demand for this is inferred by the goal of attaining "Quiet neighborhoods" in the Metro 2040 plan (Metro 2017, p. 4).

8.3.1.4 Improve Ecological Environment

Sustainable causes and conditions for natural plants and animals to thrive are imperative for the long-term health of local ecology. Thus, protecting natural resources by public demand is another goal that is consistently mentioned in the Metro 2040 plan (Metro 2017).

8.3.1.5 Improve Safety

The topic of safety for a public transportation agency includes traffic safety, physical, and cybersecurity, and injury prevention. TriMet's plan to replace old infrastructure such as rail safety switches and tracks (TriMet Business Plan for Fiscal Years 2018–2022 2014d, p. 3) and the constant emphasis from the agency's General Manager's office on creating a culture of safety for employees and the general public (TriMet Safety and Service Excellence Task Force 2017, p. 14) show its commitment to safety. Safety is primarily expressed by the agency firstly by emphasizing that everyone in society plays a role in safety, such as expressed in the "See something say something" campaign. This emphasizes a directive that society and culture must invest in maintaining safety and that self-responsibility should be actively attended to while living in a region of 1.5 million people where safety-related incidents are bound to occur and can be greatly minimized by attending to safety protocols (Crumley 2017).

8.3.1.6 Land Utilization

Land utilization is a primary factor in determining the distance between destinations after construction in the area is complete. Recently, the effect of zoning laws on the walkability of area has been studied. For example, a neighborhood will become less walkable if the grocery stores are restricted to a neighboring commercial area that is far away from most houses (McKinsey and Company 2017, p. 14). However, these zoning laws also preserve environmental serenity by removing factories and commercial activity from residential areas. Portland is particularly known for its urban growth boundary, land-use planning laws, and protection of rural lands from urban sprawl as shown in the Metro 2040 plan (Metro 2017).

8.3.1.7 Improve Walkability

Walkability improves the financial health of the consumer and increases "...efficient land use, community livability, improved fitness and public health, economic development, and support for equity objectives" (Litman 2003, p. 1) Although this tends to be underemphasized in urban planning (Litman 2003, p. 1), it connects to many of the market drivers in this roadmap and thus plays a central role because "...Many beneficial activities (socializing, waiting, shopping and eating) occur in pedestrian environments..." (Litman 2003, p. 2).

In the Metro 2014 plan, the priority to "...reduce the need to drive or travel long distances by making jobs and shopping more convenient to where people live" (Metro 2017, p. 12) shows that walkability is a priority for the region. Also, when evaluating walkability, "...factoring in the number of places between origin and destination, rather than pure physical distance" (Noulas et al. 2012, p. 1) allows us to evaluate which bus routes provide the most walkability. Like the tables shown above, the Portland Metropolitan area benefits economically from improved walkability, for example, half of the trips on TriMet are for shopping and recreation (TriMet Business Plan for Fiscal Years 2018–2022 2014d, p. 8). Also, The MAX light rail system has helped "spur more than $13 billion in development within walking distance of MAX station (TriMet Business Plan for Fiscal Years 2018–2022 2014d, p. 8).

8.3.1.8 Lower Operating Costs

Decreasing the cost of TriMet operations is a function of removing costs through improving the utilization of assets such as buses and employee time. The best examples of this include better decision making with the creation and scheduling of bus routes.

8.3.1.9 Saving Time

Riders of public transit commonly compare their commute time with that of privately owned cars, which by their nature, will always require less time than public transit for a commute, although their use may be cost prohibitive in certain areas, such as in New York City or Tokyo. Options to reduce the commute times for public transit include incentivizing not driving, building a bridge exclusively for public transit use to help riders avoid traffic congestion, and collaboration with other transportation organizations for inter-regional travel (Crumley 2017).

8.3.1.10 Traffic Footprint

This is the amount of road used to transport a person. The form factor of vehicle is key in reducing the traffic footprint, for example, individual drivers of cars that occupy a city block of congested road can instead ride a single bus.

8.3.1.11 Vehicle Occupation Rate

This is the population density of people inside of a bus and infers how much utility a bus route is providing. TriMet's measure of this is the "…number of boardings per vehicle hour" which in turn infers that "The higher the productivity for any service, the lower the cost per ride – and the more riders can be served with the same budget."; thus, "Higher productivity lines are stronger candidates for increases in service." while "Lines that fall below 15 boarding rides per vehicle hour will receive special focus" (TriMet's Service Guidelines Framework 2015, p. 5).

Excellent decisions regarding the changing nature of demand for bus service across the region is the primary factor for the success of this market driver.

8.3.2 Products and Product Features

Our team came up with four categories that the implementation of autonomous buses should aim to achieve as a way to overcome the limitation faced by current bus systems: connectivity, communications, safety, and efficiency. After performing research, our team settled on six products that are important to successfully implement autonomous vehicles and meet commuter needs. These are smart grid, intelligent transportation systems, localized autonomous systems, modular vehicles, remote parking garages, and extra services.

8.3.2.1 Smart Grid

Smart grid is a hive of artificial intelligence (AI)-controlled transportation systems working together to handle traffic operations. It aims to address some of the critical issues of autonomous buses such as data management, reliability, communications between buses and road infrastructure, etc. It also enables the efficient movement of goods and people in a coordinated, reliable, and environmentally friendly way (Graham et al. 2017). The product features identified under smart grid are data network/integration and communication networks. These are described in more detail below.

8.3.2.2 Data Network/Integration System

For autonomous buses to acquire real-time information about their environments, they need to use a number of sensing technologies such as lasers, radar, and lidar. The information from the sensors are not always inaccurate; therefore, it is important for autonomous buses to share information with other buses through the use of vehicle-to-vehicle communications and vehicle-to-infrastructure communications which would allow them to make decisions such as detecting pedestrians, traffic levels on the road, and accidents on the bus route (Chowdhury et al. 2017).

8.3.2.3 Communication Networks

Communication is important in the implementation of autonomous buses as it enables them to perform cooperative driving maneuvers and improve their safety and efficiency (Furda et al. 2010). The two types of communication technologies available to autonomous vehicles are vehicle-to-vehicle (V2V) communication and vehicle-to-infrastructure (V2F) communication. Networks are important in the development of autonomous vehicles. In as much as each car is an individual, it becomes part of a complex ecosystem. Communication with other cars, road-side infrastructure, and data centers are important for the successful implementation of autonomous vehicles. Communication networks will be able to allow nearby buses to exchange relevant information about their traffic environments such as accidents, potential hazards, their position, etc. Vehicle-to-infrastructure communication is important for the success of autonomous buses because this communication can be used to improve road network efficiency as well as reducing pollution by sending notifications to the autonomous vehicles regarding traffic conditions (such as traffic congestion, alternative routes).

8.3.2.4 Intelligent Transportation System (ITS)

These are systems that utilize synergistic technologies and engineering concepts to develop and improve different transportation modes. The product features identified under intelligent transportation systems are discussed in more detail below.

8.3.2.5 Automatic Fare Payment

Automatic fare payment or convenient payment systems are important as it helps people travel without delay especially when they undertake multi-legged trips that require different payment methods. Singapore recently proposed a hands-free fare gate which allows commuters to automatically pay for their fares via their mobile device using Bluetooth technology or radio frequency identification (RFID) (Lim 2017).

8.3.2.6 Monitoring Rider Presence and Real-Time Capacity Reporting

In current bus systems, it is a challenge for riders on buses to be able to see free seats due to seats being too far from the entry or hidden by obstacles. However, technology in autonomous vehicles will be able detect and display seat occupancy and provide this information in real time. Information regarding status of every seat can be displayed in the entry area of the bus, in advance at a bus stop or on a smartphone app. This information is important for transit companies to help them modify and update their timetable according to traffic congestion, and to inform passengers about the congestion aboard the buses about to arrive. This would help passengers to make decisions on whether to get on a heavily loaded bus or wait a few minutes for a lightly loaded (Oransirikul et al. 2014). The well-being of passengers on the bus will also be monitored through these systems.

8.3.2.7 Computer Vision and Computation-Rich Processing

The computer vision software uses images and videos to automate tasks that the human visual system can do. Computer vision plays an important role in understanding capabilities that autonomous buses require to be able to correctly operate not only under normal conditions but also during unexpected situations. Computer vision is important for the successful implementation of autonomous vehicles as it enables detection of traffic lights and objects (Janai et al. 2017). Currently, the car crash avoidance systems in experimental autonomous vehicles rely on radar and other sensors to detect objects and pedestrians on the roads. However, the development of object detection that can perform near-real-time detection based on video clues will make object or pedestrian detection more effective.

8.3.2.8 Dynamic Routing

Dynamic routing enables autonomous buses to dynamically change their routes according to passenger destinations and road conditions. The information from Wireless communication among the autonomous buses on a road network can be used to predict the optimal route for origin–destination pairs. This in turn has the potential to enhance vehicle safety and improve traffic flow.

8.3.2.9 AI Traffic Feature (Radar Signal Sensing)

AI traffic features such as radar and lidar signal sensing will enable autonomous buses to determine where obstacles and other cars are positioned and at what speed they are traveling at. One of the advantages of radar signal sensing is that they are not affected by weather conditions.

8.3.2.10 Weather Navigation/Adoption

It is important for autonomous buses to be equipped with weather navigation systems that are able to function during adverse weather conditions. Bad weather conditions make it complex for autonomous vehicles to depend on cameras and sensors to see the road and make correct decisions, especially at high speeds. Ford announced the creation of a high-fidelity 3D maps of the roads its autonomous vehicles will travel on with details such as the exact position of curbs and lane lines, trees and signs, speed limits and other relevant road rules. The assumption behind the creation of this 3D maps is that the more a car knows about an area, the more it can focus its sensors and computing power to detect people and other obstacles in real time and under adverse conditions (Davies 2016).

8.3.2.11 Localized Autonomous Systems

These are the autonomous systems and subsystems localized to an individual bus. The product features identified under localized autonomous systems are discussed in greater detail below.

8.3.2.12 Identify Empty Parking Space

Using information from sensors and other communications systems, autonomous vehicles should be able to communicate with the environmental infrastructure and detect a free parking space and obstacles in the parking lot.

8.3.2.13 Lane Detection and Object Tracking

Lane detection is an important component of autonomous buses. Lane detection features provide means of warning the autonomous buses of potential dangers; this leads to car-crash avoidance and increases safety. The lane detection systems consist of the localization of specific road features such as road markings on the surface of painted roads. The three common methods of lane detection and object tracking are radar, laser scanner, and computer vision.

8.3.2.14 Tracking Nearby Objects

Tracking nearby objects and object recognition will enable autonomous buses to recognize hailing passengers at bus stops, intersections and also be able to differentiate different types of objects. The systems should also be able to detect construction zones present in the area, construction workers holding stop signs, or construction equipment.

8.3.2.15 Blind Spot Control

Autonomous buses will need to be equipped with corner cameras that are able to capture the images that are in the bus's blind spot.

8.3.2.16 Adaptive Cruise Control (ACC)

The adaptive cruise control will allow autonomous vehicles to control speed intelligently which will reduce the amount of braking the buses will perform. It would also help reduce fuel consumption because fuel use increases when buses vary their speed unnecessarily.

8.3.2.17 Modular Vehicles

A modular bus is assembled from modular components with "standard interfaces and minimal interdependencies between the modules" (Wegner et al. 2008). Modular vehicles are designed to be an efficient, coordinated network of buses that shuttle people door to door. The design features a series of "modular, self-driving electric pods that pick passengers up on demand and link together in a bus like form for the aim of getting each passenger from point A to point B as efficiently as possible" (Weiss 2015). The modular vehicles will attach and detach while in motion, while the automatic doors connect the interiors so that drivers can freely move between the pods. This will help in saving time and operating costs because the system would send individual modules where they are needed and only links them in high-traffic areas.

8.3.2.18 Remote Parking Garages

Remote parking garages will enable relocation of bus parking facilities to areas outside the city where land is cheaper. Autonomous buses will be able to depart the parking facility at its assigned time for it to meet passengers.

8.3.2.19 Extra Services

The autonomous buses will be equipped with extra services such as restaurants on wheels and coffee vending so that commuters do not have to worry about disembarking before their final destination to get their coffee, which will save them time.

8.3.3 Technologies

8.3.3.1 VANET (Vehicular Ad Hoc Network) and WAVE 802.11p

A VANET is an ad hoc network established between multiple cars. These vehicles can then establish connections and share their processing and sensor data with each other to uniformly accelerate and brake allowing for vehicles to "platoon," or travel a few inches from each other, without risk of colliding (Ieeexplore.ieee.org 2017). VANETs can similarly be used to form an autonomous vehicular cloud in which "the vehicles themselves will pool their computational resources together creating the effect of a powerful supercomputer" (Olariu et al. 2011). When this powerful technology is integrated to work with a given city infrastructure, it becomes possible to give real-time obstacle reports, road safety warnings, and disseminate information.

The VANET is able to do this by using the IEEE 1609 WAVE (Wireless access in Vehicular Environments) protocol with the IEEE 802.11p standard. This standard sends and receives data on seven reserved channels on the 5.9-GHz band. Because it is fixed on given channels, there is almost no significant latency to connect or transmit data between vehicles. This is what allows for platooning vehicles, but it also works at distances of up to 300 meters with data rates exceeding 8 Mbits/s (Teixeira et al. 2014).

The IEEE Communications Society has a technical subcommittee on Vehicular Networks & Telematics Applications and is actively working to establish this technology as a cornerstone of ITS-based services.

8.3.3.2 Modular Vehicles

Modular vehicles are still in their developmental phases, but, given the capabilities of autonomous vehicles and VANETs, Next Future Transportation Inc. is working to create the first swarm modular car. A rider will be expected to be able to use the app to call a modular vehicle to their location and drive them where they want to go. "When joined, they create an open, bus-like area among modules, allowing passengers to stand and walk from one module to another" (Next Future Transportation 2017).

8.3.3.3 Video Image Processing

Video image processing typically utilizes multiple cameras, a microprocessor, and software for interpreting the images and converting them into traffic data. Many cameras can be used together to cover a larger area. These can replace in-road detection systems at a reduced cost and can classify vehicles by their length and report vehicle presence, flow rate, occupancy, and speed for each class (Tewolde 2012).

8.3.3.4 Artificial Intelligence (AI)

AI has already been shown to be used in intelligent transportation systems as it can be used to convert traffic sensors into intelligent agents that can automatically detect and report traffic accidents or predict traffic conditions. It can be more reliable in assessing and predicting traffic conditions based on microscopic traffic data collected from vehicles that pass through its path. It is expected to be applied to the areas of traffic operations, travel demand modeling, transportation safety and security, public transportation, and infrastructure design and construction (Artificial Intelligence Applications to Critical Transportation Issues 2012).

8.3.3.5 Computation-Rich Processing (Transit Cloud Computing)

Computing, sensing, communication, and physical resources of autonomour vehicles can be coordinated and dynamically allocated to authorized users for the purpose of processing traffic management scenarios like synchronizing traffic lights after clearing an accident. Transit cloud computing can also be used to augment the capabilities of businesses by utilizing the computing power of its employees' vehicles sitting in a parking garage (Olariu et al. 2011).

8.3.3.6 Autonomous Driving

Autonomous driving systems are currently being developed by several major companies. They generally fall into one of two spaces. They are either smaller packaged systems that can be installed in partnership with vehicle manufacturers such as the Intel Mobileye and the nVidia Drive PX2, or they are an entire vehicle built from the ground up for the express purpose of providing an entire autonomous driving solution like the Google Waymo (Waymo 2017).

8.3.3.7 Passive Acoustic Array

These two-dimensional array microphones operate by detecting acoustic energy produced by vehicular traffic from a variety of sources within each vehicle and from the interaction of a vehicle's tires with the road. An increase in sound energy recognized as a vehicle passing through the detection zone is processed by the signal-processing algorithm to generate a vehicle presence. To minimize interference, sounds from locations outside the detection zone are attenuated. These sensors can measure vehicle passage, presence, and speed (Tewolde 2012).

8.3.3.8 Data Security

Autonomous vehicles processing data and directly communicating with the infrastructure of a city requires extreme levels of data security. There is a strong argument to be made for AES 256-bit data encryption as it is nearly uncrackable. But, in a system made up of millions of vehicles, the weak aspects of blockchain encryption can be turned into a significant strength as it becomes very difficult to compromise 51% of the vehicles in a city. Blockchain can also provide a trust network between otherwise anonymous autonomous vehicles using visible light and acoustic side channels (Rowan et al. 2017).

8.3.3.9 Big Data

Perhaps one of the most important aspects of autonomous vehicles and artificial intelligence is access to Big Data. TriMet has been collecting lots of data that it can feed into an AI to help make decisions, and autonomous vehicles can read a database to determine routes, but smart and fast Big Data are necessary for TriMet to fully utilize the potential of autonomous buses.

MapR has showcased its ability to provide a platform that enables customers to harness the power of Big Data by combining analytics in real time to operational applications to improve business outcomes (Mapr.com 2017).

8.3.3.10 RFID

RFID uses radio waves to read and capture information on an object with a unique identifier. Integrating sensors that activate when max train doors open would effectively create an RFID sensor web that could read the HOP card on any given rider. When someone passes through the web without an active HOP card, the train could easily signal fare inspectors to come run an inspection.

Identiv has used this technology to manage attraction access control of five million visitors per year on a resort island in Asia (Files.identiv.com 2017).

8.3.3.11 Active GPS

GPS comes in two forms, passive and active. Passive GPS does not track information in real time. It can only track it on the device which must then be downloaded onto a computer. Active GPS is a real-time stream of an object's pinpointed location. With active GPS, a bus can transmit its exact location to a centralized server at all times. This has the advantage of telling TriMet exactly where their buses are and being able to communicate how late they may be and can be used to send police and other emergency services to the exact location of a bus.

The active GPS is an integral part of autonomous buses that are currently being used in Taiwan (Horton 2017), Japan (The Japan Times 2017), and the USA (Lavars 2017). Companies that make the GPS for autonomous vehicles like NovaTel use active GPS technology in conjunction with other sensors to create a multi-frequency, multi-constellation absolute positioning system (NovAtel 2017).

8.3.3.12 Vibration-Resistant Cameras

This technology has existed for a long time but is necessary to apply on autonomous vehicles both inside and outside. Kappa Optronics produces IP cameras used by rail vehicles to monitor its riders (Cameras 2017). Rail vehicles operate at extreme speeds, but buses must also deal with all manners of road conditions so their cameras must be able to stabilize its video feed and safeguard its sensor technology from vibration fatigue.

It is just as important to have vibration-resistant cameras on the outside in order for traffic sign recognition software to work. This prevents a bus from seeing a sign that says 85 mph instead of 35 mph.

8.3.3.13 Radar/Lidar

Velodyne is creating lidar technology that provides accurate distance and reflectivity measurements at highway speeds, enabling critical requirements such as road sign and lane marking detection (Velodynelidar.com 2017a, 2017b). Current long-range lidar can scan in 360 degrees, up to 200 meters, and about 600,000 points per second. Other models exist which can scan in 360 degrees, up to 120 meters, 2.2 million points per second, and are accurate within less than 2 centimeters (Velodynelidar.com 2017a, 2017b).

8.4 Conclusion

Although the purpose of this project is to create a roadmap for implementing the use of autonomous buses, we believe the QFDs infer that this project should be included in a larger project to create infrastructure for economic development within TriMet's region of service due to Metro's economic development plans and TriMet's support for this through creation of transportation routes that facilitate economic opportunities for its riders. It is also inferred that environmentally focused projects will have better results when handled directly by Metro rather than TriMet, as the operation of products and technologies for public transportation cannot significantly eliminate damage to the local environment. Electric vehicles may be an exception to this; however, their damage to other environments because of mining and disposing of battery materials is due for consideration.

Due to autonomous vehicle product features receiving mid-level ratings, their adoption should require the use of highly rated product feature(s), which in this case, is modular vehicles, due in the year 2042. Since the year 2042 is far in the future and has been proposed as the year to release autonomous modular vehicles, it is recommended that this roadmap be split into phases, with autonomous vehicles being a later phase. Also, "Autonomous vehicle (AV) technology will not dramatically shift TriMet's business activities during the five-year period, though some opportunities may be available to assist operators with safety technology using some AV technology; e.g., crash avoidance technology" (TriMet Business Plan for Fiscal Years 2018–2022 2014d, p. 6).

Product features related to smart cities received the highest ratings; therefore, improvement in computation and monitoring infrastructure is a top priority. Since these will be the first features available for completion, it is recommended that they be assigned to their own phase, early in the roadmap.

With the possible release dates of product features and their ratings taken into consideration, we recommend that implementation of the roadmap be divided into two phases:

Phase 1
Invest in intelligent transportation systems to improve accessibility and safety.

Phase 2
Postpone autonomous vehicles until modular vehicles are complete in 2042. Use autonomous modular vehicles to improve walkability, accessibility, saving time, and increasing the vehicle occupation rate. Electrically powered modular vehicles will also have a significant effect on the quality of the local environment and will need autonomous capabilities to keep their operational cost reasonable because they may only seat around a maximum of 10 people, which is financially unsustainable when employing a human as the vehicle operator.

Appendix – Technology Roadmap for Implementing Autonomous Bus Service

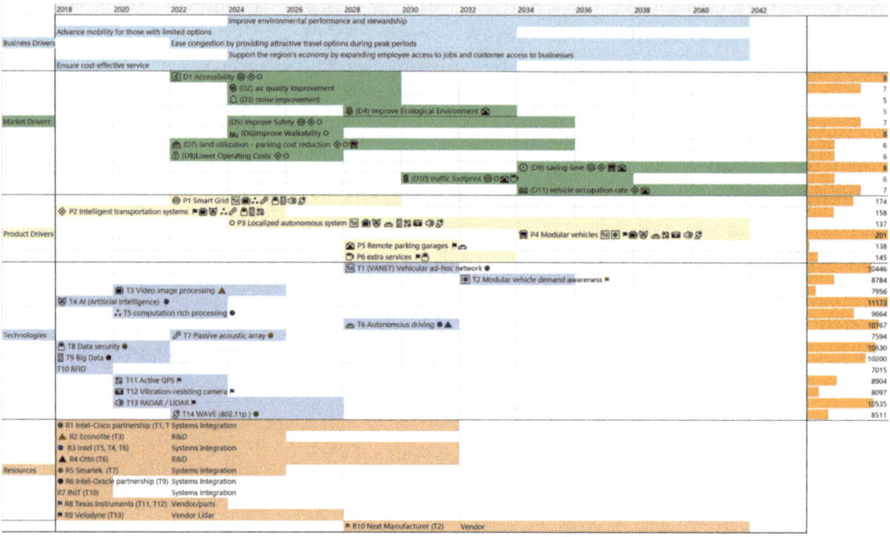

References

Artificial Intelligence Applications to Critical Transportation Issues. (2012). Available at: http://TriMet.org/access. Accessed 11 Dec 2017.

Cameras, R. (2017). Vibration-resistant rail traffic cameras - kappa optronics. Kappa-optronics.com. Available at: https://www.kappa-optronics.com/en/experts/cameras-for-rail-road-offroad/rail-traffic-cameras.cfm. Accessed 10 Dec 2017.

Chowdhury, M., Gawande, A., & Wang, L. (2017). Secure information sharing among autonomous vehicles in NDN. EEE/ACM Second International Conference, 15–26. https://doi.org/10.1145/3054977.3054994.

Crumley, M. (2017). Integrating new and autonomous technologies in bus service.

Davies, A. (2016). The Clever Way Ford's Self-Driving Cars Navigate in Snow. Retrieved from https://www.wired.com/2016/01/the-clever-way-fords-self-driving-cars-navigate-in-snow/

Fagnant, D. J., & Kockelman, K. M. (2013), Preparing a Nation for Autonomous Vehicles: Opportunities, Barriers and Policy Recommendations, Eno Foundation.

Files.identiv.com. (2017). Cite a website - cite this for me. Available at: http://files.identiv.com/products/rfid-nfc-inlays/rfid-tickets/RFID_Tickets_DS.pdf. Accessed 10 Dec 2017.

Furda, A., Bouraoui, L., Parent, M., & Vlacic, L. (2010). The Role and Future Challenges of Wireless Communication Networks for Cooperative Autonomous City Vehicles, 241–251.

Garcia, M. L., & Bray, O. H., Fundamentals of Technology Roadmapping. Sandia National Laboratory, 34, 1997.

Graham, R. L., Francis, J., Bogacz, R. J., Ee-, D. O. E., Berube, M., Farrow, B., et al. (2017). Challenges and Opportunities of Grid Modernization and Electric Transportation.

Groenveld, P. (1997). Roadmapping integrates business and technology. *Research-Technology Management, 40*(5), 48–55. https://doi.org/10.1080/08956308.1997.11671157.

Hansen, C., Daim, T., Ernst, H., Herstatt, C.. (2016). The future of rail automation: A scenario-based technology roadmap for the rail automation market, Technological Forecasting and Social Change, Available online 12 January 2016.

Horton, C. (2017). In Taiwan, modest test of driverless bus may hint at big things to come. Nytimes.com. Available at: https://www.nytimes.com/2017/09/28/automobiles/wheels/taiwan-autonomous-bus-test.html. Accessed 10 Dec 2017.

Ieeexplore.ieee.org. (2017). Performance Evaluation of Vehicle-Based Mobile Sensor Networks for Traffic Monitoring - IEEE Journals & Magazine. Available at: http://ieeexplore.ieee.org/document/4627421/. Accessed 10 Dec 2017.

Janai, J., Behl, A., & Geiger, A. (2017). Computer vision for autonomous vehicles: problems, datasets and state-of-the-art. ISPRS Journal of Photogrammetry and Remote Sensing.

KPMG (2012), Self-driving cars: the next revolution, KPMG and the center for automotive research.

Lavars, N. (2017). University of Michigan's self-driving buses will shuttle students across campus. Newatlas.com. Available at: https://newatlas.com/university-michigan-self-driving-bus/50164/. Accessed 10 Dec 2017.

Lim, A. (2017). Hands-free fare gates to be tested at selected MRT stations in 2018. Retrieved from http://www.straitstimes.com/singapore/hands-free-fare-gates-autonomous-food-delivery-trolleys-on-show-at-future-of-transport.

Litman, T. (2003). Economic value of walkability. *Transportation Research Record., 1828*, 3–11. https://doi.org/10.3141/1828-01.

Litman, T. (2014) Autonomous vehicle implementation predictions. Victoria Transport Policy Institute.

Mapr.com. (2017). MapR Showcases Big Data technologies at connected cars and autonomous vehicles 2017 | MapR. Available at: https://mapr.com/company/press-releases/mapr-showcases-big-data-at-connected-cars-autonomous-vehicles/. Accessed 10 Dec 2017.

Marinakis, Y., Walsh, S., & Chavez, V. Here there be dragons: The TSensors systems technology roadmap, IEEE, 2017.

McDermott, S. (2017). Forget self-driving cars. Automated public transportation is coming. *Cnet, 29*.

McKinsey & Company. (2017). Urban mobility at a tipping point. Available at: https://www.mckinsey.com/business-functions/sustainability-and-resource-productivity/our-insights/urban-mobility-at-a-tipping-point. Accessed 8 Dec 2017.

Metro (2017). The Nature of 2040. Portland: Metro, pp. 3,4,12,14.

National Research Council. (2012). Transportation research board. E-C168. In *Transportation research circular*. Washington, DC: Transportation Research Board.

Next Future Transportation. (2017). Next Future Transportation. Available at: http://www.next-future-mobility.com/analysis. Accessed 10 Dec 2017.

Noulas, A., Scellato, S., Lambiotte, R., Pontil, M., Mascolo, C., & Añel, J. (2012). A tale of many cities: Universal patterns in human urban mobility (universal patterns in human urban mobility). *PLoS One, 7*(5), E37027.

NovAtel. (2017). High-precision GPS for autonomous vehicles. Available at: https://www.novatel.com/industries/autonomous-vehicles/. Accessed 10 Dec 2017.

Olariu, S., Eltoweissy, M., & Younis, M. (2011). Towards autonomous vehicular clouds. *ICST Transactions on Mobile Communications and Applications, 11*(7–9), e2.

Oransirikul, T., Nishide, R., Piumarta, I., & Takada, H. (2014). Measuring bus passenger load by monitoring Wi-Fi transmissions from mobile devices. International Workshop on Innovations in Information and Communication Science and Technology.

Patterson, D. (1990). *Introduction to artificial intelligence and expert systems*. Prentice Hall.

Phaal, R., Farrukh, C., & Probert, D. (2004). Technology roadmapping—A planning framework for evolution and revolution. *Technological Forecasting and Social Change, 71*(1–2), 5–26. https://doi.org/10.1016/s0040-1625(03)00072-6.

Rowan, S., Clear, M., Gerla, M., Huggard, M., & Goldrick, C. (2017). Securing Vehicle to vehicle communications using blockchain through visible light and acoustic side-channels. Available at: https://arxiv.org/abs/1704.02553. Accessed 10 Dec 2017.

Scherer, M. (2017). NHTSA and autonomous vehicles (Part 1): The 5 levels of automation. Law and AI, 11 Dec. 2017.

Schleiffer, R. (2002). Intelligent agents in traffic and transportation. *Transportation Research Part C, 10*, 325–329.

Teixeira, F., e Silva, V., Leoni, J., Macedo, D., & Nogueira, J. (2014). Vehicular networks using the IEEE 802.11p standard: An experimental analysis. *Vehicular Communications, 1*(2), 91–96.

Tewolde, G. (2012). Sensor and network technology for intelligent transportation systems. 2012 IEEE International Conference on Electro/Information Technology.

The Japan Times. (2017). Test of autonomous bus system starts in Tochigi | The Japan Times. Available at: https://www.japantimes.co.jp/news/2017/09/03/national/autonomous-bus-test-starts-tochigi-prefecture-roadside-rest-area/. Accessed 10 Dec 2017.

Tri-County Metropolitan Transportation District of Oregon (TriMet). (2014d). TriMet Business Plan for Fiscal Years 2018–2022. Portland: Tri-County Metropolitan Transportation District of Oregon.

Tri-County Metropolitan Transportation District of Oregon (TriMet). (2015). TriMet's service guidelines framework. resolution 14–12-60 exhibit a. Portland: tri-county metropolitan transportation district of Oregon.

TriMet Safety & Service Excellence Task Force. (2017). Portland, OR, USA: Tri-County Metropolitan Transportation District of Oregon (TriMet), pp. 14, 21.

Velodynelidar.com. (2017a). HDL-64E. Available at: http://velodynelidar.com/hdl-64e.html. Accessed 10 Dec 2017.

Velodynelidar.com. (2017b). VLP-32C. Available at: http://velodynelidar.com/vlp-32c.html. Accessed 10 Dec 2017.

Waymo. (2017). Technology – Waymo. Available at: https://waymo.com/tech/. Accessed 10 Dec 2017.

Wegner, P. M., Blower, P., & Wilkenfeld, J. (2008). standard buses, modular buses, and plug-and-play modular buses; what is the difference and why does it matter? Small Satellites Systems and Services.

Weiss, C. C. (2015). Next Future modular transportation swarms the commuting hordes. Retrieved from https://newatlas.com/next-future-modular-transportation/39952/.

Chapter 9
A Technology Roadmap for a Standardized Platform for Autonomous Vehicle Systems

Thimo Bielsky and Tuğrul U. Daim

9.1 Introduction

The year 1925 can be stated as the beginning of the era of autonomous driving, when a radio-controlled vehicle drove along Broadway in New York City (Seif and Hu 2016). Since then, a small number of prototypes have been developed as part of research about radio-controlled and driverless transportation. Besides the curiosity of the possible and the devotion to develop new technologies, the topic of autonomous driving comes along with several advantages and new opportunities for our daily life. Because every autonomous vehicle will be part of an optimized system, it is claimed that autonomous driving technology has the potential to decrease traffic congestion, energy use, and land use. The fact that it is claimed to significantly increase safety for road users and pedestrians can be moreover counted as one of the most important factors (Anderson et al. 2016).

Since the turn of the millennium, the research in the field of autonomous vehicles has significantly increased (Anderson et al. 2016). Before that, autonomous transportation was mainly established in fields within controlled systems such as, for example, indoor transportation systems in warehouses, or outdoor transportation systems in differentiated container areas (Maurer et al. 2016). These systems have in common that the sequence of the system process is regulated in a closed environment, which decreases the complexity of the system. Autonomous vehicles outside of such plants have been established to, for example, access dangerous areas for disarming munitions or difficult-to-access areas in the deep sea or on mountains. An interaction with other vehicles can be excluded (Maurer et al. 2016). Even airplanes,

T. U. Daim (✉)
Department of Engineering and Technology Management, Portland State University, Portland, OR, USA
e-mail: tugrul.u.daim@pdx.edu

T. Bielsky
Northern Institute of Technology, Hamburg, Germany

T. U. Daim (ed.), *Roadmapping Future*, Applied Innovation and Technology Management, https://doi.org/10.1007/978-3-030-50502-8_9

which are highly automated due to safety reasons, cannot be operated fully autonomous yet, especially during flight phases with a potential interaction of several other airplanes such as take-off and landing.

Today, the technology has progressed to a level at which autonomous vehicles can coexist within a complex system. This is mainly due to technologies like sensors, cameras, and advanced microelectronics (Anderson et al. 2016). With these technologies, the vehicles can "see" and progress the current situation in real time. The aim is that the vehicle can act accordingly in uncontrolled situations, for example, when the road is blocked due to rock slides or pedestrians cross a street at a position where they are not supposed to cross. Especially the development of the remaining 10% of these technologies is proving to be very complex and is still part of the research today (Herrmann and Brenner 2018). Deep learning algorithms, for instance, are used to teach the vehicle how to handle situations which can daily occur. Companies equip their vehicles with cameras and sensors and drive with them around to optimize the algorithms. The vehicle learns how the driver reacts in specific situations. The more the vehicle learns, the better will be the resulting algorithm (Kong et al. 2017).

The technology progress has led to the current situation today, in which several established companies and start-up projects develop technologies for autonomous driving. This trend has been encouraged by governmental objectives and projects as from the European Union such as the *Vision Zero* for no traffic fatalities by 2050 (Maurer et al. 2016), *PROMETHEUS*, *AutoNet2030*, and *i-Game* (Hobert et al. 2015). Conventional car manufacturers are following this trend for both improving their product and because they are facing new competitors. These new competitors are mostly technology companies such as Google, which try to get into a new autonomous driving market with their capability for software development (Herrmann and Brenner 2018).

With all the mentioned new and established players that try to acquire a large market share, a situation occurs in which companies try to develop intellectual property to obtain a competitive advantage (Herrmann and Brenner 2018). In this situation, technologies are mainly developed by the companies itself without cooperation within the industry. Governmental institutions such as the European Union define standards as a result of the beforementioned projects to build interfaces between the technology which is developed by different companies (Herrmann and Brenner 2018). However, recent developments have also shown that established vehicle manufacturers are merging for the development of autonomous driving to be able to withstand the new competitors on the market. One example is the agreement about this between the Daimler AG and the BMW Group in 2019 (Weldemann and Becker 2019).

The aim of this chapter is to propose a solution for a standardized development platform for autonomous vehicles based on the standards released by governmental institutions. The methodology which is used to acquire this aim is a technology roadmap. A pre-existing technology roadmap that addresses this issue is used and updated to the current state of technology based on literature research (Alramadan et al. 2016). Based on the technology roadmap, it can be shown at what time which technologies have to be introduced due to a specific reason based on expected market drivers of autonomous transportation.

9.2 Literature Review

This section contains the literature review with different aspects in the field of autonomous transportation. The focus of the literature review lies, according to the aim of the developed technology roadmap, on a standardized platform for autonomous vehicles and currently discussed important topics in this field.

In addition to fundamental topics of autonomous transportation, the focus in this section is set on the standardization of communication because this topic is still under development and the industry currently puts a lot of effort into its development (Johanning and Mildner 2015).

Based on the above, this section is separated into three parts. In the first part, a definition of autonomous driving is established which elaborates the different stages of autonomous transportation by the Society of Automotive Engineers (SAE) (SAE International 2014). An overview about current established systems for communication and its future trends based on current literature is given in the second part. In the third part, the conflict and new competition between conventional car manufacturers and the technology companies which try to get into the market of autonomous on-road transportation is discussed.

9.2.1 Definition and Stages of Autonomous Driving

Autonomous vehicles in general have been used in many different areas so far, but mostly to fulfill a specific purpose within a company or an industry. Indoor and outdoor transport systems are an example for that. Indoor transport systems are mainly used in warehouses to distribute and store goods while outdoor transport systems are used in, for example, container terminals (Maurer et al. 2016). Other purposes for autonomous vehicles are their usage in areas which are difficult to access, for instance, in deep-sea explorations, to work on mountainsides, or to work in the mining sector (Maurer et al. 2016). All these transport systems have in common that they work in a closed system with a clearly defined system boundary. Even if in a warehouse operate other non-autonomous vehicles together with autonomous transport vehicles, both vehicles would be separated from each other by using different paths within the warehouse (Maurer et al. 2016). Today, these autonomous transport vehicles use different approaches for orientation within the closed system. Indoor transportation systems use cameras and laser combined with digital environment maps to be able to observe the surrounding area. With Wi-Fi and Radio-Frequency Identification (RFID) chips which are distributed in the warehouse, an Indoor Positioning System (IPS) is established for the vehicles to determine their position within the warehouse (Maurer et al. 2016). Autonomous outdoor transport vehicles work in a similar way. Instead of laser, they use radar technology due to safety reasons. In addition, markings on the ground are used as orientation for the vehicles, mostly to make sure that the vehicle cannot leave the closed system (Maurer et al. 2016). The same applies for autonomous vehicles used in areas which are difficult to access. These vehicles typically use radar, laser, and a satellite positioning system such as the Global Positioning System (GPS) for orientation (Maurer et al. 2016).

The main differences between the autonomous vehicles which are described in the paragraph above and autonomous on-road vehicles are its purpose and its system boundary. The purpose of autonomous on-road vehicles is to transport people or goods from a departure point to a destination point. These vehicles operate in an open system with potential interactions with people, objects, or other vehicles, which are not part of the autonomous system. These autonomous on-road vehicles shall substitute cars and trucks within the next 30 years (Litman 2020). The technology that autonomous on-road vehicles use at the state of current industrial development and research is comparable to what has been described in the previous paragraph about technologies for indoor and outdoor transport vehicles operating in closed systems. Autonomous on-road vehicles, which are from now on described as autonomous vehicles, use fundamental technologies at the basis of radar, laser, cameras, and GPS for orientation (Kong et al. 2017).

However, the substitution of human-driven vehicles to autonomous vehicles will take several decades (Litman 2020). In this time, the system is even more complex because both non-autonomous vehicles and autonomous vehicles will share the same roads. Autonomous vehicles are with no significant effort able to predict how other autonomous vehicles behave, but it is very difficult for them to predict how a human-driven vehicle behaves (Maurer et al. 2016). In 2017, the Autonomous Rail Rapid Transit (ART) was unveiled in China (Rastogi 2017). It is a mixture of an omnibus and an electric tramway. The ART is primarily guided by painted lines on the road. However, it is as well equipped with different technologies for orientation and it still contains a human driver. This is necessary because the ART operates in a system with most vehicles being non-autonomous. With the equipped technologies, the driver can decide to do detours because of, for example, traffic congestions. In this case, the ART has to leave its guidance system and change its orientation supported by the driver (Rastogi 2017).

On-road vehicles at the current state are significantly different from each other by their implemented technologies. There are human-driven vehicles which can operate autonomously in specific cases, for example, for driving on the highway or supporting the driver while driving in the city. These vehicles are typically equipped with radar technologies as it can simultaneously detect both the range to an object and the velocity independently whether it is dark or during bad weather conditions (Gamba 2020). Short-range radar is in this case used for blind-spot warnings, and warnings for pedestrians or cyclists in cities while long-range radar is used for the automatic cruise control and a collision warning system on highways (Gamba 2020). Autonomous vehicles, however, which can operate fully autonomous, additionally contain the Light Detection and Ranging (LiDAR) sensor (Johanning and Mildner 2015). LiDAR sensors work similar to radar, but instead of radio waves, laser beams are used (Kong et al. 2017). This technology is primarily used by autonomous vehicles for orientation. Nevertheless, LiDAR sensors are disadvantageous when being used under bad weather conditions. It is also blocked by obstacles, for example, a truck, and it has problems recognizing smaller objects (Kong et al. 2017). To decrease these weaknesses, communication systems are used in autonomous vehicles which are further elaborated in Sect. 2.2.

As it has been indicated in the paragraph above, there are different autonomous stages of a vehicle. A partial autonomous vehicle might be human-driven and might be able to operate autonomously only on highways, while a full autonomous vehicle can be operated driverless in all kinds of situations. These different stages are divided into six levels defined by the SAE, which are further elaborated in the following.

The six levels of autonomous driving are based on the past and expected evolution of human on-road transportation. At first, systems as the anti-skid system or the electronic stability program were introduced. These systems were safety-focused and are mandatory in vehicles today (Herrmann and Brenner 2018). With the development of new technologies, further systems which support the driver were introduced with new vehicle generations. With each new vehicle generation, the possibility is created that more and more tasks of the driver can be handed to the vehicle system (Herrmann and Brenner 2018). In the future vehicle generations, these supporting systems will increase until the driver is only a back-up and can take over the driving task when needed (Herrmann and Brenner 2018). This situation is somehow comparable to pilots flying passenger airplanes. It is the last step before the vehicle can handle problematic situations without a human back-up (Herrmann and Brenner 2018). Furthermore, without the need of a human back-up, the full potential of autonomous vehicles would be utilized. For example, it will be possible for disabled and blind people to use vehicles as well (Anderson et al. 2016). Another point in this context is that the vehicle interior can be focused on creating an environment for entertainment, consumption, work, or rest (Janssen and Kenemans 2015). In Table 9.1, the development stages for autonomous vehicles which are defined by the SAE in six levels are displayed together with the characteristics of each level.

As shown in Table 9.1, *Level 0*, which stands for *No Automation*, means that the human driver performs all necessary tasks. This includes that the human driver executes the driving task for both the lateral and longitudinal direction, and the human driver monitors the driving environment around the vehicle. The human driver is at the same time the back-up because there are no supporting systems included in the vehicle (SAE International 2014).

The next defined level is *Level 1* which is called *Driver Assistance*. In this level, systems are implemented to support the human driver with the driving task. The system is taking over the execution of driving in either the lateral direction or the longitudinal direction (SAE International 2014). Examples are the beforementioned anti-skid system and the electronic stability program (Herrmann and Brenner 2018).

Table 9.1 Summary of SAE levels for autonomous vehicles (SAE International 2014)

SAE level	SAE name	Execution of driving	Monitoring of environment	Fallback of driving task	System capability
0	No automation	Human	Human	Human	n/a
1	Driver assistance	Human and system	Human	Human	Some driving modes
2	Partial automation	System	Human	Human	Some driving modes
3	Conditional automation	System	System	Human	Some driving modes
4	High automation	System	System	System	Some driving modes
5	Full automation	System	System	System	All driving modes

The succeeding *Level 2*, called *Partial Automation*, is defined as that the vehicle system can drive the vehicle in both the lateral and longitudinal direction. However, the human driver must monitor the vehicle and the surrounding traffic at all times (SAE International 2014). Examples are the parking assistant and the distance assistant in which the vehicle has a defined speed limit but keeps a constant distance to the vehicle ahead (Herrmann and Brenner 2018).

In *Level 3*, the *Conditional Automation*, the human driver does not necessarily need to pay attention to the vehicle and the traffic when the vehicle system is performing the driving task. However, the human driver needs to be able to take over the driving task whenever it is required (SAE International 2014). One example is the highway assistant. This assistant allows the vehicle, for instance, to pass other vehicles by itself (Herrmann and Brenner 2018).

The next level is *Level 4*, which is called *High Automation*. At this level, the vehicle system can perform the driving task in various situations and the human driver does not need to pay attention to the vehicle and the traffic. But the human driver still has the possibility to intervene and can overrule the vehicle system (SAE International 2014).

In the last defined level, which is *Level 5* and called *Full Automation*, the vehicle system can perform every driving task. In this stage, the human driver is not able to intervene in the system anymore (SAE International 2014).

9.2.2 Current Established Systems for Vehicle Communication

Communication systems are fundamental for autonomous vehicles because they increase safety and optimize the entire transportation system. This topic is elaborated in this section of the literature review chapter because it is still part of current research and the associated companies have not agreed yet which technologies will be used for different communication channels (Johanning and Mildner 2015).

The communication systems of autonomous vehicles are divided into three categories. One category is the communication between vehicles, the so-called vehicle to vehicle communication (V2V). Another category is the communication between the vehicle and any other imaginable object (V2X) and the third discussed category is the communication between the vehicle and pedestrians. The succeeding structure of this section is based on these three categories.

Vehicle to Vehicle Communication (V2V)

With V2V vehicles can communicate and exchange data with each other. Vehicles which can communicate with each other are described as Intervehicle Communication (IVC) enabled vehicles (Fernandes and Nunes 2012). The IVC system needs partly to be capable of transmitting a high amount of data. Especially in areas with a high traffic density, the communication must not be terminated (Zheng et al. 2015). This challenge can most likely be solved with communication standards based on millimeter waves which is comparable to the 5G cellular network standard (Heddebaut et al. 2005). One discussed option for IVC is a mixture of the 5G technology and Wi-Fi and is called Dedicated Short-Range Communication (DSRC) (Fernandes and Nunes 2012). The main challenge in research about this topic is to make IVC reliable (Wang et al. 2017).

While the sensors and cameras on a vehicle can only process the environment directly around the vehicle (Heddebaut et al. 2005), a communication platform allows vehicles to “see” other vehicles which are visually blocked for the sensors and cameras (Litman 2013). That is to say a redundancy for increased system safety. If a sensor fails, the vehicle still can “see” by using the data of the sensor from a vehicle nearby (Anderson et al. 2016). Within cities, V2V can be used for increasing the vehicle system safety as well. By transmitting data of the vehicle status information, for example, the current velocity, the acceleration, and the relative position to other vehicles, other vehicles are able to estimate their own position on the road (Vlachos et al. 2017). This information complements the visual information from the cameras and sensors and can compensate the bad signal quality of GPS in cities (Parra et al. 2017).

Furthermore, V2V can be used to transmit further information of the vehicle. This information is called event information and contains data about the planned driving characteristics of the vehicle and information which are received about the traffic (Vlachos et al. 2017). The vehicle can, for example, be warned about an accident or a traffic jam ahead. A sudden weather change with icy roads can be transmitted as well (Johanning and Mildner 2015). Vehicles with special rights such as ambulances, police cars, or fire trucks can transmit the signal via V2V to clear the road (Maurer et al. 2016). Fully autonomous vehicles can optimize the traffic flow with V2V by adapting to the current traffic and road situation (Herrmann and Brenner 2018).

However, V2V can also be implemented in vehicles which operate not fully autonomously. While in this case this system would be another supporting system for the driver of the non-autonomous vehicle, fully autonomous vehicles can use this information to be able to adapt to the driving behavior of human drivers

(Herrmann and Brenner 2018). The autonomous vehicle can communicate via V2V with the human driver in situations such as an intersection crossing or during overtaking (Hobert et al. 2015).

Truck manufacturing companies are currently collaborating with logistic companies to test the transport of goods in partial autonomous convoys. Such a convoy consists of several trucks. The first truck of the convoy is driven by a human driver while the other trucks follow the first truck autonomously via V2V. This concept is also referenced as Platoon (Fernandes and Nunes 2012). Besides decreasing human labor cost, this concept increases the efficiency of transporting goods. Because the reaction time of a human driver is not relevant for an autonomous vehicle, the distance to other autonomous vehicles can be significantly reduced. Transporting goods in such a convoy reduces the drag to the following trucks within the convoy which at the same time improves the road capacity (Fernandes and Nunes 2012).

This concept can also be transferred to the concept of fully autonomous vehicles. With the exchanged information by V2V, vehicles can build convoys by themselves for shared route sections to save fuel and to increase the road capacity (Herrmann and Brenner 2018).

Vehicle to Object Communication (V2X)

Vehicle to object communication (V2X) is emerging from the collective term Internet of Things (IoT), which describes the connection of, among others, devices, machines, and items to the internet. In terms of autonomous vehicle communication, these objects are traffic lights, traffic signs, parking spaces, and so forth (Herrmann and Brenner 2018).

The meaning behind V2X can be divided into further sub-categories to separate the purpose of each communication channel. Four chosen sub-categories are further elaborated in the following (Johanning and Mildner 2015; Herrmann and Brenner 2018):

- Vehicle to infrastructure (V2I)
- Vehicle to home (V2H)
- Vehicle to enterprise (V2E)
- Vehicle to cloud (V2C)

The communication between vehicle and infrastructure (V2I) is defined as the receiving of information from, for example, traffic lights, traffic signs, parking spaces, intersections, or curbs (Herrmann and Brenner 2018). Furthermore, smaller objects are recognized as well and saved in a public cloud space, so other vehicles have access to the information about the recognized objects as well (Kong et al. 2017). In general, V2I is used to further optimize the operating system of autonomous vehicles. With V2I, both the vehicles can adapt their driving behavior according to information from the V2I network and the infrastructure objects itself can adapt according to the current traffic situation (Mishra et al. 2018).

The sub-category vehicle to home (V2H) is defined as the possibility to access media from home. This includes accessing music, videos, and pictures from the home network, or personal cloud storages (Johanning and Mildner 2015).

A definition of the sub-category vehicle to enterprise (V2E) communication is a communication channel for non-public facilities. This includes the communication to car repair shops, gas stations, hotels, doctor offices, restaurants, and entertainment facilities such as movie theaters (Johanning and Mildner 2015). These facilities can be contacted for information such as opening hours, reservations, fees, and location data via this communication channel. Parking decks can send information about available parking spaces and its fees (Johanning and Mildner 2015). The definition of V2E can be extended to include the communication to car manufacturers as well. This communication channel is necessary for, among others, software updates (Herrmann and Brenner 2018).

The meaning behind the communication channel vehicle to cloud (V2C) is that via this channel, information from online services such as weather data, news, or information about the local area can be transmitted to the vehicle (Herrmann and Brenner 2018).

The development of different V2X communication channels is still part of current research and technology development. This applies especially to V2I because new objects have to be implemented into the infrastructure while other communication channels can be configured by new apps or digital service concepts. In this context, the V2I concept is implemented in prospective models for smart cities to provide fundamental interfaces for mobility concepts (Herrmann and Brenner 2018).

Currently, a high effort is performed in digitalizing parking lots and providing them to the V2I communication network. The traffic in central business districts which is created by people who search a parking space is about 30% of the total traffic in this area (Shoup 2007). Knowing the location of available parking spaces would reduce the traffic within cities.

The German car manufacturers AUDI and BMW, for example, developed concepts for connected parking spaces as part of V2I. The technology of AUDI shows parking spaces around a specific location. If possible, available parking spaces and the parking fee can be accessed from the vehicle as well (Johanning and Mildner 2015). BMW developed a network called *ParkNow* that can be joined by operators of parking decks which are equipped with the necessary technology. In this network, parking spaces can be viewed, reserved, and payed from the vehicle (Johanning and Mildner 2015). Another example is the concept *Valet Park4U* developed by the supplier for car manufacturers VALEO. This concept is still under development, but the aim is to provide parking decks for autonomous vehicles. After the passengers left the vehicle in front of the parking deck, the vehicle searches a free parking space and parks by itself (Johanning and Mildner 2015).

AUDI introduced a traffic light information system in their new vehicle generation. In the USA, the vehicle can access the *Traffic Light Information* network which informs the vehicle about the time until the traffic light turns green (Volkswagen 2019). In the German city Ingolstadt, AUDI is enhancing this technology with implementing smart traffic lights for further test purposes in everyday traffic situations (Volkswagen 2019).

Another safety-related aspect is the emergency call (eCall) and the breakdown call (bCall). The eCall can be seen as a V2E communication channel and is a service that detects an accident of the vehicle. It automatically informs the car manufacturer and the emergency call about, among other information, the vehicle model, the amount of passengers, time of the accident, the last known position of the vehicle, the direction into which the vehicle was driving, and which Airbags expanded (Johanning and Mildner 2015). DAIMLER has introduced the eCall in their vehicles since 2012 (Daimler, n.d.), while the EUROPEAN UNION made it mandatory for new vehicles built from 2018 onward (European Commission 2017). The bCall can be seen as a V2E communication channel as well. If the vehicle detects a problem by itself, it can contact the manufacturer and near located repair shops via this communication channel (Johanning and Mildner 2015).

As it is assumed based on the descriptions and examples in the paragraphs above, an autonomous vehicle needs to be able to send a lot of data for both safety and network optimizing reasons. The cellular phone network is used for the transfer of this data (Vlachos et al. 2017). The current standard of 4G, also known as *Long Term Evolution* (LTE), cannot be used for this amount of data, especially due to its problems with high latency times when many users are connected to the same LTE network (Zheng et al. 2015). The next generation of the cellular phone network, 5G, is capable for the expected amount of data of autonomous vehicles. This expectation also applies to cities, in which many devices are in a network (Vlachos et al. 2017). However, 5G is more sensitive of weather conditions and the signal has a shorter range compared to LTE. To provide a fully covered network, more radio towers have to be built for 5G so that autonomous vehicles have always access to the network (Heddebaut et al. 2005).

Vehicle to Pedestrian Communication

A point that should not be neglected is the interface between vehicles and pedestrians and the necessary communication between them. The pedestrian and the driver usually communicate in a non-verbal way with each other (Maurer et al. 2016). This non-verbal communication can be seen as informal rules to help to guide the traffic. If a pedestrian approaches a crosswalk, the driver assumes that the pedestrian wants to cross. Both of them use the facial expression and eye contact as the communication channel (Maurer et al. 2016). If the pedestrian stands at a crosswalk and looks at the driver, the pedestrian usually gets the signal to cross when the driver returns the eye contact (Maurer et al. 2016).

Gestures and body movements can be seen as a communication channel between a pedestrian and a driver as well. However, these are dependent on the local culture (Maurer et al. 2016). Typical examples for Europe are that nodding stands for agreement, moving the arm up and down requests the driver to stop, moving the arm down signals the driver to slow down, and a sweeping hand movement or the palm facing upward signals the driver to go ahead (Maurer et al. 2016).

However, if such a situation occurs between a pedestrian and a fully autonomous car, the non-verbal communication between humans cannot be applied the same way. A so-called Intent Communication System (ICS) needs to be integrated into the autonomous vehicle that replaces the human driver in the aspect of non-verbal communication. But so far, there are many potential solutions to establish an ICS but no consensus yet between different companies (Matthews et al. 2017). In general, the ICS needs to tell the pedestrian whether he is recognized by the vehicle and whether it is save, for instance, to cross the street (Herrmann and Brenner 2018).

Looking at research between humans and machines, it becomes clear that pedestrians at the current stage are still looking for a human driver, although they might know that the vehicle is operating fully autonomous (Matthews et al. 2017). Human beings need a time to adapt to that during which an increase of accidents could occur due to miscommunication (Matthews et al. 2017).

So, the ICS needs to be designed in a way that it represents the face of the autonomous vehicle. The ICS should not only be able to see what the pedestrian is about to do, the system should also be capable of interacting with the pedestrian. This includes gesture identification and visual feedback, or audio feedback (Matthews et al. 2017). With strobe lights, the vehicle can get the attention of the pedestrian, while a display and speaker can tell the pedestrian what the vehicle is about to do. If a pedestrian crosses a street, moving headlights can track the movement of the human as the "eyes" of the autonomous vehicle (Matthews et al. 2017).

Daimler tried to test potential possibilities for a distinct communication between pedestrians and autonomous vehicles. In the futuristic model *F015*, Daimler implemented a projector that projects signs and words on the street which signal the pedestrian whether they are recognized by the vehicle and what the vehicle is about to do, or rather tells the pedestrian what to do (Herrmann and Brenner 2018). Nissan implemented a screen that says, "Please wait" or "Please go". The vehicle sends a clear signal to the pedestrian whether the vehicle recognized the pedestrian and what the pedestrian is supposed to do according to the status of the vehicle (Herrmann and Brenner 2018).

Other concepts have suggested to use the V2I communication channel to contact pedestrians which are located near the road segment that the autonomous vehicle is about to pass on their smartphone to warn and inform them about the arriving vehicle (Herrmann and Brenner 2018). In case a crash cannot be prevented anymore, either due to a miscommunication between the pedestrian and the vehicle or because pedestrians overestimate the safety of autonomous vehicles, exterior airbags are discussed as a further safety feature for pedestrians (Keferböck and Riener 2015).

9.2.3 Approaches of Car Manufacturers and Technology Companies

A new market is created with autonomous vehicles. They will step-by-step replace the market of human-driven on-road vehicles within the next 30–40 years (Litman 2020). The main players of the market for human-driven vehicles are the conventional car manufacturers such as Mercedes-Benz, Ford, or Toyota. However, these companies share the new autonomous vehicle market with new players. These new players are technology companies such as Nvidia or Alphabet (Google) and start-ups, which try to develop disruptive concepts for autonomous vehicle technologies (Urmson 2015).

Conventional car manufacturers introduce new technologies in every new vehicle generation. Many decades ago, the vehicles did not have any systems to support the driver. The first supporting systems were the anti-skid system and the electronic stability program which are today mandatory for new vehicles in the EUROPEAN UNION (Herrmann and Brenner 2018). Newer vehicle generations have further supporting systems, for example, a parking assistant, a highway driving assistant, or a city assistant, which takes over the acceleration and deceleration of the vehicle (Herrmann and Brenner 2018). So far, conventional car manufacturers have followed the defined levels for autonomous driving that are defined by the SAE, step by step. In these development stages, every car manufacturer tries to develop their own technologies to acquire a competitive advantage. However, looking at new technologies of different car manufacturers, it can be seen that it always goes into the direction to support the human driver with the driving task (Herrmann and Brenner 2018).

Technology companies which try to get into the market of autonomous transportation do not go through all SAE levels for autonomous driving to develop autonomous vehicles. These companies start at the highest level, full automation, with their first vehicle to be released (Herrmann and Brenner 2018). But there are also cases of established collaborations. GOOGLE with their subsidiary WAYMO collaborates with conventional car manufacturers to implement their technology for autonomous driving (Lambert 2016).

In the last years, the market of on-road transportation has already been disrupted by, among others, TESLA. At the current state, the market is facing a change in the drive technology from combustion engines to a fully electric engine and drive chain (Herrmann and Brenner 2018). This means for conventional car manufacturers that new knowledge has to be acquired. Instead of specialists for gear transmissions, specialists for electric drive chains and batteries are needed. Another challenge for conventional car manufacturers is the purchase behavior of their customers. Today, younger generations are less willing to own a vehicle. They rather use it as part of a whole mobility concept. According to this, many conventional car manufacturers see themselves as mobility provider instead of a car manufacturer by now by offering the car as a mobility service. DAIMLER and BMW, for example, established a concept in Germany with *Car2Go* and *DriveNow*. In these concepts, the customer is only paying for the usage of the vehicle (Hepe 2019).

With the development of fully autonomous vehicles, conventional car manufacturers are facing the problem that they are lacking intellectual property and experience with information technologies (Johanning and Mildner 2015). This can lead to a threat for conventional car manufacturers, because the handling of large amounts of data for deep learning processes which teaches autonomous vehicles to handle different situations is fundamental for its development (Johanning and Mildner 2015). Also security risks can lead to a threat because technology companies are more experienced in handling software security issues (Khaliq 2015). Furthermore, the concept of the vehicle itself will change. The interior at a level 5 autonomous vehicle does not need an interface to a human driver anymore and the crash design of vehicles within a system of only autonomously operated vehicles can be changed which results in a lighter vehicle design (Anderson et al. 2016).

One industry that has not been mentioned before is the industry for providing the cellular phone network. This industry is about to play a more important role for the autonomous vehicle transportation due to the high necessity of communication possibilities. Due to a high need of availability and a high amount of transferred data, the expansion of the 5G cellular network plays a key role for the development of an infrastructure for autonomous vehicles (Kong et al. 2017).

9.3 Methodology

The technology roadmap methodology is the applied methodology to achieve the aim of this chapter, which is to propose a solution for a standardized development platform for autonomous vehicles.

In literature, it is often stated that MOTOROLA applied a technology roadmap for the first time in 1984. MOTOROLA used this methodology as a planning tool for its strategic department with the aim to improve the positioning of its products in the market. According to MOTOROLA, applying the technology roadmap methodology allowed a cross-functional cooperation during the strategic planning process between different departments of the company, such as design, marketing, and development (Cho et al. 2016). In general, the technology roadmap methodology is mainly used as a visual-based strategic planning tool for technologies with the aim to determine the necessary efforts in research and development in the future (Geum et al. 2015). This methodology is used in companies, in government agencies, and in research facilities (Kostoff and Schaller 2001). Especially in complex and uncertain environments, the technology roadmap is used to visualize and coordinate activities and resources for the planning of future products and its technologies (Kostoff and Schaller 2001).

The topic of autonomous transportation is both complex due to many different involved players and still in an early development stage with the necessity of a lot of research to be conducted. A technology roadmap shall propose a development path of this industry for the next years.

In this chapter, the theory of the technology roadmap methodology is described as a preparation for its application in the succeeding chapter. This chapter is further divided into three sections. The first section is about fundamental definitions for technology development approaches. In the second section, the idea of the technology roadmap methodology is further discussed and illustrated. In the third section, the fundamental structure of the technology roadmap which is used to apply the proposed development of autonomous transportation in the succeeding chapter is presented.

9.3.1 Fundamentals of Technology Development in a Business Context

The competitiveness of many companies relies on their technology-based intellectual properties and products. For these companies, *technology* is defined as a key resource. Investments in technologies are part of their strategic planning (Moehrle et al. 2013). The previous years have shown in many industries that on the one hand, the development cycles of technologies are decreasing. This goes in hand with a necessary increase of innovative work to develop new technologies. On the other hand, the complexity of technologies is increasing. This development increases the need for strategic planning tools which supports the decision-making process in companies (Moehrle et al. 2013).

These topics together with defining the purpose of the company and its market situation are usually covered in a *technology management* department. The technology management department is responsible for the technologies developed by a company. This includes the identification of potential opportunities of new technologies, the selection and acquisition of technology relevant to the company, and the protection of technology (Moehrle et al. 2013). Since the term technology is above defined as a resource of a company, the meaning behind acquisition can also be understood as the recruitment of specialists for a field which is relevant for the development of a new technology. Thereby, the technology management department is playing a key role for core business processes such as strategic planning and operations, which create the value for the company (Moehrle et al. 2013).

The strategic planning of the technology management department can be described as *technology forecasting* of technologies which are relevant for the future business of the company. It projects the current knowledge about technologies to the future (Haddad and Uriona Maldonado 2017). However, the term technology forecast can be enhanced to *technology foresight*, which, by definition, not only includes the technology forecasting of the future, but also the social vision of the future, which affects the future business environment due to changing market trends (Moehrle et al. 2013).

The technology management department uses different methodologies and tools for technology foresight. One among them is the technology roadmap which is closer defined in the following.

The term "roadmap" is well known in the context of travel. It is used among travelers to find a path between a departure point and a destination point. The traveler is to a specific degree free to decide which path to take among the available alternatives. However, the traveler can also see on the roadmap what equipment he needs to travel along the chosen path. The roadmap provides the traveler with information about the environment, directions, and it provides certainty because the traveler can both rely on the information of the roadmap to a specific degree, and in case something unexpected happens during the journey, the traveler can decide for alternative paths according to the information of the roadmap (Kostoff and Schaller 2001).

In the business context, a *technology roadmap* is a visualization of the future strategic planning of technologies. It covers external and internal factors which might influence the development of technologies and it visualizes new opportunities due to interrelations which are not likely to be recognized without such a visualization (Moehrle et al. 2013). Furthermore, the technology roadmap provides information about the path of the company to develop technologies which are required by external factors, such as expected market requirements. It also shows the resources which are needed to achieve the planning of new technologies (Kostoff and Schaller 2001). The process of creating a technology roadmap in the business context is called *roadmapping*.

The technology roadmap is a framework which is not only used by the technology department of companies, but also for the technology foresight for an industry, for cross-industry applications, and on national and international level. A technology roadmap of a company focuses rather on a particular technology or product while a technology roadmap of an industry covers a larger field of research and focuses on technologies and products at higher level (Cho et al. 2016). The roadmapping at industry level is usually performed by the participation of several companies which are affected by the resulting technology roadmap. The participation of government agencies is possible as well, especially when the purpose of the technology roadmap is to secure the competitiveness of an industry of a whole country (Kostoff and Schaller 2001). Advantages of creating a technology roadmap on industry level are, for example, to increase the effectiveness of strategic planning and decision making, because the information is more credible as more stakeholders participate in the roadmapping. Furthermore, the research and development activities are more coordinated and align the companies of the particular industry (Moehrle et al. 2013).

In Canada, for example, small- and medium-sized companies within an industry section create a technology roadmap in collaboration. The aim of this strategy is that smaller companies can still compete with larger companies with innovative products. This process is supported by government agencies (Moehrle et al. 2013).

The Ministry of Economy Trade and Industry (METI) of Japan, for instance, develops technology roadmaps on industry level as well. The aim is to boost the economy with potential research projects from different perspectives (Yasunaga et al. 2009)

Another example for international industry-wide roadmapping is the International Technology Roadmap for Semiconductors (ITRS). Companies of the semiconductor industry from the USA, Europe, Japan, Korea, and Taiwan are developing a technology roadmap in collaboration every 2 years to set short- and long-term requirements for this industry sector. International universities, research institutes, and about 800 experts are involved in the roadmapping (Kajikawa et al. 2008).

So far, the basic concept of technology management and of the technology roadmap process has been elaborated in this section. In the following section, the methodology of the technology roadmap is presented in more detail.

9.3.2 Theory and Applications of Technology Roadmaps

As it is described in the previous section, the technology roadmap is used as a path for companies for future strategic planning of products and the associated technologies. It is a visualized methodology that illustrates the relationships among business planning variables such as markets, technologies, and products over a chosen time. Furthermore, it supports the cooperation between research programs, development programs, capability targets, and requirements (Lee and Park 2005).

However, there is no fundamental definition of how a technology roadmap is presented. The methodology itself needs to be adapted according to the situation it is applied on. The composition of a technology roadmap differs between companies and industries (Lee et al. 2012). The definition of a technology roadmap is rather originating from a functional approach which relies on the purpose and targets of such a technology roadmap, as discussed in the previous section. But besides the function of a technology roadmap, it can be said that the result is usually visually presented (Lee and Park 2005).

An example for the basic structure of a common composition of a technology roadmap is presented in Fig. 9.1. As it is shown in Fig. 9.1, a technology roadmap is a two-dimensional model. The horizontal axis represents the time. A time interval between now to 10 or 15 years in the future is usually chosen (Moehrle et al. 2013). The vertical axis consists of different layers with different topics which are assessed during the roadmapping. Potential layers are shown in Fig. 9.1 as well. Layers represent either the external environment, for example, the market, which represents the demand side and sets the requirements for the roadmapping, or matters which belong to the internal business, such as products, services, or technologies. Connections between the layers indicate dependencies between the defined events on each layer.

The layers that represent the internal business indicate the current and planned capabilities of the company and indicate the supply side of the methodology (Moehrle et al. 2013). Drivers for the demand side can be described as capturing value because the purpose of the technology roadmap is to comply with the projected customer needs which are represented by the market requirements. Value

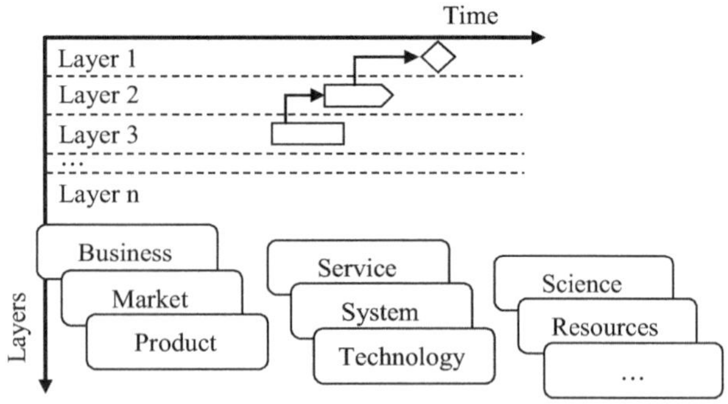

Fig. 9.1 Exemplary design of a technology roadmap with common layers (Moehrle et al. 2013)

creation, however, belongs to supply-side drivers and describes the built-up of intellectual property with resulting innovative products which increase the capabilities of a company (Phaal et al. 2011).

The most simplistic technology roadmap which is often described in literature consists of the layers market, products, and technology (Kajikawa et al. 2008).

The roadmapping is usually performed either in a top-down approach or in a bottom-up approach (Yasunaga et al. 2009).

The top-down approach is a synonym for a market-pull approach. In this case, the external environment represented by the market layer has to be created as a starting point. In case the potential market is unknown, a preceding internal step of defining the business layer needs to be conducted to define the strategy and capabilities of the company. The outcome of such a business analysis is the targeted market (Lee and Geum 2017). Drivers such as, among others, social, economic, environmental, and political, of the relevant market are predicted for the timeline of the roadmap (Phaal et al. 2011). Having defined the predicted market needs, the products or services need to be defined which would comply with the market needs and fit into the business context. Based on this, the layers are defined which are essential to satisfy the needs of the products or services. These layers are usually categorized as systems or technologies. The last step is to find out the science and resources which are necessary for research and development efforts to develop the technologies and systems that are defined in the preceding layer (Hansen et al. 2016). In conclusion, the market-pull approach is a requirement-based approach that analyzes the necessary steps to be able to accomplish another step (Moehrle et al. 2013).

In comparison to the top-down approach, the bottom-up approach can be described as a technology-push approach. This approach turns the procedure of the market-pull approach around. The first step is to define the resources and technology layers to analyze the current and estimated capabilities of the company (Moehrle et al. 2013). Based on that, products are predicted which require suitable market drivers for being sold. This approach focuses more on opportunities of the company, while the top-down approach focuses more on the needs of the customer (Hansen et al. 2016).

The creation of a technology roadmap is in general conducted with a group of experts. The people of this group should have different backgrounds for a "diverse composition" (Kostoff and Schaller 2001). Within a company, this group usually includes members from the strategy department, from the research and development department, and as well from external consultants. For creating technology roadmaps of an industry sector, this group consists usually of employees from participating companies, of people working for affected governments, and of researchers from different institutes of the specific industry field (Kajikawa et al. 2008). This expert-based roadmapping establishes a technology roadmap that is customized to the context which is analyzed (Lee and Park 2005). However, the technology roadmap which is created in such an environment is difficult to update and maintain (Lee and Park 2005). The assumptions which are made for the technology roadmap are valid for the time the roadmap is created. At a different point in time, these assumptions might not be valid anymore. The same might apply for the opinions of the experts who created the assumptions based on their current knowledge. If the roadmap is updated, many of the correlations and dependencies have to be updated as well, which might lead to a point at which it is more feasible to create a new technology roadmap. On the contrary, computer-based approaches to create a technology roadmap are used to increase the maintainability of the technology roadmap (Lee and Park 2005). This approach is practical for forecasting products within an industry of short product life cycles, duo to the expected regular changes of external factors (Lee and Park 2005). The algorithms of the software replace the different opinions that were stated before by the experts which might also lead to a limited result of the technology roadmap (Lee and Park 2005). Furthermore, the software for computer-based roadmapping needs to be adapted when the context of the technology roadmap changes.

As it is mentioned before, the roadmapping itself is a rather qualitative process. This becomes challenging when a diverse group of experts is trying to include the opinions of each expert based on their targets and experiences. To encounter this, tools from strategy and decision making are included in the roadmapping with the aim to increase the quantification of the process.

The planning for the business and marketing layers can be conducted by management tools such as the Porter's five forces (Lee and Geum 2017), the SWOT (strengths, weaknesses, opportunities, threats) analysis (Hansen et al. 2016), and the STEEP (social, technological, economic, environmental, political) analysis (Cho et al. 2016). The purpose of these tools is mainly to support the analysis of the expected business development with the relevant market including external factors which cannot be influenced by the company.

A gap analysis is applied to analyze the technology and system layer in a more quantitative way (Cho et al. 2016). The gap analysis indicates the difference of the current state of the technology or the system and the development state they need to be in to comply with the requirements (Lee et al. 2012).

Furthermore, decision-making tools can be included in the roadmapping to prioritize items. For example, each individual of the diverse group of experts that is involved in the planning of the technology roadmap might have a different idea for products to satisfy the market needs and for technologies to comply with the product requirements. Tools such as Hierarchical Decision Model (HDM) (Daim et al. 2018) and Quality Function Deployment (QFD) (Lee et al. 2013) are used to determine the most relevant items. These tools consider the different opinions of the experts and evaluate them according to the specified weighting.

As mention before, technical roadmaps are usually adapted to the specific context on which they are applied to. One possibility is, for instance, to adapt the layers. Furthermore, the methodology of the technology roadmap itself can be extended by different methods. The aim of this is to enhance the roadmapping by context-related methods, because the less general a technology roadmap is developed, the more strategic and the more credible the result is (Moehrle et al. 2013). The credibility of a roadmap in general is difficult to determine. Factors such as an existing cooperation between the team which develops the technology roadmap and the team which uses it (Lee et al. 2012) or that relevant experts have participated in the roadmapping can be taken into account (McDowall 2012). However, the credibility is significantly increased by stating reasonable assumptions and applying methods to the process (McDowall 2012).

One such method is described by RINNE which is about looking at the past (Rinne 2004). The aim of this strategy is to discover which approaches have been successful in the past (Rinne 2004). PHAAL proposes a similar approach which includes that the time axis of the technology roadmap is extended to the past and uses it as a reference point to make the future strategy compatible (Phaal et al. 2011).

JEONG developed a method that includes a patent analysis to identify the trajectory of technology development. This method is based on neural network and forecast patents that are likely to be registered in the near future (Jeong et al. 2015).

The method which is presented by GEUM is called Association Rule Mining (ARM) and is based on data mining. The ARM method identifies relationships between the items of different layers of the technology roadmap by analyzing keywords from business documents. This method enhances quantifying decision models such as the QFD in prioritizing the items of one layer due to their influence to the items of another layer (Geum et al. 2015).

LEE presents a method which consists of a Cross Impact Analysis (CIA) and an Analytical Hierarchy Process (AHP). The CIA is used to measure the impact of external environments for the market requirements by calculating the likelihood of the occurrence of specific future events. AHP is rather comparable to a decision-making process and used for the product and technology layer to decide which products and technologies to be developed (Lee and Geum 2017).

9.3.3 The Structure of the Adapted Technology Roadmap for Autonomous Driving

The structure of the technology roadmap to propose a standardized platform for autonomous driving within the scope of this thesis in Chap. 4 is based on the fundamental structure shown in Fig. 9.1. The technology roadmap consists of four layers, which are market, product, technology, and resources and is displayed as a schematic illustration in Fig. 9.2. This structure of a technology roadmap is in literature

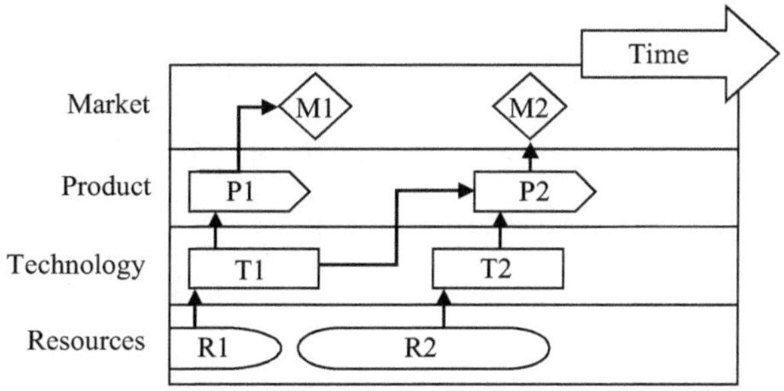

Fig. 9.2 Structure of the adopted technology roadmap

defined as a "common framework" (Kajikawa et al. 2008). The purpose of this technology roadmap is to give an overview of the market and of the current available and predicted technologies and products.

The technology roadmap is developed by using the top-down approach. After analyzing the expected market drivers for the relevant timeline, the products are derived which satisfy the expected market drivers. Based on this, the technologies are derived, which are needed for the functionalities of the product features of the products. The last step is to analyze the resources. The resources include efforts from companies, research institutes, and the government to develop the required technologies (Phaal et al. 2011).

As it is indicated in Fig. 9.2, connections are used to show the correlations between the layers. Each technology is connected with the resources that are needed to develop the specific technology. The product features, which are part of the products, are connected with the technologies which are needed to ensure the required functionalities. The products with their product features need to comply with the relevant market requirements and are connected accordingly. The most important aspect considering the correlation between the layers which represent internal business matters and the market layer is the available capability of developing required technologies and the timing (Moehrle et al. 2013). Market drivers might change after some time because it loses its relevance due to other market drivers that evolve. This leads to a situation in which products cannot satisfy the market drivers anymore because the market either does not need these products anymore or perceive these products for granted. For this reason, the research and development of technologies needs to be finished early enough to be able to implement the derived technologies into the products, so the products itself are available when required by the market (Moehrle et al. 2013).

The quality function deployment (QFD) tool is used to quantify the content of the technology roadmap to prioritize product features and technologies. A theoretical model of a QFD is shown in Fig. 9.3. The QFD tool which is integrated into the

Fig. 9.3 Fundamental structure of a matrix for a QFD

		Correlating parameters				
		AA	AB	AC	AD	
Weight		k_1	k_2	k_3	k_4	Priority
Parameters to be evaluated	BA	h	l			$h \cdot k_1 + l \cdot k_2$
	BB		m	h		$m \cdot k_2 + h \cdot k_3$
	BC	m	m			$m \cdot k_1 + m \cdot k_2$
	BD			l	h	$l \cdot k_3 + h \cdot k_4$

roadmapping helps to prioritize the content of different layers which relate to each other.

The QFD helps prioritize one set of parameters, in Fig. 9.3 described as *parameters to be evaluated*, by correlating it to another set of parameters, the *correlating parameters*. The correlated parameters are weighted according to their importance. Afterward, every parameter of both sets is compared with each other to determine whether there is no, a low, a medium, or a high correlation between them. Multiplying the correlation factors with the weight of the relevant parameter and summing these values for each parameter up results in the priority value of the parameter to be evaluated. According to the priority value, the parameter set which is to be evaluated can be structured in groups of high, medium, and low priority.

One application, for example, is to evaluate the priority of technologies in correlation with the product features. The result of this analysis is to prioritize the technologies to focus on for the development.

9.4 Application of the Methodology

The application of the introduced technology roadmap in Sect. 3.3, for proposing a standardized platform for autonomous transportation, is presented in this chapter. The procedure is based on the pre-existing technology roadmap about this topic (Alramadan et al. 2016). However, since this technology roadmap has been created in the year 2016, the presented technology roadmap in this chapter is an updated version of the existing technology roadmap.

Many product features and technologies which are presented in the technology roadmap of 2016 have been established by now. The updated technology roadmap focuses on the remaining technology gaps and the current focus of research in this topic as it is introduced in Chap. 2. According to this, the timeframe of the technology roadmap has been adapted as well. The duration of the timeframe is between the current year, 2020, and the year 2030.

The technology roadmap itself is representing the industry sector of autonomous on-road transportation. It can be seen as a proposal for the industry and a forecast based on literature research about the future requirements. The technology roadmap is only based on literature research and on the input of the student group which developed the technology roadmap in 2016 (Alramadan et al. 2016).

This chapter is structured according to the introduced technology roadmap in Sect. 3.3. The following sections will follow the procedure of the roadmapping which is structured according to the creation of each layer according to the top-down approach. This means that the following section starts with establishing the expected market drivers, followed by the section in which the product analysis is presented. In the third section, the creation of the technology layer is discussed and in the fourth section, the resource layer is elaborated. In the last section of this chapter, the roadmapping is summarized and discussed with the created technology roadmap. The finalized technology roadmap is shown in Appendix B.

9.4.1 Market Analysis

The roadmapping is based on the top-down approach. Since the targeted market sector is already known, which is the market for autonomous on-road transportation, the first step is to find the relevant market drivers for the upcoming 10 years.

Table 9.2 shows these market drivers which are resumed from the technology roadmap from 2016 (Alramadan et al. 2016). The market drivers are allocated to specific categories which are *saving time on the road*, *safety*, *environmental drivers*, *economical drivers*, and *consumer interest*. They summarize the market drivers according to business models such as STEEP (social, technological, economic, environmental, political) (Cho et al. 2016) and context-related drivers which represent the expectations of autonomous transportation (Anderson et al. 2016). The latter aspect mainly applies to the categories *saving time on the road*, *safety*, and *consumer interest*.

The listed market drivers describe aspects which are expected to influence the market demand in the future which results in the requirement of a product or a service that can satisfy the market demand. The market driver *population growth* (D1), for example, is a driver that is based on the assumption of both the general growth of a population, and a continuously increased movement of people moving to urban areas and increasing the population density in such areas (Maurer et al. 2016). This market driver evolves in the development of other market drivers which are also listed in Table 9.2, such as *traffic jam reduction* (D2), *reduce parking search time* (D4), and *space management for vehicles and pedestrians* (D5). With an increase of the population density in specific areas, these factors are expected to worsen which creates the need for products and technologies to improve that situation.

Table 9.2 Market drivers sorted in categories and weighted

Category	Code	Market driver	Weight
Saving time on the road	D1	Population growth	4
	D2	Traffic jam reduction	4
	D3	Managing accidents effectively	2
	D4	Reduce parking search time	4
	D5	Space management for vehicles and pedestrians	2
Safety	D6	Accidents – prevention and report	4
	D7	Traffic prediction	2
	D8	Pedestrian safety	4
Environmental drivers	D9	Reduce CO_2	2
	D10	Reduce number of cars on the road	1
	D11	Reduce environmental destruction	2
Economical drivers	D12	Fuel price increasing	2
	D13	Insurance cost	1
Consumer interest	D14	Driver's ability to multitask	4
	D15	Security	4

Another point is that the development of autonomous transportation needs to convince the customers by its capabilities to buy an autonomous vehicle over a non-autonomous vehicle. In this context, this could be due to safety reasons, because a machine prevents human errors, and the incentive to increase the efficiency during driving. While driving an autonomous vehicle, for instance, the driver can focus on other topics such as work or entertainment (*driver's ability to multitask* (D14)) (Anderson et al. 2016). However, if at one point, the number of autonomous vehicles increases because more customers are able and willing to buy such a vehicle, interfaces of autonomous vehicles become more attractive for hackers to manipulate the system. This creates the market driver for *security* (D15) at a specific point in time (Johanning and Mildner 2015).

Other market drivers such as *reduce* CO_2 (D9), *reduce environmental destruction* (D11), or *fuel price increase* (D12) describe expected developments of the market as a projection from today's perspective. There are products and technologies that fit into the concept of autonomous transportation and can counteract the beforementioned drivers as an improvement of the environmental situation. This might be, for example, that autonomous vehicles can drive together in close convoys with a short distance between them. The environmental drivers are expected to significantly benefit from the usage of the wind shadow in such settings (Fernandes and Nunes 2012). Furthermore, the beforementioned setting decreases the necessary space on the road which leads to a decreased need for further development of traffic infrastructure (Fernandes and Nunes 2012).

As it is shown in Table 9.2, a weight is assigned to each market driver to describe its significance of the market driver. A value of *one* describes a low significance, a value of *two* describes a medium significance, and a value of *four* describes a high significance. The weights are used for the further roadmapping to quantify the process and prioritize the market drivers to focus on.

The expected temporal occurrence of the relevant market drivers is presented in the technology roadmap in Appendix B. The temporal arrangement is further explained in Sect. 4.5.

9.4.2 Product Analysis

After deriving the expected market drivers, the next step in the chosen top-down approach is to develop ideas for product features which comply with the market drivers.

First of all, product categories are defined which are used as a generic term for product features to cover the required functionalities based on the market drivers in Table 9.2. The defined product categories are based on the technology roadmap from 2016 and are listed in the following:

- P1: Connectivity
- P2: Smart

- P3: Safety
- P4: Usability

In the next step, product features are defined which directly comply with the market drivers in Table 9.2. Table 9.3 lists the product features with its associated product categories. Product features can cover more than one market driver. For example, the product feature *auto parking searching function* (PF4) seems to be a direct fit to the market driver *reduce parking search time* (D4). However, this product feature can as well impact the market driver *traffic jam reduction* (D2), because it is proven that a lot of traffic in the city center is created from people searching free parking spaces (Shoup 2007). Adapting to that, the market driver *fuel price increasing* (D12) is affected by this product feature as well. Since the autonomous vehicle does not have to cruise and search a free parking space, fuel is saved. It has to be emphasized at this point that the product features only try to satisfy related market drivers without further analysis. The fact that autonomous vehicles most likely have an electric power train is neglected at this point. Furthermore, this product feature can also impact the market driver *space management for vehicles and pedestrians* (D5) because the autonomous vehicle can because of its programming algorithm only park in assigned parking spaces.

To structure the derived product features, a Quality Function Deployment (QFD) analysis is set up with the aim to prioritize the product features. The QFD itself is displayed in Fig. 9.4 in Appendix A. The results of the QFD are displayed in Table 9.3 as well. The QFD has been updated from one of the pre-existing technology roadmaps (Alramadan et al. 2016). The update is performed based on relevant topics, which are about safety, security, and communication, that are elaborated in the literature review in Chap. 2. As it is indicated in the previous paragraph, the product features are compared with the weighted market drivers. If a product feature satisfies a market driver, a correlation is established between the product feature and the specific market driver. The correlation itself is evaluated by its assigned value

Table 9.3 Product features with related product categories (Connectivity (P1), Smart (P2), Safety (P3), and Usability (P4)) and calculated priorities from the QFD in Fig. 9.4

Code	Product feature	Product category	Priority (QFD)
PF1	Intelligent merging	P1, P3	20
PF2	Reduced distance between vehicles	P1, P2, P3	16
PF3	Auto accident reporting and tracking system	P1, P3	19
PF4	Auto parking searching function	P1, P2	*26*
PF5	Energy saving mode	P2, P4	8
PF6	Road elements sensors	P1, P2, P3	*28*
PF7	Sense traffic lights and calculate quickest route	P1, P2	*28*
PF8	Automatic recognition of road laws	P1, P2, P3, P4	17
PF9	Protection against hacking	P1, P2, P3	*32*
PF10	Passengers health monitoring	P1, P2, P3, P4	4
PF11	Feature to adapt to weather condition	P2, P3, P4	*27*
PF12	Object identification/avoidance	P2, P3, P4	*41*

between the two items. The value *one* represents a weak correlation, the value *two* represents a medium correlation, and the value *four* stands for a strong correlation. No value means that there is no correlation or a negligible correlation between the product feature and the specific market driver. With the correlation values and the weights, the priority of each market feature is calculated according to the explanation in Sect. 3.3.

The prioritized product features are divided into three groups. This is conducted according to the resulting priority values. Product features with a priority value of less than ten are low prioritized and values above 25 are assigned to a high priority. The product features with a high priority are further used for the roadmapping, while the other product features are neglected. The product features with the high priority are displayed in bold font in Table 9.3.

Furthermore, based on the QFD and the prioritization of the product features, the market drivers can be sorted as well. Market drivers, that according to Fig. 9.4 only correlate with product features which are not assigned a high priority, can be neglected because it is assumed that the market drivers are not satisfied by the remaining product features due to the missing correlation. The same is assumed for the correlation between market drivers and product features with a low correlation value assigned.

9.4.3 Technology Analysis

The technology analysis is performed as the next step after determining and prioritizing the product features. During the technology analysis, the technologies are derived which are needed to comply with the functionalities of the product features with a high priority.

Therefore, similar to the product analysis, technology categories are derived which are a generic term of technologies needed to comply with the functionalities of the relevant product features. The technology categories are based on the ones from the pre-existing technology roadmap as well (Alramadan et al. 2016) and are listed in the following:

- TC1: Software systems
- TC2: Objects recognition technologies
- TC3: Artificial Intelligence (AI)
- TC4: Infrastructure
- TC5: Road elements technologies

According to the defined technology categories, technologies are derived to satisfy the needs of the product features. The derived technologies with the corresponding technology categories are listed in Table 9.4. These technologies are resumed from the pre-existing technology roadmap as well. Some technologies are assigned to more than one category. For example, the technology *smart broadcasting road objects* (T9), which represents the needed technology for smart traffic lights, smart traffic signs, or smart parking spaces, is assigned to the technology category *infrastructure* (TC4) and *road elements technologies* (TC5). The impact on the infrastructure is that it has to be adapted to work digitally.

Hereinafter, the listed technologies are analyzed by conducting a gap analysis. The purpose of the gap analysis is to determine the current status of the technology and the requirement of the technology to cover the functionalities of the product features (Cho et al. 2016). The difference between the required state of the technology and the current state of the technology is the so-called gap. The information of the gap is needed to determine the necessary resources to develop the technology or to improve the level of development of the technologies. The gap analysis has been updated compared to the one from the pre-existing technology roadmaps based on the literature review in Chap. 2.

For example, the technology for communication platforms between vehicles (*V2V*) or between a vehicle and an object (*V2X*) (T2) is currently developed to a state at which it can be integrated into vehicles. But one current restriction for its usage is the reliability of the connection (Wang et al. 2017). However, this technology is not only needed for optimizing the driving characteristics by, for instance, adapting the velocity according to the traffic lights signals ahead, but also for safety purposes. This technology shall be used, for example, to communicate with other vehicles about planned maneuvers so the surrounding vehicles can adjust accordingly (Johanning and Mildner 2015). For this reason, the reliability of this technology needs to be increased as it is stated as the gap for T2 in Table 9.4.

A QFD is applied to structure the derived technologies in Table 9.4 and to prioritize the technologies to be able to allocate the resources accordingly. The QFD is displayed in Fig. 9.5 in Appendix A. The target of the prioritization is different to the one conducted for the product features in the previous step of the roadmapping. The target is not to sort out technologies with low priorities but to know which technologies should be developed with a higher focus.

Therefore, weights are assigned to the product features. The approach is the same as prioritizing the market drivers with the assigned weights in Table 9.2. As it is presented in Fig. 9.5, the product features *object identification and avoidance* (PF12) and *protection against hacking* (PF9) are prioritized. This decision is mainly based on the high amount of effort to be performed to develop these product features according to literature research. The light detection and ranging (LiDAR) sensors, for instance, have difficulties to identify smaller objects, such as plastic bags. To solve this problem, the V2X is currently developed to store such information about objects on the street in databases which can be accessed by other vehicles (Kong et al. 2017).

The correlation values between the technologies and the product features are divided into three categories for a low, medium, and high correlation. The difference to the QFD for prioritizing the product features against the market drivers is that all technologies are to a certain extent correlated with the product features in the technology QFD. However, the technologies which are not essential for the product features but would support the functionalities of the products are in general marked with a low correlation. The technology *smart broadcasting road objects* (T9), for

Table 9.4 Necessary technologies for product features with defining the gap, the associated technology groups, and the priorities calculated in the QFD in Fig. 9.5

Code	Technologies	Need	Current status	Gap	Cat.	Priority (QFD)
T1	Accurate sensors	Reliable and affordable sensors	Sensors are not 100% accurate	100% accuracy	TC2	28
T2	V2V; V2X	A robust standard system for cars to communicate	System developed, but not 100% reliable	100% reliability	TC1	28
T3	LiDAR sensors	Ability to identify small/moving objects	LiDAR technology is not 100% accurate	100% accuracy	TC2	23
T4	AI (road decision-making system)	Analyze the feed from various sources and make decisions	Under development (learning)	More road testing for deep learning algorithm	TC1, TC3	26
T5	Element identification system	Analyzing the sensors around the vehicle (cameras, radar, etc.)	Under development (learning)	More road testing for deep learning algorithm	TC2	30
T6	Hotspot wireless (5G)	Cars always connected to fast internet	No 100% coverage	Enhance availability of hotspot wireless (radio tower)	TC4	21
T7	Central road technology system	A system that contains information about the roads, vehicles, law, etc.	No such system	Develop the system and collect data	TC1, TC5	17
T8	Night vision (laser)	Ability to identify objects accurately in the night	Current night vision technology is good but not 100% accurate	100% accuracy	TC2	28
T9	Smart broadcasting road objects	A broadcasting object that says: "I am a curb" or "I am a traffic light"	A few traffic lights are already connected to the internet	Digitalize all road objects and traffic lights	TC4, TC5	29
T10	Accident assessment system	A system to estimate and react to accidents	A few traffic lights are already connected to the internet	Add more features and connect it to emergency center	TC1, TC3	12

example, is assumed to have a high correlation with the product features *road elements sensors* (PF6), *feature to adapt to weather condition* (PF11), and *object identification and avoidance* (PF12). Smart broadcasting road objects are part of smart traffic lights and of smart objects that, for instance, mark road works, and have to be avoided by the vehicle. Furthermore, these sensors can analyze the condition of the road and warn about dangerous conditions such as icing. It is decided that T9 has a medium correlation with the product feature *sense traffic lights and calculate quickest route* (PF7) because it is assumed that this product feature is mainly driven by other technologies which provide a central supervision of the traffic and its flow, such as the *central road technology system* (T7). The correlation to the other product features is assumed to be low. For example, there is no direct correlation with the product feature *protection against hacking* (PF9) visible. However, it could be possible that the sensor influences the behavior of the vehicle via an emergency communication channel. This might be advantageous if the vehicle cannot control itself anymore because the system is manipulated. If the sensor realizes that, it warns the vehicle system when it gets too close with a high velocity to an object that should be avoided. Via this emergency communication channel, the vehicle can be forced to stop and to be shut off.

The prioritization of the technologies is divided into three categories. Low prioritized technologies have a priority value of less than 18, while technologies with a high prioritization have a priority value of more than 27. Technologies with a high priority are *accurate sensors* (T1), *V2V and V2X* (T2), *element identification system* (T5), *night vision* (T8), and *smart broadcasting road objects* (T9).

9.4.4 Resource Analysis

The last step of the roadmapping is, according to Sect. 3.3, the resource analysis. The resource analysis gives an overview about the required efforts which are essential to close the gaps of the technology analysis. Some suggested and common resources based on the pre-existing technology roadmap (Alramadan et al. 2016) for this technology roadmap are listed in Table 9.5.

As it is shown in Table 9.5, the resources are structured into categories. One of these categories is the field of *research and development*. This category is mainly focused on the creation of something new or the improvement of something existing. One basic part of research and development is the collaboration with research institutes at universities. Especially institutes of university with a focus on autonomous transportation are involved in fundamental research work for technologies which are newly developed in this area. In the context of autonomous transportation, computer hardware and computer software companies are important resources in research and development because they have a high potential to develop technologies which are based on computer technologies such as the V2V and V2X or smart broadcasting road objects (Herrmann and Brenner 2018). Furthermore, conventional car manufacturers play a key role as well. Their task as integrators is to verify these technologies when they are integrated into the actual system and applied to everyday scenarios. Besides that, knowledge of other fields such as the space industry, for example, can support the research and development of technologies due to their experience with autonomous vehicles on other planets.

Table 9.5 Resources with associated categories and technologies

Category	Code	Resource	Example	Tech. code
R&D	R1	Partnership with university	Tsinghua Univ., Stanford Univ., Oxford Univ.	T2/T4
	R2	Computer hardware	Intel, Nvidia	T2/T4/T6/T10
	R3	Computer software	Google, Apple	T2/T4/T5/T7/ T9/ T10
	R4	Car manufacturer	Tesla, Mercedes, BMW, Audi, GM	T1/T2/T3/T5/ T6/ T8/T10
	R5	Space research institutes	NASA	T1/T3/T5/T8
	R6	Cabinet departments	Dep. of Energy, Dep. of Transportation	T6
Incentives	R7	Government	Federal, State, Local	T5/T6/T7/T9/T10
	R8	Industries	Google, Tesla, Baidu Apollo	T1/T2/T3/T4/T9
Standards	R9	Standards	SAE	T1/T2/T3/T4/T5/ T7/T9/T10

Another category of the resources is *incentives*. Incentives are external influences which might affect the development of technologies. The government, for example, can release laws with the motivation to shift the research to specific fields. For instance, if the government increases the tax for oil, the incentive to buy an electric vehicle increases. With an expected increase of customers buying electric vehicles, the industry is shifting toward this field. Incentives, however, can be created within the industry as well. For example, TESLA disrupted the market of on-road vehicles with their electric vehicle fleet and their strong efforts to push autonomous transportation (Herrmann and Brenner 2018). This might create the incentive for other car manufacturers to adapt to this development.

The last resource category is standards. Standards within the transportation industry are set up, among others, by the SOCIETY OF AUTOMOTIVE ENGINEERS (SAE). They provide guidelines which are referred to when approving a new technology or a new system within a vehicle for the everyday usage.

9.4.5 Roadmapping

In this section, the creation of the technology roadmap based on the creation of each layer in the sections before is further elaborated and visually assembled to the template of the technology roadmap derived in Sect. 3.3. This technology roadmap is displayed in Appendix B.

The technology roadmap is structured into the four topics represented by market drivers, product features, technologies, and resources. The introduced categories for each topic are displayed as well and are either structured by using different colors for the categories within one layer or by structuring it spatial. The items of each topic are displayed according to the displayed timeframe. All the boxes which represent the considered market drivers, product features, technologies, and resources are marked with the corresponding code, which is shown in Tables 9.2, 9.3, 9.4, and 9.5 for further reference.

The market layer of the technology roadmap is structured according to the categories listed in Table 9.2. However, not all market drivers are displayed in the technology roadmap. Only the market drivers, which are defined as relevant, are displayed. This is decided according to the QFD in Fig. 9.4. Market drivers that share a medium or high correlation with high prioritized product features are assumed to be relevant and are displayed in the technology roadmap.

The arrangement of the market drivers in the technology roadmap is based on different assumptions. Some requirements of the market are always assumed to be relevant. The required technologies serve in this case the purpose to optimize the product features which comply with these market drivers. For example, the market drivers *traffic jam reduction* (D2) and *reduce* CO_2 (D9) are assumed to be always relevant and can never be fully satisfied. Other market drivers, for instance, *reduce parking search time* (D4) are requirements, which are satisfied after the corresponding product feature has been introduced. The chosen time is a compromise between the expected time the market requires this feature and the readiness of the technology. For this feature, the market demand can already be assumed to be significant, but it is not visible yet because the technology itself is not ready to provide the functionalities for the product features to cover this demand. This demand has to be created by further research and technological development. Other market drivers such as *pedestrian safety* (D8) and *driver's ability to multitask* (D14) can be counted to the beforementioned first group of market drivers, which are always assumed to be relevant. However, these market drivers are rather counted to so-called hygiene factors. In the beginning, these market drivers are relevant, but after some time, the technologies and products are assumed to be taken for granted. Considering D14, it is expected that it is possible to do different kinds of activities while the autonomous vehicle is performing the driving task. However, if this is not possible anymore, this market driver may become relevant again.

The development of the product layer is based on the assumptions made for the market layer. The product features need to be available when they become relevant due to the expected requirements of the market.

The displayed product features in the technology roadmap in Appendix B are the ones which result in a high priority after performing the QFD in Fig. 9.4. Each product category is assigned to the product feature by the associated color. The product features are connected with correlating market drivers. Based on Fig. 9.4, market drivers and product features with a high correlation are connected with a solid arrow. The ones with a medium correlation are connected with a dotted arrow. The connections of the ones with a low correlation are not displayed to not overload the technology roadmap. As mentioned before, the product features are supposed to be available at the time the market requires it.

However, the product feature *sense traffic lights and calculate quickest route* (PF7), for instance, is expected to be introduced in the second half of the upcoming decade. This product feature has a high correlation with the market driver *traffic jam reduction* (D2), which is, as mentioned before, a market driver which is assumed to be always relevant. Furthermore, different approaches can be used to further improve this market driver. But the approach represented by PF7 is not assumed to be introduced earlier. Other product features such as *object identification and avoidance* (PF12) are available now, but they have a correlation with the market driver *driver's ability to multitask* (D14), which is assumed to be relevant in 2030. PF12 is a fundamental feature for autonomous driving. Without that, the market would not develop the demand that the driver can focus on other tasks while the vehicle is taking over the driving task. Another point to be mentioned is the later introduction of product features. *Protection against hacking* (PF9), for example, is introduced at a point while one of the correlating market drivers, *pedestrian safety* (D8), is already considered to be relevant. In this case, the product feature would further improve the situation and adds features to comply with the requirements.

The technology layer considers all technologies listed in Table 9.4. In the technology roadmap in Appendix B, the items representing the technologies are assigned to the categories by the associated color. The boxes representing the technologies are connected with correlating product features to visualize the associations between them. Solid arrows are used to visualize the association between technologies and product features which have, based on the QFD in Fig. 9.5, a high correlation and dotted arrows are used to visualize the associations with a medium correlation. The visualized association of low correlations is not included in the technology roadmap to establish a clear arrangement.

Based on the temporal arrangement of the technologies, many technologies are available today or are assumed to be implemented soon. This is mainly due to the further development of relevant technologies in this sector. In comparison to the technology roadmap from 2016 (Alramadan et al. 2016), some of the technologies which are available today were predicted in that roadmap to be introduced between 2016 and 2020.

The length of the boxes in the technology analysis layer of the technology roadmap in Appendix B represents the relevant timeframe for the development of each specific technology. At the time the box starts, it is assumed that this specific technology is introduced. During the temporal duration of the box, the technology is further developed and significantly improved to satisfy all the requirements for the needed functionalities of the product features. For example, the technology *road decision-making system based on AI* (T4) is assumed to be introduced in the next years and is significantly developed until around 2030. One relevant project feature that correlates is the *auto parking searching function* (PF4) which is assumed to be introduced in a similar time. T4 can provide PF4 with information for the vehicle to choose and assign a free parking space nearby according to the information of the other vehicles which search a free parking space in that specific area as well. Furthermore, T4 is also correlating with the product feature *sense traffic lights and calculate quickest route* (PF7) which is assumed to be required in the end of the upcoming decade. T4 is supposed to be further developed to a stage where it cannot only statically assign free parking spaces but also dynamically can influence the whole transportation system including traffic lights, traffic signs, and the driving behavior of the vehicles itself.

Lastly, the resource layer is created. The resources are arranged according to the categories listed in Table 9.5. The connection between the relevant resources and the corresponding technologies is visualized with arrows in the technology roadmap in Appendix B.

Since the technologies of autonomous transportation itself have evolved to a certain state and are getting more and more present in everyday life, all the considered resources are already involved in the development process today. Research and development activities are performed for all considered technologies, incentives are created and discussed, and standards are proposed, such as the SAE J3016 (SAE International 2014).

The technology roadmap, which is developed and described in this chapter and displayed in Appendix B, is based on the methodology which is derived in detail in Sect. 3.3. The technology roadmap proposes the further development of the autonomous on-road transportation from the point of view of the year 2020. It is not a completely new field anymore, but the development of autonomous transportation is still in an early stage. The advantage is that many boundary conditions are known at this development stage which are included in this technology roadmap.

9.5 Potential Risks of Implementing the Technology Roadmap

After creating the technology roadmap for a standardized platform for autonomous on-road vehicles based on literature research, potential drivers need to be discussed which might be a risk to the occurrence of the assumed events in the technology roadmap.

The discussions about such drivers are in this chapter focused on external factors which influence the market drivers. Hansen (Hansen et al. 2016) and Lee (Lee and Geum 2017) write about an enhancement of the technology roadmap methodology by adding different scenarios to the process. These scenarios consider different future events with a likelihood to occur. Especially market drivers are difficult to predict in the long-term. For this reason, scenarios are created which predict different market situations with an assigned probability of occurrence. Each scenario influences the resulting required product features and technologies, so the technology roadmap has to be prepared and adapted for each of the considered scenarios.

In the scope of this chapter, scenario-based roadmapping is not performed. Three risk factors affecting the market drivers are chosen and their potential impact on the technology roadmap and on the development of the industry is discussed. Therefore, this chapter is further structured into three sections.

The first section includes the responsibility of the government. It is elaborated which impact the government has if no incentives for autonomous transportation are created. This includes, for example, the required cellular network expansion of 5G. The second section focuses on collaborations within the industry. It is elaborated what the advantages and disadvantages are if companies work together on specific topics. The last section is about the impact of the customers. It is discussed how the public perception of autonomous driving is influencing the development of this technology.

9.5.1 Responsibilities of the Government

The government has several responsibilities when considering the development of autonomous transportation. One of them is to regulate safety by introducing regulations and laws. The European Union, for example, published the strategy *Vision Zero* with the aim to have no fatalities in traffic in 2050 anymore (Maurer et al. 2016). This strategy recommends projects to perform research to develop technologies which help to achieve this target. If specific companies fail to achieve this, they can get penalized which might be defined in further regulations based on this strategy.

One common aspect which is based on this is the moral and judicial handling of pedestrian safety. For example, a group of children runs on the street in front of a vehicle. The vehicle would have to head toward the sidewalk to avoid an accident, but at the same time, an elderly couple is walking on the sidewalk. It has to be discussed and defined, whether it is morally applicable that the vehicle hits either of the pedestrians and whether the vehicle should be allowed to trade human life against human life. Based on such an ethical discussion, the government can propose regulations for standardization.

Another aspect which is related to that is the handling of data. The proposed technologies such as vehicle to vehicle communication (V2V), vehicle to object communication (V2X), a road decision-making system, and a central road technology system will require the exchange of a lot of data. However, some of this data can be traced back to the vehicle and to its owner. That is to say that the data security needs to be regulated by the government. However, one aspect that has to be taken into account is the different perception of data security in different countries. China, for example, is known to be much more liberal about handling personal data compared to Germany.

Furthermore, the government is obliged to continue to decrease the allowed CO_2 production for specific industries. The implementation of such restrictions is not only conducted by new technologies, but also by the development of new concepts.

The concept of Platooning, which is introduced in Chap. 2, is becoming more interesting. If vehicles drive together with a minimum distance between them, the energy consumption and the CO_2 production will decrease. However, Platooning is only possible with autonomously driven vehicles due to a required minimum reaction time which is not ensured in human drivers (Fernandes and Nunes 2012).

Lastly, the government needs to create incentives for an area-wide coverage of the 5G cellular network service, which is fundamental for the V2X technology and with that for the whole concept of autonomous transportation (Johanning and Mildner 2015). If the 5G network is only available in specific areas, autonomous transportation is only feasible in such areas. This may lead to a shift of the whole technology roadmap to a later point in time.

In conclusion, the government needs to step in to create incentives and to introduce new regulations which are fundamental for the development of technologies and products for autonomous transportation. The developed technology roadmap displayed in Appendix B is based on the assumption that the government is acting as proposed.

9.5.2 Collaboration within the Industry

The purpose of the technology roadmap is to introduce a standardized platform for autonomous transportation. This means that the proposed steps in the technology roadmap apply for all companies which are part of this industry. One fundamental aspect is that interfaces between technologies are created and published, so they can be accessed and used by technologies developed by competing companies.

For instance, the car manufacturers Audi and BMW currently develop their own product as a basis for an auto parking searching function. This is further elaborated in Chap. 2. However, after developing such product features, it should be made available to the whole industry. If every car manufacturer that develops such a product sticks to their own product, the market driver for reducing the parking search time might be satisfied from their point of view. But the customer might not see it as satisfied and would be facing different systems which might partly not be compatible to each other. Some operators of parking decks, for instance, might only work with the system from Audi, while others only collaborate with BMW. The owners of the parking deck have an interest that not only vehicles from one car manufacturer can use their parking decks but collaborating with all available systems might be too costly. A solution which is compatible for everyone would be in the interest for the whole industry sector.

Projecting this to other aspects such as safety, the outcome might be even more severe. As part of the target to achieve the *Vision Zero* as described in the previous section, car manufacturers have an interest in eliminating the fatalities in traffic. However, if one car manufacturer develops safety features to achieve this target, it most likely protects the passengers inside the vehicle. If this vehicle is safe, another vehicle produced by another car manufacturer might not have the same safety technologies and can still crash into the vehicle which is considered to be safe. That is to say that car manufacturers should not use safety technologies as a competitive advantage, because the system for autonomous transportation might fail based on that. Also, to create the ability that players in the market can compete with fundamental technologies of existent car manufacturers, the industry needs to collaborate in these fields to achieve the targets set by the government. All the market drivers which are assigned to the category safety are depending on such decisions. Safety is a crucial topic for the evolvement of the market of autonomous transportation which is further discussed in the succeeding section.

9.5.3 Public Perception of Autonomous Driving

The public perception of a product is one of the main influences of the market drivers. The customers who decide to buy a certain product do this according to their willingness and ableness. However, the willingness is influenced by many criteria, with one of them being the perception of the product in general. If the product performs not as predicted, the willingness of the customers to buy this product may decrease. This progress is fostered by product reviews which are easily accessible via social media channels today.

In the transportation industry, this effect has been recently provoked by the *Volkswagen emission scandal* (Amelang and Wehrmann 2020). After some time, it has been proved that not only VOLKSWAGEN, but also many other German car manufacturers were involved in this scandal. This scandal weakened the German car industry because customers lost their trust in the involved car manufacturers. Due to the public perception of this scandal, the market for German cars has decreased (Amelang and Wehrmann 2020).

Autonomous on-road transportation is a new product which is established. Public perception plays a key role in the evolvement of this new product. It is sensitive and can change with small impacts.

One topic which can have a high impact on the public perception of autonomous driving is safety. If an autonomous vehicle is not perceived as safe, people will not trust the technology of the vehicle. Even if it is statistically proven that autonomous vehicles might conduct less mistakes than human drivers, people still might not trust the vehicles if the general perception is that these vehicles are not safe (Herrmann and Brenner 2018).

To avoid a bad public perception due to deficits in safety, companies increase their focus on the topic of safety and reliability. This topic has been highly focused on in the aviation industry (Maurer et al. 2016). In aviation, this is necessary because flying is not a natural way of movement for human beings. With a high focus on safety and reliability, humans are trusting and using the product because it has been proven that the technology can be considered as safe. However, it is still allowed that an airplane can crash once every trillion flight hours with the assumption that none of the passengers survive. This is necessary, because, apart from unknown risks, safety cannot be guaranteed in general. Furthermore, a further improvement of safety aspects might not be feasible anymore. Such a compromise has to be found for autonomous transportation as well.

The system of autonomous transportation is more complex compared to aircraft systems because it is even more difficult to predict all possible failure situations that can occur in such a system. The complexity of the system is so high because it is an open system which interferes with unplanned incidents. One example is the non-verbal communication between pedestrians and human car drivers as it is described in Chap. 2. This way of communication is for an autonomous vehicle not as clear as the signals of a traffic light or of road signs. The vehicle needs to be able to interpret these signals, which brings the possibility for misunderstandings along (Maurer et al. 2016).

However, the compromise between a safe product and a product that is not completely safe has to be determined carefully for autonomous vehicles. On the one hand, a too safe product might lead to a perception that the product is excessively trusted. This might lead to an environment in which pedestrians feel too safe. For example, pedestrians are more likely to cross a road without paying attention to the traffic anymore (Maurer et al. 2016). On the other hand, autonomous vehicles which are not considered as safe might lead to insufficient trust by the people. This results in a general bad public perception of these vehicles (Maurer et al. 2016).

In the first time after introducing autonomous vehicles commercially, the public perception of these vehicles has to be observed and the industry has to act accordingly, because it is more likely that safety incidents occur. The incident of a TESLA with a malfunction of the autopilot which led to an accident is an example of this topic. However, representatives of the industry reassured that these incidents could happen, especially in the beginning of introducing new technologies and that in the end, autonomous vehicles are still safer than human-driven vehicles (Chokshi 2020).

However, if such incidents occur more often, the public perception might suffer from that. The technology roadmap would be affected that the predicted events might not occur or are delayed until the general trust in this product is retrieved.

9.6 Conclusion and Outlook

The topic of autonomous on-road transportation is becoming increasingly present in our everyday life. In the recent years, computer technologies and data transfer technologies have been evolved to a status, at which safe autonomous transportation is feasible.

Besides conventional car manufacturers, new players, for example, technology companies such as GOOGLE and start-ups with new mobility concepts, join the new established market for autonomous transportation. Each of these players aims to maximize their market share. They try to develop new technologies to establish intellectual property to resume a competitive advantage.

This situation created a need to overview the current situation of the autonomous transportation industry sector. For this purpose, the technology roadmap methodology is applied because it visualizes the current status of the development and shows potential development paths in the upcoming years. The aim of the created technology roadmap is to predict the occurrence of events within the autonomous transportation industry in the next years based on relevant assumed market drivers. Furthermore, the aim is to propose a standardized development platform for the industry sector.

The technology roadmap is based on a technology roadmap created in 2016 which is updated accordingly. It indicates the product features that need to be developed to comply with the predicted future market requirements of autonomous transportation. Based on the product features, technologies are indicated to provide the needed functionalities for the product features. Lastly, resources are elaborated as well, which are necessary to develop the indicated technologies. Based on this approach, it could be displayed that important topics of the future developments are communication, reliability, safety, and security.

The technology roadmap is developed with current knowledge based on literature research. The occurrence of the predicted events is dependent on external factors. One of them is the government that needs to introduce incentives and rules which are used as a baseline to establish autonomous transportation. Another important point is the public perception of autonomous vehicles. If autonomous vehicles are not considered as safe, customers will be hesitating to buy them. Collaborations between different players of the markets are proposed to develop an optimized product on a standardized technological platform that runs within an optimized system.

For further research, the aforementioned external drivers, which are a risk for the predicted events on the technology roadmap, should be elaborated in more detail. Different methods which enhance the technology roadmap methodology can be considered. One of them is to define different future scenarios for the market development based on external conditions. The events in the technology roadmap are changed according to these scenario-based market drivers. This would increase the general credibility of the proposed technology roadmap.

Appendix A: QFDs

Correlation: None, Low, Medium, High

Priority: Low, Medium, High

Product Features		D1 Population growth	D2 Traffic jam reduction	D3 Managing accidents effectively	D4 Reduce parking search time	D5 Space management for vehicles and people	D6 Accidents – prevention and report	D7 Traffic prediction	D8 Pedestrian safety	D9 Reduce CO_2	D10 Reduce number of cars on the road	D11 Reduce environmental destruction	D12 Fuel price increasing	D13 Insurance cost	D14 Driver's ability to multitask	D15 Security	Priority
		4	4	2	4	2	4	2	4	2	1	2	2	1	4	4	
Intelligent merging	PF1		2				2								1		20
Reduced distance between vehicles	PF2	1	1			1	1					1					16
Auto accident reporting and tracking system	PF3			4			2	1						1			19
Auto parking searching function	PF4		1		4	2							1				26
Energy saving mode	PF5									2			2				8
Road elements sensors	PF6		1			2	4		1								28
Sense traffic lights, calculate quickest route	PF7		4			1		2		2			1				28
Automatic recognition of road laws	PF8						2		2					1			17
Protection against hacking	PF9						2		2							4	32
Passengers health monitoring	PF10														1		4
Feature to adapt to weather condition	PF11						4	1	2					1			27
Object identification/avoidance	PF12						4		4					1	2		41

Fig. 9.4 QFD to prioritize product features correlations weighted market drivers

Correlation: None, Low, Medium, High

Priority: Low, Medium, High

Technologies		Auto parking searching function	Road elements sensors	Sense traffic lights,calculate quickest route	Protection against hacking	Feature to adapt to weather condition	Object identification/avoidance	
		PF4	PF6	PF7	PF9	PF11	PF12	Priority
		1	1	1	2	1	4	
Accurate Sensors	T1	1	4	1	1	4	4	28
Vehicle to Vehicle (V2V) communication	T2	4	2	2	4	4	2	28
Radar technology (LiDAR sensors)	T3	1	1	1	1	2	4	23
A.I. (Road decision making system)	T4	4	4	4	2	2	2	26
Element identification system	T5	2	4	2	1	4	4	30
Hotspot wireless (G5)	T6	4	1	2	2	2	2	21
Central road technology system	T7	2	1	4	1	4	1	17
Night vision (laser)	T8	1	4	1	1	4	4	28
Smart broadcasting road objects	T9	1	4	2	1	4	4	29
Accident assessment system	T10	1	1	1	2	1	1	12

(Product Features spans columns PF4–PF12.)

Fig. 9.5 QFD to prioritize technologies correlations product features with high priority

Appendix B: Technology Roadmap

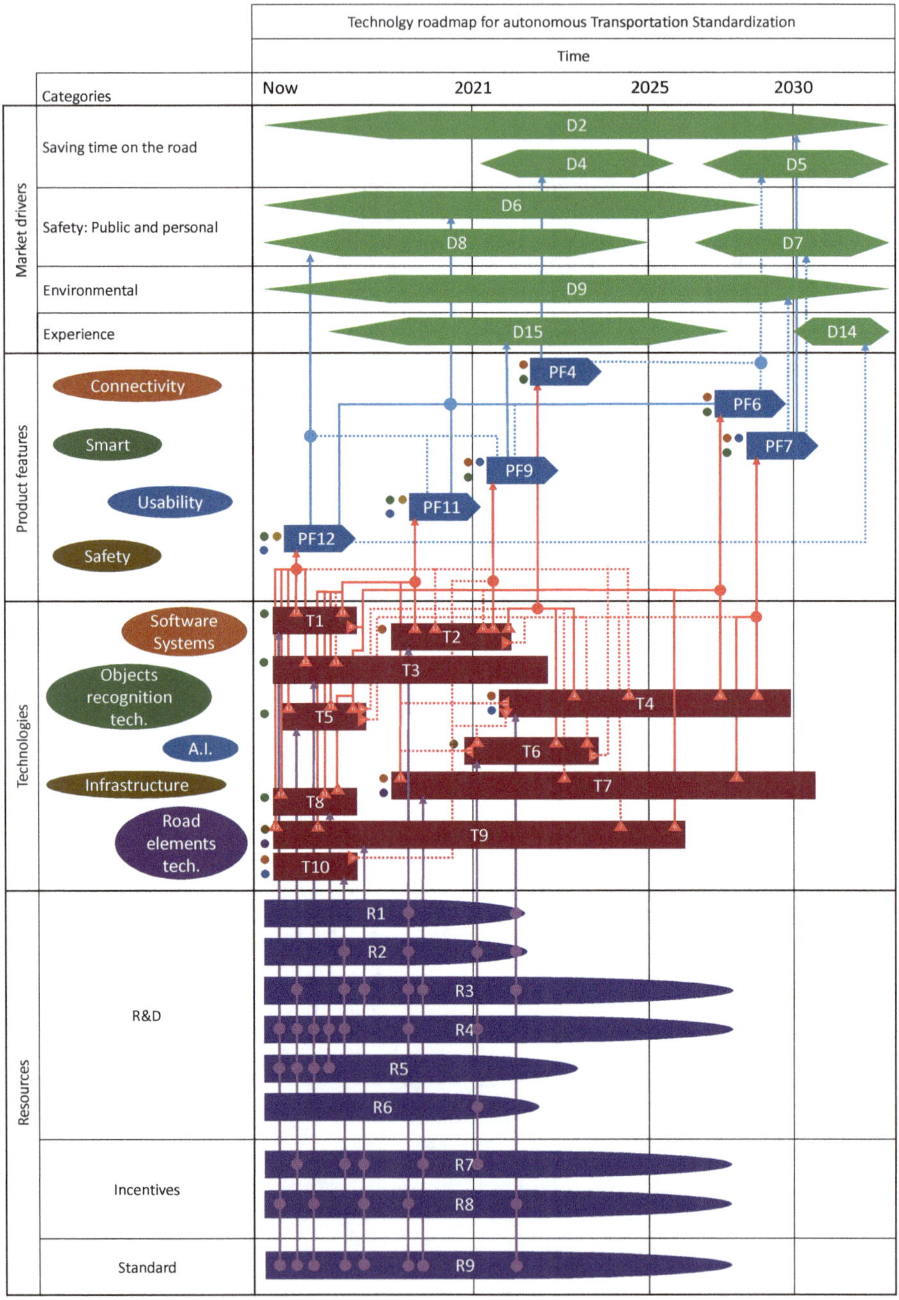

References

Alramadan, G., Barham, H., Shirasaki, M., Guo, Q., Patibanda, S. and Valentine, T. (2016). *Technology Roadmap for Autonomous Vehicles Systems: Standardization*, Portland State University.

Amelang, S. and Wehrmann, B. (2020). *"Dieselgate" - a timeline of the car emissions fraud scandal in Germany*. Available at https://www.cleanenergywire.org/factsheets/dieselgate-timeline-car-emissions-fraud-scandal-germany. Accessed 2 May 2020.

Anderson, J. M., Kalra, N., Stanley, K. D., Sorensen, P., Samaras, C., & Oluwatola, O. A. (2016). *Autonomous vehicle technology: A guide for policymakers*. Rand Corporation.

Cho, Y., Yoon, S.-P., & Kim, K.-S. (2016). An industrial technology roadmap for supporting public R&D planning. *Technological Forecasting and Social Change, 107*, 1–12.

Chokshi, N. (2020). *Tesla Autopilot System Found Probably at Fault in 2018 Crash*. Available at https://www.nytimes.com/2020/02/25/business/tesla-autopilot-ntsb.html. Accessed 2 May 2020.

Daim, T. U., Yoon, B.-S., Lindenberg, J., Grizzi, R., Estep, J., & Oliver, T. (2018). Strategic roadmapping of robotics technologies for the power industry: A multicriteria technology assessment. *Technological Forecasting and Social Change, 131*, 49–66.

Daimler, A. G. (n.d.). *Faster help at the scene of an accident.: Mercedes-Benz emergency call system*. Available at https://www.daimler.com/innovation/case/shared-services/mercedes-benz-emergency-call.html. Accessed 14 Apr 2020.

European Commission. (2017). *Intelligent transport systems: The interoperable EU-wide eCall*. Available at https://ec.europa.eu/transport/themes/its/road/action_plan/ecall_en. Accessed 14 Apr 2020.

Fernandes, P., & Nunes, U. (2012). Platooning with IVC-enabled autonomous vehicles: Strategies to mitigate communication delays, improve safety and traffic flow. *IEEE Transactions on Intelligent Transportation Systems, 13*(1), 91–106.

Gamba, J. (2020). *Radar signal processing for autonomous driving*. Springer Singapore: Singapore.

Geum, Y., Lee, H., Lee, Y., & Park, Y. (2015). Development of data-driven technology roadmap considering dependency: An ARM-based technology roadmapping. *Technological Forecasting and Social Change, 91*, 264–279.

Haddad, C. R., & Uriona Maldonado, M. (2017). A functions approach to improve sectoral technology roadmaps. *Technological Forecasting and Social Change, 115*, 251–260.

Hansen, C., Daim, T., Ernst, H., & Herstatt, C. (2016). The future of rail automation: A scenario-based technology roadmap for the rail automation market. *Technological Forecasting and Social Change, 110*, 196–212.

Heddebaut, M., Rioult, J., Ghys, J. P., Gransart, C., & Ambellouis, S. (2005). Broadband vehicle-to-vehicle communication using an extended autonomous cruise control sensor. *Measurement Science and Technology, 16*(6), 1363–1373.

Hepe, C. (2019). *BMW Group and Daimler AG invest more than €1 billion in joint mobility services provider*, BMW Group. Available at https://www.press.bmwgroup.com/global/article/detail/T0292204EN/bmw-group-and-daimler-ag-invest-more-than-%E2%82%AC1-billion-in-joint-mobility-services-provider?language=en. Accessed 15 Apr 2020.

Herrmann, A., & Brenner, W. (2018). *Autonomous driving: How the driverless revolution will change the world*. Bingley: Emerald Publishing Limited.

Hobert, L., Festag, A., Llatser, I., Altomare, L., Visintainer, F., & Kovacs, A. (2015). Enhancements of V2X communication in support of cooperative autonomous driving. *IEEE Communications Magazine, 53*(12), 64–70.

Janssen, C. P. and Kenemans, J. L. (2015). Multitasking in autonomous vehicles: Ready to go?

Jeong, Y., Lee, K., Yoon, B., & Phaal, R. (2015). Development of a patent roadmap through the generative topographic mapping and bass diffusion model. *Journal of Engineering and Technology Management, 38*, 53–70.

Johanning, V., & Mildner, R. (2015). *Car IT kompakt*. Springer Wiesbaden: Wiesbaden.

Kajikawa, Y., Usui, O., Hakata, K., Yasunaga, Y., & Matsushima, K. (2008). Structure of knowledge in the science and technology roadmaps. *Technological Forecasting and Social Change, 75*, 1–11.

Keferböck, F. & Riener, A. (2015). Strategies for negotiation between autonomous vehicles and pedestrians.

Khaliq, A. (2015). Electric and autonomous vehicle security concerns.

Kong, L., Khan, M. K., Wu, F., Chen, G., & Zeng, P. (2017). Millimeter-wave wireless communications for IoT-cloud supported autonomous vehicles: Overview, design, and challenges. *IEEE Communications Magazine, 55*, 62–68.

Kostoff, R. N., & Schaller, R. R. (2001). *Science and technology roadmaps.*

Lambert, F. (2016). *Google starts deploying its self-driving Chrysler Pacifica minivans: first prototypes spotted.* Available at https://electrek.co/2016/10/09/google-self-driving-chrysler-pacifica-minivans-first-prototypes-spotted/. Accessed 13 Apr 2020.

Lee, H., & Geum, Y. (2017). Development of the scenario-based technology roadmap considering layer heterogeneity: An approach using CIA and AHP. *Technological Forecasting and Social Change, 117*, 12–24.

Lee, S., & Park, Y. (2005). Customization of technology roadmaps according to roadmapping purposes: Overall process and detailed modules. *Technological Forecasting and Social Change, 72*, 567–583.

Lee, J. H., Kim, H.-i., & Phaal, R. (2012). An analysis of factors improving technology roadmap credibility: A communications theory assessment of roadmapping processes. *Technological Forecasting and Social Change, 79*, 263–280.

Lee, J. H., Phaal, R., & Lee, S.-H. (2013). An integrated service-device-technology roadmap for smart city development. *Technological Forecasting and Social Change, 80*, 286–306.

Litman, T. (2013). *Autonomous vehicle implementation predictions: Implications for transport planning*. Victoria Transport Policy Institute.

Litman, T. (2020). *Autonomous vehicle implementation predictions: Implications for transport planning*. Victoria Transport Policy Institute.

Matthews, M., Chowdhary, G. and Kieson, E. (2017). Intent communication between autonomous vehicles and pedestrians.

Maurer, M., Gerdes, J. C., Lenz, B., & Winner, H. (2016). *Autonomous driving*. Berlin, Heidelberg: Springer Berlin Heidelberg.

McDowall, W. (2012). Technology roadmaps for transition management: The case of hydrogen energy. *Technological Forecasting and Social Change, 79*, 530–542.

Mishra, S., Bhattacharya, D., Gupta, A. & Singh, V. R. (2018). Adaptive Traffic Light Cycle Time Controller Using Microcontrollers And Crowdsource Data Of Google APIs For Developing Countries, *ISPRS Annals of Photogrammetry, Remote Sensing and Spatial Information Sciences*, IV-4/W7, pp. 83–90.

Moehrle, M. G., Isenmann, R., & Phaal, R. (2013). *Technology Roadmapping for strategy and innovation*. Heidelberg: Springer.

Parra, I., Garcia-Morcillo, A., Izquierdo, R., Alonso, J., Fernandez-Llorca, D. & Sotelo, M. A. (2017). *Analysis of ITS-G5A V2X communications performance in autonomous cooperative driving experiments*.

Phaal, R., O'Sullivan, E., Routley, M., Ford, S., & Probert, D. (2011). A framework for mapping industrial emergence. *Technological Forecasting and Social Change, 78*, 217–230.

Rastogi, N. (2017). *China Unwraps World's first Driverless Rail Transit System with Autonomous Technology*. Available at https://www.engineersgarage.com/egblog/china-unwraps-worlds-first-driverless-rail-transit-system-with-autonomous-technology/. Accessed 12 Apr 2020.

Rinne, M. (2004). Technology roadmaps: Infrastructure for innovation. *Technological Forecasting and Social Change, 71*, 67–80.

SAE International. (2014). Surface vehicle information report.

Seif, H. G., & Hu, X. (2016). Autonomous driving in the iCity—HD maps as a key challenge of the automotive industry. *Engineering, 2*, 159–162.

Shoup, D. (2007). Cruising for parking.

Urmson, C. (2015). Green lights for our self-driving vehicle prototypes, Alphabet. Available at https://www.blog.google/inside-google/alphabet/self-driving-vehicle-prototypes-on-road/. Accessed 15 Apr 2020.

Vlachos, E., Lalos, A. S., Berberidis, K. & Tselios, C. (2017). *Autonomous Driving in 5G: Mitigating Interference in OFDM-Based Vehicular Communications.*

Volkswagen AG (ed). (2019). *Audi networks with traffic lights in Europe*. Available at https://www.volkswagenag.com/en/news/2019/05/Audi_networks_with_traffic_lights_in_Europe.html. Accessed 14 Apr 2020.

Wang, C., Gong, S., Zhou, A., Li, T. & Peeta, S. (2017). Cooperative adaptive cruise control for connected autonomous vehicles by factoring communication-related constraints.

Weldemann, B. & Becker, K. (2019). *Daimler AG and BMW Group to jointly develop next-generation technologies for automated driving*, Daimler AG. Available at https://media.daimler.com/marsMediaSite/en/instance/ko.xhtml?oid=42640196. Accessed 20 Mar 2020.

Yasunaga, Y., Watanabe, M., & Korenaga, M. (2009). Application of technology roadmaps to governmental innovation policy for promoting technology convergence. *Technological Forecasting and Social Change, 76*, 61–79.

Zheng, K., Zheng, Q., Yang, H., Zhao, L., Hou, L. & Chatzimisios, P. (2015). *Reliable and Efficient Autonomous Driving: the Need for Heterogeneous Vehicular Networks.*

Part II
Technology Intelligence Analysis

Chapter 10
Technology Intelligence Map: Finance Machine Learning

Mohammadsaleh Saadatmand and Tuğrul U. Daim

Although the terms of machine learning and deep learning have been widely used in the financial press and media, lack of agreement in the scientific and professional community about a holistic view of best practices, use cases, and trends still exists. Considering the need for filling this gap, the main aim of this study is to investigate and map the literature at the intersection of machine and deep learning as a subset, and finance and investment. This research proposes the use of bibliometric analysis of the literature that highlights the most important articles for this area of research. Specifically, this technique is applied to the literature about machine learning applications in investment and finance, resulting in a bibliographical review of the significant studies about the topic. The author evaluates papers indexed in the Scopus database. This study opens avenues for further research by concentrating on the importance of artificial intelligence and, specifically, machine learning in investment research and practice. Additionally, this review contributes by showing scholars and investment professionals in the areas in which machine learning can add value to investment research.

10.1 Introduction

Market efficiency, forecasting capability, and finding anomalies in the markets are the most important and challenging areas in finance. The efficient market hypothesis has been widely researched and tested (Jensen 1978; Schwert 2003; Chan et al. 1997; Fama 1998; Merton 1980; Lo and Mackinlay 1987; Shiller and Perron 1985; Fama 1991; Malkiel 2003; Malkiel and Fama 1970). In spite of extensive research

M. Saadatmand · T. U. Daim (✉)
Department of Engineering and Technology Management, Portland State University, Portland, OR, USA
e-mail: tugrul.u.daim@pdx.edu

T. U. Daim (ed.), *Roadmapping Future*, Applied Innovation and Technology Management, https://doi.org/10.1007/978-3-030-50502-8_10

on efficient market hypothesis that proposes financial markets follow random walks and therefore are essentially unpredictable, seeking anomalies and active portfolio management is still a big piece of research in academia and asset management industry. Moreover, predictive alpha-generating models and active portfolio management that can consistently outperform the market indices over time is evidence against the efficient market hypothesis and has tremendous importance for asset and investment management companies (Malkiel 2003; Atsalakis and Valavanis 2009; Ang 2013; Lo 2007; Cremers and Petajisto 2009; Fama and French 2009; Cohen et al. 2010).

Artificial intelligence and machine learning are changing virtually every aspect of the financial services industry. Investment research and practice are benefiting from the rise of machine learning accomplishments that until recently only professional human experts could perform. Price prediction, hedging, portfolio construction and optimization, alpha capture, and sentiment analysis, to name a few, are areas that machine learning has already impacted (Lopez de Prado 2018a; Lopez de Prado 2017; Lopez de Prado 2018b; Lopez de Prado 2016). As an example, due to the complex and chaotic nature of financial market prices, forecasting is a challenging problem that happens in a dynamic environment. To address this problem, many studies from various areas of research have used machine learning to provide some sort of prediction. These methods have resulted in promising outcomes (Gu et al. 2019; Nelson et al. 2017; Chen et al. 2015; Aadith Narayan et al. 2019; Akansu et al. 2016; Baldominos et al. 2018; Kim and Won 2018; Hasan et al. 2017; Lachiheb and Gouider 2018; Manurung et al. 2018; Ni et al. 2019; Rodrigues et al. 2019). These studies show how financial research has started to consider the value of machine learning in price forecasting given that the data are chaotic and relationships are non-linear.

Technological advances in big data analytics and our increased computational capacity have made it possible to analyze a large volume of structured and unstructured data in a short period of time. One example is the application of machine learning in portfolio construction and optimization which requires high computational power. Shen et al., for example, have proposed an orthogonal portfolio framework that represents the combining effects of passive and active investment styles based on a risk-adjusted function. Results demonstrate outperformance in both risk-adjusted return and cumulative wealth (Shen et al. 2015). Gu et al. suggested an empirical asset pricing framework for portfolio construction based on asset risk premia, which highlights the value of machine learning in both empirical studies and financial innovation (Gu et al. 2019).

Machine learning demonstrates great promise for empirical asset pricing as well. At the holistic level, machine learning has shown the potential to improve empirical testing and understanding of expected asset returns (Gu et al. 2019). The capability of crunching a massive amount of big data with high variety, velocity, and volume and feeding into predictive models enables researchers and professionals to empirically dig deeper into the empirical analysis that goes beyond traditional econometric models. Rapach et al., for instance, apply Lasso to predict global stock market returns (Rapach and Zhou 2013). Several papers apply artificial neural networks and decision trees to forecast derivative prices and credit card defaults. In addition, other types of investment applications of machine learning include the study of a cross-section of stock returns, factor pricing models, portfolio sorting, and selection. All recently applied machine learning techniques in such studies represent a promising future (Hutchinson et al. 1994; Yao et al. 2000; Khandani et al. 2010; Butaru et al. 2016; Deep Learning in Finance 2020).

As Marco Lopez de Prado stated, machine learning provides the opportunity to gain insights from: "(a) new datasets that cannot be modeled with econometric methods; and (b) old datasets that incorporate complex relationships still unexplored. Key strengths of ML methodologies include (i) focus on out-of-sample predictability over variance adjudication; (ii) usage of computational methods to avoid relying on (potentially unrealistic) assumptions; (iii) ability to 'learn' complex specifications, including non-linear, hierarchical, and non-continuous interaction effects in a high-dimensional space; and (iv) feature importance analysis robust to multicollinearity."

In addition to this, asset pricing and factor models that have been a rich area of research in the last four decades are recognized potentially be another area of research that machine learning and deep learning can shed light on (Ang 2013; Fama and French 2015; Asness 2016; Asness et al. 2017; Bean and Singer 2009; Bender et al. 2019; Bektić et al. 2019; Blitz 2015; Blitz and Vidojevic 2019; Cerniglia and Fabozzi 2018; Cong and Xu 2016). There is an abundance of empirical research on this area based on the econometric analysis. As machine learning and deep learning are very good at absorbing large datasets from a variety of sources and finding reliable patterns, they are essentially a good fit to the empirical study of asset pricing models. The main problem, in this case, is prediction and empirical testing of which machine learning is very capable in prediction problems and its empirical testing methods have been developing over time.

Notably, the current studies of data science, analytics, and machine learning in the investment landscape are still sporadic and fragmented. Furthermore, the implementation of machine learning and the application of advanced data science techniques are also in the initial step among practitioners. Therefore, by using a bibliographic technique, this study seeks to shed light on the avenues of research in the intersection of investment and machine learning and analytics to gain a broad view of analytics applications in this area. By categorizing and visualization of current literature, this study provides future direction to researchers and investment professionals. The potential of machine learning in the financial context is significant. However, to benefit from such capability, investment companies should first understand and identify value drivers of such trends. This knowledge essentially will provide grounds for the proper implementation of analytics and machine learning in investment decision making that effectively unlock value for organizations. This knowledge bridges the gap between econometricians and machine learners (Hsu et al. 2016; Lopez de Prado 2019; Cerniglia et al. 2016; Jang and Lee 2019; Mullainathan and Spiess 2017).

In addressing the issues, this study is organized as follows: first, we provide a brief review of machine learning methods used in financial applications. Second, the methods used in identifying the related papers and literature available on the Scopus scientific database that focus on the applications of machine learning and deep learning in the investment landscape are described. Subsequently, the presentation of the results of the bibliometric research on the literature is presented, revealing the most cited papers and topics in the field. Then, based on the reviewed literature, the conclusion and future research directions are outlined.

10.2 Brief Review of Financial Machine Learning

Machine learning as a field is considered differently from multiple perspectives. Some categorize it on the intersection of computer science and statistics. On the other hand, some people just simply classify it as a subsection of the broad field of artificial intelligence. In whatever way we see it, its goal is the same: "enable machines to learn from their experience and improve performance as their experience grows" (Kolanovic and Krishnamachari 2017; Mondal 2019). Machine learning techniques seek to extract complex patterns embedded in the data which are not evident to human analysis.

At the broadest level, machine learning models are categorized into three main classes: supervised learning, unsupervised learning, and deep/reinforcement learning. Each one of these classes has its own specific techniques and is suitable for different kinds of tasks (Lopez de Prado 2018b; Aadith Narayan et al. 2019; Lopez de Prado 2019; Kolanovic and Krishnamachari 2017; Arnott et al. 2018). In supervised learning, models are provided with both input and output variables and historical data in which algorithms learn the mapping function between inputs and outputs. Regression and classification methods are in this category of study. However, in unsupervised learning, algorithms are fed with the entire dataset and without any predetermined target variable. The goal, in this case, is typically clustering and finding determining factors. Principal component analysis (PCA) is one of the most used techniques in this class. Deep learning is the third set of algorithms that are inspired by the working of the human brain. This method is specifically designed to process data through multiple layers of non-linear computational units. This multilayer process gives deep learning the capability of learning complex concepts that arise from simple ones (Heaton et al. 2017; LeCun et al. 2015; Nahil and Abdelouahid 2019; Fomin et al. 2019; Chong et al. 2017a; Li et al. 2016).

The investment landscape and financial services have always been interested in applying forecasting methods to have financial predictions. Numerous studies have been published using a wide variety of forecasting models including machine learning and deep learning (Heaton et al. 2017; Abe and Nakayama 2018; Arévalo et al. 2018; Bao et al. 2017; Bhanja and Das 2019; Blažiunas and Raudys 2019; Choi and Renelle 2019; Deng et al. 2017). One of the focused areas of research and implementations is financial times series forecasting which has been the realm of financial econometrics in the last four decades and the current field of machine learning (Tsai and Wu 2008; Patel et al. 2015a; Tsai 2009). Bruno et al., for instance, applied bibliography techniques to demonstrate and review the literature of machine learning techniques used to financial market forecasting. Given machine learning capabilities in identifying complex patterns, they found out machine learning techniques are among the latest approaches that are most researched and applied in financial market prediction. They also revealed the most studies used US market data, and support vector machines (SVMs) and neural networks (NNs) are the most commonly used prediction models (Henrique et al. 2019).

Additionally, Sezer et al. apply a systematic literature review technique to analyze the related literature of financial time series forecasting using deep learning models. They demonstrated that deep learning techniques have been widely applied in different asset classes including equity, index, forex, commodity, and so on. Their findings indicate despite the long research history of financial forecasting, with the rise of new datasets and investment products utilizing deep learning models open the door of opportunities for the academic and professional communities (Sezer et al. 2019).

The range of machine learning models applied in financial research has been wide in the last couple of years. The diverse variety of techniques including support vector machines, artificial neural networks, principal component analysis, and decision trees, to name a few, have been used by researchers. Stock value prediction, credit analysis via classification, genetic algorithm optimization, and sentiment analytics are just some of the applications (Aadith Narayan et al. 2019; Bayraktar et al. 2018; Chung and Shin 2018; Aggarwal et al. 2019; Chong et al. 2017b). In fact, machine learning and specifically deep learning have shown the value and it seems it is still the beginning of an industry transformation in investment and finance. These technologies potentially have the capability of disruption that can come from different streams.

10.3 Research Methodology

The main method used in this literature review is the bibliometric analysis which is commonly applied in depicting the scientific fields of study. This method is intended to provide answers related to the evolution of the literature over time and demonstrates the influential areas of the field such as papers, authors, and journals (Henrique et al. 2019; Milian et al. 2019). The main scientific database searched in this study to find the most relevant papers about machine learning and deep learning in finance and investment is Scopus. In order to review the literature systematically, we have used a protocol to filter out the associated research (see Fig.10.1). This research aimed to include the most relevant papers and the date of search is from 1970 through 2019. Moreover, to restrict the scope of search on the database, the following keywords are used to obtain the papers that are mostly related: "machine learning*" AND "finance*," "machine learning*" AND "investment*," "deep learning*" AND "finance*," "deep learning*" AND "investment*." This procedure resulted in a total of 901, 874, 170, and 190 papers respectively. Then all papers are

Fig. 10.1 Search protocol to filter out the most relevant research

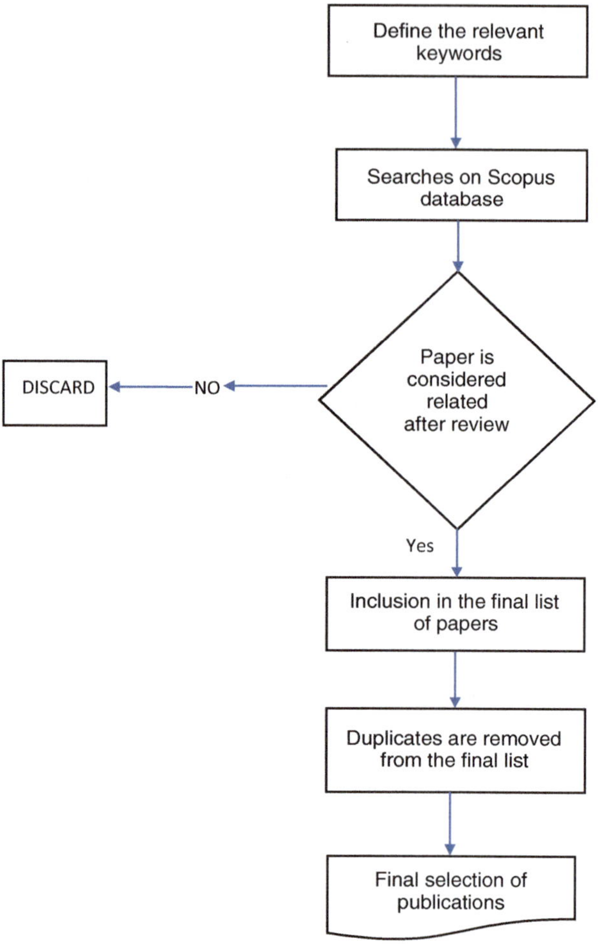

reviewed manually to select the most related papers in the field of study. A total of 833 (see Appendix 1) out of 2135 papers are found relevant and constitute the final list of papers under study.

In order to concentrate on the objectives of this study, the following are the main research questions of this research:

- RQ1: How has the literature in general on this subject evolved over time?
- RQ2: What are the most influential papers by the number of citations?
- RQ3: What are the most influential authors in this research area?
- RQ4: What are the main journals for this subject and how has the number of publications evolved over time?
- RQ5: What are the countries with the highest number of publications and cited papers?

Products of bibliometric analysis of the literature could be the list of the most influential authors, the most cited papers, and the most important journals. Additionally, furthermore, this research intends to demonstrate the relationship between the target list of papers by depicting bibliographic coupling and co-citation networks as well. Bibliometric analysis is applied to recognize the spectrum of scientific literature in the intersection of machine learning, deep learning, finance, and investment landscape, presenting the trends over time and main topics addressed. This method helps to find the most productive authors, tracking the evolution of the field of study, and the most cited research (Ramos-Rodríguez and Ruíz-Navarro 2004; Mariano et al. 2015). This approach, in general, helps researchers and practitioners to identify the relationship between machine learning and investment. Citation analysis and bibliographic coupling are efficient techniques to assess the degree of recognition of academic papers and commonality of references they share.

10.4 Results

Following the protocol demonstrated in the research method section, 833 papers were selected for this bibliometric study. The frequency of publications is shown in Fig. 10.2, in which it can be observed that there is a growing trend of publications starting from 2009 to 2019 that is continuing. This trend suggests potential opportunities for further research on machine learning and deep learning applications in finance. Also, the rate of growth is higher in recent years and many papers are published recently that show the interest of the academic community on this topic and

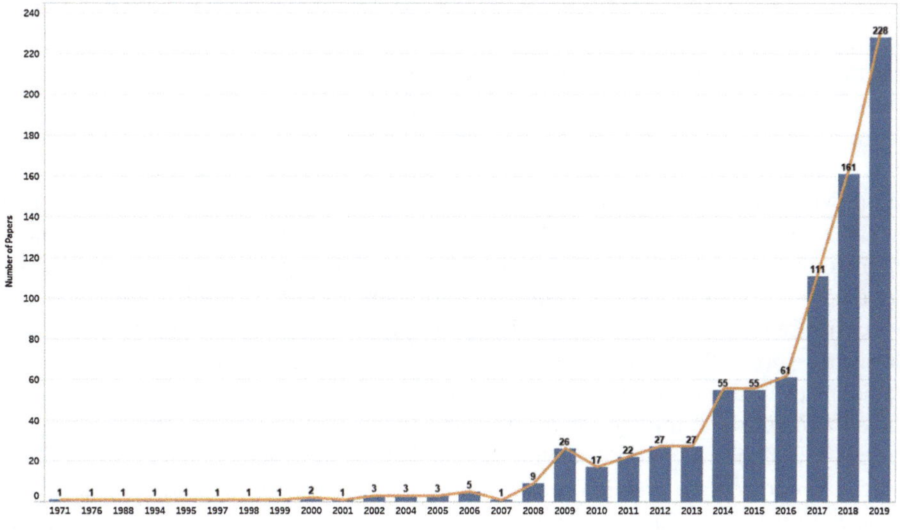

Fig. 10.2 Number of publications over time

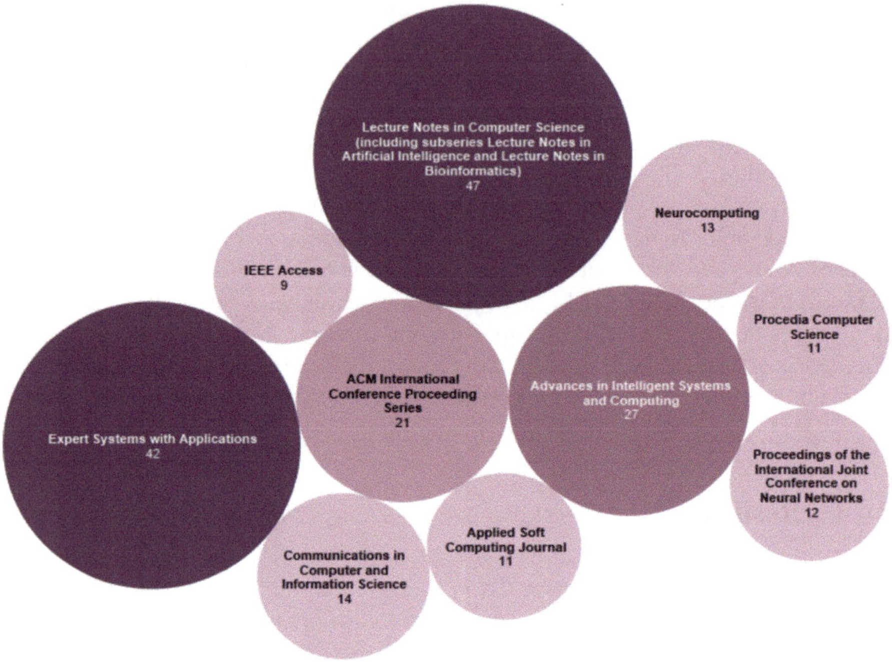

Fig. 10.3 Top 10 journals by the number of publications

its importance. In fact, starting in 2013, the number of publications is approximately doubled each year.

The 10 journals with the highest number of publications in the final list of papers are represented in Fig. 10.3. These journals constitute approximately 25% of the papers in this study and can be considered the most sources of research in the intersection of machine learning and finance. Digging deeper into the list shows *Lecture Notes in Computer Science* and *Expert Systems with Applications*, far from the rest of resources, have 47 and 42 number of papers respectively. Following the top 2 journals, *Advance in Intelligent Systems and Computing* and *ACM International Conference Proceeding Series* have had a great contribution in publishing papers with 27 and 21 publications. These journals thus would be the potential avenues of future submission of new research. Furthermore, the co-occurrence of keywords by researchers is depicted in Fig. 10.4. This network relationship shows the importance and trend of keywords by size and color. It is evident from the visualization that *investments, commerce, neural networks, and deep learning* are among the most used keywords. In addition, color coding of terms reveals the keywords related to machine learning and deep learning including *long-short term memory, convolutional neural network, sentiment analysis, and recurrent neural network* are the most currently used terms after 2017 in this field of research. This indicates the importance of applying machine learning and

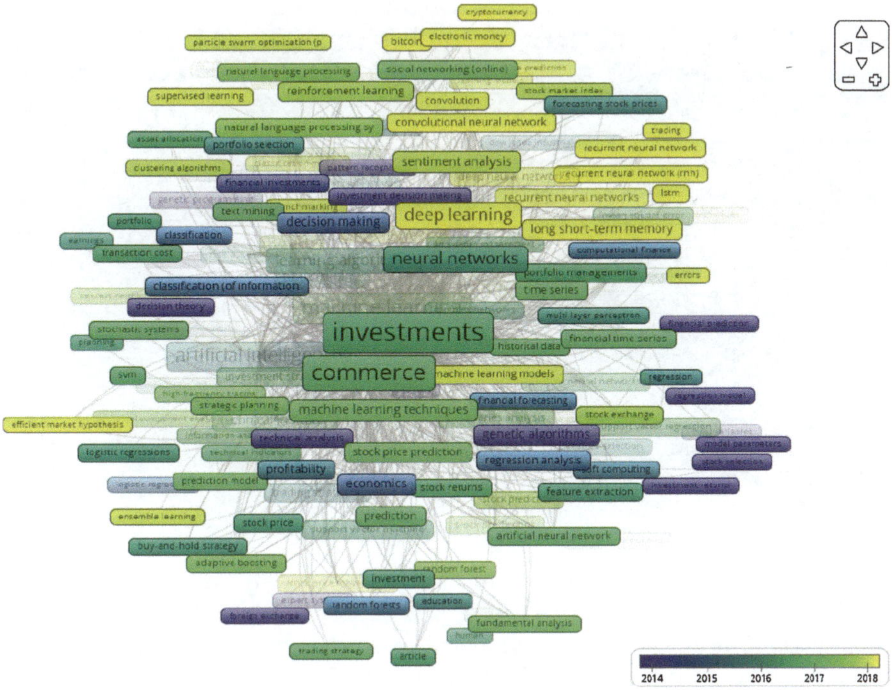

Fig. 10.4 Co-occurrence of keywords

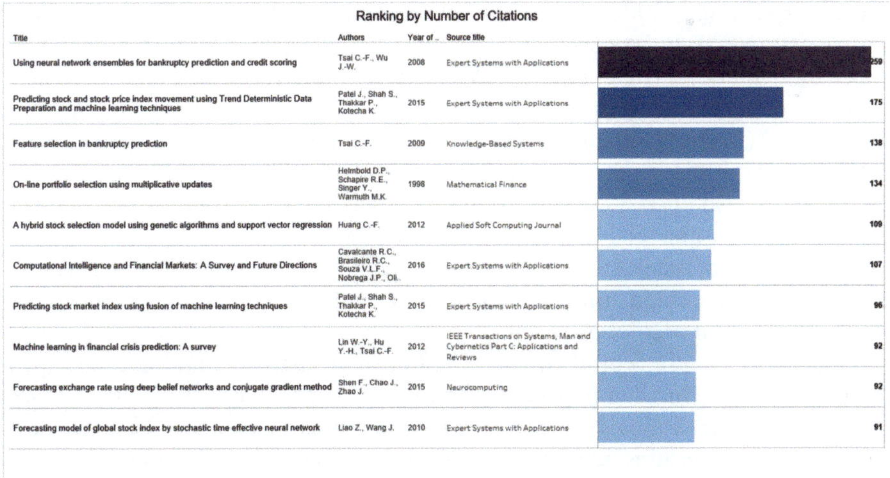

Ranking by Number of Citations

Title	Authors	Year of ..	Source title	
Using neural network ensembles for bankruptcy prediction and credit scoring	Tsai C.-F., Wu J.-W.	2008	Expert Systems with Applications	259
Predicting stock and stock price index movement using Trend Deterministic Data Preparation and machine learning techniques	Patel J., Shah S., Thakkar P., Kotecha K.	2015	Expert Systems with Applications	175
Feature selection in bankruptcy prediction	Tsai C.-F.	2009	Knowledge-Based Systems	138
On-line portfolio selection using multiplicative updates	Helmbold D.P., Schapire R.E., Singer Y., Warmuth M.K.	1998	Mathematical Finance	134
A hybrid stock selection model using genetic algorithms and support vector regression	Huang C.-F.	2012	Applied Soft Computing Journal	109
Computational Intelligence and Financial Markets: A Survey and Future Directions	Cavalcante R.C., Brasileiro R.C., Souza V.L.F., Nobrega J.P., Oli..	2016	Expert Systems with Applications	107
Predicting stock market index using fusion of machine learning techniques	Patel J., Shah S., Thakkar P., Kotecha K.	2015	Expert Systems with Applications	96
Machine learning in financial crisis prediction: A survey	Lin W.-Y., Hu Y.-H., Tsai C.-F.	2012	IEEE Transactions on Systems, Man and Cybernetics Part C: Applications and Reviews	92
Forecasting exchange rate using deep belief networks and conjugate gradient method	Shen F., Chao J., Zhao J.	2015	Neurocomputing	92
Forecasting model of global stock index by stochastic time effective neural network	Liao Z., Wang J.	2010	Expert Systems with Applications	91

Fig. 10.5 Ranking of papers by the number of citations

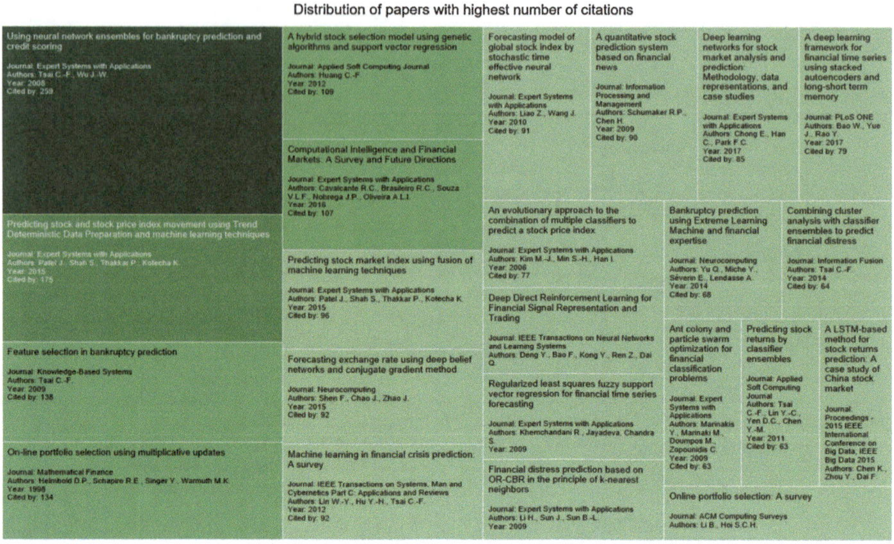

Fig. 10.6 Distribution of papers by the number of citations

deep learning in financial and investment research. The top 10 papers with the highest number of citations are listed in Figs. 10.5 and 10.6. With respect to the ranking of papers, the order of papers illustrates the high concentration of citations among the top 4 papers with a total of 706 citations. Paper by Tsai et al. (Tsai and Wu 2008) entitled "Using neural network ensembles for bankruptcy prediction and credit scoring" (2008) with 259 citations distinguishes itself from the rest of the papers and might be considered the most popular papers among researchers. Paper by Patel et al. (Patel et al. 2015b) called "predicting stock and stock price index movement using trend deterministic data preparation and machine learning techniques" (2018) stands in second place with total citations of 175. It is worth noting that both these papers are published in *Expert Systems with Applications* journal. It is also crystal clear from the bar chart that the range of machine and deep learning in financial applications is wide including stock prediction, bankruptcy prediction, credit scoring, portfolio selection, and so on. However, general themes of papers demonstrate that machine learning techniques are mostly applied for the problem of prediction.

Figure 10.7 shows the bibliography coupling of the network of the chosen papers. It means the relatedness of papers is determined based on the number of references they share (Kessler 1963). In fact, each node represents a paper, and the edges demonstrate the relationships between them. Therefore, the relationships show the same reference papers share with each other in the intersection of machine and deep learning in finance and investment. These bibliographic networks direct researchers to the highly shared work in the specific field of study and make them capable of tracking the evolution of the literature. In this study, research papers of Helmbold

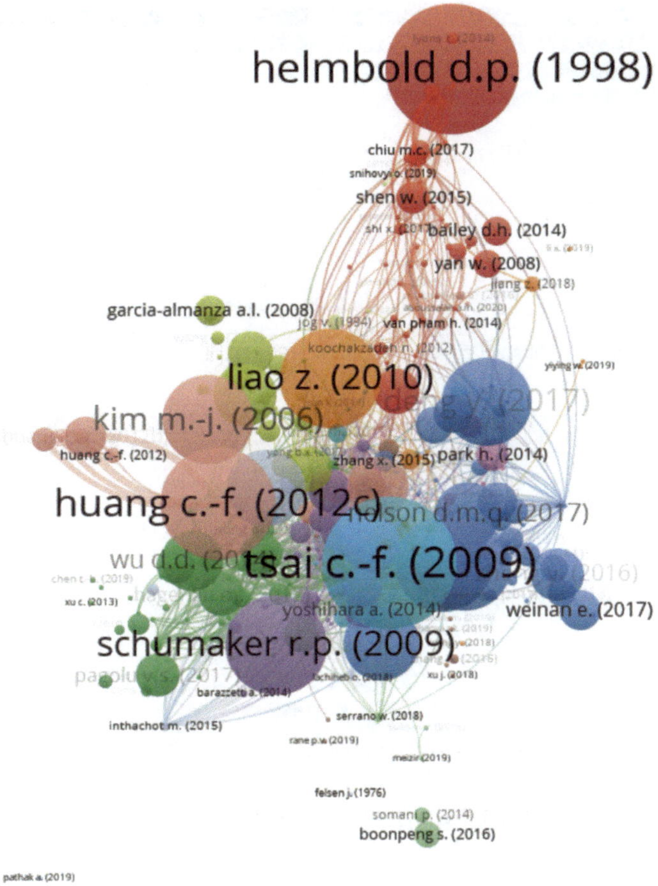

Fig. 10.7 Bibliographic coupling network of papers

d.p. (1998), Tsai c.f. (2009), and Huang c.f. (2012) stand out as the highly influential papers (Tsai 2009; Helmbold et al. 1998; Huang 2012).

In terms of a number of publications and citations by country, the ranking of the countries shows that China, the United States, and India are the top 3 countries in the number of publications respectively. Furthermore, the United States, China, and Taiwan have had most citations and stand out as the top 3 countries. These results indicate that, except for the United States, the interdisciplinary research at the intersection of machine learning and finance has found much attention among the scholars of eastern Asian countries. Taiwan is an interesting case as this country has a top ranking based on both the number of citations and publications. The United Kingdom with a rich academic environment and a long history of financial research has roughly the same number of publications but Taiwan's citations are approximately two times the United Kingdom's (Figs. 10.8 and 10.9).

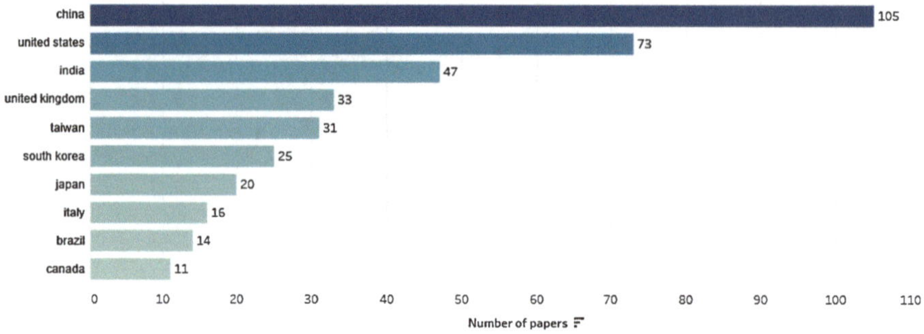

Fig. 10.8 Ranking of published papers by country

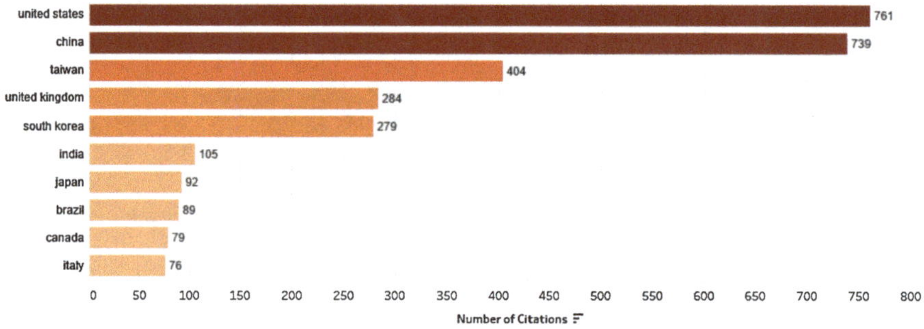

Fig. 10.9 Ranking of number of citations by country

10.5 Discussion and Future Research

This research presents a review of scientific publications on applications of machine learning and deep learning in finance and investment through the lens of descriptive analysis and bibliographic analysis of the most influential papers. In order to pull out the most relevant papers, the authors have applied a step-by-step protocol to search for papers in the Scopus database. After the initial screen of the pulled papers, the authors found 833 papers directly related to the field which constituted the final list. In the assessment process, we reviewed the abstract, conclusion, and general content of the papers to come up with papers that are directly related to the area. In turn, by the means of bibliographic technique and descriptive statistics, we found the most cited papers, main journals that have published the associated publications, the most influential authors, countries with the highest contribution, and finally the general trend of the literature over time from 1970 through 2019. Therefore, this study contributes to the overview of the literature in the area of machine learning and finance.

In response to the main research questions of this study, we demonstrated results by the means of descriptive statistics and bibliographic as follows. Starting from 2016, the number of published papers has skyrocketed and experienced a growth of approximately 274% until 2019. This result indicates too many scholars have concentrated on this subject area. Based on the number of citations, papers of Tsai et al. (2008), Patel et al. (2015), and Tsai (2009) are the top three papers in the scientific community (Tsai and Wu 2008; Tsai 2009; Patel et al. 2015b). Regarding the authors with a high contribution, a network of papers reveals that Helmbold d.p. (1998), Tsai c.f. (2009), and Huang c.f. (2012) have been highly influential authors in this field of research. In terms of valued journals in the field, *Lecture Notes in Computer Science* and *Expert Systems with Applications* are, by far, the most related journals for this research area. Finally, at the global level, the United States, China, and Taiwan are three countries with the most active researchers and a number of citations in this interdisciplinary realm.

Through the lens of descriptive and bibliographic analysis, we illustrated the vast applications of the machine and deep learning in the financial landscape in which investment has a key role. These areas include credit default prediction, portfolio analysis and construction, assessment of semantics and news, and asset pricing, to name a few (Lopez de Prado 2016; Tsai and Wu 2008; Bayraktar et al. 2018; Ahmadi et al. 2019; Batra and Daudpota 2018; Almahdi and Yang 2019). Furthermore, in each of these subjects, much innovation is observed in the means researchers have applied machine learning techniques. This helps and shows future researchers, interdisciplinary areas of machine learning and investment in particular, to narrow down and categorize the current literature and concentrate on the most important studies. Thus, this study generates insight and demonstrates different avenues for future research.

Based on the descriptive statistics shown in this paper, one of the interesting routes of future research that would incredibly provide insight would be using machine learning and deep learning in asset pricing research (Fama 1998; Malkiel 2003; Malkiel and Fama 1970; Fama and French 2009; Fama and French 2015; Fama and French 1992). In the last three decades, numerous factors are introduced in the literature and are claimed to have explanatory power in explaining the cross-section of equity returns. With the power of deep learning in the evaluation of large datasets, future research can benefit from this capability in empirically testing a large number of factors systematically and generate insight into the dynamics of factor cyclicality. Thus, researchers can focus more on the real factors that could help in explaining average equity returns.

Moreover, abundance and growing trend of alternative datasets might be another avenue of research for using machine learning capability in investment research (Kolanovic and Krishnamachari 2017; Teng and Lee 2019; Hafiz et al. 2015; Heiden et al. 2017; Wang et al. 2014; Matin et al. 2019; Cavalcante et al. 2016; Chiong et al. 2018; Day and Lee 2016; Man et al. 2019; Mishev et al. 2019). As machine learning can potentially contribute to identifying complex patterns from large datasets, researchers can benefit from adopting such technologies in alpha-generating investment strategies. Recently, we have been observing the presence of many alternative datasets that claim to have valued an edge over the classical dataset. Future research on applying machine and deep learning to test the value of these datasets can shed light on the credibility of such a hypothesis. Additionally, the results from this study reveal the importance of an interdisciplinary approach in bridging the gap between the seemingly unrelated scientific fields. Therefore, future research might further investigate the value of machine learning in detecting pure alpha strategies from alternative datasets.

This research, in general, sets the initial stage for investment scholars and professionals in observing the potential, challenges, and possible routes of future research in applications of financial machine learning techniques and technologies. This macro-level analysis helps researchers to simply find the related literature and facilitates finding the most relevant concepts and papers. As this field is still in its infancy, showing the boundaries of possible research streams would be beneficial for future academic research. Furthermore, this study demonstrates the value of a systematic and comprehensive view in dealing with multidisciplinary streams of research.

10.6 Conclusion

The contribution of this paper is understanding the general context of financial machine learning and investment in particular in different ways.

First, a graphical view of the related literature in the intersection of machine learning and finance shows the evolution of the field over time. Financial machine learning is greatly covered in the media, but such research illustrates the relevant knowledge which can decrease the fuzziness of this fairly ambiguous area. Additionally, there are just a handful of publications in mapping the literature of financial machine learning and none of them have tried to depict the field using bibliography. According to the results obtained in this research, one can see the growing trend of publications in current years that shows the interest of the academic community in the subject.

Second, this paper reveals the key subjects, journals, and the most influential authors. This attempt focuses on the relevant literature dealing with multiple applications of machine learning in investment and finance. This shows the great emphasis of researchers in financial prediction and too much opportunity for future research on areas of empirical asset pricing and risk management. These areas are not only important in academic research but also are practical issues that reflect the concerns of practitioners in the industry.

Third, the systematic method used in this study brings a comprehensive understanding of machine learning techniques in investment research. Too many investment companies have been investing in artificial intelligence and machine learning technologies and talent in recent years. Thus, this research helps them to start more robust initiatives that can result in best practices and providing insight into different use cases.

Finally, future researchers can start comparing and assessing different machine learning techniques and datasets in a range of investment problems especially empirical asset pricing and portfolio construction. The scarcity of systematic research on such areas would be too beneficial for investment companies to produce more scientifically based investment products for society.

References

Aadith Narayan, R., Dayana, B. D., Yagneshwaran, B., Vignesh Babu, M. R., & Krishna, K. A. V. (2019). Stock value prediction using machine learning. *Int. J. Recent Technol. Eng., 7*(6), 839–843.

Abe M., & Nakayama H. (2018). Deep learning for forecasting stock returns in the cross-section, *Lect. Notes Comput. Sci. Subser. Lect. Notes Artif. Intell. Lect. Notes Bioinforma.*, vol. 10937 LNAI, pp. 273–284, doi: https://doi.org/10.1007/978-3-319-93034-3_22.

Aggarwal A., Gupta I., Garg N., & Goel A. (2019). Deep learning approach to determine the impact of socio economic factors on bitcoin price prediction, in *2019 12th International Conference on Contemporary Computing, IC3 2019*, doi: 10.1109/IC3.2019.8844928.

Z. Ahmadi, P. Martens, C. Koch, T. Gottron, and S. Kramer, "Towards bankruptcy prediction: Deep sentiment mining to detect financial distress from business management reports," in Proceedings - 2018 IEEE 5th international conference on data science and advanced analytics, DSAA 2018, 2019, pp. 293–302, doi: 10.1109/DSAA.2018.00040.

Akansu, A. N., Kulkarni, S. R., & Malioutov, D. (2016). *Overview: Financial signal processing and machine learning*. Wiley-IEEE Press.

Almahdi, S., & Yang, S. Y. (2019). A constrained portfolio trading system using particle swarm algorithm and recurrent reinforcement learning. *Expert Systems with Applications, 130*, 145–156. https://doi.org/10.1016/j.eswa.2019.04.013.

Ang, A. (2013). Factor investing. *Social Science Research Network, Rochester, NY, SSRN Scholarly Paper ID, 2277397.*

Arévalo A., Nino J., León D., Hernandez G., & Sandoval J. (2018). Deep learning and wavelets for high-frequency price forecasting, *Lect. Notes Comput. Sci. Subser. Lect. Notes Artif. Intell. Lect. Notes Bioinforma.*, vol. 10861 LNCS, pp. 385–399, doi: 10.1007/978-3-319-93701-4_29.

Arnott, R. D., Harvey, C. R., & Markowitz, H. (2018). A backtesting protocol in the era of machine learning. *Social Science Research Network, Rochester, NY, SSRN Scholarly Paper ID, 3275654.*

Asness, C. S. (2016). The siren song of factor timing. *Social Science Research Network, Rochester, NY, SSRN Scholarly Paper ID, 2763956.*

Asness, C. S., Chandra, S., Ilmanen, A., & Israel, R. (2017). Contrarian factor timing is deceptively difficult. *Social Science Research Network, Rochester, NY, SSRN Scholarly Paper ID, 2928945.*

Atsalakis, G. S., & Valavanis, K. P. (2009). Surveying stock market forecasting techniques – Part II: Soft computing methods. *Expert Systems with Applications, 36*(3), 5932–5941. https://doi.org/10.1016/j.eswa.2008.07.006.

Baldominos, A., Blanco, I., Moreno, A. J., Iturrarte, R., Bernárdez, Ó., & Afonso, C. (2018). Identifying real estate opportunities using machine learning. *Applied Sciences, 8*(11). https://doi.org/10.3390/app8112321.

Bao, W., Yue, J., & Rao, Y. (2017). A deep learning framework for financial time series using stacked autoencoders and long-short term memory. *PLoS One, 12*(7). https://doi.org/10.1371/journal.pone.0180944.

Batra R., & Daudpota S. M. (2018). Integrating StockTwits with sentiment analysis for better prediction of stock price movement, in *2018 International Conference on Computing, Mathematics and Engineering Technologies: Invent, Innovate and Integrate for Socioeconomic Development, iCoMET 2018 - Proceedings*, vol. 2018-January, pp. 1–5, doi: 10.1109/ICOMET.2018.8346382.

Bayraktar, M., Aktas, M. S., Kalipsiz, O., Susuz, O., & Bayraci, S. (2018). Credit risk analysis with classification restricted Boltzmann machine [Siniflandirilmis Kisitli Boltzmann Makinesi ile

Kredi risk Analizi]. In *26th IEEE signal processing and communications applications conference, SIU 2018* (pp. 1–4). https://doi.org/10.1109/SIU.2018.8404397.

Bean, A. J., & Singer, A. C. (2009). Factor graphs for universal portfolios. *Conference Record - Asilomar Conference on Signals, Systems and Computers, Pacific Grove, CA*, 1375–1379. https://doi.org/10.1109/ACSSC.2009.5469881.

Bektić, D., Wenzler, J.-S., Wegener, M., Schiereck, D., & Spielmann, T. (2019). Extending Fama–French factors to corporate bond markets. *Journal of Portfolio Management, 45*(3), 141–158. https://doi.org/10.3905/jpm.2019.45.3.141.

Bender, J., Blackburn, T., & Sun, X. (2019). Clash of the titans: Factor portfolios versus alternative weighting schemes. *Journal of Portfolio Management, 45*(3), 38–49. https://doi.org/10.3905/jpm.2019.45.3.038.

Bhanja, S., & Das, A. (2019). Deep learning-based integrated stacked model for the stock market prediction. *Int. J. Eng. Adv. Technol., 9*(1), 5167–5174. https://doi.org/10.35940/ijeat.A1823.109119.

Blažiunas, S., & Raudys, A. (2019). Comparative study of neural networks and decision trees for application in trading financial futures. In *Proceedings - 2019 international conference on deep learning and machine learning in emerging applications, deep-ML 2019* (pp. 33–38). https://doi.org/10.1109/Deep-ML.2019.00015.

Blitz, D. (2015). Factor investing revisited. *Social Science Research Network, Rochester, NY, SSRN Scholarly Paper ID, 2626336.*

Blitz, D., & Vidojevic, M. (2019). The characteristics of factor investing. *Journal of Portfolio Management, 45*(3), 69–86. 2019. https://doi.org/10.3905/jpm.2019.45.3.069.

Butaru, F., Chen, Q., Clark, B., Das, S., Lo, A. W., & Siddique, A. (2016). Risk and risk management in the credit card industry. *J. Bank. Finance, 72*, 218–239. https://doi.org/10.1016/j.jbankfin.2016.07.015.

Cavalcante, R. C., Brasileiro, R. C., Souza, V. L. F., Nobrega, J. P., & Oliveira, A. L. I. (2016). Computational intelligence and financial markets: A survey and future directions. *Expert Systems with Applications, 55*, 194–211. https://doi.org/10.1016/j.eswa.2016.02.006.

Cerniglia, J., & Fabozzi, F. J. (2018). Academic, practitioner, and investor perspectives on factor investing. *Journal of Portfolio Management, 44*(4), 10–16. https://doi.org/10.3905/jpm.2018.44.4.010.

Cerniglia, J. A., Fabozzi, F. J., & Kolm, P. N. (2016). Best practices in research for quantitative equity strategies. *Journal of Portfolio Management, 42*(5), 135–143. https://doi.org/10.3905/jpm.2016.42.5.135.

Chan, K. C., Gup, B. E., & Pan, M.-S. (1997). International stock market efficiency and integration: A study of eighteen nations. *J. Bus. Finance Account., 24*(6), 803–813. https://doi.org/10.1111/1468-5957.00134.

Chen, K., Zhou, Y., & Dai, F. (2015). A LSTM-based method for stock returns prediction: A case study of China stock market. In *2015 IEEE international conference on big data (big data)* (pp. 2823–2824). https://doi.org/10.1109/BigData.2015.7364089.

Chiong, R., Adam, M. T. P., Fan, Z., Lutz, B., Hu, Z., & Neumann, D. (2018). A sentiment analysis-based machine learning approach for financial market prediction via news disclosures. In *GECCO 2018 companion - proceedings of the 2018 genetic and evolutionary computation conference companion* (pp. 278–279). https://doi.org/10.1145/3205651.3205682.

Choi S., & Renelle T. (2019). Deep learning price momentum in US equities, in *Proceedings of the International Joint Conference on Neural Networks*, vol. 2019-July, doi: 10.1109/IJCNN.2019.8852067.

Chong, E., Han, C., & Park, F. C. (2017a). Deep learning networks for stock market analysis and prediction: Methodology, data representations, and case studies. *Expert Systems with Applications, 83*, 187–205. https://doi.org/10.1016/j.eswa.2017.04.030.

Chong, E., Han, C., & Park, F. C. (2017b). Deep learning networks for stock market analysis and prediction: Methodology, data representations, and case studies. *Expert Systems with Applications, 83*, 187–205. https://doi.org/10.1016/j.eswa.2017.04.030.

Chung, H., & Shin, K.-S. (2018). Genetic algorithm-optimized long short-term memory network for stock market prediction. *Sustainable Switzerland, 10*(10). https://doi.org/10.3390/su10103765.

Cohen, R. B., Polk, C., & Silli, B. (2010). Best ideas. *Social Science Research Network, Rochester, NY, SSRN Scholarly Paper ID, 1364827.*

Cong, L. W., & Xu, D. (2016). Rise of factor investing: Asset prices, informational efficiency, and security design. *Social Science Research Network, Rochester, NY, SSRN Scholarly Paper ID, 2800590.*

Cremers, M., & Petajisto, A. (2009). How active is your fund manager? A new measure that predicts performance. *Social Science Research Network, Rochester, NY, SSRN Scholarly Paper ID, 891719.*

Day, M.-Y., & Lee, C.-C. (2016). Deep learning for financial sentiment analysis on finance news providers. In *Proceedings of the 2016 IEEE/ACM international conference on advances in social networks analysis and mining, ASONAM 2016* (pp. 1127–1134). https://doi.org/10.1109/ASONAM.2016.7752381.

Deep Learning in Finance [Online]. Available: https://arxiv.org/abs/1602.06561. Accessed 9 Jan 2020.

Deng, Y., Bao, F., Kong, Y., Ren, Z., & Dai, Q. (2017). Deep direct reinforcement learning for financial signal representation and trading. *IEEE Trans. Neural Netw. Learn. Syst., 28*(3), 653–664. https://doi.org/10.1109/TNNLS.2016.2522401.

Fama, E. F. (1991). Efficient capital markets: II. *J. Finance, 46*(5), 1575–1617. https://doi.org/10.1111/j.1540-6261.1991.tb04636.x.

Fama, E. F. (1998). Market efficiency, long-term returns, and behavioral finance1The comments of Brad Barber, David Hirshleifer, S.P. Kothari, Owen Lamont, Mark Mitchell, Hersh Shefrin, Robert Shiller, Rex Sinquefield, Richard Thaler, Theo Vermaelen, Robert Vishny, Ivo Welch, and a referee have been helpful. Kenneth French and Jay Ritter get special thanks.1. *Journal of Financial Economics, 49*(3), 283–306. https://doi.org/10.1016/S0304-405X(98)00026-9.

Fama, E. F., & French, K. R. (1992). The cross-section of expected stock returns. *J. Finance, 47*(2), 427–465. https://doi.org/10.1111/j.1540-6261.1992.tb04398.x.

Fama, E. F., & French, K. R. (2009). Luck versus skill in the cross section of mutual fund returns. *Social Science Research Network, Rochester, NY, SSRN Scholarly Paper ID, 1356021.*

Fama, E. F., & French, K. R. (2015). Dissecting anomalies with a five-factor model. *Social Science Research Network, Rochester, NY, SSRN Scholarly Paper ID, 2503174.*

Fomin F., Golovach P., Panolan F., & Simonov K. (2019). Refined complexity of PCA with outliers, in *36th International Conference on Machine Learning, ICML 2019*, vol. 2019-June, pp. 10204–10213.

Gu, S., Kelly, B. T., & Xiu, D. (2019). Empirical asset pricing via machine learning. *Social Science Research Network, Rochester, NY, SSRN Scholarly Paper ID, 3159577.*

Hafiz, A., Lukumon, O., Muhammad, B., Olugbenga, A., Hakeem, O., & Saheed, A. (2015). Bankruptcy prediction of construction businesses: Towards a big data analytics approach. In *Proceedings - 2015 IEEE 1st international conference on big data computing service and applications, BigDataService 2015* (pp. 347–352). https://doi.org/10.1109/BigDataService.2015.30.

Hasan A., Kalipsiz O., & Akyokuş S. (2017). Predicting financial market in big data: Deep learning [Büyük Verilerde Finansal Piyasa Tahmini: Derin Öğrenme], in *2nd International Conference on Computer Science and Engineering, UBMK*, 2017; 510–515, doi: 10.1109/UBMK.2017.8093449.

Heaton, J. B., Polson, N. G., & Witte, J. H. (2017). Deep learning for finance: Deep portfolios. *Appl. Stoch. Models Bus. Ind., 33*(1), 3–12. https://doi.org/10.1002/asmb.2209.

Heiden E. et al. (2017). Web text-based network industry classifications: Preliminary results, in *Proceedings of the 3rd International Workshop on Data Science for Macro-Modeling with Financial and Economic Datasets, DSMM 2017 - In conjunction with the ACM SIGMOD/PODS Conference*, doi: 10.1145/3077240.3077245.

Helmbold, D. P., Schapire, R. E., Singer, Y., & Warmuth, M. K. (1998). On-line portfolio selection using multiplicative updates. *Mathematical Finance, 8*(4), 325–347. https://doi.org/10.1111/1467-9965.00058.

Henrique, B. M., Sobreiro, V. A., & Kimura, H. (2019). Literature review: Machine learning techniques applied to financial market prediction. *Expert Systems with Applications, 124*, 226–251. https://doi.org/10.1016/j.eswa.2019.01.012.

Hsu, M.-W., Lessmann, S., Sung, M.-C., Ma, T., & Johnson, J. E. V. (2016). Bridging the divide in financial market forecasting: Machine learners vs. financial economists. *Expert Systems with Applications, 61*, 215–234. https://doi.org/10.1016/j.eswa.2016.05.033.

Huang, C.-F. (2012). A hybrid stock selection model using genetic algorithms and support vector regression. *Appl. Soft Comput. J., 12*(2), 807–818. https://doi.org/10.1016/j.asoc.2011.10.009.

Hutchinson, J. M., Lo, A. W., & Poggio, T. (1994). A nonparametric approach to pricing and hedging derivative securities via learning networks. *J. Finance, 49*(3), 851–889. https://doi.org/10.1111/j.1540-6261.1994.tb00081.x.

Jang, H., & Lee, J. (2019). Machine learning versus econometric jump models in predictability and domain adaptability of index options. *Phys. Stat. Mech. Its Appl., 513*, 74–86. https://doi.org/10.1016/j.physa.2018.08.091.

Jensen, M. C. (1978). Some anomalous evidence regarding market efficiency. *Social Science Research Network, Rochester, NY, SSRN Scholarly Paper ID, 244159.*

Kessler, M. M. (1963). Bibliographic coupling between scientific papers. *American Documentation, 14*(1), 10–25. https://doi.org/10.1002/asi.5090140103.

Khandani, A. E., Kim, A. J., & Lo, A. W. (2010). Consumer credit-risk models via machine-learning algorithms. *J. Bank. Finance, 34*(11), 2767–2787. https://doi.org/10.1016/j.jbankfin.2010.06.001.

Kim, H. Y., & Won, C. H. (2018). Forecasting the volatility of stock price index: A hybrid model integrating LSTM with multiple GARCH-type models. *Expert Systems with Applications, 103*, 25–37. https://doi.org/10.1016/j.eswa.2018.03.002.

Kolanovic, M., & Krishnamachari, R. T. (2017). *Machine Learning and Alternative Data Approach to Investing* (p. 280).

Lachiheb, O., & Gouider, M. S. (2018). A hierarchical deep neural network design for stock returns prediction. *Procedia Computer Science, 126*, 264–272. https://doi.org/10.1016/j.procS.2018.07.260.

LeCun, Y., Bengio, Y., & Hinton, G. (2015). Deep learning. *Nature, 521*(7553), 436–444. https://doi.org/10.1038/nature14539.

Li, Z., Tam, V., & Yeung, L. (2016). Combining cloud computing, machine learning and heuristic optimization for investment opportunities forecasting. In *2016 IEEE congress on evolutionary computation, CEC 2016* (pp. 3469–3476). https://doi.org/10.1109/CEC.2016.7744229.

Lo, A. W. (2007). Where do alphas come from?: A new measure of the value of active investment management. *Social Science Research Network, Rochester, NY, SSRN Scholarly Paper ID, 985127.*

Lo, A. W., & Mackinlay, A. C. (1987). Stock market prices do not follow random walks: Evidence from a simple specification test. *Social Science Research Network, Rochester, NY, SSRN Scholarly Paper ID, 346975.*

Lopez de Prado, M. (2016). Building diversified portfolios that outperform out-of-sample. *Social Science Research Network, Rochester, NY, SSRN Scholarly Paper ID, 2708678.*

Lopez de Prado, M. (2017). The 7 reasons most machine learning funds fail (presentation slides). *Social Science Research Network, Rochester, NY, SSRN Scholarly Paper ID, 3031282.*

Lopez de Prado, M. (2018a). Ten financial applications of machine learning (presentation slides). *Social Science Research Network, Rochester, NY, SSRN Scholarly Paper ID, 3197726.*

Lopez de Prado, M. (2018b). Advances in financial machine learning (chapter 1). *Social Science Research Network, Rochester, NY, SSRN Scholarly Paper ID, 3104847.*

Lopez de Prado, M. (2019). Beyond econometrics: A roadmap towards financial machine learning. *Social Science Research Network, Rochester, NY, SSRN Scholarly Paper ID, 3365282.*

Malkiel, B. G. (2003). The efficient market hypothesis and its critics. *The Journal of Economic Perspectives, 17*(1), 59–82. https://doi.org/10.1257/089533003321164958.

Malkiel, B. G., & Fama, E. F. (1970). Efficient capital markets: A review of theory and empirical work. *J. Finance, 25*(2), 383–417. https://doi.org/10.1111/j.1540-6261.1970.tb00518.x.

Man, X., Luo, T., & Lin, J. (2019). Financial sentiment analysis (FSA): A survey. In *Proceedings - 2019 IEEE international conference on industrial cyber physical systems, ICPS 2019* (pp. 617–622). https://doi.org/10.1109/ICPHYS.2019.8780312.

Manurung, A. H., Budiharto, W., & Prabowo, H. (2018). Algorithm and modeling of stock prices forecasting based on long short-term memory (LSTM). *ICIC Express Lett., 12*(12), 1277–1283. https://doi.org/10.24507/icicel.12.12.1277.

Mariano, E. B., Sobreiro, V. A., & Rebelatto, D. A. d. N. (2015). Human development and data envelopment analysis: A structured literature review. *Omega, 54*, 33–49. https://doi.org/10.1016/j.omega.2015.01.002.

Matin, R., Hansen, C., Hansen, C., & Mølgaard, P. (2019). Predicting distresses using deep learning of text segments in annual reports. *Expert Systems with Applications, 132*, 199–208. https://doi.org/10.1016/j.eswa.2019.04.071.

Merton, R. C. (1980). On estimating the expected return on the market: An exploratory investigation. *Social Science Research Network, Rochester, NY, SSRN Scholarly Paper ID, 262067.*

Milian, E. Z., Spinola, M. d. M., & de Carvalho, M. M. (2019). Fintechs: A literature review and research agenda. *Electronic Commerce Research and Applications, 34*, 100833. https://doi.org/10.1016/j.elerap.2019.100833.

Mishev, K., et al. (2019). Performance evaluation of word and sentence embeddings for finance headlines sentiment analysis. *Commun. Comput. Inf. Sci., 1110*, 161–172. https://doi.org/10.1007/978-3-030-33110-8_14.

Mondal, B. (2019). Artificial intelligence: State of the art. *Intell. Syst. Ref. Libr., 172*, 389–425. https://doi.org/10.1007/978-3-030-32644-9_32.

Mullainathan, S., & Spiess, J. (2017). Machine learning: An applied econometric approach. *The Journal of Economic Perspectives, 31*(2), 87–106. https://doi.org/10.1257/jep.31.2.87.

Nahil, A., & Abdelouahid, L. (2019). Portfolio construction using KPCA and SVM: Application to Casablanca stock exchange. *Adv. Intell. Syst. Comput., 915*, 885–895. https://doi.org/10.1007/978-3-030-11928-7_80.

Nelson, D. M. Q., Pereira, A. C. M., & de Oliveira, R. A. (2017). Stock market's price movement prediction with LSTM neural networks. In *2017 international joint conference on neural networks (IJCNN)* (pp. 1419–1426). https://doi.org/10.1109/IJCNN.2017.7966019.

Ni, L., Li, Y., Wang, X., Zhang, J., Yu, J., & Qi, C. (2019). Forecasting of forex time series data based on deep learning. *Procedia Computer Science, 147*, 647–652. https://doi.org/10.1016/j.procs.2019.01.189.

Patel, J., Shah, S., Thakkar, P., & Kotecha, K. (2015a). Predicting stock market index using fusion of machine learning techniques. *Expert Systems with Applications, 42*(4), 2162–2172. https://doi.org/10.1016/j.eswa.2014.10.031.

Patel, J., Shah, S., Thakkar, P., & Kotecha, K. (2015b). Predicting stock and stock price index movement using trend deterministic data preparation and machine learning techniques. *Expert Systems with Applications, 42*(1), 259–268. https://doi.org/10.1016/j.eswa.2014.07.040.

Ramos-Rodríguez, A.-R., & Ruíz-Navarro, J. (2004). Changes in the intellectual structure of strategic management research: A bibliometric study of the strategic management journal, 1980–2000. *Strategic Management Journal, 25*(10), 981–1004. https://doi.org/10.1002/smj.397.

Rapach, D., & Zhou, G. (2013). Forecasting stock returns. In *Handbook of economic forecasting* (Vol. 2, pp. 328–383). G. Elliott and A. Timmermann: Eds. Elsevier.

Rodrigues, F., Markou, I., & Pereira, F. C. (2019). Combining time-series and textual data for taxi demand prediction in event areas: A deep learning approach. *Inf. Fusion, 49*, 120–129. https://doi.org/10.1016/j.inffus.2018.07.007.

Schwert, G. W. (2003). Anomalies and market efficiency. In *Handbook of the economics of finance* (Vol. 1, pp. 939–974). Elsevier.

Sezer O. B., Gudelek M. U., & Ozbayoglu A. M. (2019). Financial time series forecasting with deep learning: A systematic literature review: 2005–2019, *ArXiv191113288 Cs Q-Fin Stat.*

Shen, W., Wang, J., Jiang, Y.-G., & Zha, H. (2015). Portfolio choices with orthogonal bandit learning. In *Twenty-fourth international joint conference on artificial intelligence.*

Shiller, R. J., & Perron, P. (1985). Testing the random walk hypothesis: Power versus frequency of observation. *Social Science Research Network, Rochester, NY, SSRN Scholarly Paper ID, 294073.*

Teng, H.-W., & Lee, M. (2019). Estimation procedures of using five alternative machine learning methods for predicting credit card default. *Review of Pacific Basin Financial Markets and Policies, 22*(3). https://doi.org/10.1142/S0219091519500218.

Tsai, C.-F. (2009). Feature selection in bankruptcy prediction. *Knowledge-Based Systems, 22*(2), 120–127. https://doi.org/10.1016/j.knosys.2008.08.002.

Tsai, C.-F., & Wu, J.-W. (2008). Using neural network ensembles for bankruptcy prediction and credit scoring. *Expert Systems with Applications, 34*(4), 2639–2649. https://doi.org/10.1016/j.eswa.2007.05.019.

Wang X., Liu X., Matwin S., & Japkowicz N. (2014). Applying instance-weighted support vector machines to class imbalanced datasets, in *Proceedings - 2014 IEEE International Conference on Big Data, IEEE Big Data 2014*, pp. 112–118, doi: 10.1109/BigData.2014.7004364.

Yao, J., Li, Y., & Tan, C. L. (2000). Option price forecasting using neural networks. *Omega, 28*(4), 455–466. https://doi.org/10.1016/S0305-0483(99)00066-3.

Chapter 11
Technology Intelligence Map: Nanotubes

Sercan Ozcan, Nazrul Islam, and Tuğrul U. Daim

This chapter develops and demonstrates a technology mapping process, which explores collaborative innovation mechanisms. The framework is applied to the case of nanotubes. Through the use of extensive patent analysis followed by expert interviews, this chapter maps the collaborations in real time. The results indicate that Asian organizations are leading the nanotube field by leveraging small to medium enterprises (SMEs) and the electronics industry. We found that mono-linkage collaborations are an effective model which considers development progress of both parties. Furthermore, firms in this sector seem to be at the initial stages of developing robust collaboration and innovation mechanisms, while they leverage new funding systems to work together with others right from the start of the collaboration.

11.1 Introduction

This chapter presents an exploratory process to map out mechanisms of collaborative innovation. The process is applied to the case of nanotubes. Specifically, the objectives of the chapter are to analyze the nanotube innovation system with respect to the linkages of actors, the innovation system processes, the type of networks or clusters, and national technology capabilities. This chapter focuses on nanotube

S. Ozcan
University of Portsmouth, Portsmouth, UK

N. Islam
University of Exeter Business School, Exeter, UK

T. U. Daim (✉)
Department of Engineering and Technology Management, Portland State University, Portland, OR, USA
e-mail: tugrul.u.daim@pdx.edu

T. U. Daim (ed.), *Roadmapping Future*, Applied Innovation and Technology Management, https://doi.org/10.1007/978-3-030-50502-8_11

innovation systems with regard to their network structure and collaboration mechanisms in different stages of the innovation process. It illustrates the collaboration system with a case study. This chapter attempts to answer following research questions:

- How are the leading actors linked to each other and how effective is their network?
- What is the trend with respect to the dominant actors in the nanotube case?
- What are the current network structures in terms of the linkages between and among organizations?
- What kind of collaboration mechanisms do exist at different stages of innovation process?

The process uses both quantitative and qualitative analyses to answer the above questions.

A collaborative patent generation system is compatible with the idea of changing a centralized approach (e.g., joint funds, research networks, and research centers) on organizational research that is more and more reliant on knowledge networks and markets rather than individual dominant players (Breschi et al. 2009; Huang et al. 2012; Fraunhofer 2009; Nam and Barnett 2011; Ma et al. 2009; Guellec and van Pottelsberghe de la Potterie 2001; Guan and Chen 2012b; Busom and Fernández-Ribas 2008). Patenting activities and technology diffusion in high-tech regions are being increasingly driven by collaborative, international, and technology-based new entrants, such as spinoffs and SMEs (Gredel et al. 2012; Qian and Chen 2011). Technology diffusion is highly dependent on a market structure. The demand for new inventions, and thus for patents, is increasing greatly. However, this may not be the case with emerging technologies, as the demand may need to be created or the process needs to be supported by other actors so that technology diffusion is efficient.

This study specifically focuses on collaborative patenting activities in nanotechnology, as strong relationships between private and public actors have gained importance in improving the efficiency of innovation systems. This research analyzes the innovation systems of nanotube-related activities in three different ways. First, it analyzes the network of patenting activities in the field with regard to different linkages of actors, as well as the structure of the networks. Second, it explores different collaboration mechanisms at different stages of the innovation process. Third, it presents a nanotube-related case study where it shows the linkages between an academic and an industrial organization. Furthermore, this research provides a comparative study with regard to regional and organizational differences of these innovation systems. Thus, governments, universities, and private sector companies can benefit from these research findings.

The nanotube is one of the most mature nanostructures available today, so an analysis of the patents in this field is significant, as there are more applications for nanotubes and the technology is closer to commercialization than almost any other form of nanotechnology. This study attempts to find potential emerging technologies by analyzing patent documents and mapping technology trends.

Although there have been previous studies that used similar methods patent analysis and focused on implementing patent networks, these studies did not particularly focus on the nanotube case and did not look at how nano-patents are linked to each other in a quantitative and qualitative manner. This study also supports and extends its findings with interview data that is collected from various nanotechnology experts. Furthermore, the study improves upon patent data collection methods with regard to the reliability of analyzed patent documents.

The results indicate the following:

- Asian organizations are the leaders in the nanotube field.
- SMEs are the key type of organizations.
- Applications are concentrated in the electronics industry.
- Mono-linkage collaborations were found to be an effective model by considering the development of both parties.

Organizations in this sector seem to be at the initial stages of developing robust collaboration and innovation mechanisms. New funding systems were identified as ways to force companies to work together right from the start of the collaboration.

11.2 Literature Review and Theoretical Background

Innovation Systems

Innovation management theories have evolved over time. Following Schumpeter (1934), many management scholars have studied innovation. These include the technology push theory (Souder 1989; Brem and Voigt 2009; Herstatt and Letti 2004; Walsh 1984) and the market pull theory (Walsh 1984; Walsh et al. 2002; Chidamber and Kon 1994; Scherer 1982; Nemet 2009). Building on these prior studies, Lundvall (1992) introduced a new model which he termed systems of innovation. The innovation system comprises the linkages and flow of information among actors, such as inventors and organizations, that are engaged in innovation processes (Lundvall 1992; Liu and White 2001; Guan and Chen 2012a; Doloreux 2002; Yim and Kang 2008). An innovation system model aims to describe the processes and interactions between actors to facilitate the value chain from the beginning of an invention to a commercialized innovation stage (Yim and Kang 2008; Roper et al. 2008). Various studies have been based on the innovation system concept, including national systems of innovation (Lundvall 1992; Freeman 1995; Nelson 1993; Mahroum and Al-Saleh 2013; Samara et al. 2012; Hobday et al. 2012; Kramer et al. 2011), regional innovation systems (Cooke and Morgan 1994; Cooke and Morgan 1998), sectoral systems of innovation (Malerba 2004), technological innovation systems (Carlson and Stankiewicz 1991), and functions in innovation systems (Johnson 1998).

The technological innovation system appears to be the most suitable innovation system theory for nanotechnology field. A sectoral system of innovation is not a suitable model, as the nanotechnology field cannot be identified as a specific sector

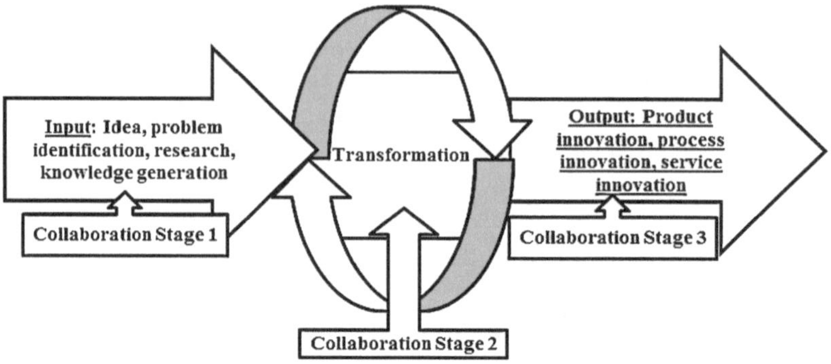

Fig. 11.1 Proposed framework for the stages of collaboration

due to its dispersed nature. The nanotechnology field is highly diffused into other major fields, such as chemistry, physics, and biology. Thus, nanotubes are used in various areas at the industrial level, such as in chemical components, electronic devices, and cosmetic products. It is also not possible to limit nanotechnology innovation systems to national or regional boundaries, since currently the main financial resources are provided by common sources where various organizations from different countries apply for funding. Furthermore, nowadays the research approach of global organizations is highly differentiated at the national, international, and even global level where they have research facilities. The framework illustrates the general model (see Fig. 11.1). It would be expected that the highest level of collaboration would occur at the transformation stage. General linkages between academia and industry will be presented by patent analysis. Types of collaborations for nanotube technology at different stages will be explained by interview analysis that is gathered from nanotechnology experts in the field.

Innovation Network Studies

Existing inter-organizational collaboration models and network studies regarding innovation can be divided into five subcategories: (1) impact of collaboration mechanisms; (2) determinants that affect collaboration mechanisms; (3) types of inter-organizational collaboration for the innovation generation process; (4) inter-organizational collaboration related to comparative studies; (5) and network studies. Each of these subcategories will be discussed in more detail in the following section.

The first subcategory is related to the impact of collaboration mechanisms in innovation processes and outcomes. In general, this subcategory has three research motivations: (1) identifying benefits (Beaudry and Schiffauerova 2011; Cuijpers et al. 2011; Patrakosol and Olson 2007; Petzel et al. 2010; Liyanage 1995), (2) identifying negative effects (Lokshin et al. 2011; Baba et al. 2009), and (3) identifying risks of collaboration mechanisms (Mata and Woerter 2013; Kloyer and Scholderer 2012).

Some studies identified benefits of collaboration mechanisms pointed to cost reduction (Cuijpers et al. 2011), innovation performance (Cuijpers et al. 2011), evolutionary improvement (Patrakosol and Olson 2007), patent quality (Beaudry and Schiffauerova 2011), and sustainability of innovation (Petzel et al. 2010). On the other side, a number of studies identified the negative effects and risks of collaboration, such as prolonging research time frame (Cuijpers et al. 2011), and negative effects of repeated collaboration on patent quality (Beaudry and Schiffauerova 2011), and risks of outsourcing (Kloyer and Scholderer 2012).

The second subcategory focuses on determinants that affect collaboration mechanisms. This section covers various determinants such as geographical (Smith and Dickson 2003; Ben Letaifa and Rabeau 2013; McKelvey et al. 2003; de Jong and Freel 2010; Petruzzelli 2011; Montobbio and Sterzi 2013), technological (Vaccaro et al. 2010; Tsai and Wang 2009; Santamaría et al. 2009), organizational (Katila and Mang 2003; Mora-Valentin et al. 2004; Cowan and Daim 2011), economic (Salter and Martin 2001; Kim et al. 2012), societal/individual factors (Calamel et al. 2012; Rycroft 2003; Blomqvist et al. 2005), and barriers/challenges (Tripsas et al. 1995; Lakitan 2013) for inter-organizational collaboration. Furthermore, there are large numbers of studies that focus on how involvement of different types of actors (academia, industry, intermediaries, suppliers, etc.) affects collaboration mechanisms (Kang and Park 2012; Guan and Zhao 2013; Baba et al. 2009; Tether and Tajar 2008; Drejer and Jørgensen 2005). This is due to the changing nature of research practices as current funding systems show significant influence from governmental funding systems to force academia and industry to work together.

The third subcategory is types of inter-organization collaboration for the innovation generation process. Different collaboration mechanism occurs when there are differences in types of actors (NGOs, intermediaries, suppliers, etc.) (Wright et al. 2008; Fawcett et al. 2012; Fitjar and Rodríguez-Pose 2013; Nieto and Santamaría 2007), sizes of actors (large organization, SMEs, global firms, etc.) (Gnyawali and Park 2011; Bougrain and Haudeville 2002; Leiponen and Byma 2009), types of partnering (R&D consortia, acquisitions, alliances, etc.) (De Man and Duysters 2005), geographical boundaries (regional, national, international, or global) (Solleiro and Gaona 2012; Ma et al. 2009; Ma and Lee 2008; Nam and Barnett 2011), sectorial boundaries (electronics, pharmaceutical, automobile, etc.) (Malerba 2004; McKelvey et al. 2003; Muscio and Nardone 2012), types of products (tangible and intangible) (Wu 2012; Nieto and Santamaría 2007; Tsai and Wang 2009; Ray and Kanta Ray 2011; Sakata et al. 2013), cross/inter-sectorial (Murphy et al. 2012; Adams et al. 2013) or cross/inter-discipline (Alzarooni et al. 2011; Swan et al. 2007), and types of innovation paradigm (open innovation) (Wi et al. 2011).

The fourth subcategory involves inter-organizational collaboration related to comparative studies. A number of studies have compared geographical locations (nations or regions) (Van Beers et al. 2008; Chang and Shih 2004) and types of sectors (i.e., nanotechnology vs. biotechnology) (Genet et al. 2012).

The final subcategory is related to network studies. This subcategory involves network studies that are related to knowledge flow analyses (i.e., co-citations) (Guan and Chen 2012b), inter-organizational linkage analyses (i.e., co-authorship) (Velden et al. 2010; Beaudry and Schiffauerova 2011), and gathering visual information (key actors, clusters, network structure, etc.) (Shapira et al. 2011). Metrics also varied. Some papers focused on the involvement of scientists in this field in terms of number of publications (Kostoff et al. 2007), citations (Motoyama and Eisler 2011), and diffusion of this field into other fields (Miyazaki and Islam 2007). With regard to patent analysis, there were similar approaches followed by various academics. Porter and Youtie (2009) looked at nanotechnology positions in relation to other disciplines by considering its multidisciplinary nature. Miyazaki and Islam (2007) focused on cross-country comparisons, actors, and institutions. They used similar quantitative methods (bibliometrics and technology mining) to understand the sectorial innovation systems in nanotechnology from a global perspective. Shapira et al. (2011) focused on an overview of corporate entry into nanotechnology through patents and publications and nanotechnology innovation factors in the shift to commercialization. Huang et al. (2003) completed similar work by presenting a longitudinal patent analysis on nanotechnology patents between 1976 and 2002. Their work included content map analysis and citation network analysis by obtaining the required data from individual countries, institutions, and technology fields.

Nanotechnology Innovation Systems

Studies related to nanotechnology innovation systems use a number of different innovation models, such as national (Chamas 2008; Ghazinoory et al. 2009; Pardo-Guerra 2011) and sectoral (Miyazaki and Islam 2007; Rampersad et al. 2010). There are also studies focused on the interactions of actors and collaboration mechanism within nanotechnology innovation systems (Beaudry and Schiffauerova 2011; Pandza et al. 2011; Raesfeld et al. 2012). Van Merkerk and van Lente (2005) focused on analyzing emerging irreversibilities underlying the dynamics of ongoing activities of key nanotube actors. Their aim was to trace technology in socio-cognitive patterns by considering three interrelated levels of analysis: research groups, the technological field, and society. Gupta and Pangannaya (2000) performed a bibliometric analysis of patents to gather information on the inventive activity in nanotube technology from key regions and organizations within the field. They identified the most active companies and the technological trends. Wry and Lounsbury (2013) studied the nanotube field in terms of patent classifications and the size of investments within this field by looking at start-up ventures in nanotube technology. Their study examines the implications for which firms received the largest investments. Köhler and Som (2013) studied the emergence of nanotextiles. Baglieri et al. (2012) studied nanocluster evolution. Lee and Song (2007) studied clusters in the nanotechnology field. Wonglimpiyarat (2005) explored the general emergence of nanotechnology.

Looking at previous studies with regard to method used, such as innovation system studies and nanotube case studies, a number of gaps can be identified. First, there are not many studies that are related to the collaboration mechanism of nanotechnology innovation systems, and there is not any previous literature that focuses on collaboration mechanism that is specific to the nanotube field. Therefore, the purpose of this study is to fill these gaps and to clarify important determinants with regard to the linkages of actors within this technology domain, as well as to present a case study about how collaboration of different actors resulted in carbon nanotube

(CNT)-related products. Furthermore, as this is a rapidly developing field, an updated study is required to present current trends and changes in nanotube technology. Current research gaps in related fields are listed below, and they can be referred to the nanotechnology field and nanotube technology as well:

- Nanotechnology is a highly dispersed field, so it is difficult to define and categorize where the technology is diffused into and the relationship between various actors.
- Due to externalities of this technology, it is difficult to have preaccepted borders, such as at national level or sectoral level for nanotechnology-related studies.
- There are difficulties to identify how knowledge flow occurs between interdisciplinary scientists and between organizations from different fields.
- There are few studies that explain successful collaboration mechanisms stage by stage and how nanotechnology-related product has been commercialized.
- There are very limited numbers of nanotube-related patent analyses available with up-to-date data.
- There is no study focused on collaboration mechanisms that are specific to nanotube research and products.

This study aims to analyze the nanotube innovation system with respect to the linkages of actors, the innovation system process, the type of networks or clusters, and national technology capabilities. This study looks at the nanotube innovation system with regard to its network structure and collaboration mechanisms in different stages of its innovation process and uses a case study to illustrate the collaboration system. This study attempts to answer the following fundamental issues:

- How the leading actors are linked to each other and how effective their networks are?
- What the trend is with respect to the dominant actors in the nanotube field?
- What the current network structures are in terms of the linkages between and among organizations?
- What kind of collaboration mechanisms exist at different stages of innovation process?

11.3 Research Methodology and Data Analysis

This study uses a technology mining methodology proposed by Porter and Cunningham (2005), combining bibliometrics using patent abstracts from patent databases. Patent analyses have been used extensively in research for a number of purposes, including identification of new technology opportunities (Lee et al. 2009), identification of partners (Geum et al. 2013), assessment of innovations (Abraham and Moitra 2001), evaluating innovation networks (Goetze 2010), and forecasting technologies (Daim et al. 2006, 2007, 2008; Harell and Daim 2009).

Table 11.1 The outline of research methods

Quantitative work	Qualitative work
Patent database selection Patent search Patent data optimization Patent data analysis The outcome of patent analysis	Selection of interviewees according to the patent analysis The design of interview questions according to the quantitative work Case study illustration Gathering key factors for collaboration mechanism

Technology mining can be used to determine relations between actors and technologies within a given innovation system. This study used specialist keywords derived publications by the Nano Science and Technology Institute. Subsequent analysis was performed using dedicated technology mining software called Thomson Data Analyser (TDA). The software automates mining and clustering of terms occurring in article abstracts and article descriptors, such as authors, affiliations, or keywords that it recommends. Table 11.1 shows an outline of research methods used in this study. Gathering valid patent data, efficient analysis of large data sets, and handling and interpreting the outcomes of the analysis are crucial for the accuracy of the results.

Considering the limitations and drawbacks of previous patent data collection approaches in related studies, it was thought that the best nanotechnology search practice would be to use all available nanotechnology classifications to gather all relevant patents. These classifications include 977 by USPTO, B82 by IPC, Y01N by ECLA, and 3C082 by Japanese F-Terms. All irrelevant patents classified within these categories could be eliminated by using Boolean search logic with very broad nanotechnology-related terms, such as 'nano*', 'quantum*', and 'fullerene*'. Afterward, Derwent Patent Index (DWPI) is used to exclude patents that appeared more than once in the search results, since patents are often granted more than once in various patenting authorities to secure inventions in those respective countries or regions. As a result, 49,544 individual nanotechnology patents were obtained for the period from 1970 to 2012. The obtained results were imported into the Thomson Data Analyser (TDA) and, to validate results further, the duplicate results were eliminated and variations of company, inventor, institute, and university names were unified where they appeared as separate patent assignees. After the dataset was prepared, various functions were utilized using the same tool, Thomson Data Analyser, to generate the required analysis.

11.3.1 Leading Organizations and Their Research Focus in CNTs

As shown in Table 11.2, the leading organizations in the nanotube field are Foxconn (Hon Hai Precision), and Samsung and Tsinghua University (Qinghua University). At the national level, the United States and Japan are the leading countries, but

Chinese and Korean organizations appear to dominate nanotube technology overall. Interestingly, all the Asian organizations became involved in nanotube patenting activity after the millennium. The most significant patent increases occurred for Foxconn and Tsinghua University, which grew by over 26%. Table 11.2 shows that the key application for nanotubes is in the electronics industry, as the main patent holders are companies in this field.

Among the top patent holders, there are only two academic organizations: Tsinghua University and the University of California. A large number of private organizations involved in nanotube technology can be seen in Table 11.2. The

Table 11.2 Profile of nanotube organizations

Fujitsu Ltd	134	JP [113];US [29];WO [20]	Kondo Daiyu [31];Awano Yuji [28];Kawabata Akio [21];Nihei Mizuhisa [21]	1998–2010	9% of 134
University California	135	US [133];WO [73]	Zettl Alexander Karlwalter [14];Jin Sungho [11];Zhan Guodong [7]	1995–2010	7% of 135
Sony Corp	141	JP [124];WO [33];US [15]	Kajiura Hisashi [47];Ata Masafumi [26];Shiraishi Seiji [17];Yagi Takao [17]	1993–2009	3% of 141
Toray Industries Inc	145	JP [139];WO [18];KR [3]	Tsukamoto Jun [34];Sato Kenichi [28];Ozeki Yuji [27]	2000–2010	8% of 145
Nantero Inc	150	US [146];WO [32]	Rueckes Thomas [118];Segal Brent M. [77];Bertin Claude L. [64]	2001–2010	4% of 150
Nec Corp	157	JP [157];WO [60];US [6]	Iijima Sumio [37];Yudasaka Masako [30];Miyamoto Yoshiyuki [20]	1992–2010	15% of 157
Dokuritsu Gyosei Hojin Sangyo Gijutsu	229	JP [226];WO [57];US [7]	Yumura Morio [48];Hata Kenji [44];Kataura Hiromichi [38]	1999–2011	14% of 229
University Qinghua	397	CN [391];US [307];JP [25]	Fan Shou Shan [286];Jiang Kai Li [199];Liu Liang [168]	2002–2010	28% of 397
Samsung Electronics Co Ltd	485	KR [470];US [325];EP [52]	Han In Taek [33];Choi Jae Young [28];Lee Young Hee [27]	1999–2010	10% of 485
Hon Hai Precision Ind Co Ltd	557	CN [429];US [371];TW [117]	Fan Shou Shan [369];Jiang Kai Li [268];Liu Liang [197]	2002–2010	27% of 557
Organization names	*Number of records*	*Top countries*	*Top people*	*Year range*	*Percentage of records in last 3 years*

University of California is the leading academic player in the United States, further strengthening its country's dominance furthering the field. The involvement of a large number of private organizations plays a vital role in the technology diffusion process, as well as in the commercialization process, given of the large number of patents they hold and their role within technology transfer activity networks.

11.3.2 Organizational Network of Patent Co-ownership

One of the strongest linkages in the nanotube technology field appears to be between Hon Hai Precision (Foxconn) and Qinghua University (Tsinghua University) (see Fig. 11.2). These two organizations they jointly own 376 patent documents on nanotube technology. This is the largest number of patent collaborations in the nanotube technology field. Beijing Funate Innovation is a spin-off organization that shares 20 patents with Foxconn and Tsinghua University. The second highest number patent collaborations is between two Japanese organizations they share 24 nanotech patents between them. The third highest degree of collaboration is between South Korean players, Samsung and Seoul National University. They have a strong linkage, sharing eight nanotube patents. Seoul National University (SNU) is one of the leading players in graphene as well, and Samsung and SNU collaborate in various other nanotechnology fields. Looking at the general picture for nanotube technology, the strongest cluster occurs in the Japanese nanotechnology innovation system. This is a good example of a decentralized network as there are many central players, such as Dokuritsu Gyosei Hojin Sangyo Gijutsu, NEC, and the Japan Science and Technology Agency. The Japanese innovation system appears to have a great network in terms of linking public and private organizations. Furthermore, the national innovation system is linked to international organizations. This shows how nanotechnology has diffused from a national level to an international level. The Japanese cluster appears to be decentralized, and this type of cluster has the best characteristics in terms of its stability and efficiency.

South Korea appears to have a highly centralized network around Samsung, but it appears that the network is growing and there are some international linkages with other networks. This is due to Samsung's multinational conglomerate size and their international research collaborations. The reason why the Korean innovation system appears centered around Samsung is because they fund a variety of academic organizations and have agreements with them. Samsung is also the largest South Korean chaebol. Considering the leading position of the United States in terms of number of nanotube patents, it would be expected that US players would be at the center of patent activity collaboration. However, Japan and South Korea have a much greater degree of collaborative involvement. Another interesting result that can be seen in Fig. 11.2 is that even though there are large numbers of patents in China between three organizations – Foxconn, Tsinghua University, and Beijing Funate Innovation – the Chinese nanotube cluster does not look very effective when the numbers of linkages are considered. Cross-country collaboration occurs between various

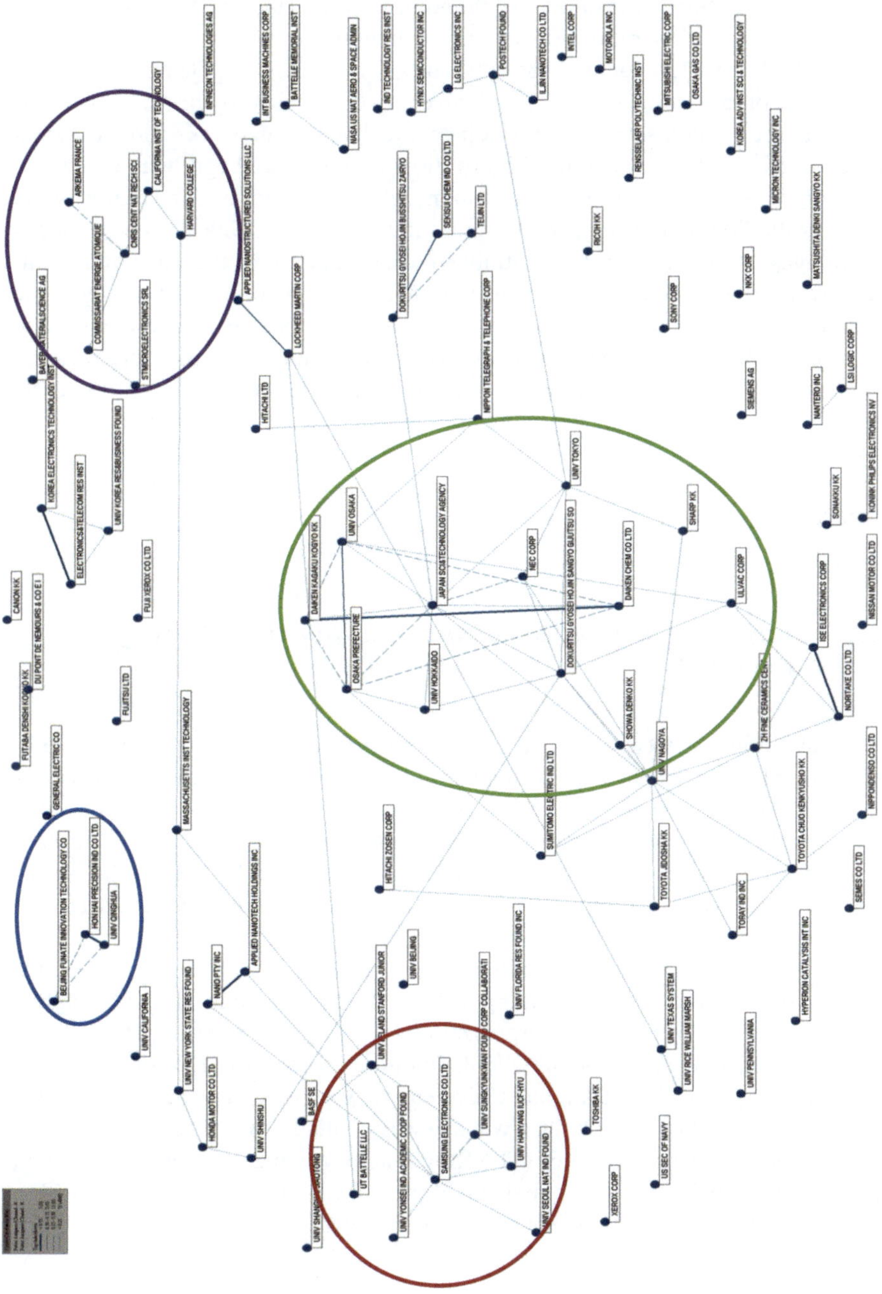

Fig. 11.2 Institutional network of nanotubes

countries. The strongest international collaboration as a whole appears between the United States and France. Specifically, the strongest linkage appears to be between CNRS and the California Institute of Technology. Some large organizations are not involved in any collaboration in nanotube patenting activity, such as Sony and Toshiba. IBM owns 124 nanotube patents, and only one of their patents is co-owned with another organization, namely with University Cambridge Technology Services. Looking at the general structure of nanopatent technology networks and clusters, it can be inferred that the structure of innovation systems may begin with a key collaboration between two or more organizations which agree to form the bidirectional linkage or the first narrow-scoped cluster, as in the case of China. This new formation enlarges and establishes the centralized cluster due to the presence of a dominant player in the system, such as Samsung. After centralized clusters develop, the structure evolves to a decentralized cluster model, such as the Japanese case. The next stage is the international connection of organizations that takes place as the cluster moves to the stage where there is an established network. For this case, Samsung is a great example, considering how they created their network of national and international linkages. It is also interesting to see the development of the Korean innovation system and how a marketing-oriented network moves toward being a complete innovation network.

11.3.3 Case Illustration: Academic and Industrial Linkage for CNTs

The collaboration between Foxconn and Tsinghua University is the strongest linkage identified by the patent analysis in this study. For that reason, it was very important to interview somebody at the managerial level in the Tsinghua-Foxconn Nanotechnology Research Center (TFNRC). Therefore, Professor Fan Shoushan[1] was interviewed to help analyze the findings of this research. The results of this interview were then used to find key determinants for such successful collaborations.

The collaboration between Tsinghua University and Foxconn started in 2000 with Foxconn's target of participating with a strong academic player. Tsinghua University was one of the largest organizations with high number of patents in the nanotechnology field. With the initial collaboration mechanism, Foxconn offered a grant of $5 million USD to Tsinghua University to work together with it for 5 years. A new department was built, which was called the Tsinghua-Foxconn Nanotechnology Research Center (TFNRC). One of the main targets for Foxconn was to increase their amount of intellectual property in this field and work with a partner who was doing fundamental research. Many patents were applied for

[1] Prof. Fan Shoushan is the Director of the Tsinghua-Foxconn Nanotechnology Research Center, and he is the second highest number of patent holder in the nanotechnology field (398 patents).

together by Foxconn and Tsinghua University. Currently, all of their nanotechnology patents are shared by both sides. According to their agreement, both parties can use each others' patents. But, if there is a third party that would like to collaborate, then both parties have to agree on it. Most of their collaborations are related to CNTs, such as CNT sensors and field emission displays (FEDs). From 2000 to 2011, $15 million USD have been spent on this collaboration. Another contract for $150 million USD will continue this collaboration through 2020.

The collaboration between Foxconn and Tsinghua University shows how the needs of an academic and industrial actor can be linked. Foxconn needed an organization with expertise in the fundamental research, and Tsinghua University needed an industrial player who could invest fund and had expertise in the commercialization and mass production of the technology.

Tsinghua University has transferred many nanotechnology-related patents to Foxconn. One of their technology transfers was related to the nanopolymer materials. This new material that Tsinghua University developed is a fire-resistant material that is not toxic, and it will be used in various applications, such as electronic cables. The fire-resistant material is the result of joint research by Tsinghua's Chemical Engineering Department and Professor Fan Shoushan. Multiple departments at Tsinghua University have been involved in work with TFNRC. Tsinghua University and Foxconn have a committee composed of member from both organizations that decides what projects should be funded. The current structure of the internal and external collaboration mechanism can be seen in Fig. 11.3.

Tsinghua University provides their various inventions and ideas in the nanotechnology field to Foxconn, and then, they agree on which of these inventions are transferable to the market. Foxconn's focus is to decide which of Tsinghua University's inventions have the most commercialization potential. Foxconn reviews research proposals twice a year to see which ones have the highest potential. Figure 11.4 presents the relationship between Tsinghua University and Foxconn. There are a number of factors that encourage Foxconn to continue investing in Tsinghua University. One of the key factors is the number of shared patents.

One of the main problems that has been identified between academia and industry that academic researchers are unaware of the needs of industry and mainly focus on fundamental research where the outcome of their work is not always possible to commercialize. Conversely, most companies do not engage in fundamental research and focus mainly on commercial applications. Furthermore, industries function on different time scales, as some have very rapid development cycles, and research has to be profitable and beneficial in a short-term. So, both industry and academia need to work better together to commercialize new inventions. Foxconn has expertise in commercialization and mass production of nanotechnology. Tsinghua University has a strong research capability for working in this highly dispersed field. Therefore, this mono-structured collaboration linkage has worked well.

There can be barriers for collaboration mechanisms involving third parties. This is why both Foxconn and Tsinghua University must agree when third-party collaborations are involved. So far, neither side has focused on these types of relationships, so their collaboration mechanism has remained highly effective.

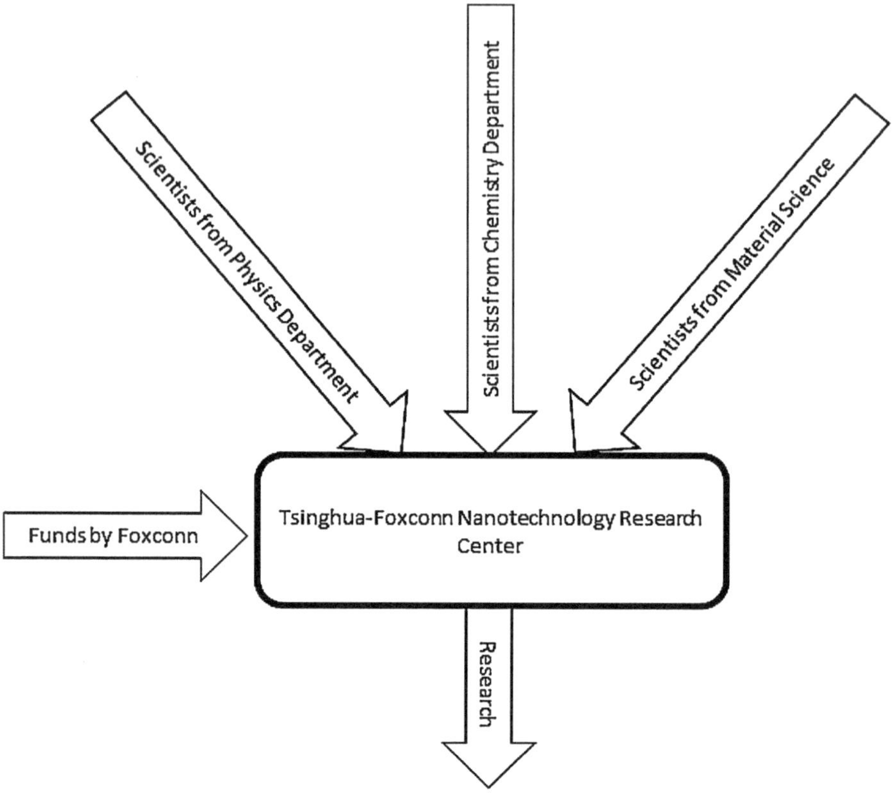

Fig. 11.3 Linkage of internal collaboration with external party

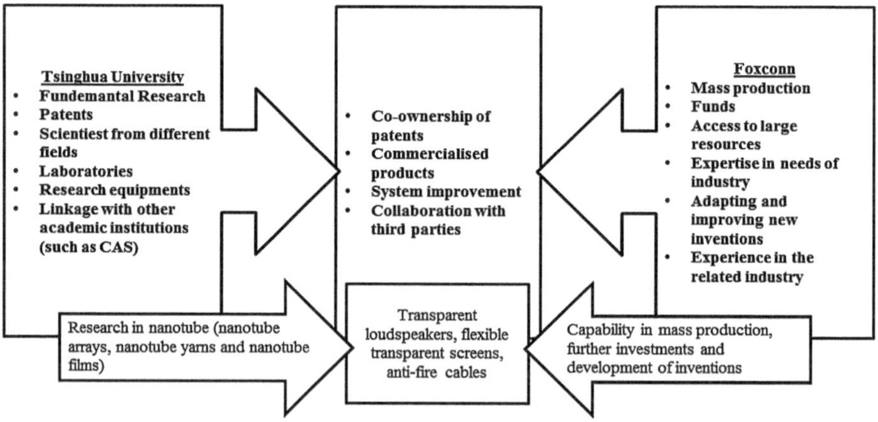

Fig. 11.4 The collaboration between Tsinghua University and Foxconn

Therefore,mono-linkages between large academic and industrial players appear to be effective when considering their own technological development and innovation processes, but it does not appear to be a very effective model for broader innovation system, including other SMEs and academic institutions.

In the relationship between Foxconn and Tsinghua University, it was interesting to note that there were successful commercializations of nanotechnologies from the beginning of the commercialization stages to the end. This is not common, since many nanotechnology-related products are still at an early stage of commercialization. Both industry and academia are currently focused on the development of their own processes. Since nanotubes were first invented, Professor Fan Shoushan has made great progress. The first major breakthrough was controlling the growth of CNTs in the production process. The second step was the mass production of CNTs, becoming the first organization in the nanotechnology field to achieve this. Tsinghua University is capable of transferring CNTs into long films in multi-walled carbon nanotubes (MWCNTs), and it is possible to produce single-walled carbon nanotubes (SWCNTs) and multi-walled carbon nanotubes (MWCNTs). The first products using transparent conductive films were touchscreens for mobile phones. Now, carbon nanotube thin films are used to produce touch screen appliances that replace formerly used materials, like indium tin oxide (ITO). Furthermore, this technology can be used for organic light-emitting diode (OLEDs) to make screens flexible. Flexible displays have the potential to change many current technologies. These are some of examples from their fundamental research findings. The core finding was how to make nanotubes into film. Two key factors were extremely important for this technology. The first is that it has unique characteristics that other materials do not have, and the second is that it can be mass-produced. It can then be used by many types of companies to apply it to a wide range of technologies. Tsinghua University now focuses on nanotube yarns, nanotube arrays, and nanotube films, which have potential applications in many different areas.

To manage such linkages between academia and industry, linkages and support at the CEO and president level were found to be a key factor. Personal relationships between parties and trust between them were vital element as well. Most importantly, there had to be a mutual benefit and both parties had to have a need for each other. For the academic side of these collaborations, it was important to know the types of laboratories, equipment, patents, and researchers available at these research institutions. It appears that some collaborations progressed were the result of highly personal initiatives, so personal relationships and trust was important for these types of collaborations. Also, collaborations among academic institutions were found to be easier than industrial collaborations, since scientists from different departments and universities frequently already share labs for their academic work with other universities. For example, Tsinghua University collaborates with other academic organizations, since they are a member of the National Center for Nanoscience and Technology, along with the Chinese Academy of Science and Peking University. Mono-level industrial-academic collaborations allow for academic linkages, but they create barriers for involvement of other industry players. Furthermore, nanotechnology centers in universities have important role considering the dispersed nature of the nanotechnology field. Many researches in this field require scientists from different fields, so universities serve an important function in enabling and encouraging these researchers to come together.

Tsinghua University has made some remarkable achievements in the nanotechnology field, both in fundamental and applied research. Examples of these achievements range from the discovery of super-aligned CNT arrays to the commercialization of real CNT products. Some additional applications include electrodes for batteries, supercapacitors, transparent loudspeakers, and flexible displays. Many products such as transmission electron microscopy grids and touch screens are ready to be commercialized. Super-aligned CNT products appear to have wide potential and are expected to be available in the market in the near future. Most importantly, the mass production problem of many of these products has been solved, which gives them the potential to be mass marketed in coming years. Furthermore, these new CNT materials have three advantages over previous materials used for these purposes: First, these materials are cheaper than the previous materials; second, they have a longer useful life, or duty cycle; and third, they are flexible materials that can be bent into a variety of shapes.

Figure 11.4 shows how collaborations between an academic organization and a corporate organization can increase the commercial potential of nanotechnology. As we have seen in previous examples, these kinds of collaborations and technology transfer processes have allowed many ideas from academic nanotechnology research to be made into viable commercial products. Tsinghua University generated many patent on nanotube technology since 1995, but only after the investment from Foxconn, and the expertise they shared about the electronics industry was it able to turn these inventions into commercialized products. Furthermore, Foxconn had access to the market channels where these technologies could be offered to the end-users, including large clients, such as Apple, Sony, and Acer.

11.3.4 Collaboration Practices in the Innovation Process

Based on the interviews with nanotechnology-related companies, a number of stages were identified in their innovation processes, from the beginning of idea creation or problem recognition to the last phase of innovation. Figure 11.5 shows five different stages of collaboration mechanisms.

In the input phase, collaborations are mainly internal between scientists. However, there are two current approaches that encourage organizations to collaborate at this level. The first is a collaborative funds application, such as the FP7 program. The aim of this program is to merge various nanotechnology organizations onto the same project. Using this mechanism, it is possible to align the work of academia and industry from the beginning, such as in the example the production mechanism of graphene. Another approach, at very early stages, is direct investment by one large organizations in another. This mainly occurs between industry players and academia. For example, large organizations like Samsung, Toyota, and Foxconn

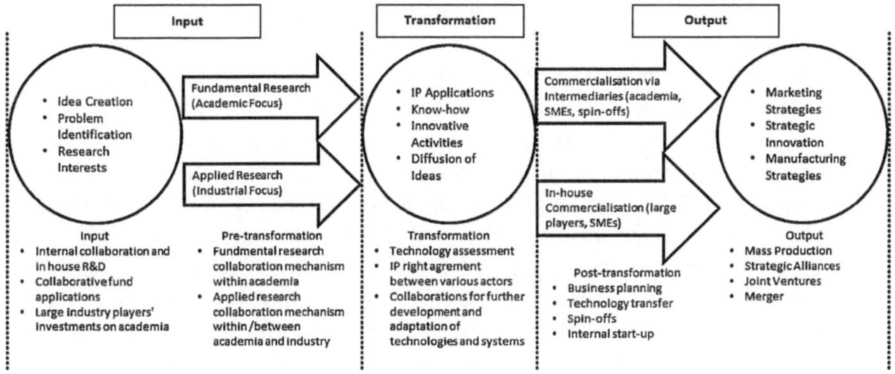

Fig. 11.5 Stages of collaboration mechanisms

invest in universities to establish research facilities that work to meet industry requirements from the very first stages of their research work.

The next stage is called pretransformation. This stage and its related collaborations are often difficult to distinguish from previous one, but it is the collaboration stage that starts the process of conducting research. It is not truly the beginning of the research process, since many preliminary steps have already been taken. For example, many scientists identify the requirement for support from different parties during the early research work. Other preliminary considerations include the requirement for expertise from different fields, the requirement for different research tools, or the requirement for tests and measurements of their samples by industry players. Also, due to the large investments at stake, many SMEs will want to send their samples to be tested, be measured, and have characterizations of their materials completed before they finalize their research and move on to the next stage. Academic institution often approaches industrial players at this stage to learn from them about applied research needs.

Collaborations in the transformation stage occur mainly when an organization has IP rights to a technology, but may not have the applied knowledge to commercialize it. It is often difficult to establish trust relationships in these types of situations, since the IP owner tends to want to maintain the secrecy of the technology and often is unsure of which organizations it should share its technology with to achieve the best commercialization results. Although some cases were found where collaborations started without the IP rights, these kinds of collaboration structures would be included in collaboration mechanisms 1 and 2. This is one of the stages where there is usually high involvement by different types of organizations other than the main industry players. Some intermediary actors are also highly involved at this stage, such as consultants, advisors, patent authorities, and IP lawyers. There were some collaboration mechanisms that are found where industrial organizations collaborated with academic organizations to further improve their existing patents. Therefore, IP rights are not used only for commercialization or as a barrier for competitors, but it also allows organizations to collaborate on their existing technology to further improve or protect it.

Collaborations in the posttransformation stage fall into two broad categories: (1) where organizations transfer the technology to other parties; or (2) where they try to commercialize it themselves. Technology transfer models are the most collaborative structures of the mechanisms discussed so far. This level usually has the highest involvement for SMEs, as they rely on academia rather than their in-house R&D. However, large organizations also sometimes collaborate at this level, such as in the case of Foxconn and Tsinghua University. But, these are rare cases, since significant investments are required. Furthermore, actors who would like to commercialize products themselves still rely on other parties for mass production. This is a key issue in the nanotechnology field that needs to be addressed.

In the output stage, collaborations occur mainly between industrial players. An example of this is that there are some organizations that have the expertise to produce certain nanoparticles, and they have access to the market, but they cannot fulfil the needs of industry, so they collaborate with some other industrial players to produce the required materials for them. Then, they process these materials to produce their own products and provide it to their clients. At this level, it was found that many nanotechnology industrial players are not willing to collaborate with each other. But, while this is not generally a highly collaborative stage, some industrial players were found to be collaborating through strategic alliances. SMEs often have higher motivations to do this, especially if their collaborative funds are provided by governments or other networks.

Based on all the previous research up to this point, several positive factors were found that support collaborations and several negative factors were found that deter collaboration in the nanotechnology field. Key determinants for collaboration are listed in Table 11.3.

11.4 Conclusions

This research provided the opportunity to construct an international profile of nanotube technology. This provided many type of valuable information, including identification of key regions where nanotube patents are being produced. This research also identified the dominant countries in key nanotechnology domains and the dominant players within those countries. An interesting outcome was to see the changing trends of involvement by different countries in nanotube technology. Asian players appear to have very high involvement in this area. It appears that South Korea and China are now catching up with Japan and close to the United States in terms of the number of nanotube patent granted.

Networks and clusters for nanotube technology vary greatly from one country to another. The largest network that was found is Samsung's centralized network in South Korea. This network has international linkages with many other countries, including several US-based organizations. This is due to the international

Table 11.3 Key determinants for collaboration

Supportive factors for collaboration	Deferent factors for collaboration
IP rights as a positive factor Joint funds Industry expertise Research capability Organizational learning Requirement for different disciplines Financial benefits Risk minimization Technology transfer Difficulties in mass production Reputation of the company Expertise of the organization Personal issues, such as trust Industrial problems To avoid IP restrictions To access other organizations Previous students or colleagues from the same organization	IP rights as a negative factor Secrecy of information Legal agreements and restrictions Competitor companies Lack of innovation potential for a product or service Relying on know-how Shortness of deadlines Negative structure of a network or cluster Requirement for large investments Required radical changes in traditional systems Negative perceptions related to cosmetics, medicines, etc.

externalities of multinational companies, such as Samsung. At the opposite extreme regarding international externalities, the strongest collaboration in nanotube patenting activities was found to exist between the United States and France. There was a high degree of co-ownership by French and US organizations, both in the academic and in private spheres. However, it was found that the main focus of these relationships was within the electronics sector.

According to network theory, three structures exist: centralized, decentralized, and distributed networks. The general structure of nanotube networks was found to be somewhere between centralized and decentralized, and it was very far from being a distributed network structure. This means that the network relies greatly on organizations such as Samsung, which is an organization that dominates Korea's centralized network. It would be expected that the United States would have the most efficient network or cluster structure, but looking at its patent activities, the United States has a national cluster rather than a network, and the number of collaborating organizations is lower than in the Korean case. Another surprising emerging from this research is that the Chinese collaboration system is not very strong in terms of linkages between private and public organizations. The key linkage in the Chinese context is between Tsinghua University and Foxconn – an organization is headquartered in Taiwan, but has most of its production assets in China. In China, the number of collaborative organizations would need to be increased to move it to the stage where there is an innovative cluster that can increase the technology diffusion process. This research suggests that the government should take action to bring this about.

With respect to the key actors in the nanotube field, the electronics industry's ownership of patents is dominated mostly by large organizations. There are two main reasons why there is considerable heterogeneity in nanotube patenting activity. First, large organizations have the capability to provide the large investment necessary for R&D activities, and they are aware of the benefits of nanotube technology in terms of its efficiency and its nature for bringing about incremental innovation characteristics. Second, they collaborate with academic organizations, such as universities, to benefit from their inventions as well. The second point is not found in every national innovation system, but of those that do, Korea, the United States, and Japan appear to have the most effective systems.

The following points summarize the most important implications of this study:

- Asian organizations, especially in South Korea and the Chinese region, appear to be having a great impact in the nanotube field.
- There is a sector concentration in the nanotube field (mainly the electronics industry). However, key capabilities of each country and various organizations were identified, and the application of nanotube technology is highly varied.
- Mono-linkage collaborations are effective models, such as in Tsinghua-Foxconn example. However, this model appears to be as a barrier for further collaboration with other industrial organizations.
- In terms of collaboration and innovation models, nanotube technology was found to be in its initial stage, where various centralized clusters or networks exist. However, some nations, such as South Korea, Japan, and the United States, are far ahead in terms of the stability and efficiency of their networks in the nanotube field.
- Looking at the various stages of collaborations, new funding systems appear to be a good way to force companies to work together from the beginning collaborations.
- In most of the stages of collaboration, it appears that SMEs are currently the key organizations in the nanotube technology, and they are the key types of organizations that link industry and academia.

Acknowledgments The authors gratefully acknowledge the valuable advice of Prof. Fred Phillips, Prof. Philip Shapira, and Prof. Scott W. Cunningham. The authors also wish to express their gratitude for the financial support given by Rowland's Foundation and the Aberystwyth University Research Fund. They are grateful to nanotechnology experts Prof. Fan Shoushan, Prof. Yonghai Chen, Dr. Bizhong LI, and Prof. Chun-Ru Wang for giving up their time for the interviews. The paper was presented at the IM2012 Beijing Conference and thanks for suggestion made by those experts who were at the conference.

References

Abraham, B. P., & Moitra, S. D. (2001). Innovation assessment through patent analysis. *Technovation, 21*(4), 245–252.

Adams, P., Fontana, R., & Malerba, F. (2013). The magnitude of innovation by demand in a sectoral system: The role of industrial users in semiconductors. *Research Policy, 42*(1), 1–14.

Alzarooni, S. A., Campbell, R. W., Wang, Y., & Miller, C. Z. (2011). Exploring the strategic value of interdisciplinary collaboration: COINs in the creation of business. *Procedia - Soc. Behav. Sci., 26*, 130–135.

Baba, Y., Shichijo, N., & Sedita, S. R. (2009). How do collaborations with universities affect firms' innovative performance? The role of "Pasteur scientists" in the advanced materials field. *Research Policy, 38*(5), 756–764.

Baglieri, D., Cinici, M. C., & Mangematin, V. (2012). Rejuvenating clusters with "sleeping anchors": The case of nanoclusters. *Technovation, 32*(3), 245–256.

Beaudry, C., & Schiffauerova, A. (2011). Impacts of collaboration and network indicators on patent quality: The case of Canadian nanotechnology innovation. *European Management Journal, 29*(5), 362–376.

Ben Letaifa, S., & Rabeau, Y. (2013). Too close to collaborate? How geographic proximity could impede entrepreneurship and innovation. *Journal of Business Research, 66*(10), 2071–2078.

Blomqvist, K., Hurmelinna, P., & Seppänen, R. (2005). Playing the collaboration game right—Balancing trust and contracting. *Technovation, 25*(5), 497–504.

Bougrain, F., & Haudeville, B. (2002). Innovation, collaboration and SMEs internal research capacities. *Research Policy, 31*(5), 735–747.

Brem, A., & Voigt, K.-I. (2009). Integration of market pull and technology push in the corporate front end and innovation management—Insights from the German software industry. *Technovation, 29*(5), 351–367.

Breschi, S., Cassi, L., Malerba, F., & Vonortas, N. (2009). Networked research: European policy intervention in ICTs. *Technology Analysis and Strategic Management, 21*(7), 833–857.

Busom, I., & Fernández-Ribas, A. (2008). The impact of firm participation in R&D programmes on R&D partnerships. *Research Policy, 37*(2), 240–257.

Calamel, L., Defélix, C., Picq, T., & Retour, D. (2012). Inter-organizational projects in French innovation clusters: The construction of collaboration. *International Journal of Project Management, 30*(1), 48–59.

Carlson, B., & Stankiewicz, R. (1991). On the nature, function and composition of technological systems. *Journal of Evolutionary Economics, 1*(2), 93–118.

Chamas, C. (2008). Nanotechnology intellectual property in Brazil: Preliminary research note. *World Pat. Inf., 30*(2), 146–149.

Chang, P.-L., & Shih, H.-Y. (2004). The innovation systems of Taiwan and China: A comparative analysis. *Technovation, 24*(7), 529–539.

Chidamber, S. R., & Kon, H. B. (1994). A research retrospective of innovation inception and success: The technology-push, demand-pull question. *International Journal of Technology Management, 9*(1), 19.

Cooke, P., & Morgan, K. (1994). The creative milieu: A regional perspective on innovation. In M. Dodgson & R. Rothwell (Eds.), *The handbook of industrial innovation* (pp. 57–89). Aldershot, England; Brookfield, Vt., USA: E. Elgar.

Cooke, P., & Morgan, K. (1998). *The associational economy: Firms, regions, and innovation*. Oxford, England; New York, USA: Oxford University Press.

Cowan, K. R., & Daim, T. U. (2011). Review of technology acquisition and adoption research in the energy sector. *Technology in Society, 33*(3–4), 183–199.

Cuijpers, M., Guenter, H., & Hussinger, K. (2011). Costs and benefits of inter-departmental innovation collaboration. *Research Policy, 40*(4), 565–575.

Daim, T., Iskin, I., Li, X., Zielsdorff, C., Bayraktaroglu, A. E., Dereli, T., & Durmusoglu, A. (2012). Patent analysis of wind energy technology using the patent alert system. *World Patent Information, 34*(1), 37–47.

Daim, T., Monalisa, M., Dash, P., & Brown, N. (2007). Time lag assessment between research funding and output in emerging technologies. *Foresight, 9*(4), 33–44.

Daim, T. U., Ploykitikoon, P., Kennedy, E., & Choothian, W. (2008). Forecasting the future of data storage: Case of hard disk drive and flash memory. *Foresight, 10*(5), 34–49.

Daim, T. U., Rueda, G., Martin, H., & Gerdsri, P. (2006). Forecasting emerging technologies: Use of bibliometrics and patent analysis. *Technological Forecasting and Social Change, 73*(8), 981–1012.
De Jong, J. P. J., & Freel, M. (2010). Absorptive capacity and the reach of collaboration in high technology small firms. *Research Policy, 39*(1), 47–54.
De Man, A.-P., & Duysters, G. (2005). Collaboration and innovation: A review of the effects of mergers, acquisitions and alliances on innovation. *Technovation, 25*(12), 1377–1387.
Doloreux, D. (2002). What we should know about regional systems of innovation. *Technology in Society, 24*(3), 243–263.
Drejer, I., & Jørgensen, B. H. (2005). The dynamic creation of knowledge: Analysing public–private collaborations. *Technovation, 25*(2), 83–94.
Fawcett, S. E., Jones, S. L., & Fawcett, A. M. (2012). Supply chain trust: The catalyst for collaborative innovation. *Business Horizons, 55*(2), 163–178.
Fitjar, R. D., & Rodríguez-Pose, A. (2013). Firm collaboration and modes of innovation in Norway. *Research Policy, 42*(1), 128–138.
Fraunhofer, I. (2009). The impact of collaboration on Europe's scientific and technological performance. Final Report to the European Commission, DG Research.
Freeman, C. (1995). The "National System of innovation" in historical perspective. *Cambridge Journal of Economics, 19*(1), 5–24.
Genet, C., Errabi, K., & Gauthier, C. (2012). Which model of technology transfer for nanotechnology? A comparison with biotech and microelectronics. *Technovation, 32*(3), 205–215.
Geum, Y., Lee, S., Yoon, B., & Park, Y. (2013). Identifying and evaluating strategic partners for collaborative R&D: Index-based approach using patents and publications. *Technovation, 33*(6), 211–224.
Ghazinoory, S., Divsalar, A., & Soofi, A. S. (2009). A new definition and framework for the development of a national technology strategy: The case of nanotechnology for Iran. *Technological Forecasting and Social Change, 76*(6), 835–848.
Gnyawali, D. R., & Park, B.-J. (2011). Co-opetition between giants: Collaboration with competitors for technological innovation. *Research Policy, 40*(5), 650–663.
Goetze, C. (2010). An empirical enquiry into co-patent networks and their stars: The case of cardiac pacemaker technology. *Technovation, 30*(7), 436–446.
Gredel, D., Kramer, M., & Bend, B. (2012). Patent-based investment funds as innovation intermediaries for SMEs: In-depth analysis of reciprocal interactions, motives and fallacies. *Technovation, 32*(9), 536–549.
Guan, J., & Chen, K. (2012a). Modeling the relative efficiency of national innovation systems. *Research Policy, 41*(1), 102–115.
Guan, J., & Chen, Z. (2012b). Patent collaboration and international knowledge flow. *Information Processing and Management, 48*(1), 170–181.
Guan, J., & Zhao, Q. (2013). The impact of university–industry collaboration networks on innovation in nanobiopharmaceuticals. *Technological Forecasting and Social Change, 80*(7), 1271–1286.
Guellec, D., & van Pottelsberghe de la Potterie, B. (2001). The internationalisation of technology analysed with patent data. *Research Policy, 30*(8), 1253–1266.
Gupta, V. K., & Pangannaya, N. B. (2000). Carbon nanotubes: Bibliometric analysis of patents. *World Patent Information, 22*(3), 185–189.
Harell, G., & Daim, T. U. (2009). Forecasting energy storage technologies. *Foresight, 11*(6), 74–85.
Herstatt, C., & Letti, C. (2004). Management of `technology push' development projects. *International Journal of Technology Management, 27*(2/3), 155–175.
Hobday, M., Boddington, A., & Grantham, A. (2012). Policies for design and policies for innovation: Contrasting perspectives and remaining challenges. *Technovation, 32*(5), 272–281.
Huang, M.-H., Dong, H.-R., & Chen, D.-Z. (2012). Globalization of collaborative creativity through cross-border patent activities. *Journal of Informetrics, 6*(2), 226–236.

Huang, Z., Chen, H., Yip, A., Ng, G., Guo, F., Chen, Z.-K., & Roco, M. C. (2003). Longitudinal patent analysis for nanoscale science and engineering: Country, institution and technology field. *Journal of Nanoparticle Research, 5*(3–4), 333–363.

Johnson, A. (1998). *Functions in innovation system approaches*. Chalmers University of Technology.

Kang, K.-N., & Park, H. (2012). Influence of government R&D support and inter-firm collaborations on innovation in Korean biotechnology SMEs. *Technovation, 32*(1), 68–78.

Katila, R., & Mang, P. Y. (2003). Exploiting technological opportunities: The timing of collaborations. *Research Policy, 32*(2), 317–332.

Kim, Y., Kim, W., & Yang, T. (2012). The effect of the triple helix system and habitat on regional entrepreneurship: Empirical evidence from the U.S. *Research Policy, 41*(1), 154–166.

Kloyer, M., & Scholderer, J. (2012). Effective incomplete contracts and milestones in market-distant R&D collaboration. *Research Policy, 41*(2), 346–357.

Köhler, A. R., & Som, C. (2013). Risk preventative innovation strategies for emerging technologies the cases of nano-textiles and smart textiles. *Technovation.*

Kostoff, R. N., Koytcheff, R. G., & Lau, C. G. Y. (2007). Global nanotechnology research literature overview. *Technological Forecasting and Social Change, 74*(9), 1733–1747.

Kramer, J.-P., Marinelli, E., Iammarino, S., & Diez, J. R. (2011). Intangible assets as drivers of innovation: Empirical evidence on multinational enterprises in German and UK regional systems of innovation. *Technovation, 31*(9), 447–458.

Lakitan, B. (2013). Connecting all the dots: Identifying the "actor level" challenges in establishing effective innovation system in Indonesia. *Technology in Society, 35*(1), 41–54.

Lee, S., Yoon, B., & Park, Y. (2009). An approach to discovering new technology opportunities: Keyword-based patent map approach. *Technovation, 29*(6), 481–497.

Lee, Y.-G., & Song, Y.-I. (2007). Selecting the key research areas in nano-technology field using technology cluster analysis: A case study based on National R&D Programs in South Korea. *Technovation, 27*(1), 57–64.

Leiponen, A., & Byma, J. (2009). If you cannot block, you better run: Small firms, cooperative innovation, and appropriation strategies. *Research Policy, 38*(9), 1478–1488.

Liu, X., & White, S. (2001). Comparing innovation systems: A framework and application to China's transitional context. *Research Policy, 30*(7), 1091–1114.

Liyanage, S. (1995). Breeding innovation clusters through collaborative research networks. *Technovation, 15*(9), 553–567.

Lokshin, B., Hagedoorn, J., & Letterie, W. (2011). The bumpy road of technology partnerships: Understanding causes and consequences of partnership mal-functioning. *Research Policy, 40*(2), 297–308.

Lundvall, B. (1992). *National innovation system: Towards a theory of innovation and interactive learning*. London: Pinter.

Ma, Z., & Lee, Y. (2008). Patent application and technological collaboration in inventive activities: 1980–2005. *Technovation, 28*(6), 379–390.

Ma, Z., Lee, Y., & Chen, C.-F. P. (2009). Booming or emerging? China's technological capability and international collaboration in patent activities. *Technological Forecasting and Social Change, 76*(6), 787–796.

Mahroum, S., & Al-Saleh, Y. (2013). Towards a functional framework for measuring national innovation efficacy. *Technovation, 33*(10), 320–332.

Malerba, F. (2004). *Sectoral systems of innovation: Concepts, issues and analyses of six major sectors in Europe*. New York: Cambridge University Press.

Mata, J., & Woerter, M. (2013). Risky innovation: The impact of internal and external R&D strategies upon the distribution of returns. *Research Policy, 42*(2), 495–501.

McKelvey, M., Alm, H., & Riccaboni, M. (2003). Does co-location matter for formal knowledge collaboration in the Swedish biotechnology–pharmaceutical sector? *Research Policy, 32*(3), 483–501.

Miyazaki, K., & Islam, N. (2007). Nanotechnology systems of innovation—An analysis of industry and academia research activities. *Technovation, 27*(11), 661–675.

Montobbio, F., & Sterzi, V. (2013). The globalization of technology in emerging markets: A gravity model on the determinants of international patent collaborations. *World Development, 44*, 281–299.

Mora-Valentin, E. M., Montoro-Sanchez, A., & Guerras-Martin, L. A. (2004). Determining factors in the success of R&D cooperative agreements between firms and research organizations. *Research Policy, 33*(1), 17–40.

Motoyama, Y., & Eisler, M. N. (2011). Bibliometry and nanotechnology: A meta-analysis. *Technological Forecasting and Social Change, 78*(7), 1174–1182.

Murphy, M., Perrot, F., & Rivera-Santos, M. (2012). New perspectives on learning and innovation in cross-sector collaborations. *Journal of Business Research, 65*(12), 1700–1709.

Muscio, A., & Nardone, G. (2012). The determinants of university–industry collaboration in food science in Italy. *Food Policy, 37*(6), 710–718.

Nam, Y., & Barnett, G. A. (2011). Globalization of technology: Network analysis of global patents and trademarks. *Technological Forecasting and Social Change, 78*(8), 1471–1485.

Nelson, R. (1993). *National innovation systems : A comparative analysis*. New York: Oxford University Press.

Nemet, G. (2009). Demand-pull, technology-push, and government-led incentives for non-incremental technical change. *Research Policy, 38*(5), 700–709.

Nieto, M. J., & Santamaría, L. (2007). The importance of diverse collaborative networks for the novelty of product innovation. *Technovation, 27*(6), 367–377.

Pandza, K., Wilkins, T. A., & Alfoldi, E. A. (2011). Collaborative diversity in a nanotechnology innovation system: Evidence from the EU framework Programme. *Technovation, 31*(9), 476–489.

Pardo-Guerra, J. P. (2011). Mapping emergence across the Atlantic: Some (tentative) lessons on nanotechnology in Latin America. *Technology in Society, 33*(1), 94–108.

Patrakosol, B., & Olson, D. L. (2007). How interfirm collaboration benefits IT innovation. *Information Management, 44*(1), 53–62.

Petruzzelli, A. M. (2011). The impact of technological relatedness, prior ties, and geographical distance on university–industry collaborations: A joint-patent analysis. *Technovation, 31*(7), 309–319.

Petzel, R., Archer, A.-M., & Fei, R. (2010). Collaboration for sustainability in a networked world. *Procedia - Social and Behavioral Sciences, 2*(4), 6597–6609.

Porter, A. L., & Cunningham, S. W. (2005). *Tech mining: Exploiting new Technologies for Competitive Advantage*. New Jersey: John Wiley & Sons.

Porter, A. L., & Youtie, J. (2009). Where does nanotechnology belong in the map of science? *Nature Nanotechnology, 4*(9), 534–536.

Qian, Q., & Chen, Y. (2011). SME, technological innovation and regional environment: The case of Guangdong, China. *Procedia Earth Planet Science, 2*, 327–333.

Raesfeld, A., von Geurts, P., Jansen, M., Boshuizen, J., & Luttge, R. (2012). Influence of partner diversity on collaborative public R&D project outcomes: A study of application and commercialization of nanotechnologies in the Netherlands. *Technovation, 32*(3), 227–233.

Rampersad, G., Quester, P., & Troshani, I. (2010). Managing innovation networks: Exploratory evidence from ICT, biotechnology and nanotechnology networks. *Industrial Marketing Management, 39*(5), 793–805.

Ray, S., & Kanta Ray, P. (2011). Product innovation for the people's car in an emerging economy. *Technovation, 31*(5), 216–227.

Roper, S., Du, J., & Love, J. H. (2008). Modelling the innovation value chain. *Research Policy, 37*(6), 961–977.

Rycroft, R. W. (2003). Technology-based globalization indicators: The centrality of innovation network data. *Technology in Society, 25*(3), 299–317.

Sakata, I., Sasaki, H., Akiyama, M., Sawatani, Y., Shibata, N., & Kajikawa, Y. (2013). Bibliometric analysis of service innovation research: Identifying knowledge domain and global network of knowledge. *Technological Forecasting and Social Change, 80*(6), 1085–1093.

Salter, A. J., & Martin, B. R. (2001). The economic benefits of publicly funded basic research: A critical review. *Research Policy, 30*(3), 509–532.

Samara, E., Georgiadis, P., & Bakouros, I. (2012). The impact of innovation policies on the performance of national innovation systems: A system dynamics analysis. *Technovation, 32*(11), 624–638.

Santamaría, L., Nieto, M. J., & Barge-Gil, A. (2009). Beyond formal R&D: Taking advantage of other sources of innovation in low- and medium-technology industries. *Research Policy, 38*(3), 507–517.

Scherer, F. (1982). *Demand-pull and technological invention: Schmookler revisited.* Washington, DC: U.S. Federal Trade Commission Bureau of Economics.

Schumpeter, J. (1934). *The theory of economic development an inquiry into profits, capital, credit, interest, and the business cycle*. Cambridge Mass: Harvard University Press.

Shapira, P., Youtie, J., & Kay, L. (2011). National innovation systems and the globalization of nanotechnology innovation. *The Journal of Technology Transfer, 36*(6), 587–604.

Smith, H., & Dickson, K. (2003). Geo-cultural influences and critical factors in inter-firm collaboration. *International Journal of Technology Management, 25*(1), 34–50.

Solleiro, J. L., & Gaona, C. (2012). Promotion of a regional innovation system: The case of the state of Mexico. *Procedia - Social and Behavioral Sciences, 52*, 110–119.

Souder, W. (1989). Improving productivity through technology push. *Research - Technology Management, 32*(2), 19–24.

Swan, J., Goussevskaia, A., Newell, S., Robertson, M., Bresnen, M., & Obembe, A. (2007). Modes of organizing biomedical innovation in the UK and US and the role of integrative and relational capabilities. *Research Policy, 36*(4), 529–547.

Tether, B. S., & Tajar, A. (2008). Beyond industry–university links: Sourcing knowledge for innovation from consultants, private research organizations and the public science-base. *Research Policy, 37*(6), 1079–1095.

Tripsas, M., Schrader, S., & Sobrero, M. (1995). Discouraging opportunistic behavior in collaborative R & D: A new role for government. *Research Policy, 24*(3), 367–389.

Tsai, K.-H., & Wang, J.-C. (2009). External technology sourcing and innovation performance in LMT sectors: An analysis based on the Taiwanese technological innovation survey. *Research Policy, 38*(3), 518–526.

Vaccaro, A., Parente, R., & Veloso, F. M. (2010). Knowledge management tools, inter-organizational relationships, innovation and firm performance. *Technological Forecasting and Social Change, 77*(7), 1076–1089.

Van Beers, C., Berghäll, E., & Poot, T. (2008). R&D internationalization, R&D collaboration and public knowledge institutions in small economies: Evidence from Finland and the Netherlands. *Research Policy, 37*(2), 294–308.

Van Merkerk, R. O., & van Lente, H. (2005). Tracing emerging irreversibilities in emerging technologies: The case of nanotubes. *Technological Forecasting and Social Change, 72*(9), 1094–1111.

Velden, T., Haque, A., & Lagoze, C. (2010). A new approach to analyzing patterns of collaboration in co-authorship networks: Mesoscopic analysis and interpretation. *Scientometrics, 85*(1), 219–242.

Walsh, S. T., Kirchhoff, B. A., & Newbert, S. (2002). Differentiating market strategies for disruptive technologies. *IEEE Transactions on Engineering Management, 49*(4), 341–351.

Walsh, V. (1984). Invention and innovation in the chemical industry: Demand-pull or discovery-push? *Research Policy, 13*(4), 211–234.

Wi, H., Oh, S., & Jung, M. (2011). Virtual organization for open innovation: Semantic web based inter-organizational team formation. *Expert Systems with Applications, 38*(7), 8466–8476.

Wonglimpiyarat, J. (2005). The nano-revolution of Schumpeter's Kondratieff cycle. *Technovation, 25*(11), 1349–1354.

Wright, M., Clarysse, B., Lockett, A., & Knockaert, M. (2008). Mid-range universities' linkages with industry: Knowledge types and the role of intermediaries. *Research Policy, 37*(8), 1205–1223.

Wry, T., & Lounsbury, M. (2013). Contextualizing the categorical imperative: Category linkages, technology focus, and resource acquisition in nanotechnology entrepreneurship. *Journal of Business Venturing, 28*(1), 117–133.

Wu, J. (2012). Technological collaboration in product innovation: The role of market competition and sectoral technological intensity. *Research Policy, 41*(2), 489–496.

Yim, D., & Kang, B. (2008). Policy options for establishing effective subnational innovation systems and technological capacity-building. *Asia-Pacific Trade Invest. Rev., 4*, 115–137.

Chapter 12
Technology Intelligence Map: Autonomous Car

Shuying Lee, Fayez Alsoubie, and Tuğrul U. Daim

Autonomous vehicle technology affects the traditional automotive industry and other related industries. However, it is much more difficult to achieve this transformation than electric car. We investigated different perspectives regarding the impact of autonomous driving through a Strengths Weaknesses Opportunities Threats (SWOT) analysis: Strength, Weakness, Opportunities, and Threats. Patent analysis and Social Network Analysis (SNA) are combined as a major tool for strategic planning to reveal the implicit R&D partnerships and explicit strategies at the company level. The case study of Google's Autonomous Vehicle Technology shows that competitive and complementary interactions influence the formation of partnerships in the market. This study shows that SNA in a complex network setting will provide the abundant and unbiased analysis to high-quality decision-making.

12.1 Introduction

Not every emerging technology becomes an advanced and mature industry. Emerging technologies should be accepted, preferred, and adapted by people. Therefore, it is necessary to develop a new product through continuous innovation, and to create the supply capacity to meet this demand in order to change the emerging technology into a mature industry. There is a general agreement in the car industry that autonomous vehicles are the future as evidenced by the significant research and development (R&D) investments undergone by automotive players and tech giants to examine the potential success of Autonomous Vehicle (AVs). Autonomous vehicle technology affects the traditional automotive industry and other related

S. Lee · F. Alsoubie · T. U. Daim (✉)
Department of Engineering and Technology Management, Portland State University, Portland, OR, USA
e-mail: tugrul.u.daim@pdx.edu

T. U. Daim (ed.), *Roadmapping Future*, Applied Innovation and Technology Management, https://doi.org/10.1007/978-3-030-50502-8_12

industries. However, it is much more difficult to achieve this transformation than electric car. We investigated the impact of autonomous driving from the perspectives of technology, the marketplace, legal issues, and infrastructure. In order to analyze the overview of the technologies, to identify potential competitors and partnerships, and to track the evolution of a landscape relative to possible strategy, we used a quantitative analysis as the major tool for strategic planning at the company level.

The case study of Google's self-driving company Waymo showed that Waymo's car should keep the best R&D for the customer-oriented technology so that it can maintain continuous cooperation with partners. Second, niche areas need to be identified before a large-scale switch, such as public transport; access for those with mobility limitations; or for campus or indoor-closed area use. Thirdly, Waymo needs to provide users and collaborators self-driving car solutions for sustaining the ecosystem to expand and realize economies of scale. Last but most important strategy is focusing on the R&D activities in critical technologies to achieve more core patents.

12.2 Literature Review

12.2.1 Different Perspective in Self-Driving Technology

Autonomous driving is recognized as the inevitable future of transportation systems. Google is leading the innovation to bring fully autonomous and commercially feasible vehicles to the market. Many vehicle companies including Ford and Tesla are racing to be the first to bring viable self-driving cars, but it is Waymo that has made significant milestones toward the goal. However, various factors have a substantial effect on the technology. Therefore, this study examines the SWOT analysis for self-driving technology and the factors that influence autonomous driving.

The strength of autonomous driving technology lies within the provision of safer transportation. Autonomous technology focuses on the elimination of human error in driving, which increases the ability to control traffic flow thus promoting a higher level of safety (Möller and Haas 2019). It seeks to reduce the fatalities associated with the current mode of transportation, which are attributable to negligence or distraction. Additionally, it offers convenience, easy operation and less stressful driving, access to transportation for all, and provides time for other activities. It also enables the providers to create an attractive user experience thus increasing operation and comfort.

The opportunities presented by the technology seem limitless. Foremost, it offers chances to obtain higher levels of road safety, fewer speed violations, and reduction in road congestion thereby facilitating the efficient flow of traffic reducing road accidents and deaths (Möller and Haas 2019). It will promote better use and shaping of scarce land in cities by reducing congested parking and eliminating parking spaces. The development of unused space will support the gradual but drastic growth of cities. The technology requires advertisements for linking the customer with the service, and thus creates new mobility and business models (Ploeger and Lacroix 2015). The provision of better experiences will enable providers like Google to obtain direct relationships with customers by creating brand loyalty. Most importantly, it will offer an opportunity for reducing carbon emissions that affect the climate. It will provide a chance for ride sharing, which will possibly dominate transportation turning it into a pay-as-you-go-service.

Nonetheless, the technology presents weaknesses and risks. The weaknesses include concerns about the reliability of the technology. Placing driving entirely on the control of automatic systems is risky and raises safety concerns (Möller and Haas 2019). Safety is contingent on the program's infrastructure and software, and there are uncertainties around the potential for failure. Furthermore, there are concerns about the ability of the systems to react and make tough decisions to road uncertainties. There are uncertainties about completion dates for the technology along with the modifications needed for implementation in some places. The vehicles are expensive for individual owners to purchase and maintain.

As with the opportunities for AV technology, the risks are also unlimited. The main danger is the uncertainty around and possibility of unforeseeable scenarios that can cause harm. The tampering of the technology is a significant threat due to possible access by unauthorized persons. The related software and apps that link with customers pose a risk of illegal access to personal data, hence the concerns about the guarantee of anonymity. Cybersecurity is the main threat that elicits privacy concerns (Ploeger and Lacroix 2015). Also, the efficient operation of self-driving vehicles requires a single network for operation and communication making it susceptible to cybercrime. Government intervention and regulation affect the potential market for self-driving cars in an effort to ensure the provision of high-level safety standards. Political factors and uncoordinated legislation can impede its progress (Ploeger and Lacroix 2015).

12.2.2 Strategic Planning Based on Patent Analysis

Patents are useful documents for competitive analysis and technical information (Liu and Shyu 1997; Tsuji 2012; Ma and Porter 2014). Patent analysis is designed to understand the IP for products and technologies important to an enterprise present and future with many implications for R&D management, competitive intelligence, and businesses strategic planning (Lee et al. 2009a). Under this kind of scenario, patent analysis has two major applications.

First, patent map analysis is used as a patent intelligence tool or system to discover the relationship between patent assignee, inventor, and classification or text in various data analytics platforms, such as Derwent Innovation, PatSnap, and Relecura (Park et al. 2013). Patent technology clustering is usually helpful in monitoring competitive environments and innovation trends at the macro level to assist companies in gaining a competitive advantage for R&D strategic planning (Lee et al. 2009b; Hsu et al. 2006). Although patent maps are widely used for their simplicity and ease of use based on patent platforms, they are limited in technical clustering based on frequency and occurrence of words or text in a document; therefore, they cannot be integrated into microlevel strategic planning.

Second, patent network analysis has been used to explore and identify R&D relationships in open innovation to reveal inside business strategy and outside challenges (Wang et al. 2009; Park et al. 2014; Pereira and Soares 2007). Cooperation networks between inventors and applicants are frequently helpful in competitor analysis, and for explaining the behavior of actors in the marketplace by emphasizing R&D activity, technology trajectories, and to what extent they build on each other's knowledge (Sternitzke et al. 2008). According to Agrawal et al. (2008), the spatial and social proximity of actors increases the probability of knowledge flows. Therefore, we believe that both the R&D cooperation network and strategic partnerships in business will be considered as factors that influence a company's strategic planning. However, there are only a few studies that explore the link between the two.

To fill the gaps, we used patent landscape and Social Network Analysis (SNA) as a tool to analyze technology clusters, to identify network and partnerships in business, to assess the competitors' strengths, weaknesses, and most importantly business strategy, and to track the evolution of a landscape relative to possible strategy.

12.3 Methodology

The overall research process is illustrated with several steps as shown in Fig.12.1.

Step1: SWOT analysis is a strategic planning tool used to specify the internal and external factors that are favorable and unfavorable to self-driving technology, and that attempts to generate the environmental impact of the Google's self-driving technology. The different perspectives, such as the market, technology, legal issues, social impact, and infrastructure, are identified through the strengths, weaknesses, opportunities, and risks (SWOT) of Google's self-driving technology.

Step2: The case study is focused on Google's self-driving company Waymo, and discusses the possible strategy it should pursue.

Step3: A patent map or patent landscape is used as a major tool to analyze the overview of the technologies, to identify potential competitors and partnerships, to assess the competitors' strengths, weaknesses and most importantly business strategy, and to track the evolution of a landscape relative to possible strategy. We used themeScape in Derwent Innovation (Möller and Haas 2019) as the patent analysis tool to show data topographically, and to identify common themes, text clustering groups' records into related sets in order to pinpoint a clear picture of the company's key technologies.

Step4: The three kinds of networks are built based on patent data, including patentee collaboration network, transfer network and citation network. Social Network

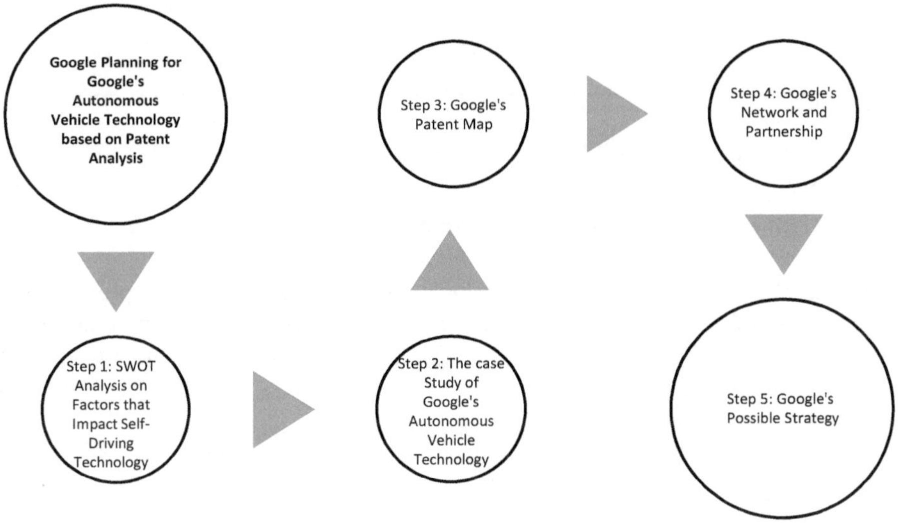

Fig. 12.1 Overall process of the research

Analysis (SNA) is used to visualize the various R&D relationships between companies. On the other hand, by analyzing the current partnerships and alliances in the press release and sorting out the competition and cooperation between companies, it is helpful to identify potential competitors and partners, in order to foresee the strategy for the development of Google driverless technology.

Step5: Based on the above conclusion, the internal and external strategies were developed comprehensively.

12.4 Factors That Impact Self-Driving Technology Based on SWOT

Driverless technology is viewed as the next phase of transportation and is important to the reduction of road accidents and deaths. However, some people perceive it as a disruptive technology, which threatens the future of the current modes of transportation. Different factors influence autonomous driving, but the main issues as shown in Fig. 12.2 include safety, reliability, trust, legal issues, and performance expectations (Benleulmi and Blecker 1970). Safety is the primary factor influencing autonomous driving and forms the basis for the development of self-driving cars. The rising rates of accidents and deaths on the roads necessitated the idea of autonomous driving to help solve the problem. Safety is also a significant concern since the vehicles depend on programs or software, which can have possible glitches or suffer cyberattacks. Autonomous vehicles present safety risks like serious crashes mainly during their first deployment (Kalra 2017). As such, the criteria for evaluating safety

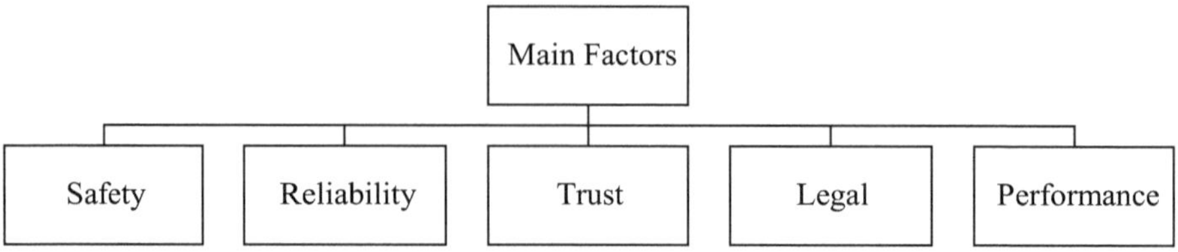

Fig. 12.2 Different factors influence autonomous driving

entails the safeguards placed in the vehicles to ensure safety, and to prevent, and respond to unforeseeable scenarios (Kaur and Rampersad 2018). The other measures include provider assurance of safety in the case of problems, and government protection to establish the liability of the use of driverless cars. Road testing does not prove or guarantee user safety.

Performance expectancy also influences autonomous driving, which affects the adoption of the technology. One of the requirements of autonomous driving is the need for high performance since failure in a single component of the driverless cars can result in fatalities. In order to encourage its adoption and usage, the first two criteria used to promote AV technology includes its high performance over traditional vehicles, and the fact that driverless cars outperform the safety records made by ordinary vehicles. There are nonautonomous cars with good safety records, and thus driverless ones must perform better. The third criterion is the associated efficiency where the use of these vehicles should improve work/life efficiency through easier and better transportation. The reliability that the autonomous cars must continuously demonstrate is another factor. Testing is one of the measures under reliability along with the cars' ability to cope and effectively react to unlikely occurrences.

Trust is fundamental to the acceptance of autonomous driving. The development of trust from the user is essential for the success of the driverless cars whose occurrence depends on the handling of concerns such as privacy and security. Self-driving cars will have associated security and privacy risks such as hacking, remote access and control, and privacy invasion through the unauthorized acquisition of personal data (Kaur and Rampersad 2018). As such, the criteria involve the likelihood of software error or failure and the guarantee of privacy and security. Further data protection and personal autonomy assurance are essential to its adoption and success. The other approach is the assurance of acceptance of liability by the manufacturers together with government assurance. Finally, the certainty of the manufacturer's technology, reliability, and safety assurance will influence the acceptance of driverless driving. All these measures promote confidence in the technology and hence promote its usage.

Legal factors have a significant impact on the technology. The reality of driverless vehicles and their feasibility for commercial use raises red flags for government intervention to primarily ensure high safety standards. Government regulations influence the making and even adoption of innovations to ensure public protection hence affecting the adoption of the autonomous technology. Regulations can either promote or cripple further developments by allowing or impede testing and deployment of the driverless technology (Schreurs and Steuwer 2015; Walker 2019). Legal standards such a high level of safety influences the costs manufacturers face in improvements to meet the standards which translated into high prices for the vehicles which affect their purchase and utilization. As such, legal factors can present challenges such as regulatory restrictions and political blowback since self-driving technology depends mostly on political and legal factors. Others factors that can affect autonomous driving include consumer behaviors due to fear of the unknown and change, road infrastructure, liability issues, and political problems like lobbying groups opposing the technology.

Concisely, there has been a massive and aggressive investment into self-driving technologies. Many companies are racing to create the technology that will lead to the inevitable transformative change. The SWOT analysis of the technology shows huge potential mainly in the provision of road safety, convenience, and the elimination of human error.

However, there are threats such as safety and privacy concerns. Autonomous driving will undoubtedly dominate the roads despite the factors influencing it adoption such as legal and safety issues. The inception of the technology hinges on political factors that will likely affect its reception.

12.5 The Case Study of Google's Autonomous Vehicle Technology

12.5.1 Data Collection

The patent race of autonomous vehicles has become more and more competitive among both automakers and tech companies. The R&D budget rose 20% for tech players and 5% for automakers. According to research conducted by Oliver Wyman and WIPO (2017), between 2012 and 2016, there were slightly more than 5000 mobility patents filed by 12 leading automakers and global tech companies. The six giant car companies—Audi, Daimler, General Motors, Volkswagen, BMW, and Tesla are focusing more on the hardware for mobility services related to green car technology, including electric cars, batteries, fuel cells, and alternative fuel. Tech companies like Google, Facebook, and Apple are pouring enormous resources into software with a 50% increase over 5 years, while the six automakers decreased their resources (Wyman 2017). Therefore, alliances and concentration of R&D resources are needed in the self-driving industry. For auto manufacturers, a more focused patent strategy is needed. There is great potential to partnership with tech companies in traffic monitoring and navigation services. For the tech companies, they focused on areas with a high level of customer interaction and engagement, which will establish a new customer interface.

The case study is focused on Google's autonomous vehicle technology for their self-driving project. Google began the self-driving car project in 2009. Waymo is an autonomous car development company and a subsidiary of Google's parent company Alphabet. Waymo is a self-driving technology company with the mission to make it safe and easy for everyone to get around—without the need for anyone in the driver's seat. Although Google began the self-driving project in 2009 and is racking up more patents than most automakers on connected and self-driving cars, it still lacks experience in building, maintaining, and engineering of cars. The purpose of case study is to discuss the future of autonomous automotive industry in detail, and what kind of strategy Google should pursue, which has significant investments in this technology.

In our case study, we used patent landscape as major tool to analyze the overview of the technologies, to identify potential competitors and partnerships, to assess the competitors' strengths, weaknesses, and most importantly business strategies, and to track the evolution of a landscape relative to possible strategy. We retrieved patent data in the U.S. from 2013 to 2017 on Derwent Innovation (DI) on May 8th in 2019. In line with patent analysis, the technology strategy will be made for Google's related investments. The news data on Waymo were retrieved from the website Autoblog on February 26, 2018 (https://www.autoblog.com/tag/waymo/).

12.5.2 *Data Analysis and Results*

12.5.2.1 Patent Race in Self-Driving Technology

According to patent analysis for automated driving vehicles (Fig. 12.3.), the top 20 companies were investigated for assessment of strategy. Toyota, Ford, Hyundai, GM, and Google are the aggressive leaders in this field. Through the analysis of a self-driving patent maps (Fig. 12.4). The self-driving car industry chain should be divided into three main participators: the yellow area is producers and suppliers of auto parts, accessories, and coating, such as Bosch, Dense, ZF Friedrichshafen, and Aisin; the red area is automotive manufacturers and suppliers, such as Ford, GM,

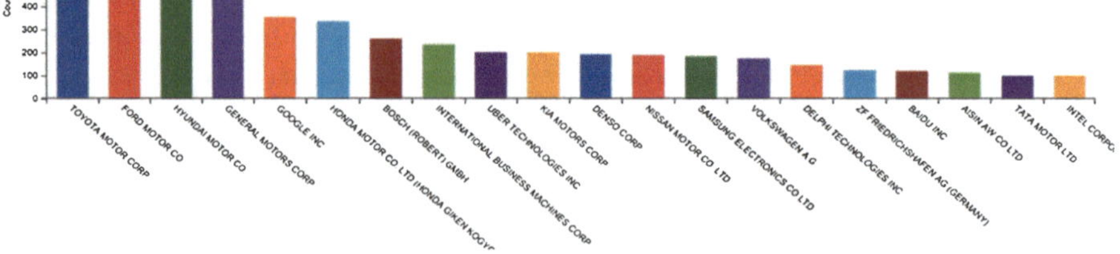

Fig. 12.3 Top 20 patent assignee/applicants in self-driving technology

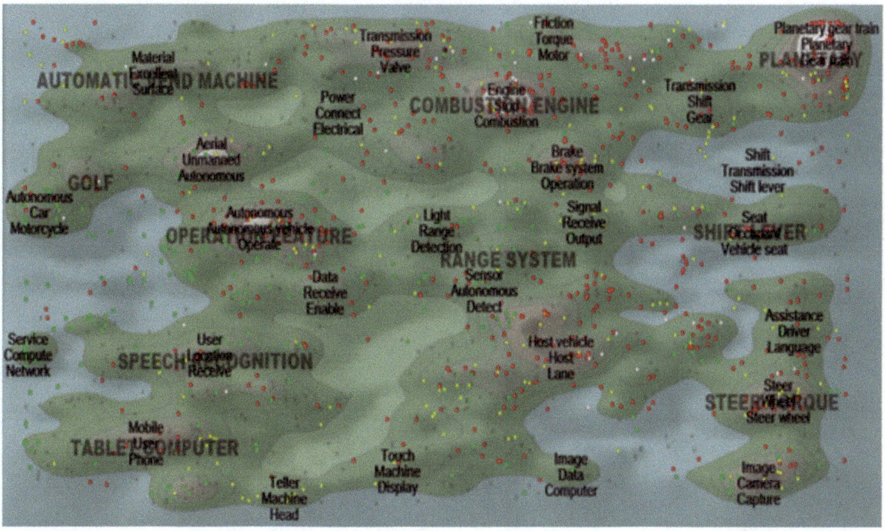

Fig. 12.4 Patent map for self-driving technology

Toyota, and Honda; the green area is emerging high-tech enterprises (including Internet companies and IT companies), such as Google, IBM, and Uber. Both carmakers and high-technology companies are involved with automated driving system.

For a high-technology company like Google, it is more likely that they will be a self-driving technology company rather than becoming a manufacturing car company. Google enjoys a strong position to leverage their patents in self-driving.

As shown in Fig. 12.5, patent maps for Google's self-driving technology, Google is already entering Level 4 in the autonomous car industry. Google patents demonstrated three features, which is safety-critical driving functions, monitoring roadway conditions, and also the option for human driving. The patent strategies of Google are as follows: First, it makes a full layout on the software-based technology for the self-driving system, such as a vision system, light detection, LiDAR (ranging radar), sensors, and objects detection. Second, it is developing the key components of Self Driving Vehicle (SDVs), such as a cameras, radars, sensor devices, safety mechanisms, windowsills, pulley shafts, and rotary joint devices. Thirdly, it is working on the different roadway conditions, for instance, different weather conditions, stop signs, signal lights, height dimensions, map data, and fields of views. Fourthly, there is trend for car-sharing services, such as predetermined locations, mobility services, navigation, pickup destination, and suggested location. Recently, minimizing injuries and delivering freight were considered as new R&D direction. There are special attractive patent filing technologies for "softer" cars, autonomous trucks, and minivans.

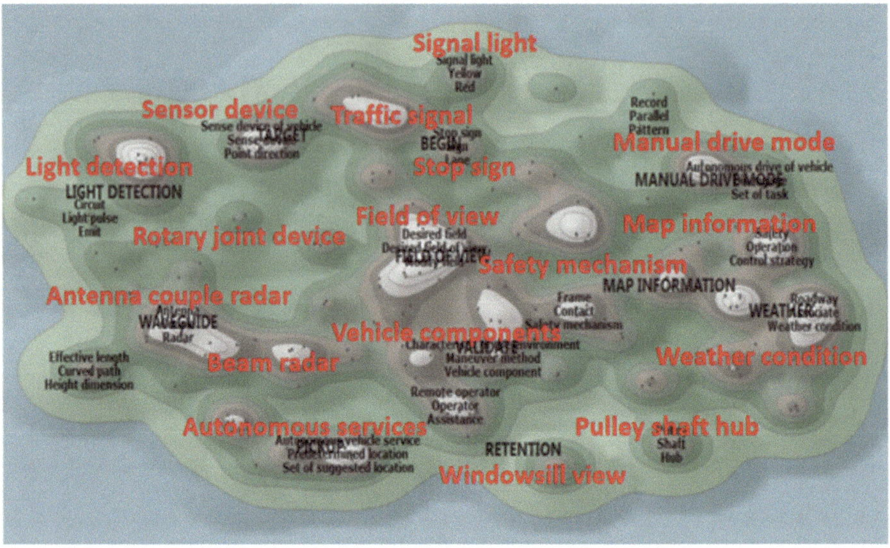

Fig. 12.5 Google's patent map

Fig. 12.6 Patentee collaboration network for Google(2013–2017)

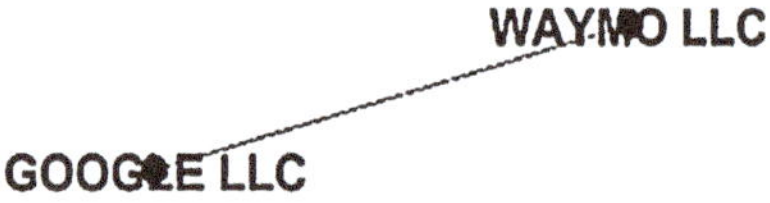

12.5.2.2 Google's Network and Partnership in Self-Driving Technology

The patent document with its legal, technical, and economic value has become the main source of measurement of R&D development and knowledge flow. Patent analysis and Social Network Analysis (SNA) are used to build Google's various networks and partnerships, including patentee cooperation network, patentee transfer network, patentee citation network, and partnerships built from 2013 to 2017. Therefore, the comparison among networks provide important information for us to comprehensively understand Google's strategies in R&D development, technology influence, and patent transaction ,etc., in order to help with better planning of the company's future development direction and strategy.

Figures 12.6 and 12.7 show Google's patentee cooperation network and transfer network. Waymo is an autonomous car development company and subsidiary of Google's parent company. Thus, Google has made a great many patent transfers with Waymo to help Waymo improve its patent layout.

Figure 12.8 shows Google's citation network based on VOSviewer. Compared with other networks, it is the largest one with relative abundance of information, which means Google's driverless technology has a big impact on other companies in the industry. As shown in Fig. Error! Reference source not found. IBM, Toyota, Ford, and Uber have been citing Google's patent extensively. In other words,

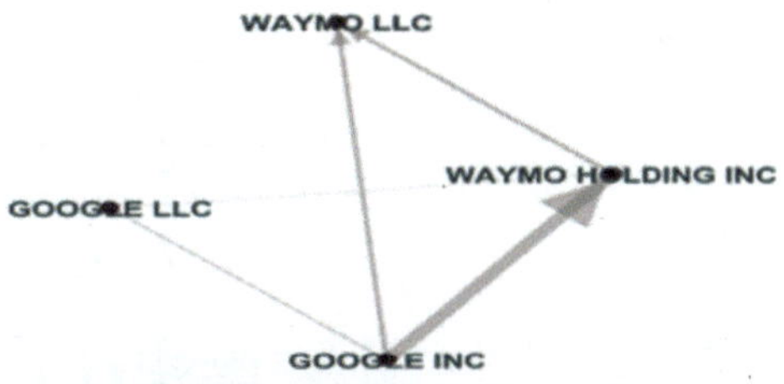

Fig. 12.7 Patentee transfer network for Google (2013–2017)

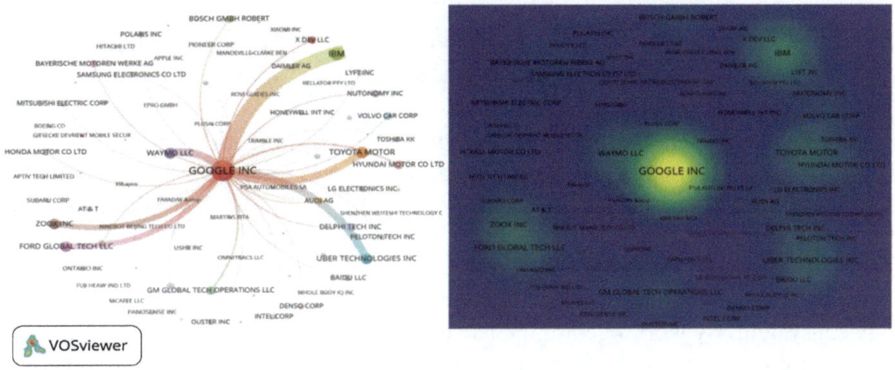

Fig. 12.8 Patentee citation network (2013–2017)

Google's patented technology has received the continuous attention of patent applicants in this field, and may contain groundbreaking technical contributions. Second, automobile companies take the initiative to lay out patents. By focusing on Google-related technologies, they lay out a large number of peripheral patents with similar technologies, in order to curb the expansion of Google's driverless technology and to obtain market and technical advantages for themselves. Google has a high rate of self-citation, which means that Google is also developing its own technology and targeting at patent protection, and has put forward a large number of follow-up patent applications for improvement, so as to achieve technology continuity and improve the layout

Figure 12.9 shows us the partnerships of Google's self-driving company Waymo in Autoblog from 2013 to 2017. Both car manufacturers and high-tech companies are working together now. Waymo has accumulated impressive allies with Lyft, Chrysler, Avis, and Intel. Waymo works with automotive partners to provide service support for companies such as Chrysler and Honda. Waymo has also used Intel-produced technology from connectivity to sensor-data processing since 2009. In this instance, the Intel–Waymo collaboration can help make self-driving vehicles become a reality during the design stage (Hawkins 2017). At the same time, Intel is continuing to position itself at the forefront of autonomy and self-driving by teaming up with other companies (Intel and Waymo team up to build self driving cars (2017). Waymo has also announced its partnership with Trov, the world's leading on-demand insurance technology company and with car rental giant Avis in 2017

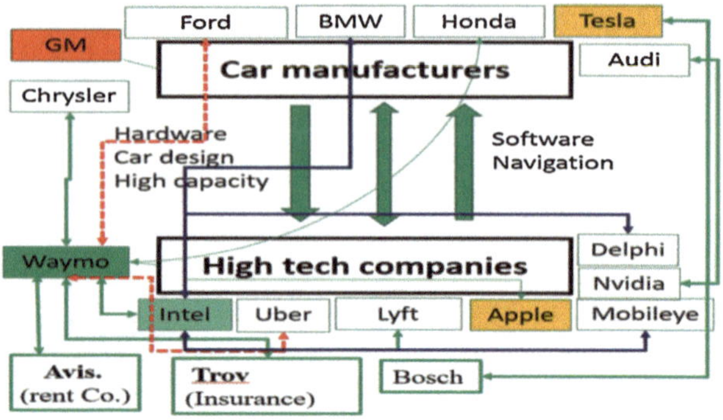

Fig. 12.9 Partnerships in self-driving technology

(Condliffe 2017). Another powerful Graphics Processing Unit (GPUs) company, NVIDIA has partnerships with Audi, Tesla, Honda, BMW, and Bosch, which makes it more formidable in the market.

Through these partnerships and patent analysis, we believe that the biggest competitors of Waymo might be Toyota and GM. Intel, as a new entrant, could be a potential competitor, and also the most profitable company in self-driving technology ecosystem. By working with intensive collaboration, Intel can provide car chips and sensor-data processing services to both carmakers and high-tech companies. According to Zacks Equity Research (2017), "…Intel will continue its leading role in helping realize the promise of autonomous driving and a safer, collision-free future," which implies Intel's ambitions for self-driving systems. Second, Tesla is embarking on Level 5 to produce fully its own autonomous cars, and continues to grow successfully. Recently, Tesla stated that its autopilot is semiautonomous, and that it had tested the tech well beforehand. The new partnership for Waymo could be expanding to Ford.

12.5.2.3 Waymo's Possible Strategy

Waymo aims to be the market leader in integrating self-driving technology into carmakers' vehicles. At the company level, Waymo is not advanced in developing AVs as it is not a mass car manufacturer, so it is dependent on others to partner, acquire, or license its technology. Therefore, Waymo's possible internal and external strategies will be needed as follows:

First, internal strategies will need to be based on building capabilities for the development of customer-oriented technology. Four directions should be considered as part of their main business strategy: autonomous mobility services ("robottaxi"), technology for monitoring systems to reduce congestion, delivery and logistics services, and autonomous trucks and "softer" cars.

Second, external strategies will be need to be based on the business model and will need to be explored and adjusted. There are two possible and potential directions as external strategies: shared mobility and/or driverless services. As Waymo expands their partnerships, self-driving ecosystems will be of great importance in the future. At the same time, Waymo should work to link hardware with software, and to integrate this into cars through partnerships with automakers.

12.6 Discussion

Electric car technology has significantly changed traditional car technology. While fuel changes are still the big concern for car companies, electric cars will not transform the entire market. The technology is still expensive, production takes place at a remote location, and more importantly, the change is toward reducing the energy footprint of the transportation industry. Autonomous vehicle technology also seems to affect the traditional automotive industry and other related industries. But, it is much more difficult to make a complete transformation of this effect than just moving away from the electric car. In addition, the conversion is not on the energy side, but on the data (IoT, Big Data) part. On the other hand, in the choice of electric vehicles, people still prefer the design of the vehicle and do a price analysis accordingly. In this context, high-tech companies in the self-driving industry cannot ignore the necessity to keep the design perception. This is a very serious handicap. Nowadays, it is very risky for people to completely abandon their control over technology in a period in which technology–personal interaction in the person-specific configuration is rapidly developing.

The change in this auto sector can be achieved as a result of the "low hanging fruit." It can be predicted that technology tends to be in areas that are easily accessible and yield quick results. In this context, Waymo and Google should determine areas that can be transformed by reaching this easily accessible and critical mass. One is for public transport, another is for accessible transportation for the other-abled, and another for campus or indoor-closed area use. These uses will lead the industry if there is a niche in the market, and a strategy in which the design perception and price for this purpose is balanced. This should be the strategy until reaching the threshold of critical mass, which would provide the technology with an overall transformation.

The critical mass access strategy should focus on the development of sufficient capacity to meet production priority and demand. Waymo is not a car manufacturer, so creating a serious production capacity from scratch and achieving economies of scale requires serious resources. No matter how much hope there is in the industry, it is unlikely that Google has the funding for many production sectors, Strategic cooperation in production at this point is very important. However, traditional car manufacturers like to be in the electric car technology, like to partner with the company, to buy or to benefit the most with limited cooperation. In this context, Waymo skates and the process of converting any car into an automatic car is critical.

The paper provides insight into the place and role of different companies in the self-drive car industry. It analyzes the emerging trends and events in this technology. For instance, we suggest that the self-driving industry chain ought to be divided into three main participators, including suppliers of vehicle accessories, automotive manufacturers, and high-tech enterprises. Vehicle accessories suppliers include Bosch and Dense, who will produce and avail auto parts for self-driving cars. Automotive manufacturers, such as General Motors, Ford, and Honda will develop self-driving cars for the market. Finally, high-tech enterprises, including IBM, Uber, and Google, will pioneer self-driving technological tools in the industry. Therefore, the article allocates specific duties to particular sectors of the economy.

Our research is useful to engineering managers, who are responsible for supervising projects, delegating duties to employees, and determining the need for particular skills and talents in an organization. Engineering managers are also required to set goals for projects and lead research and development activities. The paper helps these individuals to discern the particular needs of their organizations in line with their firm's ultimate goals within the self-drive car industry. For instance, engineering managers in Google will prioritize the hiring of IT experts in an effort to develop patents and innovative solutions. Currently, the company has established a subsidiary company to consolidate the organization's high-tech products and tools in the industry. Engineering managers will use the information contained in this paper to facilitate efficient coordination between Waymo and Google. Such a coordinated effort will help the company beat other competing firms in the industry. Most importantly, the article encourages Google to continue investing in this emerging self-driving technology.

12.7 Conclusions

The strategic planning of patent analysis and a company's networks, based on both patent data and news releases, play a crucial role in revealing the implicit R&D partnerships and explicit strategies at the company level. The case study shows that competitive and complementary interactions influence the formation of partnerships in the market. We believe SNA in a complex network setting will provide the abundant and unbiased analysis to high-quality decision-making.

The Waymo car is one of the new generations of self-driving technology. It is expected in the near future to be one of the most innovative companies. The technology adoption will vary since it is a new product and new way of driving. Hence, Waymo should focus on R&D for customer-oriented technology so that it can maintain continuous cooperation with other related industries. In this context, Waymo is already well situated as a company that develops AVs. The advanced technology will be attractive for others to partner with, acquire, or license Waymo's technology. Therefore, partnerships will help Waymo to extensively expand into the automobile industry. It will be necessary to make a differentiation strategy in this emerging area, and to develop many different business models to share the same technology. This strategy will reduce R&D costs and allow economies of scale. The combined ecosystem of both the hardware and software in self-driving transportation along with users and partners will be an ideal network for Waymo, which will provide the company the irreplaceable position.

Briefly, the complete transformation of autonomous automotive technology into the entire sector is a long and difficult road because consumers are not willing to pay a lot of money for self-driving cars or have different perceptions about how time-consuming driving can be or their personal driving experience. However, the technology is still in development and will eventually create its own market, which will make a huge impact on the future of transportation. The important thing is to construct the strategy in two stages. The first stage is to focus on pre-conversion, which are niche areas that need to be identified before the large-scale transformation. The second stage is to provide a self-driving car solution for sustaining an ecosystem of users and collaborators, such as building (not construct) at capacity of car manufacturers on a permanent basis such as a long-term company marriage or joint venture that realizes economies of scale. The last but the most important is focusing on the R&D activities in critical technologies to achieve more core patents.

Our study has some limitations, which might be seen as future research. First, from the technology adoption perspective, we've only explored and analyzed five key perspectives of self-driving technological impact, but we did not reveal the potentials of self-driving to create a profound and lasting reshaping of the future transportation system. Also, the key factors that could improve user adoption of autonomous vehicles need to be identified to help form policy and regulation governing self-driving technology. Secondly, from a business and market perspective, the business models to generate profits based on self-driving car technology are not clear, and there is a lack of focused commercialization and value chain. Thus, the interactive relationship between technology, business models, and the market should be part of a company's strategies. Thirdly, from a technology convergence point of view, the available technologies, such as sensor-based solutions and commutation application, do not completely enable self-driving, and are not robust enough to achieve Level 5.

References

Agrawal, A., Kapur, D., & Mchale, J. (2008). How do spatial and social proximity influence knowledge flows? Evidence from patent data. *Journal of Urban Economics, 64*(11), 258–269.

Benleulmi, A. Z., & Blecker, T. (1970). Investigating the factors influencing the acceptance of fully autonomous cars. Retrieved from https://www.econstor.eu/handle/10419/209304.

Condliffe, J. (2017). Why Waymo's partnership with Avis makes sense. Retrieved from https://www.technologyreview.com/s/608176/why-waymos-partnership-with-avis-makes-sense.

Hawkins,A.J.(2017).InteliisworkingwithWaymotobuildfullyself-drivingcars.Retrievedfromhttps://www.theverge.com/2017/9/18/16328284/intel-waymo-partnership-self-driving-car-chrysler.

Hsu, F. C., Trappey, A. J., Trappey, C. V., Hou, J. L., & Liu, S. J. (2006). Technology and knowledge document cluster analysis for enterprise R&D strategic planning. *International Journal of Technology Management*, (17), 36.

Intel and Waymo team up to build self driving cars. (2017). Retrieved from https://www.ciol.com/intel-and-waymo-team-up-to-build-self-driving-cars/.

Kalra, N. (2017). Challenges and approaches to realizing autonomous vehicle safety.
Kaur, K., & Rampersad, G. (2018). Trust in driverless cars: Investigating key factors influencing the adoption of driverless cars. *Journal of Engineering and Technology Management, 48*, 87–96.
Lee, S., Yoon, B., Lee, C., & Park, J. (2009a). Business planning based on technological capabilities: Patent analysis for technology-driven roadmapping. *Technological Forecasting and Social Change, 76*(6), 769–786.
Lee, S., Yoon, B., & Park, Y. (2009b). An approach to discovering new technology opportunities: Keyword-based patent map approach. *Technovation, 29*(6–7), 481–497.
Liu, S. J., & Shyu, J. (1997). Strategic planning for technology development with patent analysis. *International Journal of Technology Management, 13*(5/6), 661–680.
Ma, J., & Porter, A. L. (2014). Analyzing patent topical information to identify technology pathways and potential opportunities. *Scientometrics, 102*(1), 811–827.
Möller, D. P. F., & Haas, R. E. (2019). Automotive E/E and automotive software technology. Guide to automotive connectivity and cybersecurity computer communications and networks, 83–169.
Park, H., Kim, K., Choi, S., & Yoon, J. (2013). A patent intelligence system for strategic technology planning. *Expert Systems with Applications, 40*(7), 2373–2390.
Park, I., Jeong, Y., Yoon, B., & Mortara, L. (2014). Exploring potential R&D collaboration partners through patent analysis based on bibliographic coupling and latent semantic analysis. *Technology Analysis & Strategic Management, 27*(7), 759–781.
Pereira, C. S., & Soares, A. L. (2007). Improving the quality of collaboration requirements for information management through social networks analysis. *International Journal of Information Management, 27*(2), 86–103.
Ploeger, J., Lacroix, L. (2015). Opportunities and challenges for automated vehicles in the Zuidvleugel.
Schreurs, M. A., & Steuwer, S. D. (2015). Autonomous driving - political, legal, social, and sustainability dimensions. In *Autonomes fahren* (pp. 151–173).
Sternitzke, C., Bartkowski, A., & Schramm, R. (2008). Visualizing patent statistics by means of social network analysis tools. *World Patent Information, 30*(16), 115–131.
Tsuji, Y. S. (2012). Profiling technology development process using patent data analysis: A case study. *Technology Analysis & Strategic Management, 24*(3), 299–310.
Walker, J. (2019). Autonomous Vehicle Regulations – Near-Term Challenges and Consequences. Retrieved from https://emerj.com/ai-sector-overviews/autonomous-vehicle-regulations-near-term/.
Wang, M.-Y., Chiu, T.-F., & Chen, W.-Y. (2009). Exploring potential R&D collaborators based on patent portfolio analysis: The case of biosensors. In *PICMET 09–2009 Portland international conference on management of engineering & technology*.
Wyman, O. (2017). Google racking up more patents than most automakers on connected and self-driving cars. Retrieved from http://www.oliverwyman.com/our-expertise/insights/2017/may/Google-Racking-Up-More-Patents.html.
Zacks Equity Research. (2017). 3 Stocks in Focus Post Intel - Waymo Autonomous Vehicle Deal. Retrieved from http://www.zacks.com/stock/news/276345/3-stocks-in-focus-post-intel-waymo-autonomous-vehicle-deal.

Chapter 13
Technology Intelligence Map: Fast Charging for Electric Vehicles

Sultan AlGhamdi, Bret Hunley, Angel Contreras Cruz, Roland Richards, Gwendolyn Jester, and Tuğrul U. Daim

As part of technological forecasting the electrical vehicle (EV) ecosystem, there are specific technologies that need to be developed to reach the desirable extreme fast charging technology (XFC) to propel commercial use and passenger markets. There are economic, environmental, and industry efficiencies of EV charging station technologies necessary for success. From what was gathered during the research, results have shown that XFC stations will be able to recharge battery electric vehicles (BEVs) in less than 10 min and provide roughly 200 miles of driving time. Data collected do support and identify the critical component technologies for the enablement and widespread adoption of XFC technology and infrastructure, to evaluate current and projected progress of critical component technologies, and to detect the potential gating technologies and recommend areas that will most benefit from accelerated development. The authors of this chapter used an evaluative bibliometric approach and literature review to gather data sets and industry findings to lead their results. The data sets are represented in four S-curves elaborated using the Pearl method. The authors' goal is to identify which technology of the EV charging ecosystem component engineering, specifically XFC batteries to forecast the evolution of this new technology.

13.1 Introduction

The world continues to evaluate the dependency on fossil fuels and has made exciting technological strides to cause less impact on our planet. Of great global interest is the development and move toward clean energy powered modes of transportation.

S. AlGhamdi · B. Hunley · A. C. Cruz · R. Richards · G. Jester · T. U. Daim (✉)
Department of Engineering and Technology Management, Portland State University, Portland, OR, USA
e-mail: tugrul.u.daim@pdx.edu

T. U. Daim (ed.), *Roadmapping Future*, Applied Innovation and Technology Management, https://doi.org/10.1007/978-3-030-50502-8_13

The focus of this chapter is the electrification of the world's vehicles and the necessary ecosystem as it becomes more globally deployed. To enable this path forward, two main items are needed. Firstly, the cost of energy storage needs to continue to decline while efficiencies need to continue to increase. Secondly, the ability to charge electric vehicles (EV) at faster rates to enable efficient travel and commerce will be imperative for successful transition to EV adoption.

This chapter aims to identify critical component technologies for the enablement and widespread adoption of extreme fast charging technology and the necessary infrastructure. This chapter will employ forecasting research techniques to extrapolate the progress – based on the available data of the technological development of the critical component technologies to determine a projection of widespread adoption of extreme fast charging technology. Thus, the data analysis in this chapter hopes to identify the areas whose accelerated development enables faster adoption of advanced charging methods for consumer and industrial EVs.

The expansion of the electric vehicle market is now beyond the passenger car market and is entering the age of the commercial vehicle. As this transition happens, the capabilities of batteries and chargers will continue to rise. With the introduction of heavy and medium duty trucks/vehicles, the batteries will need to continue to grow in size and capacity. As this capacity continues to grow, the ability of light-duty chargers to provide adequate charging power will quickly diminish. This will make way for a new set of charging technology known as "extreme fast chargers" or heavy-duty chargers where charging rates will "need to reach the megawatt scale." These chargers will be able to charge these batteries in as little as 30 min (Walkowicz et al. 2020).

According to the Department of Energy's National Renewable Energy Laboratory (NREL), "Research and development on safe, efficient, and cost-effective [extreme fast charging] is needed now to mitigate technical barriers in time to coincide with the anticipated large-scale adoption of [medium and heavy-duty] electrified vehicles…" "Technological areas of interest include connectors, contactors, solid-state transformers, grid interface devices, power transfer mechanisms, charging control systems, XFC-capable energy storage, and automated charging" (Walkowicz et al. 2020). As extreme fast charging (XFC) technology continues to grow and evolve, there will be an increasing need to determine the maturity of this upcoming technology to make sure that it aligns with the development of the battery technology it supports. The following chapter will provide a method and forecast as to when extreme fast charging will reach economic maturity.

We spend a brief part of the reading to discuss the political, environmental, and sustainability of EV Fast Charging. First, politically speaking the vast financial backing of Big Oil wants to slow the adoption of electric vehicles. However, countries are demanding action to curb fossil fuel engines (Hussien 2017). Emissions vary dependent on the geography (Energysage.com 2019). Thus, as we present in the chapter, we research methods and data that help illuminate our two goals: (1) the cost of batteries needs to continue to decline while efficiencies need to continue to increase, and (2) the ability to charge at faster rates to enable efficient commerce will be needed.

13.2 Literature Review

According to the United States transportation sector, transportation consumes roughly 30% of the total amount of energy used in the country and is also responsible for 92% of the petroleum-based energy required to meet ever-growing transportation demands, which is approximately 70% of the total oil consumption in the United States (Onat et al. 2019). Due to these data, alternative technologies, including electronic vehicles (EVs), have been recognized as a sustainable solution (Onat et al. 2019).

Electric vehicles were invented in the early 1900s, and for a short time, actually outnumbered the number of gasoline-powered vehicles (Young et al. 2013). As is public knowledge today, that trend is no longer intact. However, over the past few years, electric vehicles (EVs) have begun to increase in popularity again. While this growth has been slow, it has also been steady, and the trend of increasing volumes has only accelerated as customer's social preferences have started to lean more toward clean power vehicles instead of the traditional internal combustion engine. To support this trend, government agencies have started to support public and private initiatives to improve the EV technologies in hopes that the technology base becomes more self-sustaining, and in an attempt, to reduce the overall CO_2 emissions (Young et al. 2013). Even with all of the public and government support, EVs are still struggling to gain a mainstream hold and still have some technical hurdles to overcome in the areas of purchase price, range, charging time, and lifetime of components (Elizabeth et al. 2017).

13.2.1 Understanding the System

To understand the higher level ecosystem of the power distribution system you have to understand that there are many parts before that charge is able to get to your vehicle – all the way back to the utility supply – such as hydroelectric power through the switchers, transformers, and load balancing center to the EV charging systems. There are many needs on regional utility systems and municipal load balancing. EV charging roll out has to be done in a way that supports the needs and demand of the community with the power allotment for a given area. Areas such as Fremont and Alameda California have the most EVs sold so it stands that they have the most EV stations at 82 for the region. Next in the chapter, we will break down the particular elements of EV charging.

13.2.2 *Electric Vehicle Charging*

Charging an electric car is known as a simple process, because it involves simply plugging in your car into a charger that is connected to the electric grid. An electric vehicle charging station, also known as an EV charging station, electric re-charging point, charging point, charge point, electronic charging station (ECS), and electric vehicle supply equipment (EVSE), is an element in an infrastructure that can supply electric energy for the recharging of plug-in electric vehicles (Onat et al. 2019). The most commonly used term used to describe these charging methods is electric vehicle supply equipment, or EVSE (energy.gov 2020).

13.2.2.1 Types of Chargers

The types of EV chargers available in the market are not federally regulated, but there are certain standards that have been put in place to promote consistency within the market. One such standard is published by the Society of Automotive Engineers (SAE) and is called J1772. This standard defines how a charger should work and connect to the vehicle, as well as the power loads that can be passed, also known as levels (energy.gov 2020).

Level 1 Charging

Level 1 charging is the most basic of the charging systems as it simply plugs into the standard 120V outlet found in most homes or businesses. This level of charger does not require the installation of specialized equipment or infrastructure and is capable of charging at a rate of "two to file miles of range per hour" (energy.gov 2020). This type of charger is typically used in the home for charging vehicles overnight as they take a long time to charge the battery. The advantage of this system is that it is by far the least expensive of the options available (energy.gov 2020).

Level 2 Charging

Level 2 charging uses a higher voltage to charge the vehicle faster than level 1 technology. For residential, the systems use 240V and for commercial they use 208V. These systems cannot be connected to a standard wall socket and are usually installed by a licensed electrician. This type of charger is capable of delivering "ten to sixty miles of range per hour" and is used in homes and commercial spaces where charging in 2–6 h is desired (energy.gov 2020).

Level 3 Charging

Level 3 charging is the most advanced, commercially available charging system. This system uses higher voltages and current to deliver energy to the vehicle at an accelerated rate. This system is capable of delivering "sixty to a hundred miles of range" per hour and is typically used in commercial settings. However, some passenger cars have started to implement this technology. The reason for the limited spread of this level of charging is the specialized equipment and infrastructure needed to make it work properly (energy.gov 2020).

Extreme Fast Charging

Extreme fast charging (XFC) is considered the next level of charging technology. XFC is defined as any charging system where energy transfer is 350 kW or higher. This technology is currently in its infancy and is currently seen as the way to make electric vehicles the new normal within society. At these charge rates, a user can expect to charge their vehicle in as little as 10 min. However, as with level 3 chargers, this requires specialized equipment and infrastructure (Ahmed et al. 2017).

XFC requires thicker electrodes and they present several barriers to faster charging; as electrode thickness increases, the charge times also increase to avoid lithium plating. In relation to infrastructure, the optimization of XFC charging station locations are needed within cities and across highway structures in order to properly account for convenience of the user and also availability of power from the current utility (Howell et al. 2017).

13.2.3 Electric Vehicle Trends

Electric vehicles play a significant and crucial role in the reduction of fuel consumption and pollutant emissions that will result in a more sustainable transportation (Hu et al. 2020). Throughout the past two decades, different EVs including battery electric vehicles (BEVs) and hybrid electric vehicles (HEVs) have been developed to replace more traditional internal combustion engine vehicles (Hu et al. 2020).

Since their invention in the early 1900s, electric vehicles have not always been a common vehicle in the market. They spent many decades as the "one-off" or "kit-car" that only the rich and famous could afford. However, as technology has increased, the prevalence of the electric vehicle has also grown. This growth has been driven by increasing social acceptance, carbon consciousness, increasing oil prices, and government incentives (Edis 2019). According to the Bloomberg, NEF 'Electric Vehicle Outlook 2019', they expect EVs to account for more than half of all vehicles produced worldwide and more than 30% of passenger cars in operation by 2040 (Bloomberg Finance L.P. 2019).

This is an impressive trend when you consider that in 2010, only a few thousand EVs were produced, and in 2018, it is estimated that EV production hit approximately 2 million vehicles. This growth is driven primarily by cost improvements in battery technology and it is expected that EV prices will match that of internal combustion engines (ICEs) by the mid-to-late 2020s. This expected parity is forecasted to lead to a decrease in the overall ICE production over the same 20-year time period (Bloomberg Finance L.P. 2019).

13.2.4 Parallel Technologies

In order for EVs to reach their full potential within the mainstream of 2020, then battery performance must improve at an increased rate (Elizabeth et al. 2017). According to data results by the Technology Forecasting with Data Envelopment Analysis (TFDEA) analysis, it was found that incremental improvements showcased that battery technologies that are studied will leave EVs with a much shorter performance specification for vehicle ranges (Elizabeth et al. 2017). Some recommendations for the evolving technology of EVs involve analyzing renewable energy integration challenges, including best practices among charging utilities, as well as developing tools that provide users with stakeholder-specific estimates for long-term electrification costs and benefits in the long-run (Walkowicz et al. 2020). These recommendations will help enhance the enhancement and usability of EV charging stations so that consumers, businesses, and additional users will contribute to the ever-growing demand of electric vehicles.

13.2.4.1 Identification of Enabling Technologies

Among the electrical vehicle manufacturers, there are specific technologies that need to be developed to reach the desirable extreme fast charging technology (XFC). As more people are using electrical vehicles, improvements in current batteries are needed in order to meet the range of the conventional vehicles. Such characteristics are storage capacity, power, cost, density, and so on (Gao et al. 2018). Manufacturers are working in the development of level 2 and level 3 batteries that are going to reduce time charging (Gao et al. 2018).

In addition, manufacturers are working toward the same goals as National Labs are too. According to the National Renewable Energy Laboratory as shown in Fig. 13.1, the development of new electrical vehicle batteries must meet in the medium- and long-term high voltage, high capacity, new materials systems, reduction of the time charging, and reduction of cost (Onat et al. 2019).

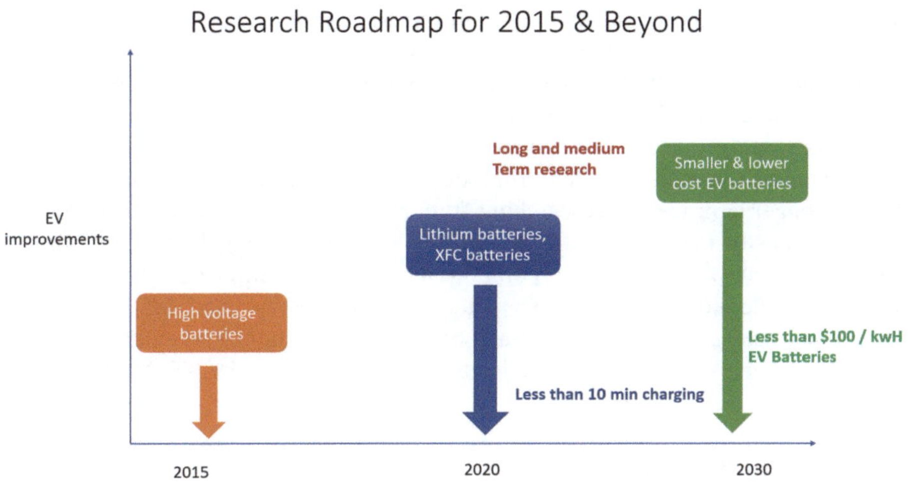

Fig. 13.1 Research roadmap. (Source: National Renewable Energy Laboratory)

13.2.4.2 Batteries

One of the major parallel technologies to charging is EV battery technology. While this technology does not have a direct impact on the charging technology itself, as the storage medium of the output of the charger, it does have a major impact on the implementation of the technology. Based on a literary review of a battery forecast from 2017 on battery technology and its expected technology trends, it was determined that battery technology itself is at a crossroads. Based on the findings, the current technology in the market is not capable of hitting the Department of Energy's cost or performance goals: "This tells us that if current incremental advancements continue then the DOE goals cannot be met unless a new disruptive technology is introduced" (Elizabeth et al. 2017).

The chapter went on to say, "battery performance must improve at a faster rate. While researchers are actively working towards achieving the battery characteristic targets...analysis find[s] that incremental improvements realized in the battery technologies studied will leave EVs considerably short of the performance specification for vehicle range. Based on the data mining results, this study identified a focus in the basic research stage on metal-air battery technologies. This makes sense because research shows that metal-air battery technologies have a higher specific energy rating than the current lithium-ion batteries. This study found that more emphasis should be placed on finding new candidate technologies" (Elizabeth et al. 2017).

This conclusion presents an issue for extreme fast charging technology development, as this will limit the capability of the charger itself. Before extreme fast charging can become truly useful, batteries that are capable of "3–4C-rate charging with good energy density, low cost, and the ability to support 4000 charge cycles," are needed (Walkowicz et al. 2020). Based on the literature review on Department of

Energy (DOE) goals and the findings of the NREL, battery technology will continue to be a limitation on the capabilities of charging.

Taking into account the evaluative bibliometric approach of this chapter, this referenced analysis of battery technology development is pertinent. While patent-based data analysis may show heavy attention being paid by industry to gain an edge on competitors, the shorter term focus of most corporate project horizons versus the longer term view of research institutions may overstate the impact corporate patent expansion may have. In other words, patents may be signaling large efforts in incremental innovation, while corporations may not be incentivized to fund the deep research needed for truly disruptive battery technologies. As such, bibliometrics of overall research publication trends may be a better indicator of whether or not battery technology is advancing as needed to effectively enable EV XFC technology.

13.2.4.3 Infrastructure

An electric vehicle charging station, also known as an EV charging station, electric re-charging point, charging point, charge point, electronic charging station (ECS), and electric vehicle supply equipment (EVSE), is an element in an infrastructure that can supply electric energy for the recharging of plug-in electric vehicles. The goal of this literature review was to compare three general components in electric vehicle charging stations. The three components were (a) Enablement of XFC Technology and Infrastructure, (b) Projected Progress of Critical Component Technologies, and (c) Potentials of Gating Technologies and Recommendations. The reviewed literature suggests that there are advantages to the use of electric vehicle charging stations when utilized properly, due to the ever-changing evolution of technological advances. Current research supports the use, construction, and implementation of electric vehicle charging stations, as discussed above; however, a continuation of current research with consistent methodologies will assist in the justification of its use and benefits.

The charging infrastructure will also play a major role in how extreme fast charging is adopted. For this, there are three major items: (1) available power, (2) physical layout, and (3) location availability.

Power Availability

The available power to charge is a function of the electrical grid itself. This is where all mainstream users source their power today, including that of our businesses and homes. The connection of large amounts of electric vehicles to this grid can cause large spikes in power demand which are undesirable for the grid as a whole (Clement-Nyns et al. 2010). For example, one chapter estimates that approximately 5% of the total power consumption of Belgium in 2030 will be from electric vehicles alone (Clement et al. 2007). This additional load will have to be minimized for the grid to stay functional. This comes with a risk that not only losing grid power

(power outages) will become a major problem, but also power quality. While losing power is a major concern, the quality of the power should not be understated. Without voltage control and noise suppression, the power being delivered is useless and general operation of the appliances will become hindered (Clement-Nyns et al. 2010).

The impact to the grid also extends to the charger level as well. Depending on if the vehicle uses AC or DC charging methods, the energy impact to the grid will change. In order to enable extreme fast charging, it will likely mean that future charging schemes will need to be DC based due to the high current draws and the high cost of putting AC converters on the vehicles themselves (Walkowicz et al. 2020).

One option to ease the implementation of extreme fast charging in a way that has minimal grid impact would be to have the utility companies themselves be the direct providers of the service. This is known as Direct Current as a Service (DCaaS). In this business model, utilities own the power-conversion hardware, power lines, and power generation, similar to the normal grid. Due to these similarities, the existing business models and practices should apply and be more easily integrated, rather than the fleet managers or individuals trying to manage it themselves. However, there are regulatory and technological hurdles that must be overcome before a system like this can be implemented. The good news is that this is currently being worked on by a working group at the Electric Power Research Institute (Walkowicz et al. 2020).

Power Layout

For the layout, there are several major considerations that should be accounted for:

- Cable management and parking configuration – account for cord lengths, sizes, port locations, and cooling requirements.
- Security – account for the need for private charging areas where maintenance can be safely performed, and competitors can be excluded from using the services.
- Footprint – the facility must be reasonable, manageable, and capable of providing the necessary power needs (Walkowicz et al. 2020).

If all of these items can be accounted for, the addition of fast charging stations should not be an issue. If one of these items is missed, however, the implementation of a system with extreme fast charging capabilities will face significant challenges (Walkowicz et al. 2020).

13.2.4.4 Location Availability

The number of locations capable of charging vehicles will also play a key role in the adoption of EVs. In 2018, there were approximately 630,000 electric chargers installed worldwide, growing at a compound annual growth rate of 44% since 2013

(Bloomberg Finance L.P. 2019). Adoption requires availability: while faster charging technology may exist, without available locations, adoption will suffer. As such, locational availability is a potential gating factor for XFC adoption.

13.2.4.5 Fuel Cell Technology

Fuel cell electric vehicles (FCEVs) are powered through the use of hydrogenous hydrogen, which is stored onboard in high-pressure tanks that are converted into electricity by multiple cells (Fernández et al. 2015). For the nature of EVs to work correctly, a critical need for consistent and accurate security is needed to ensure that FCEVs remain safe, secure, and reliable for DC charging (Howell et al. 2017). In order to properly enhance the energy efficiency and market penetration for FCEVs, then extensive work needs to be implemented on all FCEV components, powertrain control, and energy management (Gao et al. 2018). Even with the public's and government's support, EVs are still struggling with gaining a mainstream hold while still having some technical hurdles to overcome in the areas of purchase price, range, changing time, and lifetime of components. The motor and battery are two of the main components in an EV (Gao et al. 2018). Three major pillars that are considered in EVs are connectivity, automation, and electrification; these three pillars support a more efficient future transportation system as well as a cleaner environment, therefore boosting energy efficiency, driving range, and market adoption (Gao et al. 2018).

13.3 Methodology

A technological forecast is a prediction of the future characteristics of useful machines, procedures, or techniques (Howell et al. 2017). Organizations, companies, and governments use the results to make plans for future developments. According to experts, the future growth of electrical vehicles will not depend on new vehicle technology inventions, but more likely on electrical components innovations, extreme fast charging technologies, and charging stations (Gao et al. 2018; Neubauer et al. 2014). To forecast the timing and market adoption of XFC technology, as is the focus of this chapter, supporting technologies were examined to determine their stage of development and market adoption with data retrieved by Technology Forecasting and Social Change Journal. By analyzing the evolution of these individual technologies, a forecast is produced that projects the development and adoption of XFC technology.

Using technology adoption extrapolation techniques as outlined by Amol Adamuthea and Gopakumaran Thampib (Adamuthe and Gopakumaran 2019), S-curves are generated. Among the technological forecasting methods, analog methods are used to compare technologies using historical data for a past technology (Fernández et al. 2015). The approach used in this chapter is growth curves, which falls into the category of analog methods. The growth pattern shows an

S-Shape curve to explain the technological and social impact. Pearl curve and Gompertz curve are the most recognized among experts (Fernández et al. 2015).

The formula for Pearl (logistic) curve is:

$$y_t = \frac{L}{1 + ae^{bt}}$$

The formula for Gompertz growth curve is:

$$y_t = Le^{ae^{bt}}$$

The data used to create the S-curves were obtained using patent and bibliometric analysis. Patent analysis provides technological information that can be used by researchers and inventors to find new solutions to technical problems or to identify trends in the industry. Meanwhile, bibliometric analysis may explain the influence of a single researcher, or to describe the relationship between two or more works (Hu et al. 2020). The data obtained were classified into four categories: (1) Fast Charging for EV, (2) Fast Charging Batteries EV. (3) EV legislation, and (4) EV Charging. Using Pearl and Gompertz curves, we obtained the resultant curves for XFC technology for electrical vehicles.

13.3.1 What Will It Produce?

We obtained data from the journal *Technology* Forecasting and Social Change following the method used by Amol Adamuthea, and Gopakumaran Thampib (Adamuthe and Gopakumaran 2019). We expected to present four S-curves elaborated using the Pearl method. We want to identify which would be the technology life cycle on which XFC batteries are right now and we want to forecast the evolution of this new technology.

Data acquisition

- Publications on science direct (per keyword, per year)
- Patents (per keyword, per year) *Disc: application date versus issue date*

Application

- S-curve extrapolation to forecast XFC batteries for Electric Vehicles.

13.4 XFC Forecast Result

Using the methodology outlined previously, a predictive maturity curve was developed to determine when we can expect XFC to reach technological maturity. This process was done in steps and is outlined here:

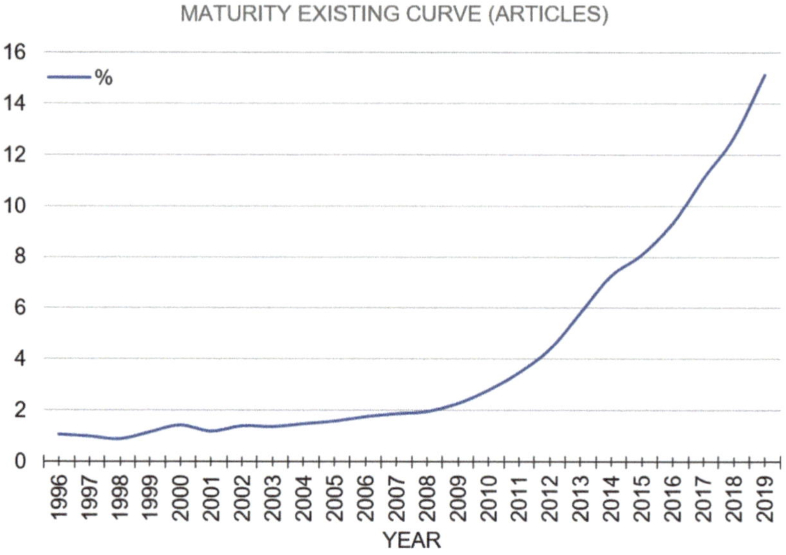

Fig. 13.2 The existing articles in the journals

13.4.1 Step 1: Baseline Data

The first step was to gather a data baseline. To do this, an article search was performed to determine the number of articles published as it pertains specifically to XFC. These results showed that there were no measurable data prior to 1996 and has steadily progressed until today. The data from 2020 were excluded due to it only being a partial year. The curve as shown in Fig. 13.2 was generated to represent the maturity to date. This also shows that the growth to date has begun to accelerate in recent years, and while still relatively new from a technological standpoint, has begun to garner some attention from the literary field.

13.4.2 Step 2: The Predictive Curve

The next step was to extrapolate the baseline curve into a full predictive curve. The curve was extrapolated out until a maturity level close to 100 was reached. Full technological maturity is expected in 2076, Appendix A, per the prediction curve as shown in Fig. 13.3. This does not mean that the technology will not be viable until then, but simply that the literature on the technology will be robust and considered "complete."

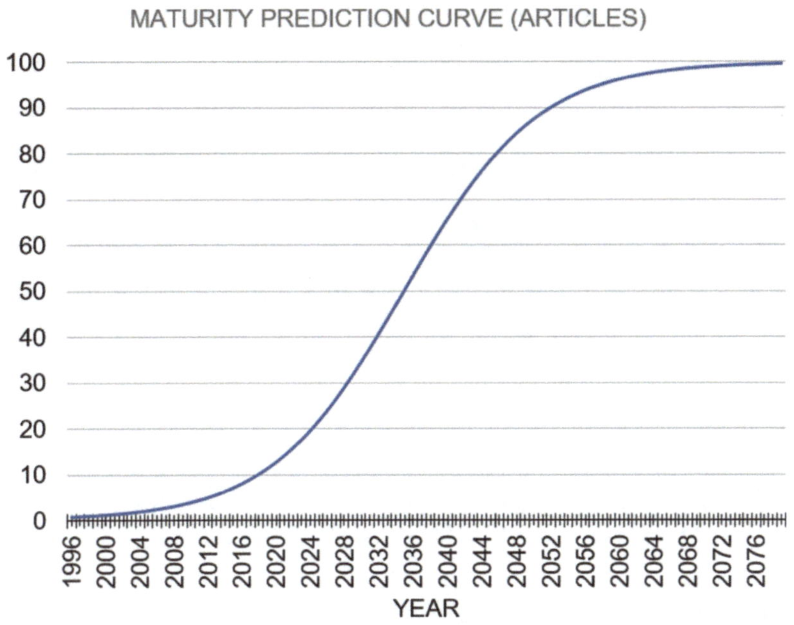

Fig. 13.3 The predictive curve of the maturity level

13.4.3 Step 3: Growth Curve

The last step was to generate a predictive growth curve. This curve will identify the inflection point where growth in literature will peak and begin to fall. This point is considered to be the point of technological maturity. The point of technological maturity is tied to this point as at this point, the literature is considered well known and it starts to decline as research begins to fall. This is a sign that the technology has reached a point where it is mainstream and predictively mature in the market. In our predictive growth curve, the year of technological viability and maximum article output is 2035 as shown in Fig. 13.4.

13.5 Technology Enablement Tipping Point

For the purposes of this chapter, a technology will be considered 'adopted' and enabled by other technologies once an S-curve exceeds 16%. This is the level of market adoption referred to by Moore as 'crossing the chasm' in the market adoption cycle (Moore 1991), where the adopters of an innovation are no longer "innovators" or "early adopters," but part of the majority of the obtainable market. Many innovations fail to 'cross the chasm' and be adopted by the majority, and Gladwell refers to this as the 'tipping' point (Gladwell 2000), at which the market is primed

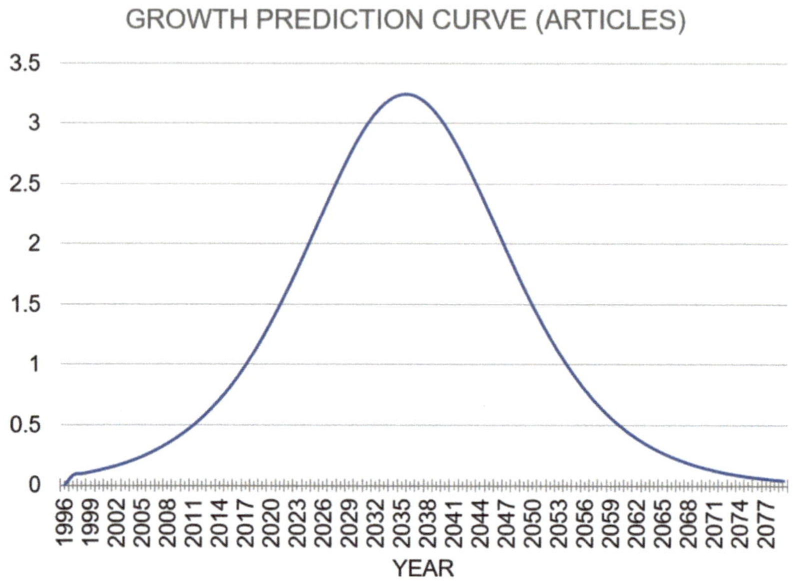

Fig. 13.4 The inflection point in the article

to complete the adoption cycle. Rogers in 2003, relates the adoption cycle directly to S-curves, correlating the 'early majority' as occurring at 16% of the way through the S-curve, with innovations that do not pass this tipping point 'failing to diffuse' to mainstream adoption (Rogers 2003).

While the ecosystems of consumers in markets and industry players differ in many ways, the authors of this chapter consider the incentives and disincentives inherent in the adoption of new technologies to have many similar properties across these environments. Both may see potential benefit in new technology, but refrain from investing resources on an unproven new method to solve a problem. Both will be more likely to adopt once others in their space have shown the innovation to be successful, though parties with more resources may be less risk averse.

If a consumer sees a technology, they are interested in being discarded by other consumers, they will likely choose not to invest in that technology themselves. Similarly, if a company in an industry is considering moving toward a particular technology to solve an industry problem, and peer companies are choosing different paths for the same problem, the company in question will take a harder, more skeptical look at the proposed technology.

This is an area that could benefit from further research and validation, though these factors lead the authors to believe both consumers and industry players will behave similarly in terms of 'following the trends' of emerging technologies, and that the 'tipping point' of industry adoption and research will fall upon similar 'chasm' dynamics near 16% of the complete curve. Relating this back to the predictive curves, the 16% threshold should be hit in 2020 and be the springboard for the rapid growth of the technology and its implementation.

13.6 Conclusion

Upon conducting the literature review, the overall benefits of electric vehicle charging station technologies have become a staple in modern day society with the interconnected advancement of technological solutions for everyday life. From what was gathered during the research, results have shown that XFC stations will be able to recharge BEVs in less than 10 min and provide roughly 200 miles of driving time (Howell et al. 2017). Evidence from an agnostic approach showed that battery performance will enable vehicle level performances that are required of a commercially successful, mass market BEV (Neubauer et al. 2014). Before extreme fast charging can become truly useful, batteries that are capable of "3–4C-rate charging with good energy density, low cost, and the ability to support 4,000 charge cycles," are needed (Walkowicz et al. 2020). The overall enablement and impact of XFCs will provide a large positive result for consumers and organizations as technological advances grow in popularity. BEVs offer significant potential when it comes to reducing the nation's petroleum consumption-based products and the production of greenhouse gases (Neubauer et al. 2014).

The data collected do support and identify the critical component technologies for the enablement and widespread adoption of XFC technology and infrastructure, to evaluate current and projected progress of critical component technologies, and to detect the potential gating technologies and recommend areas that will most benefit from accelerated development. The data collected from this literature review detail the significance of fast charging for electric vehicles and its prolonged impact upon adoption and implementation. However, there are still limitations to XFCs, with the batteries being the major hurdle. On the other hand, continuous technological advancements will allow XFCs to reach its peak and perform at its best and maximize user ability.

Based on the predictive curves used in this review, the year of 2020 will be a critical year as it is the year of market "adoption." This is a key milestone for the technology and should result in an acceleration in the development and expansion of the technology. This key milestone is followed by the technological maturity in 2035. This is the point when the technology should be in wide use and should warrant minimal improvements. The last milestone is met in 2076 with full technical maturity and the sundown of all further development.

The key takeaway from this review of the XFC technology is that its time is now. This is key in that EV batteries are also under heavy development at that same time, and these two technologies will gate to each other. While technological maturity is still many years away, XFC is needed now to meet the market needs of today. Therefore, it is likely that an acceleration within the technological space is needed on both EV batteries and XFC to meet the future market demands.

13.7 Future Research

This chapter implies the need for an in-depth investigation into the relation and potential time lags between the development of a technology and the advancements that enable them. Further research into this relationship, especially the timings of the adoption (s-curve) of technologies within this relationship, would provide a grounded basis for the projection of developing advancements across the vehicle and provide a full picture of where additional pressure should be applied to meet the technological demands of the future EV market.

Additionally, the 'technology enablement tipping point' described in this chapter is primarily conjecture and could benefit from further study. In particular, it could be very useful to take a retrospective look at the relationship between the adoption and failure of past technologies that were considered necessary to enable specific future innovations. If the conjecture in this chapter is accurate, that consumers and industry players will behave similarly in terms of adoption trends, such retrospective examinations should show similar 'tipping point' behavior in technology development as that observed in the adoption of innovation in consumer markets.

Appendix A Raw Data for Prediction Curve

Table 13.1 Raw data for the prediction curve

Year	Number of chapters	%	y/(L-y)	log{(y/(L-y)}	Predicted by
1996	522	0.0104	0.000104265	−3.981861809	0.006530375
1997	487	0.0097	9.72733E-05	−4.012006387	0.00739383
1998	434	0.0087	8.66862E-05	−4.062050216	0.008371442
1999	568	0.0113	0.000113454	−3.945179986	0.009478302
2000	704	0.0141	0.000140623	−3.851943865	0.010731494
2001	587	0.0117	0.00011725	−3.930888572	0.012150358
2002	691	0.0138	0.000138026	−3.860039604	0.013756792
2003	676	0.0135	0.000135029	−3.869572257	0.015575584
2004	732	0.0146	0.000146217	−3.835003014	0.017634796
2005	783	0.0156	0.000156406	−3.805747909	0.019966199
2006	870	0.0174	0.000173787	−3.759982871	0.022605754
2007	928	0.0185	0.000185375	−3.731949115	0.025594172
2008	973	0.0194	0.000194366	−3.711380347	0.028977536
2009	1129	0.0225	0.000225535	−3.646785712	0.032808009
2010	1388	0.0277	0.000277289	−3.557067717	0.037144636
2011	1731	0.0346	0.000345836	−3.461130355	0.042054245
2012	2186	0.0437	0.000436779	−3.359737784	0.047612475
2013	2899	0.0579	0.000579325	−3.237077851	0.053904929
2014	3630	0.0725	0.000725511	−3.139355995	0.061028486
2015	4041	0.0807	0.000807722	−3.092738093	0.069092772

Year	Number of chapters	%	y/(L-y)	log{(y/(L-y)}	Predicted by
2016	4656	0.0930	0.000930764	−3.031160582	0.078221836
2017	5542	0.1107	0.001108077	−2.955430105	0.088556034
2018	6336	0.1265	0.001267032	−2.897212506	0.100254153
2019	7577	0.1513	0.001515575	−2.819422597	0.113495821
2020					0.128484212
2021					0.145449105
2022					0.164650328
2023					0.186381638
2024					0.210975085
2025					0.238805925
2026					0.270298134
2027					0.305930599
2028					0.346244061
2029					0.391848878
2030					0.44343371
2031					0.501775203
2031					0.501775203
2032					0.567748771
2033					0.642340565
2034					0.726660737
2035					0.821958066
2036					0.929636052
2037					1.051270527
2038					1.18862883
2039					1.343690561
2040					1.518669861
2041					1.716039128
2042					1.938553969

Source: Self-elaborated with data retrieved by Technology Forecasting and Social Change and predicted values using Pearl method

References

Adamuthe, C., & Gopakumaran, T. (2019). *Technology forecasting: A case study of computational technologies, technological forecasting and social change 2019*. ScienceDirect, 143, 181–189.

Ahmed, S. et al. (2017). Enabling fast charging - a battery technology gap assessment. *Journal of Power Sources*, 1 Nov 2017. Journal of Power Sources: ScienceDirect.

Bloomberg Finance L.P. (2019). *Electric vehicle outlook 2019: BNEF's annual forecast of electric vehicles, shared mobility and road transport to 2040, New York, 2019*. New York: BloombergNEF.

Clement, K., Van Reusel, K., & Driesen, J. (2007). The consumption of electrical energy of plug-in hybrid electric vehicles in Belgium. In *Proc. 2nd Eur. Ele-drive transportation Conf*, Brussels, May. Belgium: Belgium.

Clement-Nyns, K., Haesen, E., & Driesen, J. (2010). The impact of charging plug-in hybrid electric vehicles on a residential distribution grid. *IEEE Transactions on Power Systems, 25*(1), 371–380.

Edis, O. (2019). *What you need to know about batteries for electric vehicles*, engineering.com, Inc., Mississauga, Ontario, Sept 25. Online: engineering.com, Inc.

Elizabeth, G., Kevin, B., Jisun, K., Daim, T., & Garces, E. (2017). Forecasting the electric transformation in transportation: The role of battery technology performance. *Technology Analysis & Strategic Management, 29*(10), 1103–1120. https://doi.org/10.1080/09537325.2016.1269886.

energy.gov. (2020). *Vehicle charging* (p. 2019). Washington DC: Office of Energy Efficiency& Renewable Energy.

Energysage.com. (2019). Electrical vehicles and the environment, electric vehicles, United State, December 13. United State: energysage.com.

Fernández, R. Á., Cilleruelo, F. B., & Martínez, I. V. (2015). A new approach to battery powered electric vehicles: A hydrogen fuel-cell-based range extender system. *International Journal of Hydrogen Energy, 41*, 4808–4819.

Gao, Z., LaClair, T., Ou, S., Huff, S., Wu, G., Hao, P., Boriboonsonsin, K., & Barth, M. (2018). Evaluation of electric vehicle component performance over eco-driving cycles. *Energy, 172*(2019), 823–839.

Gladwell, M. (2000). *The tipping point: How little things can make a big difference*. Boston: Little, Brown, and Company.

Howell, D., Boyd, S., Cunningham, B., Gillard, S., Slezak, L., (2017). *Enabling fast charging: A technology gap assessment*. U.S. Department of Energy: Office of Energy Efficiency & Renewable Energy.

Hu, X., Zheng, Y., Howey, D. A., Perez, H., Foley, A., & Pecht, M. (2020). Battery warm-up methodologies at subzero temperatures for automotive applications: Recent advances and perspectives. *Progress in Energy and Combustion Science, 77*, 100806.

Hussien, D. (2017). *Jobs, tax and politics: Three ways electric vehicles will change our world.* Boston: The Conversation US, Inc.: The Conversation.

Moore, G. A. (1991). *Crossing the chasm: Marketing and selling technology products to mainstream customers*. New York: HarperBusiness.

Neubauer, J., Pesaran, A., Bae, C., Elder, R., & Cunningham, B. (2014). Updating United States Advanced Battery Consortium and Department of Energy battery technology targets for battery electric vehicles. *The Journal of Power Sources, 271*, 614–621.

Onat, N. C., Kucukvar, M., & Afshar, S. (2019). Eco-efficiency of electric vehicles in the United States: A life cycle assessment based principal component analysis. *Journal of Cleaner Production, 212*(2019), 515–526.

Rogers, E. (2003). *Diffusion of innovations* (5th ed). Simon and Schuster. ISBN: 978-0-7432-5823-4. New York, NY

Walkowicz, K., Meintz, A., & Farrell, J. (2020). *R&D insights for extreme fast charging of medium- and heavy-duty vehicles: Insights from the NREL commercial vehicles and extreme fast charging research needs workshop*. Golden: National Renewable Energy Laboratory. NREL/TP-5400-75705.

Young, K., Wang, L., & Strunz, K. (2013). *Electric vehicle integration into modern power networks*. New York: Springer New York.

Chapter 14
Technology Intelligence Map: Space Tourism

Bhavana Ramesh, Rita Snodgrass, Bharat Verma, Moise Degan, and Tuğrul U. Daim

Since the Apollo 11 launch to the moon in 1969, there have been numerous space missions. However, this space travel has always been conducted by NASA alone or NASA in collaboration with the US military and other country's space agencies. Now, through research and development, private companies like Virgin Galactic, Blue Origin, and SpaceX are racing to fulfill the dreams of many, by proposing to take civilian customers on space excursions. But space tourism is in its infancy. This chapter looks at seven technologies that are necessary for space tourism to succeed; the vessel, fuel, navigation, health and safety technology, physics, communication systems, orbit control systems, and sensors, examining each through a literature review. Our seven key technologies are then examined using a technology forecasting technique – patent analysis – to reveal their growth curve. Our study shows that the last of the technology necessary to support safe space tourism will not reach maturity for another century.

14.1 Introduction

Space travel is a fascinating, but very complicated subject, where nothing is constant. Every object present in space is constantly moving at a specific speed and time, due to its gravity and mass. Scientists around the world, who continually monitor these space objects, are discovering something new almost every day. Our solar system, formed four billion years ago, has been stabilized in such a way that there are not enough objects, debris, dust, gas or rocks left out in space to form new planets or to become satellite moons that could collide with another planet. This is

B. Ramesh · R. Snodgrass · B. Verma · M. Degan · T. U. Daim (✉)
Department of Engineering and Technology Management, Portland State University, Portland, OR, USA
e-mail: tugrul.u.daim@pdx.edu

T. U. Daim (ed.), *Roadmapping Future*, Applied Innovation and Technology Management, https://doi.org/10.1007/978-3-030-50502-8_14

known as a stabilized system. Hence, it provides a perfect opportunity to study our solar system as a revolving body in the Milky Way galaxy.

To understand the potential prospects for space tourism, we need to look at how our solar system works. In our solar system, the moons or satellites revolve around the planets, the planets revolve around the Sun, and the Sun revolves around the massive black hole present in the middle of our galaxy. As humans have gained a greater understanding of the galaxy's geography, scientists studied the inner and outer solar system in greater detail. Continuous study and research over decades have revealed that Venus is the twin planet of Earth (Swaminathan 2007) and that Mars is the most likely candidate for life in the inner solar system (Perez 2019), because it falls in the habitable zone, along with Earth (Oceans, beaches, cosmic shorelines: our changing views of habitable planets 2019). It has also been discovered that each of these three planets once had an abundance of water (Dockrill 2020) and Mars may still have water to this date (Anderson 2015). Over time only Earth retained all of its water, due to its strong magnetic field, dense atmosphere, and because it falls in the habitable zone, which is the distance from the Sun where water is in the form of liquid and not in ice or gas. It has been discovered that in every ten billion years, the Sun increases its brightness by 10% and keeps getting bigger and bigger as it becomes a "Giant Dwarf" star, on its course toward its eventual death. It has also been found that in the next seven to eight billion years, the Sun will become so big that it will swallow Mercury, Venus, Earth, and possibly Mars (Appell 2008). Therefore, for humankind to avoid extinction, we need to move toward the outer solar system. Currently, the proposed idea is to shift humans to Mars followed by Jupiter's moon, Europa, or Saturn's moon, Titan, which contain the building blocks of life, in some shape or form, and are geologically active. Another idea proposes moving humankind out of the solar system to another star system, Proxima Centauri, to escape the death of our solar system.

For these reasons, space tourism is more than a leisure concept. Researching and developing space tourism develop these ideas and the technologies needed to support them so that humans can safely travel out into space as they race for future survival. Our very existence depends on this.

In this chapter, we explore the technologies that must be developed to full maturity to support successful space tourism in the near future.

14.2 Literature Review

Our research looks at space tourism using a technology forecasting process; patent analysis. Several private companies are working toward building a new industry, space tourism, flying ordinary civilians on excursions outside of the earth's atmosphere. For space tourism to succeed, there is a confluence of technologies that must all be developed to full, stable maturity. To forecast when this technology will reach this level of maturity, we will look at related patents. Through an examination of these patents, we can forecast when all of the technological pieces necessary for safe, successful space tourism will take place.

14.2.1 Space Tourism

Ever since Apollo 11 made it to the moon and back in 1969, people have dreamt about what it would be like to be in space. Space travels have always been reserved for government missions. The general public could never embark on this adventure, no matter their social or economic status. SpaceShipOne's prize-winning suborbital jaunt in 2004 is a step toward opening space travel to the general public and helped to advance congressional legislation to establish a space travel industry in the United States (Reddy et al. 2012). In April 2001, an American Businessman, Denis Tito, became the first space tourist, (Reddy et al. 2012) when he boarded a Russian Soyuz spacecraft at a cost of $20 million. Since then, we have seen breakthroughs in the new space tourism industry. Virgin Galactic, Blue Origin, and SpaceX are racing to make space tourism a reality.

Virgin Galactic is one of the three main competitors to offer space tourism commercially. The company is run by Richard Branson. The company's goal is to have a spaceship able to carry six passengers and two crew members, 62 miles above Earth's surface. The passengers will experience 4–5 minutes of weightlessness. The company has reported that an estimated 700 tickets have already been sold at a price of $200,000. In April 2013, Virgin Galactic performed its first rocket-powered test flight of SpaceShipTwo, which was built in collaboration with Scaled Composites called The SpaceShip Company in July 2005. The company suffered a major setback in October 2014 when their fourth rocket test exploded, killing the co-pilot and injuring another pilot. The launch of the VSS Unity test vehicle into space on December 13, 2019, was another major achievement for Virgin Galactic (Howell 2019).

Blue Origin was founded in 2000 by Amazon founder, Jeff Bezos. The company is planning on having a booster rocket with an attached passenger chamber, carrying six passengers and two crew members, 62 miles above Earth's surface. Once the spaceship reaches suborbital, the passenger chamber will detach from the main rocket where the passengers will experience 4–5 minutes of weightlessness before re-entering Earth's atmosphere. Between 2014 and 2019, Blue Origin flew several unmanned test flights, including one with a mannequin (Blue Origin: Space Tourism Company Profile 2019). So far, no accidents have been reported during their tests.

The third major competitor in space tourism is SpaceX. The company was founded in 2002 by Tesla founder, Elon Musk. The company is giving Space Tourism a whole new meaning, where its competitors only plan on taking about six to eight passengers above Earth's atmosphere; SpaceX envisions carrying 100 passengers and taking passengers to Mars. SpaceX has not advertised the price per trip yet as they will have several trip options. "A trip with SpaceX could be from a few hours to an Entire lifetime" Elon Musk has famously stated, "I've said I want to die on Mars, just not on impact." (Blue Origin: Space Tourism Company Profile 2019) the company is still in the testing phase. In March 2019, SpaceX launched Crew Dragon, a human-related space capsule, which autonomously docked with the International Space Station (ISS) as a proof of concept (Halseide 2020). SpaceX normally uses its rockets for several launches, proving reliability and dependability of the program. However, in April 2019, the company had a major setback when their unmanned Crew Dragon capsule was destroyed during its return to Earth.

Currently, these companies are in test phases; although most of their flight tests have been successful, they also have fatal accidents. Their biggest challenge is figuring out how to give their customers a space tourism experience while maintaining a 100% success rate.

14.2.2 *Technology Forecasting*

Technology forecasting covers a wide variety of activities and research techniques. It is useful to look at each word in the term "technology forecasting" to gain a better understanding of its meaning. The Merriam–Webster dictionary defines technology as "the practical application of knowledge especially in a particular area" and "a capability given by the practical application of knowledge" (Merriam-Webster 2020a). Forecasting is defined by the Merriam–Webster dictionary as "to calculate or predict (some future event or condition) usually as a result of study and analysis of available pertinent data" and "to indicate as likely to occur" (Merriam-Webster 2020b). Therefore, technology forecasting can be seen as using the study and analysis of available pertinent data to calculate or predict a capability given by the practical application of knowledge. Cho explored the definitions of technology forecasting developed by numerous sources, starting from its inception in 1940 through the 1990s and synthesized their definitions to "the analysis and the evaluation of performance parameters, timing of advancements, new concepts, products, processes, market penetration, and sales in a given time frame with probability statements, on a relatively high confidence level, which anticipates opportunities and threats from technological changes in order to provide for more well-informed R&D decision" (Cho 2000). Technology forecasting does not guarantee a specific outcome, rather it can be thought of as a process of documenting and assessing potential outcomes in a systematic way. Phillips and Linstone state that "The real point of forecasting is not to be 'right,' but to deal with the future in a maximally positive and economic way" (Phillips and Linstone 2016).There are many different approaches to technology forecasting. Our research will utilize patent analysis to produce a technology forecast for space tourism technology.

14.2.3 Patent Analysis

Patents are a good source for gauging the state of technology development because they contain content that can be analyzed; bibliographic information about inventions, inventors' certificates, utility certificates, and use models. The US Patent office database is the most well-classified and most up-to-date source for technical documents on new and innovative technologies (Hong 2020).

Patent analysis can be used to perform competitive analysis and technology trend analysis (D'Atri et al. 2014). It is a valuable approach that derives information about specific technologies through the use of patent data. Through this analysis, you can forecast a technology's future use. A specific technology's patent growth generally follows an s-shaped growth trend. In the early stages of a technology, the number of patents issued is very limited. If the technology continues to be developed, a fast-growing period where the number of patents filed and issued increases can be observed before, finally, a plateau is reached (Liu and Shyu 2020).

This chapter analyzes technologies related to space tourism by examining related patent filings. Through our analysis of this patent data, we will determine where in the development lifecycle, the critical technologies are for supporting successful space flights for ordinary people as tourists toward different destinations within the inner or outer solar system.

14.2.4 Technologies Used in Space Tourism

We've identified seven technologies that must all be mature in their development before space tourism can become a reality; the vessel, fuel, navigation, health and safety technology, physics, communication systems, orbit control systems, and sensors. If any one of these technologies is not fully developed, space tourism cannot proceed.

14.2.4.1 The Vessel

The spacecraft is an important part of NASA's Artemis program, which includes sending the first woman and the next man to the Moon by 2024. One important aspect of a vessel or spacecraft's design is mass. If a spacecraft has more mass, it will require more energy to launch, which increases the overall burden and cost. NASA's Orion spacecraft is designed to be the safest transportation and will carry four astronauts and everything they will need to live and work in space for up to 21 days per mission.

Orion has three main parts:

1. The launch abort system (LAS) at the top
2. The crew module in the center
3. The service module at the bottom (Dunbar 2015)

Antimatter thrusters are being designed that can one day carry us into deep space and eventually, travel between stars. Antimatter rockets have several inherent limitations; the immense load of dangerous gamma radiation caused by antimatter reaction, difficulty creating enough antimatter for fuel, and doing this while limiting the size of the payload. NASA has funded research to address the gamma radiation issue and has discovered positrons, which can be used in place or antiprotons, resulting in gamma rays that are far lower in their dangerous energy output. A former Fermilab physicist, Gerald Jackson, is attempting to address the issue of antimatter creation through the use of an antimatter thruster, which could make antimatter-based propulsion a reality (Tangermann 2017).

Other types of spacecraft propulsion technologies being considered include nuclear rockets, which work either by heating fuel with a nuclear reactor or by bombing nuclear bombs behind the craft, and ion drive thrusters and plasma drive thrusters, which both provide acceleration through an electrical charge (How Interstellar Space Travel Works 2013).

14.2.4.2 Fuel

The fuel used for space exploration must be able to simultaneously meet the constraints imposed by atmospheric pressure, gravity, heat, and temperature differences. Space agencies have utilized the combination of hydrogen and liquid oxygen to form liquid hydrogen. Hydrogen has the lowest molecular weight of any known element and discharges high energy during combustion, at over 5500 degrees Fahrenheit. Handling liquid hydrogen has proven to be quite challenging, since both oxygen and liquid hydrogen can only be liquefied at an extremely low temperature. To keep liquid hydrogen fuel in a spacecraft's rockets from evaporating or boiling off, it must be insulated from all heat sources, including engine exhaust and air friction created during flight through the atmosphere and the radiant heat of the Sun must be protected against, once the vehicle reaches space. When exposed to heat, liquid hydrogen will expand rapidly which may cause an explosion of the rocket. The signature fuel of the American space program is liquid hydrogen; other countries utilize it to launch satellites (Dunbar 2020).

14.2.4.3 Navigation

Navigating in space presents unique challenges that must be met before space tourism can become a reality. Technologies we rely on for navigation on earth; for example, global positioning system (GPS) satellites and the planet's magnetic field aren't available as you travel outside of the earth's orbit. To meet this challenge, NASA's Jet Propulsion Laboratory (JPL) has developed the Deep-Space Positioning System (DPS). DPS utilizes cameras, a coelostat and X-band receiver, and a patch antenna and a central processor that performs navigation computation and controls the coelostat to determine a spacecraft's position relative to images of solar objects and optionally, via a one-way radio to earth (Deep-Space Positioning System 2020). Technologies used in DPS include Optical Navigation (OpNav), which provides the core capability for in situ target observation, and optionally, provides 1-way Doppler radiometric measurements from a known source beacon or another spacecraft. Accurate radiometric data require a very good frequency reference, which can be provided by the Deep Space Atomic Clock (DSAC) (Guinn et al. 2016). NASA scientist Zaven Arzoumanian states that using DPS would provide "a fallback, so that if a crewed mission loses contact with the Earth, they'd still have navigation systems on board that are autonomous" (NASA's 2020). Without accurate navigational technology to ensure that travelers won't get lost on the way to their destination and return to earth, the space tourism industry will never come to fruition.

14.2.4.4 Health and Safety Technology

According to NASA, 4% of people who have flown to space have lost their lives, (Bensoussan 2010) most recently with the tragic accident of Columbia Space Shuttle upon re-entering the Earth's atmosphere, killing all seven astronauts on board. Space tourism is moving toward becoming a reality, with Virgin Galactic, SpaceX, and Blue Origin making progress on delivering this exhilarating experience. These companies have already accepted applications and payment from potential customers. These customers must face the health and safety risks that are associated with traveling above the earth's atmosphere. Health and safety is a huge concern for space tourism. Astronauts spend months training and going through numerous health checks before traveling into space. Currently, there are no data supporting whether space travelers will be checked for health issues, and how long any training provided to them will last. At this time, "insurance coverage is not required for passengers to travel via commercial ground or air transport. However, that may well be different for space travel" (Bensoussan 2010). Companies like Virgin Galactic or Blue Origin may require their potential customers to sign a waiver that would relieve them of any liabilities during their space exploration. If customers sign a waiver, there may be no liability for the companies providing space tourism services. Customers will probably have to purchase their own insurance, but it is not yet clear how any insurance coverage will apply if there is an incident where their space vessel suffers a mechanical or system failure that results in a fatal accident.

14.2.4.5 Communication System

When sending human beings into space, the ability for people and equipment to communicate with earth is crucial to ensure safe passage and to return pertinent data. The current state of the art in space communication is NASA's Deep Space Network (DSN). The DSN is operated by NASA's Jet Propulsion Laboratory (JPL).

Three facilities located approximately 120 degrees apart in longitude; Goldstone, near Barstow, California; near Madrid, Spain; and near Canberra, Australia, provide a series of large antennas that can be contacted from space during all stages of the earth's rotation (About the Deep Space Network 2020). The need to transmit larger volumes of data from greater distances is necessitating an update to the DSN. NASA estimates that, over each of the next three decades, deep space communications capability will need to grow by a factor of 10 to meet these needs (Deep Space Communications 2020). NASA is currently testing a laser-based system to augment or replace the aging radio-based DSN system. Lunar Laser Communication is an optical communications system, uses optical frequencies of the magnetic spectrum instead of radio frequencies. Smaller optical frequencies, ranging from the ultraviolet or visible range of 2–10 microns, or short-wave infrared bands between 1 and 1.5 microns, are capable of high-speed modulation and can carry more data over greater distances. The component parts of the test system consist of a space terminal, the Lunar Lasercom Space Terminal (LLST), and a primary ground terminal, the Lunar Lasercom Ground Terminal (LLGT), a transportable system. In addition, several other ground terminals provided additional backup for the tests. The space terminal was tested from the Lunar Atmospheric Dust and Environment Explorer (LADEE) spacecraft and proved to be successful (Boroson and Robinson 2015). Further tests and development of the laser-based communication may meet the increased communication requirements to allow space tourism to succeed.

14.2.4.6 Orbit Control Systems

The satellite orbit control system is a combination of various sensors, actuators, and software. During any mission being executed in space, the satellite has to be pointing toward the Earth at all times to send and receive signals. This system provides three-axis stabilized earth pointing attitude control. This provides the orbital position of the spacecraft. This system initially includes a number of upgrades from an initial Gravity Recovery and Climate Experiment (GRACE) mission's design for this type of function. The parts of the system include a "GPS receiver, star tracker assembly, high-performance gyro package, coarse Earth and Sun sensor, fluxgate magnetometer, inertial measurement unit, magnetic torques, and a cold gas propulsion system" (Attitude and Orbit Control System 2020).

Actuators are often used to control space systems. These actuators have a limited output level and strictly control relative position and attitude. Autonomous spacecraft rendezvous and docking (RVD) maneuvers have been extensively studied in past years, performing maneuvers using controlled trajectories where the active vehicle tries to dock a passive target (Dentis, Matteo, Capello, Elisa, and Giorgio 2016).

14.2.4.7 Sensors

Spacecrafts are among the most advanced and sophisticated vehicles ever built, with the ability to travel above Earth's atmosphere, withstanding the pressure difference, heat, magnetic field, temperature extremes, and more. Spacecrafts are equipped with an array of sensors, along with dependent and independent sensors that enable the command center operators to monitor the spacecraft more critically during lift-off and re-entering Earth's atmosphere. "Specific areas of work include chemical species sensors, thin-film thermocouples and strain gauges, heat flux gages, fuel gages, SiC-based electronic devices and sensors, space-qualified electronics, and MicroElectroMechanical Systems (MEMS) as well as integrated and multifunctional sensor systems" (Aerospace Sensor Systems 2020). As the first commercial space tourism flight, promised by Virgin Galactic, is coming in the not too distant future, the shuttle that will take those eager consumers' needs to be equipped with the most advanced sensors that have been used and tested on NASA's previous spacecraft. Using a hierarchical design method, this flight will utilize a sophisticated system infrastructure that includes electronics, sensors, and a connected actuator, all linked together through a main controller (Aerospace Sensor Systems 2020).

14.3 Methodology

14.3.1 Framework

In this chapter, we analyzed seven categories of technologies necessary for smooth and safe operation of space tourism. Using literature review as a method, 42 technologies were identified which were then categorized into seven buckets. Table 14.1 provides the classification of 42 technologies into buckets. Using the technologies as keywords as shown in Table 14.1, Cooperative Patent Classification (CPC) codes are collected using lens.com. As the scope of this research was forecasting the growth curve of various technologies, data points such as patent published dates, geographic fields, count of published patents by year, and the titles of patents associated with selected keywords as listed in Table 14.1 are gathered. Data cleaning and manipulation activity is performed using Excel spreadsheets. The growth curve is analyzed separately for all seven categories and Appendices 1 through 7 show the growth curves and data used for analysis. All seven growth curves are then layered on a single time-series graph to identify the leading and lagging technologies based on the historical patent data from lens.com (Fig. 14.1).

Table 14.1 Categories of technologies and keywords for used for patent data extraction and analysis

Technologies categories	Vessel	Fuel	Navigation	Health and safety	Physics	Communication systems	Sensors
Technologies sub-categories (Keywords)	Heat (ablative heat shield)	Power control	GPS	Radiation	Gravity, orbits	Telemetric (communication back and forth with earth)	flight software
	Vibration	Batteries that store power	SatNav	Mental health	Space junk	Payload data	Actuators (reaction wheels, thrusters)
	Temperature, safety	Solar	Core Satellite navigation systems – GLONASS, Galileo, BeiDou	Physical health	Magnetic levitation	Transponders	Antennas
	Micrometeorite damage	Nuclear photonic rocket	Global Satellite Based Augmentation systems (SBAS)		Weather forecasting	Amplifying received radio signals	Remote sensing
	Orbit stability	Propulsion	Radio navigation			Cameras	Measure vehicle orientation
	Aerobraking Reusable launch system	Beam-powered propulsion	NAVIC				
	Flexible wings (X-53 Active Aeroelastic Wing, Adaptive Compliant Wing)	Photon rocket	QZSS				
	Fluidic flight controls	Propellant depot					
	Fusion rocket						

Technologies categories	Vessel	Fuel	Navigation	Health and safety	Physics	Communication systems	Sensors
	Pulse detonation engine						
	Booster (rocketry)						
	Gantry						
	Reusable launch system						

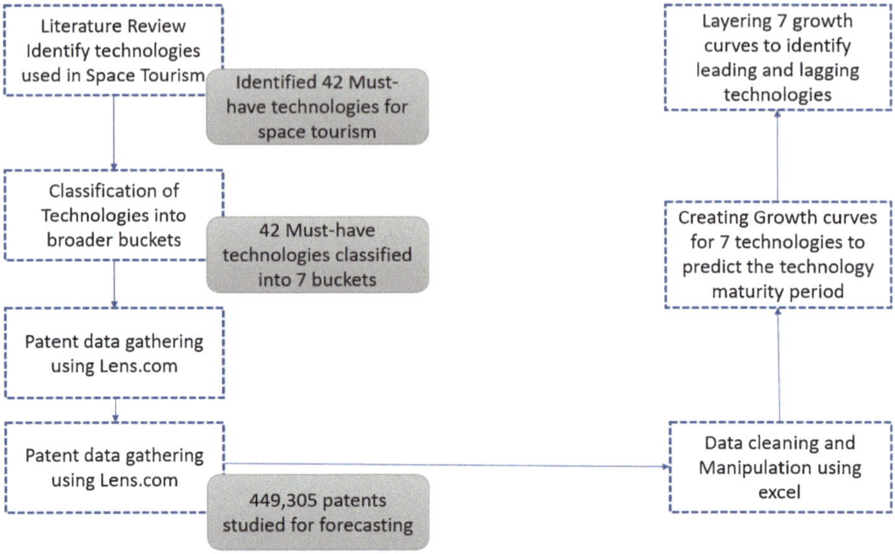

Fig. 14.1 Framework for patent analysis used in the study of space tourism technologies

Table 14.2 Patent data analyzed

Data	Number of patents analyzed
Overall patents analyzed	449,305
Vessel	55,383
Communication systems	56,309
Sensors	154,963
Physics	5597
Navigation systems	50,671
Health & Safety	41,920
Fuel	84,462

14.3.1.1 Exploratory Data Analysis

For this analysis, we gathered patent data dating from January, 1920 to March, 2020. It was noticed that some technologies such as propellers categorized under the "Fuel" bucket were researched as far back as the 1920s, where technologies such as sensors and navigation systems were relatively newer and most of the research began in early 1970s. In order to maintain data consistency, analysis has been conducted using patent data since the 1970s. Table 14.2 shows the details of the patent data analyzed in this chapter.

14.3.2 *Analysis and Results*

Figure 14.2 shows the growth curve of all the seven technologies forecasted into the future. We use a growth curve that uses a logistic function to predict the technology maturity based on the historical process made. It is important to note that this forecast is purely based on historical patent data and no expert advice is considered.

Looking at Fig. 14.2, we notice that some technologies are progressing much faster than the others. In fact, some technologies classified as laggers could fall a couple of decades behind, which could affect the implementation of space tourism. The seven technologies analyzed are vessel, sensors, orbital physics, navigation systems, health and safety in space, fuel, and communication systems.

This research believes that the above-listed technologies are absolutely necessary to launch space tourism. Of the identified technologies, with the growth phase we notice from patent analysis, space tourism seems a distant vision which could take another century to be completely safe and functioning. Below is the in-depth analysis of leading and lagging technologies.

14.3.2.1 Leading Technologies

1. *Fuel:* We considered looking at various fuel-related patents that are used in space flights such as power controls, batteries, solar, nuclear, propulsion, beam-powered propulsion, photon rockets, and propellant depot. Looking at data from the 1970s, the growth curve looks as shown in Fig. 14.3; we can see that fuel technologies will reach the maturity curve by late 2070s, making it the leading technology.
2. *Navigation:* We looked at multiple navigation patents that are used in space flights. GPS, SatNav, core satellite navigation systems – GLONASS, Galileo, BeiDou, Global Satellite Based Augmentation Systems (SBAS), Radio

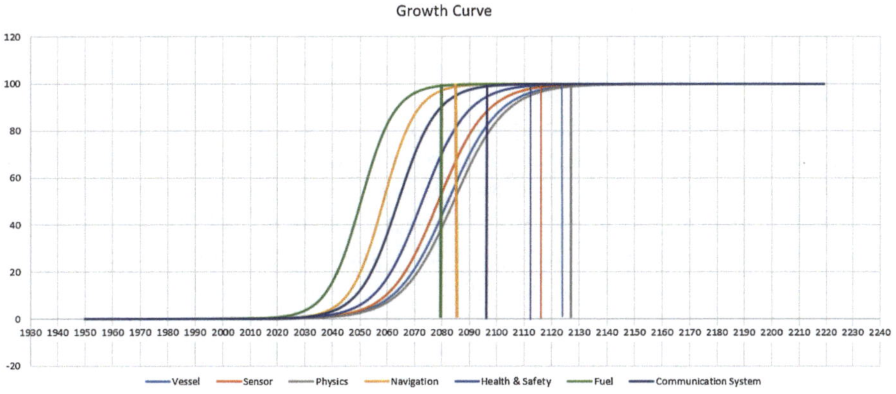

Fig. 14.2 Consolidated growth curve for space technologies

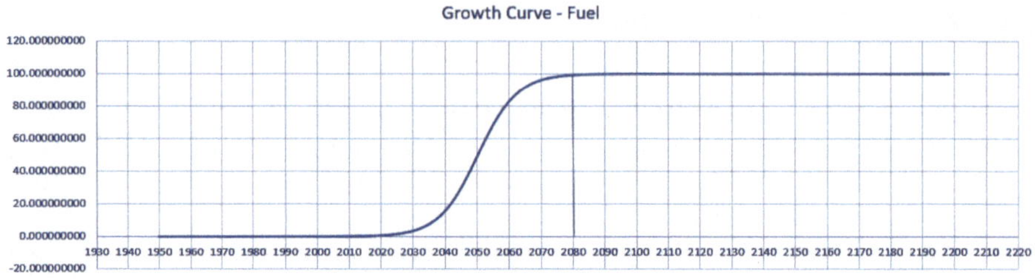

Fig. 14.3 Growth curve for fuel technologies

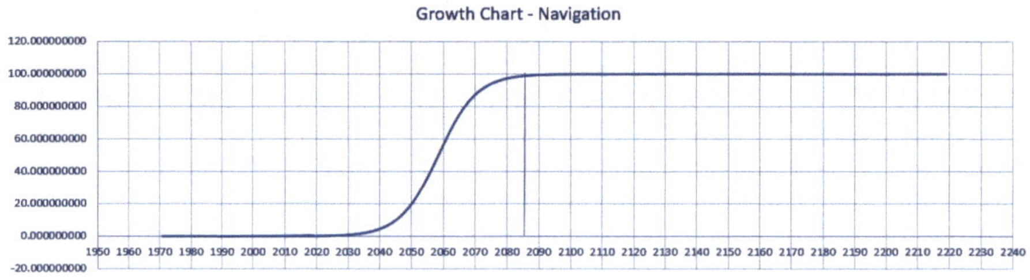

Fig. 14.4 Growth curve for navigation technologies

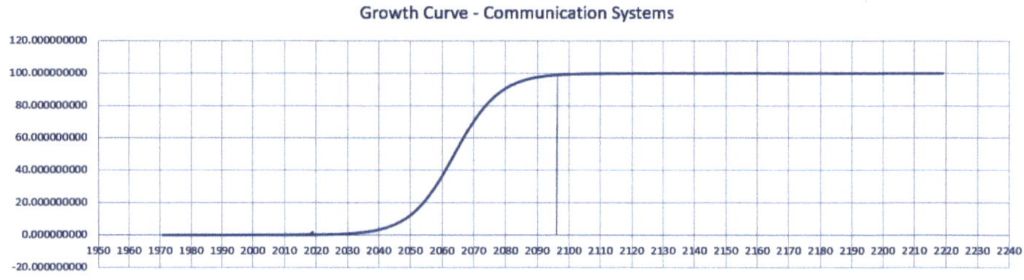

Fig. 14.5 Growth curve for communication technologies

Navigation, NAVIC, and QZSS are some of the patents we looked into. Figure 14.4 shows the growth curve of Navigation technologies and is expected to reach maturity in mid-2080.

3. *Communication Systems:* Communication systems are identified as one of the leading technologies in comparison to other space technologies. We looked into technologies such as payload data, telemetric communications, transponders, amplifying signals technologies, and cameras & imaging technologies in analyzing the communication patents. Figure 14.5 shows the growth curve for communication systems reaching technology maturity in the mid-2090s.

14.3.2.2 Lagging Technologies

1. *Vessel:* Technologies such as ablative heat shield, vibrations, temperature safety, micrometeorite damages, orbit stability, aerobraking systems, reusable launch systems, flexible wings such as X-53 Active Aeroelastic Wings and Adaptive Compliant Wings, fluidic flight controls, fusion rockets, pulse detonation engines, boosters, and gantry were considered in the analysis. Space Vessel is a new technology and according to the forecast from growth curve, the technology is expected to reach maturity sometime in the 2120s (Fig. 14.6).

1. *Sensors:* Sensors can also be categorized into lagging technology, basing our forecast on the logarithmic growth curve. Even though sensor technologies have progressed in other areas such as computation and artificial intelligence, we see that sensors in the area of space tourism might take longer, given the complexities involved in space tourism. The analysis considers patents on remote sensing, antennas, inflight software, data transmissions, actuators, and vehicle orientation sensing to forecast the sensor growth curve.
2. *Health and Safety:* Health and safety is a primary concern in space tourism; both the mental and physical health of people involved in travel needs a lot of research and trials before actual tourism can take place. From our analysis, we see that these areas are lagging behind and could create a bottleneck for launching the space flights. Figures 14.7 and 14.8 show that it might take another century for these technologies to mature.
3. *Orbital Physics:* Orbital physics studies gravity, space junk, magnetic levitation, and outer space weather forecasting accuracy. These are extremely important for successful space tourism. We observe from patent data that this is one area that still needs a lot of research. Figure 14.9 shows the research under this category takes longer than the other studied technologies and the growth curve suggests that it could take longer than 100 years for us to understand and navigate orbital zones while dealing with free-floating space junk.

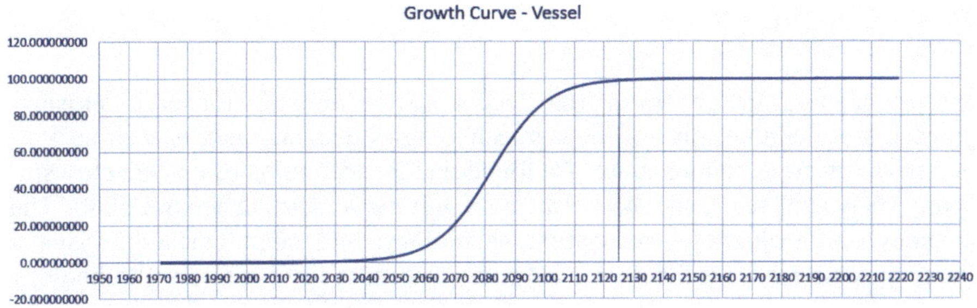

Fig. 14.6 Growth curve for spacecraft vessel technologies

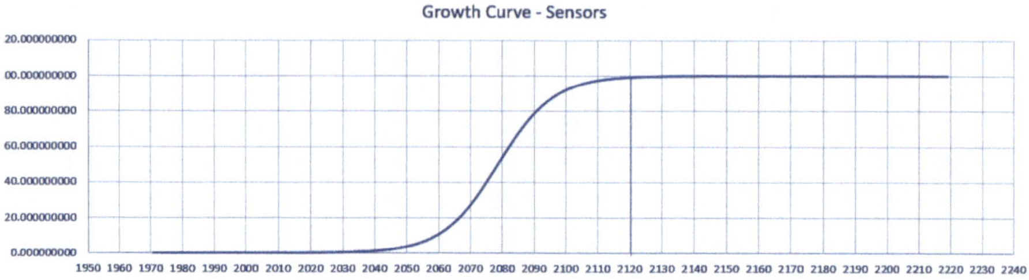

Fig. 14.7 Growth curve for spacecraft sensor technologies

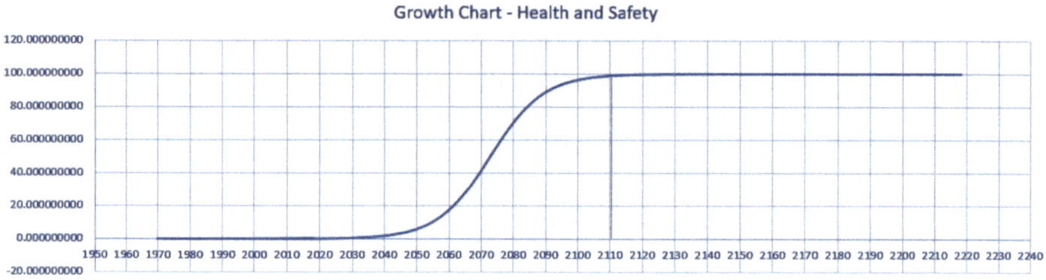

Fig. 14.8 Growth curve for spacecraft health and safety technologies

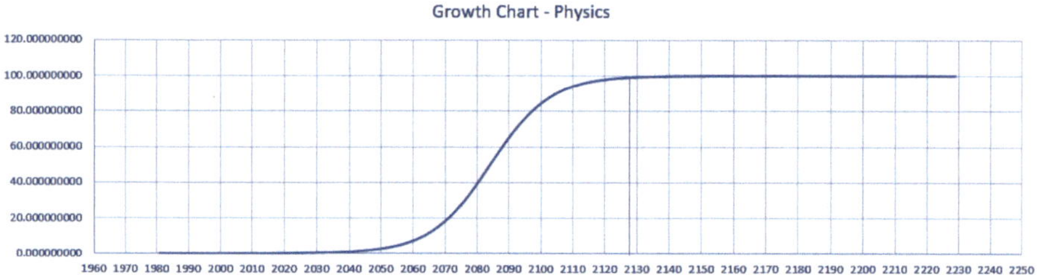

Fig. 14.9 Growth curve for spacecraft orbital physics technologies

14.4 Conclusion

Our patent analysis of the seven critical technologies necessary for space tourism to proceed revealed that some of the key technologies that can make space tourism a reality are far from being mature. We found that the technologies that are closest to being ready will reach the top of their growth curve starting around 2080. This includes fuel technology, navigation technology, and communication systems. Lagging behind these technologies in the development curve are vessel technology, sensors, health and safety related technology, and orbital physics. The last of these technologies will reach the top of the growth curve over 100 years from now, around 2130. For space tourism to advance to become a viable industry, development on all of these key technologies should be accelerated. Companies like SpaceX, Blue Origin, and Virgin Galactic are racing to become the premier providers of space tourism. Each of the technologies we examined are key to providing a safe experience for burgeoning space tourists and also serve to carry humankind into space, should Earth become uninhabitable. Without further and faster development by the companies racing to provide this service, space tourism will remain a dream for at least another century, and a safe and timely escape from planet Earth, should it become necessary, may prove to be impossible.

Appendix

1.

L=	100.000000000				
Year	%	log(y/L-y)	Predicted y		
1971	0.000091836	-6.0369876	0.0006034	Intercept	97.72450615
1972	0.000146937	-5.8328674	0.00067226	Slope	0.046933087
1973	0.000202039	-5.6945644	0.00074898		
1974	0.000220408	-5.6567758	0.00083446		
1975	0.000238773	-5.6220136	0.00092968		
1976	0.003893838	-4.4096052	0.00103578		
1977	0.003104050	-4.5080578	0.00115399		
1978	0.003508127	-4.4549094	0.00128568		
1979	0.002563035	-4.5929321	0.00143241		
1980	0.002993847	-4.5237574	0.00159587		
1981	0.003471393	-4.4594811	0.001778		
1982	0.003581596	-4.4459078	0.0019809		
1983	0.003508127	-4.4549094	0.00220696		
1984	0.004279548	-4.3685835	0.00245882		
1985	0.004206080	-4.3761043	0.00273942		
1986	0.004591790	-4.3379981	0.00305204		
1987	0.004793829	-4.3192967	0.00340034		
1988	0.006281569	-4.2019046	0.00378838		
1989	0.005767288	-4.2390033	0.0042207		
1990	0.007062989	-4.1515961	0.00470236		
1991	0.006759115	-4.1700808	0.00523898		
1992	0.008650932	-4.0628995	0.00583683		
1993	0.007897879	-4.1024552	0.0065029		
1994	0.007714207	-4.1126752	0.00724498		
1995	0.007989714	-4.097434	0.00807173		
1996	0.008485628	-4.0712792	0.00899281		
1997	[illegible]	[illegible]	[illegible]		

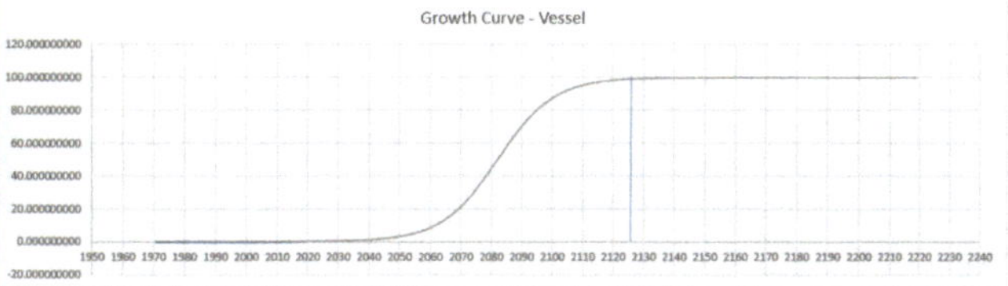

Snippet of log function, intercept, and slope used to analyze growth curve for vessel technologies

2.

L=	100.000000000				
Year	%	log(y/L-y)	Predicted y		
1971	0.000092614	-6.03332	0.000442149	A	103.324512
1972	0.000072768	-6.13806	0.000495763	B	0.04970578
1973	0.000119075	-5.92418	0.000555879		
1974	0.000165382	-5.78151	0.000623283		
1975	0.000304303	-5.51669	0.000698861		
1976	0.003565640	-4.44785	0.000783603		
1977	0.002738729	-4.56244	0.000878621		
1978	0.002778421	-4.55619	0.00098516		
1979	0.002454272	-4.61007	0.001104618		
1980	0.002672576	-4.57306	0.001238561		
1981	0.002957034	-4.52913	0.001388745		
1982	0.002864420	-4.54295	0.00155714		
1983	0.002348427	-4.62921	0.001745953		
1984	0.003234876	-4.49013	0.00195766		
1985	0.003466411	-4.4601	0.002195038		
1986	0.003711177	-4.43047	0.002461199		
1987	0.004187477	-4.37803	0.002759632		
1988	0.004101479	-4.38704	0.003094251		
1989	0.004829160	-4.31611	0.003469442		
1990	0.004822545	-4.3167	0.003890126		
1991	0.005126848	-4.29013	0.004361816		
1992	0.006152218	-4.21094	0.004890698		
1993	0.005424536	-4.26561	0.005483704		
1994	[illegible]	-4.22182	0.00614861		

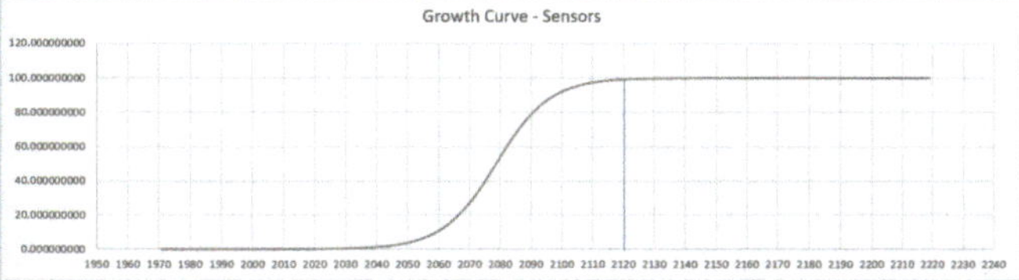

Snippet of log function, intercept, and slope used to analyze growth curve for sensors technologies

3.

L= 100.000000000

Year	%	log(y/L-y)	Predicted y
1981	0.001667284	-4.77798	0.0017
1982	0.002408299	-4.61828	0.0019
1983	0.001296777	-4.88713	0.0021
1984	0.002223046	-4.65304	0.0024
1985	0.002037792	-4.69083	0.0026
1986	0.003705076	-4.43119	0.0029
1987	0.002964061	-4.5281	0.0033
1988	0.006113375	-4.21369	0.0036
1989	0.005372360	-4.26981	0.0040
1990	0.005742868	-4.24085	0.0045
1991	0.005928122	-4.22706	0.0050
1992	0.007039644	-4.15242	0.0056
1993	0.006113375	-4.21369	0.0062
1994	0.005742868	-4.24085	0.0069
1995	0.005742868	-4.24085	0.0076
1996	0.006669137	-4.1759	0.0085
1997	0.008892182	-4.05095	0.0094
1998	0.012597258	-3.89967	0.0105
1999	0.012782512	-3.89333	0.0117
2000	0.012782512	-3.89333	0.0130
2001	0.014820304	-3.82908	0.0144
2002	0.015561319	-3.80789	0.0161
2003	0.019081141	-3.71931	0.0179

A	96.1187182
B	0.04611614

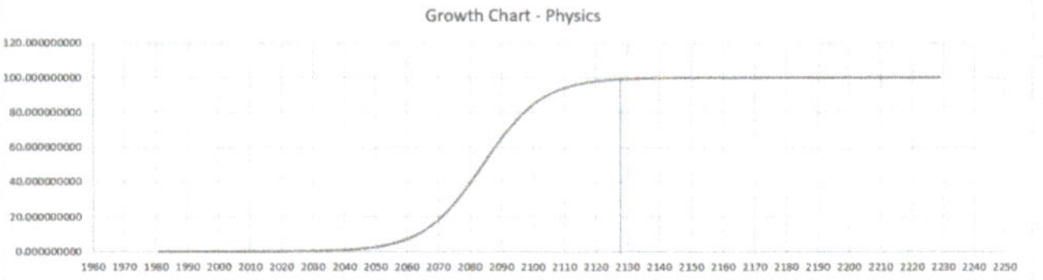

Snippet of log function, intercept, and slope used to analyze growth curve for orbital physics technologies

4.

L= 100.000000000

Year	%	log(y/L-y)	Predicted y
1971	0.000040862	-6.38868	4.96644E-05
1972	0.000020431	-6.68971	5.86212E-05
1973	0.000040862	-6.38868	6.91934E-05
1974	0.000040862	-6.38868	8.16722E-05
1975	0.000020431	-6.68971	9.64015E-05
1976	0.000531208	-5.27473	0.000113787
1977	0.000306466	-5.51362	0.000134308
1978	0.000388191	-5.41095	0.00015853
1979	0.000490346	-5.30949	0.000187121
1980	0.000551640	-5.25834	0.000220867
1981	0.000531208	-5.27473	0.0002607
1982	0.000265604	-5.57576	0.000307716
1983	0.000306466	-5.51362	0.000363211
1984	0.000347329	-5.45926	0.000428715
1985	0.000449484	-5.34728	0.000506032
1986	0.000286035	-5.54358	0.000597293
1987	0.000735519	-5.1334	0.000705012
1988	0.000551640	-5.25834	0.000832158
1989	0.001511901	-4.82047	0.000982233
1990	0.000939830	-5.02695	0.001159374
1991	0.001082848	-4.96543	0.001368461
1992	0.001532332	-4.81464	0.001615254
1993	0.001430177	-4.8446	0.001906554

A	148.234951
B	0.07200964

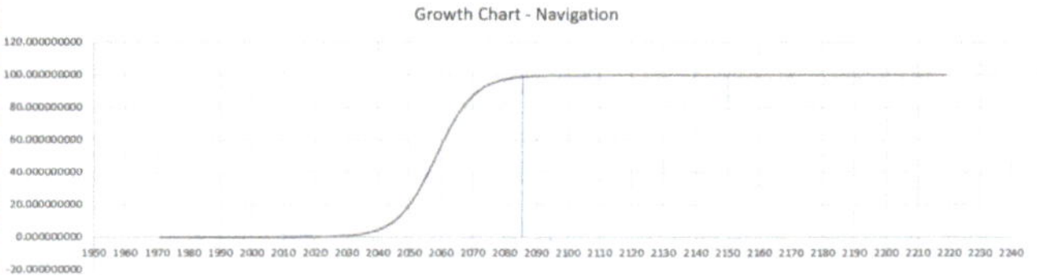

Snippet of log function, intercept, and slope used to analyze growth curve for navigational technologies

5.

L= 100.000000000

Year	%	log(y/L-y)	Predicted y
1970	0.000054057	-6.26715	0.000375556
1971	0.000027028	-6.56818	0.000424071
1972	0.000108114	-5.96612	0.000478853
1973	0.000108114	-5.96612	0.000540712
1974	0.000243256	-5.61393	0.000610562
1975	0.003108276	-4.50747	0.000689434
1976	0.002675820	-4.57253	0.000778496
1977	0.002243365	-4.64909	0.000879063
1978	0.002162279	-4.66508	0.00099262
1979	0.002837991	-4.54698	0.001120847
1980	0.002810963	-4.55113	0.001265638
1981	0.002621763	-4.5814	0.001429133
1982	0.002702849	-4.56817	0.001613748
1983	0.002919077	-4.53474	0.001822211
1984	0.003108276	-4.50747	0.002057603
1985	0.002865020	-4.54286	0.002323401
1986	0.004838099	-4.3153	0.002623535
1987	0.003892102	-4.4098	0.002962437
1988	0.006594951	-4.18076	0.003345117
1989	0.005567869	-4.25429	0.003777229
1990	0.005973296	-4.22376	0.004265157
1991	0.007189578	-4.14327	0.004816111
1992	0.007135521	-4.14654	0.00543823

A	109.370086
B	0.05276384

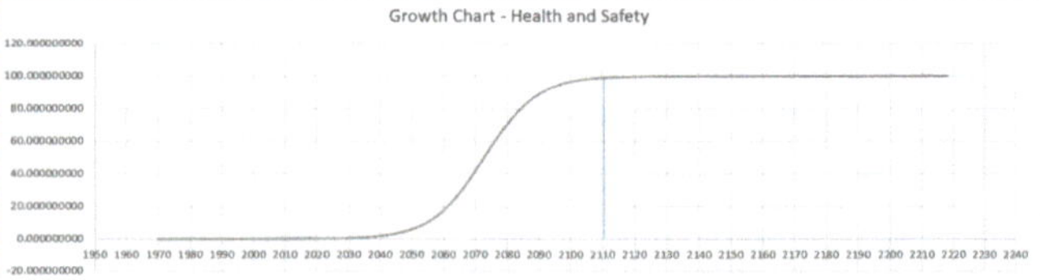

Snippet of log function, intercept, and slope used to analyze growth curve for health and safety technologies

6.

L= 100.000000000

Year	%	log(y/L-y)	Predicted y
1950	0.000024256	-6.6151816	7.66716E-06
1951	0.000024256	-6.6151816	9.02738E-06
1952	0.000024256	-6.6151816	1.06289E-05
1953	0.000012128	-6.9162117	1.25146E-05
1954	0.000024256	-6.6151816	1.47347E-05
1955	0.000012128	-6.9162117	1.73488E-05
1956	0.000048512	-6.3141515	2.04266E-05
1957	0.000024256	-6.6151816	2.40505E-05
1958	0.000048512	-6.3141515	2.83172E-05
1959	0.000072768	-6.1380602	3.33409E-05
1960	0.000012128	-6.9162117	3.92559E-05
1961	0.000072768	-6.1380602	4.62202E-05
1962	0.000072768	-6.1380602	5.442E-05
1963	0.000012128	-6.9162117	6.40746E-05
1964	0.000060640	-6.2172415	7.54419E-05
1965	0.000048512	-6.3141515	8.88259E-05
1966	0.000024256	-6.6151816	0.000104584
1967	0.000012128	-6.9162117	0.000123138
1968	0.000048512	-6.3141515	0.000144984
1969	0.000024256	-6.6151816	0.000170705
1970	0.000072768	-6.1380602	0.00020099
1971	0.000048512	-6.3141515	0.000236647
1972	0.000121280	-5.9162112	0.00027863

A	145.423082
B	0.07092703

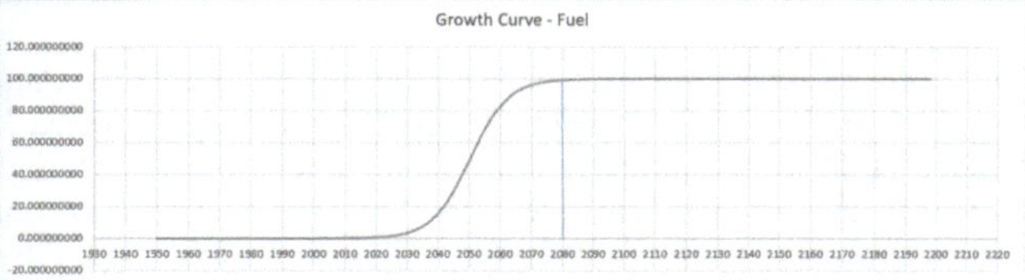

Snippet of log function, intercept, and slope used to analyze growth curve for fuel technologies

7.

L= 100.000000000

Year	%	log(y/L-y)	Predicted y
1971	0.000018291	-6.73776	0.000191813
1972	0.000018291	-6.73776	0.000220969
1973	0.000036582	-6.43673	0.000254557
1974	0.000109745	-5.95961	0.00029325
1975	0.002048581	-4.68854	0.000337824
1976	0.001298654	-4.8865	0.000389174
1977	0.001444981	-4.84013	0.000448329
1978	0.001444981	-4.84013	0.000516476
1979	0.001591308	-4.79824	0.000594981
1980	0.001225490	-4.91168	0.000685419
1981	0.001865672	-4.72916	0.000789604
1982	0.001353526	-4.86853	0.000909624
1983	0.001627890	-4.78837	0.001047888
1984	0.001627890	-4.78837	0.001207167
1985	0.001536435	-4.81348	0.001390657
1986	0.003164325	-4.49971	0.001602036
1987	0.002432689	-4.6139	0.001845545
1988	0.003274071	-4.4849	0.002126066
1989	0.003310653	-4.48007	0.002449226
1990	0.003127744	-4.50476	0.002821503
1991	0.004426397	-4.35393	0.003250364
1992	0.003987416	-4.39929	0.003744409
1993	0.004572725	-4.33981	0.004313543

A	126.841694
B	0.06145336

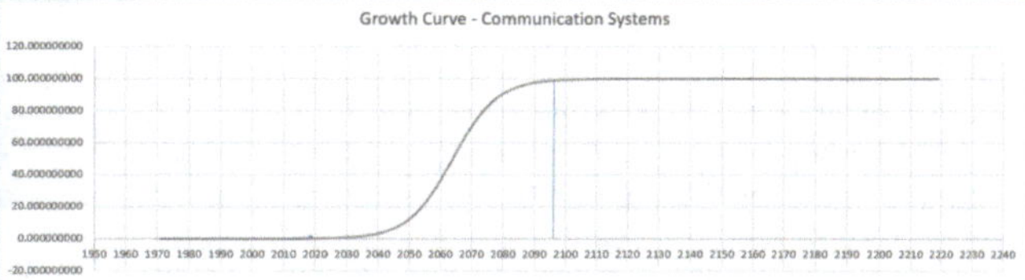

Snippet of log function, intercept, and slope used to analyze the growth curve for communication systems technologies

References

About the deep space network. NASA. [Online]. Available: https://deepspace.jpl.nasa.gov/about/. Accessed 15 Mar 2020.

Aerospace sensor systems: From sensor development to vehicle application. NASA. [Online]. Available: https://ntrs.nasa.gov/search.jsp?R=20130013129. Accessed 22 Mar 2020.

Anderson, G. NASA confirms evidence that liquid water flows on today's mars, NASA, 28-Sep-2015. [Online]. Available: https://www.nasa.gov/press-release/nasa-confirms-evidence-that-liquid-water-flows-on-today-s-mars. Accessed 22 Mar 2020.

Appell, D. The sun will eventually engulf earth—maybe. *Scientific American*, 01-Sep-2008. [Online]. Available: https://www.scientificamerican.com/article/the-sun-will-eventually-engulf-earth-maybe. Accessed 22 Mar 2020.

Attitude and orbit control system – GRACE-FO, NASA. [Online]. Available: https://gracefo.jpl.nasa.gov/attitude-and-orbit-control-system. Accessed 22 Mar 2020.

Bensoussan, D. Space tourism risks: A space insurance perspective. Acta Astronautica, 24-Feb-2010. [Online]. Available: https://www.sciencedirect.com/science/article/pii/S0094576510000111. Accessed 22 Mar 2020.

Blue origin: Space tourism company profile. *Space Tourism Guide*, 11-Dec-2019. [Online]. Available: https://spacetourismguide.com/blue-origin/. Accessed 22 Mar 2020]

Boroson, D. M. & Robinson, B. S. The lunar laser communication demonstration: NASA's first step toward very high data rate support of science and exploration missions, the lunar atmosphere and dust environment explorer Mission (LADEE), pp. 115–128, 2015.

Halseide, P. (2020). "SpaceX: Space tourism company profile. *Space tourism guide*, 07-Jan-2020. [Online]. Available: https://spacetourismguide.com/spacex/. Accessed 22 Mar 2020.

Cho, Y. (2000). Exploring technology forecasting and its implications for strategic technology Planning. https://pdxscholar.library.pdx.edu/open_access_etds/4224. https://doi.org/10.15760/etd.6108

D'Atri, A., Marco, M. D., Braccini, A. M., & Cabiddu, F. (2014). *Management of the interconnected world ItAIS: The Italian association for information systems*. Heidelberg, Neckar: Physica.

Deep space communications. NASA. [Online]. Available: https://scienceandtechnology.jpl.nasa.gov/research/research-topics-list/communications-computing-software/deep-space-communications. Accessed 15 Mar 2020.

Deep-space positioning system (DPS). *NASA*. [Online]. Available: https://technology.nasa.gov/patent/NPO-TOPS-26. Accessed 15 Mar 2020.

Dentis, Matteo, Capello, Elisa, and Giorgio. A novel concept for guidance and control of spacecraft orbital maneuvers. *International Journal of Aerospace Engineering*, 29-Nov-2016. [Online]. Available: https://www.hindawi.com/journals/ijae/2016/7695257/. Accessed 22 Mar 2020.

Dockrill, P. Venus may have been habitable until a mysterious catastrophe millions of years ago, ScienceAlert. [Online]. Available: https://www.sciencealert.com/venus-may-have-been-habitable-until-a-mysterious-catastrophe-millions-of-years-ago. Accessed 22 Mar 2020.

Dunbar, B. What is Orion?. NASA, 20-May-2015. [Online]. Available: https://www.nasa.gov/audience/forstudents/k-4/stories/nasa-knows/what-is-orion-k4.html. Accessed 22 Mar 2020.

Dunbar, B. Liquid hydrogen--the fuel of choice for space exploration. NASA. [Online]. Available: https://www.nasa.gov/topics/technology/hydrogen/hydrogen_fuel_of_choice.html. Accessed 19 Mar 2020.

Guinn, J. R., Riedel, J. E., Bhaskaran, S., Park, R. S., Vaughan, A. T., Owen, W. M., Ely, T., Abrahamsson, M., & Martin-Mur, T. (2016). The deep-space positioning system concept: Automating complex navigation operations beyond the earth. *Aiaa Space, 2016*, 5409.

S. Hong, *The magic of patent information*. [Online]. Available: https://www.wipo.int/sme/en/documents/patent_information_fulltext.html. Accessed 22 Mar 2020.

How interstellar space travel works (infographic) | Space. 1 Jul 2013. https://www.space.com/17619-how-interstellar-travel-works-infographic.html. Accessed 22 Mar 2020.

Howell, E. Virgin galactic: Richard Branson's space tourism company, Space.com, 09-Jan-2019. [Online]. Available: https://www.space.com/18993-virgin-galactic.html. Accessed 22 Mar 2020.

Liu, S. -J., & Shyu, J. Strategic planning for technology development with patent analysis. *International Journal of Technology Management*. [Online]. Available: https://www.inderscienceonline.com/doi/abs/10.1504/IJTM.1997.001689. Accessed 22 Mar 2020.

Merriam-Webster, Online ed. (2020a). Technology. [online]. Available at: https://www.merriam-webster.com/. Accessed 22 Mar 2020.

Merriam-Webster, Online ed. (2020b). Forecasting. [online]. Available at: https://www.merriam-webster.com/. Accessed 22 Mar 2020.

NASA's got a plan for a 'galactic positioning system' to save astronauts lost in space. LiveScience. [Online]. Available: https://www.livescience.com/62309-galactic-positioning-system-nasa.html. Accessed 15 Mar 2020.

Oceans, beaches, cosmic shorelines: our changing views of habitable planets, NASA, 19-Jun-2019. [Online]. Available: https://exoplanets.nasa.gov/news/1583/oceans-beaches-cosmic-shorelines-our-changing-views-of-habitable-planets. Accessed 22 Mar 2020.

Perez, J. NASA's HRP trailblazing better ways to protect astronauts, NASA, 29-Oct-2019. [Online]. Available: https://www.nasa.gov/feature/nasa-s-human-research-program-trailblazing-better-ways-to-protect-astronauts. Accessed 22 Mar 2020.

Phillips, F., & Linstone, H. (2016). Key ideas from a 25-year collaboration at technological forecasting & social change. *Technological Forecasting and Social Change, 105*, 158–166. http://www.sciencedirect.com/science/article/pii/S0040162516000081.

Reddy, M. V., Nica, M., & Wilkes, K. (2012). Space tourism: Research recommendations for the future of the industry and perspectives of potential participants. *Tourism Management, 33*(5), 1093–1102.

Swaminathan, N. Twisted sister: twin planets Earth and Venus were 'separated at birth'. *Scientific American*, 29-Nov-2007. [Online]. Available: https://www.scientificamerican.com/article/twisted-sister-venus-earth. Accessed 22 Mar 2020.

Tangermann, V.. Here is the future of interstellar spacecraft. *Futurism*, 25-Aug-2017. [Online]. Available: https://futurism.com/here-future-interstellar-spacecraft. Accessed 22 Mar 2020.

Chapter 15
Technology Intelligence Map: Lithium Metal Battery

Amee Sankhesara, Dao Dang, Erika Ogami, Igor Goulart, and Tuğrul U. Daim

Technology trends can be a useful resource for evaluating emerging technologies. Such evaluations should utilize multiple models to account for the complexity of these technologies. This paper explores how patent analysis along with Twitter data mining can be used to identify and monitor an emerging technology. The technology to be evaluated is the lithium metal battery.

15.1 Introduction

Researchers have historically used academic papers and patent data to detect and identify trends in emerging technologies. However, another untapped source of data is social media. Users' behavior is also an essential source to obtain and predict the adoption of technology in modern centuries. From social media data, we can determine the public's awareness, response, and expectation of a technology which can then be correlated to a potential for demand in the market (Xin et al. 2019). Identifying technology trends of emerging technologies and making timely decisions play an important role in establishing strategies. Through the analysis of technical and social awareness data, organizations can determine whether they need to move early in emerging technologies. This paper will demonstrate the use of patent analysis and Twitter data mining to analyze and predict the role of lithium metal battery technology. Through our results, we hope to gain insights into possible future and development trends of this technology.

A. Sankhesara · D. Dang · E. Ogami · I. Goulart · T. U. Daim (✉)
Department of Engineering and Technology Management, Portland State University, Portland, OR, USA
e-mail: tugrul.u.daim@pdx.edu

T. U. Daim (ed.), *Roadmapping Future*, Applied Innovation and Technology Management, https://doi.org/10.1007/978-3-030-50502-8_15

15.2 Literature Review

15.2.1 *Patent Analysis*

Patent analysis has been well known as a management tool for analyzing the intelligence information of technology for identifying and forecasting technological trends. This tool helps researchers and experts obtain the intelligence information about technology and product/service development processes from documents to support R&D strategies (Xin et al. 2019). It supports getting information of the relevant technology areas in order to forecast market needs. Based on this tool, researchers can obtain an overview of the technologies that the competitors are working on, understand the current state of your technology relative to the competitors, and then can assess the competitor's strengths, weaknesses, and most important strategy.

Patent analysis plays a main role in merging and acquisition of technologies and inventions because it helps researchers and experts narrow down the choices to consider and also help them find solutions to technical problems.

Patent text mining allows users access to technical information in patent documents and offers more pictures and trends of the technological emergence and development process.

Patents and other indicators can be used in many variations to predict technologies (Daim et al. 2006; Li et al. 2019), integrated social network analysis, and analogy to analyze autonomous vehicles. Lin et al. (2019) leveraged text mining to predict solar technologies. Goncalves et al. (2019) integrated patent analysis and decision modeling to evaluate biotechnologies. Madani et al. (2017, 2018) used more sophisticated patent analyses to analyze smart building technologies.

Bibliometrics are often used just alone or with patents for intelligence analysis as well. These analyses include refrigeration (Garces et al. 2017), manufacturing (Marzi et al. 2017), electric vehicles (Gibson et al. 2017), pacemakers (Daim et al. 2013), renewable energy (Daim et al. 2012), health information technology (Behkami and Daim 2012), and airplane technologies (Lamb et al. 2010).

Similar intelligence indicators were used to predict the diffusion of technologies (Daim and Suntharasaj 2009) including television technologies (Cho and Daim 2016), wind energy (Daim et al. 2012), and nanotechnology (Martin and Daim 2008; Daim et al. 2007).

15.2.2 *Twitter Data Mining*

Twitter is a global social media platform, so all tweets are public and extractable, so it helps researchers easily gather large amounts of data from users for analysis. These databases are very specific that makes them extremely good for prediction and forecasting.

Twitter data mining is a helpful tool that analyzes public opinions and comments on Twitter. It contributes to getting users' sense, response to and expectations for emerging technology to predict the trends of technologies and then researchers can make right decisions and strategies based on social perspectives and awareness.

To get a database on Twitter Application Programming Interface (API), we apply various queries, keywords, and users profiles (Xin et al. 2019). A technique widely used in data mining is sentiment analysis.

15.2.2.1 Twitter Sentiment Analysis

Twitter Sentiment Analysis uses advanced text mining techniques to automate the analysis of the sentiment of a tweet in the form of positive, negative, and neutral tweets (Bonani 2018).

As an example, take a look at the following tweet:

I hate slow internet @ATT u-verse sucks balls #worstinternet #dontgetatt

This tweet expresses a negative opinion about AT&T service U-Verse.

Sentiment analysis provides the ability to process thousands of tweets all at once. It's often used for public relations monitoring, getting insights about a brand or topic and tracking trends over time, detect potential PR crises, market research, and other useful applications.

15.2.3 Lithium Metal Battery Technology

Lithium metal batteries have lithium metal or lithium compounds as an anode. The Liquid Li electrolytes are found between the cathode and anode. Lithium metal batteries have a solid lithium anode as opposed to lithium ions that have lithium present in the ionic form in the electrolyte. Traditionally, lithium-ion and lithium polymer batteries have been used for rechargeable applications such as mobile telephones, laptop computers, tablets, power tools, and e-bikes.

15.3 Methodology

15.3.1 Step 1: Patent Analysis

The patent data were acquired using the Patent Integration web-based application. This software was used for patent database mining. We acquired the data using the following Boolean search.

Lithium Metallic OR metal battery OR Anode OR Cathode NOT "Lithium-Ion"

There were limitations to the analysis imposed by the web-based software used and several constraints implemented during the analysis. The constraints for the analysis are below.

Analysis Constraints:

1. Patent integration can only download data for up to 3000 patents
2. Looked only at US patents because of the 3000 patent limit and the technology development is mostly in the United States
3. No time for expert verification of key words
4. If too long of a search phrase was used, the software would not be able to correctly parse that database

Once the data were obtained, they were used to fit a Pearl-Reed Growth curve. The assumptions made with this model are that the upper limit to growth is known, and the historical data yield correct coefficients of the chosen curve. The L was chosen to give the closest fit to the data point for 2019. The data did not include 2020 as the year is not yet complete, but it did include all patents on the technology starting in 1972. The revised Pearl Reed growth curve is below.

$$y = \frac{L}{1 + 10^{A-Bt}}$$

It was necessary to obtain the coefficients A and B by linearizing the above equation. With the results of this linearization below

$$Y = \log\left[\frac{y}{L-y}\right] = -A + Bt$$

Coefficients A and B were gathered from plotting Y and calculating the y intercept and slope. These coefficients were plugged into the revised Pearl Curve to forecast technology. The L value was mathematically fit to produce the closest result to patents using the year 2019. This revised Pearl Reed growth curve was used to develop a technology forecast.

15.3.2 Step 2: Twitter Data Mining Using Twitter Sentiment Analysis

For twitter sentiment analysis, we have used twitter API to collect the data. From twitter API, we have collected tweets from 2010 to 2020. We have used the "Lithium Metal Battery" search keyword to collect the tweets. And for each from 2010 to 2020, we got around 3000 tweets having the keyword "Lithium Metal Battery."

We have written code in python to do sentiment analysis on collected tweets. After collecting the tweets, we did data cleaning (Preprocessing) on tweets in python, then on that cleaned data we did sentiment analysis in python using Tweepy library. After doing sentiment analysis, we got positive, negative, and neutral percentages for each year. We stored these resultant percentages for each in excel and we have created graphs from this resultant data in Excel.

In the below figure all steps for twitter sentiment analysis which explained earlier is shown:

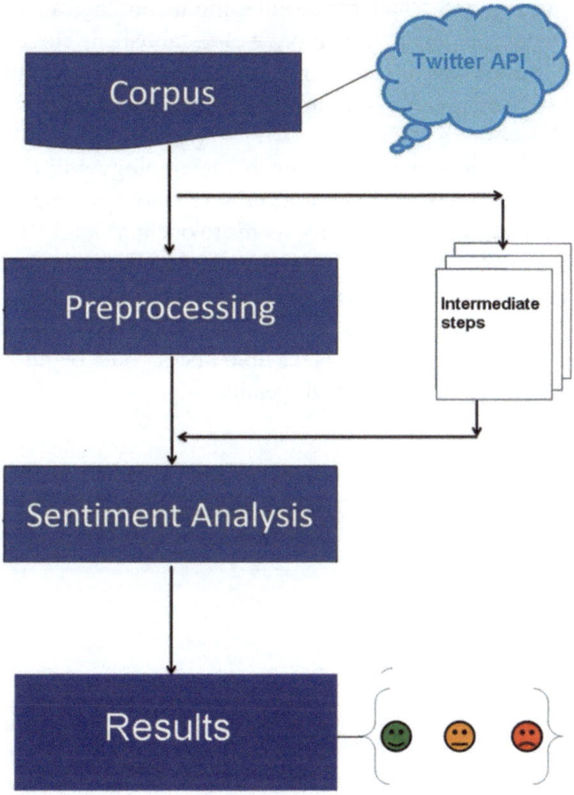

15.3.3 *Step 3: Identify the development trends of emerging technology*

This step will be shown in the results section.

15.4 Results and Discussion

15.4.1 Patent Analysis

The predicted Pearl Reed curve fit the data well. Based on the current rate of growth in patents in the Lithium metal technology, it appears the technology is in the emerging phase based on Figs. 15.1 and 15.2. Emerging technologies are harder to forecast than technologies that are already more well established. These data are a good representation of the amount of research and development funds are being fed into the development of rechargeable lithium metal batteries for use in electric cars and other electronic devices.

Based on Fig. 15.3, the development of this technology will not start to flatten out until 2060. So, we can expect a saturation of technological development at that point. The inflection point or productivity seems to occur around 2040. At this point, the technology is expected to switch from growth to maturity stages of growth. We can expect that there will be significant development in solid-state batteries in the next 20 years while the technology is growing the fastest. This technology has numerous barriers to overcome, but it is feasible that it could begin to take the place of lithium-ion batteries in the next 10–20 years.

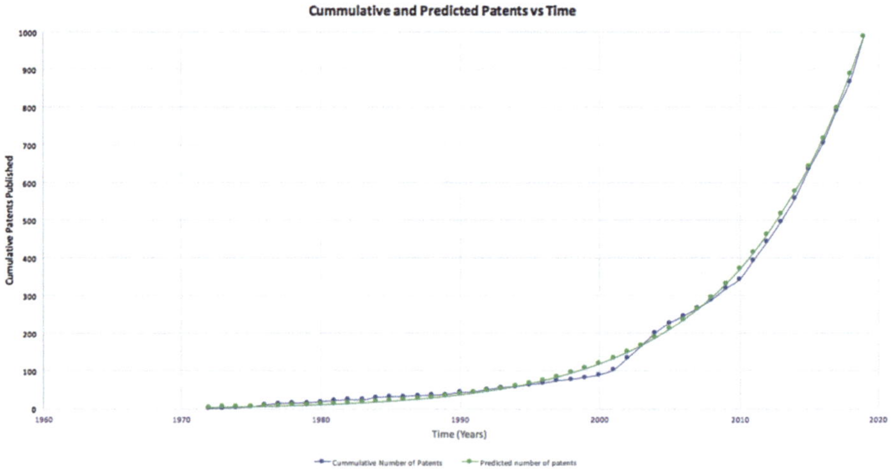

Fig. 15.1 Cumulative and Predicted Patents

Fig. 15.2 Prediction Parameters

L	A	B
11061	102.04	0.05004

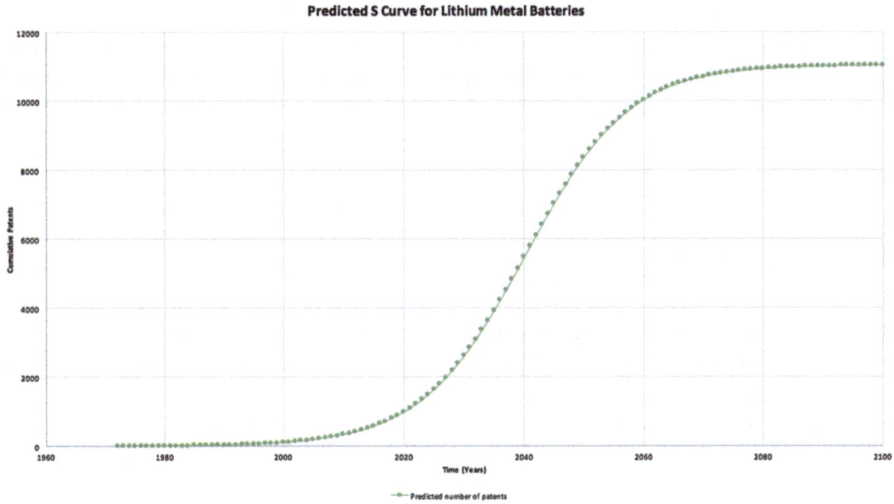

Fig. 15.3 Predicted S curve

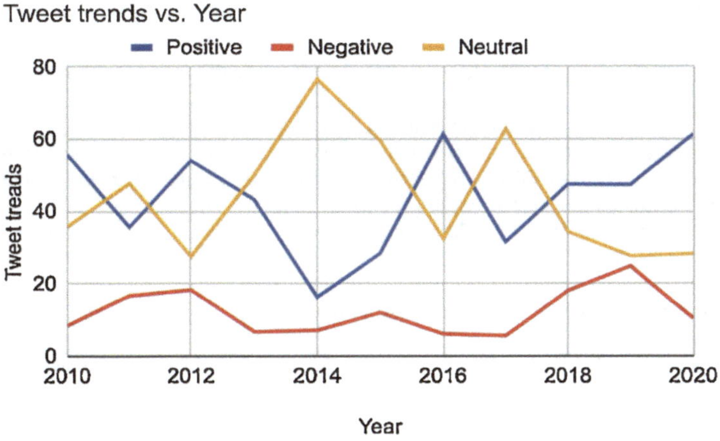

Fig. 15.4 Tweet trends vs year

15.4.2 Twitter Sentiment Analysis (Keyword – "Lithium Metal Battery")

Figure 15.4 represents the graph of twitter sentiment analysis of tweets having the keyword "Lithium Metal Battery." This graph shows analysis for one decade from 2010 to 2020.

In this graph, we can see that responses about Lithium Metal Battery from people are not uniform. For example, we can see that, in the year 2016, positive responses are very high then it goes down and then again it goes up. We can see the same for neutral and negative responses. This may be a very alternative to lithium metal batteries available in the market like lithium-ion batteries. But since the last 4 years from 2017, we can see that positive responses increase and negative and neutral responses decrease so we can predict that future usage of "Lithium Metal Battery" will be increased.

15.5 Conclusion

This paper used the complementary methods of patent analysis and twitter data mining to characterize the technology of lithium metal batteries. Through the use of patent analysis, lithium metal battery technology was identified as being in the emerging phase. The sentiment analysis results helped to validate this identification. Opportunities for future research involve using the results from other social media resources.

References

Behkami, N., & Daim, T. (2012). Research forecasting for health information technology (HIT), using technology intelligence. *Technological Forecasting and Social Change, 79*(3), 498–508.

Bonani, B. (2018). Twitter sentiment analysis – introduction and techniques. *Data Analytics*. https://www.digitalvidya.com/blog/twitter-sentiment-analysis-introduction-and-techniques/

Cho, Y., & Daim, T. (2016). OLED TV technology forecasting using technology mining and the fisher-pry diffusion model. *Foresight, 18*(2), 117–137.

Daim, T., & Suntharasaj, P. (2009). Technology diffusion: Forecasting with bibliometric analysis and bass model. *Foresight, 11*(3), 45–55.

Daim, T., Grueda, G., Martin, H., & Gerdsri, P. (2006). Forecasting emerging technologies: Use of bibliometrics and patent analysis. *Technological Forecasting and Social Change, 73*(8), 981–1012.

Daim, T., Monalisa, M., Pranabesh, D., & Brown, N. (2007). Time lag assessment between research funding and output in emerging technologies: Case of scope applications. *Foresight, 9*(4), 33–44.

Daim, T., et al. (2012). Patent analysis of wind energy technology using the patent alert system. *World Patent Information, 34*(1), 37–47.

Daim, T., Gomez, F., Martin, H., & Sheikh, N. (2013). Technology roadmap development process (TRDP): Case of pacemakers. *International Journal of Business Innovation and Research, 7*(2), 228–263.

Garces, E., van Blommestein, K., Anthony, J., Hillegas-Elting, J., Daim, T., & Yoon, B. (2017). Technology domain analysis: A case of energy-efficient advanced commercial refrigeration technologies. *Sustainable Production and Consumption, 12*, 221–233.

Gibson, E., Blommestein, K., Kim, J., Daim, T., & Garces, E. (2017). Forecasting the electric transformation in transportation. *Technology Analysis & Strategic Management, 29*(10), 1103–1120.

Gonçalves Pereira, C., Lavoie, J., Garces, E., Basso, F., Dabić, M., Porto, G., & Daim, T. (2019). Assessment of technologies: Forecasting of emerging therapeutic monoclonal antibodies patents based on a decision model. *Technological Forecasting and Social Change, 139*, 185–199.

https://monkeylearn.com/sentiment-analysis/
https://tadiranbatteries.de/pdf/IATA_Lithium_Battery_Guidance_2019.pdf
https://www.greentechmedia.com/articles/read/lithium-metal-battery-promise-challenge
https://www.marshall.usc.edu/sites/default/files/oleary/intellcont/Twitter-Mining-1-1.pdf
https://www.quora.com/What-is-Twitter-sentiment-analysis

Lamb, A., Anderson, T., & Daim, T. (2010). Forecasting airplane technologies. *Foresight, 12*(6), 38–54.

Li, S., Garces, E., & Daim, T. (2019). Technology forecasting by analogy based on social network analysis: The case of autonomous vehicles. *Technological Forecasting and Social Change, 148*, 119731.

Lin, X., Xie, Q., Daim, T., & Huang, L. (2019). Forecasting technology trends using text mining of the gaps between science and technology: The case of perovskite solar cell technology. *Technological Forecasting and Social Change, 146*, 432–449.

Madani, F., Daim, T., & Weng, C. (2017). Smart building technology network analysis: Applying core periphery structure analysis. *International Journal of Management Science and Engineering Management, 12*(1), 1–11.

Madani, F., Daim, T., & Zwick, M. (2018). Keyword-based patent citation prediction via information theory. *International Journal of General Systems, 47*(8), 821–841.

Martin, H., & Daim, T. (2008). Technology roadmapping through intelligence analysis: Nanotechnology. *International Journal of Society Systems Science, 1*(1), 49–66.

Marzi, G., Dabic, M., Daim, T., & Garces, E. (2017). Product and process innovation in manufacturing firms—a thirty-year bibliometric analysis. *Scientometrics, 113*(2), 673–704.

Xin, L., Qianqian, X., Jiaojiao, J., Yuan, Z. H., Lucheng, H. (2019). Identifying and monitoring the development trends of emerging technologies using patent analysis and Twitter data mining: The case of perovskite solar cell technology. *Technological Forecasting and Social Change*, 432–449. https://doi.org/10.1016/j.techfore.2018.06.004

Chapter 16
Technology Intelligence Map: Twitter and Currency

Adya Pandey, Aman Singh Solanki, Divya Sravani, Vinaya D. Bhat, and Tuğrul U. Daim

Social media is becoming the backbone of virtual communication in our modern society. It is widely used to spread information, voice opinions and foster relationships. It also plays a major role in politics where people are using social media as a way to make their voices heard. The vast amount of data on these platforms can be used to analyse public engagement and criticism. This chapter aims at exploring the means of social media to understand and predict the economy. We are using sentiment analysis and regression techniques to capture the wider sentiment of the public and verify if these changes in sentiment lead to fluctuations in the economy as well. To estimate the economy, we will check the historic rates of Indian National Rupee (INR) and its conversion value to the United States Dollar (USD). The results show that public sentiment might be correlated to the INR value.

16.1 Introduction

On 8 November 2016, the Government of India announced that it had decided to end the legal tender of the ₹500 and ₹1000 notes of the Mahatma Gandhi Series. These were the two largest denominations in its currency system and accounted for 86% of the country's circulating cash. Prime Minister Narendra Modi also said in his address, new notes of ₹500 and ₹2000 would be released to the public in exchange of the demonetized notes.

In the past, demonetization has been a tool for a cash-dependent developing economy to combat corruption and crime. India was trying a similar model to promote a cashless economy and to eradicate counterfeit currency, tax evasion and

A. Pandey · A. S. Solanki · D. Sravani · V. D. Bhat · T. U. Daim (✉)
Department of Engineering and Technology Management, Portland State University, Portland, OR, USA
e-mail: tugrul.u.daim@pdx.edu

T. U. Daim (ed.), *Roadmapping Future*, Applied Innovation and Technology Management, https://doi.org/10.1007/978-3-030-50502-8_16

black money gotten from money laundering and funding terrorist activities. Individuals or corporations with huge amounts of cash money would need to take it to the bank, whereby they had to provide proof of their tax payments as well. If unable to do so, a penalty of 200% of the owed amount would be imposed.

This change took place with very little warning for the general public, as a result of which chaos ensued. Since 78% of the customer transactions in India are by cash, there were staggering lines outside banks and ATMs to withdraw money. Only 60% of the ATMs out of the 200,000 were functional as they needed to be recalibrated for the size and thickness of the new notes. People and small businesses struggled to find cash even for the smaller denominations.

As a result, people had mixed opinions about this announcement. It initially received support from some bankers and international commentators. But was later criticized for being unfair and poorly executed. There were strikes against the government in many parts of India, and at the same time, processions commending the government for its efforts.

In this project, we are analysing the sentiment of people and seeing how the INR value correlates to it. Social media was a powerful tool used by people to voice their opinions about this topic. We used social media monitoring to gain an overview of the wider public opinion. It can be used by tracking conversations about certain topics so that we can shortlist the relevant discussions from diverse, independent groups. The platform we chose for our analysis is twitter.

16.2 Phase I: Finalizing Dataset

For this project, we needed two different datasets. The first one is a compilation of tweets and retweets posted by Indians during demonetization. We tried setting up the Twitter developer account to use the twitter Application Programming Interface (API) to get tweets. There were some limitations there in getting past tweets and not much information was provided by the Twitter team. Instead, we tried to find data online and found multiple comma-separated values (CSV) files that we combined together to make our dataset. The second one is the INR rate during demonetization. We used the historic conversion rate from USD to INR in 2016 for our project.

We created two separate CSV files called "demonitization_rates.csv" and "INR_rates.csv." This process involved data cleaning and data processing. We also added a date column in these two CSV files so that we can compare and contrast during the analysis.

16.3 Phase II: Sentiment Analysis

Sentiment Analysis is classifying emotions, which could be either positive, negative or neutral. We do this by reading a text line, paragraph or a whole book and applying text analysis techniques. It's used in a variety of domains like suggesting movies on Netflix, creating ratings by reading user reviews on Amazon, public opinion of a new law by reading social media posts, etc. Sentiment analysis is a powerful part of machine learning since the oncoming of popular languages like R or Python. For the purpose of this project, we used the R language due to its ease of use and simplicity.

For the purpose of this project, we have three levels of sentiments as depicted in the image:

1. Positive
2. Negative
3. Neutral

In short, this is how we are doing it. We will analyse each step in detail in the following sections:

1. Start parsing (reading) a tweet.
2. Keep a sentiment score for each tweet, initially, it would be zero.
3. For each positive word we encounter, increment the sentiment score by one.
4. For each negative word we encounter, decrement the sentiment score by one.
5. If the final sentiment score is greater than zero, then the overall sentiment of that tweet is positive.
6. If the final sentiment score is less than zero, then the overall sentiment of that tweet is negative.
7. If there are more positive sentiment tweets on a particular day, then that day is a positive day. Similarly, we determine the negative and neutral days.

We use two large text files, one each for positive words and negative words. These files have already been created by the open-source platforms and are widely used for sentiment analysis. The positive file contains a wide variety of positive English words like good, like amazing, etc. The negative file contains words such as bad, hate, disappointed etc. Although each of these files has a lot of words, they are exhaustive in nature but are sufficient for the nature of this project.

16.3.1 Data Clean-Up

We start by cleaning up the data, which here are the tweets.

1. We convert the tweet to all lowercase letters, which will make it easier to compare their words with the positive/negative files later.
2. We remove all the extra spaces and tabs from the tweets. This will help us to separate each word in the tweet in a clean manner.

3. We remove all punctuations from the tweets.
4. We remove all numbers.
5. We remove all HTML links from the tweets.

16.3.2 Processing the Tweet

Next, we break the tweet into individual words. Each time a word exists in a positive file, we increment the positive score and vice versa for the negative file.

$$\text{Totals score} = \text{Positive Score} - \text{Negative Score}$$

16.3.3 Analysis

We calculate the mean and median score of all the tweets by using the R library function "mean" and "median" respectively.

$$\text{Mean Score} = \text{Sum of scores of all the tweets} / \text{Total number of tweets}$$

$$\text{Median Score} = \text{Middle value of score if scores are ordered in increasing order}$$

Next, we plot the graph, using the library method "ggplot" in R. The below figure depicts the graph created using the above steps.

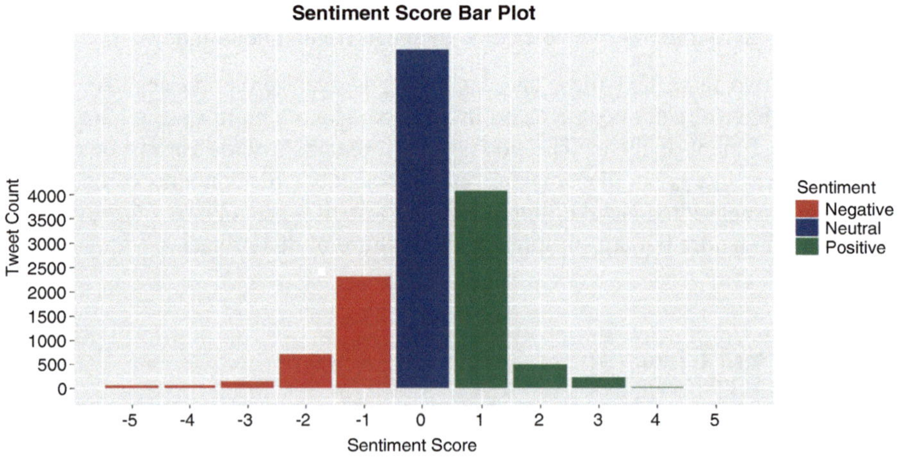

The graph suggests that most of the tweets about demonetization are neutral in nature. This might suggest that the majority of the people still could not make up their minds or are giving mixed reactions in the same tweet.

If we ignore the neutral tweets, we can see that the positive reactions (depicted in green) far outweighs the negative reactions (contrary to most media reports for that time). The lower values for absolute values greater than or equal to two suggested that people were restraining on giving their final judgment and were still open to debate. As this event was unprecedented anywhere in the world, it is quite understandable. A lack of opinion, in this case, might have gone on to suggest that the tweets could not be solely relied on predicting the future trend in the rupee (INR) value.

For visual understanding, we also created what is called a word cloud. It creates a diagram with a list of words, the size of which is proportional to how often that word is used across different tweets. "Wordcloud" is a feature provided by the R language.

On feeding our demonetization tweets to the function, we get the following diagram:

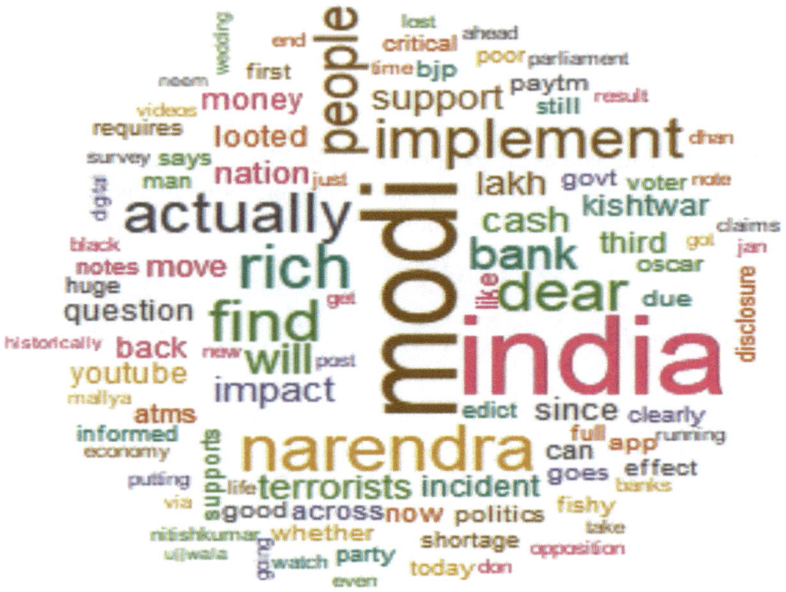

This helps us to understand on a human level, what is the general sentiment of the public on this issue and what are the most common terms used to express those sentiments.

The below graph depicts the varying sentiment with time. As portrayed earlier, the number of neutral tweets are almost always larger than either positive or negative tweets. Apart from that, the number of positive tweets and negative tweets are almost equal in number. This might go on to suggest that people's sentiments on demonetization might not affect the value of the Indian Rupee as much as we would like it to for this project.

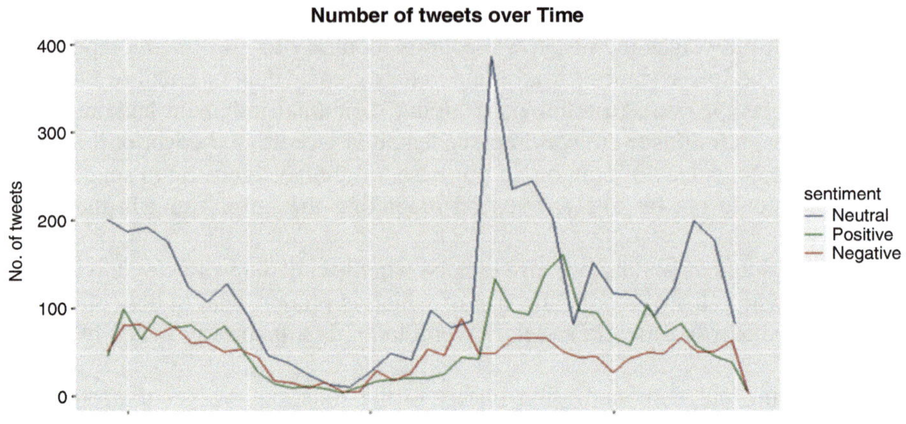

16.4 Phase III: Regression

In this experiment, we take the rupee value as the dependent variable and tweet positive and negative sentiment as independent variables. The purpose of this experiment is to see if the variation in the independent variables affects the dependent variable (https://www.statisticssolutions.com/what-is-multiple-linear-regression/).

16.4.1 Data Cleaning

Using the pandas library, we imported the twitter data and the rupee value data from the CSV files. The date and time were in one column called "Created," so we separated the data into a separate column called "Date." We retained only the Date and tweets column from the original CSV file.

	Date	text
0	11/14/2016	RT @rssurjewala: Critical question: Was PayTM ...
1	11/14/2016	RT @Hemant_80: Did you vote on #Demonetization...
2	11/14/2016	RT @roshankar: Former FinSec, RBI Dy Governor,...
3	11/14/2016	RT @ANI_news: Gurugram (Haryana): Post office ...
4	11/14/2016	RT @satishacharya: Reddy Wedding! @mail_today ...
5	11/14/2016	@DerekScissors1: India□s #demonetization: #Bla...
6	11/14/2016	RT @gauravcsawant: Rs 40 lakh looted from a ba...
7	11/14/2016	RT @Joydeep_911: Calling all Nationalists to j...
8	11/14/2016	RT @sumitbhati2002: Many opposition leaders ar...

Next, we grouped the text column based on the Date. So all the tweets on 1 day are in one cell.

	Date	text
0	01/01/2017	RT @ModiBharosa: Putting Nation over Party Pol...
1	01/02/2017	RT @ModiBharosa: Putting Nation over Party Pol...
2	01/03/2017	RT @gauravcsawant: Rs 40 lakh looted from a ba...
3	01/04/2017	RT @PIB_India: Watch briefing on #Demonetizati...
4	01/05/2017	RT @gauravcsawant: Rs 40 lakh looted from a ba...

Next, we imported the rupee data.

	rupee	Date
0	67.6375	11/14/2016
1	67.7260	11/15/2016
2	68.0314	11/16/2016
3	67.9540	11/17/2016
4	68.1251	11/18/2016

We imported the Unicode data python library and used it to normalize the tweets. The Unicode module provides access to the Unicode Character Database (UCD) which defines character properties for all Unicode characters. NFKD normalization form which will apply the compatibility decomposition, that replaces all compatibility characters with their equivalents (https://docs.python.org/3.2/library/unicodedata.html) was used.

16.4.2 Multiple Linear Regression

We used the sklearn SentimentIntensityAnalyzer. A SentimentAnalyzer is a tool to implement and facilitate Sentiment Analysis tasks using natural language tool kit (NLTK) features and classifiers. From the normalized tweet, we take the polarity scores method of SentimentIntensityAnalyzer which gives a sentiment dictionary. It gives positive, negative, neutral and compound scores of the tweets. "*VADER (Valence Aware Dictionary and sEntiment Reasoner) is a lexicon and rule-based sentiment analysis tool that is specifically attuned to sentiments expressed in social media and works well on texts from other domains. The compound score is computed by summing the valence scores of each word in the lexicon, adjusted according to the rules and then normalized to be between −1 (most extreme negative) and +1 (most extreme positive). This is the most useful metric if you want a single unidimensional measure of sentiment for a given sentence. Calling it a 'normalized, weighted composite score' is accurate."*

- *Positive sentiment: compound score >= 0.05*
- *Neutral sentiment: (compound score > −0.05) and (compound score < 0.05)*
- *Negative sentiment: compound score <= −0.05*

The positive, neutral and negative scores are ratios for proportions of text that fall in each category (so these should all add up to be 1 or close to it with float operation)" (Hutto and Gilbert 2014)

```
% of positive tweets=  69.44444444444444
% of negative tweets=  30.555555555555557
```

	Date	Prices	Comp	Negative	Neutral	Positive
0	01/01/2017	67.9409	0.9984	0.114	0.654	0.232
1	01/02/2017	68.1462	-0.9945	0.152	0.73	0.118
2	01/03/2017	68.2869	-0.9992	0.167	0.802	0.032
3	01/04/2017	67.865	-0.9982	0.132	0.832	0.036
4	01/05/2017	67.7683	-0.9976	0.16	0.74	0.1
...	...	...	...	...	...	...
175	12/27/2016	67.9822	-0.9924	0.158	0.71	0.132
176	12/28/2016	68.2183	-0.7239	0.143	0.696	0.161
177	12/29/2016	68.0048	-0.996	0.149	0.738	0.113
178	12/30/2016	67.9106	-0.9655	0.148	0.708	0.145
179	12/31/2016	67.9547	-0.9909	0.173	0.669	0.157

16.4.3 Ordinary Least Squares (https://statisticsbyjim.com/glossary/ordinary-least-squares/)

```
                            OLS Regression Results
==============================================================================
Dep. Variable:                 Prices   R-squared:                       0.166
Model:                            OLS   Adj. R-squared:                  0.156
Method:                 Least Squares   F-statistic:                     17.56
Date:                Sun, 22 Mar 2020   Prob (F-statistic):           1.10e-07
Time:                        19:47:48   Log-Likelihood:                -309.69
No. Observations:                 180   AIC:                             625.4
Df Residuals:                     177   BIC:                             635.0
Df Model:                           2
Covariance Type:            nonrobust
==============================================================================
                 coef    std err          t      P>|t|      [0.025      0.975]
------------------------------------------------------------------------------
const         66.6491      0.514    129.605      0.000      65.634      67.664
Positive      -4.4443      1.592     -2.791      0.006      -7.587      -1.302
Negative       6.9890      2.733      2.557      0.011       1.595      12.383
==============================================================================
Omnibus:                       14.003   Durbin-Watson:                   0.132
Prob(Omnibus):                  0.001   Jarque-Bera (JB):               15.660
Skew:                          -0.708   Prob(JB):                     0.000398
Kurtosis:                       2.710   Cond. No.                         29.8
==============================================================================
```

R-squared: It signifies the "percentage variation independent that is explained by independent variables." Here, a 16.6% variation in Rupee value is explained by the positive and negative score of tweets. If R-squared value is closer to 100% then it means that the model is fit well. So our model does not fit well using negative and positive values for independent variables.

Adj R-squared is less than R-squared and shows that some independent variables chosen are not the correct fit for the model.

Prob(F-Statistic) This gives the overall significance of the regression. There is no good linear relationship between positive, negative and rupee value.

AIC/BIC It stands for Akaike's Information Criteria. A lower AIC implies a better model which is not the case here.

If P-value is less than 0.05, we reject the null hypothesis. Since both the positive and negative P-value is less than 0.05, an optimal team of independent variables that can predict a rupee value with the highest statistical significance is actually composed of two independent variables - positive and negative.

Because of the R-squared values and F-statistics value, we can conclude that the independent variables chosen might not be the correct fit for the model (http://www.biostathandbook.com/multipleregression.html; https://medium.com/@jyotiyadav99111/statistics-how-should-i-interpret-results-of-ols-3bde1ebeec01).

16.4.4 Regression

We split the data into train and test sets in the ratio of 70:30. Applied the multiple linear regression on the train set and tried to predict the test set.

	Actual	Predicted
0	68.0656	66.518833
1	66.8475	67.177226
2	68.024	66.663062
3	65.3267	65.810765
4	68.4255	66.934253
5	68.0843	66.795738
6	67.9812	66.750200
7	66.458	66.963731
8	64.4391	66.802799
9	64.2386	66.525505

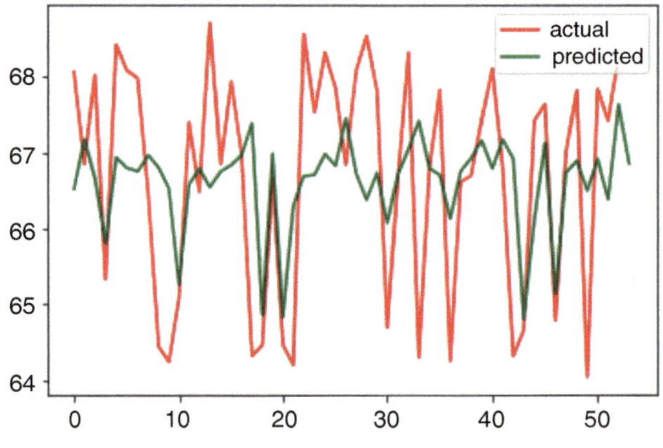

16.4.5 Results

The graph above depicts the trend in rupee value. We can see from the graph that on most of the instances the predicted value of rupee follows the overall increase or decrease of the actual rupee value. This regression model is only predicting whether the rupee value would increase or decrease based on twitter user sentiment. However, it is taking the larger absolute values in each of these cases, in a sense that it is going too much positive or too negative. Based on our understanding, this can be due to multiple reasons. One of the primary reasons can be a lack of data to train the regression model. Although we have trained the model on two-third of the total data, it might still fall short of properly training the model. The most obvious cause would be that twitter sentiment analysis alone could not single-handedly predict the rupee value. As we know, currency fluctuations depend on a lot of external factors like oil prices, dollar values, and natural calamities in a different country with public sentiment being only one of them. Another reason we suggest is the quality of the data. As already mentioned in the sentiment analysis section, the majority of users have neutral opinions on this matter. This suggests that there is a lack of positive or negative sentiment for the most part, which makes only using twitter analysis as an incomplete tool. For example, if the overall mood was positive, then twitter can weigh in on other factors and we could have seen an appropriate rise in rupee values.

Still, coming close to the overall trend in the rupee value achieves the purpose of this project. Twitter sentiment analysis can be an important part of a full-fledged study on predicting currency value, among other factors mentioned above.

16.4.6 Conclusion

The broad aim of this chapter is to see how social media has gained importance over the years and how it can be used to validate public sentiment about changes in our society. As a result, it can be used to predict currency values. Often underestimated, public opinion can have a huge impact in the business environment. All the results in this chapter lead us to believe that social media, more specifically twitter, can have a strong statistical correlation with INR values. After conducting this research that, in order to have a proper forecasting tool, other factors must also be taken into account. We speculate that these factors may include the country's monetary policy, the rate of inflation and political and economic conditions. Social media can work as predictors of the economy, but a larger range of algorithms and factors must be taken into account. Sentiment can be combined with the other traditional factors to make better decisions and to predict the economic situation of a country.

16.4.7 Future Work

- Future research could be to rank the importance of this analysis with those other factors as well.
- Future Analysis could also include seeing whether the twitter sentiment analysis affects the rupee value in the coming days rather than the present day.
- To try and compare how other regression methods like Logistic Regression, DecisionTreeRegression and RandomForestRegression would accurately predict currency value.

References

https://monkeylearn.com/sentiment-analysis/
https://www.statisticssolutions.com/what-is-multiple-linear-regression/
https://docs.python.org/3.2/library/unicodedata.html
https://www.nltk.org/api/nltk.sentiment.html
https://www.kaggle.com/nltkdata/vader-lexicon
https://www.coursehero.com/file/p6c8s53j/is-more-positive-than-I-like-America-We-will-use-the-VADER-Valence-Aware/
https://statisticsbyjim.com/glossary/ordinary-least-squares/
http://www.biostathandbook.com/multipleregression.html
https://medium.com/@jyotiyadav99111/statistics-how-should-i-interpret-results-of-ols-3bde1ebeec01
https://github.com/VinayaBhat/ETM.git
Hutto, C. J. & Gilbert, E. E. (2014). VADER: A parsimonious rule-based model for sentiment analysis of social media text. Eighth International Conference on Weblogs and Social Media (ICWSM-14). Ann Arbor, MI, June 2014.

Chapter 17
Technology Intelligence Map: Biotechnology

Cristiano Gonçalves Pereira, Joao Ricardo Lavoie, Amir Shaygan, Tuğrul U. Daim, Maria Angélica Oliveira Luqueze, and Geciane Silveira Porto

Recently, biotech companies are increasingly attracting the attention from investors. From 2016, of the top 10 performers in the NASDAQ-100 (index consisting of 100 largest non-financial companies listed on NASDAQ), six were biotech. However, the reasons behind their success, or the actions that lead them to be successful, are still ambiguous. This chapter aims to give insights into the critical success factors in the biotech industry by analyzing the strategic and market data of the 10 biotech companies included in NASDAQ-100. Thus, the enterprise value (EV) of companies from 2002 to 2016 was considered to identify fluctuations that can be associated with key indicators such as R&D expenditure, corporate deals, patent portfolio, pipelines, and additional indicators. Our results give an overview of the main critical success factors and strategic actions, in a temporal manner, which the top-valued biotech firms fall into. The factors of highest importance included constant and high investment on R&D, intense transactions related to mergers and acquisitions for big biotech companies, and a favorable regulatory environment for drug approvals. Additional relevant factors involve the company's focus and a proper management of their innovations. This chapter opens a new perspective by understanding the consolidated critical factors among the top US biotech companies, indicating what they are dependent on to steer them to failure or success.

C. G. Pereira · J. R. Lavoie · A. Shaygan · T. U. Daim (✉) · M. A. O. Luqueze · G. S. Porto
Department of Engineering and Technology Management, Portland State University, Portland, OR, USA
e-mail: tugrul.u.daim@pdx.edu

T. U. Daim (ed.), *Roadmapping Future*, Applied Innovation and Technology Management, https://doi.org/10.1007/978-3-030-50502-8_17

17.1 Introduction

Biotechnology involves the integration of biological systems, living organisms, or genetic engineering to develop technologies on a wide range of applications in major areas, which include healthcare (medical), food and agriculture, environmental, and industrial. The worldwide biotech market was estimated at USD 369.62 billion in 2016 and is projected to reach USD 727.1 billion by 2025 (Grand View Research 2017). The sector is facing an unprecedented growth lately, and it is remarkably transforming the economy of the United States, the country with the largest biotech sector. Its contribution to the US economy was estimated to represent more than 2% of the gross domestic product (GDP), and from 2007 to 2012, the biotech aggregates' revenues have grown on an annual average rate of more than 10%, much faster than the US economy as a whole (Carlson 2016). The growth of biotechnology is largely driven by its applications in healthcare, which differs from the traditional pharmaceutical companies through the nature of their drugs. While biotechnology companies develop their drugs on a biological basis, the pharmaceuticals use traditional chemical-based drugs. However, both sectors intersect and complement each other to develop the so-called biopharmaceuticals (Rader 2008).

The progress of the industry is mainly motivated by factors such as massive R&D investments, intense protection of intellectual property rights, and also the establishment of cooperations and strategic alliances (Evens and Kaitin 2014). Moreover, advances in our understanding of the disease's molecular level triggered by the OMICs have led to significant improvements in drug discovery, consisting of identification of new targets and therapeutic molecules (Bilello 2005).

The huge R&D expenditure is the hallmark for the biotechnology industry, known to be one of the most R&D-intensive industry sectors in the world. Among the US industries, biopharmaceuticals have the highest R&D reinvestment as a percentage of the revenue, a rate of 21.3%, i.e., nearly 3 percentage points ahead of the semiconductor industry. The steps in the discovery and development through market approval are known to be costly and lengthy, besides being marked by financial risks due to failure in obtaining approval by regulatory agencies. The capitalized R&D cost for a newly approved biopharmaceutical was estimated at 2558 billion in 2014 (2013 US dollars), two times more than the cost 10 years from that time (DiMasi et al. 2016). Most of the R&D expenses go to the clinical phases to develop a drug's pipeline. Each stage can be very costly due to the complexity of human health, compounding, and response to treatment (Morgan et al. 2011). The surge in R&D expenses over the years resulted in an increased number of biopharmaceuticals in clinical development. In 2001, there are 369 biopharmaceuticals in the clinical phase, compared to 901 in 2012, an increase of 245% in the number of drug candidates (PhRMA 2013). The biggest increase was seen by the monoclonal antibody class, which grew 351% over the 11 years of analysis consisting of 338 from the 901 biologics in development in 2012. Monoclonal antibodies are not only the highest biopharmaceuticals class in development, but they are also the most approved in the USA and Europe (Walsh 2014). Thus, the high R&D expenditure, as a consequence, allows companies to leverage their discoveries, submit new candidates for clinical evaluation, and if it fulfills all stages successfully, new drugs are approved, leading to revenue for companies.

Behind the steps of preclinical and clinical phases resides the strength of intellectual property rights. The healthcare biotechnology benefits the most from the patent system, which guarantees exclusivity in the commercial exploration of their technologies. The biotechnology intellectual property rights also take advantage from some peculiarities of the market, a biopharmaceutical product can carry multiple patents from the molecule involving different process and improvements (Evens 2016), also by changes in law in 2010 (by the Patient Protection and Affordable Care Act) awarded 12 years of data exclusivity from the approval date to biotech patents. On the other hand, more recently, the industry has been facing a patent cliff, with the expiry of patents of some top-selling biological drugs. This phenomenon opens up an opportunity for biosimilars, which are less costly and more long lasting than the reference biopharmaceuticals (Misra 2012).

Biopharmaceutical companies can disclose part of their knowledge stocks through patents (Erden et al. 2015). Besides the interest of biotech companies in protecting their products through patents, unexpected values from their patent portfolio can be extracted in many ways. A study led by Harlin and O'Connor (Harlin and O'Connor 2008) in 2008 discussed that biotech companies can leverage their intellectual property though nonexclusive or exclusive out-licensing, which is conducive for early-stage technology to raise capital and narrow the company's focus. By negotiating cross-licensing agreements with competitors promoting collaborations, the biotech companies can create opportunities to sell their intellectual properties or royalty streams to raise short-term capital and narrow the focus, and they can also obtain capital by using the patent assets as security for loans.

Another aspect of the biotech industry, which promotes the progress of the sector, is the frequent collaborations and mergers and acquisitions. In order to combine knowledge and resources for successfully developing a new product, biotech companies take advantage of alliances and cooperation (Haeussler et al. 2012). Historical data on R&D partnerships among pharmaceutical biotechnology companies revealed a pattern of overall growth since the mid-1970s (Roijakkers and Hagedoorn 2006). Cooperation is a common strategy of biopharmaceuticals; partnerships can arise from companies motivated by reasons related to sharing knowledge and R&D efforts to invest in a new path or drug class when the company has no expertise and to enter into a new market location through cooperation with firms that are geographically distant (Moorkens et al. 2017). Besides the high cooperation rate in biopharmaceuticals, the sector is massively influenced by its mergers and acquisitions. From Big Pharma's perspective, an interest in the market segment of biotechnology resulted in several acquisitions in order to achieve the right infrastructure and knowledge to develop their biopharmaceuticals (Moorkens et al. 2017). In 2016, the acquisition of biotech companies comprised 55% of all potential merger and acquisition value. The years of 2011, 2013, 2015, and 2016 were marked by mega deals (>US$ 5 billions) among biotech companies (Ernst and Young 2016). In

2015, mergers and acquisitions in the USA and Europe reached a high in terms of total potential value and volume records.

The biotech sector is dependent on mergers and acquisitions to accomplish their growth goals and during its life cycle, the initial public offering (IPO) is an important stage (Quintana-García and Benavides-Velasco 2016). Most of the US biopharmaceuticals, involved in the greatest deals, went public years ago on NASDAQ (National Association of Securities Dealers Automated Quotation System). NASDAQ differs from traditional stock exchanges because it is an electronic stock exchange, in which only shares of technological companies known as "New Economy" such as Apple, Google, Microsoft, Intel, and others are traded. The NASDAQ-100 Index includes the top 100 US and international nonfinancial companies listed on the stock market based on market capitalization. The index is updated quarterly and reflects companies in all major industry groups, including hardware and software, telecommunications, retail/wholesale, and biotechnology. By the end of 2016, six of the ten performers of the NASDAQ-100 index were from biotechnology. Overall, 10 biotech companies with a weight of 8.6% comprised the index by November of 2016, which includes Alexion Pharmaceuticals, Inc. (ALXN), BioMarin Pharmaceutical Inc. (BRMN), Incyte Corporation (INCY), Regeneron Pharmaceuticals Inc. (REGN), Vertex Pharmaceuticals Incorporated (VRTX), Biogen Inc. (BIIB), Celgene Corp. (CELG), Amgen Inc. (AMGN), Gilead Science Inc. (GILD), and Illumina Inc. (ILMN). As mentioned above, the NASDAQ-100 included companies based on its market capitalization; however, the metric does not represent the entire worth of the company such as the enterprise value (EV). Since the sector is marked by acquisitions, an accurate valuation of the company is very important for the negotiations.

EV is an economic metric used by economists and financial analysts to measure the value of a publicly traded organization. It is a way through which the market can determine the entire worth of a company, instead of just looking at its market capitalization (market cap) (Investopedia 2018). The EV formula includes extra information that balances out the sometimes-misleading figures provided by market capitalization. The measure of EV is given by the market value of common stock plus market value of preferred equity + market value of debt + minority interest minus cash and investments (A. H. CFA 2018). In more simple terms, EV would be the market cap plus the net debt. EV is a good indicator to measure an organization's value and it is especially useful when evaluating companies prior to a merger and acquisition (M&A), since it includes aspects such as debt and cash reserves, which could drastically change the actual value of a company in a potential deal. EV is sometimes referred to as the "takeover price" and it is considered to be more accurate than a regular market capitalization calculation (A. H. CFA 2018).

EV has also been consistently applied to value companies in academic studies. Aiming to compare US and Canadian equity markets and to assess whether those were integrated or segmented, King and Segal used statistical regressions, EV, and EBITDA (earnings before interest, tax, depreciation, and amortization) in order to compare companies' values (King and Segal 2008). Similarly, other studies also use EV and EBITDA in valuing companies for several purposes (Ribal et al. 2010;

Chullen et al. 2015; Sareewiwatthana and Janin 2017). In a study conducted to identify and analyze crucial value drivers for players in the mining sector, MacDiarmid et al. used EV as the metric, mentioning that the metric is especially useful for comparing the value of different companies (MacDiarmid et al. 2018). Mariani et al. used EV as a way to account for the performance of University spin-offs and investigated how technology transfer efforts and investments would influence the EVs, using the University of Pisa as a case (Mariani et al. 2018). Njowa and Musingwini criticized standard financial methods and statements for not accounting for future perspectives and factors, and proposed a new framework using the mining sector as a case, while admitting that EV is a more adequate measure of value than market capitalization (Njowa and Musingwini 2018).

Taking into consideration the aspects discussed above, this chapter aimed to trace a historical profile of the 10 biotech companies included on NASDAQ-100 index (November 2016) in order to identify critical success factors that lead them to outperform in the industry. The evolution of the EV between 2002 and 2016 was analyzed to get insights of the company's performance. The factors are shown and discussed with highlights on the current scenario for biotech during the period and each company's decisions and strategies which likely reflect on its EV. At the end, the identified factors are consolidated for an overview of what biopharmaceuticals depend on and what lead the sector to fail and succeed.

17.2 Data Collection

The criteria for selecting the companies were based on the NASDAQ stock market considering the biotechnology firms listed on NASDAQ-100 index (NDX). Ten biotech companies comprised the index by November 2013 including Alexion Pharmaceuticals, BioMarin, Incyte Corporation, Regeneron Pharmaceuticals, Vertex Pharmaceuticals, Biogen Inc, Celgene Corp, Amgen Inc, Gilead Science, and Illumina.

Information regarding the companies was retrieved from three main sources.

17.2.1 Financial Data

The database Thomson EIKON™, from Thomson Reuters, was used to gather financial data of the companies from 2002 to 2016. Thomson Reuters Eikon™ has the broadest and deepest financial data and offers a comprehensive solution for establishing customizable benchmarks for the assessment of corporate performance (Reuters 2018). Data related to company's income statements, operating metrics, balance sheet, deals, merge and acquisitions, major costumers, ratio key metrics, ratios profile value, and risk were extracted from the database. Some key information including EV, net income, revenue, R&D expenditure, company's pipeline and deals' description were compared among companies. Additional information available through the database was collected when needed.

17.2.2 Patent Data

The intellectual property assets from companies were collected from the Derwent Innovation (formerly Thomson Innovation) of *Clarivate Analytics*. The database provides access to global patent data covering more than 50 patent issuing authorities, with English translations from 30 languages (Clarivate Analytics 2018). The patent portfolio from all corporate tree members of each biotech company was collected based on the earliest priority data ranging from 2001 to 2016 and grouped by INPADOC (International Patent Documentations) families.

17.2.3 Company's Annual Reports

The annual reports from the 10 US biotech companies listed on NASDAQ-100 were manually downloaded from the company's website available to investors. The reports were used to gather detailed information regarding the company's strategies, decisions, purposes behind their deals, drug's pipeline performance, and company's future perspectives.

Additional information available online related to financial news, opinions from financial analysts and experts, reports from consulting firms to the sector was also used for supplementary support to our findings.

17.3 Results and Discussion

The EV for the ten companies analyzed in this chapter are shown in Fig. 17.1 for the years 2001–2017.

On the individual firm analysis level, Amgen started out as the most valuable company with an EV of $60 billion in 2002, while the remaining nine companies started out with EVs less than $10 billion. It is interesting to notice the evolution of Gilead overtime – the company's curve shows major spikes and ends up with an even higher EV than Amgen (around $110 billion). Celgene finishes the period with an EV of approximately $95 billion, while Biogen comes in fourth place with an EV of approximately $55 billion, both detaching themselves form the lower tier group of companies. After these four companies, Regeneron and Alexion show up with approximately $35 and $30 billion, respectively. The remaining four companies finish 2016 with EVs equal to or less than $15 billion. Although there are different patterns followed by each individual firm, one can notice that some patterns are

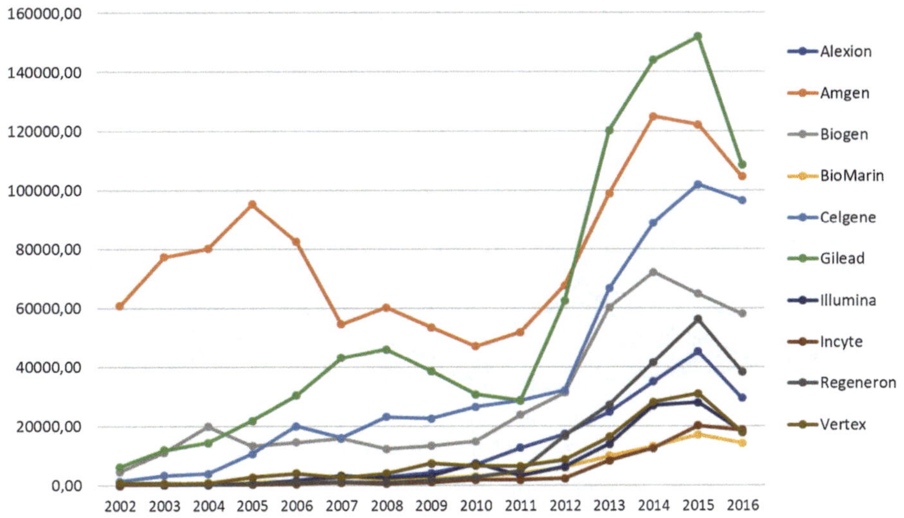

Fig. 17.1 Enterprise Values from 2002 to 2016 of US biotech companies

observed in the curves of all ten companies. For instance, it is noticeable that in the period 2011–2012, all ten companies experienced a spike in their EVs (to a greater or lesser extent, depending on each case). Additionally, precisely, the opposite is observed around 2015, when nearly all ten companies see their EVs fall down drastically and rapidly. All of these points are discussed in subsequent sections of this chapter.

A study published in 2010 showed that survival of a company on a stock market such as NASDAQ is directly dependent of the patents they hold (Wagner and Cockburn 2010). Since NASDAQ include high-technology companies and the biotech sector is absolutely dependent of intellectual property rights, the patent portfolio from each biotech company studied in the present research was plotted for further analysis. Figure 17.2 brings the patent family count the companies first filed from the period of 2001 to 2016. The chart shows that Incyte starting invested deeply in intellectual property in 1997 (almost 250 patents – data not shown) but quickly losing momentum and being surpassed by Amgen, which files the highest number of patents from that year until 2012, when Celgene starts catching up. Gilead, Biogen, and Vertex also show strong intellectual property output throughout those years, and Illumina and Regeneron get close to 50 patents in 2014. The remaining companies stay, for the most part of the reported period, below the line of 20 patents.

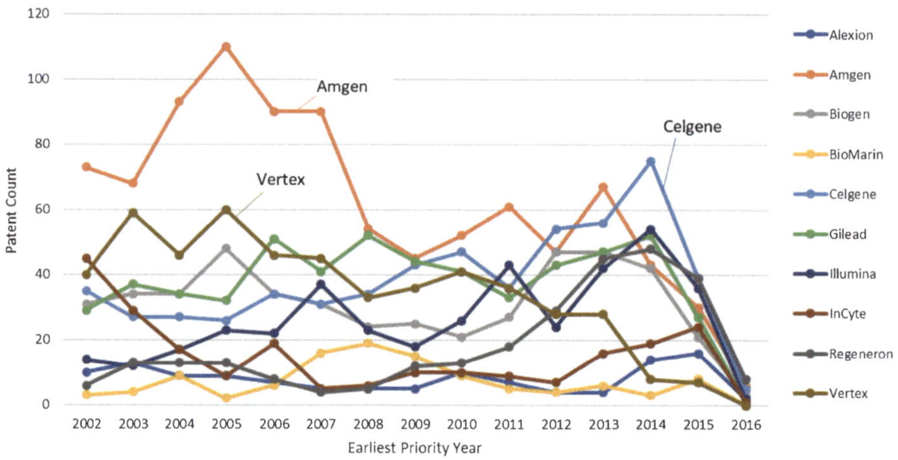

Fig. 17.2 Patent count from 2001 to 2016

17.3.1 Company's Features and EV Performance

17.3.1.1 Alexion

Alexion Pharmaceuticals Inc. is a Connecticut-based (moving to Boston in 2018) pharmaceutical company, founded in 1992 (IPO in 1996), best known for its development of Soliris (eculizumab), a drug used to treat rare disorders such as atypical hemolytic uremic syndrome (aHUS) and paroxysmal nocturnal hemoglobinuria (PNH). Soliris is considered one of the world's most expensive drug (Z. I. Research 2018) approved by FDA in 2007 which subsequently reflected the company's EV's historical rise since then. The hemolytic–uremic syndrome epidemic in 2010 also helped to boost Soliris sales impacting on Alexion's EV from about $10 billion to more than $40 billion in less than 5 years and landing it in NASDAQ-100 in 2011. Alexion has completed three major acquisitions to date, which are antibody, hypophosphatasia treatment, and rare disease-related drug. The company has been raising its R&D spending every year with bigger leaps starting after 2007 (going from $10 million to $757 million from 2002 to 2016). The company has done a total number of 13 deals (including M&A (7), Equity (2), and loans (4)) with the biggest acquisition being Synageva Biopharma Corp for $7.6 billion in 2015. Alexion's main strategy is to target rare diseases. With this purpose, Alexion shortened FDA approval times and cut trial costs, leading to a more productive pipeline in addition to the fact that rare diseases are normally ignored by big pharma companies. Although Alexion's drug prices are high, it approaches public relations firms to help families institute campaigns to pressure their governments to pay for the drug ("How a pharmaceutical firm priced its life-saving drug at $500K a year" CBC News. [Online].). Alexion's second drug Strensig received FDA approval in 2015, which was part of their 2011 acquisition of Enobia. Also, Kanuma drug was

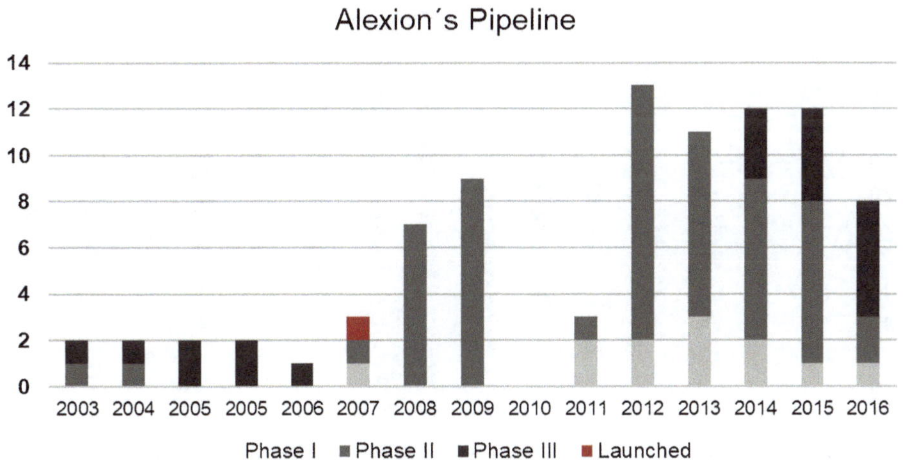

Fig. 17.3 Alexion pipeline from 2003 to 2016

approved in 2015 and was connected to their Synageva acquisition (with $1 billion in projected annual sales ("Biotech blastoff: Synageva up 112% on $8B Alexion deal," USA TODAY. [Online].)). In October 2017, the FDA approved the use of Soliris to treat adult patients with generalized myasthenia gravis (gMG) and in November 2017, the company received a patent for Soliris from the Japanese Patent Office. Figure 17.3 brings the company's pipeline evolution in terms of drug candidates in phases I, II, and III.

17.3.1.2 Amgen

Amgen is one of the world's leading biotechnology companies focused on the treatment of serious disease through the discovery, development, and manufacturing of innovative treatments. Amgen is among the biopharma companies which spend the most in R&D. Besides the expense being very high, the percentage of the R&D of the revenue was constant between 2002 and 2016, around 20% of the revenue; recently, the ratio decreased in 2015 (18.7% of the revenue) and 2016 (16.7%).

The company's therapies are very diverse, consisting of sixteen drugs including six recombinant peptides, five monoclonal antibodies, and one oncolytic virus. The company is very intense in protecting their discoveries through patents, mostly international deposit (as PCT), besides new inventions falling dramatically after 2007 (Fig. 17.2). The drug discovery activity of Amgen is focused on drugs that hit a target or has a mechanism of action that no existing drugs address. The company is constantly testing new products in clinical trials as seen by the high number of late-stage drugs. After 2011, Amgen was with 10 (or more) candidates undergoing testing in phase III – Fig. 17.4.

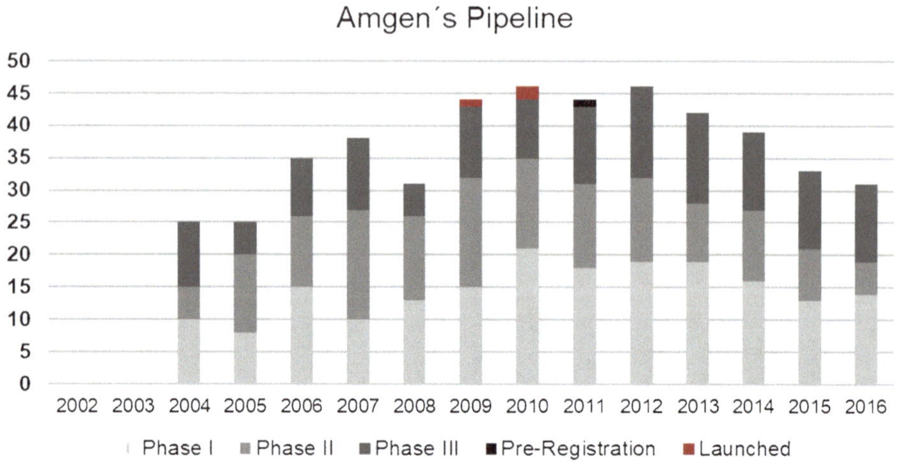

Fig. 17.4 Amgen pipeline from 2002 to 2016

The company had an expressive growth between 2002 and 2005 driven by significant merge and acquisitions (M&A). As in 2002, it was the world's largest biotechnology company, after completing the acquisition of Immunex Corporation, a leader in inflammation and one of biotechnology's premier companies. The acquisition on a $16 billion deal was the largest so far of one biotech company acquiring another. Two years later, Amgen acquired Tularik, a pioneer company in drug discovery related to cell signaling and the control of gene expression, for $1.3 billion. At the time, Amgen did not have any exciting new drugs under development to sustain its rapid growth, and with the acquisition, Amgen aimed at the Talariḱs research talents besides any specific molecule on its pipeline. In 2005, Amgen agreed to acquire Abgenix for approximately $2.2 billion. The acquisition gave Amgen full ownership of one of its most important advanced pipeline products, panitumumab, and consequently eliminated the need of royalty payment on denosumab, another important pipeline drug.

During 2006–2010, the company experienced a dramatic fall in EV for many reasons. One was about the concerns for the use of Aranesp, a flagship drug, in cancer patients, which showed on clinical trials that it could make cancer worse and shorten patient survival. As a result, in 2007, the FDA issued a black-box warning about the potential side effects of Aranesp, which led to sales decline in the following years. In subsequent years, Amgen faced a lawsuit about the Aranesp scheme, accused of boosting drug sales along with the International Nephrology Network and ASD Healthcare, a wholesaler owned by AmerisourceBergen Corp. There were some other issues that Amgen had to face from 2007 to 2009, such as FDA approval delays on its most promising drug at the time, the monoclonal antibody denosumab, followed by some failures on clinical trials for other drug candidates.

Amgen started rising in EV after 2011, thanks to the numerous mergers and acquisitions. At total, it was 8 M&A in 2011 totaling $7 billion, 4 in 2012 totaling

$2.6 billion, and 2, which included the acquisition of Onyx Pharmaceuticals in 2013, totaling $9.1 billion. Onix Pharmaceuticals developed carfilzomib, one of the major drugs on sales of Amgen. In 2011, Amgen acquired for $1 billion a venture-funded, biotechnology company named BioVex Group, Inc., owner of OncoVEX (GM-CSF), a novel oncolytic vaccine in Phase 3 clinical development to treat melanoma and head and neck cancer, which were approved by FDA in 2015.

In 2012, Amgen acquired Micromet ($1.1 Billion), a biotech company focused on the discovery, development, and commercialization of antibody-based therapies for the treatment of cancer. At the same year, Amgen acquired Mustafa Nevzat (700MM) to expand its presence in Turkey and the surrounding region. At the end of the year, the company acquired Decode Genetics ($415 million), a global leader in human genetics, to strengthen the identification and validation of human disease targets.

17.3.1.3 Biogen

Biogen Idec Inc. (NASDAQ: BIIB) is an American multinational biotechnology leader in the discovery, development, manufacturing, and commercialization of innovative therapies for the treatment of neurodegenerative, hematologic, and autoimmune diseases. The company R&D expenses turned around an average of 25% of the revenue from 2002 to 2016, and the ratio was nearly constant each year, but it dropped to an average of 18% of the revenue in 2014–2016.

Currently, the company has 12 therapies including three biosimilars and many other drugs under development on its pipeline. From the current therapies of Biogen, eight are immunotherapy-based, being six monoclonal antibodies (natalizumab, daclizumab, obinutuzumab, rituximab, and the biosimilars infliximab and adalimumab) and two interferons: one interferon beta1-a and its pegylated form. Before 2002, the company had two FDA approvals (Avonex and rituximab) and two approvals for other companies of drugs licensed from Biogen. It is worth mentioning that in-licensing of drugs is also a key part of building Biogen's pipeline. There are also three phase III drugs on its pipeline, a monoclonal antibody (aducanumab) and a BACE1 inhibitor for Alzheimer disease, and a monomethyl fumarate prodrug for multiple sclerosis. The company had a powerful drug development engine which resulted in several FDA approvals of drugs independently commercialized, in cooperation with biotech and pharmaceutical companies and also drugs which were sold/transferred to other companies, resulting in royalties for their commercialization. Its pipeline consisted of several new candidates being tested, late-stage drugs, and many on pre-registration phase since 2011, which could possibly lead to an increase in EV after this year – Fig. 17.5.

The second largest acquisition involving two biotech companies was the merging of Biogen with Idec Pharmaceuticals on a $6.8 billion deal in 2003; the first was Amgen's acquisition of Immunex which has been described previously. With the merger, two blockbuster drugs were commercialized by the newly Biogen Idec company, rituximab originally from Idec (U.S. sales shared with Genentech) and

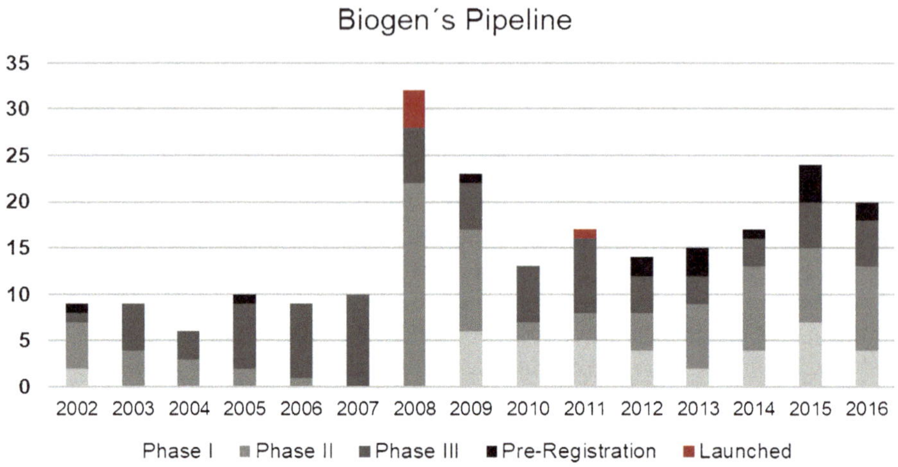

Fig. 17.5 Biogen pipeline from 2002 to 2016

Avonex from Biogen. Idec also had another drug named Zevalin, which was sold to Cell Therapeutics in 2007 for up to $30 million. The merger resulted in a growth of 78% of EV in 2004 driven by significant achievements: one was the first full year as a combined organization and also the outstanding performance of Avonex and rituximab. Another major acquisition by Biogen was the European Fumapharm AG, which develops therapeutics derived from fumaric acid esters. The acquisition provided support to the company's interest in treating multiple sclerosis, since Fumapharm had a co-developed drug with Biogen named at the time BG-12 (later Tecfidera), which would become one of the top sellers for Biogen. Also in 2007, Biogen acquired Syntonix Pharmaceuticals, which at the time had two drugs to treat hemophilia, Eloctate and Alprolix, and the acquisition resulted in intense sales for Biogen and in 2016, Biogen announced the spin-off Bioverativ for the its hemophilia business to keep the focus on its core multiple sclerosis portfolio. Biogen did also some small acquisition after 2007, and the largest was the Convergence Pharmaceuticals in 2015 for $675 million.

The year 2012 was very successful for the company, driven by the progress to late-stage pipeline in clinical trials and upcoming product launches (Tecfidera and hemophilia drugs). The revenue growth of natalizumab and Avonex was also a significant factor for the EV growth in 2012.

17.3.1.4 Biomarin

BioMarin is a world leader in developing and commercializing innovative biopharmaceuticals for rare diseases driven by genetic causes. Its mission is to bring new treatments to market that will make a big impact on small patient population. A leading source for global clinical trial information, *Center Watch,* named Biomarin as one of the fastest drug developers in the industry in its monthly report in September 2014.

The company's strength to develop new drugs is explained by its massive R&D investments each year. In 2002, the company had no revenue, but even then, it invested around 27 million in R&D. After the approval of Aldurazyme in 2003, the company started earning revenue and invested massively in R&D, reaching 445% of the revenue in this year. In 2004, 268% of the revenue was invested in R&D and 219% in 2005. After 2005, the company started increasing its revenue each year after drugs approval by FDA, and the expressive investment in R&D was still high. From 2012 to 2016, the company invested an average on 64% of revenue in R&D.

The giant R&D investments resulted in a series of successful therapies and FDA approvals initiated in 2003. The first drug to treat mucopolysaccharidosis type I (MPS I), Aldurazyme (which was commercialized by Genzyme Corporation), was approved in Europe and by the FDA in 2003. Two years later, the FDA approved Naglazyme, making the second drug to treat MPS and the first drug independently commercialized by the company. In 2007, the FDA approved Kuvan for the treatment of phenylketonuria (PKU); it was the first medication to treat this condition. Kuvan was also approved in Europe, Japan, and Canada in the following years. In 2014, Vimizim received FDA approval for Morquio A syndrome, the first treatment for this condition. At the same year, Vimizim was also approved in Europe, Canada, and Australia. Lastly, in 2017, the FDA and the European Commission approved Brineura to treat patients with CLN2 disease, a form of Batten disease. It is worth mentioning that Biomarin has no direct competitors for their products.

Biomarin did some major acquisitions during its history. The first one was Glyko Biomedical, a leading provider of carbohydrate-related research reagents, processing enzymes, and analytic products, for $228 million in 2002. In 2009, Biomarin acquired Huxley Pharmaceuticals, focusing to get access to a proprietary therapy to treat a rare autoimmune disease, the Lambert Eaton Myasthenic Syndrome (LEMS). The therapy, named Firdapse by Biomarin, was approved in Europe in 2010. The next major acquisition was Lead Therapeutics, an early-stage development company, with a key asset, a PARP inhibitor. Biomarin acquired Lead for 97 million in 2010, then sold the PARP inhibitor (BMN673), already in phase III to treat advanced breast cancer, 5 years later for 570 million, keeping its efforts on rare diseases. Still, in 2010, Biomarin also acquired ZyStor Therapeutics, a biotech company developing a novel class of targeted protein therapeutics enhanced with the company's proprietary glycosylation independent lysosomal targeting (GILT) technology, which was incorporated further by many drug candidates in clinical trials. The biggest acquisition was the Dutch Prosensa, which had as the key asset an experimental Duchenne muscular dystrophy treatment named Drisapersen (Kydrisa), which failed in phase 3 and was rejected by the FDA in 2016. Recently, there are some speculations about Biomarin being acquired by giant pharmaceuticals and biotech companies such as Gilead and Sanofi. Figure 17.6 brings the company's pipeline evolution in terms of drug candidates in phases I, II, and III.

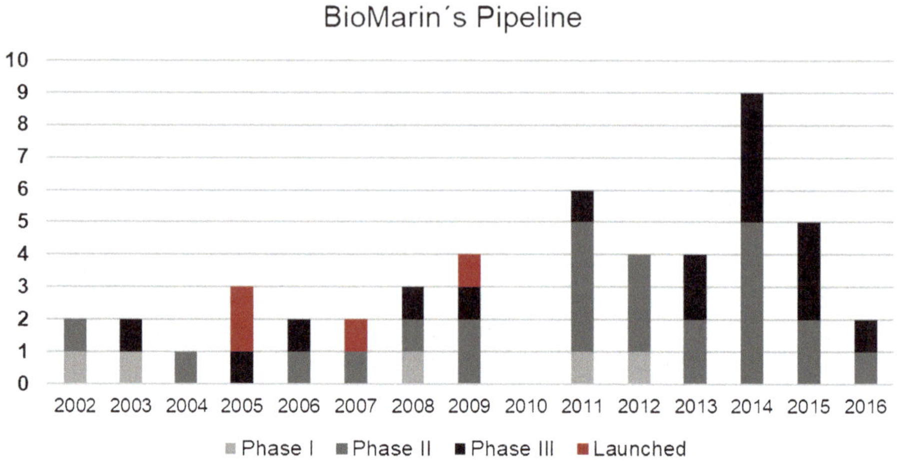

Fig. 17.6 Biomarin pipeline from 2002 to 2016

17.3.1.5 Celgene

Celgene, headquartered in Summit, NJ, was founded in 1986 with its initial public offering (IPO) 1 year later. Having received several recognitions and awards, including Fortune 500's fastest-growing pharmaceutical companies (Awards & Recognition 2016), the company focuses on oncology medicines and has multiple products in the pipeline, in the areas of multiple myeloma, leukemia, lymphoma, and others. The company has successfully developed several medicines in different areas and has three major centers for research and development: The Celgene Institute for Translational Research Europe (CITRE) in Seville, Spain; the Celgene Translational Development center in San Francisco; the Drug Discovery & Alliance Development center in San Diego.

The company is actively making deals, having completed 35 M&A in the period 2008–2017, totaling $45 billion. In 2010, Celgene acquired Abraxis BioScience in a transaction of almost $3,5 billion, in an attempt to enlarge their portfolio of oncology-related products. Abraxis had in its portfolio drugs that attack solid tumors, and although the transaction amount was considered very high, Celgene was confident that Abraxis was a perfect fit and also would be able to boost sales of Abraxis' existing drugs (Celgene to buy Abraxis BioScience for $2.9 billion 2010). In 2016, Celgene finalized a deal that had started in 2013 and acquired part of the assets of Acetyon Pharmaceuticals – a deal totaling $1.7 billion – bringing more cancer-related drug candidates to its pipeline (Xconomy 2016). In 2015, Celgene acquired Receptos for more than $7 billion. Expecting to attack markets with very large patient populations, Celgene yielded from Receptos medicines (candidates in the late stage of development) that address multiple sclerosis and inflammatory bowel disease.

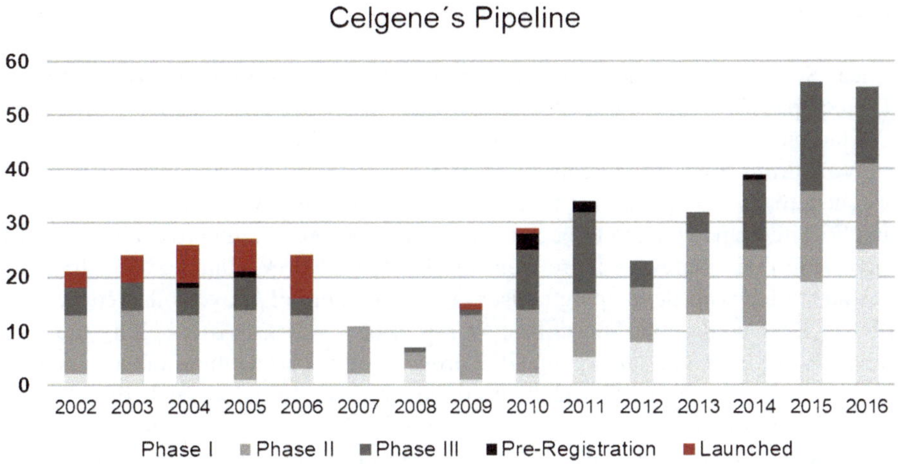

Fig. 17.7 Celgene pipeline from 2002 to 2016

Celgene's pipeline is large and had 14 products in clinical phase III as of 2016 as Fig. 17.7 shows. The company's portfolio is also large and it grows larger with every acquisition Celgene completes. Among the drugs within the portfolio, Revlimid, Abraxane, and Otezla could be listed as the most important ones – but some of their drug candidates are also considered to have an extremely high potential. Revlimid is a drug used to fight multiple myeloma (MM), myelodysplastic syndromes (MDS), and mantle cell lymphoma (MCL), and was first approved by the FDA in 2005, having had several subsequent approvals for different applications. Abraxane, originally developed by Abraxis BioScience, is a drug used to treat metastatic breast cancer, advanced non-small cell lung cancer and metastatic pancreatic cancer – this drug was approved by the FDA in 2005 for usage in metastatic breast cancer, but received other approvals in 2012 and 2013 for lung and pancreatic cancer. Otezla, the first pill to treat psoriasis, is prescribed to treat plaque psoriasis and active psoriatic arthritis, and it was approved by the FDA in 2014.

The income of Celgene has been steadily on the rise, going from \$136M in that year to \$11B in 2016. As opposed to other biotechnology companies, Celgene has not only achieved high income but also has enjoyed profits for most years since 2002 – profits have risen to almost \$2 billion in 2016. In fact, between 2002 and 2016, the only year for which the company reported losses was in 2008 (a loss of \$1,5 billion), and that is attributed to fees associated with the acquisition of Pharmion in the same year (Celgene swung back to profits in the first quarter 2009). Similar to the income, R&D investments have also been steadily increasing, going from \$85 million in 2002 to \$4,4 billion in 2016.

17.3.1.6 Gilead

Gilead Sciences is an American biopharma company based in Foster City, California. The company's main area of success and focus has been the treatment of Hepatitis C (Harvoni and Sovaldi), but allocates a great deal of R&D investment in treating antiviral drugs used in the treatment of HIV, hepatitis B, and influenza.

The already strong pipeline base and on top of that innovative drugs like Sovaldi and Stribild helped Gilead to become a gargantuan force in the biotech industry by increasing their shares by more than 400% since 2009 (Williams 2015). Initially, Gilead started with developing antisense with the help of Glaxo followed by their IPO in 1992 with about $86 million in raised proceedings. After selling the antisense IP, they started focusing on AIDS drugs through partnerships and acquisitions (1996–2002). However, in 2002, they shifted their concentration solely on antivirals and jettisoned their oncology knowledge to OSI Pharma for $200 million dollars. Gilead stayed focused on antivirals until 2006 where they added cardiovascular and respiratory drugs to their portfolio by acquiring Myogen Inc. and Corus Pharma Inc.. This helped them in having a steady rise in the EV from 2006 to 2008. In the meantime, acquisition of Raylo Chemicals Inc. helped them to increase their profits margin by bolstering their supply chain in terms of ingredients and intermediates. Gilead strengthened its position in the cardiovascular and respiratory fields by acquisitions in 2007 and 2009. In the meantime, they kept strengthening their portfolio through acquisitions of companies like Arresto Biosciences, CGI Pharmaceuticals, and Calistoga Pharmaceuticals. An important point in Gilead's success is definitely the acquisition of Pharmasset Inc. for about $10,8 billion. As a part of this acquisition, they got Sovaldi in their hands, which turned out to be phenomenal blockbuster drug for the treatment of Hepatitis C. This drug minimized the side effects of the previous drugs and raised the success rate to over 94% of the patients (Keating 2015). This acquisition helped their EV to rocket to over $140 billion from 2011 to 2015. The revenue generated from these drugs helped them to bolster their R&D spending from $142 million in 2002 to $4.2 billion in 2016. The M&A consequently led to the increase in the products in their pipeline as well (from 3 products in 2002 to 22 in 2011 and 25 in 2016). In 2015, the three acquisitions made by Gilead strengthened their explorations into nonalcoholic steatohepatitis, cancer therapy, and anti-inflammatory drugs, as they sensed the need to broaden their pipeline horizons. This action was taken as Gilead realized that they have more competition in the Hepatitis C field as Merck & Co. marketed Harvoni, which is the combination of Sovaldi and Ledipasvir, leading to litigations with Gilead. AbbVie launched a competing treatment for hepatitis C as well, which forced Gilead to sell at more competitive prices. They slowed down on the acquisitions in the recent years, missing out on the acquisitions of Pharmacyclics and Medivation in 2016 (Campbell 2018) and this chain of events added to other reasons decreased Gilead's EV to less than $110 billion in 2016.

Although the hepatitis C market seems to be leveling off, it should be taken into account that the addressable market for that is still huge. They finally made another big acquisition in the oncology field in 2017 with Kite Pharma ($11 billion), which

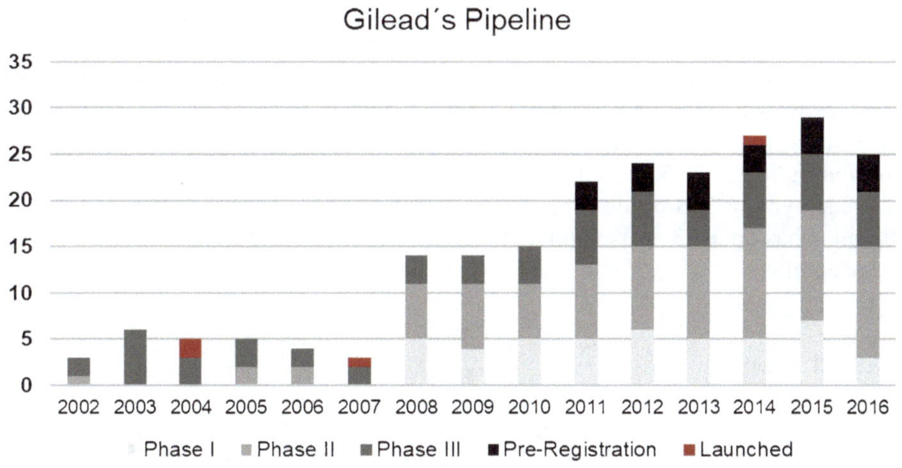

Fig. 17.8 Gilead pipeline from 2002 to 2016

has 14 clinical trials ongoing or planned targeting 10 cancer indications which can potentially boost Gilead's EV and maintain their sales with its drugs. All these new acquisitions and industry climate mean that hepatitis C is no longer going to be the sole focus of Gilead and they need to go after (already going) fields such as hepatitis B, cancer, and HIV more aggressively on a quest of finding their new blockbuster drug while their current cash cows are still standing. Figure 17.8 brings the company's pipeline evolution in terms of drug candidates in phases I, II, and III.

17.3.1.7 Illumina

Illumina Inc. is a San Diego-based company established in April 1998 (IPO 2000) that develops, manufactures, and markets integrated systems for the analysis of genetic variation and biological function. It produces a line of products and services revolving around sequencing, genotyping, and gene expression and proteomics markets. Illumina's biggest achievement is reducing Personal Full Genome Sequencing price from around $48,000 (June 2009) to around $4000 (May 2011) serving the academic, government, pharma, biotech, and other industries. Illumina was in the business of selling research instruments until 2010 before getting the FDA approval for BeadXpress for multiplex genetic analysis. This was the start of their EV increase going from $8 billion in 2010 to $28 billion in 2016. Illumina went through acquiring Epicentre Biotechnologies in 2011. The move was followed by the introduction of HiSeq X with cheaper and faster sequencing capabilities, helping them to own about three-fourths of the genome sequencing machines market by 2014 and producing more than 90% of the produced genome data (Regalado 2018). Since 2011, Illumina has been offering gene sequencing management and analysis platform (BaseSpace). The acquisition of GenoLogics in 2015 helped them to bolster their lab info management capabilities. Illumina has been selling NovaSeq and HiSeq in a pretty stable way. Illumina has been following the goal of achieving $100 gene sequencing making its addressable market way bigger as discussed by its CEO de Souza. They have acquired 10 companies in total (biggest being Verinata Health Inc for $450 million in 2013). Their R&D spending has risen from $26.8 million in 2002 to $504.4 million in 2016 concurrent with widening the increase gap each year ($103 million increase in R&D spending between 2015 and 2016). Illumina has done 23 deals in total in which 20 of them are M&As ($9 billion), and 3 are equities.

17.3.1.8 Incyte

Incyte, founded in California (Palo Alto) in the early 1990s, is headquartered in Wilmington, Delaware. Focusing on oncology-related medicines, the company has several molecules under development and dozens of discoveries being further developed, some of which are in the clinical proof-of-concept stage. Being keen on the importance of collaborative research and development, the company has several license agreements and collaborations with important players in the industry, for example, Novartis and Eli Lilly, and has two medicines approved in 2016, one in the USA and one in Europe. Later in 2016, with the acquisition of ARIAD Pharms, a third drug – Iclusig – was added to the European portion of the company's portfolio.

In order to strengthen its pipeline and portfolio of oncology drugs, Incyte has reached a deal with Merus NV in 2016, a Dutch pharmaceutical company. The deal indicates that Incyte has exclusive rights for 11 of Merus antibody discovery initiatives, and Incyte has also bought $83 million worth of Merus shares – a total of 3.2 million shares. Also, in 2016, Incyte acquired ARIAD Pharmaceuticals, in a $274 million agreement, strengthening even further its presence in the market for oncology medicines. The deal gives Incyte exclusive rights for the commercialization of a kinase inhibitor called Iclusig (ponatinib) to fight chronic myeloid leukemia.

Analyzing Incyte's pipeline, it is very interesting to note that the company had, as of 2016, 21 drug candidates in the clinical phase II, as shown in Fig. 17.9. Provided the company is able to bring some of those drugs to the market, it might translate into a huge growth potential. However, it is also noteworthy that the number of early-stage development cases has been very low in the last few years, and the company has no drug candidates in phase I as of 2016. Incyte's portfolio has three drugs: Jakafi (ruxolitinib); Iclusig (ponatinib); Olumiant (baricitinib). Jakafi is used to treat myelofibrosis and polycythemia Vera, and was FDA approved for the first application in 2011 and for the second application in 2014. According to Incyte's annual report for 2016, Jakafi's sales have largely driven the company's spike in revenues in the past few years. Iclusig is indicated for adults suffering from different kinds of leukemia and also a specific type of abnormal gene (T315l-positive) – this drug is approved for markets in Europe, but has not received FDA approval. Olumiant, also approved for markets in Europe, is used to treat severe rheumatoid arthritis when previous therapies are not tolerated by the patient. Interestingly, FDA

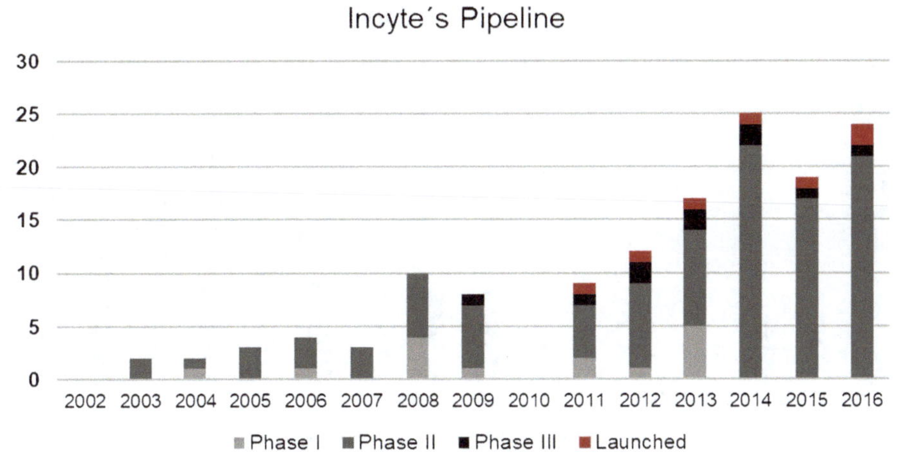

Fig. 17.9 Incyte pipeline from 2002 to 2016

did not approve this drug, rather asking for more data, which can severely delay the approval (U.S. FDA 2017).

In the period 2004–2009, the income has varied from \$9M to \$14M, and after 2010, the income has steadily increased, reaching \$1,1 billion in 2016. In the period 2004–2016, the R&D investments have also increased steadily, reaching \$581 million in 2016. Regarding profits, the company has presented losses until 2014. In 2015, Incyte has had small profits (\$6,5 million) and in 2016, the company enjoyed over \$100 million.

17.3.1.9 Regeneron

Regeneron was founded in 1988 and it is headquartered in Tarrytown, NY. The company employs more than 5,000 people – more than 700 of whom are PhD, MD, or PharmD candidates – and currently has 6 FDA-approved medicines in its portfolio, while also working on more than 15 different antibodies currently undergoing clinical trials. Regeneron has invented a series of technologies that aim to accelerate and improve the success rates of drug's R&D, with collaborations with big pharmas such as Sanofi and Bayer. Regeneron focuses on seven research areas for its drug development: cardiovascular and metabolism, infectious diseases, inflammation and immunology, oncology, ophthalmology, pain, and rare diseases.

In the period 2008–2017, Regeneron was part of eight deals, totaling \$770 million dollars. In 2013, Sanofi acquired Regeneron shares, and in 2016, Regeneron acquired two companies: Intellia Therapeutics and Biomed Ritty. Intellia has developed treatments for numerous applications such as liver diseases, blood diseases, and cancer, and also is one of the pioneers in the Crisp-Cas9 technology, which focuses on gene identification and selection as a way to treat diseases.

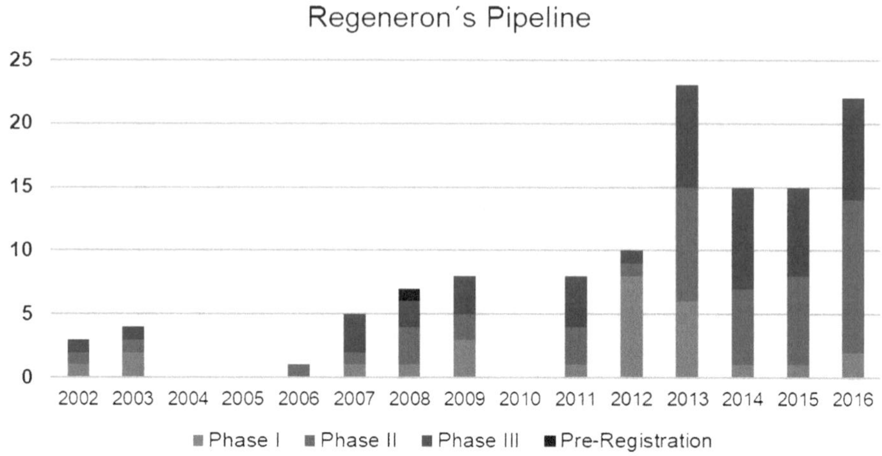

Fig. 17.10 Regeneron pipeline from 2002 to 2016

As demonstrated in Fig. 17.10, the company's pipeline was fairly modest until around 2011, when more candidates were introduced (in 2013, there were 23 candidates in the pipeline) and as of 2016, 22 products are either in phases I, II, or III. The portfolio of Regeneron is composed of five drugs, namely Arcalyst, Dupixent, Eylea, Kevzara, and Praluent. According to Regeneron's annual report for 2016, Eylea was responsible for the biggest share of the company revenues – $3.32 billion dollars. The drug is used to treat several types of macular degeneration and macular edema and it received FDA approval in 2011 for age-related macular degeneration.

Regeneron's income had been steadily increasing from 2007 to 2011, year in which the company had $445 million as revenue. In 2012, however, after the launch of Eylea partnered with Bayer, revenues skyrocketed to $1.3 billion dollars and it kept on growing – in 2016, the company had $4.8 billion in revenues. The investments in research and development have also been very strong and constantly growing. In 2002, the company invested $22 million, whereas in 2016, the amount of dollars spent in R&D was over $2 billion. With regard to profit, Regeneron has experienced losses in the period 2002–2011 (with the exception of 2004 when the company had a profit of a little over $40 million). In 2012, however, also after launching Eylea, profits came back and rose to $750 million in that year. In 2016, the net income was almost $900 million.

17.3.1.10 Vertex

Vertex was founded in 1989 and it is headquartered in Boston, Massachusetts. The company has employed more than 2000 people – two-thirds of whom actively work in research and development activities. Vertex has two approved medicines and several others in the pipeline, in the areas of cystic fibrosis, acute spinal cord injury,

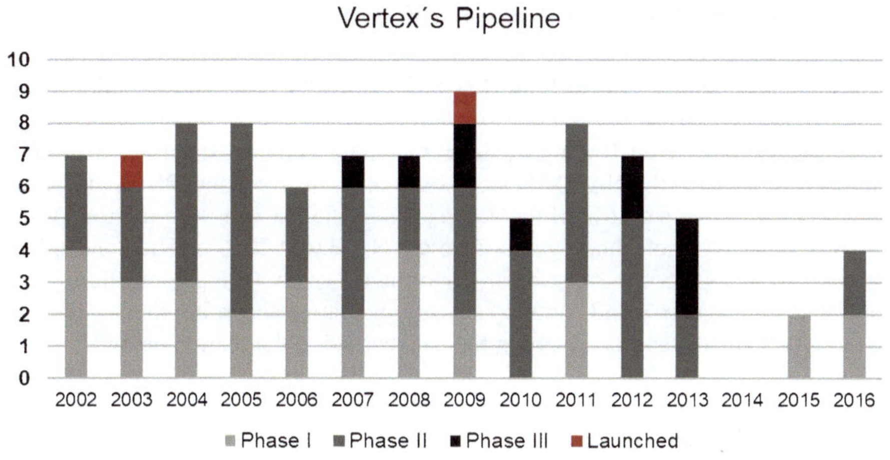

Fig. 17.11 Vertex pipeline from 2002 to 2016

oncology, pain and influenza. In 2009, Vertex had its drug Telaprevir undergoing development in Phase III, and in the same year, following their strategy of focusing on hepatitis C medicines, the company acquired ViroChem Pharma. ViroChem, another US-based biotech company, had at the time two promising drug candidates that tackled the hepatitis C virus, HCV (VCH-222 and VCH-759).

The pipeline variation in terms of clinical phases 1–3 for the years 2002 until 2016 is depicted in the Fig. 17.11. Vertex has developed a drug to tackle HCV called Telaprevir, in partnership with Johnson & Johnson and other companies, and is marketed by different companies and under different brands across the globe. Telaprevir received approval by the FDA in 2011. The company currently has two products in their portfolio: Orkambi (lumacaftor/ivacaftor) and Kalydeco (ivacaftor). Orkambi is a drug developed to target the root causes of cystic fibrosis and is administered to people who are 6 years of age and older and received approval from the FDA in 2015. Kalydeco, also developed to treat patients with cystic fibrosis, is administered to people who are 2 years of age and older and received approval from FDA in 2012.

Although the company has been obtaining increasing revenues (that jumped to more than $1 billion in 2011), the R&D investments are also massive ($1 billion in 2016) and the company has not enjoyed profits, except a small profit in 2011. New drug prospects and increased sales from existing drugs in their portfolio could change this scenario, however – the company's net income in 2014 was -$738 million in 2014, but has advance to -$112 million in 2016.

17.3.2 Companies' Key Success Factors

This chapter intended to identify the key factors that led biotech companies to success (NASDAQ 2016a). Our sample encompasses biopharmaceutical public companies listed on NASDAQ-100 index, since it aggregates the most actively traded US companies listed on the NASDAQ stock exchange. In November 2016, the healthcare sector represented 16% of the companies in the list and from the top 10 performers, six were biotechnology companies (NASDAQ 2016b). The consolidated descriptive analysis of the historical performance of the companies allows us to highlight some patterns and common success indicators according to their EV fluctuations. The main aspects are going to be discussed in the section below.

17.3.2.1 R&D Expenditure

High R&D expenditure seems to be one of the common strategies among these successful biotech companies. As an example, for Vertex, Incyte, and Biomarin, key success factors come from their massive and constant investments in R&D. They are the leaders among companies that spend the most on R&D relative to their revenue in the last 5 years of analysis (2012–2019) (Fig. 17.12). The biotech sector is highly demanding of R&D investment; thus, many of the selected companies were investing millions of dollars even before they had no revenues.

Heavy investments in research led to the development of some very important drugs for companies (such as Soliris for Alexion). Additionally, one of the reasons

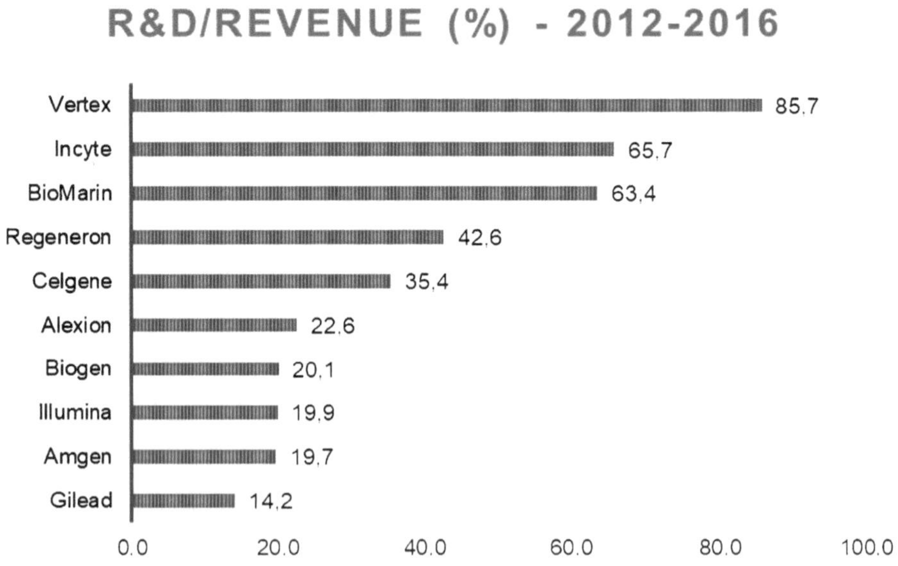

Fig. 17.12 Companies' R&D expenditure and revenues from 2002 to 2016

behind Gilead's success has been its massive R&D investment in different fields with hopes of expanding its pipeline horizons and penetrating different addressable markets for different diseases, which leads to a constant increase in the allocated budget for research (from $142 million in 2002 to over $4 billion in 2016). Along with R&D investments, Celgene's income has been increasing in a rapid and steady pace, and the company has enjoyed profits in most years since 2002.

Our data showed that most companies start with a high ratio of R&D expenditure compared to their revenue, but as time goes by, there is a changing point where the revenue would exceed the R&D expenditure. The overcoming point for revenue occurred post-2007 for Alexion, Biomarin, Illumina, Incyte, Regeneron, and Vertex. Also, as it can be seen in Fig. 17.13, the R&D expenditures for some companies are dramatically higher based on the ratio of the revenue. As an example, Alexion spent massive amount in R&D through consecutive years from 2002 to 2006, an average of 60 times more than the revenue, which consequentially may have led to the development of Soliris and its FDA approval in 2007 and the EV increase in the following years. From the other side, Biogen has a powerful drug development strategy and a diverse business core consisting of constant percentage of R&D expenditure of the revenue. In summary, R&D expenditure are crucial for companies to develop their products and launch them on market when approval occurs (Morgan et al. 2011; PhRMA 2013).

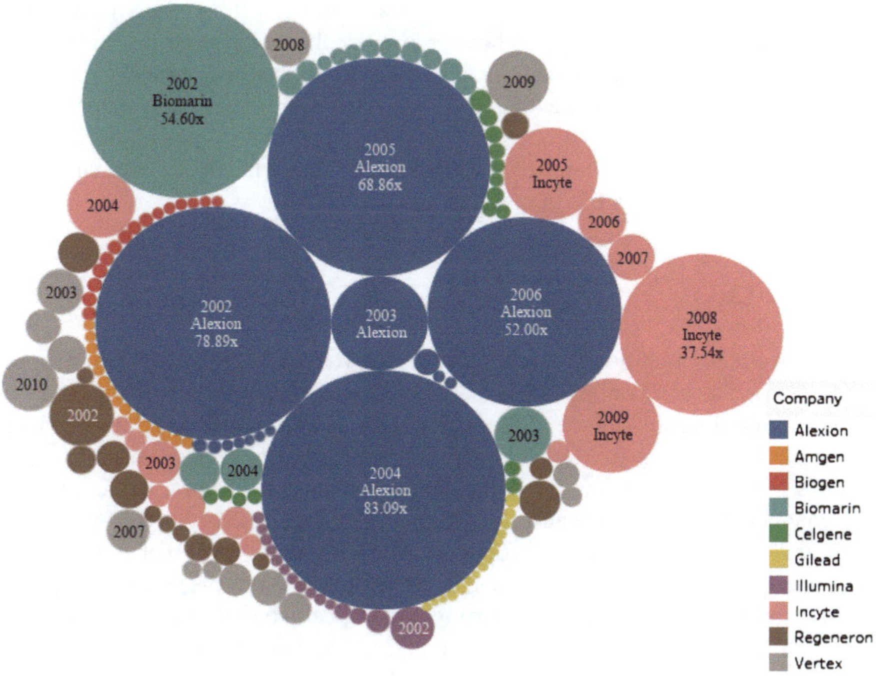

Fig. 17.13 R&D expenditure as a percentage of revenues (circle sizes show higher percentages)

17.3.2.2 FDA Approvals

The second theme should have naturally been common between companies, as getting the FDA approval would allow them to start selling their drugs and generating revenues. These drugs reflect the R&D and clinical trial results as being fruit of successful and strategic acquisitions. Thus, mergers and acquisitions focused on drugs with advanced pipeline should be positive to reach an FDA approval and consequently, revenue. This was the case for Alexion's second drug, Strensig, which received FDA approval in 2015 that resulted from the 2011 acquisition of Enobia; similarly, with its third drug, Kanuma, which was also approved in 2015 previously connected to their Synageva acquisition (with $1 billion in projected annual sales). Another example of this was the spike in the Regeneron's EV curve from 2011 to 2015, which probably was leveraged after their blockbuster drug, Eylea, was approved by the FDA in 2011. As for Incyte, Jakafi was approved by FDA for two different applications in 2011 and 2014, which indicates the spike we note in the EV curve for the period 2012–2015 due to the drug's launch leading to recurring revenue since then. Celgene also seems to have enjoyed an FDA approval-related success. The company's EV has had a big spike in the period 2012–2015, during which at least three of their drugs received approvals from the FDA, some of which for the first time (Otezla) and some for different kinds of treatment (revlimid and abraxane). Some of the rare disease-focused companies have been benefiting from FDA's faster approval times for orphan drugs. Biogen's strategy to focus on rare disease helped the company to get faster FDA approvals through orphan drug designations and a constant pipeline of drugs on advanced clinical development. It is also important to mention that Vertex had an increase in their EV value after 2011 due to the fact that two of their medicines received approval from the FDA – telaprevir and ivacaftor. The regulatory environment was very supportive to biotech during 2014 and 2015, but in 2016, the number of approvals fell dramatically (Moyer and Efram 2017), leading to a fall in EV in majority of companies.

17.3.2.3 M&As

Beside the already-mentioned impact of strategic M&A of companies with advanced pipelines on the companies' EV, the selected companies tend to have different behavior regarding this theme. Companies such as Vertex and Alexion tend to be less aggressive in terms of M&A compared to giant pharma Gilead. As an example, Vertex's M&A is rather focused and timely. The only acquisition made, in 2009, was conducted precisely because the company acquired had two very promising drug candidates in one of Vertex's focus areas – HCV infection. It seems that instead of growing out of acquisitions and mergers, the company prefers to selectively invest in their expertise areas and grow more organically based on blockbuster drugs that might come out of research in those areas. Alexion also was not a very busy player in the M&A area, but a very successful one, as two of their acquired companies resulted in fast FDA approved marketable drugs. Conversely, companies like Amgen, Gilead, and Celgene (which are the top three companies in terms of EV) have taken a more aggressive approach in M&As. Amgen became a huge biotech company and the key success factor for Amgen EV is its strategy for mergers and acquisitions. The reasons behind were diverse but mainly driven by taking control of blockbuster drugs developed from other companies, to take advantage of a R&D team and expertise from the acquired company, improve its pipeline on a novel therapeutic class, expand its territorial presence, and strengthen the support on its own R&D team. Gilead also had their big break after acquiring Pharmasset in 2015 which handed them Sovaldi, which turned out to be their most successful drug. Gilead continues on its tradition of acquiring potential successful companies in order to delve into new domains and expand its currently gargantuan territory. A great part of Biogen's revenue also comes from joint business, royalties and partnership, which showed that cooperation is the key factor for Biogen success. According to our data, some M&As such as Illumina's acquisition of Genologics are not necessarily a drug company buying another drug company. Genologics is developer of industry leading laboratory information management systems that would intend to help Illumina to develop an analytic platform for their sequencing machines. In summary, a strategic M&A, even focused on a certain drug category or expertise, contributes strongly to the dynamicity of the sector.

17.3.2.4 Others (Innovations, Dynamic Capabilities, Analytic Services)

Some of the other spotted themes which were common among these companies were related to their innovation management. With the rise of Genomics, companies like Alexion, Incyte, Biomarin, and Illumina were leveraging this technology to target rare diseases while taking advantage of FDA's less strict and faster approval times for orphan drugs designation. In terms of managing the innovations, there have been instances of strategic shift in the history of these successful companies. Vertex was focused on cystic fibrosis and HCV infection, but started focusing also on pain relief, oncology, and acute spinal cord injury from 2015 while intensifying their strategic partnership initiatives. Another example can be Gilead's shift to focus only in antivirals in 2002 followed by jettisoning their oncology knowledge to OSI Pharma for $200 million followed by their move in 2006, when they added cardiovascular and respiratory drugs to their portfolio.

The trend seems to be that bigger companies pursue a more aggressive M&A strategy and as they get bigger, need to maintain their flexibility. Since it is harder for bigger companies to change, acquisitions facilitate these shifts for them. They also tend to occur after bigger addressable markets (Gilead's hepatitis B, hepatitis C, cancer, and HIV). Smaller companies, however, are leveraging the new technologies such as Genomics and focusing on smaller patient populations and rare diseases. One factor that can be seen in both types of companies is that biotech companies need to maintain a prolific and strong pipeline to survive. They can maintain it through great R&D expenditures or acquiring companies and talents. As a high-risk, high-reward industry with a high-velocity market, companies need to have different drugs and different stages (from discovery and preclinical to phase III) in order to be more successful and even survive. Even bigger companies' (like

Gilead) time is limited, as they need to maintain their multi-billion dollars of revenues by coming up with new blockbuster drugs before the cash cows run dry.

Despite the fact that these identified themes affect the success of the biotech companies, it should be taken into account that extrinsic events can have great impact on the performance of biotech companies that are studied in this paper. Some events related to that subject must be highlighted. The financial crisis in 2008 greatly affected the pharmaceutical and especially biotech industry as companies were faced with less budgets and many of them became weaker as a result of that. Moreover, many biotech firms had to reprioritize their R&D programs (cutting many drugs which were in the earlier phases of development) so as to cope with the economic crisis, as many had to go through big employee layoffs affecting the investors' interests in them. As a result and the aftermath of the crisis, biotech investors and founders were no longer looking for building the next biotech giant and were looking for establishing companies with hopes of getting acquired by bigger players in the industry (Russo 2008). Another event was in 2016 when Hillary Clinton slammed high price increase in the biotech industry (Egan 2015). Her tweet caused the NASDAQ Biotechnology Index to fall 5% in its ETF (IBB), as it scared the investors concerning the future of the sector Although she did not end up as the president of US, the pharma and biotech industries took a hard hit as result of her remarks. The nine biggest losers on the NASDAQ 100 were all biotech stocks, led by BioMarin and Biogen, which fell 6% each. Bigger biotech companies such as Regeneron, Gilead, and Celgene were not safe either and took big hits in their numbers (Egan 2015). However, as many companies survived the 2008 crisis, there are signs of their survival past the Clinton remarks as well. Despite the fact that the biotech industry has had a couple of rough years, there are signs of revival on the horizon and biotechnology had a solid year in 2017 (rise of about 20% in the December in NASDAQ biotechnology index) (Chandra 2017a, b). In January 2018, Sanofi and Celgene spent $20 billion on adding oncology and hemophilia-based drugs to their portfolio through acquisitions. These actions may encourage other biotech companies to become more active in 2018 as well. As an example, Gilead had become less active through years, but in 2015, the company raised US$10 billion in debt, increasing analysts' expectations that the big biotech would pursue an acquisition, but in 2016, they did not announce any major transaction. Consequentially, the US biotech sector dropped by 22%, led by Gilead (down 35%) due to investors' lack of trust on the company finding a new growth engine (Spence and Giovannetti 2017).

17.4 Conclusion

This chapter sheds a new light on the factors that lead the biotechnology industries to success. The methodology differs from other studies concerning the same subject through two major points: first, we use the enterprise value (EV) as a measure of reference to compare the US biotech companies, and second, the sample chosen consists of biotech companies listed on NASDAQ-100 index – the 100 largest, most actively traded US companies on NASDAQ. The biotechnology companies contributed massively to the index by the end of 2016 through 10 companies, which were the object of this chapter.

It has been noted that the investment on intellectual property rights is an essential step as inventions take longer to achieve the developmental phase. Otherwise, companies can use that to raise capital in the early stage of development. Another interesting point for early-stage companies is the massive investment in R&D, even when there is no revenue. After a drug's approval, companies start earning revenue that is massively reinvested in R&D to develop more drugs from its pipeline, culminating in a rising cycle. The merger and acquisition are an essential step for biotech companies in whatever stage. Otherwise, the big acquisitions of companies with valuable assets are predominantly a transaction to big biotech companies, and it is also necessary to guarantee the investors' trust. In summary, the results discussed in this paper will contribute to R&D strategies for biotechnology industry, not only restricted to those with IPOs, but also start-ups and smaller companies, in order to drive their innovation management decisions that will consequently result in its growth.

Acknowledgments This research was partially funded by São Paulo Research Foundation (FAPESP) grant number 2014/22500-8 and by Coordination for the Improvement of Higher Education Personnel (CAPES) grant number BEX13297/13-9.

References

A. H. CFA. Enterprise Value (EV), *Investopedia*, 18-Nov-2003. [Online]. Available: https://www.investopedia.com/terms/e/enterprisevalue.asp. Accessed 23 Jan 2018.

Awards & Recognition, *Celgene*, 26 Apr 2016. [Online]. Available: http://www.celgene.com/about/awards-recognition/. Accessed 16 Jan 2018.

Bilello, J. A. (2005). The agony and ecstasy of 'OMIC' technologies in drug development. *Current Molecular Medicine, 5*(1), 39–52.

Biotech blastoff: Synageva up 112% on $8B Alexion deal, USA TODAY. [Online]. Available: https://www.usatoday.com/story/money/business/2015/05/06/alexion-synageva/70909976/. Accessed 26 Jan 2018.

Campbell, T. Is this Gilead sciences' biggest mistake in 2016?, *The Motley Fool*, 16-Dec-2016. [Online]. Available: https://www.fool.com/investing/2016/12/16/is-this-gilead-sciences-biggest-mistake-in-2016.aspx. Accessed 26 Jan 2018.

Carlson, R. (2016). Estimating the biotech sector's contribution to the US economy. *Nature Publishing Group, 34*(3), 247–255.

Celgene swings back to profits in 1st qtr, *The Pharma Letter*, 05 Apr 2009. [Online]. Available: https://www.thepharmaletter.com/article/celgene-swings-back-to-profit-in-1st-qtr. Accessed 15 Jan 2018.

Celgene to buy Abraxis BioScience for $2.9 billion, *Reuters*, 30 Jun 2010.

Chandra, T. (2017a). Biotech bonanza: Biotech stocks to keep shining in 2018, Seeking Alpha.

Chandra, T. (2017b). Biotech: Opportunity ahead!, Seeking Alpha.

Chullen, A., Kaltenbrunner, H., & Schwetzler, B. (2015). Does consistency improve accuracy in multiple—based valuation? *Journal of Business Economics, 85*(6), 635–662.

Clarivate Analytics. (2018). Derwent innovation. [Online]. Available: https://clarivate.com/products/derwent-innovation/. Accessed 16 Jan 2018.

DiMasi, J. A., Grabowski, H. G., & Hansen, R. W. (2016). Innovation in the pharmaceutical industry: New estimates of R&D costs. *Journal of Health Economics, 47*, 20–33.

Egan, M. (2015). Stockswatch: Hillary Clinton tweet crushes biotech stocks, CNN Money.

Erden, Z., Klang, D., Sydler, R., & von Krogh, G. (2015). 'How can we signal the value of our knowledge?' Knowledge-based reputation and its impact on firm performance in science-based industries. *Long Range Planning, 48*(4), 252–264.

Ernst & Young. (2016). "Beyond borders: Returning to Earth," *Biotechnoogy Reports Beyond*. pp. 1–86.

Evens, R. P. (2016). Pharma success in product development—does biotechnology change the paradigm in product development and attrition. *The AAPS Journal, 18*(1), 281–285.

Evens, R. P., & Kaitin, K. I. (2014). The biotechnology innovation machine: A source of intelligent biopharmaceuticals for the pharma industry—mapping biotechnology's success. *Clinical Pharmacology and Therapeutics, 95*(5), 528–532.

Grand View Research. (2017). Biotechnology market analysis by application (health, food & agriculture, natural resources & environment, industrial processing bioinformatics), by technology, and segment forecasts, 2014–2025, San Francisco.

Haeussler, C., Patzelt, H., & Zahra, S. A. (2012). Strategic alliances and product development in high technology new firms: The moderating effect of technological capabilities. *Journal of Business Venturing, 27*(2), 217–233.

Harlin, M. B., & O'Connor, K. A. (2008). Leveraging your biotech intellectual property. *Nature Biotechnology, 26*(6), 607–609.

How a pharmaceutical firm priced its life-saving drug at $500K a year, CBC News. [Online]. Available: http://www.cbc.ca/news/health/how-pharmaceutical-company-alexion-set-the-price-of-the-world-s-most-expensive-drug-1.3125251. Accessed 26 Jan 2018.

Investopedia. Using enterprise value to compare companies, *Forbes*. [Online]. Available: https://www.forbes.com/sites/investopedia/2012/11/15/using-enterprise-value-to-compare-companies/. Accessed 23 Jan 2018.

Keating, G. M. (2015). Ledipasvir/Sofosbuvir: A review of its use in chronic hepatitis C. *Drugs, 75*(6), 675–685.

King, M. R., & Segal, D. (2008). Market segmentation and equity valuation: Comparing Canada and the United States. *Journal of International Financial Markets Institutions and Money, 18*(3), 245–258.

MacDiarmid, J., Tholana, T., & Musingwini, C. (2018). Analysis of key value drivers for major mining companies for the period 2006–2015. *Resources Policy, 56*, 16–30.

Mariani, G., Carlesi, A., & Scarfò, A. A. (2018). Academic spinoffs as a value driver for intellectual capital: The case of the University of Pisa. *Journal of Intellectual Capital, 19*(1), 202–226.

Misra, M. (2012). Biosimilars: Current perspectives and future implications. *Indian Journal of Pharmacology, 44*(1), 12.

Moorkens, E., Meuwissen, N., Huys, I., Declerck, P., Vulto, A. G., & Simoens, S. (2017). The market of biopharmaceutical medicines: A snapshot of a diverse industrial landscape. *Frontiers in Pharmacology, 8*(JUN), 314.

Morgan, S., Grootendorst, P., Lexchin, J., Cunningham, C., & Greyson, D. (2011). The cost of drug development: A systematic review. *Health Policy, 100*(1), 4–17.

Moyer, C. & Efram, S. (2017). NASDAQ index research: Biotec is booming - Nasdaq's Biotech index shows impressive performace, *Nasdaq Global Information Services*, [Online]. Available: http://business.nasdaq.com/marketinsite/2017/Nasdaq-Index-Research-Biotech-is-Booming.html. Accessed 11 Dec 2017.

Nasdaq. (2016a). In: "Nasdaq 100 – Nasdaq.com," *nasdaq.com*. [Online]. Available: http://www.nasdaq.com/markets/indices/nasdaq-100.aspx. Accessed 21 Nov 2016.

Nasdaq. (2016b). In: Biotech leading the NASDAQ-100 index performance, *Nasdaq*. [Online]. Available: http://business.nasdaq.com/marketinsite/2016/biotech-leading-the-Nasdaq-100-index-performance.html. Accessed 12 Jan 2016.

Njowa, G., & Musingwini, C. (2018). A framework for interfacing mineral asset valuation and financial reporting. *Resources Policy, 56*, 3–15.

PhRMA. (2013). *Medicines in development: Biologics*, Washington, DC.

Quintana-García, C., & Benavides-Velasco, C. A. (2016). Gender diversity in top management teams and innovation capabilities: The initial public offerings of biotechnology firms. *Long Range Planning, 49*(4), 507–518.

Rader, R. A. (2008). (Re)defining biopharmaceutical. *Nature Biotechnology, 26*(7), 743–751.

Regalado, A. EmTech: Record number of genomes sequenced in 2014, *MIT Technology Review*. [Online]. Available: https://www.technologyreview.com/s/531091/emtech-illumina-says-228000-human-genomes-will-be-sequenced-this-year/. Accessed 26 Jan 2018.

Ribal, J., Blasco, A., & Segura, B. (2010). Estimation of valuation multiples of Spanish unlisted food companies. *Spanish Journal of Agricultural Research, 8*(3), 547–558.

Roijakkers, N., & Hagedoorn, J. (2006). Inter-firm R&D partnering in pharmaceutical biotechnology since 1975: Trends, patterns, and networks. *Research Policy, 35*(3), 431–446.

Russo, G. (2008). Prospects: The ongoing financial crisis will affect the biotechnology sector — but the sector's used to it. *Nature, 455*(7217), 1273–1273.

Sareewiwatthana, P., & Janin, P. (2017). Tests of quantitative investing strategies of famous investors: Case of Thailand. *Investment Management and Financial Innovations, 14*, 218–226.

Spence, P. & Giovannetti, G. T. (2017). Biotechnology report 2017 – beyond borders – staying the course.

Thomson Reuters. (2018). Thomson Reuters EIKON. [Online]. Available: financial.thomsonreuters.com/Thomson/Eikon%0A. Accessed 16 Jan 2018.

U.S. FDA declines to approve Eli Lilly and Incyte arthritis drug Olumi, *Reuters*, 17 Apr 2017.

Wagner, S., & Cockburn, I. (2010). Patents and the survival of Internet-related IPOs. *Research Policy, 39*(2), 214–228.

Walsh, G. (2014). Biopharmaceutical benchmarks 2014. *Nature Biotechnology, 32*(10), 992–1000.

Williams, S. The 3 underlying reasons Why Gilead Sciences, Inc. Could remain wildly successful," *The Motley Fool*, 03 Jan 2015. [Online]. Available: https://www.fool.com/investing/general/2015/01/03/the-3-underlying-reasons-why-gilead-sciences-could.aspx. Accessed 26 Jan 2018.

Xconomy: Acetylon crafts new buyout deal with Celgene, spins out startup regenacy," *Xconomy*, 02 Dec 2016. [Online]. Available: https://www.xconomy.com/boston/2016/12/02/acetylon-crafts-new-buyout-deal-with-celgene-spins-out-startup-regenacy/. Accessed 19 Jan 2018.

Z. I. Research. Alexion seeks label expansion, *Zacks Investment Research*. [Online]. Available: https://www.zacks.com/stock/news/50969/alexion-seeks-label-expansion. Accessed 26 Jan 2018.

Appendix 1

Drone Package Deliveries

Amazon Prime Air Technology Roadmap

Corban, Horatiu
Tuite, Nicole
Singireddy, Shiva Ram Reddy
Rapolu, Kranthi Kumar Reddy
Sathappan, Sivagamasundari

Prime Air Proposal / Application

- Drone delivery to customers in 30 minutes or less
- Warehouse to home package delivery
- Home delivery proposed
- Amazon owned drones
- Vertical integration

T. U. Daim (ed.), *Roadmapping Future*, Applied Innovation and Technology Management, https://doi.org/10.1007/978-3-030-50502-8

Prime Air Proposal

- In July 2015, Amazon met at NASA's Ames Research Center to lay out their first public proposal for safely operating a drone highway.
- Amazon drones would operate in a proposed "high speed" transit lane between 200 & 400 feet.
- 100 foot buffer zone between high speed transit and integrated air space.
- Below that corridor, amateur or hobby sUAS would operate freely.

Drones proposed would have:

- advanced GPS systems
- online flight planning
- Ability to communicate with other drones.

PrimeAir Opportunities

- Developing aerial vehicles that travel over 50 miles per hour & carry 5- pound payloads, which **cover 86% of products sold on Amazon**
- Amazon ships approx. 608 million packages per year. **Drone market at 86% extends to 522 million packages**
- Shipping revenue is $4.48 billion, and shipping costs $8.7 billion, so the **opportunity to vertically integrate their supply chain is over $4 billion dollars**

http://www.scdigest.com/ontarget/14-04-30-1.php?cid=8012

August 2015- Amazon Executive TRM Workshops

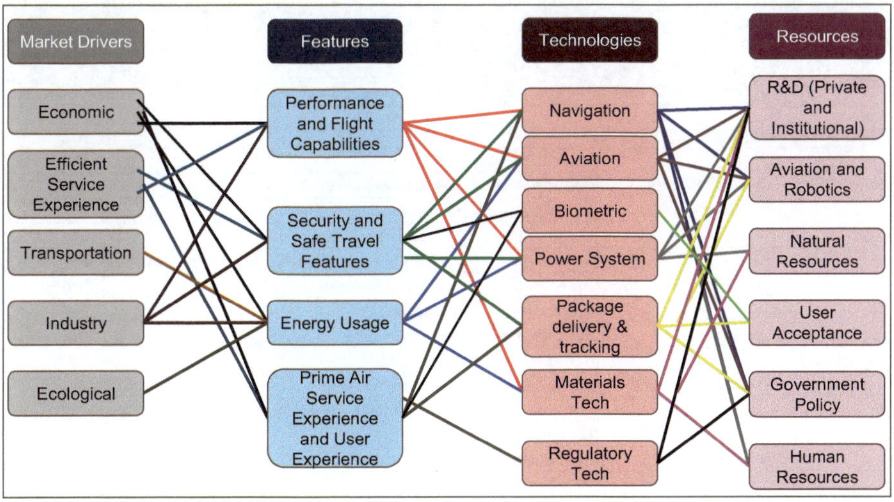

Market Drivers and Product Features

Product Feature Vs Market Drivers	
Strong Relationship	(dark green)
Medium Relationship	(light green)
Weak Relationship	(purple)
Priority	
High	(red)
Medium	(blue)
Low	(orange)

Product Features			Market Drivers													
			Efficiency				Economic		Transportation		Ecological		Industry			
			30- minute delivery	Real-time information of delivery	Last mile problem	Delivery possible in remote locations	Vertically integrate supply chain ($4 billion dollar loss on shipping)	Deliver 86% of Amazon products (5 lbs or less)	Commercial drone highway	Reduced road traffic	Reduced carbon emmisions	Potential for eco-fuel drones	Commercial needs (agriculture, medical) will drive UAV usage	Untapped Market	Expansion of business	
			D1	D2	D3	D4	D5	D6	D7	D8	D9	D10	D11	D12	D13	Priority
Drone Performance and Flight Capabilities	Motor and Blade	F1	2		3	2	3	2	2	2			2			18
	Lightweight materials	F2	2				2	3								7
	Package weight	F3						3						3	3	9
	Delivery range	F4	3		3	3	3	3		2	2		2	2	3	26
Secure and Safe Travel Features	UAV rescue	F5				2			2							4
	UAV to UAV communication	F6		2					2				2			6
	Guidance system	F7			2				3							5
Energy Usage	Power module	F8	2		3	2		2	3			2				14
Prime Air Service and User Experience	Prime Air Membership/Prime Air Standard	F9					3	3						3	3	12
	Smart connectivity	F10	2	3												5
	Delivery location	F11	3	3	3	3	2							2	2	18

Market Drivers

MARKET DRIVERS		CODE
Speed and Efficiency of Deliveries	30- minute delivery	D1
	Real-time information of delivery	D2
	Last mile problem- convoluted routes	D3
	Delivery possible in remote locations	D4
Economic	Vertically integrate supply chain ($4 billion dollar loss on shippin	D5
	Deliver 86% of Amazon products (5 lbs or less)	D6
Transportation	Commercial drone highway	D7
	Reduced road traffic	D8
Ecological	Reduced carbon emmisions	D9
	Potential for eco-fuel drones	D10
Industry	Commercial needs (agriculture, medical) will drive UAV usage	D11
	Untapped Market	D12
	Expansion of business	D13

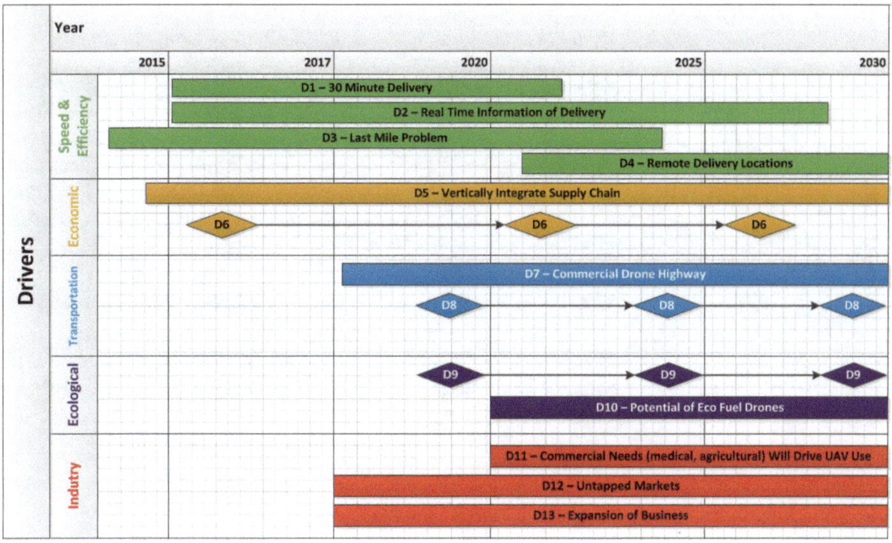

Product Features

PRODUCT FEATURES		CODE	2015	2017	2020	2025	2030
Drone Performance and Flight Capabilities	Motor&Blade	F1		Evolving			
	Lightweight Materials	F2		Evolving			
	Package Weight	F3	5lbs		7lbs	9lbs	11lbs
	Delivery Range	F4	10Miles		15Miles	20Miles	25Miles
Secure and Safe Travel Features	UAV to UAV Communication	F5			Evolving		
	UAV Rescue	F6			Evolving		
	Guidance System	F7		Evolving			
Energy Usage	Power Module	F8		Evolving			
Prime Air Service and User Experience	Prime Air Membership/Prime Air Standard	F9		Available for Prime Air Members Only	Drone delivery Method-Standard, Traditional Delivery(UPS, USPS, FedEX, OnTrac, etc.) - Paid		
	Smart Connectivity	F10		Evolving			
	Delivery Location	F11		Evolving			

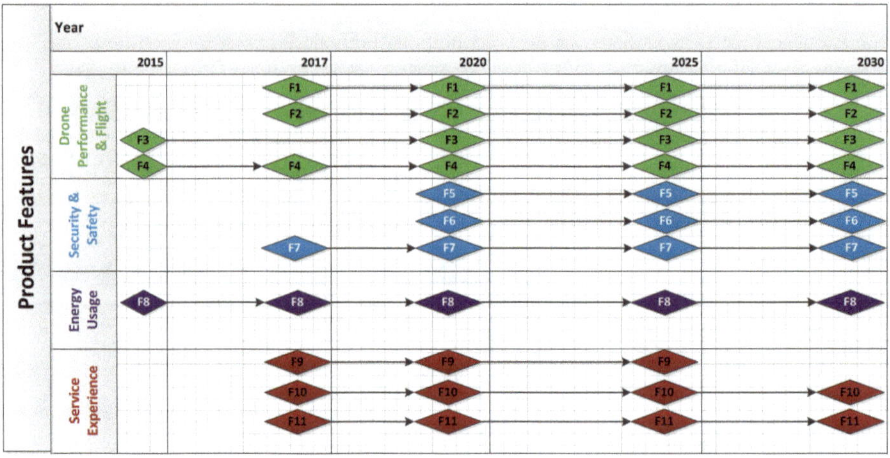

Technologies

TECHNOLOGY		CODE	2015	2017	2020	2025	2030
Regulatory Technologies	Geofencing and geographical boundaries	T1		Evolving			
	Encryption of data	T2		Evolving			
	Noise reduction technology	T3		Evolving			
	Operating system	T4		Evolving			
Navigation Technologies	Onboard Navigation	T5		Evolving			
	Lasers	T6			Evolving		
	Video Cameras	T7		Evolving			
	Infared cameras	T8		Evolving			
Aviation Technologies	Higher intensity LED lights- orientation during flight	T9		Evolving			
	Robotics (Improved arms, legs)	T10			Evolving		
	Remote Monitoring	T11	Evolving				

Technologies

TECHNOLOGY		CODE	2015	2017	2020	2025	2030
Power System	Li-ion	T12		Evolving			
	Li-po batteries	T13				Evolving	
	Solar power	T14				Evolving	
	Hydrogen fuel cell	T15					Available
Package Delivery Technologies	Barcodes and QR codes	T16		Evolving			
	Logistics	T17			Evolving		
	Inventory management (SAP)	T18			Evolving		
	RFID	T19				Evolving	
Biometric Technologies	Face recognition tech	T20				Available	
	Retina Scan	T21					Available
	Fingerprint	T22			Available		
Materials Technologies	Aluminum	T23		Available			
	Titanium	T24			Available		
	Magnesium	T25				Available	
	Graphite	T26					Available

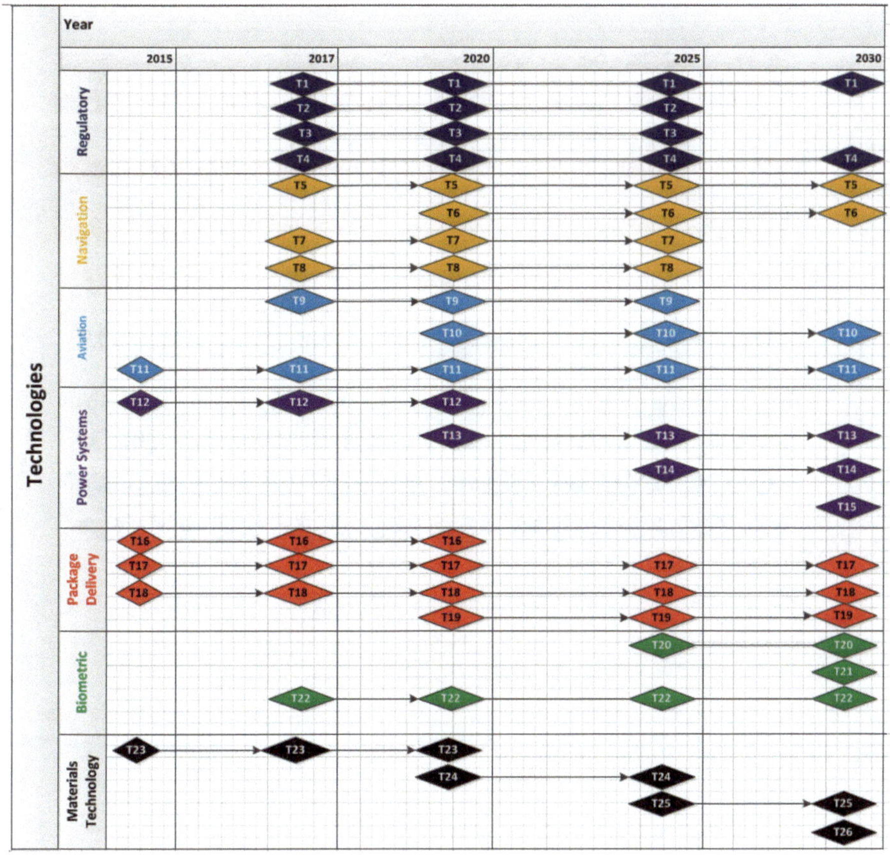

Resources

RESOURCE	CODE
Finance	R1
Research and Development - Amazon Seattle Laboratory	R2
Universities	R3
Government Organization and Policies - FAA	R4
Aviation and Robotics Industry - Boeing	R5
Natural Resources	R6
User acceptance/Customers	R7
Human Resource	R8
Material Scientist	R9
Astronaut - NASA	R10

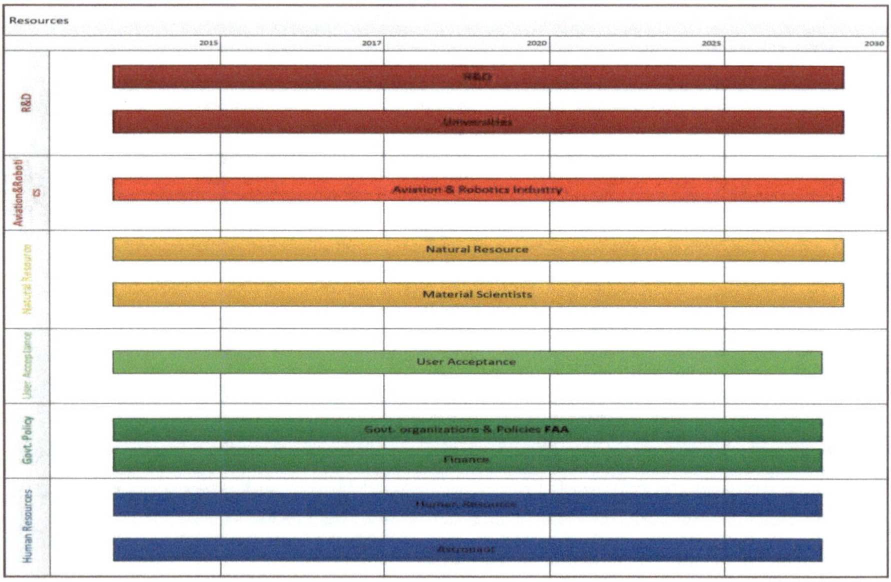

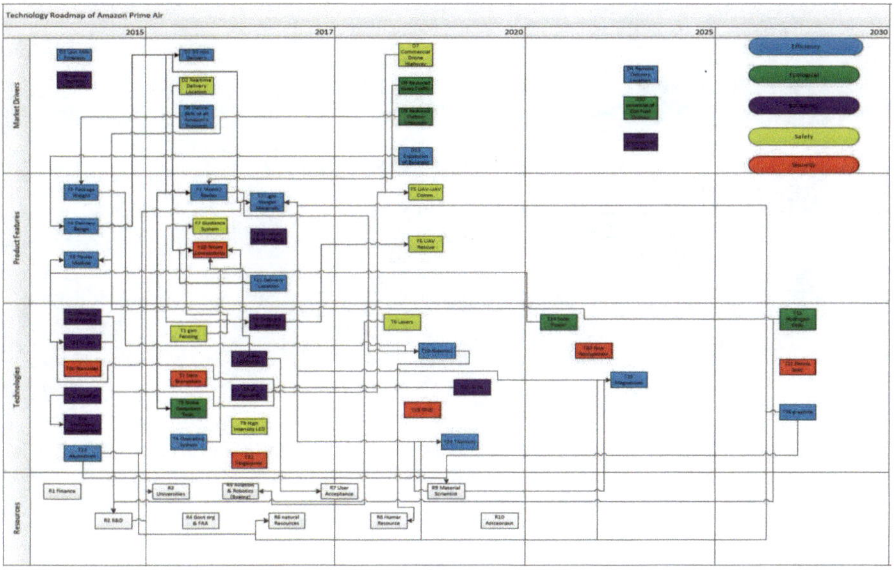
Technology Roadmap of Amazon Prime Air
2015
2017
2020
2025
2030
Market Drivers
Product Features
Technologies
Resources
Safety

Questions?

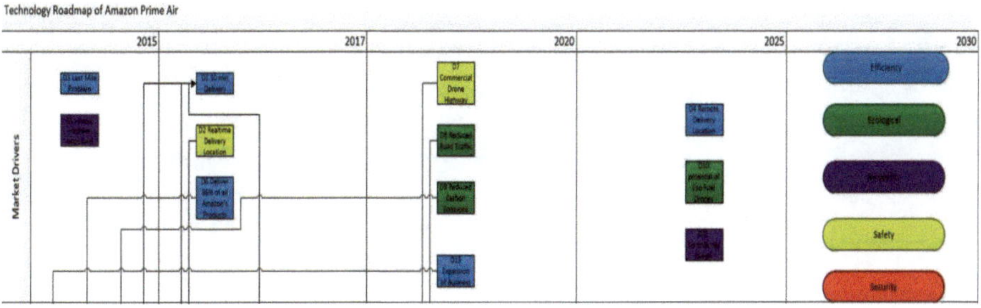
Technology Roadmap of Amazon Prime Air
2015
2017
2020
2025
2030
Market Drivers
D7 Commercial Drone Highway
D2 Realtime Delivery Location
Efficiency
Safety

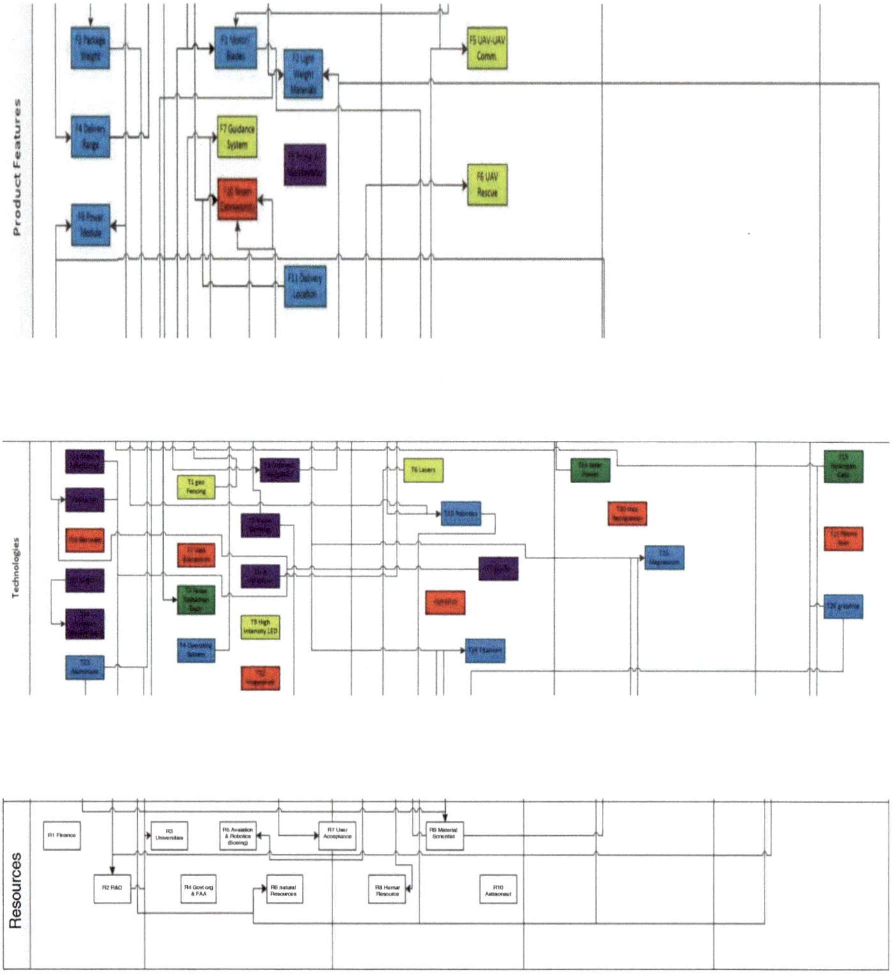
Product Features
F3 Package Weight
F1 Motor/ Blades
F2 Light Weight Materials
F5 UAV-UAV Comm.
F4 Delivery Range
F7 Guidance System
F6 UAV Rescue
F8 Power Module
F11 Delivery Location
Technologies
T1 geo Fencing
T6 Lasers
T9 High Intensity LED
Resources
R1 Finance
R3 Universities
R5 Avaiation & Robotics (Boeing)
R7 User Acceptance
R9 Material Scientist
R2 R&D
R4 Govt org & FAA
R6 natural Resources
R8 Human Resource
R10 Astronaut

Appendix

PRODUCT FEATURES	DEFINITION
Motor&Blade	Propeller spinning at 10,000 revolutions per minute.Propellers driven by 8 motors to lift the UAV. Funtional even if one fails.
Lightweight Materials	Drones will be made of lightweight materials which are strong enough to handle flight and payload. Idea is to maximize load capacity and minimize drone weight.
Package Weight	Limited to small and five pound package
Delivery Range	Delivery range is restricted to 10miles form the distribution centre.
UAV to UAV Communication	Provide information on weather, landing condition, traffic. This information is used for route planning to modify the actual route.
UAV Rescue	Drones packages can be rescued/ picked up by a replacement drone in the event of a malfunction. Drones can be picked up by replacement drone and brought back to shop.
Guidance System	Guidance is a revolutionary visual sensing system that has a powerful processing core, integrated visual cameras and ultrasonic sensors to detect object up to 65 feet(20 meters) away.
Power Module	Provide power for UAV control system and propeller motor. One or more power modules configured such that it is autonomously removed/replaced with another. Container may also include power module which can be utilized on engagement. When the container is disengaged, it used the power from UAV power module.
Prime Air Membership/Prime Air Standard	PrimeAir to be offered on an additional fee basis to "early adopters". X amount per year.
Smart Connectivity	Re-route to find package recipient through their smart phone.
Delivery Location	Products are delivered to the preferred location(Home,Work place, Current location - Deteremined by GPS data,location of the wireless network)

Appendix

TECHNOLOGY	DEFINITION
Geofencing and geographical boundaries	The use of GPS or RFID technology to create a virtual geographic boundary, enabling software to trigger a response when a mobile device enters or leaves a particular area
Encryption of data	The process of converting information or data into a code, especially to prevent unauthorized access.
Noise reduction technology	Noise control or noise mitigation is a set of strategies to reduce noise pollution or to reduce the impact of that noise, whether outdoors or indoors
Operating system	Software that supports a computer's basic functions, such as scheduling tasks, executing applications, and controlling peripherals
Onboard Navigation	Navigation system which is situated on a ship, craft, or other vehicle.
Lasers	Device that generates an intense beam of coherent monochromatic light (or other electromagnetic radiation) by stimulated emission of photons from excited atoms or molecules. Can be used in alignment or guidance.
Video Cameras	A camera for recording images on videotape or for transmitting them to a monitor screen.
Infared cameras	A camera which can view and observe light beyond the red end of the visbile spectrum. Able to identify heated objects.
Higher intensity LED lights- orientation during flight	Light-emitting diode, a semiconductor diode that glows when a voltage is applied

Appendix

TECHNOLOGY	DEFINITION
Robotics (Improved arms, legs)	The branch of technology that deals with the design, construction, operation, and application ofrobots
Remote Monitoring	Monitoring and controlling of disparate devices from a network operations center or control room and the ability to change the operation of these devices from that central office
Li-ion	A member of a family of rechargeable battery types in which lithium ions move from the negative electrode to the positive electrode during discharge and back when charging
Li-po batteries	A rechargeable battery of lithium-ion technology in a pouch format
Solar power	Power obtained by harnessing the energy of the sun's rays
Hydrogen fuel cell	A device that converts the chemical energy from a fuel (hydrogen) into electricity through a chemical reaction of positively charged hydrogen ions with oxygen or another oxidizing agent
Barcodes and QR codes	A machine-readable code in the form of numbers and a pattern of parallel lines of varying widths, printed on and identifying a product
Logistics	The detailed coordination of a complex operation involving many people, facilities, or supplies
Inventory management (SAP)	The overseeing and controlling of the ordering, storage and use of components that a company will sell

Appendix

TECHNOLOGY	DEFINITION
RFID	Radio Frequency Identificaion tags. Would be used to scan packages (near-field) and ensure the drone handles the right package.
Face recognition tech	A computer application for automatically identifying or verifying a person from a digital image or a video frame from a video source
Retina Scan	A biometric technique that uses the unique patterns on a person's retina blood vessels.
Fingerprint	A system which automatically compares a fingerprint to those stored on a database, to determine if there is one that matches
Aluminum	A strong, light, and corrosion resistant metal used in engineering applications. It is the most abundant metal in the earth's crust.
Titanium	The chemical element of atomic number 22, a hard silver-gray metal of the transition series, used in strong, light, corrosion-resistant alloys.
Magnesium	The chemical element of atomic number 12, a silver-white metal of the alkaline earth series. It is used to make strong lightweight alloys, especially for the aerospace industry
Graphite	A gray, crystalline, allotropic form of carbon that occurs as a mineral in some rocks and can be made from coke. It is used as a solid lubricant, in pencils, and as a moderator in nuclear reactors.

Appendix 2

Human Space Exploration Privatization

Anna Mathew

Daragh Finn

David Wigen

Joao Lavoie

Monticha Khammuang

Rafaa Khalifa

Slide 1

Engineering and Technology Management
MASEEH COLLEGE OF ENGINEERING AND COMPUTER SCIENCE

Portland State UNIVERSITY

T. U. Daim (ed.), *Roadmapping Future*, Applied Innovation and Technology Management, https://doi.org/10.1007/978-3-030-50502-8

Introduction

Problem Statement: Government space agencies are heavily burdened with the costs of developing and operating "space" programs

Objective: Provide Technology Roadmap for the privatization and commercialization of ***human space exploration and colonization of planets (HSE)*** within our solar system

Methodology: We have developed TRM for HSE by using analysis tools.

- A SWOT analysis
- A mkt drivers, c/s needs & pdt feature analysis grid
- A product feature & technology analysis grid
- A gap analysis to understand the product gap, technology gap, and resources gap.

Slide 2

Engineering and Technology Management
MASEEH COLLEGE OF ENGINEERING AND COMPUTER SCIENCE

Portland State UNIVERSITY

TRM Development Process

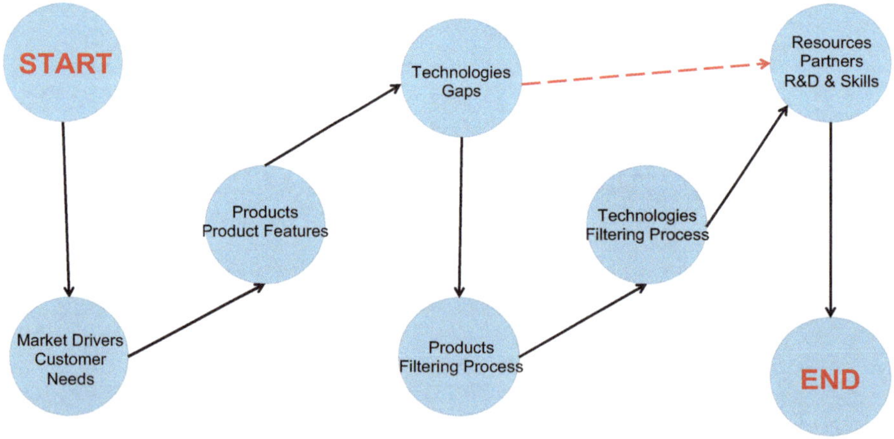

Slide 3

Engineering and Technology Management
MASEEH COLLEGE OF ENGINEERING AND COMPUTER SCIENCE

Portland State UNIVERSITY

Our TRM project targeted…..

Cargo

Crew Vehicle Systems

Space Habitate

Slide 4

Engineering and Technology Management
MASEEH COLLEGE OF ENGINEERING AND COMPUTER SCIENCE

Portland State UNIVERSITY

Products and Technologies after Filtering

- Products
  - P5: **Deep-Space Habitat**
  - P6: **Space Exploration Vehicle**
- Technologies
  - T6: **Improved thrust and in-space propulsion systems**
  - T9: **Ascent/Entry TPS**
  - T10: **Data Analysis for Decision Making**
  - T11: **Water Recovery and Management**
  - T12: **Radiation Mitigation and Biological Countermeasures**

Slide 5

Engineering and Technology Management
MASEEH COLLEGE OF ENGINEERING AND COMPUTER SCIENCE

Portland State UNIVERSITY

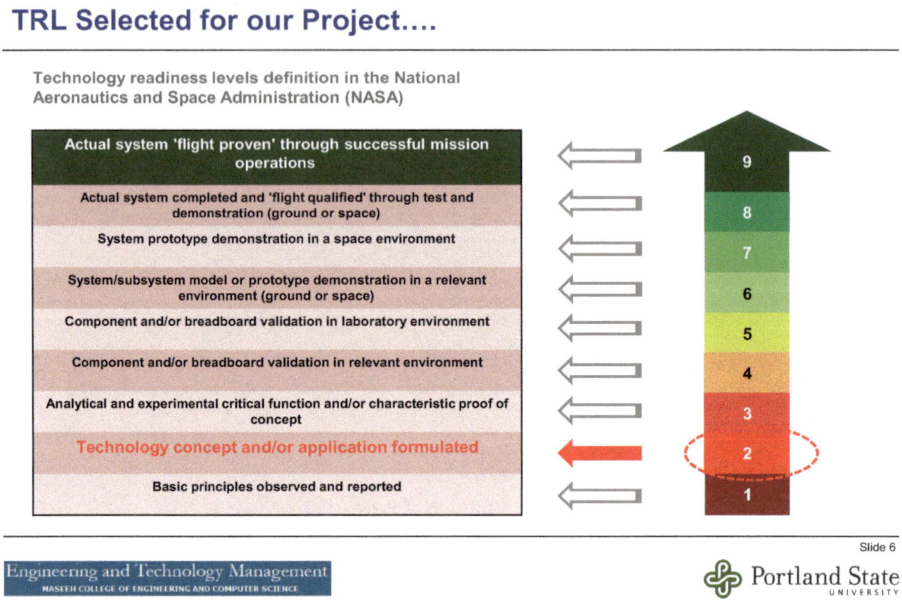
TRL Selected for our Project….
Technology readiness levels definition in the National Aeronautics and Space Administration (NASA)
Actual system 'flight proven' through successful mission operations
Actual system completed and 'flight qualified' through test and demonstration (ground or space)
System prototype demonstration in a space environment
System/subsystem model or prototype demonstration in a relevant environment (ground or space)
Component and/or breadboard validation in laboratory environment
Component and/or breadboard validation in relevant environment
Analytical and experimental critical function and/or characteristic proof of concept
Technology concept and/or application formulated
Basic principles observed and reported
9
8
7
6
5
4
3
2
1
Slide 6
Engineering and Technology Management
MASEEH COLLEGE OF ENGINEERING AND COMPUTER SCIENCE
Portland State
UNIVERSITY

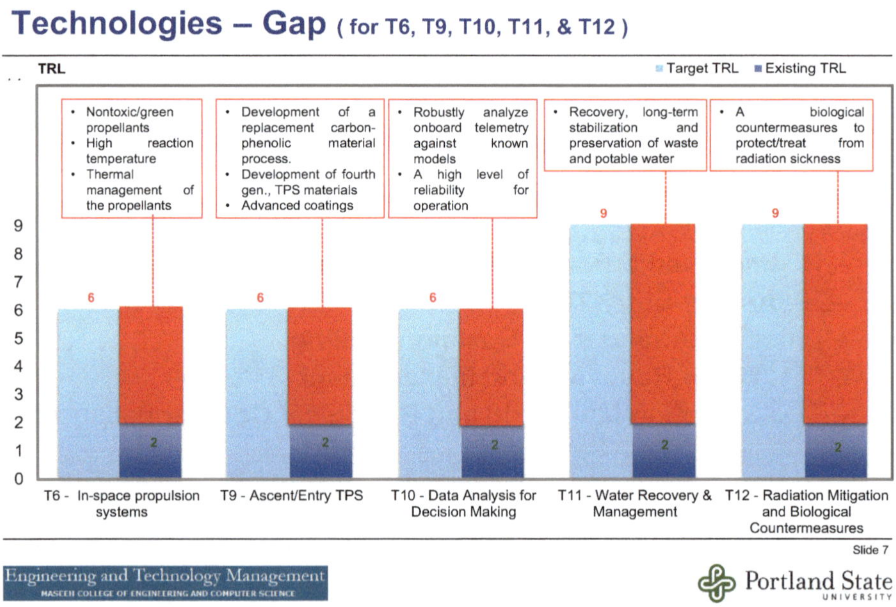
Technologies – Gap (for T6, T9, T10, T11, & T12)
TRL
Target TRL
Existing TRL
• Nontoxic/green propellants
• High reaction temperature
• Thermal management of the propellants
• Development of a replacement carbon-phenolic material process.
• Development of fourth gen., TPS materials
• Advanced coatings
• Robustly analyze onboard telemetry against known models
• A high level of reliability for operation
• Recovery, long-term stabilization and preservation of waste and potable water
• A biological countermeasures to protect/treat from radiation sickness
9
8
7
6
5
4
3
2
1
0
6
2
6
2
6
2
9
2
9
2
T6 - In-space propulsion systems
T9 - Ascent/Entry TPS
T10 - Data Analysis for Decision Making
T11 - Water Recovery & Management
T12 - Radiation Mitigation and Biological Countermeasures
Slide 7
Engineering and Technology Management
MASEEH COLLEGE OF ENGINEERING AND COMPUTER SCIENCE
Portland State
UNIVERSITY

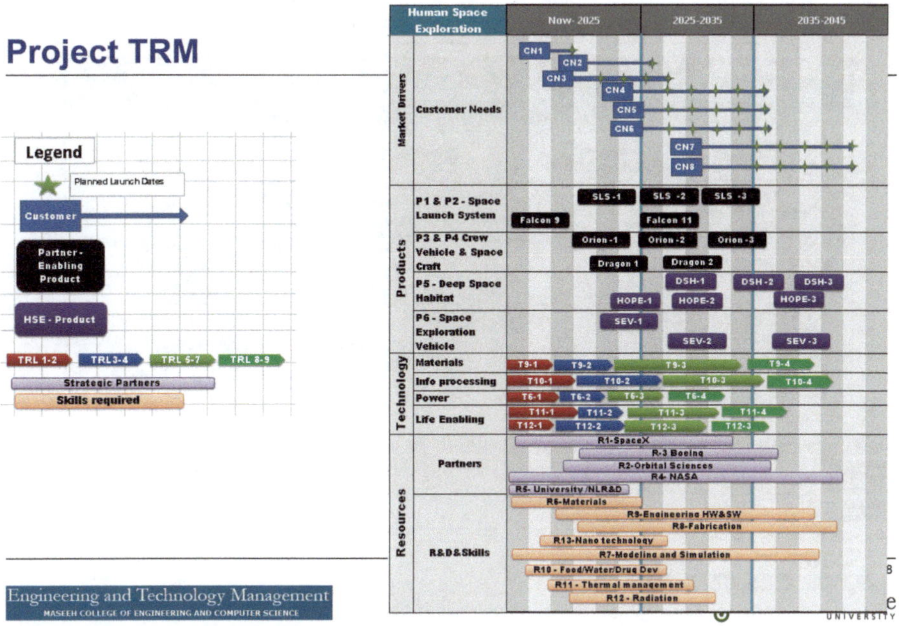

This slide shows the definition of product features

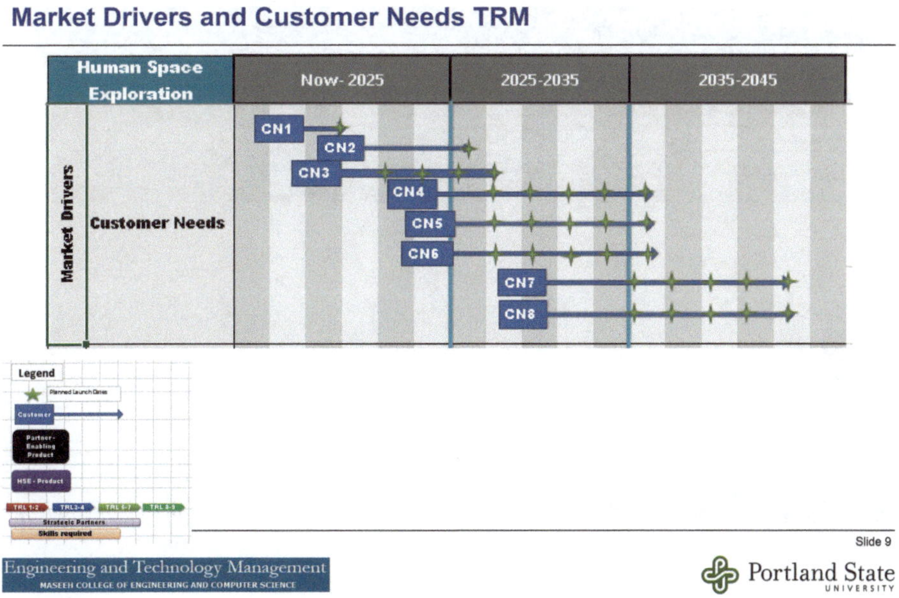

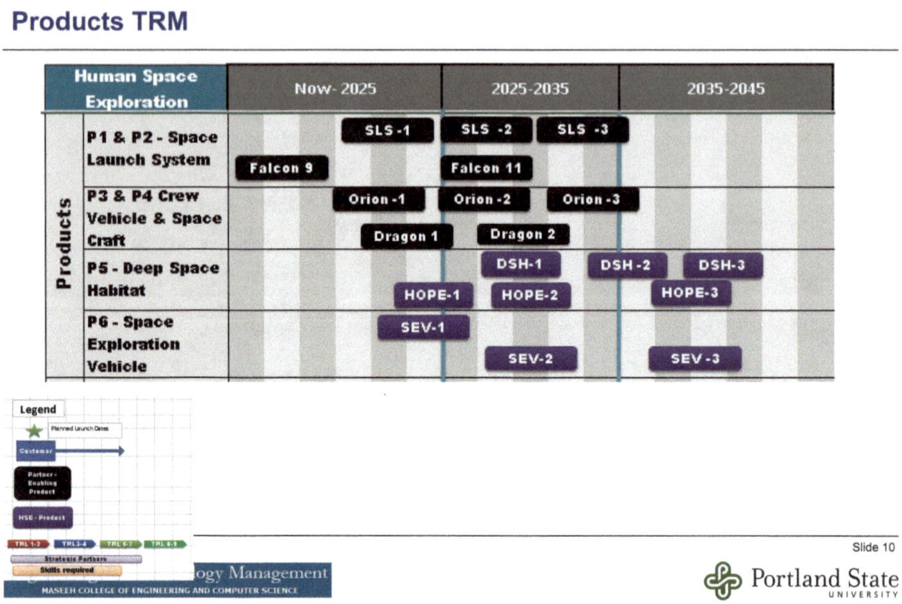
Products TRM
Human Space Exploration
Now- 2025
2025-2035
2035-2045
Products
P1 & P2 - Space Launch System
SLS -1
SLS -2
SLS -3
Falcon 9
Falcon 11
P3 & P4 Crew Vehicle & Space Craft
Orion -1
Orion -2
Orion -3
Dragon 1
Dragon 2
P5 - Deep Space Habitat
DSH-1
DSH-2
DSH-3
HOPE-1
HOPE-2
HOPE-3
P6 - Space Exploration Vehicle
SEV-1
SEV-2
SEV -3
Legend
Customer
HSE - Product
Skills required
Slide 10
Portland State
UNIVERSITY

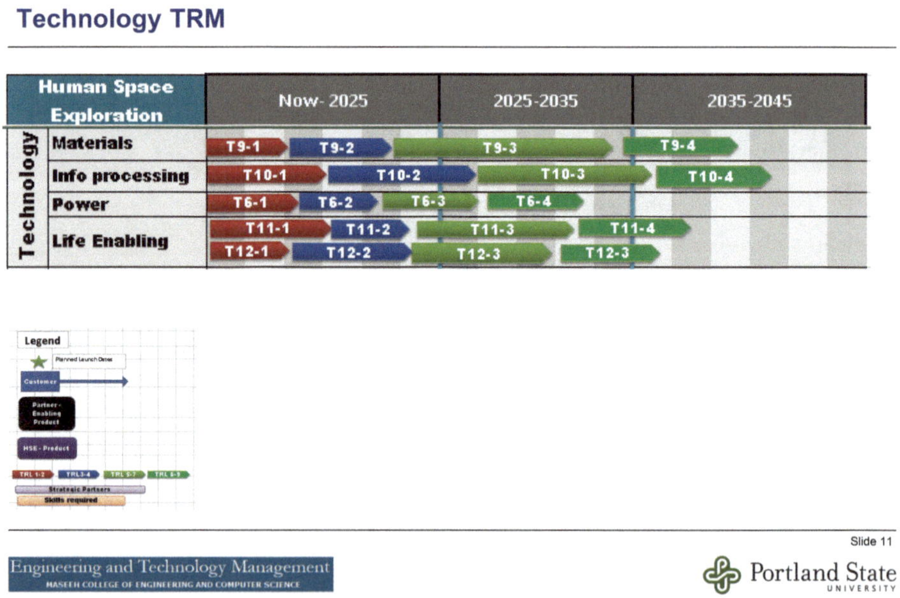
Technology TRM
Human Space Exploration
Now- 2025
2025-2035
2035-2045
Technology
Materials
T9-1
T9-2
T9-3
T9-4
Info processing
T10-1
T10-2
T10-3
T10-4
Power
T6-1
T6-2
T6-3
T6-4
Life Enabling
T11-1
T11-2
T11-3
T11-4
T12-1
T12-2
T12-3
T12-3
Legend
Customer
HSE - Product
Skills required
Slide 11
Engineering and Technology Management
MASEEH COLLEGE OF ENGINEERING AND COMPUTER SCIENCE
Portland State
UNIVERSITY

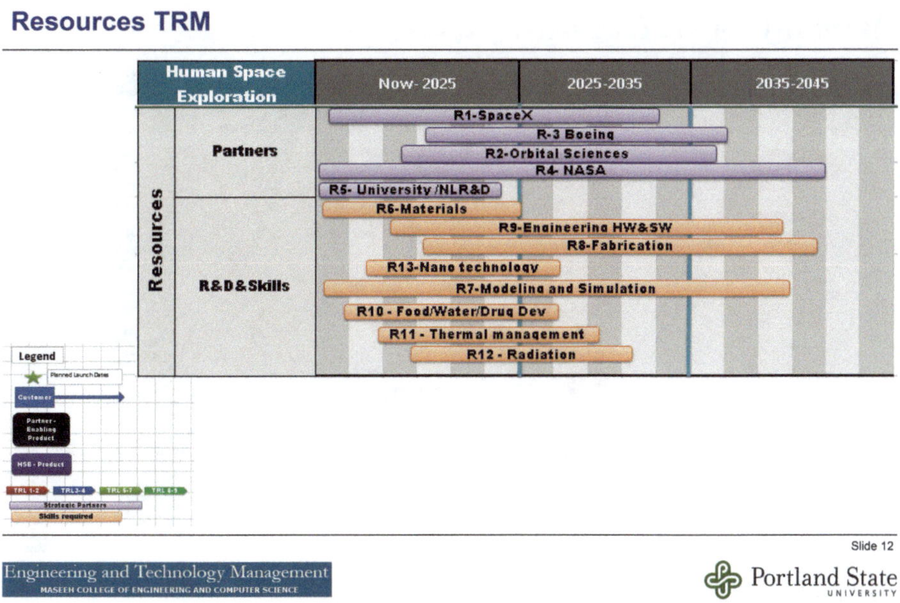

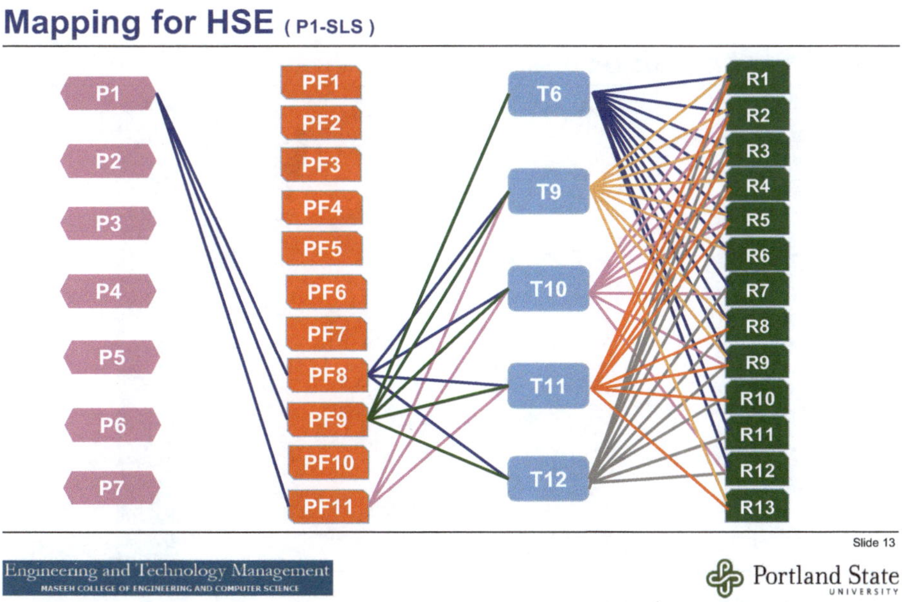

When explain to class, we have to start from Technology and expand to R, PF and finally with CN

Mapping for HSE (P2-Falcon)

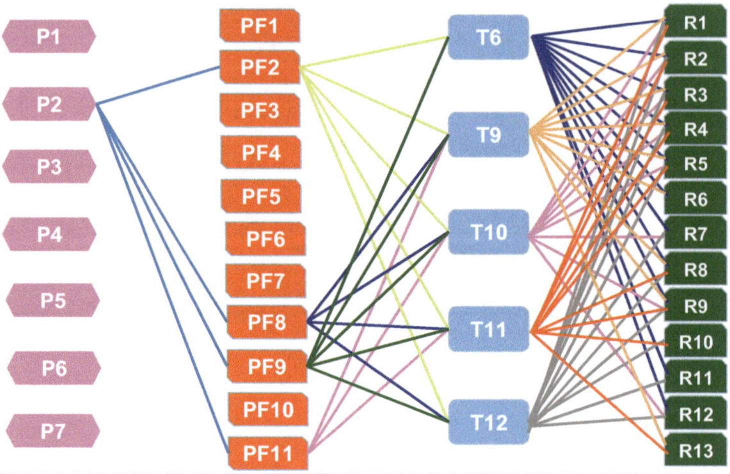

Slide 14

Engineering and Technology Management
MASEEH COLLEGE OF ENGINEERING AND COMPUTER SCIENCE

Portland State UNIVERSITY

When explain to class, we have to start from Technology and expand to R, PF and finally with CN

Mapping for HSE (P3-Orion)

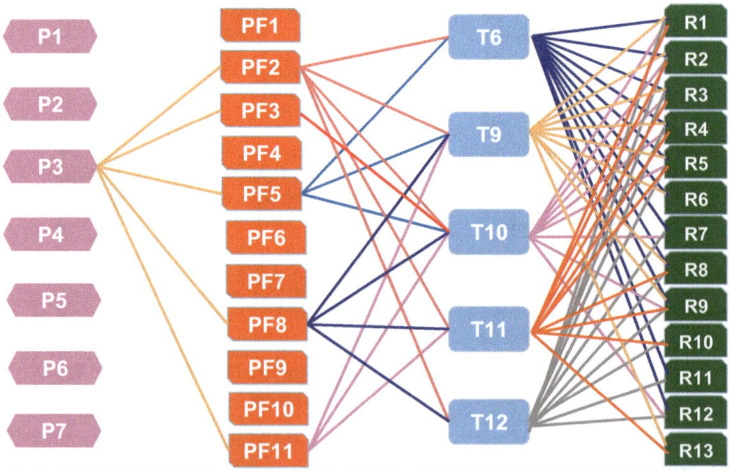

Slide 15

Engineering and Technology Management
MASEEH COLLEGE OF ENGINEERING AND COMPUTER SCIENCE

Portland State UNIVERSITY

When explain to class, we have to start from Technology and expand to R, PF and finally with CN

Mapping for HSE (P4-Dragon)

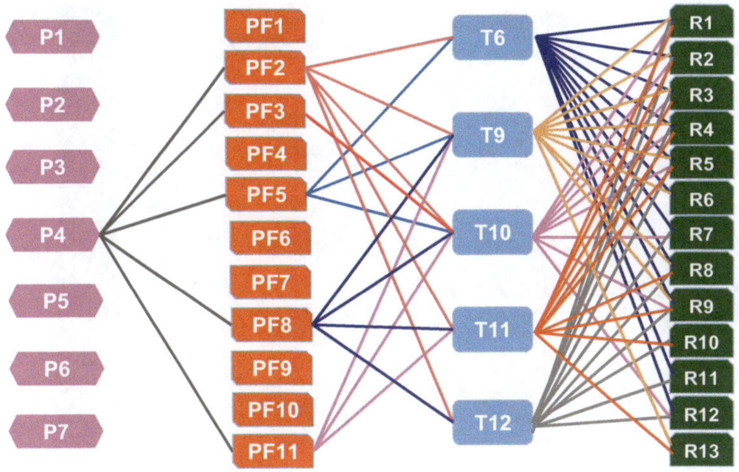

Slide 16

Engineering and Technology Management
MASEEH COLLEGE OF ENGINEERING AND COMPUTER SCIENCE

Portland State UNIVERSITY

When explain to class, we have to start from Technology and expand to R, PF and finally with CN

Mapping for HSE (P5-DSH)

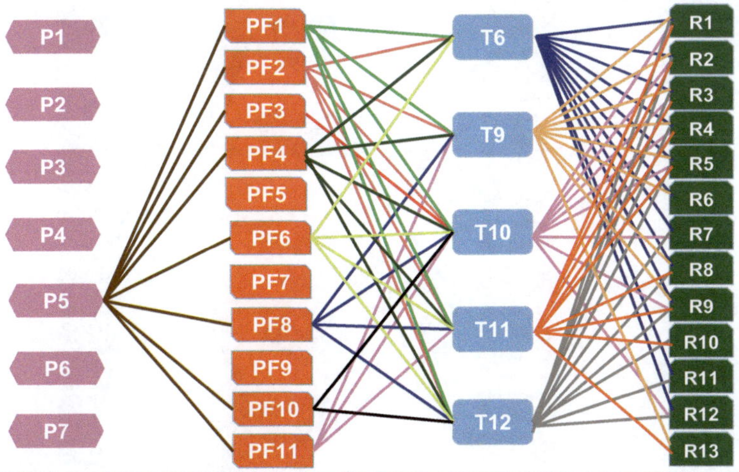

Slide 17

Engineering and Technology Management
MASEEH COLLEGE OF ENGINEERING AND COMPUTER SCIENCE

Portland State UNIVERSITY

When explain to class, we have to start from Technology and expand to R, PF and finally with CN

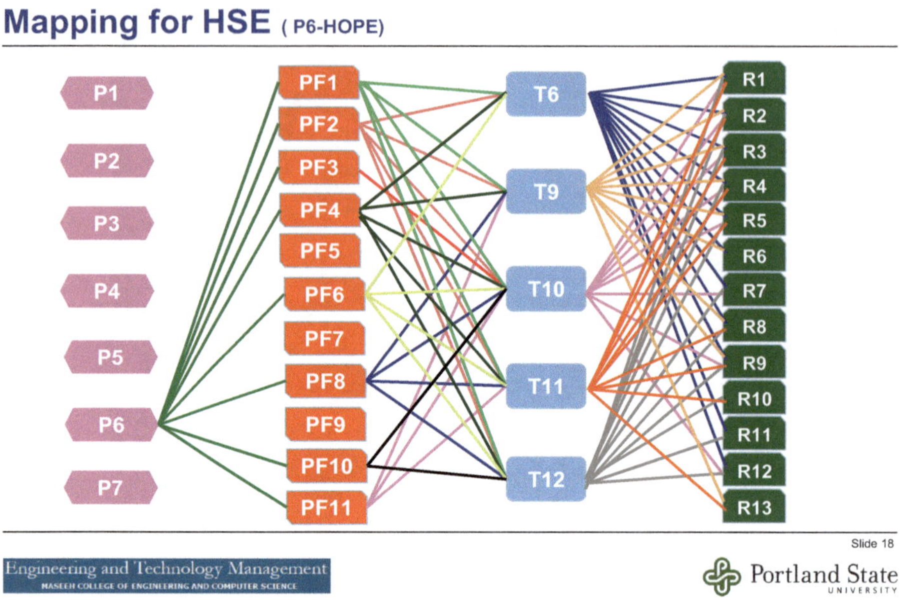

When explain to class, we have to start from Technology and expand to R, PF and finally with CN

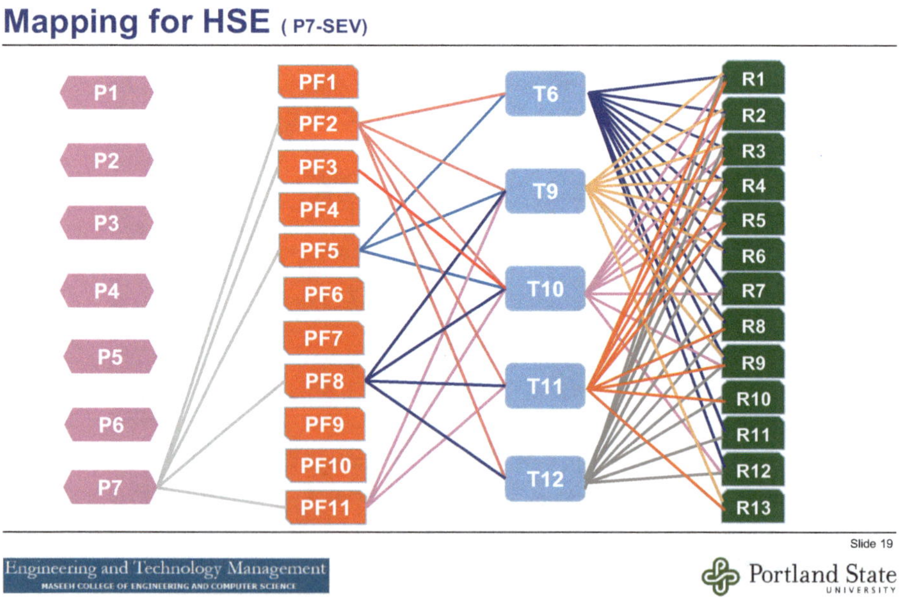

When explain to class, we have to start from Technology and expand to R, PF and finally with CN

Conclusion

TRM process is awesome!

Our team learn a lot about the process and the application of TRM

We also learned a little about HSE too!

Thank You….

Engineering and Technology Management
MASEEH COLLEGE OF ENGINEERING AND COMPUTER SCIENCE
Portland State UNIVERSITY

Market Driver Definition

Code	Market drivers
D1	Scientific activities & technology developing
D2	Quality of living
D3	Advanced data exploration
D4	Money / Economy
D5	Natural and energy resources
D6	Population growth
D7	Protect the planet /Extinction event
D8	Fulfill species
D9	New place to live / New earth asteroid

Slide 22

Engineering and Technology Management
MASEEH COLLEGE OF ENGINEERING AND COMPUTER SCIENCE

Portland State UNIVERSITY

This slide shows the definition of product features

Customer needs Definition

Code	Customer needs
CN1	The ability to lift and transport light and heavy cargo
CN2	The ability to return samples from Mars
CN3	The ability to redirect asteroids when necessary
CN4	The ability to land a crew on an asteroid
CN5	The ability to land a crew on the lunar surface
CN6	The ability to crew a flight to one of Mars moons
CN7	The ability to crew a flight that orbits Mars
CN8	The ability to land a crew on the surface of Mars

Slide 23

Engineering and Technology Management
MASEEH COLLEGE OF ENGINEERING AND COMPUTER SCIENCE

Portland State UNIVERSITY

This slide shows the definition of product features

Relation between Market Drivers & Customer Needs

Customer Needs	CN Code	Market Driver Code
The ability to lift and transport light and heavy cargo	CN1	D1, D4, D6, D9
The ability to return samples from Mars	CN2	D3, D8
The ability to redirect asteroids when necessary	CN3	D6, D7
The ability to land a crew on an asteroid	CN4	D1, D3, D4, D5, D8
The ability to land a crew on the lunar surface	CN5	D1, D2, D3, D6, D9
The ability to crew a flight to one of Mars moons	CN6	D3, D4, D5, D6, D8
The ability to crew a flight that orbits Mars	CN7	D3, D6, D8
The ability to land a crew on the surface of Mars	CN8	D1, D2, D3, D4, D5, D6, D8, D9

Slide 24

Engineering and Technology Management
MASEEH COLLEGE OF ENGINEERING AND COMPUTER SCIENCE

Portland State UNIVERSITY

This slide shows the relationship between product features and products

Products

	Products	Code	Provider
Cargo	Space Launch System (SLS)	P1	Space launch platform/NASA
	Falcon 9 heavy lift rocket Falcon 11 super heavy lift rocket	P2	SpaceX
Crew Vehicle Systems	Multi-purpose Crew Vehicle (MPCV) Orion-1,2,3	P3	Boeing/Orbital Sciences
	Dragon-1,2	P4	SpaceX
	Space Exploration vehicle SEV -1, 2, 3	P7	HSE
Space Habitate	Deep Space habitation vehicle DSH -1, 2, 3	P5	HSE
	Human off Planet Environment HOPE-1,2,3	P6	HSE

Slide 25

Engineering and Technology Management
MASEEH COLLEGE OF ENGINEERING AND COMPUTER SCIENCE

Portland State UNIVERSITY

This slide shows the information of existing and future products (Cargo and Advance creware existing/planned products, Space habitate is the future product)

Product Features

Code	Group	Product Features
PF1	Facility	Sustainability and Supportability
PF2	Performance	Powerful & sustainable energy sources
PF3	Information	Advance space communication and Navigation
PF4	Feature	Advance Space station and deep space habitation
PF5	Performance	Launching- Landing module/Advanced in-space propulsion
PF6	Information	In-situ resource utilization & Human-robotic systems
PF7	Information	Mobile Extravehicular activity & robotic platform
PF8	Safety	Safe for human
PF9	Feature	Orbital & deep space exploration
PF10	Feature	Space living community
PF11	Security	Advance security systems

Slide 26

Engineering and Technology Management
MASEEH COLLEGE OF ENGINEERING AND COMPUTER SCIENCE

Portland State UNIVERSITY

This slide shows the definition of product features

Products & Product Features

Product Features	PF Code	P Code
Sustainability and Supportability	PF1	P5, P6
Powerful & sustainable energy sources	PF2	P1, ..., P7
Advance space communication and Navigation	PF3	P3,P4,P5,P6,P7
Advance Space station and deep space habitation	PF4	P5,P6
Launching- Landing module/Advanced in-space propulsion	PF5	P3,P4,P7
In-situ resource utilization & Human-robotic systems	PF6	P5,P6,P7
Mobile Extravehicular activity & robotic platform	PF7	P3,P4,P5,P6,P7
Safe for human	PF8	P1,...,P7
Orbital exploration	PF9	P1,P2,P7
Space living community	PF10	P5,P6
Advance security systems	PF11	P1,...,P7

Slide 27

Engineering and Technology Management
MASEEH COLLEGE OF ENGINEERING AND COMPUTER SCIENCE

Portland State UNIVERSITY

This slide shows the relationship between product features and products

Customer Needs & Product Features Analysis Grid

Product Features VS Customer Needs (H=4, M=2, L=1)			Customer Needs								Total Score	Normalize
			Light/Heavy lift cargo	Return Sample from	Asteroid Redirection	Crew to Asteroid	Crew to Lunar Surface	Crew to Mars Moons	Crewed Mars Orbital	Crewed Mars Surface		
			CN(9)	CN (6)	CN (8)	CN (5)	CN (9)	CN (8)	CN (7)	CN (7)		
Product Features	Sustainability and Supportability	PF1				H	H	H	H	H	144.00	7.44
	Powerful & sustainable energy sources	PF2	H	M							48.00	2.11
	Advanced Space Communication and navigation	PF3		M	M	H	H	H	H	H	172.00	9.00
	Advance Space station and deep space habitation	PF4				M	M	M	M	H	86.00	4.22
	Launching- Landing module/Advanced in-space propulsion	PF5	H			M	M	M	M	M	108.00	5.44
	In-situ resource utilization & Human-robotic systems	PF6		M	M	H	H	H	H	H	172.00	9.00
	Mobile Extravehicular activity & robotic platform	PF7		M	M	M	M	M	M	M	100.00	5.00
	Safe for human	PF8		M		H	H	H	H	H	156.00	8.11
	Orbital exploration	PF9		M	M						28.00	1.00
	Space living community	PF10	M			M	M	M	M	H	104.00	5.22
	Advance security systems	PF11	M	M	M	H	H	H	H	H	190.00	10.00

Slide 28

Engineering and Technology Management
MASEEH COLLEGE OF ENGINEERING AND COMPUTER SCIENCE

Portland State UNIVERSITY

Technology Definition

Code	Group	Technology
T1	Materials	Lightweight structural/composite materials
T2	Information Processing	Human system integration
T3	Information Processing	Cyber security
T4	Life Enabling	Reliability, life assessment and health monitoring
T5	Materials	New fuel sources and generation process
T6	Power	Improved thrust and in-space propulsion systems
T7	Power	Energy storage and distribution
T8	Mechanical	Launch Infrastructure
T9	Materials	Ascent/Entry TPS
T10	Information Processing	Data Analysis for Decision Making
T11	Life Enabling	Water Recovery and Management
T12	Life Enabling	Radiation Mitigation and Biological Countermeasures
T13	Mechanical	Supersonic Retro-propulsion
T14	Mechanical	Repair Systems
T15	Life Enabling	Food Production, Processing and Preservation

Slide 29

Engineering and Technology Management
MASEEH COLLEGE OF ENGINEERING AND COMPUTER SCIENCE

Portland State UNIVERSITY

This slide shows the definition of product features

Technology & Product Features Analysis Grid

Technologies VS Product Features (H =4, M=2, L = 1)		Product Features: Sustainability and Supportability PF1 (8)	Powerful & sustainable energy sources PF2 (9)	Advanced Space Communication and navigation PF3 (7)	Advance Space station and deep space habitation PF4 (7)	Launching-Landing module/Advanced in-space propulsion PF5 (8)	In-situ resource utilization & Human-robotic systems PF6 (6)	Mobile Extravehicular activity & robotic platform PF7 (7)	Safe for human PF8 (10)	Orbital exploration PF9 (7)	Space living community PF10 (7)	Advance security systems PF11 (10)	Total Score	Normalized
Lightweight structural/composite materials	T1				2	4	2	4		1	1		100	**1.34**
Human system integration	T2	4		2	2				4		4		128	**2.91**
Cyber security	T3	2		2	1				1		1	4	94	**1.00**
Reliability, life assessment and health monitoring	T4	4			4				4		4		128	**2.91**
New fuel sources and generation process	T5	2	4		2	4	4	4		4			178	**5.73**
Improved thrust and in-space propulsion systems	T6	2	4		1	4	2	4		4			159	**4.66**
Energy storage and distribution	T7	4	4	1	2	2	4	2	4	4	4	1	249	**9.72**
Launch Infrastructure	T8	2	2		1	4			2	1		1	110	**1.90**
Ascent/Entry TPS	T9	2	2		1	4			2	1		1	110	**1.90**
Data Analysis for Decision Making	T10	2		4	4	4	1	4	2	4	4	4	254	**10.00**
Water Recovery and Management	T11	4	2		4		1		4	4	4		180	**5.84**
Radiation Mitigation and Biological Countermeasures	T12	4	2		4		1		4	4	4		180	**5.84**
Supersonic Retropropulsion	T13	1	2	1	2	4			2	4	1	1	144	**3.81**
Repair Systems	T14	1			2	1	4	4		2	4		124	**2.69**
Food Production, Processing and Preservation	T15	4			4		2		4		4		140	**3.59**

(Row group label: Technologies)

For HSE, related technologies are T6, T9, T10, T11, and T12

Technology TRM

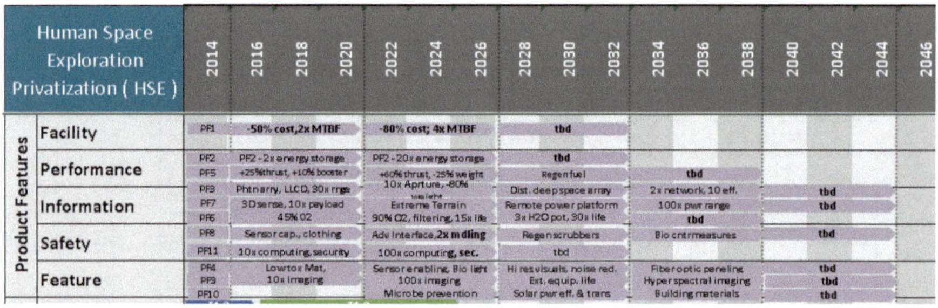

SWOT Analysis

Strengths	Weaknesses
• High number of brilliant people involved in projects • Competitive Industry and organizations • Government support • Entrepreneur visionaries • Growth industry 5-6% CAGR • $300B industry	• High cost - Cap investment • Long research and development time • Require high and complicated technology • Expensive supporting infrastructure • Low level of collaboration • High quality and safety standards ○ >6 sigma
Opportunities	**Threats**
• Technology improvement • Economic improvement • Discovery of new resources and materials • Advanced data exploration	• Impact of international regulations and government regulations • Technology limitation • Security Policies • Competition - existing companies

Slide 32

Engineering and Technology Management
MASEEH COLLEGE OF ENGINEERING AND COMPUTER SCIENCE

Portland State
UNIVERSITY

Added by Monticha

Mapping for HSE (focusing on T6, T9, T10, T11, T12)

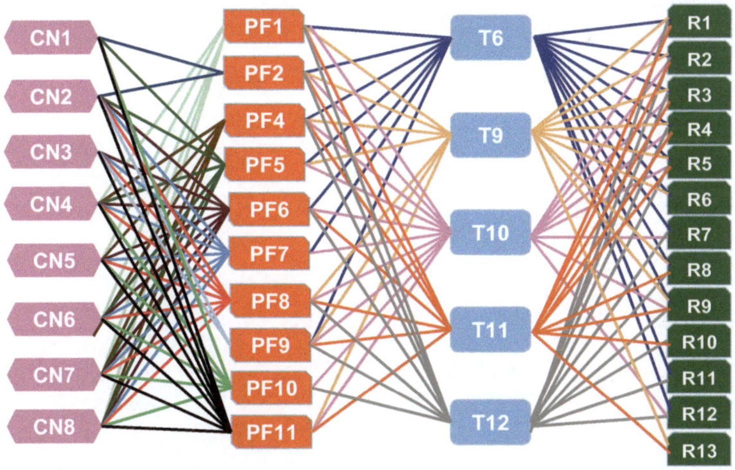

Slide 33

Engineering and Technology Management
MASEEH COLLEGE OF ENGINEERING AND COMPUTER SCIENCE

Portland State
UNIVERSITY

When explain to class, we have to start from Technology and expand to R, PF and finally with CN

Appendix 3

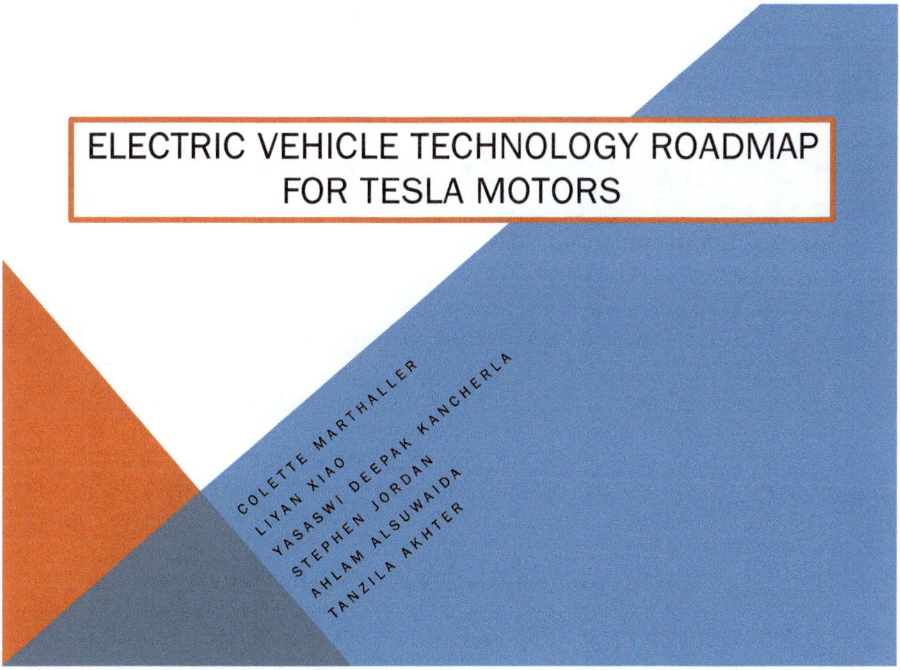

T. U. Daim (ed.), *Roadmapping Future*, Applied Innovation and Technology Management, https://doi.org/10.1007/978-3-030-50502-8

OBJECTIVES

Research Goal:

- To developa roadmap for Tesla's Electric VehicleTechnology in the coming 10 years

Research Method:

- Review Tesla's current Electric Vehicles Technologies
- Review Electrical Vehicles Industry trend
- Review university and independent research programs that are looking into new EV technology
- Look at government programs that subsidize research and give tax breaks to EV manufactures and consumers

Purpose of Research:

- Help Tesla making strategic decisions on where to focus on R&D

SWOT ANALYSIS

STRENGTHS	WEAKNESSES
• High Brand Recognition • Luxury Style • High Performance • Superior Battery Technology • Eco friendly	• High Car Price starts at $70k • Low Profitability -$4k/car • Low production amount ½ of 50,000 unit/yr sold • Range 230-270 (mi/single charge) • Only 2810 Super chargers in US and Canada • Long Charging Time 40 – 70 mins
OPPORTUNITIES	**THREATS**
• Increasing government incentive and tax breaks • Decreasing one's carbon footprint • Increasing global awareness of global warming • Lots of research on EV technology • Unstable oil prices	• Increasing competition in EV market • Other alternative fuels and technologies

MARKET DRIVER DEFINITIONS

Category	Label	Drivers	Definitions	Weight
Business Needs	D1	Market position	Maintain top marketposition/brand namein electric car market	10
	D2	Higher capacity	Develop family cars and utility vehicles withhigher power and capacity	8
	D3	Higher performance	Continuously develop highperformance electric motors forhigher speed, longer range, faster acceleration, Tesla's P Law.	7
	D4	Longer range	Continuously developelectric vehicles that drive farther on a single charge. Tesla's R Law	8
	D10	Lower production cost	Lower production cost to increase profit margin	7
	D11	Lesscharging time	Decrease time needed to fully charge the battery	9
	D12	Lower charging service cost	Low cost charging service (battery swap program)	5
	D13	Attractive to "Everyday Joe"	Make EVs more attractive to general populatioto gain more market share	6
Technology Advancements	D6	Research assistance	University research in need of monetary assistance and real life applications	6
	D7	Move away from Lithium	Lithium which can be costly to dispose of. Explore other alternative high performance ,low cost and low environmental impacbattery materials	6
	D8	Charge by renewable sources	Charge by renewable sources (solar)	6
	D9	Pioneer in innovations	First to go away from bulky cube batteries, need to continue the innovations	9
Social Trends	D5	Low carbon footprint	Emerging market/customers are becoming more aware of their carbon footprint	6
	D14	Greener everything	Apply green concept on all parts of an EV, including battery Lithium ion battery production are environmentally destructive.	6
Regulations	D15	Stringent regulations	Stringent regulations on the manufacture and recycling of batteries	7

PRODUCT FEATURE DEFINITIONS

Category	Label	Features	Definitions
Comfort	P16	Health Monitoring	Active Health Monitoring Device embedded in steering wheel
	P8	Family friendly	Family friendly models, including wagons, vans, suvs, etc
	P17	Wifi Enabled	Wifi hotspot inside car
	P18	Reconfigurable body panels	Reconfigurable body panels (transformer)
Convenience	P12	Wireless charging stations	Wireless charging facilities (future)
	P3	Portable charging equipment	Portable charging devices
	P10	Expand service network	Need more Roadside assistance andservice shops
	P1	Faster charging batteries	Charge in less time than it takes to put gas in a fossil fuel car
Safety	P13	Facial recognition	Facial recognition as anti theft protection
	P14	Hack Proof Technology	Safeguard user's identification and data stored in the EV
	P15	Night Vision	Use night vision device to help with drivingand parking
	P19	Remote vehicle control	User could remotely start or stop the vehicles
Technology Innovation	P4	Solar charging panels	Solar charging panels installed on EVor at EV charging station
	P5	Batteries	Low cost, high capacityYearly 5% increase in battery capacity
	P9	Hybrid with clean fuel	Electric and hydrogen hybridvehicle
	P2	Lithium free batteries	Aluminum Graphite battery (future) issafe won't catch fire
	P7	High efficiencymotors	Single or dual induction motor configurations
	P11	Low cost battery	Low cost battery
	P6	Battery recycling	Tesla battery recycling program, reuses Lithium reduces cost

DRIVER AND PRODUCTS ANALYSIS GRID

Product Features		Health Monitoring	Family Friendly	Wifi Enabled	Reconfigurable Body Panels	Wireless Charging Stations	Portable Charging Equipment	Expanded Service Network	Faster Charging Batteries	Facial Recognition	Hack Proof Technology	Night Vision	Remote Vehicle Control	Solar Charging Panels	High Capacity Batteries	Hybrid With Clean Fuel	Lithium Free Batteries	High Efficiency Motors	Low Cost Battery	Refurbish-able Batteries
Drivers		P16	P8	P17	P18	P12	P3	P10	P1	P13	P14	P15	P19	P4	P5	P9	P2	P7	P11	P6
Market Position	D1		×			×	×	×	×		×				×			×	×	
Higher Capacity	D2	×	×	×	×				×			×	×				×		×	
Higher Performance	D3					×	×		×				×		×			×		
Longer Range	D4		×			×	×	×	×					×	×	×		×		×
Low Carbon Footprint	D5		×			×	×		×					×	×	×	×	×		×
Research Assistance	D6				×	×	×		×	×	×	×	×	×	×	×	×	×	×	×
Move Away From Lithium	D7		×													×	×			×
Charge By Renewable Sources	D8													×						
Pioneer In Innovations	D9				×	×	×		×	×	×	×		×	×	×	×	×	×	×
Lower Production Cost	D10		×					×											×	×
Less Charge Time	D11					×	×		×						×		×	×		
Lower Charging Service Cost	D12							×												×
Attractive to "Everyday Joe"	D13	×	×	×	×	×	×	×	×	×	×	×	×	×	×		×	×	×	×
Greener Everything	D14													×		×	×			
Stringent Regulations	D15							×						×	×	×	×			×

DRIVER ROADMAP

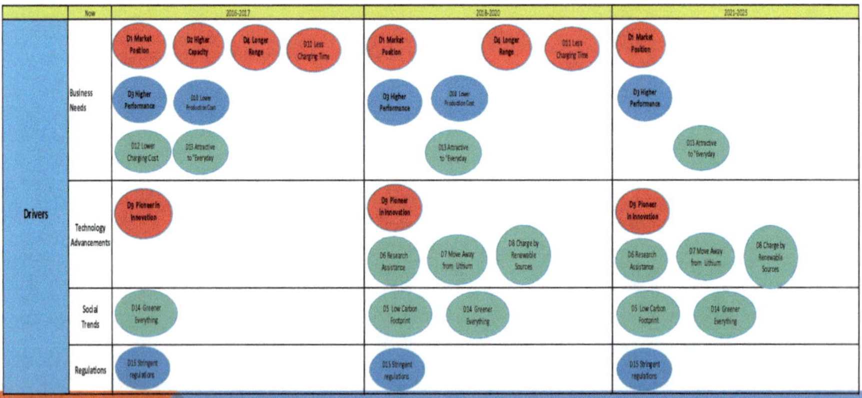

2016-2017 PRODUCT GAP ANALYSIS

Products

Lable	Drivers	WT	Active Health Monitoring (P16)	family friendly models (wagons, vans, suvs) (P8)	wifi (P17)	reconfigurable body panels (transformer) (P18)	wireless charging stations (P12)	portable charging devices (P3)	expanded service network (P10)	faster charging batteries (P1)	Facial recognition (theft protection) (P13)	Hack Proof Technology (P14)	Night Vision (P15)	remote vehicle control (P19)	solar charging panels (P4)	high capacity batteries (P5)	electric hybrid vehicle (P9)	lithium free batteries (P2)	high efficency motors (P7)	low cost battery (P11)	refurbishable batteries (P6)
D1	Maintain top market position in electric car market	10		10			10	10	10	10		10				10			10	10	
D2	Target families	8	8	8	8	8				8			8	8				8		8	
D3	Maintain the high performance appeal	7					7	7		7				7		7			7		
D4	Reduce range anxiety	8		8			8	8	8	8					8	8	8		8		8
D6	University research in need of monetary assistance and real life	6				6	6	6		6	6	6	6	6	6	6	6	6	6	6	6
D9	first to go away from bulky cube batteries, need to continue the	9				9	9	9		9	9	9	9		9	9	9	9	9	9	9
D10	increase profit margin	7		7					7											7	7
D11	decrease charge time	9					9	9		9						9		9	9		
D12	low cost charging service (battery swap program)	5							5												5
D14	Green living (lithiom ion battery production are environmentally	6													6		6	6			
D15	Stringent regulations on the manufacture and recycling of batteries	7							7						7	7	7	7			7
Total			8	33	8	23	49	49	37	57	15	25	23	21	36	56	36	45	49	40	42
Rank							2	2		1						1		3	2	3	3

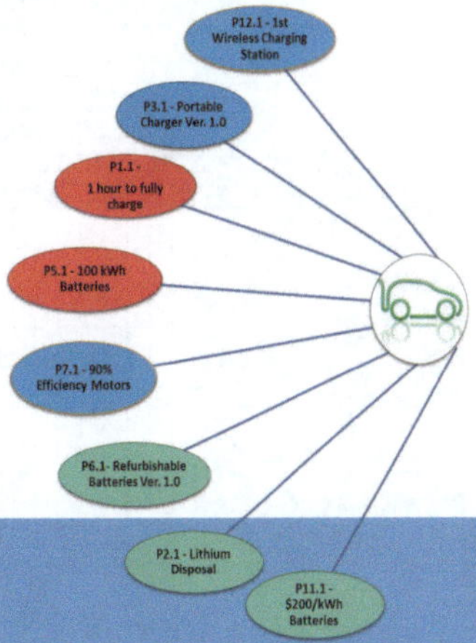

2018-2020 PRODUCT GAP ANALYSIS

Products

Label	Drivers	WT	Active Health Monitoring (P16)	family friendly models (wagons, vans, suvs) (P8)	wifi (P17)	reconfigurable body panels (transformer) (P18)	wireless charging stations (P12)	portable charging stations (P3)	expanded service network (P10)	faster charging batteries (P1)	Facial recognition (theft protection) (P13)	Hack Proof Technology (P14)	Night Vision (P15)	remote vehicle control (P19)	solar charging panels (P4)	high capacity batteries (P5)	electric hybrid vehicle (P9)	lithium free batteries (P2)	high efficency motors (P7)	low cost battery (P11)	refurbishable batteries (P6)
D1	Maintain top market position in electric car market	10		10			10	10	10	10		10				10			10	10	
D3	Maintain the high performance appeal	7					7	7		7				7		7			7		0
D4	Reduce range anxiety	8		8			8	8	8	8					8	8	8		8		8
D5	Emerging market/customers are becoming more aware of their carbon footprint	6		6			6	6		6					6	6	6	6	6		6
D6	University research in need of monetary assistance and real life applications	6				6	6	6		6	6	6	6	6	6	6	6	6	6	6	6
D7	move away from Lithium which can be costly to dispose of	6		6													6	6			6
D8	charge by renewable sources (solar)	6													6						0
D9	first to go away from bulky cube batteries, need to continue the innovations	9				9	9	9		9	9	9	9		9	9	9	9	9	9	9
D10	increase profit margin	7		7					7											7	7
D11	decrease charge time	9					9	9		9						9		9	9		0
D13	make EVs more attractive to "everyday joe"	6	6	6	6	6	6	6	6	6	6	6	6	6	6	6		6	6	6	6
D14	Green living (lithiom ion battery production are environmentally	6													6		6	6			0
D15	Stringent regulations on the manufacture and recycling of batteries	7							7						7	7	7	7			7
Total			6	43	6	21	61	61	38	61	21	31	21	19	54	68	48	55	61	38	55
Rank							2	2		2					3	1		3	2		3

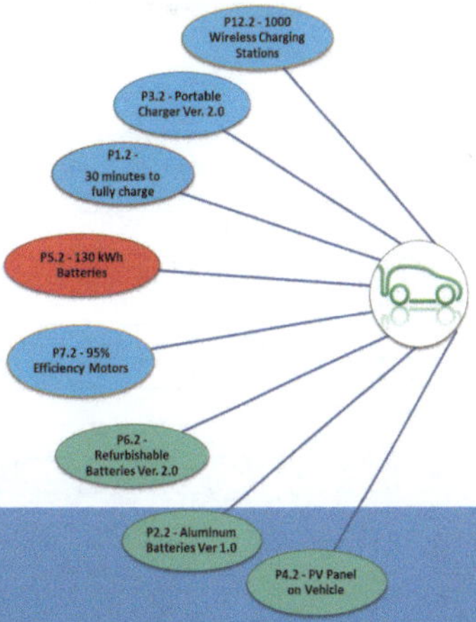

2020-2025 PRODUCT GAP ANALYSIS

Products			Active Health Monitoring	family friendly models (wagons, vans, suvs)	wifi	reconfigurable body panels (transformer)	wireless charging stations	portable charging stations	expanded service network	faster charging batteries	Facial recognition (theft protection)	Hack Proof Technology	Night Vision	remote vehicle control	solar charging panels	high capacity batteries	electric hybrid vehicle	lithium free batteries	high efficency motors	low cost battery	refurbishable batteries
Label	**Drivers**	**WT**	P16	P8	P17	P18	P12	P3	P10	P1	P13	P14	P15	P19	P4	P5	P9	P2	P7	P11	P6
D1	Maintain top market position in electric car market	10		10			10	10	10	10		10				10			10	10	
D3	Maintain the high performance appeal	7					7	7		7				7		7			7		
D5	Emerging market/customers are becoming more aware of their carbon footprint	6		6			6	6		6					6	6	6	6	6		6
D6	University research in need of monetary assistance and real life applications	6				6	6	6		6	6	6	6	6	6	6	6	6	6	6	6
D7	move away from Lithium which can be costly to dispose of	6		6													6	6			6
D8	charge by renewable sources (solar)	6													6						
D9	first to go away from bulky cube batteries, need to continue the innovations	9				9	9	9		9	9	9	9		9	9	9	9	9	9	9
D13	make EVs more attractive to "everyday joe"	6	6	6	6	6	6	6	6	6	6	6	6	6	6	6		6	6	6	6
D14	Green living (lithiom ion battery production are environmentally desctructive)	6													6		6	6			
D15	Stringent regulations on the manufacture and recycling of batteries	7							7						7	7	7	7			7
Total			6	28	6	21	44	44	23	44	21	31	21	19	46	51	40	46	44	31	40
Rank							3	3		3					2	1	3	2	3		

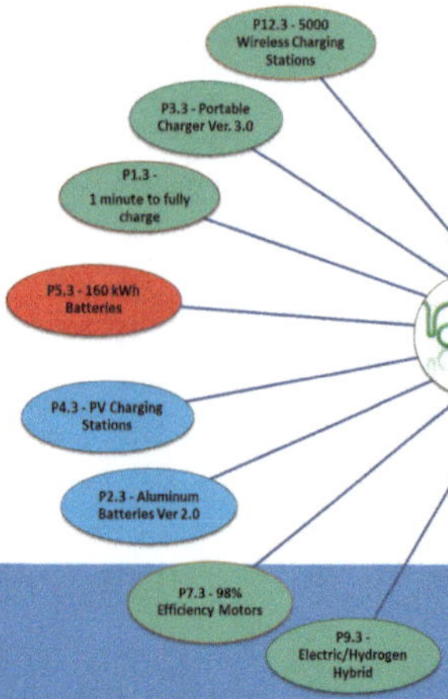

PRODUCT GAP ANALYSIS SUMMARY

Label	Features	2016-2017	2018-2020	2021-2025
P16	Health Monitoring	-	-	-
P8	Family friendly	-	-	-
P17	Wifi Enabled	-	-	-
P18	Reconfigurable body panels	-	-	-
P12	Wireless charging stations	1st wirless charging station (P12.1)	1000 wireless charging station nationwide (P12.2)	5000 wireless charging station nationwide (P12.3)
P3	Portable charging equipment	portable charger V1 (P3.1)	portable charger V2 (P3.2)	portable charger V3 (P3.3)
P10	Expanded service network	-	-	-
P1	Faster charging batteries	1 hour to fully charge (P1.1)	30 minutes to fully charge (P1.2)	1 minute to fully charge (P1.3)
P13	Facial recognition	-	-	-
P14	Hack Proof Technology	-	-	-
P15	Night Vision	-	-	-
P19	Remote vehicle control	-	-	-
P4	Solar charging panels	-	solar charging panel on vehicle for backup (P4.2)	solar charging stations nationwide (P4.3)
P5	High capacity batteries	100 kWh battery (P5.1)	130 kWh battery (P5.2)	160 kWh battery (P5.3)
P9	Hybrid with clean fuel	-	-	battery/hydrogen hybrid (P9.3)
P2	Lithium free batteries	lithium disposal (P2.1)	aluminum batteries V1 (P2.2)	aluminum betteries V2 (P2.3)
P7	High efficiency motors	90% (P7.1)	95% (P7.2)	98% (P7.3)
P11	Low cost battery	$200 dollars/kWh (P11.1)	-	-
P6	Refurbishable batteries	Refurbishable battery V1 (P6.1)	Refurbishable battery V2 (P6.2)	-

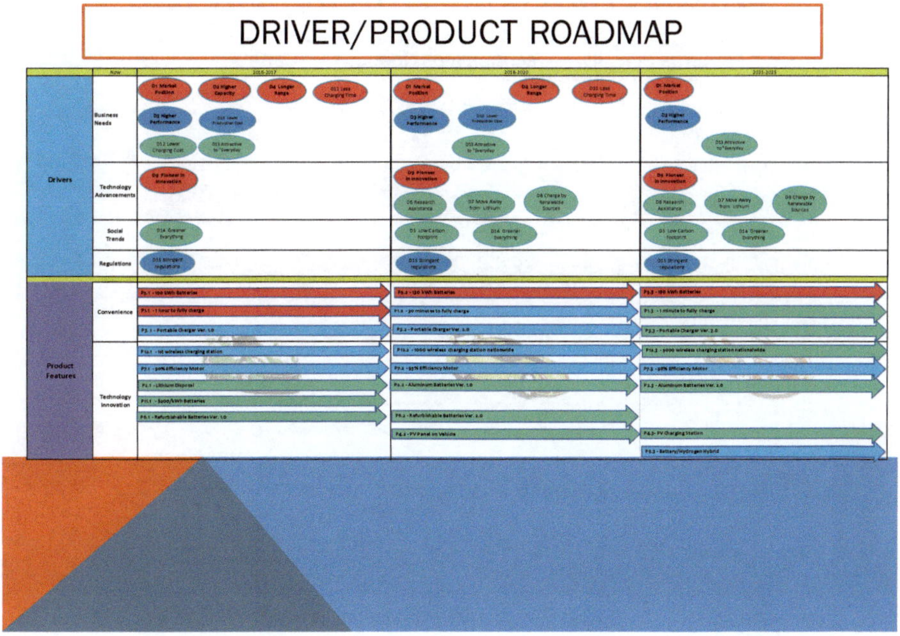

2016-2017 TECHNOLOGY GAP ANALYSIS

Product Features		ID	Technologies Needed	Gaps
P1.1	1 hour to fullycharge	T1a.1 T1b.1	Fast charging electronics Battery charging controller	Currently Available Currently Available
P2.1	Lithium Disposition	T2a.1 T2b.1	Chemical disposal Waste management	Currently Available Currently Available
P3.1	Portable Charger Ver. 1.0	T3a.1 T3b.1	Portable Charger Portable charger protection	Currently Available Currently Available
P5.1	100 kWh batteries	T5a.1 T5b.1	Battery storage capacity Battery protection	High powerdensity technology Currently Available
P6.1	Refurbish-able Batteries Ver. 1.0	T6a.1 T6b.1	Chemical recycling Refurbish process	Currently Available Reuse of recycledlithium
P7.1	90% Efficiency Motors	T7a.1 T7b.1 T7c.1	Electronic motor design Motor control electronics Motor overheating protection	Currently Available Currently Available Currently Available
P11	$250/kWh Batteries	T11a T11.b	Battery manufacturing Battery materialcost	Battery assembly process improvement Currently Available
P12.1	1st Wireless ChargingStation	T12a.1 T12b.1	Wireless charging Charging station design	High power wirelesscharging High Efficiency/low noise charging station design

2018-2020 TECHNOLOGY GAP ANALYSIS

Product Features		ID	Technologies Needed	Gaps
P1.2	30 minutes to FullyCharge	T1a.2 T1b.2	Fast charging electronics Battery charging controller	Lithium battery has a limit in charging rate. Unless change battery material to be lithium free, 30 minute to fully charge is not possible.
P2.2	Lithium Free Batteries	T2c.2	Aluminum battery	In research phase, not in production yet
P3.2	Portable Charger Ver. 2.0	T3a.2	Fast Portable charger	Increased charging capacity
P4.2	PV Panel on Vehicle	T4a.2	Integrated PV panel	Light-weight, durable PV material
P5.2	130 kWh Batteries	T5a.2 T5b.2	Battery storage capacity Battery protection	High power density technology Overheating and overchargingprotection
P6.2	Refurbish-able Batteries Ver. 2.0	T6a.2 T6b.2	Chemical recycling Refurbish process	Currently Available Cost efficient refurbish process
P7.2	95% Efficiency Motors	T7a.2 T7b.2 T7c.2	Electronic motor design Motor control electronics Motor overheating protection	High performance magnet Currently Available Currently Available
P12.2	1000 Wireless Stations	T12a.2 T12b.2	Wireless charging Wireless charging station design	High power wirelesscharging High Efficiency/low noise charging station design

2021-2025 TECHNOLOGY GAP ANALYSIS

Product Features		ID	Technologies Needed	Gaps
P1.3	1 minute to Fully Charge	T1a.3	Fast charging electronics	This is possible only with aluminum battery. It is a gap for lithium ion battery.
P2.3	Lithium Free Batteries	T2c.3	Aluminum battery	We anticipate the aluminum battery technology will be available in production by then
P3.3	Portable Charger Ver. 3.0	T3a.3	High Capacity portable charger	Higher power storage
P4.3	PV Stations Nationwide	T4a.3 T4b.3	High-production PV panel Efficient energy storage	High efficiency power conversion Currently available
P5.3	160 kWh Batteries	T5a.3 T5b.3	Battery storage capacity Battery protection	High power density technology Overheating and overchargingprotection
P7.3	98% Efficiency Motor	T7a.3	Electronic motor design	High performance magnet
P9.3	Battery/Hydrogen Hybrid	T9a.3 T9b.3 T9c.3	Hydrogen process Hydrogen storage Hydrogen fueled engine	Hydrogen fuel cell technology might not be possible to be used on Electric Vehicles
P12.3	5000 Wireless Stations	T12a.3 T12b.3	Wireless charging Wireless charging station design	High power wirelesscharging High Efficiency/low noise charging station design

PRODUCT/TECHNOLOGYROADMAP

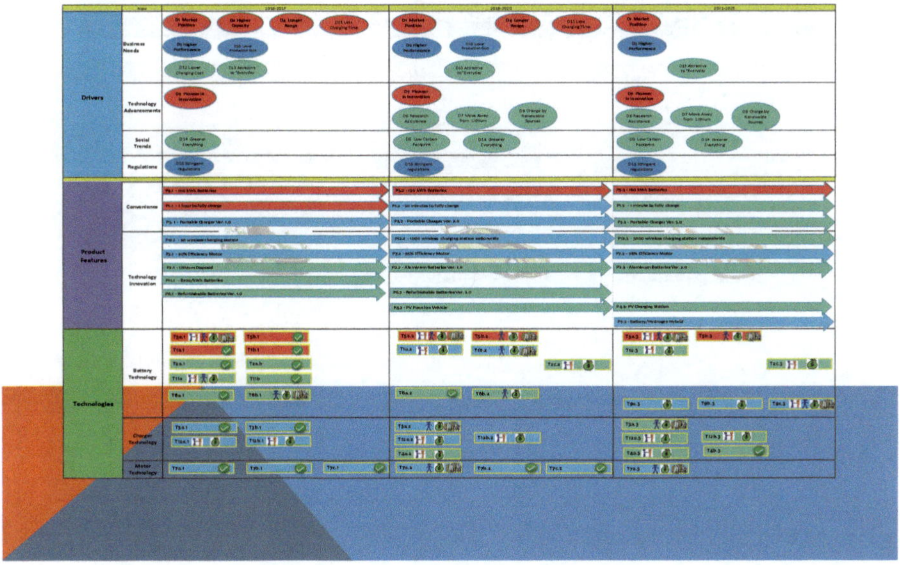

RESOURCES NEEDED– BATTERY TECHNOLOGY

Technology	2016-2017	2018-2020	2020-2025
T1a. Fast Charging Electronics	-	$1.5M 5 E.E./5 Tech Panasonic Lab	$5M 5 E.E./5 Tech Panasonic Lab
T1b. Charging Controller	-	$1M 10 E.E./5 Tech 1 General Lab	-
T2a. Chemical Disposal	-	-	-
T2b. Waste Management	-	-	-
T2c. Aluminum Battery	-	$3M Stanford University	$10M Stanford University
T5a. Battery Storage Capacity	$5M 10 E.E./6 C.E./5 Tech Panasonic Lab 1 High Power Lab	$8M 10 E.E./5 C.E./5 Tech Panasonic Lab 1 High Power Lab	$10M 10 E.E./5 C.E./5 Tech Panasonic Lab 1 High Power Lab
T5b. Battery Protection	-	$1.5M 5 E.E./5 Tech 1 General Lab	$6M 5 E.E./5 Tech 1 General Lab
T6a. Chemical Recycling	-	-	-
T6b. Refurbish Process		$1M 3 E.E./2 C.E./2Tech	-
T9a. Hydrogen Process	-	-	$7M
T9b. HydrogenStorage	-	-	$10M
T9c. Hydrogen Fueled Engine	-	-	$10M 10 M.E./5 E.E./5 C.E./15 Tech IntelligentEnergy Lab 1 High Power Lab, 1 General Lab, 2 Machinery Workshops
T11a. Battery Manufacturing	$1M 3 E.E./3 C.E./3 M.E. Panasonic Lab	-	-
T11b. Battery Material Cost	-	-	-

RESOURCES NEEDED– CHARGER TECHNOLOGY

Technology	2016-2017	2018-2020	2020-2025
T3a. High Capacity Portable Charger	-	$1.5M 10 E.E./5 Tech 1 General Lab, 1 Machinery Shop	$7M 10 E.E./5 Tech 1 General Lab, 1 Machinery Shop
T3b. PortableCharger Protection	-	-	-
T4a. PVPanel	-	$2M 4 E.E./2M.E. SolarCityPV Lab	$15M 5 E.E./3 M.E. SolarCityPV Lab
T4b. Energy Storage	-	-	-
T12a. Wireless Charging	$1.5M 5 E.E./2M.E. Plugless Inc.	$2M 5 E.E./2M.E. Plugless Inc.	$5M 5 E.E./2M.E. Plugless Inc.
T12b. Wireless Charging Station Design	$2M 4 E.E. Plugless Inc.	$3M 4 E.E Plugless Inc.	$8M 4 E.E Plugless Inc.

RESOURCES NEEDED– MOTOR TECHNOLOGY

Technology	2016-2017	2018-2020	2021-2025
T7a. Electronic Motor Design	-	$2M 10 E.E./10 M.E./10 Tech 1 General Lab, 1 Machinery workshop	$8M 10 E.E./10 M.E./10 Tech 1 General Lab, 1 Machinery workshop
T7b. Motor Control Electronics	-	-	-
T7c. Motor Overheating Protection	-	-	-

TESLA'S EV PRODUCT ROADMAP

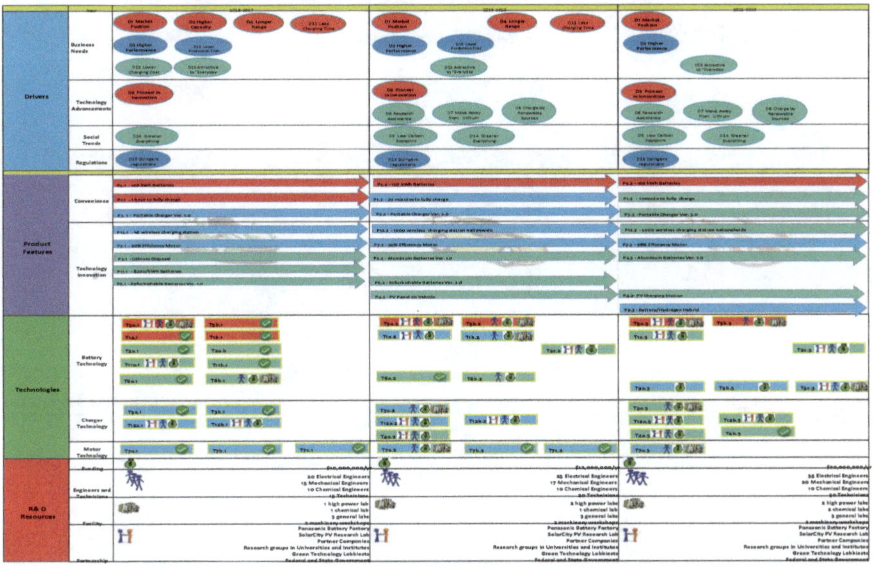

2016-2017 RESOURCE ALLOCATION

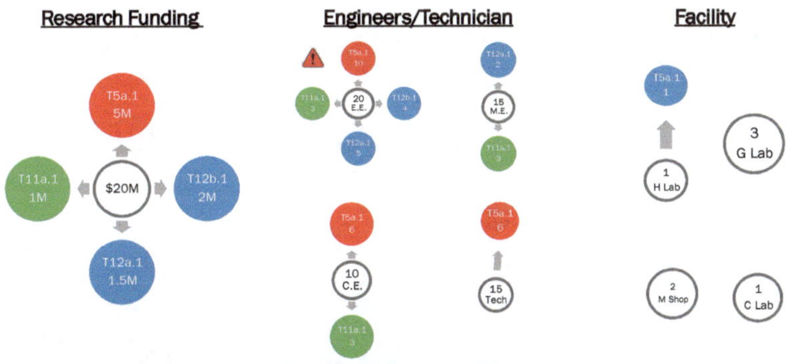

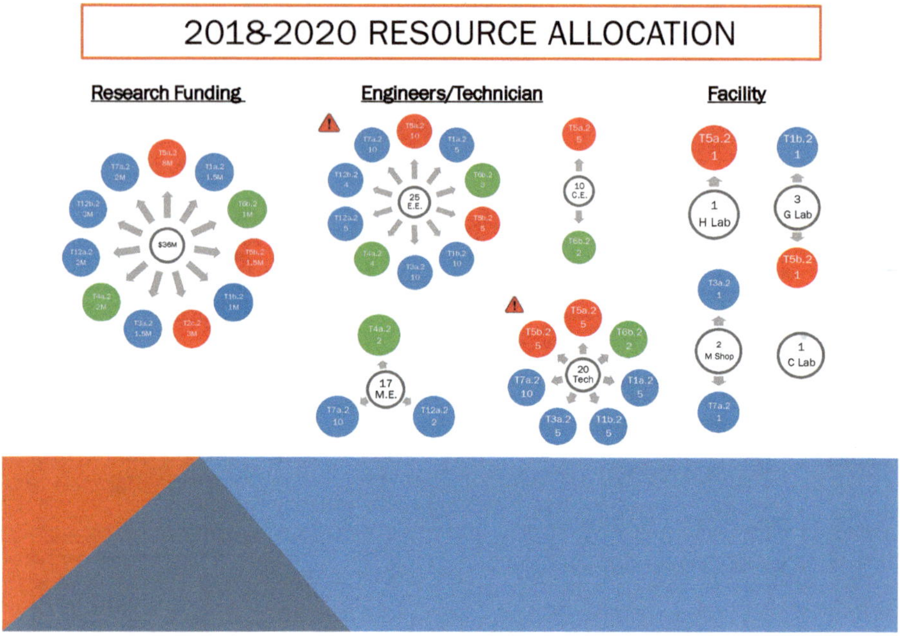
2018-2020 RESOURCE ALLOCATION
Research Funding
Engineers/Technician
Facility
25
E.E.
10
C.E.
17
M.E.
20
Tech
T5a.2
5
T5b.2
5
T6b.2
2
T7a.2
10
T1a.2
5
T3a.2
5
T1b.2
5
T4a.2
2
T12a.2
2
T5a.2
1
T1b.2
1
1
H Lab
3
G Lab
T5b.2
1
T3a.2
1
2
M Shop
1
C Lab
T7a.2
1

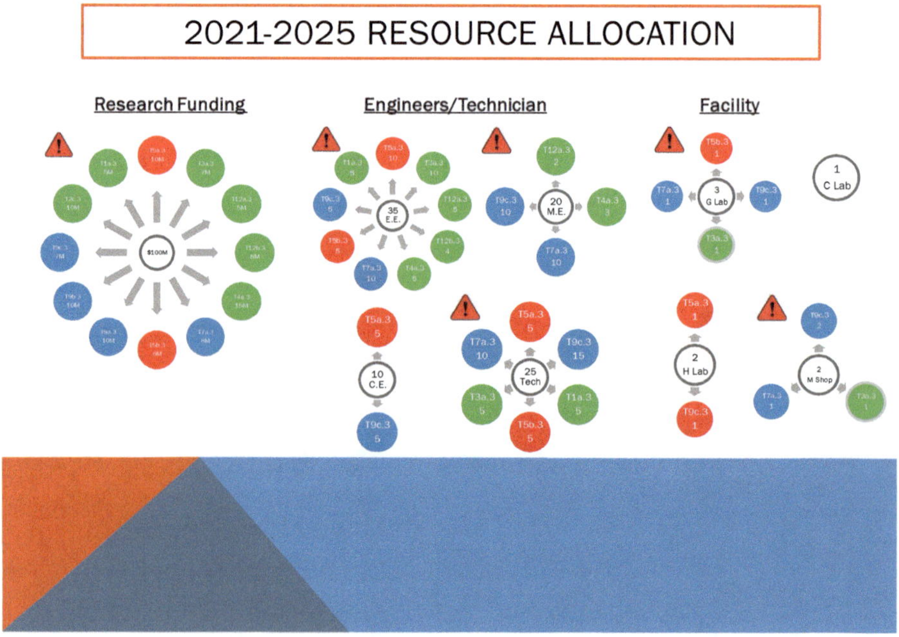
2021-2025 RESOURCE ALLOCATION
Research Funding
Engineers/Technician
Facility
$100M
35
E.E.
20
M.E.
T9c.3
10
T4a.3
3
T7a.3
10
T12a.3
2
10
C.E.
T5a.3
5
T9c.3
5
25
Tech
T5a.3
5
T7a.3
10
T9c.3
15
T3a.3
5
T1a.3
5
T5b.3
5
3
G Lab
1
C Lab
2
H Lab
T5a.3
1
T9c.3
1
2
M Shop

Appendix 4

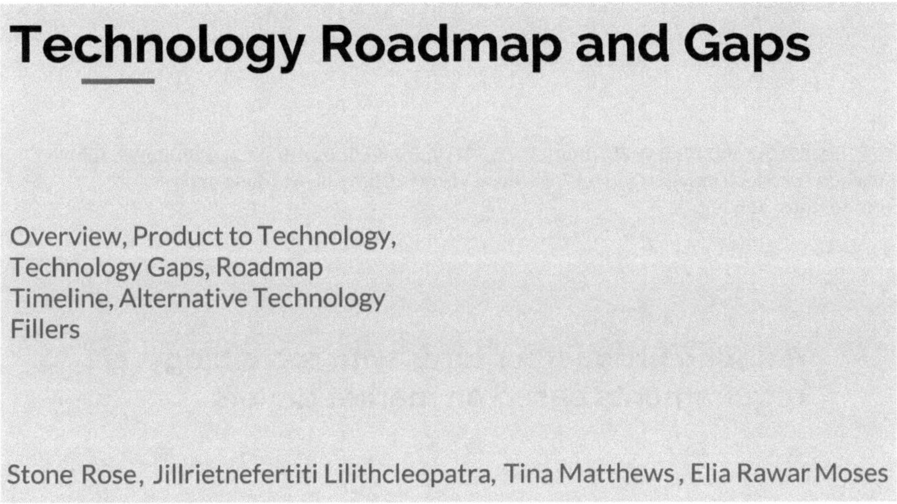

T. U. Daim (ed.), *Roadmapping Future*, Applied Innovation and Technology Management, https://doi.org/10.1007/978-3-030-50502-8

Robo Cab Overview

- Market Drivers
  - Eliminate requirement for car ownership, driver license, and insurance, save money on taxi trips
  - Reduce traffic congestion, optimize transit speed and traffic flow, minimize stop & go
  - Communicate with other vehicles, eliminate human error, reduce crashes and vehicle breakdowns
- Product Features
  - Vehicle to vehicle communication
  - Smart pick up, smart drop off, smart parking, predictive maintenance
  - Customized user experience using passenger profiles built with Machine Learning and modular design
- Technology Required
  - Sensors - camera, radar, lidar, ultrasonic
  - AI - artificial intelligence will enable vehicles to respond to data gathered from sensors
  - IoT - connectivity will enable vehicles to exchange data gathered from sensors
- Resources
  - ISO 26262 - international automotive electrical/electronic systems standard
  - HIPAA - 1996 U.S. legislation that provides data privacy for safeguarding medical information
  - PII - personally identifiable information
  - Ethics

Jill

https://www.sensorsmag.com/components/three-sensor-types-drive-autonomous-vehicles

https://www.electronicdesign.com/test-measurement/how-ai-will-help-pave-way autonomous-driving

Relating product features with technology requirements based on market drivers

	56	72	64	
	Design	**Intelligence**	**Safety**	
A.I.	User focused design with AI integration does not score as high since the features and technology relationship between the two are not as important as the customer's safety.	The focus of intelligence in term of A.I scores high because it's important to build the feature of connectivity within the driver.	As most important factors of the product features relate to safety. It scores high since A.I is the primary driver behind safety	656
Sensors	Sensors are important because their design is going to help to get an efficiency sensors product.	Intelligence is a good feature to combine with sensors for high technology that is why the score is high.	The role of sensors are very important for safety because when we have good sensors it increases the sensibility of the vehicle.	768
I.O.T.	The use of design in I.O.T does not score as high but provides good connectivity.	This scores high since the combination of intelligence with I.O.T is very important to have in smart apps.	Use of I.O.T will be needed in safety for quick response through the internet in an emergency. That is why it scores high.	656

Moses

SToJM - Process

Product Features & Technology overlap

Product Feature Areas

Technology Areas	Design (56)	Intelligence (72)	Safety (64)	
A.I.	2 / 112	4 / 288	4 / 256	656
Sensors	4 / 224	4 / 288	4 / 256	768
I.O.T.	2 / 112	4 / 288	4 / 256	656

Moses

SToJM - Process

Technology Gaps

Product Feature Areas

Technology Areas	Design	Intelligence	Safety
AI	**Current Gap:** Predictive user interface to help with user transit patterns for pickup and drop-off of commuters. (15 years) **Alternative/Filler:** Level 2 Autonomous vehicle with user transit pattern recognition (with GPS).	**Current Gap:** Lack of interconnected sensors for monitoring and tracking. (5-10 years) **Current Solution:** Standard platform for connected devices and sensors.	**Current Gap:** Cognitive Neural Network in order to omit the driver and minimize the reaction time. (5-10 years) **Alternative/Filler:** BITA
Sensors	**Current Gap:** Too few sensors available currently. Requirement of a common sensor platform with at least 1 trillion sensors. (10 years) **Alternative/Filler:** Utilize current sensors of about 20 billion.	**Current Gap:** Sensors to build user profiles based on customer transit patterns. (5 years) **Current Solution:** Multiple sensors in the car with Machine Learning for profiles.	**Current Gap:** No vehicle to vehicle communication. (10+ years) **Alternative/Filler:** Location-based road traffic control app.
IoT	**Current Gap:** Adapting to situational user needs e.g. disability needs with IoT integration. (5-10 years) **Alternative/Filler:** App allowing customer to request Robo Cab customization.	**Current Gap:** The use of IoT for device-to-device communication to enable conscious, self-aware, intuitive decision-making. (10 + years) **Alternative/Filler:** Location-based IP app for all IoT devices.	**Current Gap:** Device data is too scattered for cognitive neural networks to be able to function. (5-7 years) **Current Solution:** Device data on a common cloud platform.

Tina

Notes

IoT and Design: Custom app from robo cabs that directly links user calendar and application services

SToJM - Process

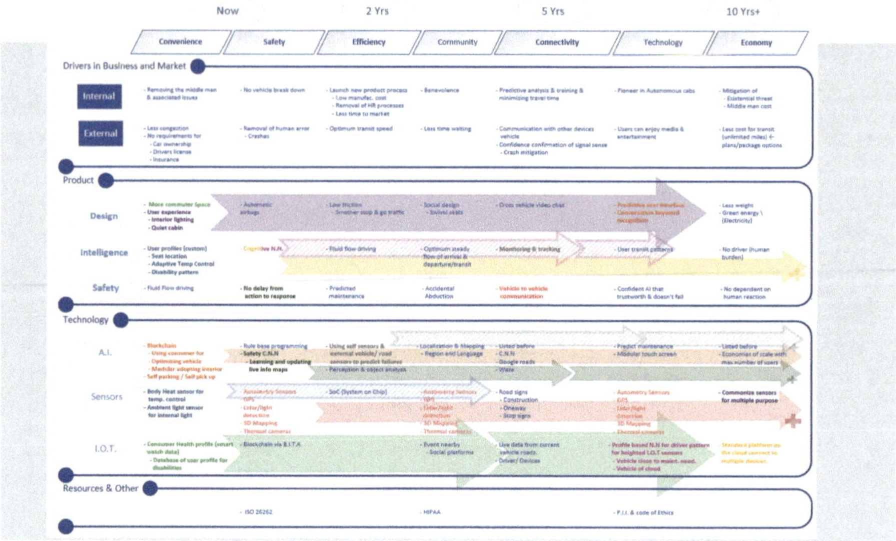

Stone

Alternative GAP Filler Implementation

AFFECTED INFRASTRUCTURE 2021 IMPACT Hwy only
2045 IMPACT Hwy+City

Insurance 20%
80%

DOT/DMV 10%
90%

Taxes
30% 70%

Utilities 40%
60%

Auto Dealers 5%
95%

Jill

Key technology Gaps with challenges as seen in the Timeline for the company and the concept that does not/will not have technological solutions in place for alignment with the rest of the timeline.

Alternative Technology Fillers - Jill

1. Key technology Gaps with challenges as seen in the Timeline for the company and the concept **that does not/will not** have technological solutions in place for alignment with the rest of the timeline. *(reference: T-plan and the gaps that we have talked about/ presented on that do not fit with the evolution of technology and the idea of Robo cabs.)*

SToJM - Process

Backup

Infrastructure: Gaps in the resources
Relate the gaps with the numbers and features that have gaps missing with the company

Business Drivers - Internal

- **Convenience**
  - Removal of associated issues and requirements of "middlemen" players and human drivers.
- **Safety**
  - Decrease in vehicle breakdowns and servicing through predictive maintenance crash mitigation.
- **Efficiency**
  - Launch of new products and processes through lowered manufacturing cost, less time to market.
- **Connectivity**
  - Predictive analysis/optimizing travel time.
- **Economy**
  - Mitigation of Existential threats and "middleman" costs.

Jill

WHAT IS GOOD FOR UBE

Market Drivers - External

- **Convenience**
  - Less congestion, No requirements for car ownership, driving license, or Insurance
- **Safety**
  - Removal of human error, crashes
- **Efficiency**
  - Optimum transit speed and free flow of traffic with minimal stop & go.
- **Connectivity**
  - Communication with other devices/vehicles
    - Confidence confirmation of signal sense
    - Crash mitigation
- **Economy**
  - Lower cost of transit (unlimited miles) - commuters can add unlimited mileage plans/packages to economize

Jill

WHAT IS GOOD FOR CUSTOMERS

Timeline

		Now		2 yrs.	5 yrs.		10 yrs.	10+ yrs.
		Convenience (H)	Safety (H)	Efficiency (H)	Community (L)	Connectivity (H)	Technology (M)	Economy (H)
Drivers Business Market	External (What the market requires)	- less congestion - NO requirements for - Car ownership - Drivers license - Insurance	- Removal of human error - Crashes	- Optimum transit speed	- Less time waiting	- Communication with other devices/vehicle - Confidence confirmation of signal sense - Crash mitigation	- Users can enjoy media & entertainment	- Less cost for transit (unlimited miles) — plans/package options
	Internal (what the company gets out of it)	- Removing the middle man & associated issues	- No vehicle break down	- Launch new product process - Low manufac. cost - Removal of HR processes - Less time to market	- Benevolence	- Predictive analysis & training & minimizing travel time	- Pioneer in Autonomous cabs	- Mitigation of - Existential threat - Middle man cost
Product	Design	- More commuter Space - User experience - Interior lighting - Quiet cabin - HVAC	- Automatic airbags	- Low friction - Smother stop & go traffic	- Social design - Swivel seats	- Cross vehicle video chat	- Predictive user Interface - Conversation keyword recognition	- Less weight - Green energy (Electricity)
	Intelligence	- User profiles (custom) - Seat location - Adaptive Temp Control - Disability pattern	- Cognitive N.N	- Fluid flow driving	- Optimum steady flow of arrival & departure/transit	- Monitoring & tracking	- User transit patterns	- No driver (human burden)
	Safety	* Fluid Flow driving	- No delay from action to response	- Predicted maintenance	- Accidental Abduction	- Vehicle to vehicle communication	- Confident AI that trustworth & doesn't fail	- No dependent on human reaction
Technology	AI	- Blockchain - Using consumer for - Optimizing vehicle - Modular adopting interior - Self parking / Self pick up	- Rule base programming - Safety C.N.N - Learning and updating live info maps	- Using self sensors & external vehicle/ road sensors to predict failures - Perception & object analysis	- Localization & Mapping - Region and Language	- Listed before - C.N.N - Google roads - Waze	- Predict maintenance - Modular touch screen	- Listed before - Economies of scale with max number of users
	Sensors	- Body Heat sensor for temp.control - Ambient light sensor for internal light	- Autometry Sensors - GPS - Lidar/light detection - 3D Mapping - Thermal cameras	- SoC (System on Chip)	- Autometry Sensors - GPS - Lidar/light detection - 3D Mapping - Thermal cameras	- Road signs - Construction - Oneway - Stop signs	- Autometry Sensors - GPS - Lidar/light detection - 3D Mapping - Thermal cameras	- Commonize sensors for multiple purpose
	I.o.T	- Consumer Health profile (smart watch data) - Database of user profile for Disabilities	- Blockchain via B.I.T.A		- Event nearby - Social platforms	- Live data from current vehicle roads. - Driver/ Devices	- Profile based N.N for driver pattern for heighted I.O.T sensors - Vehicle close to maintenance need. - Vehicle of cloud	- Standard platform on the cloud connect to multiple devices.
Resources/ Other			- ISO 26262		- HIPAA		- P.I.I. & code of Ethics	

Moses

Notes: Need to add 5 years, 10 years, 15, and 20 years

Product Features - Design

- Modular Design (space can be utilized for luggage or seating)
  - User experience -lighting, quiet cabin, HVAC.
- Low road friction - less stop & go
- Social design
  - swivel seats
  - cross vehicle video chat
- Predictive user Interface
  - Conversation keyword recognition
- Low overall weight - better fuel efficiency, less wear and tear
- Green energy (Electric)

Tina:
Notes: Advanced frontal airbags, also called dual-stage airbags, were the next-generation system. They can detect whether they need to deploy at full force, reduced force or not at all, depending on the situation.
https://www.dw.com/en/daimler-and-bosch-team-up-to-build-robo-taxis/a-38293298

Product Features - Intelligence

- User profiling (customized user-centered interiors)
  - Seat location
  - Temp control
  - Disability requirements
- Cognitive Neural Network
- Fluid flow driving
- Optimum steady flow of arrival /departure/ transit
- Monitoring & tracking
- Recognition and machine learning of user transit patterns

Tina:
https://www.marketwatch.com/video/sectorwatch/what-going-to-happen-inside-your-car-once-you-dont-have-to-drive/506C9A0B-D700-43C4-823A-0728B665ABF2.html

Product Features - Safety

- Fluid Flow driving
- No delay in reaction and response - No dependendence on human reaction
- Predictive maintenance
- Vehicle to vehicle communication (IOT)
- Confident AI that is trustworth & doesn't fail
- Emergency Braking (in the mechanics of the car connected to its sensors as well as one for passengers to use)
- Smart Airbags - Advanced Airbags and inflatable seat belts

Tina:

Technology Areas - AI

- Blockchain
    - Utilizing consumer profiles for optimizing vehicle scalable usage
    - Modular adopting interior
    - Smart parking / Smart pick up
- Rule based programming (Regulation/Safety)
- C.N.N for Safety and Connectivity
    - Learning and Updating live info maps
        - Google Roads & Waze
- Using self sensors & external vehicle/ road sensors to predict vehicle maintenance
- Perception & object analysis
- Localization & Mapping
    - Region and Language
- Economy of Scale

Stone

Technology Areas - Sensors

- Heat sensors for internal temperature control
- Ambient light sensor for internal lighting
- Odometry Sensors
- GPS
- Lidar - Light Detection and Ranging
- 3D Mapping
- Thermal cameras
- Radar
- DSRC - Dedicated Short Range Communication (V to V)
- Inertial Navigation System
- Pre-built maps
- IR Sensors
- Ultrasonic sensors

Stone:
https://www.mckinsey.com/industries/automotive-and-assembly/our-insights/self-driving-car-technology-when-will-the-robots-hit-the-road

Technology Areas - IoT

- Consumer Health profile (smart watch data)
  - Database of user profile's
  - Disabilities
- Blockchain via B.I.T.A
- Energy Efficient sensors within the IoT network
- Congestion Production references by looking at events nearby
  - Social Media and platforms
  - Event Brite
- Live data form current vehicle roads
  - Driver/ Devices
- Profile based N.N for driver pattern for heighted I.o.T sensors
  - Bad driver in area
    - Heighted sensors data usage around user
  - Vehicle close to maintenance need
  - Vehicle of cloud
- Standard platform on the cloud connect to multiple devices

Stone

Market (Customer) Driver Areas

Product Feature Areas

Importance M = 2 H = 4	Convenience	Safety	Efficiency	Connectivity	Economy	~~Technology~~	
	4	*4*	*4*	*4*	*4*	*N/A*	
Design	2 / 8	4 / 16	2 / 8	2 / 8	4 / 16	-	**56**
Intelligence	4 / 16	4 / 16	4 / 16	4 / 16	2 / 8	-	**72**
Safety	2 / 8	4 / 16	4 / 16	4 / 16	2 / 8	-	**64**
~~**Sensors**~~	-	-	-	-	-	-	-

Moses

Gaps

Return on Investment for consumer usage

- High manufacturing cost due to advanced technology
  - AI, Sensors, and IoT
- R&D for guaranteed ROI
  - Reliable neural network
- Customer trust and adoption of technology
- Affected Infrastructure: DMV, Insurance Companies, Taxes, Electric Service Companies
- Modular vehicle design for economic benefit for internal and external market
  - Axle config (economy/load bearing)
  - Modular vehicle designs for transportation of goods and people

Jill GAP with respect to the ROI for consumer usage

Technology Gaps

Product Feature Areas

Technology Areas	Design	Intelligence	Safety
AI	**Current Gap:**Predictive user interface to help with user transit patterns for pickup and drop-off of commuters (15 years) **Alternative/Filler:** Level 2 Autonomous vehicle with user transit pattern recognition (with GPS)	**Current Gap:** Lack of interconnected sensors for monitoring and tracking (5-10 years) **Current Solution:**Standard platform for connected devices and sensors	**Current Gap:** Cognitive Neural Network in order to omit the driver and minimize the reaction time. (5-10 years) Alternative/Filler: BITA
Sensors	**Current Gap:** Too many types of sensors are required currently. Requirement of a common sensor platform. (10 years) **Alternative/Filler:** Utilize current sensors.	**Current Gap:** Sensors to build user profiles based on customer patterns. (5 years) **Current Solution**: Multiple sensors in the car with Machine Learning for profiles.	**Current Gap:** No vehicle to vehicle communication (10+ years) **Alternative/Filler:** Location-based road traffic control app.
IoT	**Current Gap**: Adapting to situational user needs e.g. disability needs. (5-10 years) **Filler Solution:** App allowing customer to request Robo Cab customization	**Current Gap:** The use of IoT for device-to-device communication to enable conscious, self-aware, intuitive decision-making (10 + years) **Alternative/Filler:** Location-based IP app for all IoT devices.	**Current Gap:** Device data is too scattered for cognitive neural networks to be able to function (5-7 years) **Current Solution:** Device data on a common cloud platform.

IoT and Design: Custom app from robo cabs that directly links user calendar and application services

Analysis

Why is safety intertwined with all the market drivers?

- Solution is removing the variable of human control
  - Driver reaction and errors
  - Better visibility of vehicle control
- Communication through IOT
  - Road conditions
  - Active construction zones
  - Speed zones
  - Population zones
  - Predictive maintenance

Stone

Analysis (continued)

Why did the market drivers for Design score low?

- Convenience
  - Not a major necessity for most consumers
- Efficiency
  - Can be achieved with today's technology (Auto Pilot, Cruise Control)
- Connectivity
  - It's a 'nice-to-have' but not a 'must-have'

Stone

Thank you!

Any Questions?

Jill

T-Plan

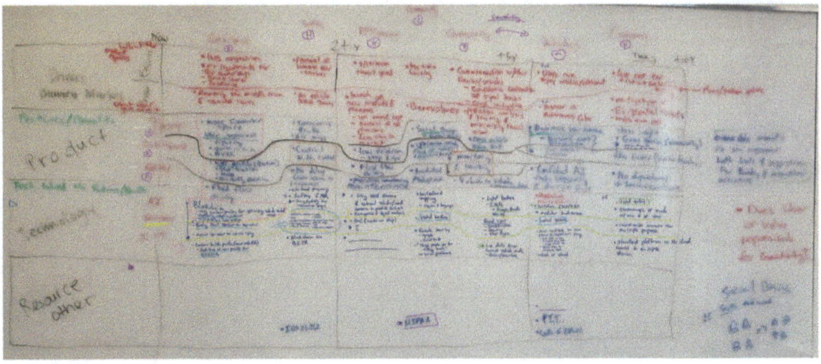

Market Drivers

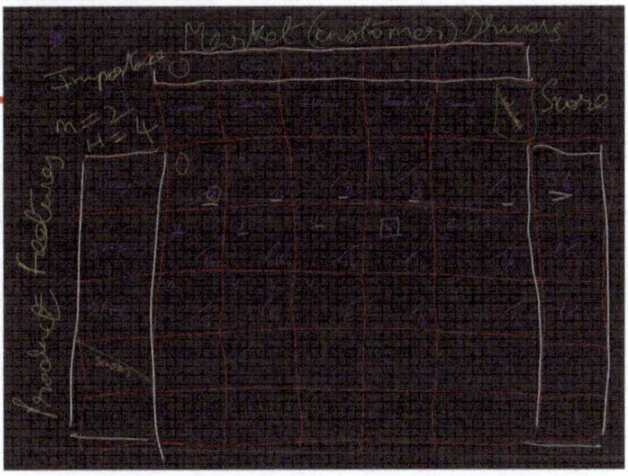

Product Feature Areas

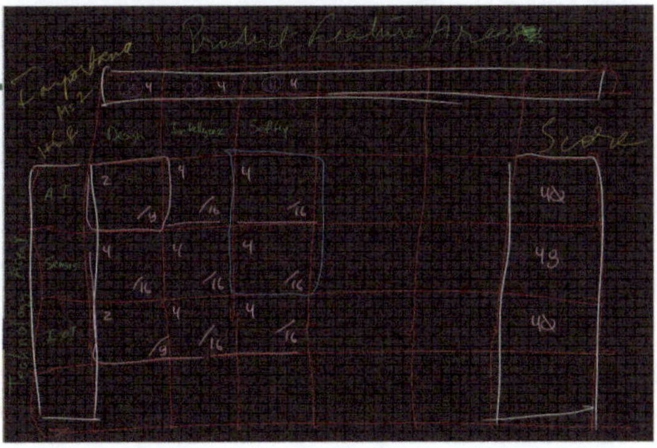

Gaps

- Customer Adoption
- Cost optimization
- Connection with other devices (IoT)
- Affected Infrastructure: DMV, Insurance Companies, Taxes, Electric Service Companies
- Modular vehicle design for economic ROI for internal and external market
  - Axle config (economy/load bearing)
  - Modular vehicle designs for transportation of goods and people

Tina

Gaps

Return on Investment for consumer usage

- High manufacturing cost due to advanced technology
  - AI, Sensors, and IOT
- R&D for guaranteed ROI
  - Reliable neural network
- Customer trust and adoption of technology

Timeline

- **1900-1980** : Level 0: Driver performs all driving tasks
- **1980-2012** : Level 1: Vehicle is controlled by the driver but some driving assist features may be included.
- **2012-2022 :** Level 2: Vehicle has some level of automation such as acceleration and steering, but driver must remain engaged and monitor the environment at all times.
- **2022-2028** : Level 3: Driver is able to completely shift "safety-critical functions" to the vehicle, under certain traffic or environmental conditions.
- **2028-2045 :** Level 4: Designed to perform all safety-critical driving functions and monitor roadway conditions for an entire trip.
- **2045-2100 :** Level 5: fully-autonomous system that expects the vehicle's performance to equal that of a human driver, in every driving scenario.

Tina
Source:
https://www.techemergence.com/self-driving-car-timeline-themselves-top-11-automakers/
https://www.techrepublic.com/article/autonomous-driving-levels-0-to-5-understanding-the-differences/
https://www.i-micronews.com/category-listing/product/sensors-and-data-management-for-autonomous-vehicles-report-2015.html

Market Drivers - External

- **Convenience**
  - Less congestion, No requirements for car ownership, driver license, or insurance
- **Safety**
  - Removal of human error, crashes
- **Efficiency**
  - Optimum transit speed and free flow of traffic with minimal stop & go.
- **Connectivity**
  - Communication with other devices/vehicles
    - Confidence confirmation of signal sense
    - Crash mitigation
- **Economy**
  - Lower cost of transit (unlimited miles) - commuters can add unlimited mileage plans/packages to economize

Jill

WHAT IS GOOD FOR CUSTOMERS

Business Drivers - Internal

- **Convenience**
  - Removal of associated issues and requirements of "middlemen" players and human drivers.
- **Safety**
  - Decrease in vehicle breakdowns and servicing through predictive maintenance crash mitigation.
- **Efficiency**
  - Launch of new products and processes through lowered manufacturing cost, less time to market.
- **Connectivity**
  - Predictive analysis/optimizing travel time.
- **Economy**
  - Mitigation of Existential threats and "middleman" costs.

Jill

WHAT IS GOOD FOR UBER

Intelligence	Sensors	Safety
User profiling (customized user-centered interiors)	Heat sensors for internal temperature control	Fluid Flow driving
Seat location	Ambient light sensor for internal lighting	No delay in reaction and response-No dependendence on human reaction
Temp control	Odometry Sensors	Predictive maintenance
Disability requirements	GPS	Vehicle to vehicle communication (IOT)
Cognitive Neural Network	Lidar -Light Detection and Ranging	Confident AI that is trustworth & doesn't fail
Fluid flow driving	3D Mapping	Emergency Braking (in the mechanics of the car connected to its sensors as well as one for passengers to use)
Optimum steady flow of arrival /departure/ transit	Thermal cameras	Smart Airbags-Advanced Airbags and inflatable seat belts
Monitoring & tracking	Radar	
Recognition and machine learning of user transit patterns	DSRC -Dedicated Short Range Communication (V to V)	
	Inertial Navigation System	
	Pre-built maps	
	IR Sensors	
	Ultrasonic sensors	

Market (Customer) Driver Areas

Product Feature Areas	*Importance* *M = 2* *H = 4*	*4* Convenience	*4* Safety	*4* Efficiency	*4* Connectivity	*4* Economy	*N/A* ~~Technology~~	
	Design	2 / 8	4 / 16	2 / 8	2 / 8	4 / 16	-	**56**
	Intelligence	4 / 16	4 / 16	4 / 16	4 / 16	2 / 8	-	**72**
	Safety	2 / 8	4 / 16	4 / 16	4 / 16	2 / 8	-	**72**
	~~**Sensors**~~	-	-	-	-	-	-	-

Appendix 5

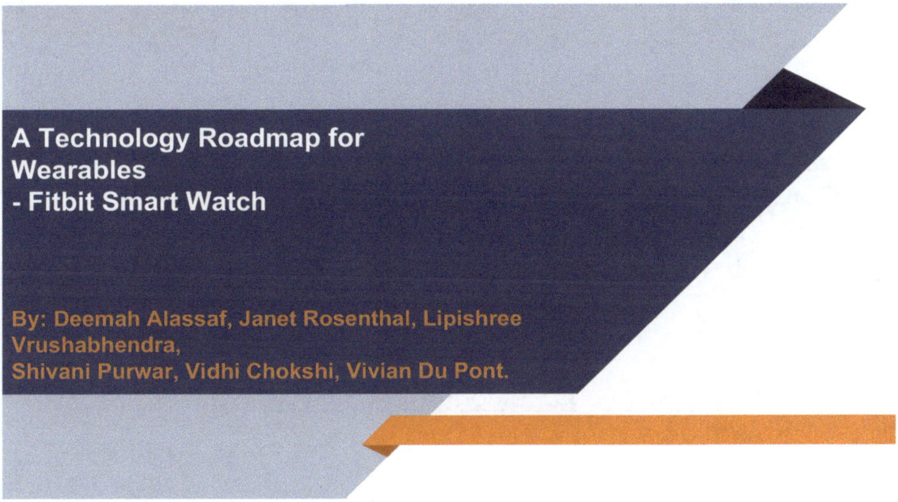

T. U. Daim (ed.), *Roadmapping Future*, Applied Innovation and Technology Management, https://doi.org/10.1007/978-3-030-50502-8

Agenda

- Introduction
- Project Proposal
- Objective
- Drivers
- Product Features
- Technology
- Resources
- Roadmap

FitBit

https://www.youtube.com/watch?v=PxpxRoWmjq8

Introduction

- Smartwatches have been around since 2012 (Mainstream market)
- Lightweight and compact size with potential to replace traditional watches and smart phones
- "The wrist is the ideal place for the power of Google to help people with their lives" (David Singleton, Google's smartwatch king)
- Health is #1 reason for wearables purchase (survey conducted by PwC in March 2016)

Project Proposal

What Is A Smartwatch?

A Smartwatch is a computerized wristwatch that has numerous functionalities similar to the ones present in a PDA or smart phones. According to Gartner, smartwatches have the "primary function of receiving electronic communications (for example, caller alerts, texts or voice calls). They provide a two-way connection via Bluetooth to a mobile connectivity hub (for example, a smartphone), or via cellular baseband or Wi-Fi." [1]

Project Objective And Customer

- To develop a Technology Roadmapping forFitBit focusing on Wellness programs for consumers in the age group of 30- 50 and seniors.
- Features need to be developed on industry level to enable effective interaction to human beings
  - Notification
  - Internet of things
  - Fitness
  - Navigation

[1] (http://www.cio.com/article/3122133/wearable-technology/smartwatches-are-big-pricey-and-ugly.html?page=2)

6

Who Are We

We Are Fitbit

- Company dedicated to create health and fitness wearable products.

Where Are We now?

- Fitbit acquired software assets from Pebble in December, 2016. Pebble created the first modern smartwatch market in 2013. [1]
- James Park, chief executive officer and co-founder of Fitbit, said "With basic wearables getting smarter and smartwatches adding health and fitness capabilities, we see an opportunity to build on our strengths and extend our leadership position in the wearables category."[1]
- Fitibit completed a second acquisition in January, 2017 - smartwatch maker Vector Watch - aimed at making a push into the smartwatch market.[2].
- Fitbit is launching their first smartwatch by the end of 2017[3].

[1] https://www.bloomberg.com/news/articles/2016-12-07/pebble-said-to-discuss-selling-software-assets-to-fitbit
[2] https://www.usatoday.com/story/tech/news/2017/01/10/fitbit-acquires-smartwatch-maker-vector-watch/96388730/
[3] https://www.digitaltrends.com/mobile/fitbit-smartwatch-2017/

7

MARKET DRIVERS

8

Ranking Analysis

Categories	Code	Market Drivers	Weight
Social/Consumer	D1	Fitness/Wellness Support Functions	4
Interest	D2	Lifestyle	3
	D3	Adaptation of Smart Technologies	4
	D4	Brand Loyalty	2
	D5	Social Media Participation	3
Health	D6	Stress Level Monitoring	2
	D7	Rise in Diabetes	3
	D8	Aging Population	2
	D9	Obesity	4
	D10	Health Industry Desire for Data	3
Technical	D11	Connectivity with Multiple Devices	4
	D12	Interest from Athletes	4
Economical	D13	Cheap Components	3
	D14	Work Productivity	3

Current Fitbit Technology Features

Fitbit's product portfolio currently offers a number of fitness trackers with multiple features that inspire its customers to live a healthier, more active life.

https://www.fitbit.com/compare

Product Categories

PRODUCT CATEGORIES	
Product	**Code**
Security	P1
Battery	P2
User Interface	P3
Connectivity	P4
Connected Wellness	P5

Product Features

PRODUCT FEATURES					
Category	**Product Feature**	**Code**	**Current Level**	**Where we want to be**	**Gap**
Security	Encrypted data	PF1	Not encrypted	Encrypt confidential data	Implement software for data encryption
	Fingerprint unlock	PF2	Unavailable product feature	Enable unlock with two fingerprints	Incorporate bio-metric recognition software
	Face recognition	PF3	Unavailable feature	Enable unlock with fface recognition	Incorporate face recognition software
Battery	Longer-life lithium batteries	PF4	Shorter-life of batteries	Retaining battery charge for longer periods	Improve efficiency of batteries
	Solar batteries	PF5	Not implemented	Use solar energy for charging	Introduce solar technology for charging
	Wireless charging	PF6	In the development phase	Only wireless charging	Develop wireless technology
User interface	Image Projection	PF7	In the development phase	Project content that user wants	Develop image processing technology
Connectivity	5G	PF8	Not implemented	With robust connectivity	Develop strong networks
	Gesture Control	PF9	At the development phase	Control connected devices using gestures	Incorporate gesture recognition software
Connected Wellness	Apps	PF10	Not achieved 100% accuracy	Achieve 100% accuracy of the apps	Improve existing software on accuracy
	IOT	PF11	In the development phase	Achieve connectivity with all smart home devices	Implement multiple device connectivity
	Seamless sync	PF12	Syncing is slow	Robust sync and updates	Improve the syncing capabilities
	Monitor body	PF13	Moderate levels of monitoring	Body monitored with utmost accuracy	Improve monitoring capabilities
	SOS button	PF14	Not yet implemented	Should be incorporated as an app	Introduce solar technology for charging
	Reverse communication with health professional	PF15	Not yet implemented	Enable communication as per the user	Introduce solar technology for charging
	Notifications	PF16	Only email/call/message notifications	Enable customizable notifications	Build the system formultiple notifications

Product Features Timeline

PRODUCT FEATURES TIMELINE				
Product Features	**Current Level**	**2 years**	**5 years**	**Where we want it to be (15 years)**
Encrypted Data	Most smartwatches send unencrypted data	Encrypt 80% of data	Data fully Encrypted	Full encryption of confidential data
Fingerprint Unlock	Only traditional passcode method is used for locking the device.	Implement fingerprint unlock		Smartwatches unlock by using fingerprints for higher confidentiality
Face Recognition			Implement face recognition system	Smatwatches enable users face recognition to unlock the device
Longer-life Litium Batteries	Lithium-ion batteries with short life, requires recharging very often	Longer-lasting batteries that can withstand plenty of recharging	Less reliance on lithium	High capacity, small, light batteries
Solar Batteries	Only lithium batteries used in current		Solar battery option	Affordable, reliable and efficient solar batteries
Wireless Charger		Charging case, sleep cycle accuracy		Verity of wireless charging cases for portable charging
Image Projection	Coin-sized displays with very difficult navigation	Projection of the user interface (UI) screen on the user's hand.	Larger and better quality projection with a camera configured to capture an image, and a processor configured to detect the commands	High quality image projection. Devices need to integrate a camera to capture user's selections on the projected image and have advanced gesture control.
5G	Limited cellular coverage. Users can connect smartwatch to their smartphones by bluetooth The cellular connection is a huge battery consumer. Some smartwatches have 3G connectivity and a few models have 4G (which makes the device bigger and heavier).	Cellular network coverage expansion	Introduction of 5G	5G compatibility Accessibility to healthcare anytime, anywhere Fast and reliable connection with large coverage to support an always-connected lifestyle and enable more appealing Fitness/Health apps and use cases. Continuous connectivity with a health professional
Gesture Control	Under Development	Single device control	Multiple devices controlling capability	

Product Features Timeline

Product Features	Current Level	2 years	5 years	Where we want it to be (15 years)
Apps	A large number of apps that provide specific medical reference, fitness/wellbeing programs.	Ability to micromonitor our behaviors and biosignals and the accumulation of massive amouts of data. Access to massive data sets through cloud		Cost-effective solutions providing apps for smartwatches able to seamlessly interact with others apps, devices and wearables and leverage the opportunity to personalize wellness programs by learning and adapting to user needs/habits.
IoT	IoT and wearables proliferation Accessible and scalable use of Artificial Intelligence		Implementation of Iot between all wearable technologies and smart devices	Allow a comprehensive lifestyle analysis and monitoring for patients by their health professionals
Seamless Syncing	Smartwatches synic and connect to limited devices as smartphones		Smartwatches able to connect and communicate with health professional for monitoring	Watch communicates with health professionals.
Monitor Body	Smartwatches have limited body monitoring		Enable smartwatches to detect wider range of vital signs and identify serious levels.	Health professionals can receive direct updates on patients for disturbing vital signals
SOS Notification	Smartwatches include SOS button for emergencies	Send direct and instant notifications to the emergency contact when serious vital signs occur		Fast responding emergency notification system

Product Features Timeline

Product Features	Current Level	2 years	5 years	Where we want it to be (15 years)
Communication with Health Professional	Smartwatches only detect some vital signs for user use only	Develop smartwatches ability to send vital signs serious levels to health professional for immediate action		Health professionals can communicate with patients to notify them about sudden serious health signals and set urgent appointments if needed
Notifications	the current notification system is not very reliable it does not show all the notifications	increase reliability of notifications	Customized notifications for healthcare, e.g. medication reminders, doctor appointments.	Integrated notification system with calendar and other health apps

QFD Market Drivers vs. Products

			Wellness	Luxury and Style	Smart Technology	Brand Loyalty	Social Media	Stress Level Monitoring	Rise in Diabetes	Senior Citizens	Oobesity	Health industry desire for data	Connectivity with multiple devices	Interest from athletes	Cheap components	Work productivity	Total
		WEIGHT	4	3	4	3	3	2	4	4	4	3	2	4	2	3	
		CODE	D1	D2	D3	D4	D5	D6	D7	D8	D9	D10	D11	D12	D13	D14	
Security	encrypted data	P1	3	0	2	0	1	1	0	1	1	3	3	0	0	2	54
	fingerprint unlock	P2	3	0	3	0	0	0	0	0	0	1	2	0	0	0	33
	face recognition	P3	3	0	3	0	0	0	0	0	0	1	2	0	0	0	31
Battery	longer life lithium batteries	P4	0	0	3	0	0	0	0	0	0	0	0	0	0	0	12
	solar batteries	P5	0	0	3	0	0	0	0	0	0	0	0	0	0	0	12
	pad charging	P6	0	0	3	0	0	0	0	0	0	0	0	0	0	0	12
User Interface	Image Projection	P7	0	0	3	0	0	0	0	0	0	0	0	0	0	0	12
Connectivity	5G	P8	3	0	3	0	2	3	3	3	3	3	2	0	0	0	85
	Gesture Control	P9	3	0	3	0	0	0	0	0	0	2	3	3	1	3	59
Connected Wellness	Apps	P10	3	0	3	1	3	2	2	2	2	3	3	0	0	1	82
	IOT	P11	3	0	3	0	0	3	3	3	3	3	3	2	0	3	98
	seamless sync	P12	3	0	3	0	3	3	3	3	3	3	3	0	0	2	96
	monitor body	P13	3	0	0	0	0	3	3	3	3	3	3	3	0	3	90
	SOS button	P14	3	0	0	0	0	2	3	3	3	3	3	0	0	0	67
	Reverse communication with health professional	P15	3	0	0	0	0	3	3	3	3	3	3	0	0	0	69
	Notifications	P16	3	0	2	0	0	2	2	3	2	2	3	0	0	1	67

Technologies

TECHNOLOGY CATEGORIES				
Code	Technology Category	Current level	Need	Gap
T1	IOT	Under development	Comprehensive and seamless connections to other devices	Build the system
T2	Accurate sensor	Not 100% accurate	Increase accuracy	100% accuracy
T3	Motion sensor	Not developed	Precision, small and cheap	Build the system
T4	Location services	Gps available	Enable emergency services	Integrate with map system
T5	Artificial Intelligence	Not developed for watch	Security	Build system for watch
T6	Battery charging	Average	More convenient charging	Reliance on lithium
T7	Cellular Connectivity	4G	Faster connectivity with more coverage	Inefficient connectivity
T8	Cloud Computing	Developed	Storage and data transmission	Advanced and integrated

QFD Technologies vs. Product Features

		encrypted data	fingerprint unlock	face recognition	longer life lithium batteries	solar batteries	pad charging	image Projection	5G	Gesture Control	Apps	IOT	seamless sync	monitor body	SOS button	Reverse communication with health professional	Notifications	Totals
	WEIGHT	3	2	2	1	1	1	1	4	3	4	4	4	4	3	3	3	
	CODE	P1	P2	P3	P4	P5	P6	P7	P8	P9	P10	P11	P12	P13	P14	P15	P16	
IOT	T1	0	0	0	0	0	0	0	0	3	2	3	3	3	3	3	2	94
Accurate sensor	T2	0	0	0	0	0	0	0	0	3	3	3	0	3	3	1	2	63
Motion sensor	T3	0	0	0	0	0	0	0	0	3	3	3	0	3	3	1	2	63
Location services	T4	0	0	0	0	0	0	0	0	0	1	3	1	1	3	2	2	45
Artificial Intelligence	T5	0	2	2	0	0	0	0	0	2	2	3	0	2	3	2	2	63
Battery charging	T6	0	0	0	3	3	3	0	0	0	0	0	2	2	1	1	1	37
Cellular Connectivity	T7	0	0	0	0	0	0	0	3	1	1	2	3	2	3	3	3	74
Cloud Computing	T8	3	0	0	0	0	0	0	2	0	1	3	3	3	3	3	3	84

Resources

Resources		
Code	**Resources**	**Examples**
R1	Sensor development	Intel
R2	Software development	Google
R3	App development	Microsoft
R4	Cloud services	AWS
R5	Battery services	Energizer
R6	Cellular Network	Sprint, Verizon
R7	Project Management	Coordinate all partners
R8	Location Services	Google
R9	Academic Institutions	PSU

Resources Analysis

Resources			
Code	Resources	Examples	Timeline
R1	Sensor development	Intel	Existing but more research needed
R2	Software development	Google	Partnership and start the development in 2 years
R3	App development	Microsoft	Immediate implementation
R4	Cloud services	AWS	Does not exist, in 5 years to be implemented
R5	Battery services	Energizer	Existing but more research needed
R6	Cellular Network	Sprint, Verizon	In 10 years the partnership should realize results
R7	Project Management	Coordinate all partners	Existing, constant monitoring
R8	Location Services	Google	Existing but more research needed in 5 years
R9	Academic Institutions	PSU	Create a partnership in next 5 years

Technologies vs. Resources

TECHNOLOGIES VS. RESOURCES			
Code	Technology	The Gap	Resources
T1	IOT	Build the system	R1,R2,R3,R4,R6, R7,R8
T2	Accurate sensor	100% accuracy	R1,
T3	Motion sensor	Build the system	R1,R8
T4	Location services	Integrate with map system	R8
T5	Artificial Intelligence	Build system for watch	R1.R2. R7, R8
T6	Battery charging	Reliance on lithium	R5, R9
T7	Cellular Connectivity	Inefficient connectivity	R6
T8	Cloud Computing	Advanced and integrated	R3,R4,R6,R7

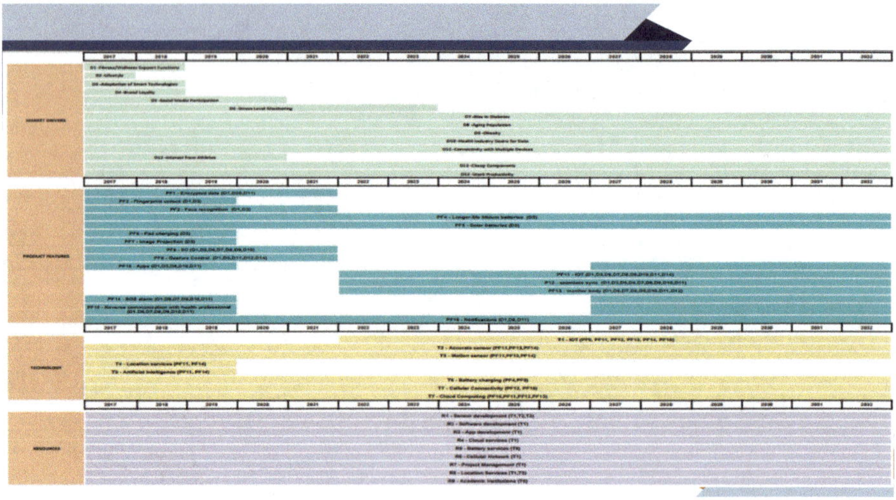

Q&A

Backup

1.Social Drivers

- Fitness/Wellness support functions
- Lifestyle
- Increased adaptation of smaller technologies
- Brand loyalty
- Social media participation via automatic upload
- Growth in biofeedback technology for sports/fitness performance

2. Health and Fitness Tracking Drivers

- Stress level monitoring
- Hazards of using smart phones
- Rise in Diabetes
- Senior citizens
- Rise in obesity

3. Technical Drivers

Health industry desire for data

- Cloud Solutions - periods of physical inactivity,recovery,duration and performance of sleep
- Consumers looking for customized health programs

Connectivity to multiple devices

- Support Wi-Fi (802.11b/g/n 2.4GHz) and Bluetooth 4.0.
- IoT - sensor technologies providing connection to multiple devices (e.g. Smart Home solutions)
- Gesture controlled

4. Economical Drivers

Sports brands entering wearable technologies market
- New Balance, Under Armour

Low manufacturing costs/ Affordability
- Less expensive components compared to smartphones

Luxury watchmakers enter the smartwatches market
- TAG Heuer, Montblanc, Movado

Predictive approach to advertising
- Customer's data analyzed for tailored advertisement

Businesses adapting new technologies to increase workers productivity
- Monitoring workers

28

Future Technology

Our goal is to...

Create a technology roadmap that will enable Fitbit to succeed in delivering what consumers are looking for in a smartwatch: stylish, well-designed devices that combine the right general purpose functionality and cutting-edge technologies with a focus on health and fitness.

Future Features:

- Track movement, such as arm gestures (biofeedback)
- Monitor health issues (blood sugar, hydration)
- Communicate data with health professionals, historically and real time
- Coaching applications
- Distress signaling, tracking outdoors
- Distress signals for at risk individuals (elderly, diabetic)
- Expansion to jewelry and clothing

29

IOT - The watch connects to the refridgerator and scale, then sends info to computer or health professional

Accurate Sensor - body monitoring

Motion Sensors- Body gestures e.g Running, Walking

Location Services - GPS

Artificial Intelligence- Predicting Patterns

Battery Charging - Solar

Cellular Connectivity

Cloud Computing - Data Collection, Reporting, Analysis

Appendix 6

Technology Roadmap for Autonomous Vehicles Systems: Standardization

Ghada Alramadan
Husam Barham
Mitsutaka Shirasaki
Qin Guo
Sriharika Patibanda
Timothy Valentine

1

T. U. Daim (ed.), *Roadmapping Future*, Applied Innovation and Technology Management, https://doi.org/10.1007/978-3-030-50502-8

Outline

- ❖ Background
- ❖ Project Proposal
- ❖ Market Drivers
- ❖ Product Features
- ❖ Drivers Vs. Products QFD
- ❖ Technologies
- ❖ Products Vs. Technologies QFD
- ❖ Resource Analysis
- ❖ Technology Analysis
- ❖ TRM
- ❖ Conclusions

ETM
2

Background

❖ According to U.S. Department of Transportation: [1] "autonomous" or "self-driving" vehicles are:

Those in which operation of the vehicle occurs without direct driver input to:

- ➢ Control the steering
- ➢ Acceleration
- ➢ Braking

❖ The driver is not expected to constantly monitor the roadway while operating in self-driving mode

ETM
3

Background

- Autonomous cars by google hit the roads in 2009, and by March 2016 reached 1.5 million miles of autonomous driving. [6]
- Many car makers working with google or on their own projects
  - Toyota, Audi, BMW, GM [7]

ETM
4

Project Proposal

- Objective is to develop a Technology Roadmap for the "autonomous vehicles standardization"
- What need to be standardized on industry level to enable **effective vehicles interaction** with:
  - Other Vehicles
  - The Road Elements
- Who are we: Consultants to the industry and government
- The industry is car makers and technology vendors.

ETM
5

Market Drivers

- **Saving Time On The Road**
  - **D1:** Population Growth
  - **D2:** Traffic Jam Reduction
  - **D3:** Managing Accidents Effectively
  - **D4:** Reduce Parking Search time
  - **D5:**Space Management for Vehicles and Pedestrians
- **Safety: Public and Personal (Driver/ Passenger)**
  - **D6:** Accidents - Prevention and Report
    - Drunk Drivers
    - Inexperienced Drivers
    - Distractions (From the Passengers - Texting)
    - Speeding
  - **D7:** Traffic Prediction
  - **D8:** Pedestrian Safety

ETM

6

Market Drivers

- **Environmental Drivers**
  - **D9:** Reduce CO2
  - **D10:** Reduce Number of Cars on the Road
  - **D11:** Reduce Environmental Destruction (e.g. Reduce Construction)
- **Economical Drivers**
  - **D12:** Fuel Price Increasing
  - **D13:** Insurance Cost (Auto Insurers)
- **Consumer Interest/Passenger Experience**
  - **D14:** Driver's Ability to Multitask
  - **D15**: Security

ETM

7

Market Drivers Table

Categories	Code	Market Drivers	Weight
Saving Time on the Road	D1	Population Growth [11]	4
	D2	Traffic Jam Reduction [12]	4
	D3	Managing Accidents Effectively [22]	2
	D4	Reduce Parking Search Time [23]	4
	D5	Space Management for Vehicles and Pedestrians [13]	2
Safety: Public and Personal	D6	Accidents - Prevention and Report [22]	4
	D7	Traffic Prediction [27]	2
	D8	Pedestrian Safety [24]	4
Environmental Drivers	D9	Reduce CO2 [13]	2
	D10	Reduce Number of Cars on the Road [29]	1
	D11	Reduce Environmental Destruction [13]	2
Economical Drivers	D12	Fuel Price Increasing [13]	2
	D13	Insurance Cost (Auto Insurers) [25]	1
Consumer Interest / Passenger Experience	D14	Driver's Ability to Multitask [28]	4
	D15	Security [26]	4

ETM

8

Product Categories

Code	Product Category
P1	Connectivity
P2	Smart
P3	Safety
P4	Usability

ETM

9

Product Features

Code	Product Feature	Product
PF1	Intelligent merging (on-ramps, exits, lane ends, free left lanes, flyovers)	P2, P3
PF2	Reduced safe follow distance	P2, P3
PF3	Auto accident reporting and tracking system [13]	P1, P3
PF4	Auto parking searching function [23]	P1, P2
PF5	Energy saving mode (Based on Road conditions)	P2, P4
PF6	Road elements sensors (I'm a curb, I'm a traffic light and I'm red now)	P1, P2, P3
PF7	Sense traffic lights and calculate quickest route	P1, P2
PF8	Auto recognition of road laws (each state has its own road laws) [15][16]	P1, P2, P3, P4
PF9	Protection against hacking [26]	P1, P2, P3
PF10	Passengers Health Monitoring [30]	P1, P2, P3, P4
PF11	Feature to adapt to weather condition	P2, P3, P4
PF12	Object Identification/Avoidance [15][16]	P2, P3, P4

ETM

10

Resource for the Product Features

QFD – Product Features Vs Market Drivers

Product Features		Population growth	Traffic jam reduction	Managing Accidents Effectively	Reduce Parking Search Time	Space Management for Vehicles and Pedestrians	Accidents - Prevention & Report	Traffic Predicition	Pedestrian Safety	Reduce CO2	Reduce Number of Cars on the Road	Reduce Environmental Destruction	Fuel Price Increasing	Insurance Cost (Auto Insurers)	Driver's Ability to Multitask	Security	
		D1	D2	D3	D4	D5	D6	D7	D8	D9	D10	D11	D12	D13	D14	D15	Priority
		4	4	2	4	2	4	2	4	2	1	2	2	1	4	4	
Inteligent Merging	PF1		2				2							1			17
Reduced safe follow distance	PF2		1			1						1					8
Tracking& Reporting System	PF3			4			2	4						1			25
Auto parking searching	PF4		1		4	2							1				26
Energy saving mode	PF5									2			2				8
Road element sensor	PF6		1				4		1					1	1		29
Traffic lights sensor	PF7		4			1		2					1				24
Road law prediction	PF8						2		4					1			25
Against hacking	PF9															4	16
Health monitor	PF10																0
Adapt to Weather condition	PF11						4							1			17
Object ID/Avoidance	PF12						4		4					1	2		41

Market Driver vs. Product Features	
H	4
M	2
L	1
None	0

Priority
High
Medium
Low

ETM

11

Priority Product Features

Code	Product Category
P1	Connectivity
P2	Smart
P3	Safety
P4	Usability

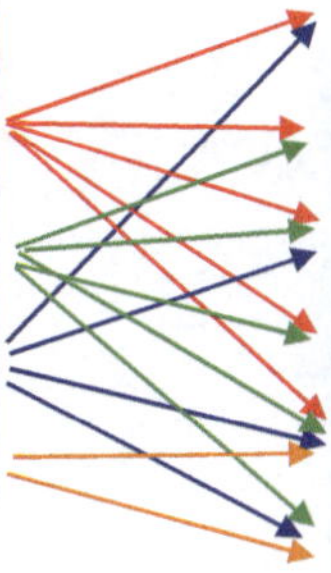

Code	Product Features	Weight
PF3	Tracking & Reporting System	25
PF4	Auto parking searching	26
PF6	Road element sensor	29
PF7	Traffic lights sensor	24
PF8	Road law prediction	25
PF12	Object ID/Avoidance	41

ETM

12

Priority Product Features: Time Horizon

	Product Category	Product Features
P1	Connectivity	PF3, PF4, PF7, PF6, PF8
P2	Smart	PF4, PF6, PF7, PF8, PF12
P3	Safety	PF3, PF6, PF8, PF12
P4	Usability	PF8, PF12

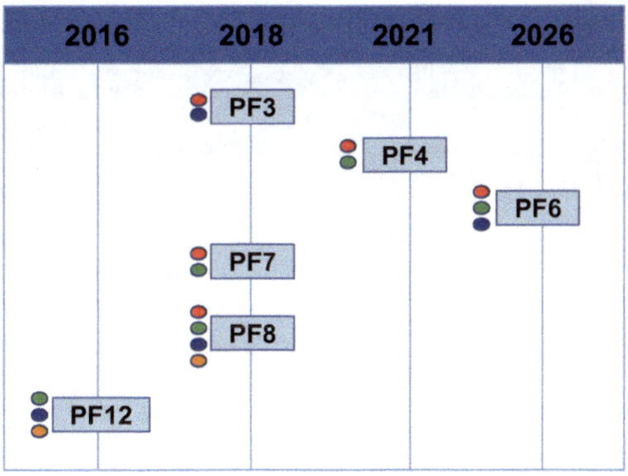

ETM

13

Technologies

Code	Technology	The Need	Current Status	The Gap
T1	Accurate Sensors [14]	Reliable and affordable sensors	Sensors are good but expensive and not 100% accurate	100% accuracy
T2	Car to Car Communication System [16][18]	A robust standard system for cars to communicate	Under Development	Finalizing the System
T3	Radar Technology [15][20]	Effective ability to identify hidden/moving objects	Radar technology is good but not 100% accurate	100% accuracy
T4	A.I. (Road Decision Making system) [19][15]	Analyze the feed from various sources and make decisions	Under Development	Finalizing the System
T5	Element Identification system [15]	Analyzing the cameras around the vehicle, sensors, radar, and laser feeds	Under Development	Build the System!

ETM
14

Technologies

Code	Technology	The Need	Current Status	The Gap
T6	Hotspot wireless [16][18]	Cars always connected to the internet	Hotspot wireless is good but not available 100% everywere	Enhancing the quality and availability of hotspot wireless
T7	Central Road Technology System [17][18]	A system that contains information about the roads, vehicles, law, etc..	No such system	Developing the system
T8	Night Vision (laser) [15][21]	Ability to identify objects accurately in the night	Current night vision Tech. is good but not 100% accurate	100% accuracy
T9	Smart broadcasting road objects	A broadcasting object that says: I'm a curb, A traffic light and so on,	No such objects	Build and install the object
T10	Accident Assessment System	A system to estimate and react to accidents	No such system	Developing the system

ETM
15

Technology Categories

Code	Technology Category	Technologies
TC1	Software Systems	T2, T4, T7, T10
TC2	Objects Recognition Technologies	T1, T3, T5, T8
TC3	A.I.	T4, T10
TC4	Infrastructure	T6, T9
TC5	Road Elements Technologies	T7, T9

16

Technologies: Time Horizon

	Technology Category	Technologies
TC1	Software Systems	T2, T4, T7, T10
TC2	Objects Recognition Technologies	T1, T3, T5, T8
TC3	A.I.	T4, T10
TC4	Infrastructure	T6, T9
TC5	Road Elements Technologies	T7, T9

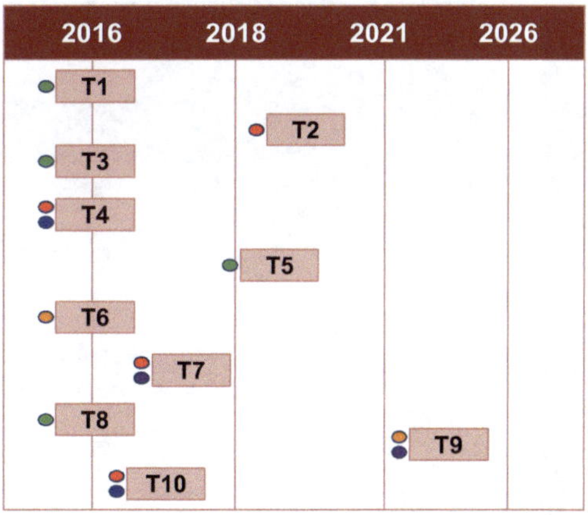

17

Technology Features vs. Product Features

			Auto accident reporting and tracking system	Auto parking searching function	Road elements identification	Sense traffic lights and calculate quickest route	Auto Recognition of Road Law	Object ID/Avoidance	
			PF3	PF4	PF6	PF7	PF8	PF12	Priority
		Weight	1	1	2	1	1	4	
Technology Features	Accurate sensors	T1	1	1	4	1	1	4	12
	Car to car communication system	T2	2	4	1	1	1	2	11
	Radar Technology	T3	1	1	4	1	1	4	12
	Artificial intelligence	T4	1	4	4	4	1	4	18
	Element identification system	T5	1	2	4	2	1	4	14
	Hotspot wireless	T6	2	4	1	2	2	2	13
	Central road technology system	T7	2	2	1	4	4	1	14
	Night vision (Laser)	T8	1	1	4	1	1	4	12
	Smart broadcasting road objects	T9	1	1	4	2	1	4	13
	Accident assessment system	T10	4	1	1	1	1	1	9

Technology features vs. Product features	
Strong relationship	
Medium relationship	
Week relationship	

Priority	
High	
Medium	
Low	

18

Resources

Resource Category	Code	Resources	Examples
R&D	R1	Partnership with University	PSU, Harvard
	R2	Computer Hardware	Intel, Cisco
	R3	Computer Software	Google
	R4	Car Manufacture	Toyota, BMW, Tesla, Audi
	R5	US Military	Army, Navy
	R6	Department of Energy	Collaboration with Department of Transportation
Incentives	R7	Government	Federal, State, Local
	R8	Industries	Google, Tesla, Toyota, BMW
Standard	R9	Standard	Transportation Dept., SAE

19

Technologies vs. Resources

Code	Technology	The Gap	Resources
T1	Accurate Sensors [14]	100% accuracy	R4,R5,R8,R9
T2	Car to Car Communication System [16][18]	Finalizing the System	R1,R2,R3,R4,R8,R9
T3	Radar Technology [15]	100% accuracy	R4,R5,R8,R9
T4	A.I. (Road Decision Making system) [19][15]	Finalizing the System	R1,R2,R3,R8,R9
T5	Element Identification system [15]	Build the System!	R3,R4,R5,R7,R9

20

Technologies vs. Resources

Code	Technology	The Gap	Resources
T6	Hotspot wireless [16][18]	Enhancing the quality and availability of hotspot wireless	R2,R4,R6,R8
T7	Central Road Technology System [17][18]	Developing the system	R3,R9
T8	Night Vision (laser) [15]	100% accuracy	R4, R5,R6
T9	Smart broadcasting road objects	Build and install the object	R3,R7,R8,R9
T10	Accident Assessment System	Developing the system	R2,R3,R4R9

21

TRM

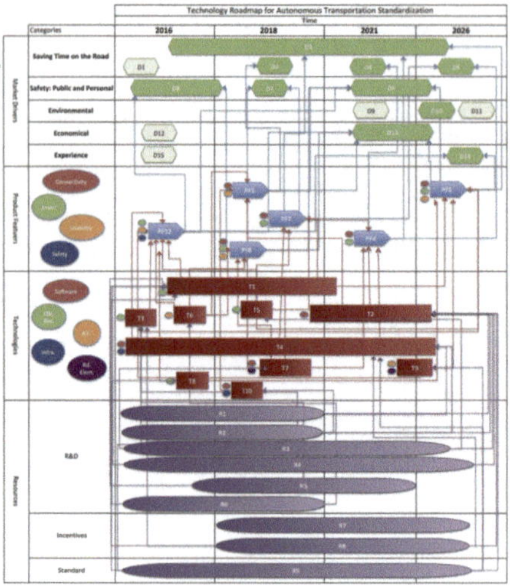

22

Conclusions

- Our TRM looked at what need to be standardized on industry level to enable effective vehicles interaction with other vehicles and road elements.
- Adoption of standardization for autonomous vehicles will require extensive collaboration across manufacturers, government agencies, and research institutions.
- Artificial Intelligence is the most critical technology for developing autonomous cars.
- While the TRM identified what standardization is needed to reduce the efforts and resources needed to enable autonomous cars, we are not sure that the timeline, car makers set, is achievable

23

Car Maker's Future Perspective [7]

- First autonomous **Toyota** to be available in **2020**
- Elon Musk now expects first fully autonomous **Tesla** by **2018**, approved by 2021
- **Volkswagen** expects first self driving cars on the market by **2019**
- **GM**: Autonomous cars could be deployed by **2020** or sooner
- **BMW** to launch autonomous iNext in **2021**
- **Ford's** head of product development: autonomous vehicle on the market by **2020**
- **Uber** fleet to be driverless by **2030**

24

References

[1] Available: http://www.nhtsa.gov/About+NHTSA/Press+Releases/U.S.+Department+of+Transportation+Releases+Policy+on+Automated+Vehicle+Development
[2] Available: http://www.motortrend.com/news/the-beginning-of-the-end-of-driving
[3] Available: http://ieeexplore.ieee.org/xpl/articleDetails.jsp?arnumber=811692
[4] Available: http://www.smart-systems-integration.org/public/documents/publications/EPoSS%20Roadmap_Smart%20Systems%20for%20Automated%20Driving_V2_April%202015.pdf
[5] R. Phaal, E. O'Sullivan, M. Routley, S. Ford, D. Probert, A framework for mapping industrial emergence, Technological Forecasting and Social Change, Volume 78, Issue 2, February 2011, Pages 217-230
[6] Available: https://static.googleusercontent.com/media/www.google.com/lt//selfdrivingcar/files/reports/report-0316.pdf
[7] Available: http://www.driverless-future.com/?page_id=384
[8] Market Drivers (Economical & Customers) Available: http://www.mckinsey.com/industries/high-tech/our-insights/disruptive-trends-that-will-transform-the-auto-industry
[9] Available: http://www.businessinsider.com/report-10-million-self-driving-cars-will-be-on-the-road-by-2020-2015-5-6
[10] Portland Autonomous car Available: http://gizmodo.com/these-7-american-cities-will-compete-for-40-million-to-1764431906
[11] Available: http://www.census.gov/popest/data/national/totals/2014/index.html,
[12] Available: http://traveltips.usatoday.com/effects-traffic-congestion-61043.html
[13] Available: http://autocaat.org/Technologies/Automated_and_Connected_Vehicles/
[14] Available: http://www.mouser.com/applications/autonomous-car-sensors-drive-performance/
[15] Available: http://gizmodo.com/how-to-teach-an-autonomous-car-to-drive-1694725874
[16] S. Biswas, T. R and D. F, Vehicle-to-vehicle wireless communication protocols for enhancing highway traffic safety, Communications Magazine, IEEE, 2006.
[17] Available: http://www.autonews.com/article/20140310/OEM06/303109959/like-evs-self-guided-cars-need-infrastructure
[18] Available: http://www.industryweek.com/datasheet/autonomous-vehicle-technology-aka-auto-iot-industry-report
[19] Available: http://www.ni.com/newsletter/50346/en/
[20] Available: http://safety.trw.com/autonomous-vehicles-will-need-3d-360-degree-sensing-and-better-object-recognition/0605/
[21] Available: http://spectrum.ieee.org/transportation/advanced-cars/bmw-laser-headlights-slice-through-the-dark/deer-spotting-diodes
[22] Available: http://www.fhwa.dot.gov/publications/publicroads/13julaug/05.cfm
[23] Available: http://shoup.bol.ucla.edu/CruisingForParkingAccess.pdf
[24] Available: https://www.cdc.gov/motorvehiclesafety/pedestrian_safety/
[25] Available: http://www.forbes.com/sites/patricklin/2016/04/25/self-driving-cars-wont-kill-insurance-industry/#28f3bcb16afa
[26] Available: http://large.stanford.edu/courses/2015/ph240/khaliq1/
[27] Available: http://orfe.princeton.edu/~alaink/SmartDrivingCars/Reports&Speaches_External/Litman_AutonomousVehicleImplementationPredictions.pdf
[28] Available: http://www.auto-ui.org/15/p/workshops/4/Multitasking%20in%20Autonomous%20Vehicles%20Ready%20to%20Go.pdf
[29] Available: http://www.driverless-future.com/?p=911
[30] Available: http://www.strategyand.pwc.com/reports/connected-car-2015-study

ETM

25

Source: http://en.hdyo.org

ETM
26

Appendix 7

TEAM: Hussain Almohameed, Maria Pat Chavarria, Sema Kirkewoog, Nirupama Mantha, Jessie Truong, Konstantin Tuv

Executive Summary

Vision:

- Create an integrated sustainable energy future - one that is sustainable, less expensive, and just better.

Product Features to focus on:

- Versatile applications
- Internet Connected
- Less Waste
- More efficient

2

T. U. Daim (ed.), *Roadmapping Future*, Applied Innovation and Technology Management, https://doi.org/10.1007/978-3-030-50502-8

Executive Summary

Technology gaps to address:

- Nanoparticle based solar cells
- Microprocessor controlled automatic sun tracker
- In-situ monitoring for wired and wireless networking systems
- Smart solar system

Future Opportunities:

- Collaborate with PG&E to install smart inverters and battery storage systems
- Software to integrate smart technologies
- Government Incentives - tax credit
- Hardware - Collaborate to design and integrate CMOS

3

SolarCity Published Roadmap

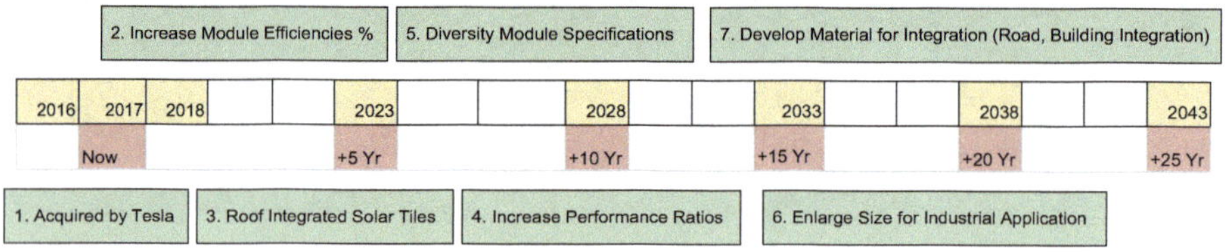

4

Market Drivers

Category	Label	Driver
Political	D1	Government Incentives
	D2	Regulations
	D3	Grid Integration
Environmental	D4	Sustainability
	D5	Global Warming
	D6	Predictable
Economic	D7	Investment Opportunity
	D8	Economic Benefits
	D9	Commercialization
Technical	D10	Performance
	D11	Storage Options
	D12	Connectivity
	D13	Innovation
	D14	IoT
Social	D15	Reduce Emissions
	D16	Limited Resources
	D17	Design

5

Product Feature

Category	Label	Feature
Design	P1	Flexible panels
	P2	Portable
	P3	Customizable
	P4	Versatile Applications
	P5	Seamless Integration
Materials	P6	Light weight
	P7	Less Waste
	P8	Recyclable Materials
Smart	P9	Self-Cleaning
	P10	Self-Maintaining
	P11	Internet Connected
	P12	Auto-Protection
	P13	Integrated Sensors
	P14	Intelligent Movement
Value	P15	Efficient
	P16	Longer life
	P17	Reliability
	P18	Affordable / Cheap

6

Driver & Product Roadmap

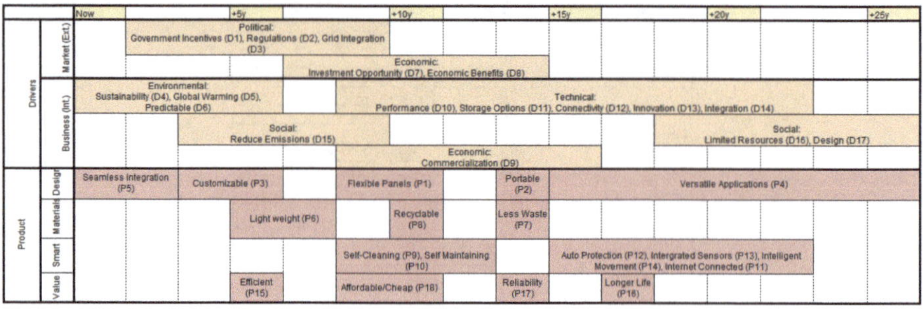

7

Technology Analysis

Category	Technology	Label	Technology Gap?
	Solar Skin	T1	No
	Micro/Nanocrystalline	T6	No
	Organic Solar Cells	T7	Yes
	Perovskite solar cells	T8	Yes
Material	Nanoparticle-based solar cell	T16	Yes
	Portable Solar Generator	T2	No
	Hybrid energy systems	T3	Yes
	Deployable Panel	T4	No
System	Solar-to-vehicle (S2V) systems	T5	No
	Electrostatic charge integration	T9	Yes
	Adaptive movement	T14	Yes
	Photovoltaics (PV), Concentrating solar power (CSP), Heating / Cooling Systems, Perovskite solar cells, Using the photons	T15	No
Integration	Smart Modules	T17	Yes
	Solar Analytics Monitor	T10	Yes
	Smart solar system	T11	Yes
	Microporcessor controlled automatic sun tracker	T12	Yes
Connectivity	In-situ solar panel monitoring system based on wired and wireless sensor network technologies	T13	Yes
Manufacturing	Dry processing	T18	Yes

8

Technology Gap Analysis

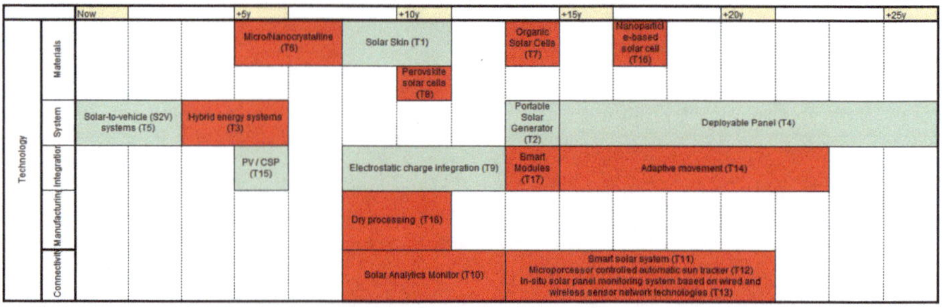

9

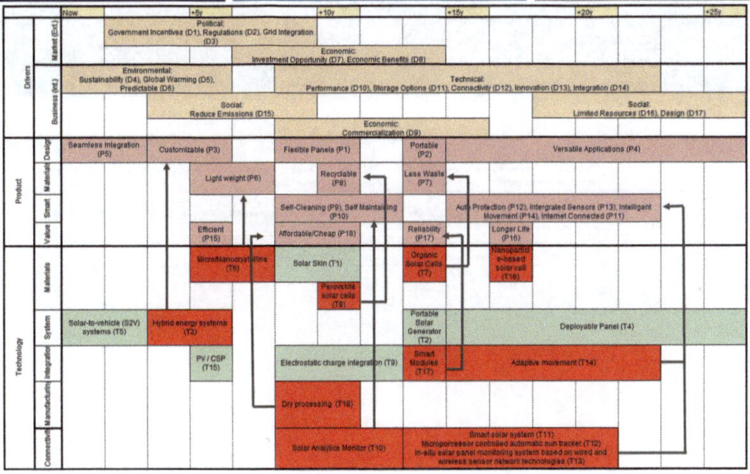

10

Resources: R&D

Category	Resources	Label	Definition	Gap	Technology
	Collaborate with PG&E	R1	PG&E is teaming up with SolarCity to install smart inverters and battery storage systems for residential rooftop solar customers	unlock additional benefits of solar power, battery storage, smart inverters for enhanced grid stability and control	T10, T11, T12, T13, T14, T15
	Computer hardware	R2	collaborate with Panasonic to design and integrate full autonomous vehicles	Develop devices for self driving systems. This could include a new piece of technology known as organic photoconductive film CMOS image sensors, which could allow for distortion-free detection of fast-moving objects. Considering Tesla prefers cameras and radar-based sensing systems to the lidar-based setup favored by most manufacturers of self-driving cars, such tech could be a boon to Elon Musk's efforts to achieve Level 5 autonomy in the near future.	T10, T11, T12, T13, T14, T15
	Computer software	R3	collaborate with software companies to design and integrate smart technologies to company products	integrate smart technology to current and future products	T10, T11, T12, T13, T14, T15
	University Partnership	R4	Collaborate with University of Michigan, University of Florida, and Princeton University to reserach and develop viable organic solar cell technology	Develop organic solar cell solution to migrate away from silicon solar cells	T7, T16
R&D	In house development	R5	Hire development Engineers to develop in house smart modules for solar panels	Integrate smart modules into existing and future product lines to enable panel-level maximum power point tracking, monitoring, and enhanced safety	T17

11

Resources: Incentive

Category	Resources	Label	Definition	Gap	Technology
	Government Tax Credit	R6	The federal government provides a solar tax credit, known as the investment tax credit (ITC), that allow homeowners and businesses to deduct a portion of their solar costs from their taxes.	Both homeowners and businesses qualify for a federal tax credit equal to 30 percent of the cost of their solar panel system minus any cash rebate. The timeline for the eventual end of the ITC in 2022.	T8, T3
	Solar renewable energy certificates (SRECs)	R7	Legislation in some states requires utilities to generate a certain percentage of their electricity from solar power	System will generate solar renewable energy certificates (SRECs) for the amount of electricity produced by the solar panel systems. Utilities will buy the SRECs so that they can count the solar power towards meeting their requirements. Selling SRECs can result in hundreds (or even thousands) of dollars more per year in income.	T8
Incentive	U.S. DOE's SunShot program	R8	With funding from the US DOE, SolarCity can dedicate more resources to efficient solar panels.	Systems will gain efficiency by > 10%.	T16

12

Resources: Partnership

Category	Resources	Label	Definition	Gap	Technology
Partnership	SpaceX	R9	Solar technology for spacecrafts already integrate Electrostatic charge to repel dust.	Gaining this IP from SpaceX will allow this technology to be applied towards industrial applications.	T9
	Partnership with Credit Suisse	R10	new fund to finance more than $300 million in solar projects	allow thousands of homeowners in the U.S. to pay less for solar electricity than they pay for utility bills.	T18
	Partnership with CSIRO	R11	Outsource printable perovskite solar cells to CSIRO, to enable hybrid solar cell product line	bridge technology from materials perspective to move from silicone solar panels to fully organic solar panel	T8

13

Resources & Technology Analysis

		Solar Skin	Portable Solar Generator	Hybrid energy systems	Deployable Panel	Solar-to-vehicle (S2V) systems	Micro/Nanocrystalline	Organic Solar Cells	Perovskite solar cells	Electrostatic charge integration	Solar Analytics Monitor	Smart solar system	Microporcessor controlled automatic sun tracker	In-situ solar panel monitoring system based on wired and wireless sensor network technologies	Adaptive movement	Perovskite solar cells	Nanoparticle-based solar cell	Smart Modules	Dry processing	
	Weight	4	1	4	2	4	4	2	2	2	2	2	2	2	2	2	2	2	2	
	Code	T1	T2	T3	T4	T5	T6	T7	T8	T9	T10	T11	T12	T13	T14	T15	T16	T17	T18	Priority
Collaborate with PG&E	R1	2	0	2	0	0	0	0	0	0	4	4	4	4	4	2	0	4	0	68
Computer hardware	R2	0	0	1	0	4	0	0	0	0	0	0	4	0	0	0	0	0	0	28
Computer software	R3	0	0	0	0	4	0	0	0	0	4	4	0	2	2	0	0	4	0	48
University Partnership	R4	0	0	0	0	0	0	4	0	0	0	0	0	0	0	0	0	0	0	8
In house development	R5	0	0	0	0	0	0	0	0	0	0	4	0	0	4	0	0	0	0	16
Government Tax Credit	R6	1	0	1	1	1	0	0	4	0	0	0	0	0	0	4	0	0	0	30
Solar renewable energy certificates (SRECs)	R7	0	0	0	0	0	0	0	4	0	0	0	0	0	0	0	0	0	0	8
U.S. DOE's SunShot program	R8	0	0	0		0	0	0	0	0	0	0	0	0	0	0	4	0	0	8
SpaceX	R9	0	0	0	0	0	0	0	0	4	0	0	0	0	0	0	0	0	0	8
Partnership with Credit Suisse	R10	0	0	0	0	0	0	0	0	0	2	2	0	0	0	0	0	0	4	16
Partnership with CSIRO	R11	0	0	0	0	0	0	4	0	0	0	0	0	0	0	0	0	0	0	8

High	>40
Medium	>20 & <40
Low	<20

14

Resources Roadmap

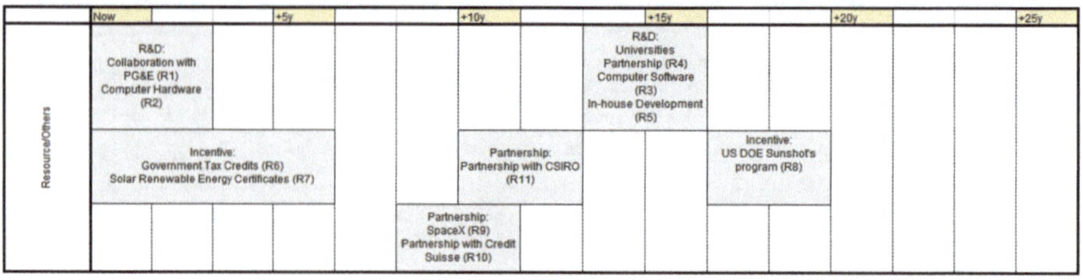

15

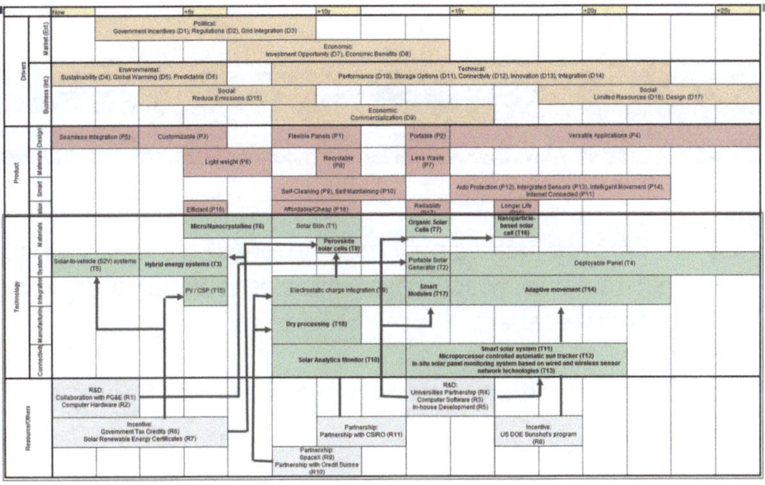

16

BACKUP: Week 3 - Week 6 Presentations

17

Objectives

Research Goal:

- Develop a roadmap for SolarCity solar energy products (25 years)

Problem Statement:

- Bridge the gap for mass user adoption

Research Method:

- Review SolarCity's current solar power technologies
- Review current solar power industry trends

Purpose of Research:

- Enable SolarCity's vision to grow market share in sustainable energy

18

Renewable Energy

Trend:

- Consumers are increasingly more interested in becoming more sustainable, which enables the renewable energy sector to grow market share.

Opportunity:

- Accelerate transition to sustainable energy by honing solar technology.

19

Porter's Five Driving Forces

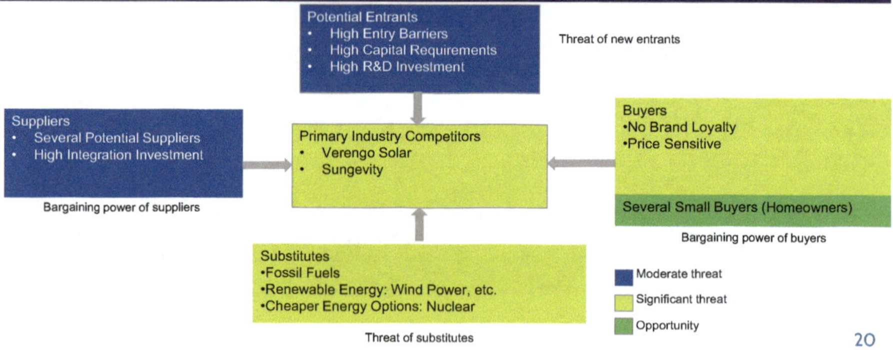

20

SWOT Analysis

Strengths	**Weakness**
Financial Resourcing Existing IP & Technology Integrated Full Value Chain Process Asset Leverage Pricing Power Economies of Scale	Dependence on Government Subsidies Profitability High Customer Acquisition Costs
Opportunities	**Threats**
Sustainability (Green) Trend Diminishing Fossil Fuels Innovation New Markets International Expansion	Brand Loyalty Other Existing Renewable Energy Sources

21

Market Drivers

Category	Label	Driver	Definition	Weight (Value)	Weight (#)
	D1	Government Incentives	Government initiatives & incentives on renewable energy - Government funding for Solar Panel research	Medium	2
	D2	Regulations	RPS regulations, Environmental regulations - Net Metering Laws (Buy back rules for energy companies)	High	4
Political	D3	Grid Integration	Solar grid - Integrated renewable energy in bulk power distribution system	Low	1
	D4	Sustainability	Sustainability, recyclable/reusable/transferable - Protect environment / ecosystem	High	4
	D5	Global Warming	Depletion of fossil fuels	Low	1
Environmental	D6	Predictable	Less Risky (compared to alternative energy)	Low	1
	D7	Investment Opportunity	Long term investment, stable, inflation-protected returns. Increase property value	Medium	2
	D8	Economic Benefits	Economic benefits, low cost, maintenance, grow market position - Protect from other energy sources increase in price	Medium	2
Economic	D9	Commercialization	Increase in applications for solar (Off Grid, Commercial, Power Plants, Residential)	Medium	2
	D10	Performance	Increase performance - Increase efficiency (reflective, thermodynamic, conductive), reliability	High	4
	D11	Storage Options	Generation of renewable energy (Storage), accessibility	Medium	2
	D12	Connectivity	Connectivity - real time data monitoring, integrated IOT	Medium	2
	D13	Innovation	First to market (pioneer in innovation), technology/innovation, Radical design enhancements, early new tech adaptor	Medium	2
Technical	D14	IoT	Radical design enhancements - Materials (integrated into roofs, building material, roads)	High	4
	D15	Reduce Emissions	support green energy - Low carbon footprint	High	4
	D16	Limited Resources	Limited land/building capacity - Floating Solar panels	Low	1
Social	D17	Design	Aesthetic	Low	1

22

Product Feature

Category	Label	Feature	Definition	Weight (Value)	Weight (#)
	P1	Flexible panels	Materials without rigid properties, moldable into several applications	Low	1
	P2	Portable	Applications beyond "one install and done," ability to move	Low	1
	P3	Customizable	Customers are allowed to customize panels size, color, design	Medium	2
	P4	Versatile Applications	Building applications (material, windows), floating unit applications, adjustable to a wide range of terrain	Medium	2
Design	P5	Seamless Integration	Easily integrated into building materials, cars, etc.	High	4
	P6	Light weight	Designs are lighter in weight with lighter materials	Low	1
	P7	Less Waste	Lower pollution rate during manufacturing process	Medium	2
Materials	P8	Recyclable Materials	Materials are easy to recycle/re-use at the end of the product life cycle	Medium	2
	P9	Self-Cleaning	Capability to remove dust, particles, etc.	Low	1
	P10	Self-Maintaining	Ability to detect possible issues preemptively and repair	Low	1
	P11	Internet Connected	Connected to IoT (i.e. Smart Home)	High	4
	P12	Auto-Protection	Damage resistant	Low	1
	P13	Integrated Sensors	Senses where the most light is coming from and adjusts position accordingly	Medium	2
Smart	P14	Intelligent Movement	Ability to move based on optimal light, ability to account for weather predictions	Medium	2
	P15	Efficient	Improved module efficiencies, conversion efficiencies	High	4
	P16	Longer life	Panels will produce at least 80% of the rated power after 30 years	Medium	2
	P17	Reliability	Long lasting, low degradation (rate of decay)	High	4
Value	P18	Affordable / Cheap	Lower manufacturing cost and more efficient use of materials, cost preference	High	4

23

SolarCity Published Roadmap

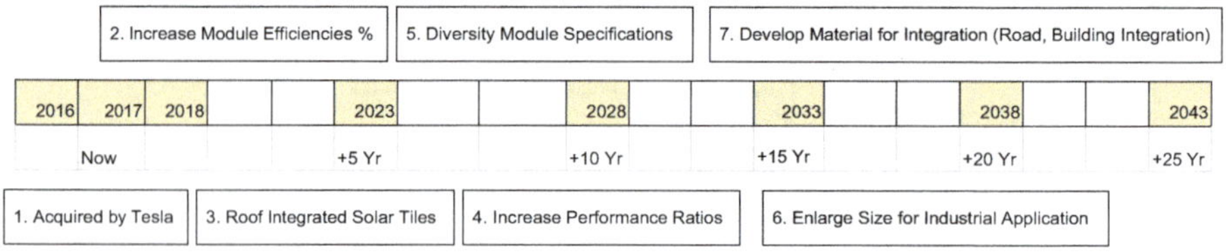

24

Driver & Product Roadmap

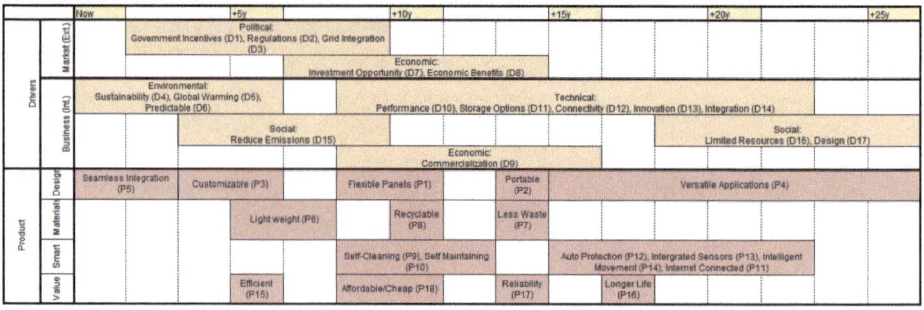

25

Driver & Product Analysis

		Government Incentives	Regulations	Grid Integration	Sustainability	Global Warming	Predictable	Investment Opportunity	Economic Benefits	Commercialization	Performance	Storage Options	Connectivity	Innovation	Integration	Reduce Emissions	Limited Resources	Design	
		D1	D2	D3	D4	D5	D6	D7	D8	D9	D10	D11	D12	D13	D14	D15	D16	D17	Priority
		2	4	1	4	1	1	2	2	2	4	2	2	2	4	4	1	1	
Flexible panels	P1	0	0	1	0	0	0	0	0	0	0	0	0	1	0	0	1	1	5
Portable	P2	0	0	1	0	0	0	0	0	0	0	0	0	1	0	0	1	1	5
Customizable	P3	0	0	0	0	0	0	0	0	0	0	0	0	1	0	0	1	2	5
Versatile Applications	P4	0	0	1	0	0	0	0	1	1	1	0	1	2	2	0	0	1	24
Seamless Integration	P5	0	0	2	1	0	0	0	0	0	1	0	1	1	1	0	0	2	20
Light weight	P6	0	0	0	0	0	0	0	0	0	1	0	0	0	0	0	0	1	5
Less Waste	P7	2	2	2	0	3	0	0	1	0	0	0	0	0	0	2	0	0	27
Recyclable Materials	P8	2	1	1	0	3	0	0	0	0	0	0	0	0	0	1	0	1	17
Self-Cleaning	P9	0	0	0	2	0	0	0	0	0	2	0	0	1	0	0	0	1	19
Self-Maintaining	P10	0	0	0	2	0	0	0	0	0	2	0	0	1	0	0	0	1	19
Internet Connected	P11	0	0	0	0	0	0	0	0	0	1	0	2	2	2	0	0	1	21
Auto-Protection	P12	0	0	0	2	0	0	0	0	0	1	0	0	1	0	0	0	1	15
Integrated Sensors	P13	0	0	0	0	0	1	0	0	0	1	0	1	2	0	0	0	1	12
Intelligent Movement	P14	0	0	0	0	0	1	0	0	0	3	0	1	2	0	0	0	1	20
Efficient	P15	1	0	2	0	1	0	2	2	0	2	0	0	0	0	0	0	0	21
Longer life	P16	0	0	0	0	0	0	2	2	0	1	0	0	0	0	0	0	0	12
Reliability	P17	1	0	0	0	0	0	1	0	0	1	0	0	0	0	0	0	0	8
Affordable / Cheap	P18	0	0	0	0	0	0	2	2	0	0	0	0	0	0	0	0	0	8

High
Medium
Low

26

Product Feature & Gap Analysis: Design

Category	Label	Feature	Current Level	Where we want to be	Gap
	P1	Flexible panels	Flexible solar panels developed, but not widely integrated into current products. Available for solar roofs.	Solar sky tower, transparent solar panels with integrated solar cells.	Transparent solar panels have been developed by competing companies through government funding. SolarCity could either acquire or partner with this program to enable a more industrial application.
	P2	Portable	Portable solar technology exists, but is not a focus for SolarCity.	Solar technology that can be installed for various purposes and moved easily.	Existing technology within SolarCity to enable mobile applications.
	P3	Customizable	Solar roof tiles are offered in 4 styles.	More customizations for the solar roof tiles as well as the solar panels so they complement the existing design.	User options.
	P4	Versatile Applications	Solar roof, solar panels for public and industrial applications.	Roll up sheets, floating panels.	Technology exists but is very expensive. Need to reduce cost so industrial applications are more common.
Design	P5	Seamless Integration	Solar panels on EV is currently capable from adding 2-20 miles per day.	Solar panel shield or roof that is capable to generate 30-50 miles of energy per day.	Need an innovative solution, such as retractable solar shield. Solar panels on the roof of car don't generate enough power.

27

Product Feature & Gap Analysis: Materials

Category	Label	Feature	Current Level	Where we want to be	Gap
	P6	Light weight	LightMount panels used in commercial installations weight 2.3lbs/Sq.Ft	Solar panels to weigh 1lb/sq. ft	Reduce solar panel weight year over year to reach 1 lb/sq. ft goal.
	P7	Less Waste	409 metric tons of CO_2, per MW installed.	Maintain 5.5%+ year over year reduction in Co_2 emissions. Organic solar cells.	Develop organic solar cell product line, and continue to optimize silicon solar panel MFG to reduce CO_2 emission.
Materials	P8	Recyclable Materials	Silicon solar panels are currently 96% recyclable, while non silicon solar panels are up to 98% recyclable.	Transition to non silicon solar panels to be best in class for recycling potential.	Develop non silicon solar panels.

28

Product Feature & Gap Analysis: Smart

Category	Label	Feature	Current Level	Where we want to be	Gap
	P9	Self-Cleaning	Annual cleaning can cost users up to hundreds of dollar	Solar panels to have "self cleaning" properties that break down organic dirt and reduce adhesion of non organic dirt.	Integrate "self cleaning glass" technology for solar panel coating.
	P10	Self-Maintaining	Users spend $100-$1000s to maintain solar panels	Low maintanance or no maintenance units. Catch pre-failures before they occur.	Solar panel self diagnostics, increased reliability.
	P11	Internet Connected	Phone app, Smart Homes (Google, Nest, etc.) integration	Full smart home integration, like Amazon Echo, etc.	Partner with smart home developers to establish partnerships, to have smart home capabilities beyond Nest.
	P12	Auto-Protection	Solar Roof is damage resistant from hail up to 100mph, lifetime warranty.	Able to withstand extreme weather conditions	wind and trauma resistant.
	P13	Integrated Sensors	NA	Integrated sensor + solar technology.	Develop integrated sensors for solar panels.
Smart	P14	Intelligent Movement	NA	Adjustable solar panels that respond to the Sun's position	Adjustable solar panels exist, but need to develop capability to integrate sensor and programmable movement.

29

Product Feature & Gap Analysis: Value

Category	Label	Feature	Current Level	Where we want to be	Gap
	P15	Efficient	Current module efficiency is 22.04%	Current best in industry is 26.3%.	Increase module efficiency by 3%+
	P16	Longer life	Panels can last 35 years, with a performance of 80% after 35 years.	Increase panel performance to 90% after 35 years.	Increase panel performance by 10%
	P17	Reliability	Current degradation is less than 0.5% per year	Module degradation is negligible, less than 0.1%	Reduce module degradation by 0.4%
Value	P18	Affordable / Cheap	Average $4.25 price per Watt for solar panels, 4% above industry average. Powerwall battery $5,500 + Supporting hardware $700. Solar roof Solar Roof will cost around $21.85 per square foot.	$4 price per Watt for solar panels, to be priced below industry average. Solar Roof and Powerwall already best value in class.	Reduce prices by $.25 per Watt for solar panels.

30

Technology Analysis

Category	Technology
Material	[T1][T6][T7][T8][T16]
System	[T2][T3][T4][T5]
Integration	[T9][T14][T15][T17]
Manufacturing	[T18]
Connectivity	[T10][T11][T12][T13]

31

Technology Analysis: Material

Category	Technology	Label	Definition	Product Gap
	Solar Skin	T1	Solar skin makes it possible for solar panels to match the appearance of a roof without interfering with panel efficiency or production.	[P1] Transparent solar panels have been developed by competing companies through government funding. SolarCity could either acquire or partner with this program to enable a more industrial application.
	Micro/Nanocrystalline	T6	Micro/Nanocrystalline or better known as Thin Film Solar Energy Panels are also one category of photovoltaic cells.	[P6] Reduce solar panel weight year over year to reach 1 lb/sq. ft goal.
	Organic Solar Cells	T7	An organic solar cell or plastic solar cell is a type of photovoltaic that uses organic electronics, a branch of electronics that deals with conductive organic polymers or small organic molecules, for light absorption and charge transport to produce electricity from sunlight by the photovoltaic effect. Most organic photovoltaic cells are polymer solar cells.	[P7] Develop organic solar cell product line, and continue to optimize silicon solar panel MFG to reduce CO2 emission
	Perovskite solar cells	T8	A perovskite solar cell is a type of non-silicon solar cell, which includes a perovskite structured compound, most commonly a hybrid organic-inorganic lead or tin halide-based material, as the light-harvesting active layer. Perovskite materials are usually cheap to produce and relatively simple to manufacture. Perovskite is relatively easy to process, and therefore cheaper to manufacture, but also has an efficiency of 22 per cent, close to silicon cells' 25 per cent.	[P8] Develop non silicon solar panels
Material	Nanoparticle-based solar cell	T16	With funding from U.S. DOE's SunShot program, the University of California, San Diego appears to have developed a revolutionary type of nanoparticle-based solar cell that can achieve a 90% efficiency.	[P16] Increase panel performance by 10%

32

Technology Analysis: System

Category	Technology	Label	Definition	Product Gap
	Portable Solar Generator	T2	Portable photovoltaic (PV) modular solar generator.	[P2] Existing technology within SolarCity to enable mobile applications.
	Hybrid energy systems	T3	The wind and solar energy are omnipresent, freely available, and environmental friendly. The wind energy systems may not be technically viable at all sites because of low wind speeds and being more unpredictable than solar energy. The combined utilization of these renewable energy sources are therefore becoming increasingly attractive and are being widely used as alternative of oil-produced energy.	[P3] User options.
	Deployable Panel	T4	A deployable panel for a spacecraft deployable solar array and other deployable structures. The panel has a number of panel sections hinged edge-to-edge for folding to a stowage configuration wherein the sections are disposed in confronting face-to-face relation and extension to a flat unfolded configuration wherein the panel sections are disposed in coplanar relation.	[P4] Technology exists but is very expensive. Need to reduce cost so industrial applications are more common.
System	Solar-to-vehicle (S2V) systems	T5	The solar power would be large enough to supply many commuters' needs. The implications for electric car design in relation to commuter range would vastly improve.	[P5] Need an innovative solution, such as retractable solar shield. Solar panels on the roof of car don't generate enough power.

33

Technology Analysis: Integration

Category	Technology	Label	Definition	Product Gap
	Electrostatic charge integration	T9	It uses electrostatic charge to repel dust and force it to the edges of the panels. It can remove 90 percent of the dust on a solar panel in a two-minute cycle.	[P9] Integrate "self cleaning glass" technology for solar panel coating.
	Adaptive movement	T14	Adaptive material greatly reduces the cost of a tracking system used in some types of solar power. It changes its reflectivity in response to heat from concentrated sunlight in a way that makes it possible capture light coming in at different angles throughout the day.	[P14] Adjustable solar panels exist, but need to develop capability to integrate sensor and programmable movement.
	Photovoltaics (PV) Concentrating solar power (CSP) Heating / Cooling Systems Perovskite solar cells Using the photons	T15	Photovoltaics (PV) : directly convert light to electricity. Concentrating solar power (CSP), which uses heat from the sun (thermal energy) to drive utility-scale, electric turbines. Heating and cooling systems, which collect thermal energy to provide hot water and air conditioning. Perovskite solar cells (as compared to the silicon cells that are used predominantly today) have seen some major breakthroughs in the past two years. The result will be a solar panel that can generate 20+ percent efficiency while still being one of the lowest cost options on the market. Using the photons, they developed a new solar cell structure for generating photocurrents. As well as demonstrating theoretical results of up to 63% conversion efficiency, it experimentally achieved up-conversion based on two photons, a mechanism unique to this solar cell.	[P15] Increase module efficiency by 3%+
Integration	Smart Modules	T17	Individual PV modules that can be enhanced/optimized for maximum production based on their location(like if they are in the shade).	[P17] Reduce module degradation by 0.4%

34

Technology Analysis: Manufacturing

Category	Technology	Label	Definition	Product Gap
Manufacturing	Dry processing	T18	PV cells are typically created via a wet chemical process that etches away layers of silicon from a crystalline wafer, leaving behind the solar cell. The SOLNOWAT process replaces the expensive and inefficient wet chemical process with the use of atmospheric pressure dry etching technology – a process that cuts costs and speeds up production. Because less silicon is removed during dry etching, the resulting cells are darker, making them very efficient at absorbing light. In fact, they are so efficient that they have been classified as having zero global warming potential.	[P18] Reduce prices by $.25 per Watt for solar panels.

35

Technology Analysis: Connectivity

Category	Technology	Label	Definition	Product Gap
	Solar Analytics Monitor	T10	Automatic diagnostic Indicator. The Solar Analytics Smart Monitor measures multiple electrical characteristics using a voltage reference frame and current measurements from the split-core current transformers. The device automatically connects to a cloud interface via 3G communications. This measured information is paired with patented smart algorithms to provide the solar PV system owner a world class monitoring and diagnostics platform.	[P10] solar panel self diagnostics, increased reliability
	Smart solar system	T11	Smart system that lets you consume your solar power intelligently. Giving you control of what appliances in the house consume solar power.	[P11] Partner with smart home developers to establish partnerships, to have smart home capabilities beyond Nest.
	Microprocessor controlled automatic sun tracker	T12	Microprocessor unit controls the movement of a solar panel that rotates and follows the motion of the sun	[P12] Wind and blunt force resistant.
Connectivity	In-situ solar panel monitoring system based on wired and wireless sensor network technologies	T13	The use of wireless sensor network technology provides low complexity and cost communication with a coordinator, while new web-based publishing technologies simplify the design, maintenance and operation of large networks of PV panels Panels with efficiency changes due to aging or other effects can be identified through the proposed in-situ characterization system and the owner will be informed in real- time for maintenance actions and possible guarantee claim.	[P13] Develop integrated sensors for solar panels.

36

Technology & Product Analysis

			Design					Materials			Smart						Value				
			Flexible panels	Portable	Customizable	Versatile Applications	Seamless Integration	Light weight	Less Waste	Recyclable Materials	Self-Cleaning	Self-Maintaining	Internet Connected	Auto-Protection	Integrated Sensors	Intelligent Movement	Efficient	Longer life	Reliability	Affordable / Cheap	Priority
		Weight	1	1	2	2	4	1	2	2	1	1	4	1	2	2	4	2	4	4	
Technology Categories	Technology	Code	P1	P2	P3	P4	P5	P6	P7	P8	P9	P10	P11	P12	P13	P14	P15	P16	P17	P18	
Material	Solar Skin	T1	4	2	2	4	2	2	0	0	0	0	0	0	0	0	1	0	1	0	36
System	Portable Solar Generator	T2	1	4	1	2	2	0	0	0	0	0	0	0	0	0	1	0	1	0	27
System	Hybrid energy systems	T3	0	0	4	4	2	0	0	0	0	0	0	0	0	0	2	0	0	0	32
System	Deployable Panel	T4	0	0	0	4	1	0	0	0	0	0	0	0	0	0	0	0	1	0	16
System	Solar-to-vehicle (S2V) systems	T5	0	0	0	2	2	0	0	0	0	0	0	0	1	0	0	0	1	0	18
Material	Micro/Nanocrystalline	T6	2	0	2	2	2	4	2	0	0	0	0	0	0	0	2	0	1	0	38
Material	Organic Solar Cells	T7	0	0	0	0	0	0	4	4	0	0	0	0	0	0	0	0	1	0	20
Material	Perovskite solar cells	T8	0	0	0	0	0	4	2	0	0	0	0	0	0	0	2	0	1	0	20
Integration	Electrostatic charge integration	T9	3	4	1	3	1	1	2	0	0	0	0	2	0	0	0	0	1	0	30
Connectivity	Solar Analytics Monitor	T10	1	0	0	0	0	0	0	0	0	0	0	0	0	0	0	4	4	4	41
Connectivity	Smart solar system	T11	0	0	1	2	0	0	0	0	0	0	4	0	4	0	4	0	1	0	58
Connectivity	Microporcessor controlled automatic sun tracker	T12	1	0	1	1	1	0	0	0	0	0	2	0	4	4	4	0	1	0	63
Connectivity	In-situ solar panel monitoring system based on wired and wireless sensor network technologies	T13	0	0	0	0	0	0	0	0	4	4	4	0	4	0	4	0	1	0	62
Integration	Adaptive movement	T14	0	0	0	0	0	0	0	0	0	0	2	2	4	4	3	0	1	0	42
Integration	Perovskite solar cells	T15	0	0	0	0	0	0	0	0	0	0	0	0	0	0	4	4	1	2	36
Material	Nanoparticle-based solar cell	T16	0	0	0	0	0	0	0	0	0	0	4	4	0	0	0	4	4	3	56
Integration	Smart Modules	T17	0	0	0	1	0	0	0	0	0	0	2	0	2	4	4	0	1	0	42
Manufacturing	Dry processing	T18	0	0	0	0	0	0	0	0	0	1	0	4	0	0	0	4	4	4	46

High	>40
Medium	>30 & <40
Low	<20

37

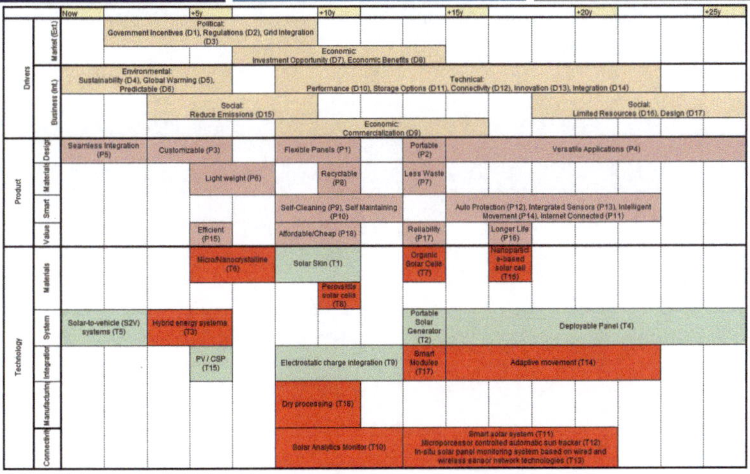

38

Appendix 8

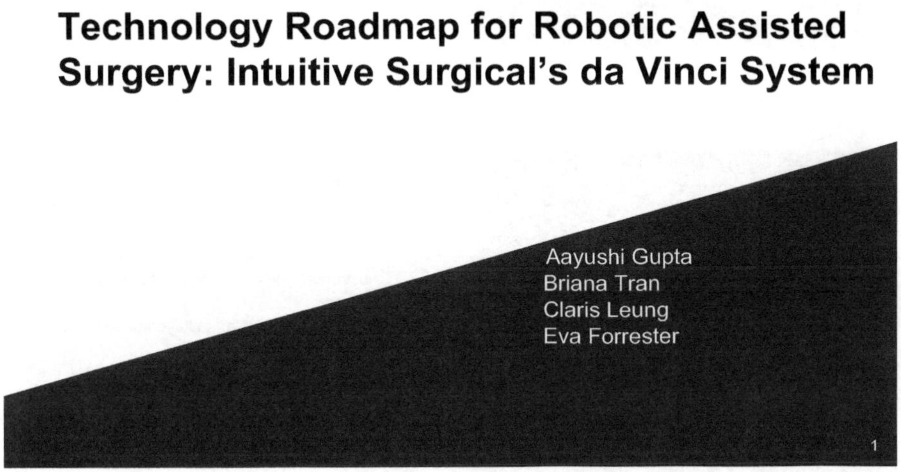

T. U. Daim (ed.), *Roadmapping Future*, Applied Innovation and Technology Management, https://doi.org/10.1007/978-3-030-50502-8

Agenda

- Intuitive Surgical Background
- Executive Summary
- SWOT Analysis
- Market Drivers
- Product features
- Technology Areas
- Resources
- QFD
- Final Roadmap
- References

2

Intuitive Surgicals

Intuitive by the Numbers

Da Vinci Intuitive Surgical HD

- An American corporation that develops, manufactures and markets robotic products designed to improve clinical outcomes of patients through minimally invasive surgery, most notably with the *da Vinci* Surgical System
- In 2017, there was an installed base of 4,271 units worldwide
- 2,770 in the United States, 719 in Europe, 561 in Asia, and 221 in the rest of the world
- corporate headquarters located in Sunnyvale, California

3

Executive Summary

Objective:

- To analyze the future possible technologies for Intuitive Surgical

Benefits:

- Less time,small incisions,better accessibility Safe and offers many potential benefits for patients, surgeons and hospitals when used appropriately and with proper training.

Current Features:

- 3D vision system
- 360° arms
- Specialist aided surgery
- Additional monitors
- Simulation and training
- Remote monitoring

4

Executive Summary

Potential Developments:

- Applicable to all surgeries
- Long-distance remote surgery
- Completely automated
- More miniaturized tools
- Faster operations
- Reduction in equipment/maintenance costs

Technology gaps to address:

- Ultra High Definition
- Integrate laser with 360° rotating arms for automated surgeries
- 8 quadrants, including underneath operating table
- Faster response from controller to surgical system. Increased CPU speeds.
- Increasingly more user-friendly interface for set-up and use.
- Increased options of surgical instruments
- More comfortable operating table with increased support of the patient's body.
- Increased number of procedures available for simulations
- Need a system to program, experts to create programs that are able to translate into movements

5

SWOT

Strengths

-Intuitive Surgical is the market leader, with 90% market share in surgical robotics.

-Intuitive Surgical is the innovation leader in the field both in terms of patents held and expenditures on R &D.

-Intuitive Surgical is operating from a position of financial strength; in a five-year span (2005-2010) revenues grew by 363%.

-The EndoWrist technology and 3D vision components enable surgeons to operate with improved precision and accuracy, resulting in improved clinical outcomes and reduced surgeon fatigue.

-Technology is able to meet the current trend of minimally invasive surgery.

Weaknesses

-Buyer's resource capability, the cost of the system is a limiting factor.

-Given the cost, only larger hospitals will be buyer's, which produces a finite market.

-Intuitive Surgical is currently dependent on one principal product, the da Vinci platform.

-The da Vinci surgical platform requires a dedicated operating room, given the bulk of the equipment and the specialized set-up.

-Not all surgeries can be performed using the da Vinci platform.

6

aayushi

SWOT

Opportunities

-Intuitive Surgical can expand further by expanding into more procedures and by expanding overseas

-As patients learn of the benefits of the da Vinci system, demand for surgeries performed using them will increase

-Intuitive Surgical also sells equipment and services related to the da Vinci system, as sales of the platform increase, so too will associated revenues.

-The trend toward minimally invasive surgery is increasing which is placing more technical demands on surgeons which the da Vinci system helps address.

Threats

-As health care reform focuses on cost containment, the expense of not only the system but the increased costs of surgeries (in increments of thousands) must be justified by Buyers

-Other companies are making heavy investments in research and development in an effort to catch up to Intuitive Surgical, for example, Titan attracted $40 million in market capitalization between 2009-2011.

-It is difficult to improve when you are the market leader; this can lead to a perception problem over time, as customers perceive the leader as staying stagnant while competitors improve

-"The characteristics of a third cohort of surgical robots are still to be defined but it is anticipated these intelligent new tools will be smaller, special purpose, lower cost, possibly disposable robots, providing alternatives to the current large, versatile and expensive systems."

7

aayushi

Market Drivers

Category	Market ID	Driver	Definition	Weight (Value)	Weight #
Technical	D1	Applicable to various procedures	Currently used for the following surgeries: Urology, Gynecology, General surgery, Thoracic surgery, Cardiac surgery	HIGH	3
	D2	FDA Approval	FDA approval for specific surgery procedures	HIGH	3
	D3	Increased surgical options	Addition of new procedures to the da Vinci system's repertoire	HIGH	3
	D4	Customer Support	Support teams are available on-site and/or in-person when needed	MEDIUM	2
Safety	D5	Minimally Invasive	Minimally invasive incisions lead to less blood loss, fewer complications, shorter hospital stays, smaller incisions for minimal scarring, and faster recovery and return to daily life.	HIGH	3
	D6	Precision	Minimize errors in the operating room.	HIGH	3
	D7	Non-Invasive Surgery	No incision needed for the surgery to occur	HIGH	3
	D8	Flexibility	Need for more flexibility than possible with just the operating surgeon's wrists and fingers	MEDIUM	2
Convienence	D9	Ease of Use - Surgeon	Easy for the surgeon to navigate the system to perform the necessary movements for the procedure	MEDIUM	2
	D10	Time	Minimizing the time needed for the procedure	MEDIUM	2

8

aayushi

Product Features

Category	Label ID	Feature	Definition
Design	P1	HD Imaging	Magnified vision system During a surgical procedure, the da Vinci Vision System displays high-definition, 3D imagery to the surgeon via the Surgeon Console and to the operating room staff via the Vision Cart.
	P2	360° Arms	Three or four robotic arms, which hold an endoscope (camera) and surgical instruments, carry out the surgeon's commands. The System seamlessly translates the surgeon's hand, wrist and finger movements into precise, real-time movement of the surgical instruments positioned inside the patient's body. These instruments can bend and rotate far greater than both traditional lap instruments and the human wrist.
Safety	P3	Specialist Aided Surgery	New da Vinci surgeons are assisted by experienced proctoring surgeons during their first da Vinci procedures. The goal is to provide direct support and help in ensuring proper technique.
Convienence	P4	Additional Monitors	A view of the operating field is available to the entire OR team on a large viewing monitor (vision cart). This widescreen view provides the surgical assistant at the patient's side with a broad perspective and visualization of the procedure.
	P5	Simulation & Training	The da Vinci Xi Skills Simulator™ allows surgeons to practice the skills required to effectively use the da Vinci Surgical System. Skills Simulator now offers patient cart simulation capabilities designed to train the perioperative staff.
	P6	Remote Monitoring	OnSite allows to the da Vinci Surgery Technical Assistance Team (dVSTAT®) remotely monitor your system status for real-time diagnostic feedback.
	P7	Remote Surgery	Allowing surgery to be performed without needing the surgeon to be in the operating room.
	P8	Complete Automation	Removing the need for a surgeon to be performing a procedure in real time. The entire procedure may be programmed in advance and optimized for the patient.

9

Claris

QFD: Market Drivers vs. Product Features

MARKET DRIVERS / PRODUCT FEATURES

	Applicable to various procedures (D1)	FDA Approval (D2)	Increased surgical options (D3)	Customer Support (D4)	Minimally Invasive (D5)	Precision (D6)	Non-Invasive (D7)	Flexibility (D8)	Ease of Use - Surgeon (D9)	Time (D10)	SCORE
	3	3	3	2	3	3	3	2	2	2	
HD Imaging (P1)	3 / 9	1 / 3	2 / 6	1 / 2	3 / 9	3 / 9	3 / 9	1 / 2	3 / 6	1 / 2	57
360° arms (P2)	3 / 9	2 / 6	3 / 9	1 / 2	3 / 9	3 / 9	1 / 3	3 / 6	3 / 6	2 / 4	63
Specialist Aided Surgery (P3)	3 / 9	1 / 3	3 / 9	3 / 6	1 / 3	3 / 9	3 / 9	1 / 2	2 / 4	1 / 2	56
Additional Monitors (P4)	1 / 3	1 / 3	1 / 3	1 / 2	1 / 3	2 / 6	1 / 3	1 / 2	3 / 6	1 / 2	33
Simulation & Training (P5)	2 / 6	1 / 3	2 / 6	1 / 2	2 / 6	2 / 6	2 / 6	1 / 2	3 / 6	1 / 2	45
Remote Monitoring (P6)	1 / 3	1 / 3	1 / 3	3 / 6	1 / 3	2 / 6	1 / 3	1 / 2	2 / 4	1 / 2	35
Remote Surgery (P7)	3 / 9	3 / 9	3 / 9	1 / 2	3 / 9	3 / 9	2 / 6	3 / 6	3 / 6	3 / 6	71
Complete Automation (P8)	3 / 9	3 / 9	3 / 9	2 / 4	2 / 6	3 / 9	3 / 9	1 / 2	1 / 2	3 / 6	65

10

Eva--very brief recall...

Market Drivers and Product Features Roadmap

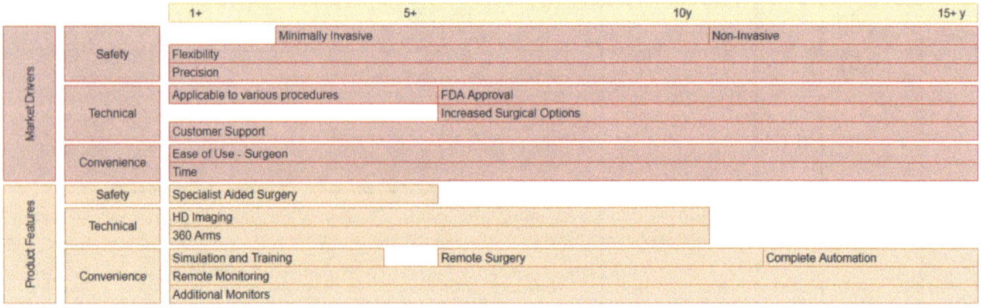

11

Eva--very brief recall...

Technology Areas

Category	Label ID	Technology	Currently	Gaps
Accuracy	T1	3DHD® Vision	Currently allows for visualizing anatomy in highly-magnified 3DHD. Crystal-clear image allows surgeons to see anatomical structures with enhanced definition and in natural color.	Ultra High Definition
	T2	Laser Targeting System	Currently, laser point at the target anatomy and the system will position the boom in an optimized configuration for the procedure	Integrate laser with 360° rotating arms for automated surgeries
Convenience	T3	Revolutionary Anatomical Access	Currently boom-mounted system with the flexibility of a mobile platform. Hybrid architecture enables placement of the surgical cart at any position around the patient while allowing four-quadrant anatomical access	access to underneath operating table
	T4	Master Controller Technology	Currently, system allows for precise, dexterous, remote control of the EndoWrist instruments.	Faster response from controller to surgical system. Increased CPU speeds.
	T5	TouchPad Technology	Comprehensive control of video, audio and system settings. Settings are stored to a unique user profile, providing automatic recall for future cases.	Increasingly more user-friendly interface for set-up and use.
Comfort	T6	EndoWrist® Instruments	Currently a set of various scissors, drivers, graspers, clip appliers, retractors etc. to provide flexibility human wrists lack	Increased options of surgical instruments
	T7	Trumpf Medical TruSystem 7000 dV Operating Table	Currently patient can be dynamically positioned while the surgeon operates	More comfortable operating table with increased support of the patient's body.
Software	T8	ERBE VIO dV Generator	Currently, integrated energy source for EndoWrist® instruments. Set desired tissue effect and the generator minimizes the energy delivered to surgical site	Higher power
	T9	da Vinci® Skills Simulator™	Currently allows surgeons to practice the skills required to effectively use the da Vinci® System	Increased number of procedures available for simulations
	T10	Programmable System	Not currently programmable	Need a system to program, experts to create programs that are able to translate into movements
	T11	Internet Capabilities	Allow either surgeon console and patient side cart to be able to connect to each other remotely. Also allow Intuitive Surgical Customer support to monitor the device and ensure all system is functioning properly.	Increased speed and function of the internet to allow surgeries across greater distances.

12

QFD: Product Features vs. Technology Areas

	Product Features								
	57	63	56	33	45	35	71	65	
Technology Areas	HD Imaging	360° Arms	Specialist Aided Surgery	Additional Monitors	Simulation and Training	Remote Monitoring	Remote Surgery	Complete Automation	Composite Score
3DHD® Vision	3	1	3	3	3	3	3	1	1019
laser targeting system	1	1	1	1	2	1	3	3	742
revolutionary anatomical access	1	3	2	1	2	1	3	3	924
Master controller technology	3	3	1	2	2	1	3	1	885
Trumpf Medical's TruSystem 7000dV Operating Table	1	1	1	1	1	1	3	3	697
TouchPad Technology	2	2	3	2	3	1	3	1	922
EndoWrist® Instruments	1	3	1	1	3	1	3	3	913
RBE VIO dV generator	1	3	1	1	1	1	3	3	823
da Vinci® Skills Simulator™	3	3	3	1	3	1	1	1	867
Programmable system	1	3	1	1	2	1	3	3	868
Internet capabilities	3	3	3	3	1	3	3	3	1185

13

Resources

Resource Code	Resources	Examples
R1	Government	FDA
R2	Software	IBM/Watson
R3	Hardware	Intel
R4	University/Company Partnerships	University/Company research labs
R5	Capital	Collabarations, Revenue stream
R6	Skills/Talent	International development
R7	Medical Knowledge	Medical Experts

14

Eva

Technology and Resources Roadmap

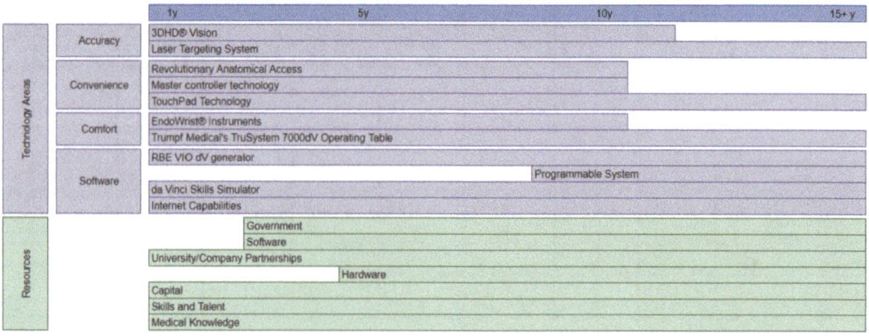

15

Briana

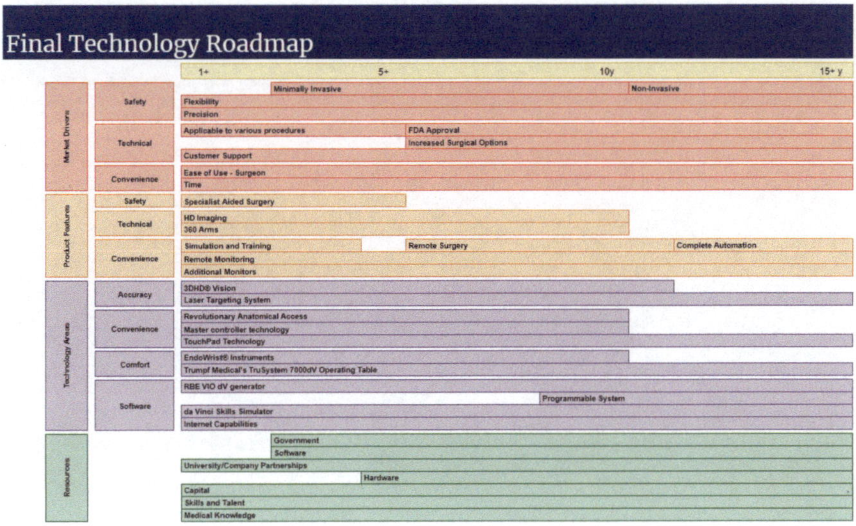

16

Briana

References

Intuitive Surgical. (2016). Frequently Asked Questions. [online] Available at: https://www.intuitivesurgical.com/company/faqs.php [Accessed 21 Jul. 2018].

17

Appendix 9

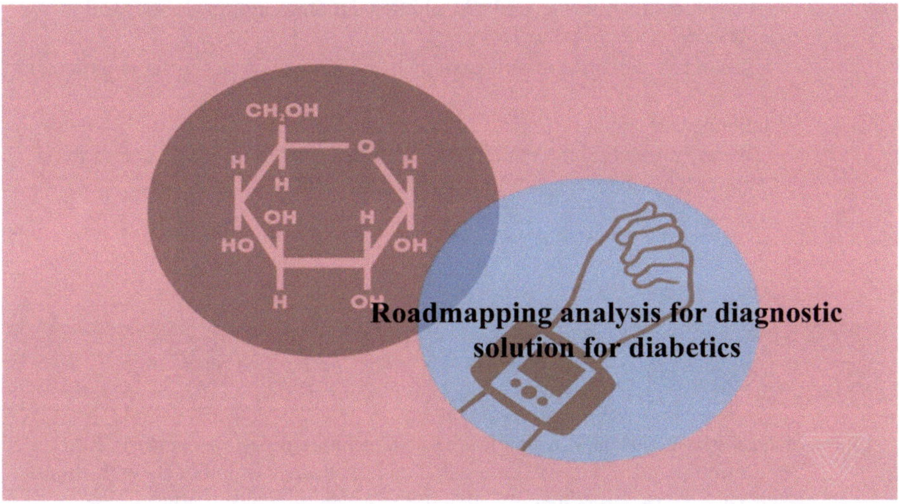

By Amir Shaygan, Bill Henry, Bobby Romanski, Buck Smith, Nouf Alswdn

Portland State UNIVERSITY

Illustration by Rosy Warren / The Verge

Vision

An electronic personal laboratory and doctor:

- Detection:
  - Chemicals
  - Light
- Data Gathering
- Data Recording
- Data Presentation
- Recommendations

The purpose of our technology:

- To enhance health outcomes and quality of life through convenience, information, and encouragement of beneficial actions.

Our goal is to make disease treatment effortless and enjoyable

T. U. Daim (ed.), *Roadmapping Future*, Applied Innovation and Technology Management, https://doi.org/10.1007/978-3-030-50502-8

Who we are / Audience

- **Who we are**:
  - Company developing monitoring solutions and analytics. We design devices, provide integrated services. Manufacturing is outsourced.
- **Market opportunity:**
  - Primary: direct to consumers. Start with glucose monitoring, heart rate + body temp monitoring, and add additional monitoring capabilities and services as identified in TRM
  - Secondary: health care providers (help pay for our service, offer to patients)
- **Audience:**
  - We are presenting the initial TRM to company decision makers, describing our process, findings, limitations, and next steps

Project Phases

Phase 1: starting with currently available wearable technology, we will perform R&D and manufacture devices to record and communicate customer data with simple analytics. Phase 1 roadmap will determine which products to offer.

Phase 2: we will perform analytics on data from our devices as well as other relevant and accessible data to achieve customer outcomes. This will include health recommendations for use in conjunction in customer healthcare provider

Phase 3: we will perform R&D for advanced monitoring solutions including passive and active. Roadmap will products to offer.

Steps taken to construct TRM

1. **Medical device/service concept development/initial proposal**
2. **Develop market drivers**
3. **Rank market drivers & STEEP analysis**
4. **Develop product features & gaps**
5. **Perform product features vs market drivers QFD**
6. **Identify technologies & availability gaps**
7. **Perform technologies vs product features QFD**
8. **'Combine / Refine' prior steps TRM**
9. **Analyze limitations and suggest next steps**

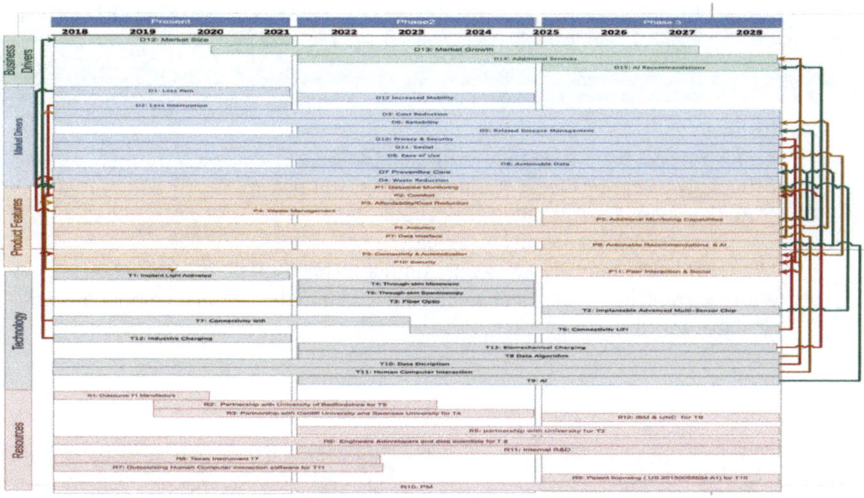
Present
Phase2
Phase 3
2018
2019
2020
2021
2022
2023
2024
2025
2026
2027
2028
Business Drivers
Market Drivers
Product Features
Technology
Resources

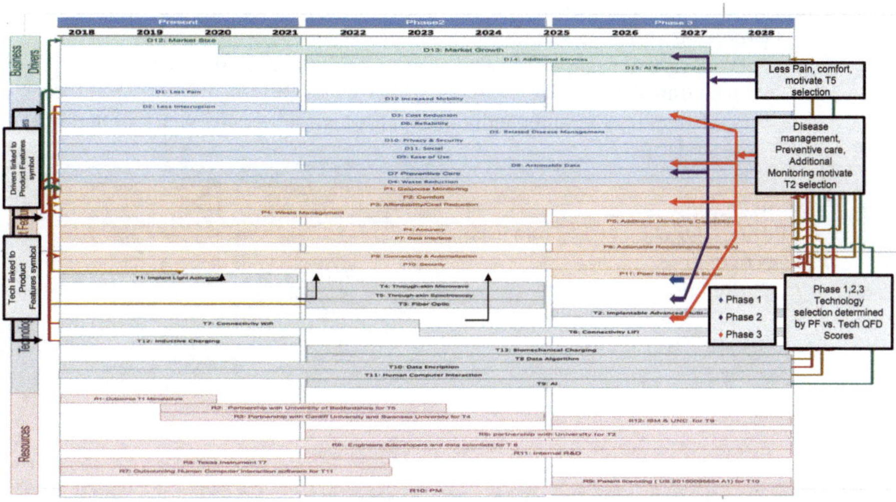
Present
Phase2
2018
2019
2020
2021
2022
2023
2024
2025
2026
2027
2028
Business Drivers
Resources
Less Pain, comfort, motivate T5 selection
Disease management, Preventive care, Additional Monitoring motivate T2 selection
Drivers linked to Product Features symbol
Tech linked to Product Features symbol
Phase 1
Phase 2
Phase 3
Phase 1,2,3 Technology selection determined by PF vs. Tech QFD Scores

Conclusions/Limitations/Next steps

- **Conclusions:**
  - Relative scores were used to prioritize and select a critical path from 13 technology categories for 3 product development phases.
  - Primary reasons for technology selection:
    - Commercial availability estimate
    - QFD score comparison against similar technologies
  - While we prioritized a critical path, many additional possible combinations are available for further analysis

Conclusions/Limitations/Next steps

- **Limitations of our analysis**
  - **Temporal:**
    - We evaluated drivers, products, technologies over a 10 year horizon including three separate phases, with a single QFD analysis
    - QFD application to the individual phases is challenged by lack of temporal granularity

- **Next Steps**
  - Perform QFD that takes into account changes over time.
    - Iterations:
      - Up & Down the TRM
      - Over time

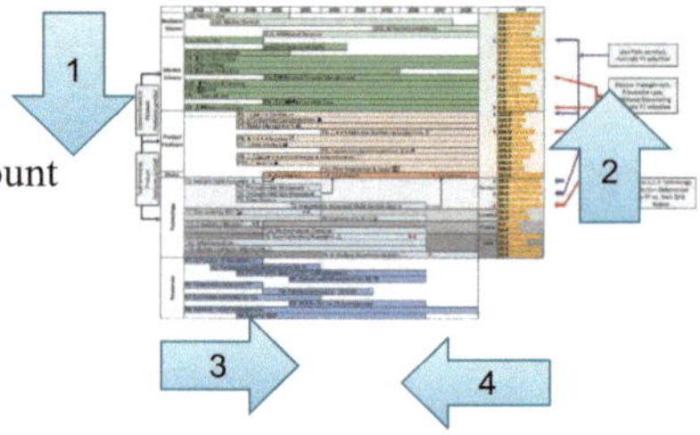

TRM Class takeaways

- Roadmapping analysis should be done in multi-disciplinary environment.
- Time management is important.
- Need for an iterative and flexible roadmapping process
- Customization of roadmap format to meet project needs

References

[1] Pitzer KR, Desai S, Dunn T, Edelman S, Jayalakshmi Y, Kennedy J, Tamada JA, Potts RO: Detection of hypoglycemia with the GlucoWatch Biographer. Diabetes Care 24:881– 885, 2001
[2] Klonoff DC: The need for separate performance goals for glucose sensors in the hypoglycemic, normoglycemic, and hyperglycemic ranges. Diabetes Care 27: 834–836, 2004
[3] Klonoff, David C. "Continuous glucose monitoring." *Diabetes care* 28.5 (2005): 1231-1239.
[4] Flamm CR: Use of intermittent or continuous interstitial glucose monitoring in patients with diabetes mellitus. Technology Evaluation Center. Blue Cross and Blue Shield Association Assessment Program. Vol. 18, No. 16. December 2003
Goldstein DE, Little RR, Lorenz RA, Malone JI, Nathan D, Peterson CM: Tests of glycemia in diabetes. Diabetes Care 18: 896–909, 1995
[5] Reach G: Which threshold to detect hypoglycemia? Value of receiver-operator curve analysis to find a compromise between sensitivity and specificity. Diabetes Care 24:803– 804, 2001
[6] Al Ameen M, Liu J, Kwak K. Security and Privacy Issues in Wireless Sensor Networks for Healthcare Applications. *Journal of Medical Systems*. 2012;36(1):93-101. doi:10.1007/s10916-010-9449-4.
[7] Dohler A. Wireless sensor networks: The biggest cross-community design exercise to-date. Recent Patents Comput. Sci. 2008;1:9–25. doi: 10.2174/1874479610801010009
[8] Munir, S. A., Yu, W. B., Ren, B., and Ma, M., Fuzzy logic based congestion estimation for qos in wireless sensor network. Wireless Communications and Networking Conference, 2007.WCNC 2007. IEEE, pp.4336–4341, 11–15 March 2007.
https://engineering.wustl.edu/news/Pages/Novel-process-to-detect-proteins-could-simplify-kidney-disease-detection.aspx

References

[9] https://www.researchgate.net/publication/280860205_A_graphene-based_affinity_glucose_nanosensor

[10] https://www.nature.com/articles/srep08311.pdf

[11] http://www.pnas.org/content/113/26/7088.full.pdf

[12] https://dspace.mit.edu/handle/1721.1/92652#files-area

[13] Heungjae Choi, Jack Naylon, Steve Luzio, Jan Beutler, James Birchall, Chris Martin, and Adrian Porch ,Design and In Vitro Interference Test of Microwave Noninvasive Blood Glucose Monitoring Sensor,IEEE TRANSACTIONS ON MICROWAVE THEORY AND TECHNIQUES, VOL. 63, NO. 10, OCTOBER 2015

[14] https://www.meddeviceonline.com/doc/needle-free-device-monitors-glucose-with-microwaves-0001

[15] Haxha, Shyqyri, and Jaspreet Jhoja. "Optical Based Noninvasive Glucose Monitoring Sensor Prototype." IEEE Photonics Journal 8.6 (2016): 1-11.

[16] https://futurism.com/ibms-watson-ai-recommends-same-treatment-as-doctors-in-99-of-cancer-cases/

https://www.technologyreview.com/s/600868/the-artificially-intelligent-doctor-will-hear-you-now/

References

[17] K. Mercer, L. Giangregorio, E. Schneider, P. Chilana, M. Li, and K. Grindrod, "Acceptance of Commercially Available Wearable Activity Trackers Among Adults Aged Over 50 and With Chronic Illness: A Mixed-Methods Evaluation.," JMIR mHealth uHealth, vol. 4, no. 1, pp. 1–24, 2016.

[18] E. C. Nelson, T. Verhagen, and M. L. Noordzij, "Health empowerment through activity trackers : An empirical smart wristband study," Comput. Human Behav., vol. 62, pp. 364–374 2016.

[19] A. Lunney, N. R. Cunningham, and M. S. Eastin, "Wearable fitness technology : A structural investigation into acceptance and perceived fi tness outcomes," Comput. Human Behav., vol. 65, pp. 114–120, 2016.

[20] C. Buenaflor and H. C. Kim, "Six human factors to acceptability of wearable computers," Int. J. Multimed. Ubiquitous Eng., vol. 8, no. 3, pp. 103–114, 2013.

[21] K. J. Kim and D.-H. Shin, "An acceptance model for smart watches," Internet Res., vol. 25, no. 4, pp. 527–541, 2015.

[22] P. Zhang and N. Li, "The Importance of Affective Quality," Commun. ACM, vol. 48, no. 9, pp. 105–108, 2005. https://www.ncbi.nlm.nih.gov/pmc/articles/PMC4701313/

[23] Heungjae Choi, Jack Naylon, Steve Luzio, Jan Beutler, James Birchall, Chris Martin, and Adrian Porch ,Design and In Vitro Interference Test of Microwave Noninvasive Blood Glucose Monitoring Sensor,IEEE TRANSACTIONS ON MICROWAVE THEORY AND TECHNIQUES, VOL. 63, NO. 10, OCTOBER 2015

[24] Wagner, Julie, Howard Tennen, and Howard Wolpert. "Continuous Glucose Monitoring: A Review for Behavioral Researchers." *Psychosomatic Medicine 74*.4 (2012): 356–365. PMC. Web. 25 July 2017

Extra Slides

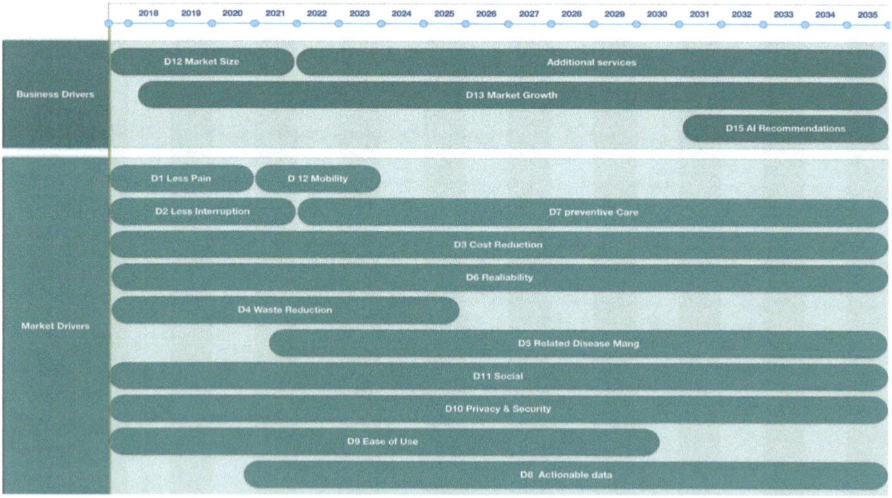

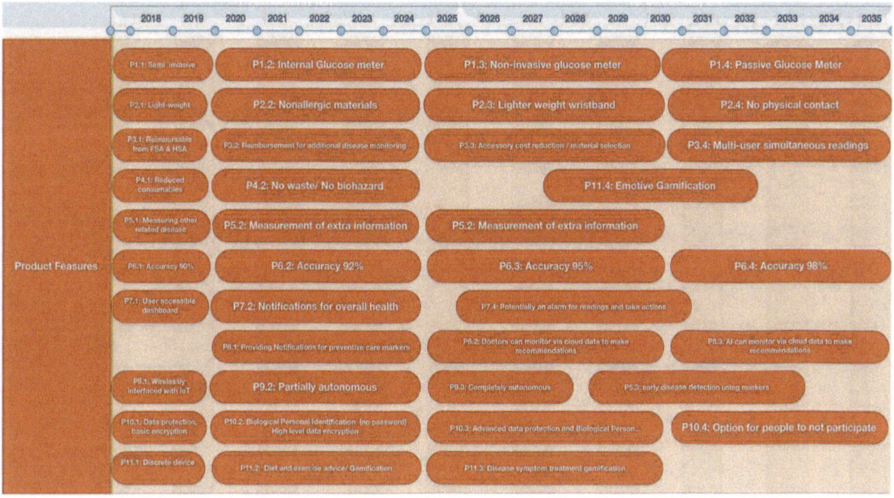
2018
2019
2020
2021
2022
2023
2024
2025
2026
2027
2028
2029
2030
2031
2032
2033
2034
2035
Product Feasures
P1.1: Semi invasive
P1.2: Internal Glucose meter
P1.3: Non-invasive glucose meter
P1.4: Passive Glucose Meter
P2.1: Light-weight
P2.2: Nonallergic materials
P2.3: Lighter weight wristband
P2.4: No physical contact
P3.1: Reimbursable from FSA & HSA
P3.2: Reimbursement for additional disease monitoring
P3.3: Accessory cost reduction / material selection
P3.4: Multi-user simultaneous readings
P4.1: Reduced consumables
P4.2: No waste/ No biohazard
P11.4: Emotive Gamification
P5.1: Measuring other related disease
P5.2: Measurement of extra information
P5.2: Measurement of extra information
P6.1: Accuracy 90%
P6.2: Accuracy 92%
P6.3: Accuracy 95%
P6.4: Accuracy 98%
P7.1: User accessible dashboard
P7.2: Notifications for overall health
P7.4: Potentially an alarm for readings and take actions
P8.1: Providing Notifications for preventive care markers
P8.2: Doctors can monitor via cloud data to make recommendations
P8.3: AI can monitor via cloud data to make recommendations
P9.1: Wirelessly interfaced with IoT
P9.2: Partially autonomous
P9.3: Completely autonomous
P5.3: early disease detection using markers
P10.1: Data protection, basic encryption
P10.2: Biological Personal Identification (no password) High level data encryption
P10.3: Advanced data protection and Biological Person...
P10.4: Option for people to not participate
P11.1: Discrete device
P11.2: Diet and exercise advice/ Gamification
P11.3: Disease symptom treatment gamification

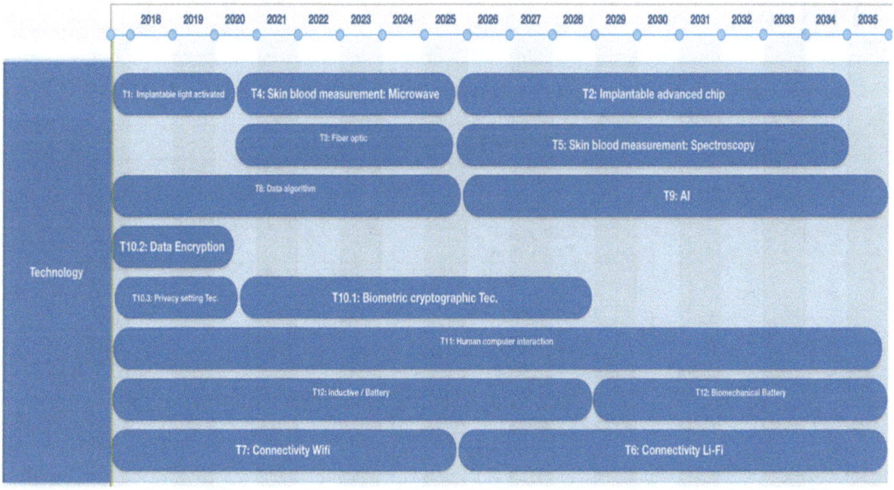
2018
2019
2020
2021
2022
2023
2024
2025
2026
2027
2028
2029
2030
2031
2032
2033
2034
2035
Technology
T1: Implantable light activated
T4: Skin blood measurement: Microwave
T2: Implantable advanced chip
T3: Fiber optic
T5: Skin blood measurement: Spectroscopy
T8: Data algorithm
T9: AI
T10.2: Data Encryption
T10.3: Privacy setting Tec.
T10.1: Biometric cryptographic Tec.
T11: Human computer interaction
T12: Inductive / Battery
T12: Biomechanical Battery
T7: Connectivity Wifi
T6: Connectivity Li-Fi

Driver Values

Drivers	Expert 1	Expert 2	Expert 3	Expert 4	Expert 5	Average
D1	9	6	8	7	9	7.80
D2	6	8	9	6	8	7.40
D3	9	9	2	9	7	7.20
D4	3	6	1	5	4	3.80
D5	5	8	7	6	7	6.60
D6	6	7	7	9	9	7.60
D7	8	9	3	6	8	6.80
D8	5	9	9	6	6	7.00
D9	6	7	4	8	8	6.60
D10	6	7	1	7	7	5.60
D11	3	6	1	6	3	3.80
D12	7	8	4	8	9	7.20
D13	7	7	6	8	8	7.20
D14	5	9	3	6	6	5.80
D15	5	7	8	7	5	6.40

Low	Med	High
Low	Med	High
Low	Med	High

STEEP

			Perspectives				
			Social	Technological	Environmental	Economical	Political
Drivers	Market (Ext)	D1- Less Pain	✓				
		D2- Less Interruption	✓				
		D3- Cost Reduction				✓	
		D4- Waste Reduction			✓		
		D5- Disease Management	✓		✓		
		D6- More Reliability		✓			
		D7- Preventive Care	✓			✓	
		D8- Actionable Data		✓			✓
		D9- Ease of Use	✓				
		D10- Privacy/Security Settings	✓	✓			✓
		D11- Social Aspect	✓				
	Business (Int)	D12- Market Size				✓	
		D13- Market Growth				✓	
		D14- Additional Services		✓		✓	
		D15- Demand for AI Recommendations	✓			✓	✓

	Market Drivers	D1	D2	D3	D4	D5	D6	D7	D8	D9	D10	D11	
	Driver Weights	7.8	7.4	7.2	3.8	6.6	7.6	6.8	7	6.6	5.6	3.8	
Product Features	***QFD***	Less pain	Less Interruption	Cost Reduction	Waste Reduction	Disease Management	Reliability	Preventive Care	Actionable Data	Ease of Use	Privacy and Security	Social	Total
P1	Glucose Monitoring	6.0	6.4	3.8	2.6	7.8	4.0	7.4	7.6	3.8	1.0	2.6	357.36
P2	Comfort	6.0	6.0	1.6	1.0	1.8	1.6	2.0	1.6	5.6	1.4	3.2	212.32
P3	Affordability	1.2	1.6	9.0	3.2	4.8	1.0	4.4	3.2	1.6	1.8	2.6	220.28
P4	Waste Management	1.4	2.2	4.4	9.0	1.8	1.8	2.2	1.8	3.6	1.4	2.6	187.68
P5	Additional Monitoring Capabilities	1.6	2.6	4.2	2.0	8.0	2.8	8.0	6.2	2.8	1.6	5.2	288.64
P6	High Accuracy	0.8	1.8	2.2	1.2	5.2	7.0	5.0	7.4	1.4	1.0	2.8	238.76
P7	Data Viewer	1.6	2.6	3.8	1.2	4.4	3.6	5.2	6.0	5.8	3.2	4.6	271.08
P8	Actionable Recommendations	1.8	1.8	4.6	2.0	5.8	2.8	6.4	8.8	4.8	2.6	3.2	291.16
P9	Connectivity	5.4	5.6	1.4	3.4	1.8	2.4	2.4	2.0	8.2	2.8	3.8	251.24
P10	Security	2.0	1.2	2.2	1.0	1.4	3.8	1.0	1.0	4.2	9.0	4.2	190.12
P11	Peer Interaction	2.2	3.0	1.2	1.0	3.6	1.0	2.6	2.2	4.6	4.8	8.6	206.16

Texas Instrument

	Product Features	P1	P2	P3	P4	P5	P6	P7	P8	P9	P10	P11	
		0.57	0.34	0.35	0.30	0.46	0.38	0.43	0.46	0.40	0.30	0.33	
Techs	QFD	Glucose Monitoring	Comfort Ergonomics	Affordability	Waste Management	Additional Monitoring Capabilities	High Accuracy	Data Viewer	Actionable Recommendations	Ergonomics/User Friendliness	Security	Peer Interaction	Total
T1	implantable light activated	8	4.6	4.8	4	6	7.2	1.25	1.8	7	1.2	1	19.23
T2	implantable advanced chip	8	4.4	4.4	3.6	8.5	7.4	1.2	3.2	6	1.4	1.4	20.54
T3	fiber optic	7	5.6	4.8	4.4	6	6.2	1.4	2.6	6.6	1.6	1	19.13
T4	Microwave	6.8	7.2	5	5.2	5.6	7	1	1.8	7.4	1.6	1.4	19.90
T5	Spectroscopy	7	6.8	5.4	5.2	5.6	7.2	1	1.8	7.8	1.6	1	20.20
T6	Li-fi	2.4	6.4	5.2	2.2	2.2	4.6	2	1	6.2	4.4	2	14.48
T7	Wifi	2.2	5	5.6	1.2	2.2	4.4	2	1	5.2	4	2	13.14
T8	Data Collection	5	1.6	2.5	1.2	5	5.6	4.8	5.6	6.2	4.75	4.8	19.10
T9	AI	3	1.6	3.6	1.2	6.4	5.4	3.8	8.8	4.4	5.4	4.8	19.44
T10	Encryption	1	2.8	2.8	1.2	2	2.6	3.4	3.6	4.4	9	3	13.29
T11	Data viewer	2.4	3.8	2.8	1.2	4.4	4.8	9	4	6.6	5.2	5.2	19.38
T12	Inductive Charging	1.6	5.8	2.6	3	1.4	1	1	1	5.4	1	1	9.33
T13	Biomechanical Charging	1.6	7.2	3.2	4.2	1.4	1	1	1	5.4	1	1	10.36

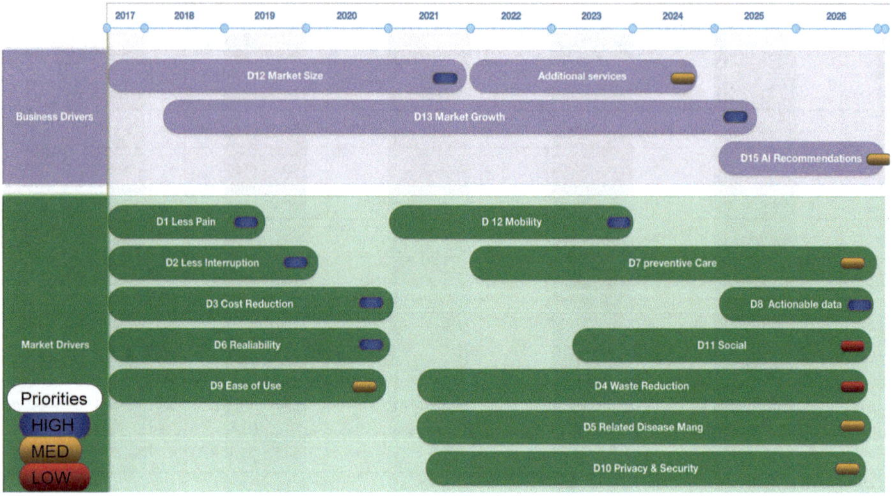

Technology Timeline

	Category	Sub-categories		Present	5 Years	10 Years	15 Years
T1	Sensor	Invasive	Implantable light activated	■			
T2			Implantable advanced chip			■	
T3			Fiber optic		■		
T4		Non-Invasive	Microwave		■		
T5			Spectroscopy			■	
T6	Connectivity/ Networking (Bobby)	Li-fi	Photons			■	
T7		Wifi	Electromagnetism	■			
T8	Software	Data Collection		■			
T9		AI					■
T10		Security and privacy settings	Encryption	■			
T11		Human Interaction	Data viewer	■			
T12	Power	Battery	Inductive charging	■			
T13			Biomechanical charging				■

D1- Less Pain

- Desire by customers to :
    - not cause any physical or psychological pain during the monitoring process.
    - allow them to monitor their blood glucose levels without the need to pierce their skin.
    - eliminate fear of needles/blood to be able to monitor their glucose levels
    - reduce finger stick blood draw for monitoring/calibration
    - eliminate finger stick blood draw for monitoring/calibration
    - eliminate all monitoring methods that require skin piercing
    - avoid soreness after test
    - eliminate probable infections from tests

http://www.gluco-wise.com/
https://cdn.diabetesdaily.com/wp-content/blogs.dir/21/files/2014/01/534049_620794216940_2076081477_n2222222222.jpg

D2 - Interruption

- Decrease in number of doctor visits by assigning appointments based on needs as opposed to routines
    - Cuts down loss time due to commuting
- Desire by customers to reduce or eliminate actions required, e.g., no blood draws or calibration activities required
    - Need to find private space multiple times a day
    - Desire by customers reduce time spent self-monitoring, e.g., don't have to constantly check glucose levels, solution will notify customer only when needed.
    - it does not require any patient action to perform the test (ex. 10 minutes per day = 60 hours per year)
- Reduce glucose management requirement, e.g., less burden for customer to constantly ingest glucose to achieve correct glucose level. (Possibly suggests active management solution)

D3 - Cost Reduction

- Co-pay
  - it reduces the health care cost including, doctors visit, regular check test.
    - Reduced requirement for doctor visits
    - Doctor visit expense (minimum copay)
  - Reduce healthcare premiums for diabetics due to lower count of doctor visits
- Supplies
  - Elimination of the need for fingerpicks/needles (supplies)
  - Band-aid, antiseptic & gauss costs
  - Device Costs + Refillable test strip costs

https://www.cdc.gov/diabetes/

- Desire to reduce lifetime healthcare costs

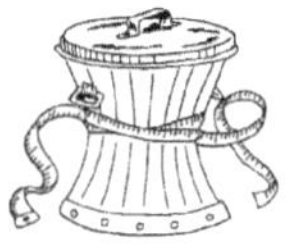

D4 - Waste Reduction

- Desire for environmentally friendly products and push for renewable resources
  - No more test strips
- Reduces biohazard waste such as bloody needles etc.
  - Reduces the chance of contagious disease transmission through blood
  - Reduced need to process ordinary waste and biohazard waste

http://rcrcommodities.com/wp-content/uploads/2012/01/mc5.png

http://rcrcommodities.com/wp-content/uploads/2012/01/mc5.png

D5 - Related Disease Management

- Desire to monitor overall health factors
- Desire by customers to achieve synergies and overlap across all healthcare activities, e.g., glucose history should be used to inform diagnosis and treatment of other conditions.
- Customer desire for total wellness solution – not simply CGM, but reduction or elimination of all conditions related to glucose levels, feeling good all day, etc.
- Desire to maximize use of medical devices, e.g., if incremental cost is low for CGM device to record other valuable metrics, overall health solution is improved. (maybe a product feature)
- Desire to prevent test site infections

https://www.cdc.gov/diabetes/

D6 - Reliability

- The desire for real time, meaningful, reliable information
- Desire to have accurate information without invasive blood tests
- Increase in technology's ability to analyze gathered health and fitness data which help users adopt the physical and health-based activities [1],[2]
- Increase in the degree to which the result of a measurement, calculation, or specification conforms to the precise value by the wearable technology [3]
- Reduction or elimination of the need to calibrate
- Desire by customer to rely on CGM solution without reverting to old methods
- Desire to have continuous monitoring for nuanced evaluation
- Desire to increase information inputs by adding multiple sensors to improve overall analysis

[1] K. Mercer, L. Giangregorio, E. Schneider, P. Chilana, M. Li, and K. Grindrod, "Acceptance of Commercially Available Wearable Activity Trackers Among Adults Aged Over 50 and With Chronic Illness: A Mixed-Methods Evaluation.," JMIR mHealth uHealth, vol. 4, no. 1, pp. 1–24, 2016.

[2] E. C. Nelson, T. Verhagen, and M. L. Noordzij, "Health empowerment through activity trackers : An empirical smart wristband study," Comput. Human Behav., vol. 62, pp. 364–374 2016.

[3] A. Lunney, N. R. Cunningham, and M. S. Eastin, "Wearable fitness technology : A structural investigation into acceptance and perceived fi tness outcomes," Comput. Human Behav., vol. 65, pp. 114–120, 2016.

D7 - Preventive Care

- Desire to receive recommendations to manage and prevent any further consequences
  - Desire by customer for increase in overall healthcare effectiveness by knowing which preventive actions have the highest likelihood of reducing future healthcare burden.
  - Desire by customer for early warning notification if at heightened risk for minor or major condition.
- Improvement in patients' motivation and proactiveness to be healthier through real time data
- Desire to receive better overall status of their health in order to be more cautious or prevent diabetes
- A bigger historical data (infographic/charts) can make data more understandable for both users and professionals
- Desire to maintain quality of life

https://static1.squarespace.com/

Avoiding amputation

D8 - Actionable Data

- Desire for a provided set of data that enable SME to make decision and real time interaction
  - Real time feedback from health monitoring systems
  - Reduce time that emergency services are notified significantly increases odds of full recovery
  - Beyond simply communicating CGM data, customer desire to receive actionable recommendations that benefit overall wellness, helping to fulfill D5 and D7
- The desire for bigger data (for patients, doctors, and analytic SMEs)
  - Meta data could be compressed and stored for analysis

http://fwdkc.co/wp-content/uploads/2017/02/actionable-data

D9 - Ease of Use

- Desire to reduce inconvenience during commuting
- Desire for the allowance of the continuous collection of data no matter where the patients are (camping,swimming, etc.)
- To be suitable, the size, weight, placement, and attachment to the body must be considered[5]
- Desire by customer for solution that is simple to use, can be easily understood, and requires low ongoing customer effort.
- Easy for children to use independently
- Easy for elderly to use independently
- Desire for no reminders or alarms to notify test times (through CGM)
- Desire for more readable screens (during different times of day and circumstances)
- Reduce the number of steps needed to produce a result (Desire for more user friendly device in general)

D10 - Privacy & Security Settings

- Consumers perception of the likelihood that device provider will protect consumers confidential information [2]
    - provides adequate security and privacy for data and connectivity
    - Concern that hackers could cause false readings, endangering lives
    - Concern that employers could read this information
- Less fear that a device may cause physical harm. [4] (safety)
- Desire for Customers to manage what notifications (and possibly recommendations they receive)
- High risk individuals want less privacy so help could easily find them in the event of an emergency

http://www.networxsecurity.org/fileadmin/user_upload/images/2015-08/internet_privacy.jpg

D11 - Social

- Patients' desire to be able to socialize and connect with each other
  - Desire by some customers to interact with other people who have similar healthcare challenges
  - Desire by some customers to share their information within their community
  - Desire by some for gamification of healthcare
  - Need for emotional support
- Desire for devices with more affective quality [The quality dimension of objects which has stimulus to create changes in people affects.][6]
- Desire for a platform to share diet and exercise advice
- Forum for parents of diabetic children
- More freedom for parents if children are continuously monitored

http://1.bp.blogspot.com/-UcjbtcixePs/VTZIAT7WDDI/AAAAAAAAA-8/YrsIbECjRZU/s1600/internet_of_things.jpg

D12 - Market Size (Business Driver)

- **9%** of American population is diabetic of which **27.8%** are undiagnosed
- **Prediabetes**: In 2012, 86 million Americans age 20 and older had prediabetes; this is up from **79** million in 2010.
- Diabetes remains the **7th** leading cause of death in the United States in 2010, with **69,071** death certificates listing it as the underlying cause of death, and a total of **234,051** death certificates listing diabetes as an underlying or contributing cause of death.
- **Prevalence in Seniors**: The percentage of Americans age 65 and older remains high, at 25.9%, or 11.2 million seniors (diagnosed and undiagnosed)
- New business model requires a large pool of potential, willing, and able customers. This could be individuals or healthcare providers.
- The CGM platform will integrate into markets beyond the affected diabetes population.
- Passive diagnostics could expand to include all illnesses.

https://www.cdc.gov/diabetes/

http://www.diabetes.org/diabetes-basics/statistics/?referrer=https://www.google.com/
https://www.cdc.gov/diabetes/data/statistics/2014statisticsreport.html

- More sensors increase amount of illnesses that can be monitored, increasing market size

D13 - Market Growth (Business Driver)

- Increase in number of diabetic people, also there is a trend of being well and monitoring the health
- Baby Boomers, who in 2030 will be aged 66 to 84—the "young old"—and will number 61 million people. In addition to the Baby Boomers, those born prior to 1946—the "oldest old"—will number 9million people in 2030.
- Growth in original CGM market
- Growth in advanced CGM market (actionable data & Recommendations)
- Growth via additional services and technologies (building off of CGM to offer other products/services)
  - Monitoring can expand to include other conditions, such as heart conditions.
  - IoT, AI
- Passive monitoring in hospitals reduces spread of diseases

https://www.cdc.gov/diabetes/

http://www.pbs.org/newshour/bb/social_issues-jan-june11-boomer_01-03/

D14 - Additional Services (Business Driver)

- Other related diseases monitoring technologies such monitoring kidneys functions, blood pressure
- Extensibility of systems developed by company. Desire for company to be able to implement new value-add product/service at lowest possible incremental cost on top of initially developed product/service offering.
- Passive monitoring – in home, car/commute, office space
- Monitoring data available to doctor
- Connection to emergency services
- Overall health monitoring
- Electronic Personal Trainer
- Nutrition Monitoring
- Open innovation, API

http://www.trainermetrics.com/

D15 - (Demand for) AI Recommendations

- Provision of recommendations based on patient's' current health status
- Doctors are very busy
- Reduction in human error
- Ultra complex correlation and decision making
- Ability to continuously monitor and make recommendations
- High risk individuals may require continuous monitoring as a condition of insurance coverage

https://cdn-images-1.medium.com/max/1200/0*ZEAL0P4wju6zg1pg.

P.Code	Product Feature	Phase 1 (-2 - +2)	Phase 2 (+5)	Phase 3 (+10)	Phase 4 (+15)	Gap
P1	Glucose Monitoring	Semi-invasive	Internal Glucose meter	Non-invasive glucose meter (skin, eye contact)	Passive Glucose meter	Blood/health analysis sensors
P2	Comfort	Lightweight wristband	Nonallergic materials	Lighter weight wristband	No physical contact	Battery size
P3	Affordability	Reimbursable from FSA & HSA	Reimbursement for additional disease monitoring	Accessory cost reduction / material selection	Multi-user simultaneous readings	Reimbursement for CGMs by insurance or government payer organizations has been limited [4]/Insurance companies are demanding rigorous scientific evidence about continuous monitoring before they will pay for this technology[5]

[4] Flamm CR: Use of intermittent or continuous interstitial glucose monitoring in patients with diabetes mellitus. Technology Evaluation Center. Blue Cross and Blue Shield Association Assessment Program. Vol. 18, No. 16. December 2003

[5] Goldstein DE, Little RR, Lorenz RA, Malone JI, Nathan D, Peterson CM: Tests of glycemia in diabetes. Diabetes Care 18: 896–909, 1995

Continuous glucose monitoring also offers the capability of expressing the mean blood glucose value in new ways. The mean blood glucose value can shift quickly with any new treatment, and it is not always practical to wait for months or weeks, respectively, for the HbA1c or fructosamine values to shift.

P.Code	Product Feature	Phase 1 (-2 - +2)	Phase 2 (+5)	Phase 3 (+10)	Phase 4 (+15)	Gap
P4	Waste Management	Reduced consumables	No waste/ No biohazard	-	-	Fingerpick and accessories waste
P5	Additional Monitoring Capabilities	Measuring (Heart beat, blood oxygen level, body temperature,Hydration, Glucose), Sleeping pattern	Measurement of extra information such as white blood cells/ vitamin levels , lactate, sodium, potassium	Measurement of kidneys functions, gangrene disease signs.	Monitoring eyes function, early disease detection using markers	High cost of materials to build biomarkers [11]
P6	Accuracy	Accuracy 90%	Accuracy 92%/High non-invasive Hypoglycemic Accuracy	95%	98%	low accurate in the hypoglycemic range[2]/ necessity of trend informations in using CGM data [3]

https://engineering.wustl.edu/news/Pages/Novel-process-to-detect-proteins-could-simplify-kidney-disease-detection.aspx[1]Pitzer KR, Desai S, Dunn T, Edelman S, Jayalakshmi Y, Kennedy J, Tamada JA, Potts RO: Detection of hypoglycemia with the GlucoWatch Biographer. Diabetes Care 24:881– 885, 2001

[2]Klonoff DC: The need for separate performance goals for glucose sensors in the hypoglycemic, normoglycemic, and hyperglycemic ranges. Diabetes Care 27: 834–836, 2004

[3] Klonoff, David C. "Continuous glucose monitoring." Diabetes care 28.5 (2005): 1231-1239.

Most currently available outpatient continuous blood glucose monitors do not actually measure the glucose concentration within whole blood, but instead measure the glucose concentration within the ISF (interstitial fluid) These ISF measurement technologies are defined as minimally invasive because they compromise the skin barrier but do not puncture any blood vessels

P.Code	Product Feature	Phase 1 (-2 - +2)	Phase 2 (+5)	Phase 3 (+10)	Phase 4 (+15)	Gap
P7	Data interface and Notifications	User accessible dashboard	Notifications for overall health (mood/stress), Health report to present to doctors	Potentially an alarm for readings outside target ranges and take actions	-	There is a trade-off between an alarm's sensitivity and specificity[6]
P8	Actionable Recommenda tions	Not yet	Providing Notifications for preventive care	Doctors can monitor via cloud data to make recommendations	AI can monitor via cloud data to make recommendat ions	FDA Approval

it is necessary to incorporate trend information in using continuous blood glucose data

In general, if the alarm is set to sound at a lower level than the hypoglycemic threshold, then the specificity will be good but the sensitivity may be poor. If the alarm is set to sound at a glucose level higher than the hypoglycemic threshold, then the sensitivity will be good but the specificity may be poor. The greater accuracy a continuous monitor can provide, the less of a trade-off is necessary [6].

[6]Reach G: Which threshold to detect hypoglycemia? Value of receiver-operator curve analysis to find a compromise between sensitivity and specificity.
Diabetes Care 24:803– 804, 2001

P.Code	Product Feature	Phase 1 (-2 - +2)	Phase 2 (+5)	Phase 3 (+10)	Phase 4 (+15)	Gap
P9	Connectivity/ Autonomization	Wirelessly interfaced with IoT	Partially autonomous	Completely autonomous	-	Communicatio n technology between the devices
P10	Security	Data protection, basic encryption	Biological Personal Identification (no password) High level data encryption	Advanced data protection and Biological Personal Identification	Option for people to not participate	Security breach in healthcare applications of sensor networks is a major concern[7]
P11	Peer interaction	Discrete device, chat rooms	Diet and exercise advice	Disease symptom treatment gamification	Emotive Gamification	complex relationships among glucose and intrapersonal, interpersonal, and contextual factors[10]

[7]Al Ameen M, Liu J, Kwak K. Security and Privacy Issues in Wireless Sensor Networks for Healthcare Applications. Journal of Medical Systems. 2012;36(1):93-101. doi:10.1007/s10916-010-9449-4.

https://www.ncbi.nlm.nih.gov/pmc/articles/PMC3279645/

T1: Invasive Light Activated Fluorescent Sensor

- Fluorescent biosensor surface coating attracts glucose molecules in presence of sensor-supplied LED light source.
- Gap: current biosensor (Eversense) degrades after 90 days
- Future sensor (Glucosense) uses sensor-supplied biosensor replenishment to last one year. (5 years to maturity)
- Gap: Factory calibration via low sensor variation (5 years to maturity)

Image Sources:
Eversensediabetes.com
glusensemedical.com

Degradation of Sensor
http://healthtechinsider.com/2017/03/01/live-cells-enable-full-year-cgm-implant/

Factory Calibrated Sensor
https://www.ncbi.nlm.nih.gov/pmc/articles/PMC4454101/

T1: Invasive Light Activated Sensor (Gaps and Potential Resources)

	Category	Subcategory	Technology	Present	Future	Gap
T1	**Sensor**	**Invasive**	Light Activated Implantable (fluorescent biosensor)	Fluorescent biosensor surface coating attracts glucose molecules in presence of sensor light source. Expected 2017 FDA Approved for 90 day measurements	Approved for 6 months to 1 year of measurements	Durability and reliability of sensor reader Gap: current biosensor (Eversense) degrades after 90 days Future sensor (Glucosense) uses sensor-supplied biosensor replenishment (5 years?) Gap: Factory calibration via low sensor variation (5 years?)

T2: Invasive Advanced Microchip Sensor (Lab on a Chip)

- Currently, chip activated implants have limited capability such as measure cardiac functions.
- New chip technology could provide significant monitoring improvements
- **Solution-Gated Graphene Transistor** [1], [2].
    - Transistor gate activates when target biomarker contacts the polymer-coated graphene.
    - Coatings developed for a range of biomarkers including glucose and DNA.
- Implanted sensors could measure single cell sequencing or DNA mutations for early cancer detection, or for cancer patients currently in remission. [3]
- Potential for low cost manufacturing. [4]
- Potential for miniaturization, reducing customer impact

Description notes:
The goal of this system is electronic measurement - presence of the analyte triggers an electrical response that be easily monitored. Low electrically charged molecules are harder to detect. The novel surface coating and the electrical properties of graphene allow for accurate detection of a wide range of analytes, including glucose. There may be significant business opportunities to offer low cost monitors for glucose, as well as many other metrics.

[1]
https://www.researchgate.net/publication/280860205_A_graphene-based_affinity_glucose_nanosensor

[2]
https://www.nature.com/articles/srep08311.pdf

[3]
http://www.pnas.org/content/113/26/7088.full.pdf

[4]
https://dspace.mit.edu/handle/1721.1/92652#files-area

T2: Invasive Chip Activated Sensor (Gaps and Potential Resources)

	Category	Subcategory	Technology	Present	Future	Gap
T2	**Sensor**	**Invasive**	Advanced Chip	Available for cardiac monitoring, not feasible for other monitoring	Reduced size and increased measurement capability, additional monitoring such as DNA mutation	• Integration of multiple sensors into one internal monitoring device • Development of nanosensor coatings for multiple analytes • Commercialization of lab-proven technology

T3: Invasive Fiber Optic Sensor

- Reliability
  - Immune to electromagnetic interference
  - Easy to implant multiple tubes for redundancy
- Ergonomics
  - Biocompatible
  - Not easily noticeable by peers
- Technology Lifecycle
  - Easy to upgrade by replacing with tubes that have enhanced coating
    - Beta Cyclodextrin coating
  - Provides a channel for future technology of additional monitoring capabilities
- Affordability
  - Inexpensive passive sensor
- Accuracy
  - Current technology is ±15% accurate

T3: Invasive Fiber Optic Sensor (Gaps and Potential Resources)

	Category	Subcategory	Technology	Present	Future	Gap
T3	**Sensor**	**Invasive**	Fiber Optic	in vivo and in vitro investigation phase (Trials with promising results)	Long distance passive reading	FDA approval / extremely light sensitive readers

T4: Skin blood measurement: Microwave

- Needle-Free Device Monitors Glucose With Microwaves [5]
- Is attached to the skin with ordinary adhesives and uses low levels of microwaves to keep tabs on glucose levels.
- Clinical testing has shown that the device's performance is comparable to commercially available continuous glucose monitors (CGMs) and finger-stick blood tests
- The device is slightly smaller than a fist and can adhere to the arm or to the side of the body. Once in place, the device sends microwaves into the skin, collects readings, and then transmits them to a smart device for interpretation. The device can achieve precise and continuous readings using only one mW of power, 1000 times less than what is used by the average smartphone [6]
- Potential resources: a partnership with Cardiff University and Swansea University

1- Heungjae Choi, Jack Naylon, Steve Luzio, Jan Beutler, James Birchall, Chris Martin, and Adrian Porch ,Design and In Vitro Interference Test of Microwave Noninvasive Blood Glucose Monitoring Sensor,IEEE TRANSACTIONS ON MICROWAVE THEORY AND TECHNIQUES, VOL. 63, NO. 10, OCTOBER 2015

2-https://www.meddeviceonline.com/doc/needle-free-device-monitors-glucose-with-microwaves-0001

T5: Skin blood measurement: Spectroscopy [7][11]

- Noninvasive measurement technique to determine the glucose levels in the human body
- Near-infrared transmission spectroscopy is used and in vitro experiments are conducted, as well as in vivo.
- Study confirms a correlation between the sensor output voltage and glucose concentration levels and reports a low-cost prototype of spectroscopy-based noninvasive glucose monitoring system that demonstrates promising results in vitro and establishes a relationship between the optical signals and the changing levels of blood–glucose concentration.

- Potential resources: a partnership with University of Bedfordshire

1- Haxha, Shyqyri, and Jaspreet Jhoja. "Optical Based Noninvasive Glucose Monitoring Sensor Prototype." *IEEE Photonics Journal* 8.6 (2016): 1-11.

Gaps

	Category	Subcategory	Technology	Present	Future	Gap
T4	**Sensor**	**Non-Invasive (through skin)**	Microwave	In vivo and in vitro investigation phase (Trials with promising results)	Non-invasive/safe version integrated into wearable devices such as smart watches	Accuracy and reliability should improve/ The device needs to be commercialized / Safety
T5			Spectroscopy	Trials with promising results (correlation between the sensor output voltage and glucose concentration levels)/ Some electromagnetic waves reader exist (Glucowise)	Non-invasive/safe version integrated into wearable devices such as smartwatches/ Shift from between the fingers to wrist for measurement	Should be integrated into a device/ Improved accuracy and refinement

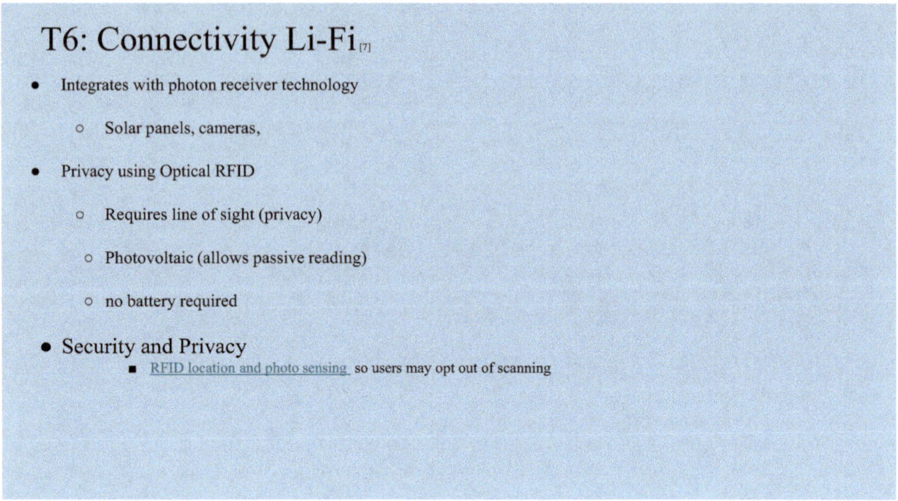

[7] Haxha, Shyqyri, and Jaspreet Jhoja. "Optical Based Noninvasive Glucose Monitoring Sensor Prototype." *IEEE Photonics Journal* 8.6 (2016): 1-11.

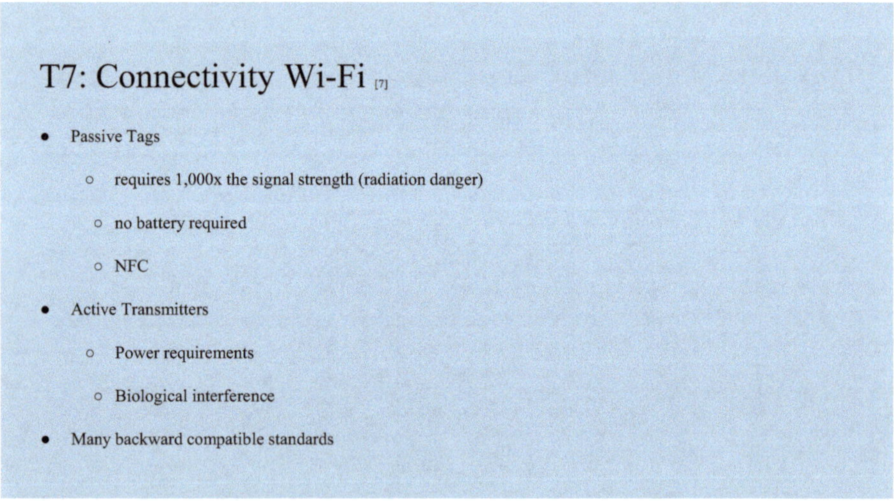

[8] Haxha, Shyqyri, and Jaspreet Jhoja. "Optical Based Noninvasive Glucose Monitoring Sensor Prototype." *IEEE Photonics Journal* 8.6 (2016): 1-11.

	Category	Subcategory	Technology	Present	Future	Gap
T6	Connectivity	**Li-Fi**	Photon	• Untested In Vivo? • Limited commercial use.	• Photon based connectivity in Vivo	• Industry standardization • Sunlight interference • Testing of light when made opaque by human body
T7		**Wi-Fi /NFC**	Electromagnetism	Currently in production	obsolescence	Waiting for photon and microwave technology

T8: Data Algorithm

- Data need to be collected, processed, validated and stored in database.

source : [13]

Gap Analysis

Technology	Present	Future	Gap
Algorithm for Data Collection, processing, validation, and storage.	• Data are collected and transferred by sensor and transistor with software. • Data processing: it would be processed by patients cellphone • Data storage needs to be addressed.	• Cloud to storage patient's data. • Data algorithm for data validation and ML for giving personalized notification. • In house data processing (full integrated data system).	Obtaining cloud storage from host company or purchase them. Integration

Resources:

Computer & Electrical engineers, software developers and data scientists.

T9: AI

- AI can analyze the collected data from patients, doctors recommendations, and the clinical trials publication and make recommendations.
- Al has proved that its diagnosis has higher accuracy than doctors diagnosis.

	Technology	Present	Future	Gaps
T9	AI	It is only on experimental stage [2] It is costly	Obtaining technology licenses and integrate the technology in the product.	Since it is not approved yet, it is not available in the market. So, more investment needed. Als, technology integration needs engineers Fund

Resources:

- IBM's Watson AI recommends which treatment plan should be chosen to treat cancer. [8]
- Also, the University of North Carolina School is involved in this project.
- U.K.-based startup Babylon, by has developed an app that listens to patient's symptoms and provides medical advice [9]
- Financial support.

[N4]
Fund
https://futurism.com/ibms-watson-ai-recommends-same-treatment-as-doctors-in-99-of-cancer-cases/
[N5] https://www.technologyreview.com/s/600868/the-artificially-intelligent-doctor-will-hear-you-now/

Source: N2

T10: Security and privacy settings

- Patients have options to select privacy settings. Meaning that, Patients select which data to be sent to whom.
- Also, recognition technologies needed to assure the wright person has the access.
- Data encryption is required.

	Technology	Present	Future	Gaps
T10	Data Encryption	Encryption software is available	-	Purchasing or partnership with encryption software company/ Better cloud security

Resources:
1- Partnership with an external company to provide security system in the first phases.
2- Biometric cryptographic needs in 2th phases to replace fingerprint with multiple biometrics recognition . Patent licensing (US 20150095654 A1) [3]
4- Privacy settings integrated in the device system
3- Engineers and developers

[N2] https://www.google.com/imgres?imgurl=https://findbiometrics.com/assets/iStock_Multi-Modal.jpg&imgrefurl=https://findbiometrics.com/un-refugee-biometrics-project-25191/&h=600&w=800&tbnid=_oTX81WjVS0o-M:&tbnh=150&tbnw=200&usg=__VocjcYdWFin9m_han-zpwl4OHiM=&vet=10ahUKEwjAgJeirrXVAhVC32MKHZu9AY8Q_B0IjQEwDg..i&docid=es9bCnhN1Z2NFM&itg=1&sa=X&ved=0ahUKEwjAgJeirrXVAhVC32MKHZu9AY8Q_B0IjQEwDg&ei=chmAWcD_B8K-jwOb-4b4CA

N3
Resources:
1- Partnership with an external company to provide security system in the first phases.
2- Biometric cryptographic needs in 2th phases to replace fingerprint with multiple biometrics recognition . Patent licensing (US 20150095654 A1) [3]
4- Privacy settings integrated in the device system
3- Engineers and developers

T11: Human computer interaction

- Effective responsive system to human interaction.

	Technology	Present	Future	Gaps
T11	Human Interaction interface	Hardware and software needed to implement interactive user interface.	In house interactive user interface	The need for software engineers to produce in house. Or outsource it.

Resources:

Partnering with a company which can provide Human Computer interaction software.

T12: Inductive / Battery

Power source options for implantable device:

- Permanently installed battery (pacemaker)
  - Gaps:
    - Battery life for long lasting devices
    - Energy demand reduction from device (chip)
- Inductive coupling (Eversense implantable CGM)
  - Uses near field communication (NFC) for exchanging information to the external device and for powering the implant.
  - Gaps:
    - Minimal (mature technology)
    - Alignment of implant
    - Health impact of electromagnetic field

Image source: eversensediabetes.com

T12: Biomechanical Charging / Energy Harvesting

Several technologies can potentially fulfill need for self-sustaining power for an implantable device:

- Thermoelectricity
- Piezoelectricity
- Electrostatic
- Electromagnetic
- Bio-fuel cells (using glucose or other sources)

Gap: All have low power output potential and are complex. Significant technology development needed -- may represent physical barrier.

Technology Category 4: Power Gaps [10]

	Categor y	Subcategory	Technology	Present	Future	Gap
T12	Power		Inductive / Battery	• Single use battery • Inductive coupling	• Long duration battery • Inductive charging of battery	• Battery technology & cost improvement (5 years) • Inductive coupling is currently commercially available
T13			Biomechanical	• No commercial availability	• Energy harvesting	• Insufficient energy potential. • Lower power devices or higher efficiency energy harvesting technology

Sources:

https://www.ncbi.nlm.nih.gov/pmc/articles/PMC4701313/

Appendix 10

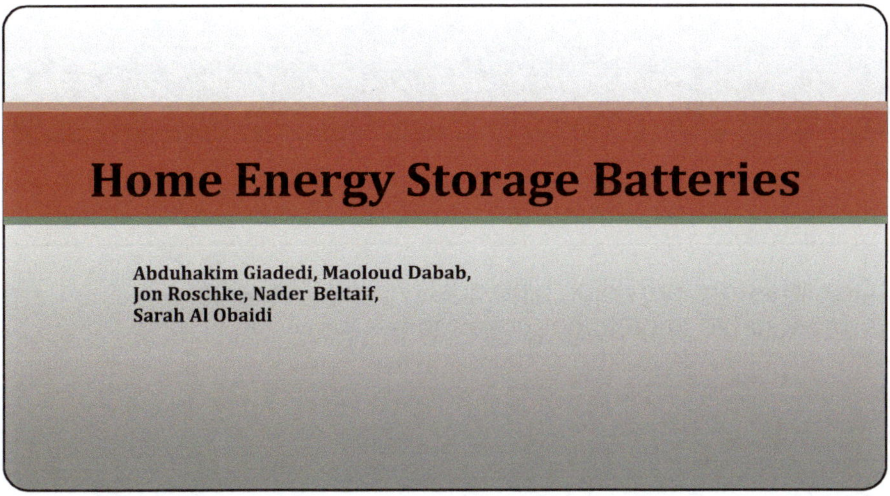

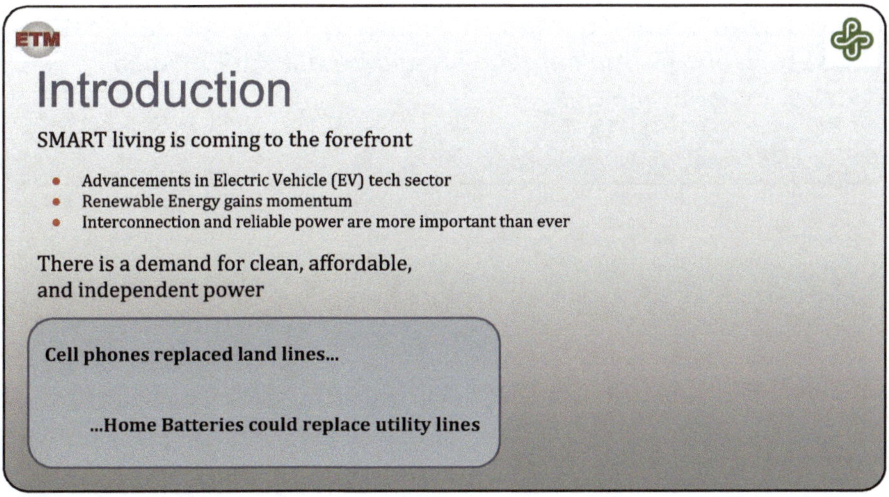

T. U. Daim (ed.), *Roadmapping Future*, Applied Innovation and Technology Management, https://doi.org/10.1007/978-3-030-50502-8

Recent News

THE WALL STREET JOURNAL.
Home World U.S. Politics Economy Business Tech Markets Opinion Arts Life Real Estate

~ Announced on August 1, 2016

"...That's why we are all doing this, to accelerate the advent of a sustainable energy world."
- Elon Musk

"For Tesla, SolarCity brings installers who will be able to install electric-car chargers, separately or together with panels and batteries, the companies said."

Our Mission: providing high quality products within a reasonable price to our customers.

Our Vision: For environmental and economic reasons, Green World Group believes the world will benefit greatly from the rapid development of renewable energy resources and that Green World Group can and should be a leader in such development.

SWOT Analysis

S — Strengths
- Charge when you want
- Use when you want
- Used with existing grid
- Smart and comprehensive controls
- Safe and reliable
- Emergency back-up

W — Weaknesses
- Short term electricity storage
- Procurement cost
- Energy lost in inefficiencies
- Additional cost and complexity
- Additional infrastructure and space requirements

O — Opportunities
- Enabling prosumers to sell power to their neighbors or other consumers
- Global markets
- Income level is at a constant increase
- Distribute renewable energy and storage technologies
- New "smart" technology can deliver greener services to customers

T — Threats
- Government regulations
- External business risks
- Price changes
- Tax code changes
- Short term Application

Market Drivers

Market Drivers	
Economic and Cost	DM1 - Decreasing Cost of RE DM2 - Decreasing Cost of Batteries DM3 - Increasing Cost of Conventional Energy
Emergency Backup and Energy Security	DM4 - Natural disasters DM5 - Terrorist threat DM6 - Energy independent
Functionality	DM7 - Perfect Design DM8 - High Quality DM9 - High Performance DM10 - Smart
Eco-friendly	DM11 - High demand for Green Energy DM12 - Non-toxic
Policies and Regulations	DM13 - Renewable Portfolio Standards (RPS) DM14 - Time of day regulation DM15 - Self consumption policy

ETM

Product Feature Definitions

Product Groups	Product Features
User friendly and easy to use	P1: WiFi-compatible
	P2: Mobile app
	P3: Digital dashboard display
	P4: Recorded system
	P5: Sleek design
Convenience	P6: Operates in wide range of temperatur
	P7: Light weight
	P8: Waterproof
	P9 Less Maintenance
	P10: Long warranty
Safety and environment	P11: No leaks
	P12: No noise
	P13: Safe to connect to grid/electrical sy
	P14: Safe to touch
System Efficiency	P15: Appropriate size
	P16: High capacity system
	P:17 Long lasting

	Definition
	Able to connect to internet and allow for SMART interaction
	Provide mobile communication to monitor and manage system
	Easy to read and understand the basics of the battery system
	Keep a digital history of system to help with troubleshooting
	Attractive look to make the system something to show off
es	Allows system to be placed in any place (cold shed, hot attic, etc)
	Easy to install and move if necessary
	Allows for outdoor placement and/or protection against flooding
	Easy for end-user to maintain the system
	Covers any malfunction of system for a reasonable period of time
	Keeps the system clean and working properly
	Prevents complaints from end-user or neighbors
stem	Follows safety standards to allow for proper connection to grid
	Prevents injury to end-user or maintenance crews
	Physical size of battery system
	Ensures the battery system is able to store as much as possible
	Keeps the batteries functional and in use as long as possible

ETM Market Drivers and Product Features Analysis Grid

	DM1 - Decreasing Cost of RE	DM2 - Decreasing Cost of Batteries	DM3 - Increasing Cost of Conventional	DM4 - Natural	DM5 - Terrorist	DM6 - Energy independent	DM7 - Perfect Design	DM8 - High Quality	DM9 - High Performance	DM10 - Smart	DM11 - High demand for Green Energy	DM12 - Non-toxic	DM13 - Renewable Portfolio Standards	DM14 - Time of day regulation	DM15 - Self consumption policy	Total
P1: WiFi-compatible	0	0	0	2	2	3	0	2	2	3	0	0	0	2	1	46
P2: Mobile app	0	0	0	1	1	2	1	1	2	3	0	0	0	3	1	39
P3: Digital dashboard display	0	0	0	0	0	1	2	1	1	3	0	0	0	1	0	25
P4: Recorded system	1	0	0	1	1	1	2	1	2	3	0	0	0	0	0	36
P5: Sleek design	0	0	0	0	0	0	3	2	1	1	1	0	0	0	0	26
P6: Operates in wide range of temperatures	2	1	0	3	3	3	2	2	3	2	0	1	0	0	0	68
P7: Light weight	0	0	0	0	0	0	3	2	0	0	0	0	0	0	0	17
P8: Waterproof	0	0	0	3	3	2	2	2	2	1	0	1	0	0	0	47
P9 Less Maintenance	2	3	1	2	2	2	3	3	2	1	1	0	0	0	0	71
P10: Long warranty	1	0	2	1	1	1	2	2	1	1	0	0	1	0	1	36
P11: No leaks	0	0	0	0	0	2	2	2	2	0	0	3	1	0	0	35
P12: No noise	0	0	0	0	0	0	3	3	1	1	1	0	1	0	0	31
P13: Safe to connect to grid/electrical system	1	1	0	3	3	3	3	2	1	1	0	0	2	0	1	58
P14: Safe to touch	0	0	0	2	2	2	3	3	2	2	2	0	2	0	2	59
P15: Appropriate size	1	1	0	2	2	3	2	2	2	1	1	0	0	0	0	54
P16: High capacity system	1	1	0	3	3	3	3	3	3	1	3	0	0	1	0	78
P:17 Long lasting	1	0	0	3	3	3	2	3	3	2	1	1	1	1	2	72

Product Feature Gap Analysis

Product Features	Current Level	Future Level		
		2020	2025	2030
P16: High capacity system	≥ 8 kWh usable capacity and 2kW output	≥ 10 kWh usable capacity and 4kW output	≥ 15 kWh usable capacity and 6kW output	≥ 25 kWh usable capacity and 8kW output
P17: Long lasting	5 Years long-life	10 years long-life	15 years long-life	20 years long-life
P9: Less Maintenance	Self Diagnostic & Required more Maintenance	Self Diagnostic & Required less Maintenance	Smart System (Self Diagnostic & Self Maintenance)	Next generation of Smart System (Self Diagnostic & Self Maintenance)
P6: Operates in wide range of temperatures	≈ -8–40°C / Discharge -10–40°C	≈ -10–50°C/ Discharge -10–45°C	≈ -12–55°C / Discharge -12–55°C	≈ -20–65°C / Discharge -20–65°C
P14: Safe to touch	Safe to touch	Safe to touch	Safe to touch	Safe to touch
P13: Safe to connect to grid/electrical system	Not totally safe	Safe System	Smart Connection System	Next generation of Smart Connection System
P15: Appropriate size	Still Big	30% Less it's Current Size	40% Less it's Current Size	50% Less it's Current Size
P8: Waterproof	70 % against flooding	80 % against flooding	100% against flooding	100% against flooding
P1: WiFi-compatible	No Wifi- compatible	Wifi- compatible	Smart Wifi- compatible	Next generation of Smart Wifi- compatible
P2: Mobile app	No mobile access	Partial Access	50% integrated with Mobile app	Fully integrated with Mobile app

9

ETM

Technology Definitions

Technology Groups	Technology	Definition
Material	T1: New Li-X battery	Best technology on the market [e.g. Lithium-Ion, Li-S]
	T2: Carbon fiber frame	Exterior frame for protection and security
	T3: Impact resistance	Used generally for water and fall resistance
Both	T4: Temperature sensor circuit	High technology for temperature control
	T5: Intelligent chipset technology	Processes info and relates to and from Material and Software
	T6: Smart switching inverter	Connects battery system smoothly with grid and household; adds protection to system by providing an individual connect point
Software	T7: Data mining analysis	Data analysis and memory to operate efficiently
	T8: Android/OS platform	Programming needed to create a user friendly mobile app
	T9: Safety alerts	Advanced warnings and self-diagnosis capabilities

ETM

Product Features and Technologies Analysis Grid

	P16: High capacity system	P17: Long lasting	P9: Less Maintenance	P6: Operates in wide range of temperatures	P14: Safe to touch	P13: Safe to connect to grid/electrical system	P15: Appropriate size	P8: Waterproof	P1: WiFi-Compatible	P2: Mobile app	Total
Score	78	72	71	68	59	58	54	47	46	39	
T1: New Li-X battery	3	3	1	2	1	2	3	1	0	0	1041
T2: Carbon fiber frame	0	2	2	2	2	1	3	1	0	0	807
T3: Impact resistance	0	1	2	2	1	0	0	3	0	0	550
T4: Temperature sensor circuit	1	1	3	3	2	2	1	0	1	0	901
T5: Intelligent chipset technology	0	1	2	1	0	3	2	0	2	2	734
T6: Smart switching invertor	1	1	3	1	1	3	2	0	2	2	942
T7: Data mining analysis	1	2	3	1	0	2	0	0	3	3	874
T8: Android/OS platform	1	0	1	0	0	0	0	0	3	3	287
T9: Safety alerts	1	3	3	2	2	3	0	2	1	2	1153

ETM

Technology Gap Analysis

Technology	Current Level	Future Level		
		2020	2025	2030
T9: Safety alerts	Available	Available	Smart technology of Safety alerts	Next generation of Smart technology of Safety alerts
T1: New Li-X battery	Lithium-ion	Advanced Lithium -ion	Lithium-sulfur	Lithium- air
T6: Smart switching inverter	Transformerless technology	Transformerless technology	Next generation of transformerless technology	Next generation of transformerless technology
T4: Temperature sensor circuit	Silicon Technology	Advanced Silicon Technology	Smart Temperature sensor circuit Technology	Next Generation of Smart Temperature sensor circuit Technology
T7: Data mining analysis	Low Capacity Memory	Low Capacity Memory	New upgraded Capacity Memory	High Efficient Capacity Memory
T2: Carbon fiber frame	Not yet	Carbon /Fiber technology	Fully Carbon Frame with High Protection	Next generation of Fully Carbon Frame with High Protection
T5: Intelligent chipset technology	Not yet	First Generation	Second Generation	Third Generation
T3: Impact resistance	Advanced nano-coatings	Advanced nano-coatings	Next generation of advanced nano-coatings	Next generation of advanced nano-material technology
T8: Android/OS platform	Trial	Adopted	Adopted	Adopted

ETM

Resources Definitions

Resources Groups	Resources Features	Definition
Raw Material	R1: Ion, Sulfur, Air, Carbon…etc	Materials used in the batteries
Funds	R2: Government	Funding from the governmental sector
	R3: Shareholders/Investors	Funding from private sector
Knowledge/Research	R4: Labs	Labs needed for the researchers
	R5: Universities	Universities to continue the research
	R6: Partners	Competitors and allies who can provide import technology or expertise
People	R7: Scientist & Researchers	Scientists and researchers that are needed in the lab
	R8: Engineers	Electrical engineers, computer engineer,…etc.
	R9: Suppliers	Suppliers that will supply us with the materials ,…etc.

Technology & Resources Analysis Grid

	T1: New Li-X battery	T2: Carbon fiber frame	T3: Impact resistance	T4: Temperature sensor circuit	T5: Intelligent chipset technology	T6: Smart switching inverter	T7: Data mining analysis	T8: Android/OS platform	T9: Safety alerts	Total
Score	1041	807	550	901	734	942	874	287	1153	
R1: Ion, Sulfur, Air, Carbon…etc	3	3	3	3	3	2	0	0	1	15136
R2: Government	2	2	2	2	2	2	2	2	2	14578
R3: Shareholders	2	2	2	2	2	2	2	2	2	14578
R4: Labs	3	3	2	3	3	2	2	1	2	17774
R5: Universities	3	3	3	3	3	2	3	1	2	19198
R6: Partners	3	1	1	2	3	2	1	2	1	12969
R7: Scientists & Researchers	3	3	1	2	3	1	1	1	1	13354
R8: Engineers	1	1	1	2	3	3	3	1	2	14443
R9: Suppliers	2	2	1	2	3	3	0	3	0	11937

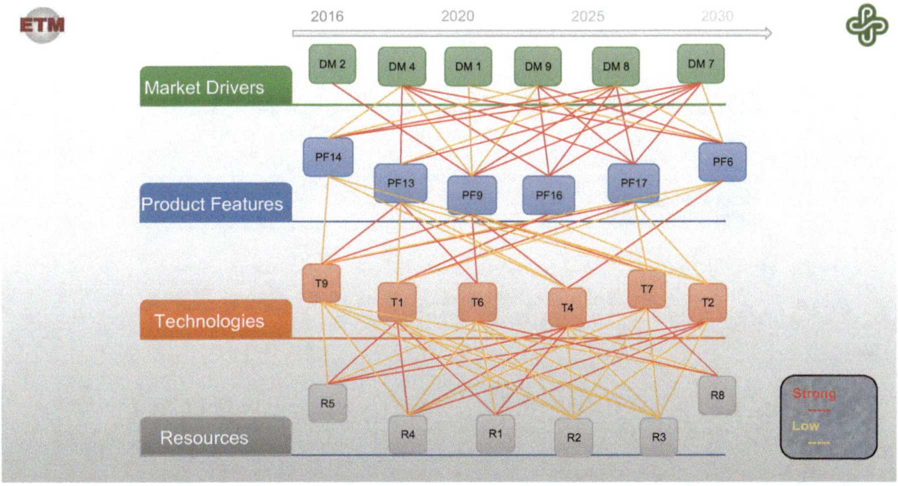

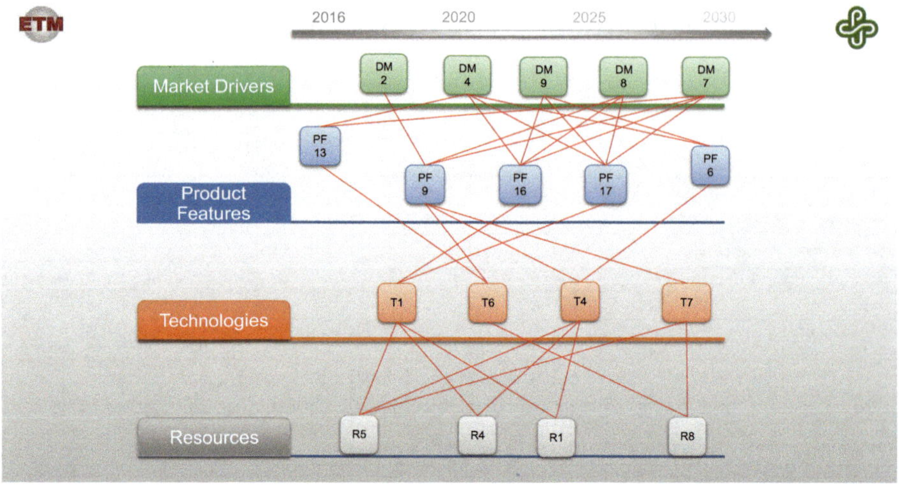
2016
2020
2025
2030
Market Drivers
DM 2
DM 4
DM 9
DM 8
DM 7
Product Features
PF 13
PF 9
PF 16
PF 17
PF 6
Technologies
T1
T6
T4
T7
Resources
R5
R4
R1
R8

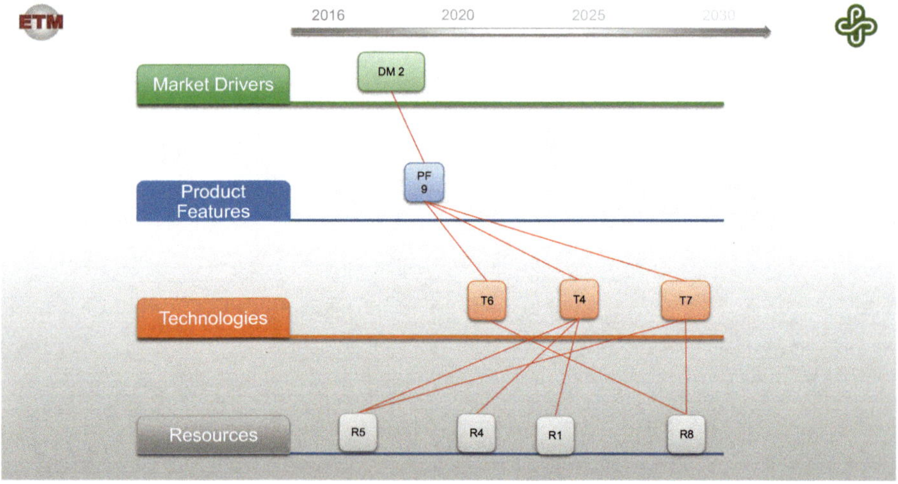
2016
2020
2025
2030
Market Drivers
DM 2
Product Features
PF 9
Technologies
T6
T4
T7
Resources
R5
R4
R1
R8

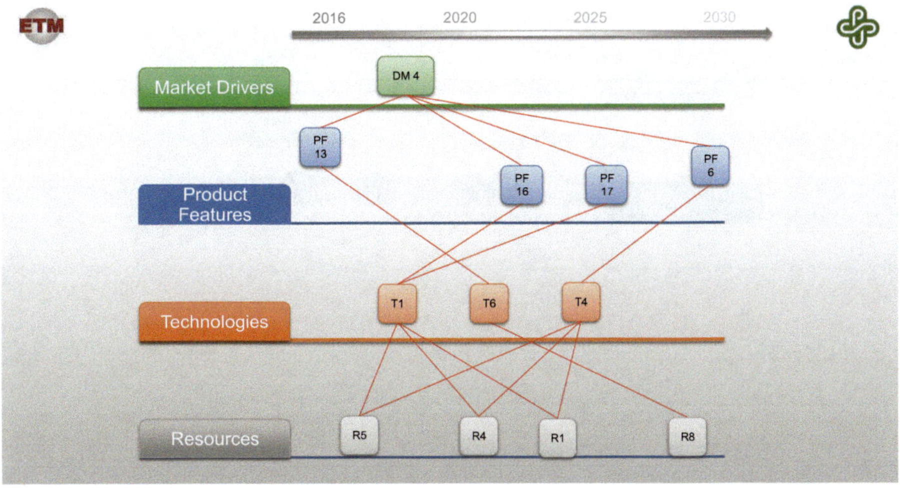
2016
2020
2025
2030
Market Drivers
DM 4
PF 13
PF 16
PF 17
PF 6
Product Features
T1
T6
T4
Technologies
Resources
R5
R4
R1
R8

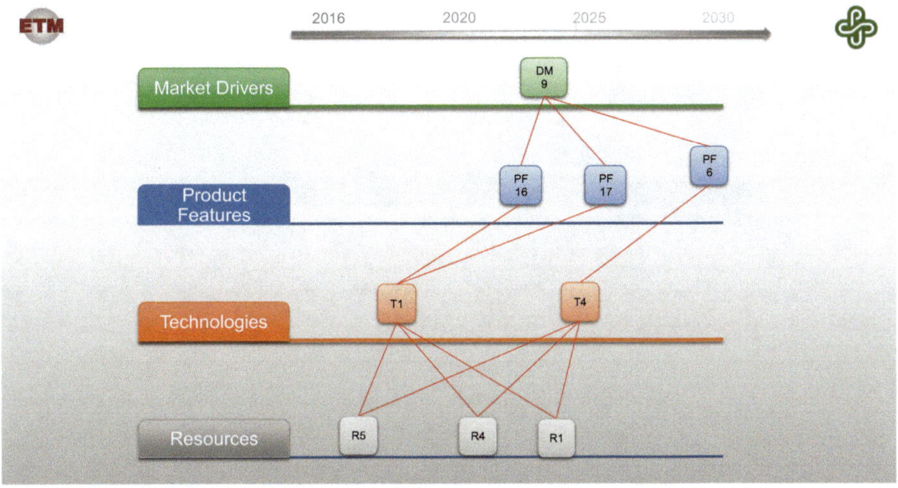
2016
2020
2025
2030
Market Drivers
DM 9
PF 16
PF 17
PF 6
Product Features
T1
T4
Technologies
Resources
R5
R4
R1

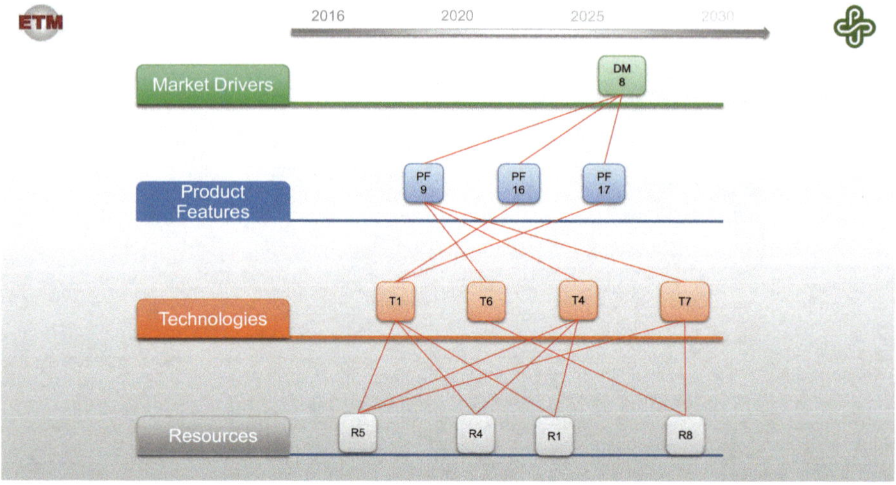
ETM
2016
2020
2025
2030
Market Drivers
DM 8
Product Features
PF 9
PF 16
PF 17
Technologies
T1
T6
T4
T7
Resources
R5
R4
R1
R8

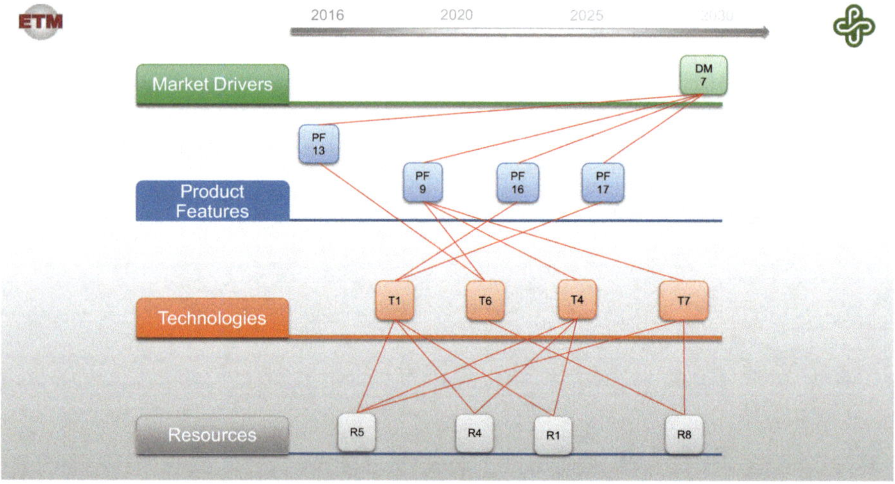
ETM
2016
2020
2025
2030
Market Drivers
DM 7
PF 13
Product Features
PF 9
PF 16
PF 17
Technologies
T1
T6
T4
T7
Resources
R5
R4
R1
R8

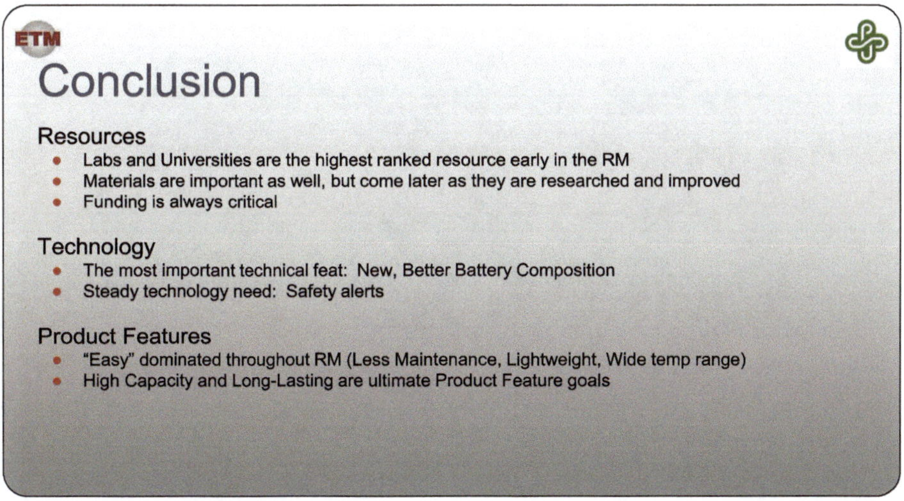
ETM
Conclusion
Resources
Labs and Universities are the highest ranked resource early in the RM
Materials are important as well, but come later as they are researched and improved
Funding is always critical
Technology
The most important technical feat: New, Better Battery Composition
Steady technology need: Safety alerts
Product Features
"Easy" dominated throughout RM (Less Maintenance, Lightweight, Wide temp range)
High Capacity and Long-Lasting are ultimate Product Feature goals

Thank You

ETM **References**

[1] http://www.ucsusa.org/clean_energy/our-energy-choices/renewable-energy/public-benefits-of-renewable.html#.V4VLYJMrLV0
[2]https://www.swellenergy.com/blog/2016/01/25/home-battery-prices-reach-an-all-time-low-and-with-sgip-your-payback-time-is-crazy-fast
[3] http://www.renewableenerg
[4] http://www.solarcity.com/residential/backup-power- supply
[5] http://dailycaller.com/2016/01/10/top-7-threats-to-americas-power-grid/
[6] http://www.wired.com/2015/05/get-tesla-home-battery-let-physics-explain/
[7] www.solarpowerworldonline.com/2015/08/whatisthebesttypeofbatteryforsolarstorage
[8] www.panasonic.com.au/energysolution
[9] http://analysis.energystorageupdate.com/homebatterystoragegrowthspursracesoftwarenewservics
[10] http://top10batteries.com/
[11] htttp://www.homepower.com/articles/solarelectricity/equipmentproducts/lithiumionbatteriesgridsystems/page/0/2?v=print
[12] http://cleantechnica.com/2016/05/31/investment-energy-storage-vital-renewable-energy-success/
[13] http://www.solarpowerworldonline.com/2015/08/what-is-the-best-type-of-battery-for-solar-storage/
[14] http://cleantechnica.com/2016/05/31/investment-energy-storage-vital-renewable-energy-success/
[15] http://www.pv-magazine.com/news/details/beitrag/hawaii- shuts-down- net-metering- to-new-customers_100021550/#ixzz4E2oS9DEw
[16] https://www.portlandgeneral.com/residential/power-choices/time-of-use/time-of-use-pricing
[17] http://iea-pvps.org
[18] www.solarpowerworldonline.com/2015/08/whatisthebesttypeofbatteryforsolarstorage
[19] www.panasonic.com.au/energysolution
[20] http://analysis.energystorageupdate.com/homebatterystoragegrowthspursracesoftwarenewservices
[21] http://top10batteries.com/
[22] htttp://www.homepower.com/articles/solarelectricity/equipmentproducts/lithiumionbatteriesgridsystems/page/0/2?v=print

ETM

[23] T7 - Youngjung Geum, HyeonJeong Lee, Youngjo Lee, Yongtae Park, Development of data-driven technology roadmap considering dependency: An ARM-based technology roadmapping, Technological Forecasting and Social Change, Volume 91, February 2015, Pages 264-279,
[24] T6 - http://www.greentechmedia.com/articles/read/Whats-Ahead-For-Inverter-Technology
[25] T3 - http://www.semblant.com/about-semblant-electronic-device-waterproofing-technology
[26] T4 - http://ntrs.nasa.gov/archive/nasa/casi.ntrs.nasa.gov/20090042373.pdf
[27] T1- Pillot, Christophe, "Battery Market Development for Consumer Electronics, Automotive, and Industrial", Batteries 2014
[28] T8- http://energydesk.greenpeace.org/2016/02/26/5-energy-apps-to-change-the-world

Appendix 11

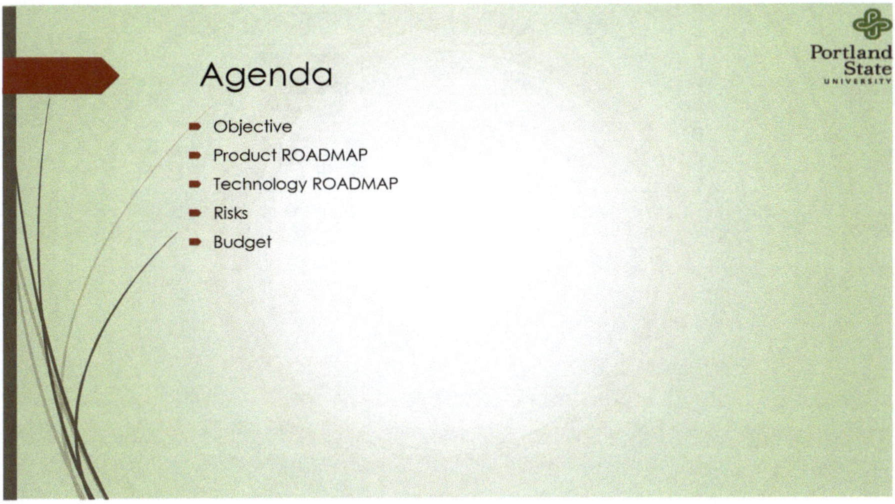

T. U. Daim (ed.), *Roadmapping Future*, Applied Innovation and Technology Management, https://doi.org/10.1007/978-3-030-50502-8

Objective:

Portland State University

- Develop a technology roadmap for Nike golf clubs and hockey sticks.

Product Roadmap

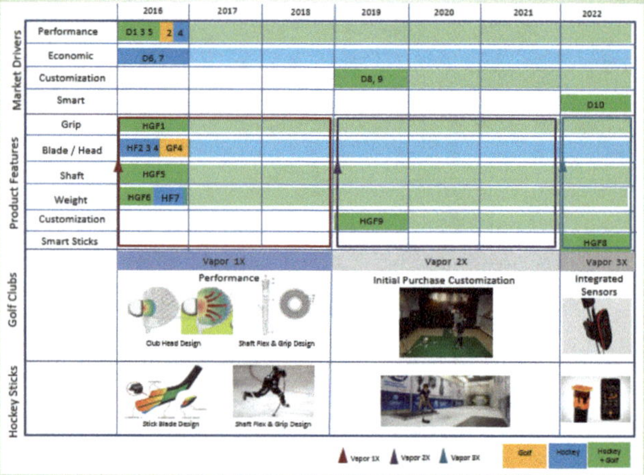

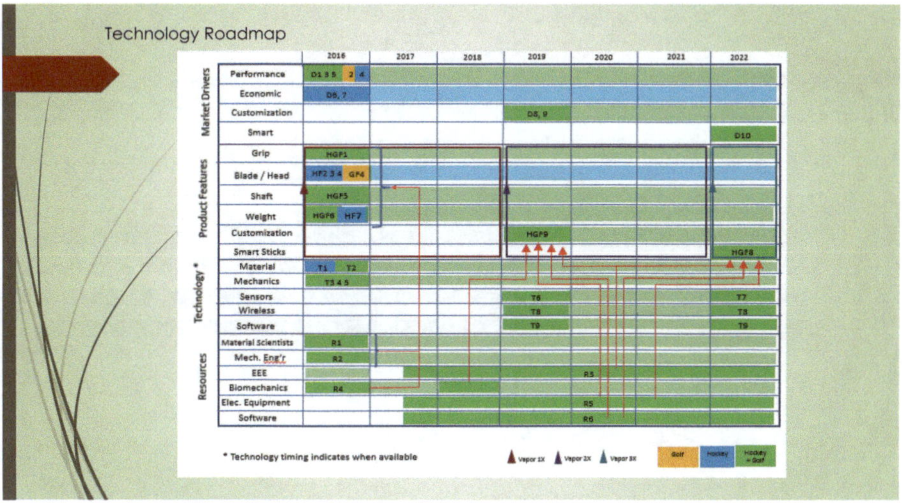

Risk Analysis

Features	Gap (1-10)	Year	Gap Description	Gap Resolution	Risk Rating
GF1 - Grip	1.4	2016	Grip that offers better control in the presence of moisture	With the help of Research/Material scientists, perforated grips can be designed which increase the surface area of the grip leading to a powerful reaction when exposed to moisture	**Low**
GF6 - Light weight/Balanced weight	0.1	2016	A lighter club without compromising the head speed	With the help of Research/Material scientists and engineers a 40 gm lighter club which can improve head speed by 4 miles/hr, otherwise difficult for a player to generate on his own.	**Low**
HF7 – Elasticity	1	2016	Durability not competitive	Incorporate cobalt nano tech	**Low**
HGF9 – Initial Purchase	10	2018	Sensor integrated on demo stick/clubs	Electronics engineers to inegrate existing sensor technology	Med
Integrated Sensors	10	2022	Miniaturization & integrating sensors to the stick/clubs and collecting the data in a cell phone	Gen 1 sensors available, bulky, develop small sensors	High

R&D Budget

		BUDGET						
		2016	2017	2018	2019	2020	2021	2022
	Head Count	18	37	37	37	37	37	37
Labor	$	4,500,000	9,250,000	9,250,000	9,250,000	9,250,000	9,250,000	9,250,000
Equipment		0	425,000	0	0	0	0	0
Software		0	75,000	0	0	0	0	0
Total		4,500,000	9,750,000	9,250,000	9,250,000	9,250,000	9,250,000	9,250,000

Q&A

Portland State UNIVERSITY

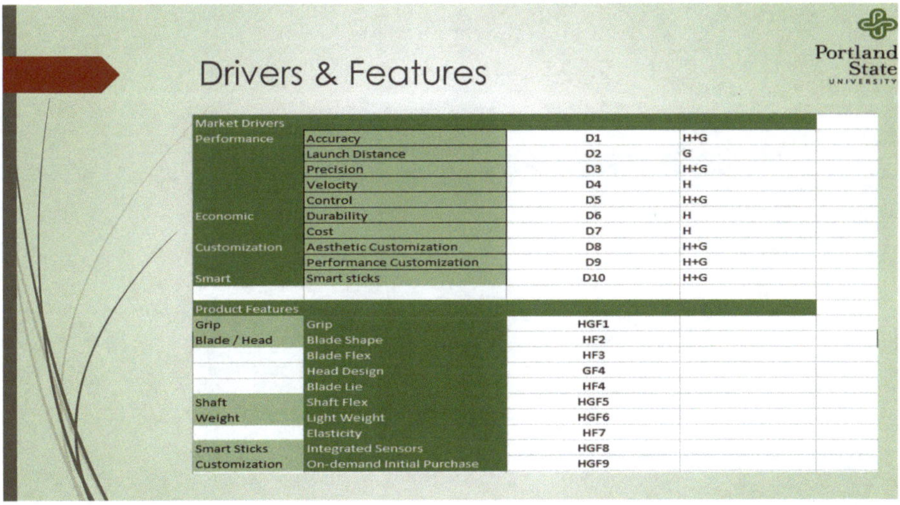

Market Drivers			
Performance	Accuracy	D1	H+G
	Launch Distance	D2	G
	Precision	D3	H+G
	Velocity	D4	H
	Control	D5	H+G
Economic	Durability	D6	H
	Cost	D7	H
Customization	Aesthetic Customization	D8	H+G
	Performance Customization	D9	H+G
Smart	Smart sticks	D10	H+G
Product Features			
Grip	Grip	HGF1	
Blade / Head	Blade Shape	HF2	
	Blade Flex	HF3	
	Head Design	GF4	
	Blade Lie	HF4	
Shaft	Shaft Flex	HGF5	
Weight	Light Weight	HGF6	
	Elasticity	HF7	
Smart Sticks	Integrated Sensors	HGF8	
Customization	On-demand Initial Purchase	HGF9	

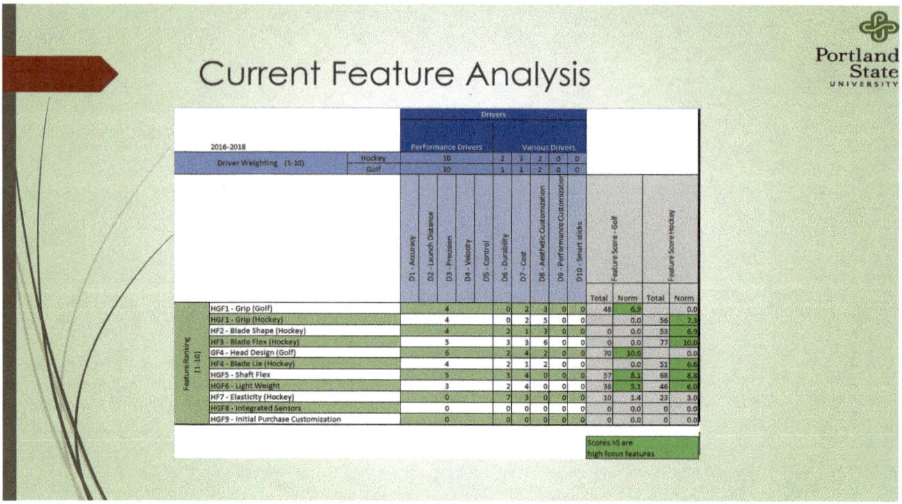

Current Feature Analysis

2016-2018

Feature Ranking (1-10)	D1 - Accuracy	D2 - Launch Distance	D3 - Precision	D4 - Velocity	D5 - Control	D6 - Durability	D7 - Cost	D8 - Aesthetic Customization	D9 - Performance Customization	D10 - Smart sticks	Feature Score - Golf Total	Feature Score - Golf Norm	Feature Score Hockey Total	Feature Score Hockey Norm
	Drivers: Performance Drivers					Drivers: Various Drivers								
Driver Weighting (1-10) Hockey	10					2	3	2	0	0				
Driver Weighting (1-10) Golf	10					1	1	2	0	0				
HGF1 - Grip (Golf)	4					0	2	3	0	0	48	6.9		0.0
HGF1 - Grip (Hockey)	4					0	2	5	0	0		0.0	56	7.3
HF2 - Blade Shape (Hockey)	4					2	1	3	0	0	0	0.0	53	6.9
HF3 - Blade Flex (Hockey)	5					3	3	6	0	0	0	0.0	77	10.0
GF4 - Head Design (Golf)	6					2	4	2	0	0	70	10.0		0.0
HF4 - Blade Lie (Hockey)	4					2	1	2	0	0		0.0	51	6.6
HGF5 - Shaft Flex	5					3	4	0	0	0	57	8.1	68	8.8
HGF6 - Light Weight	3					2	4	0	0	0	36	5.1	46	6.0
HF7 - Elasticity (Hockey)	0					7	3	0	0	0	10	1.4	23	3.0
HGF8 - Integrated Sensors	0					0	0	0	0	0	0	0.0	0	0.0
HGF9 - Initial Purchase Customization	0					0	0	0	0	0	0	0.0	0	0.0

Scores >5 are high focus features

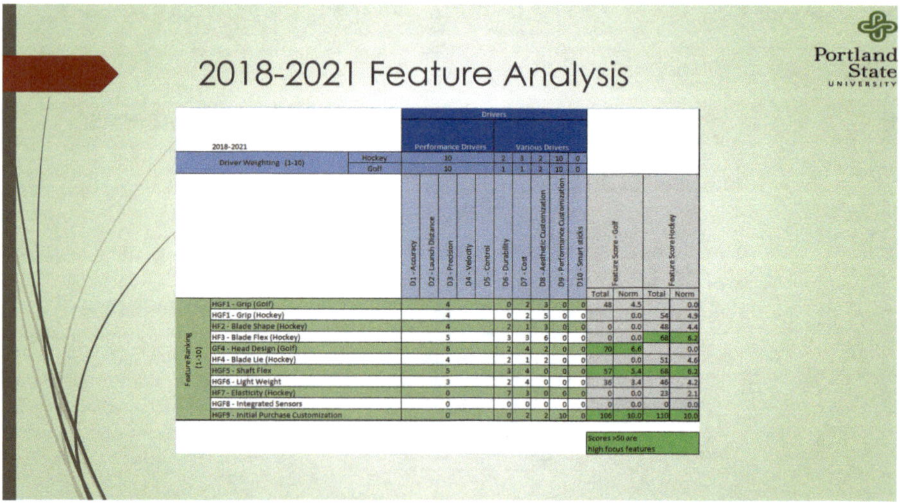

2018-2021 Feature Analysis

2018-2021

Feature Ranking (1-10)	D1 - Accuracy	D2 - Launch Distance	D3 - Precision	D4 - Velocity	D5 - Control	D6 - Durability	D7 - Cost	D8 - Aesthetic Customization	D9 - Performance Customization	D10 - Smart sticks	Feature Score - Golf Total	Feature Score - Golf Norm	Feature Score Hockey Total	Feature Score Hockey Norm
	Drivers: Performance Drivers					Drivers: Various Drivers								
Driver Weighting (1-10) Hockey	10					2	3	2	10	0				
Driver Weighting (1-10) Golf	10					1	1	2	10	0				
HGF1 - Grip (Golf)	4					0	2	3	0	0	48	4.5		0.0
HGF1 - Grip (Hockey)	4					0	2	5	0	0		0.0	54	4.9
HF2 - Blade Shape (Hockey)	4					2	1	3	0	0	0	0.0	48	4.4
HF3 - Blade Flex (Hockey)	5					3	3	6	0	0	0	0.0	68	6.2
GF4 - Head Design (Golf)	6					2	4	2	0	0	70	6.6		0.0
HF4 - Blade Lie (Hockey)	4					2	1	2	0	0		0.0	51	4.6
HGF5 - Shaft Flex	5					3	4	0	0	0	57	5.4	68	6.2
HGF6 - Light Weight	3					2	4	0	0	0	36	3.4	46	4.2
HF7 - Elasticity (Hockey)	0					7	3	0	0	0	0	0.0	23	2.1
HGF8 - Integrated Sensors	0					0	0	0	0	0	0	0.0	0	0.0
HGF9 - Initial Purchase Customization	0					0	2	2	10	0	106	10.0	110	10.0

Scores >50 are high focus features

Portland State UNIVERSITY

2022 & Beyond Feature Analysis

2022 and Beyond

Feature Ranking (1-10)	Performance Drivers: D1 - Accuracy, D2 - Launch Distance, D3 - Precision, D4 - Velocity, D5 - Control	Various Drivers: D6 - Durability	D7 - Cost	D8 - Aesthetic Customization	D9 - Performance Customization	D10 - Smart sticks	Feature Score - Golf: Total	Feature Score - Golf: Norm	Feature Score - Hockey: Total	Feature Score - Hockey: Norm
Driver Weighting (1-10): Hockey	3	2	3	2	5	10				
Driver Weighting (1-10): Golf	3	1	1	2	5	10				
HGF1 - Grip (Golf)	4	0	2	3	0	0	20	2.0		0.0
HGF1 - Grip (Hockey)	4	0	2	5	0	0		0.0	26	2.5
HF2 - Blade Shape (Hockey)	4	2	1	3	0	0	0	0.0	20	1.9
HF3 - Blade Flex (Hockey)	5	3	3	6	0	0	0	0.0	33	3.1
GF4 - Head Design (Golf)	6	2	4	2	0	0	28	2.7		0.0
HF4 - Blade Lie (Hockey)	4	2	1	2	0	0		0.0	23	2.2
HGF5 - Shaft Flex (Hockey/Golf)	5	3	4	0	0	0	22	2.2	33	3.1
HGF6 - Light Weight	3	2	4	0	0	0	15	1.5	25	2.4
HF7 - Elasticity (Hockey)	0	7	3	0	0	0	0	0.0	23	2.2
HGF8 - Integrated Sensors	0	0	2	0	0	10	102	10.0	106	10.0
HGF9 - Initial Purchase Customization	0	0	2	2	10	0	56	5.5	60	5.7

Scores >5 are high focus features

Technology GAP Analysis

Technology

Technology Group	Technology Focus	Tech Code	Aligned Driver Code	Sport Application	Definition/Need	Gaps	Gap Rating (0-5) 0= no gap
Material (Product)	Toughness	T1	D6, D7	H	Durability of hockey sticks is particularily an issue. Higher toughness materials are needed to address the issue	Competitors have developed nano cobalt technology that is superior to in-house composite polymer material	2
	Strength to weight ratio - weight reduction	T2	D5,D7	H+G	Weight is a balancing act, as weight is needed to create energy as the hockey stick or golf club is swung. This is balanced by lightness needed to control the stick or club, especially for hockey sticks	Currently on par with competitors weight to strength properties, and material properties to enable flex tuning	0
Mechanics (Development Tools)	FEA	T3	D1-D4, D5	H+G	To analyse shaft/stick and blade flex, torqueing, identify weak points, optimize weight	Leading FEA capability	0
	Kinematics	T4	D4	H+G	Analyse velocity of club and stick	Leading kinematics technology	0
	Energy	T5	D2,D4	H+G	Ultimately the kinetic and potential energy in hockey sticks and golf clubs launch pucks and golf balls to achieve velocity and distance	Leading energy analysis capability	0
Sensors (Product & Development Tools)	Strain & Acceleration	T6	D6	H+G	Measuring key characteristics during development and actual use	On par with competitors for development tools to gather strain and acceleration data	0
	Miniaturization of sensors for sensors mounted on consumer sticks & golf clubs	T7	D10	H+G	Integration of sensors on sticks and clubs for consumer use	Lagging behind independent development companies in integrating performance sensors	3
Wireless Data Transfer (Product)	Sensor signal to transmitter to receiver via wireless transmission	T8	D9,D10	H+G	Transfer sensor info to data base	Lagging behind 3rd party independents. Some in-house technology expertise with Nike FuelBand	3
Software (Product)	Data analysis and communication	T9	D9,D10	H+G	Convert sensor signal to meaningful performance information	No software is being developed to gather wireless data and create format for performance analysis	4

Market Drivers

Driver Group	Driver	Driver Coc	Sport App
Performance	Accuracy	D1	H+G
	Launch Distance	D2	G
	Precision	D3	H+G
	Velocity	D4	H
	Control	D5	H+G
Economic	Durability	D6	H
	Cost	D7	H
Customization	Aesthetic Custom	D8	H+G
	Performance Cus	D9	H+G
Smart	Smart sticks/club	D10	H+G

Product Feature GAP Analysis

Feature/ Year	2015 Target/Actual/Gap Vapor 1X	2018 Target/Actual/Gap Vapor 2X	2022 Target/Actual/Gap Vapor 3X
HGF1 - Grip	H-7.3/7.5/[illegible] G-6.9/5.5/[illegible]	H-4.9/7.5/-2.6 G-4.5/5.5/-1.0	H-2.5/7.5/-5.0 G-2/5.5/-3.5
HF2 - Blade Shape	H-6.9/7/-0.1	H-4.4/7/-2.6	H-1.9/7/-5.1
HF3 - Blade Flex	H-10/10/0	H-6.2/10/-3.8	H-3.1/10/-6.9
GF4 – Head Design	G-10/10/0	G-6.6/10/-3.4	G-2.7/10/-7.3
HF4 - Blade Lie	H-6.6/7/-0.4	H-4.6/7/-2.4	H-2.2/7/-4.8
HGF5 – Shaft Flex	H-8.8/9/-0.2 G-8.1/8.5/-0.4	H-6.2/9/-2.8 G-5.4/8.5/-3.1	H-3.1/9/-5.9 G-2.2/8.5/-6.3
HGF6 - Light Weight/Balanced Weight	H-6/6/[illegible] G-5.1/5/[illegible]	H-4.2/6/-1.8 G-3.4/5/-1.6	H-2.4/6/-3.6 G-1.5/5/-3.5
HF7 - Elasticity	H-3/2/[illegible]	H-2.1/2/[illegible]	H-2.2/2/[illegible]
HGF8 - Integrated Sensors	H-0/0/0 G-0/0/0	H-0/0/0 G-0/0/0	H-10/0/[illegible] G-10/0/[illegible]
HGF9 - Initial Purchase Customization	H-0/0/0 G-0/0/0	H-10/0/[illegible] G-10/0/[illegible]	H-5.7/0/[illegible] G-5.5/0/[illegible]

Budget Input

Position	Role Description	Manpowe	Start Year	End Year	Duration(Years)	Type of Employment	Salary(per hour	Hours/Year*	Cost to Company/Year
Head	Head Research and Department	1	2016	2022	6	Contract	150	1790	268500
Scientist	Analysis of Material Technology	2	2016	2022	6	Contract	120	1790	429600
Technician	Field work, Collection of basic data used in sticks	4	2016	2022	6	Contract	80	1790	572800
Head	HOD and Research Head	1	2017	2022	5	Contract	150	1790	268500
Scientist	Research and incorporation of technology into sticks	2	2017	2022	5	Contract	120	1790	429600
Engineer	Test the necessary Equipment and troubleshoot	4	2017	NA	Permanent	Full Time	100	1790	716000
Technician	Collect Basic data and Field Work	4	2017	NA	Permanent	Full Time	70	1790	501200
Head	HOD	1	2016	N.A	Permanent	Full Time	150	1790	268500
Developer	Develop the necessary code for sensors to capture da	4	2017	2022	5	Contract	120	1790	859200
Tester	Test the Code and Troubleshoot for Problems	4	2017	N.A	Permanent	Full Time	120	1790	859200
Head	HOD	1	2016	2022	6	Contract	150	1790	268500
Engineer	Design the stick with inputs from Metallurgy and Bio-	4	2016	2022	6	Contract	120	1790	859200
Bio-med Engine	Provide analysis on Design of sticks and Position of S	1	2016	2022	6	Contract	120	1790	214800
Technician	Collect Data as instructed by the team	4	2016	2022	6	Contract	80	1790	572800

Appendix 12

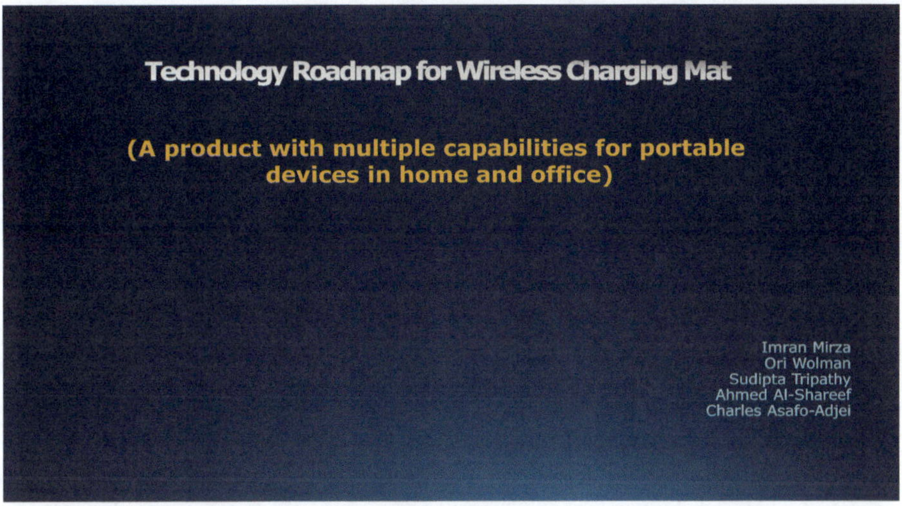

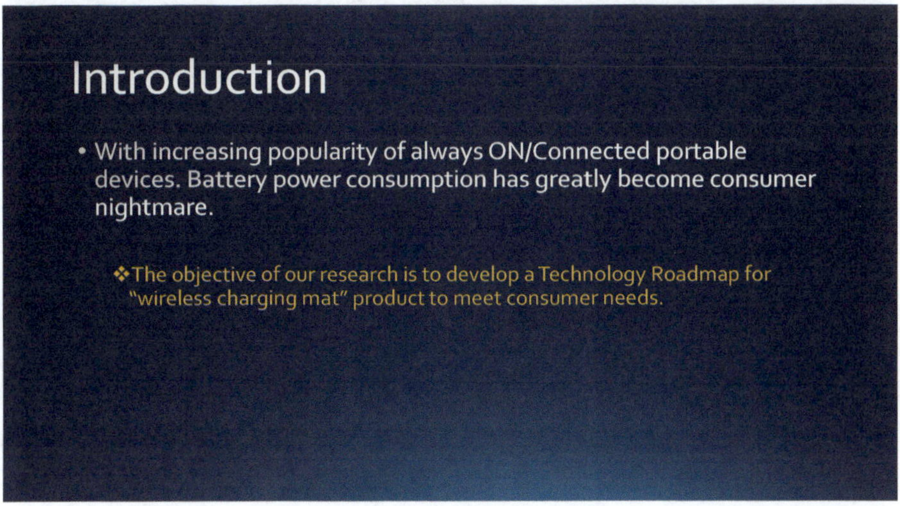

T. U. Daim (ed.), *Roadmapping Future*, Applied Innovation and Technology Management, https://doi.org/10.1007/978-3-030-50502-8

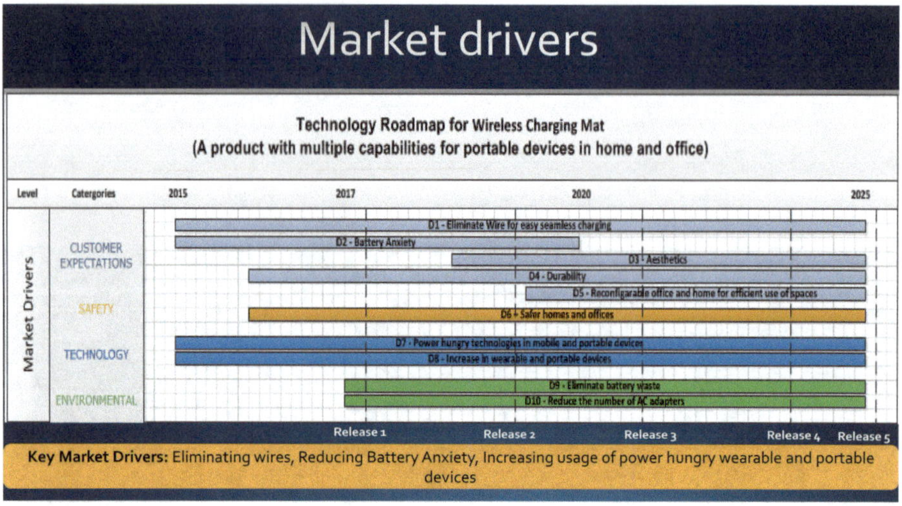

Eliminate – 10
Battery Anxiety – 10
Wearable – 9
Power hungry – 8
Durability – 7
Eliminate waste – 7
Reduce number of adapters - 7

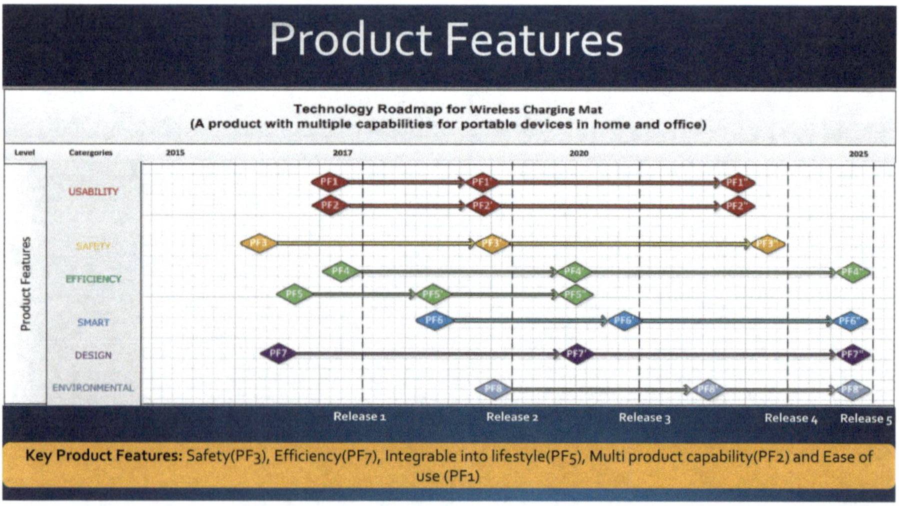

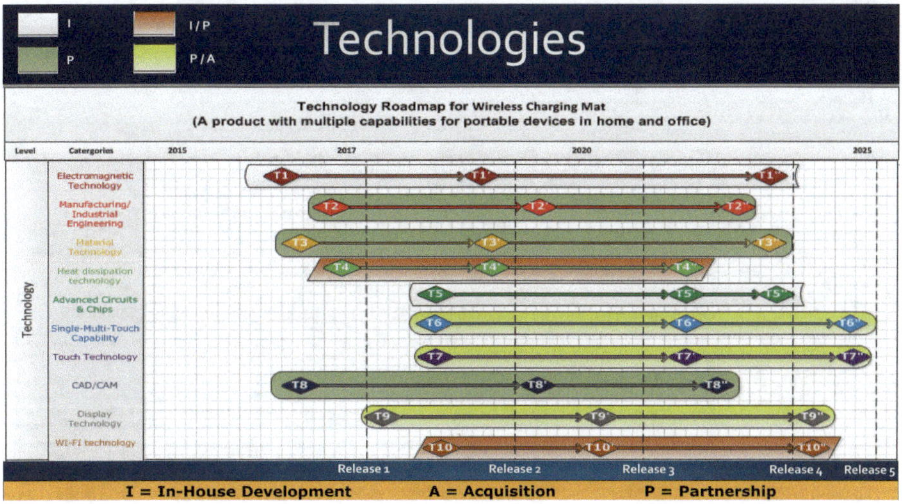
I
I/P
P
P/A
Technologies
Technology Roadmap for Wireless Charging Mat
(A product with multiple capabilities for portable devices in home and office)
Level
Catergories
2015
2017
2020
2025
Technology
Electromagnetic Technology
Manufacturing/ Industrial Engineering
Material Technology
Heat dissipation technology
Advanced Circuits & Chips
Single-Multi-Touch Capability
Touch Technology
CAD/CAM
Display Technology
WI-FI technology
Release 1
Release 2
Release 3
Release 4
Release 5
I = In-House Development
A = Acquisition
P = Partnership

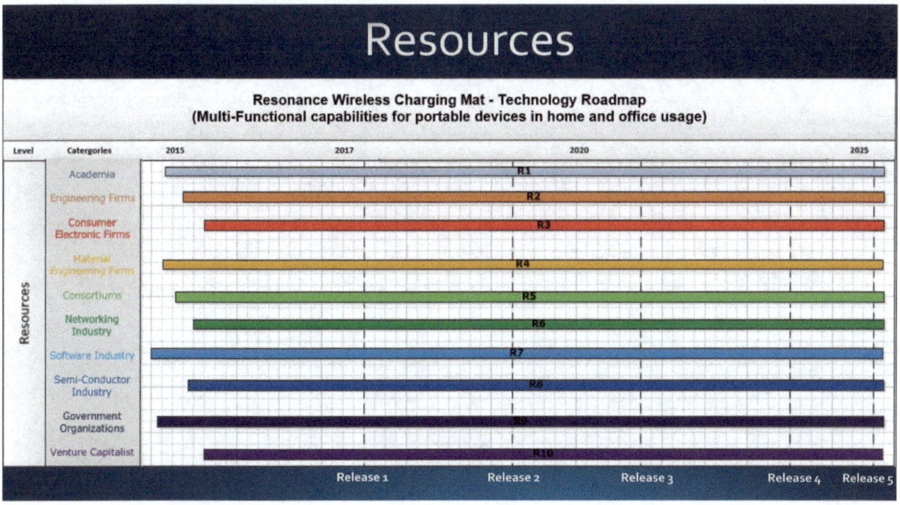
Resources
Resonance Wireless Charging Mat - Technology Roadmap
(Multi-Functional capabilities for portable devices in home and office usage)
Level
Catergories
2015
2017
2020
2025
Resources
Academia
Engineering Firms
Consumer Electronic Firms
Material Engineering Firms
Consortiums
Networking Industry
Software Industry
Semi-Conductor Industry
Government Organizations
Venture Capitalist
Release 1
Release 2
Release 3
Release 4
Release 5

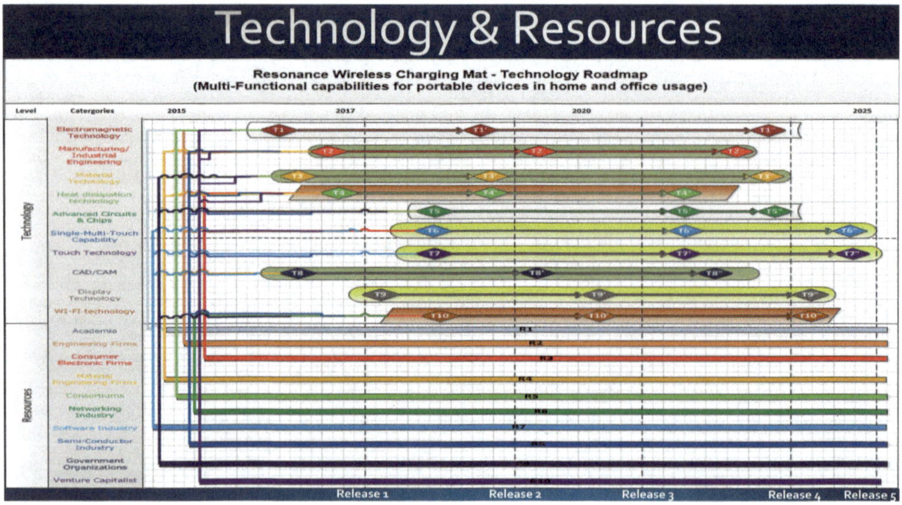

Conclusion

- Wireless Charging Technology (WCT) is gradually being realized as the solution of present and future battery anxieties for portable devices.
- Current technological advancement poses a challenges for OEMs to develop products with seamless charging capabilities. With improvements in resonance manufacturing technology and power transmission , its expected that WCT products will increase in usages within homes, offices, academia, health sectors, and government sectors.
- Resonance charging will offer unique advantages in spatial freedom, enabling the transmitter (resonance charger) to be separated from the receiver (portable device) by several inches or more.
- We aim to leverage multiple resources, and technologies across various sectors to develop WC product with multiple capabilities to enhance consumers experience and change how we live.

Appendix 13

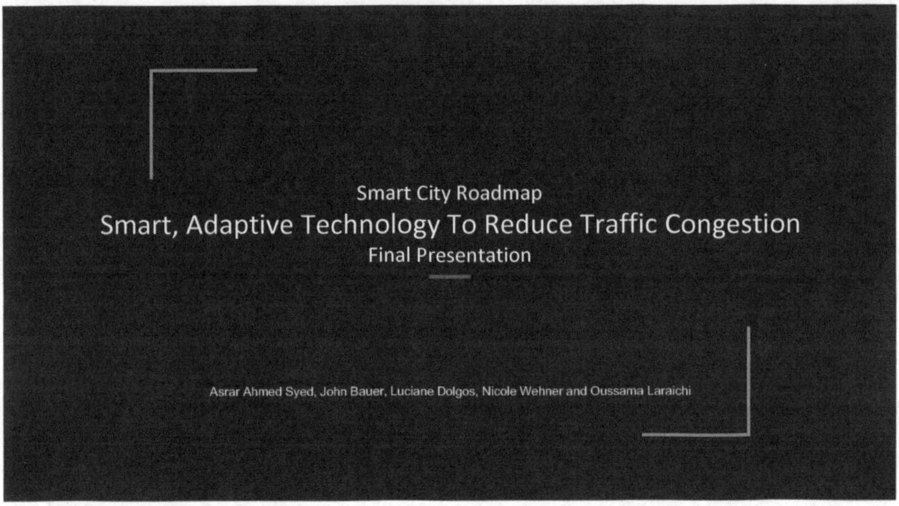

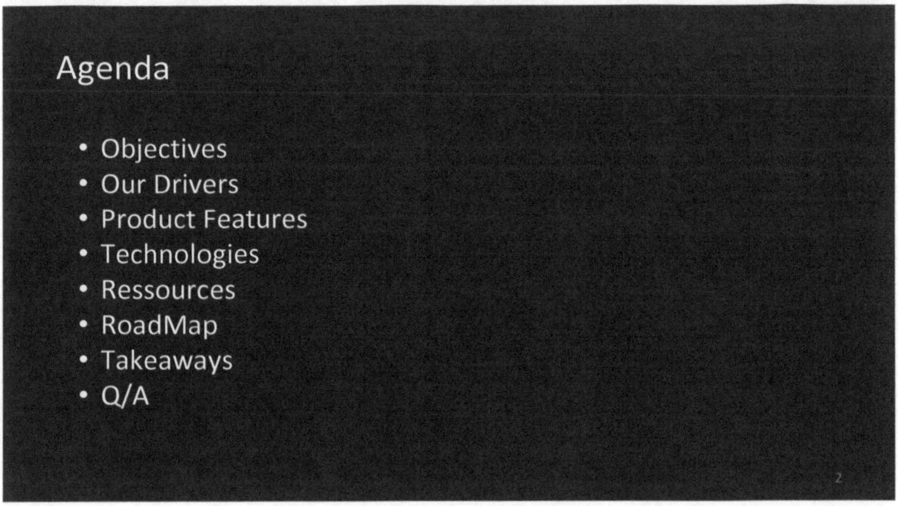

T. U. Daim (ed.), *Roadmapping Future*, Applied Innovation and Technology Management, https://doi.org/10.1007/978-3-030-50502-8

Objectives

- Who we are:
  - Consulting firm hired by PDOT
- Goal:
  - Develop a roadmap for a smart, adaptive technology for efficient urban mobility
- Purpose:
  - Find a solution to Portland's Mobility oriented problems to create a path for a winning Smart City Challenge Proposal

3

Market Driver Ranking

Market Drivers		Code	Weight
Economical	Loss of productivity	D1	9
	Disrupts supply chain	D2	7
	U.S. Impacted in the global economy	D3	5
Population	Demographic growth	D4	9
	Commuting behaviors	D5	7
	Transportations offers	D6	5
Political	Funding chasing	D7	4
	Urban Programs	D8	5
	Political accountability	D9	3
Technical	Transportation Efficiency	D10	8
	Evolution of the automotive industry	D11	7

4

Product Features and Gap Analysis

Code	Product Features	Current Level	Where we want it to be	Gap
PF1	**Consumer and commercial vehicles**	Connect through phone	Broader implementation, through car (eventuatlly), adapt routes for commercial vehicles	Select data needs to be made available in real-time. No impact study for dedicated lanes for autonomous vehicles.
PF2	**Public tranportation**	Know if it's on time or not, trimet tracking	Know how crowded it is, prediction of how late, slow it will go. GPS should be linked with app to get real time info like where is the exact location of the vehicle.	Transit tracker gives the information based on vehicle speed and last location reported. it may be faster or slower. GPS is not linked.
PF3	**Emergency vehicles**	Tracked by hospital, separate entities	Tied into single user app	Data needs to be made available in real-time, and only data which is non-sensitive.
PF4	**Combine all forms of transp. into a single user interface**	Does not exist, private companies have own app	show options for biking and public transportation, car and biking, etc, biketown, uber locations	Need algorithm for comparing different speeds of different types of transportation and optimizing what to do where based on current conditions
PF5	**Accident/road closure alerts**	User entered into Google Maps, Waze	Ambulance/police automatically send signal	Transportation API's need to update emergency vehicle status and connect information to the software
PF6	**Carpooling/connectiong people**	No way of connecting people (separate app for carpooling.)	Inform neighbors, take reservations, schedule times, update status in real time	separate app for carpooling and less awareness.
PF7	**Should adapt to new types of transportations**	Not in use	Use drones as a mode of trasportation to work place or other.	Surveillance of all the drones is not possible for illegal use
PF8	**Parking/EV charging stations**	Don't know if it's full until you get there	Before leaving, can look and see what is available and reserve spot	Real time user-sensing and updating.
PF9	**Handicap accessible**	Not currently available	Inegrated into voice commands	Voice recognition software related to tranportation modes
PF10	**Easy to use GUI**	Not in use	Reliable, high performance / efficient, intuitive / inclusive of all demographics	The user interface will depend on the speed and accuracy of real time data. Need reliable, robust infrastructure for fast data streaming.
PF11	**Emission preferences**	Not in use	rank modes / comibnations of modes in terms of time spent and emmssions, and show user impact in understanble context	need way of calculating impact if on bus/max, need to come up with context that users understand
PF12	**Cost preferences**	Some cost analysis but not complete analysis of all options	Compare more option, rank by price, $/time units	combine all options into camparative dashboard, create metric for comparing time and cost
PF13	**Time predictions**	Predict best option route under current conditions.	Predict at different times of day and year and take account user inputs on traffic conditions	Need advanced algorithms, and hardware / software platform for multiple real-time data inputs for precise predictions
PF14	**Control Traffic Signals**	control room personnel, public transportation, and emergency vehicles can change traffic signals	signals can adapt to different events and conditions	Robust SCADA system, and ITS (intelligent transportation systems) allowing adaptive signal control are currently available.

Product Feature Timeline

Category	Code	Product Features	Current Level	2 years	5 years	15 years
Intregration of vehicles and services	PF1	Consumer and commercial vehicles	PF1.1 Connect through phone	PF1.2 Broader implementation		PF1.3 Connection through car(eventually)
	PF2	Public tranportation	PF2.1Know if it's on time or not, trimet tracks	PF2.2 Know how crowded the vehicle is	PF2.3 GPS should be linked with the app to get real time info, like the exact location of the vehicle	PF2.4 More accurate
	PF3	Emergency vehicles	PF 3.1 Tracked by hospital, separate entities	PF 3.2 Real time location of emergency vehicles to users app	PF3.3 Improve accuracy	PF3.4 Separate routes for emergency vehicles
	PF4	Combine all forms of transp. into a single user interface (Biking, walking, uber, public transport, carpooling, future modes)	PF 4.1 Does not exist, private companies have own app	PF 4.2 Show options for biking and public transportation, car and biking etc	PF 4.3 Show options for bking, public transportation, car and commercial vehicles	PF 4.4 Connect the app with quasi-autonomous vehicles
	PF5	Accident/road closure alerts	PF5.1 User entered into Google Maps, Waze	PF 5.2 Traffic signals signalling different routes	PF 5.3 Ambulance/police automatically send signal	PF 5.4 Quasi-autonomous cars receiving direct signals to avoid the accident area and rerout.
	PF6	Carpooling/connectiong people	No way of connecting people(separate app for carpooling.)	PF6.1 Create communual board/forum to allow people to post and connect	PF6.2 Automatically find matches to carpool	
	PF7	Should adapt to new types of transportations (boats, drones, bike sharing programs, etc)	PF7.1 Bike sharing	PF7.2 Use boats as a way of transportation	PF7.3 Use drones for transportation	PF7.4 Monitoring drones for border protection
	PF8	Parking/EV charging stations	Don't know if it's full until you get there	PF8.1 Reserve spot	PF8.2 Pay online for fuels reduce time at stations	
Usage	PF9	Handicap accessible	Not currently available	PF9.1 Integrated into voice commands	PF 9.2 Sensors sensing wheelchairs	PF.3 Connected Wheelchair with the cloud
	PF10	Easy to use GUI	Not in use	PF10.1 Efficient, and reliable	PF10.2 Inclusive of all demographics	
	PF11	Emission preferences	Not in use	PF11.1 Rank modes	PF11. 2 Show user impact in understandable context	PF11.3 Compare more option, rank by price, $/time units
	PF12	Cost preferences	PF12.1 Some cost analysis but not complete analysis of all options		PF12.2 Compare more option, rank by price, $/time units	PF12.3 All modes of transportation included ranked by price
Movement Data	PF13	Time predictions	PF13.1 Predict best option route under current conditions.		PF13.2 Predict at different times of day and year and take account user inputs on traffic conditions	
Signal Control	PF14	Control Traffic Signals	PF14.1 Control room personnel, public transportation, and emergency vehicles can change traffic signals	PF14.2 Broad implementation of SCADA	PF14.3 Signals can adapt to different events and conditions	PF14.4 Traffic lights show differents routes during events

PF vs Drivers: QFD Matrix

High connection: 3
No connection: 0

			Economical			Population			Political			Technical			
			Productivity	Disrupts supply chain	U.S. Impacted in a the global economy	Demographic growth	Commuting behaviors	Transportations offers	Funding chasing	Urban Programs	Political accountability	Transport Efficiency	Evolution of the automotive industry		
		Weight	9	7	5	9	7	5	4	5	3	8	7		
		Code	D1	D2	D3	D4	D5	D6	D7	D8	D9	D10	D11	Total	Normalized value
Connecting Vehicles and Services	Commercial and consumer vehicles	PF1	3	3	2	3	3	0	0	0	0	2	0	122	7.1
	Public tranportation	PF2	2	0	1	3	3	3	2	3	2	3	2	153	8.9
	Emergency vehicles	PF3	3	0	1	0	0	0	0	0	1	0	1	42	2.4
	Combine all transp. types	PF4	3	0	0	2	3	3	2	2	0	3	3	144	8.4
	Accident/road closure alerts	PF5	3	3	1	0	1	0	1	0	1	1	0	75	4.4
	Carpooling/connectiong people	PF6	2	0	0	2	3	3	0	3	2	3	3	138	8.0
	Adapt to new kinds of technology	PF7	2	2	3	1	2	1	0	0	1	1	3	107	6.2
	Parking/EV charging stations	PF8	3	3	2	2	2	0	3	2	1	1	2	137	8.0
Usage	Handicap accessible	PF9	0	0	0	2	1	0	1	1	1	0	1	44	2.6
	Ease of use	PF10	1	0	0	1	2	2	0	1	0	3	2	85	4.9
	Emission preferences	PF11	0	0	2	2	2	3	3	2	3	2	2	118	6.9
	Cost preferences	PF12	0	1	0	0	3	3	0	0	1	1	1	61	3.5
Movement Data	Time/Traffic predictions	PF13	3	3	3	3	3	1	2	0	1	3	3	172	10.0
Signal Control	Traffic signal data/control	PF14	3	3	0	2	2	0	1	0	1	3	1	118	6.9

7

Technologies and Gap Analysis

Code	Technology	Current Level	Where we want it to be	Gap
T1	Data Hardware	Costly infrastructure, just enough computing power	Infrastructure, and cloud solutions that meet the performance, reliability, security and cost expectations	No Gap --> However, hardware must evolve and keep pace with more and more data, users, and long-term growth, and changing requirements
T2	Connectivity	Short range bluetooth and WiFI in all phones.	High range of bluetooth and phones equiped with Llifi for faster internet connection	LiFi still under development
T3	Motion Sense	PORTAL has loop sensors throughout city	Motion sensers at parking lots, more loop sensors in areas of bad congestion	No Gap --> However, need cheaper and wider implementation
T4	Location Services	GPS in every smart phone ~within 3.5 meters	Consistently accurate within ~1 meter	accuracy augmentation programs in service and in development
T5	Security Protocols	Cryptography, Anti-virus, Firewall, access control, on-going remote updates and monitoring. Standards: IEEE, IETF	Increased security standards for wireless networking. More robust firewall and anti-virus	IEEE 802.15.4 & 802.15.4E & 802.15.4F (5-15 years)
T6	Data Algorithms for Mode Localization and Optimization	Content-based Filtering, Collaborative Filtering, Hybrid Filtering (Combination) are current methods for developing predictive algorithms used with open source API's	Optimization through greater predicitive accuracy from dynamic patterns.	No Gap --> However, expensive infrastructure and hardware limitations. Need emerging technical platforms to provide enhanced capabilities

8

Technology Timeline

Technology Categories	Code	Current Standards/Level	> 2 years	> 5 years	> 15 years
Data Hardware	T1	T1.1- Manual migrations		T1.2 -Expanded Storage	T1.3 -Adaptive Storage
Connectivity	T2	T2.1.1 - Bluetooth: range of 10 meter T2.1.2 - Wifi: 800Mbps T2.1.3 - Lifi: 1 Gbps	T2.2.1 - Bluetooth: Range of 100 meters T2.2.2 - Wifi: 900 MHz(low power)	T2.3.1 - Broader implementation of Lifi, T2.3.2 - WiMax	T2.4.1 - Lifi: 225 Gbps
Motion Sense	T3	T3.1 - Radar: 77 GHz(Transceiver)	T3.2 - Radar: 79 GHz (improved resolution)		
Location Services	T4	T4.1 - GPS: 66 channels		T4.2 - Increased Accuracy	
Security Protocols	T5	T5.1.1 - Cryptography (IEEE P1363) T5.1.2 - Access Control (IEEE 802.1: 802.1X) T5.1.3 - WPA2 (IEEE 802.11i-2004) T5.1.4 - IPV6 (IETF) T5.1.5 - Payment Card Industry Data Security Standard (PCI DSS) T5.1.6 - LR-WPANs (IEEE 802.15.4) T5.2.7 - Intrusion Detection Systems	T5.2.1 - Consistent use of encryption between all network nodes and in all network communication. T5.2.2 - Rollout start of 5G mobile networks, designed for much larger number of devices	T5.3.1 - Consistent use of Intrusion Detection Systems and Deep Packet Inspection at large "City Scale" networks. T.5.3.2 - Wide accessibility to new 5G networks, and cheaper access to 4G networks.	T5.4.1 -- Adaptive and quasi-autonomous IDS and DPI T5.4.2 - V2V,V2I, V2P (IEEE 802.11p) T5.4.3 - CEREBRE protocol
Data Algorithms	T6	T6.1 - Collect data and organize it [P2]	T6.2 - Predictive and adaptive	T6.3 - Improved and streamlined algorithms like how re-route exactly right amount of traffic [P3]	T6.4 - Increased Quality of Predictions

9

Tech vs PF QFD Matrix

High dependence: 3
No dependence: 0

		Connecting Vehicles and Services								Usage				Movement Data	Signal Control	
		Commercial and consumer vehicles	Public tranportation	Emergency vehicles	Combine all transp. types	Accident/road closure alerts	Carpooling/connecting people	Adapt to new kinds of technology	Parking/EV charging stations	Handicap accessible	Ease of use	Emission preferences	Cost preferences	Time/Traffic predictions	Traffic signal data/control	
	Weight	7.1	8.9	2.4	8.4	4.4	8.0	6.2	8.0	2.6	4.9	6.9	3.5	10.0	6.9	
	Code	PF1	PF2	PF3	PF4	PF5	PF6	PF7	PF8	PF9	PF10	PF11	PF12	PF13	PF14	Total
Data Hardware	T1	1	1	1	1	1	2	2	2	2	2	2	2	3	3	162
Connectivity	T2	3	3	3	3	3	3	0	3	0	2	1	1	2	3	202
Motion Sense (vehicles)	T3	0	0	0	0	0	0	0	3	0	0	0	0	2	3	64
Location Services	T4	3	3	3	3	3	2	2	0	0	0	0	0	3	3	173
Security	T5	3	1	3	0	3	3	0	0	0	0	0	0	0	3	95
Data Algorithms	T6	0	2	0	3	1	1	2	0	2	2	3	3	3	3	165

10

Resources and Analysis

Code	Resource	Service	Technology	Implementation Timeline
R1	Partnership with Wimax Service providers and PureLifi	R&D for Wimax, Lifi	T2	Extended-wait for technology to mature and use R2 in the meantime
R2	Cellular phones/data networks	Connect users	T2	Existing
R3	Partnership with Intel	Software engineers, Data analysts	T6, T5	Right away
R4	Partnership with PSU-- access to PORTAL	Traffic pattern analysis, Historical Traffic Data, Data from existing loop stations	T6, T3	Partnership existing, research will expand
R5	Partnership with Amazon Cloud	Server Space	T1	Right away
R6	Partnership with Amazon Cloud	Increase server space/adaptability	T1	Extended, will grow from R5
R7	Partnership with Car2go, Uber, Lyft, biketown	Integration of private transportation companies	T6	Right away
R8	Access to publicly available GPS Augmented systems	GPS Augmentation System Development	T4	Extended--use what is available now to start
R9	Partnership with Trimet	Access to bus tracking, data, adding bluetooth/wifi to bus stops and buses	T4, T6, T2	Existing, develops over time
R10	Access to PSU and ODOT data and sensors	Traffic incident data, bike sensors around town	T6, T3	Existing
R11	Project Management staff	Communicate and coordinate all tasks and between all partners	All	Existing

11

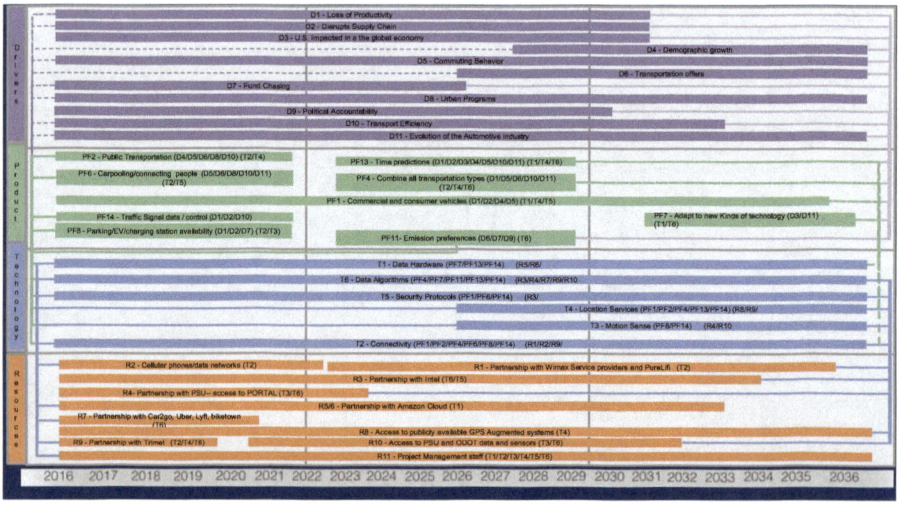

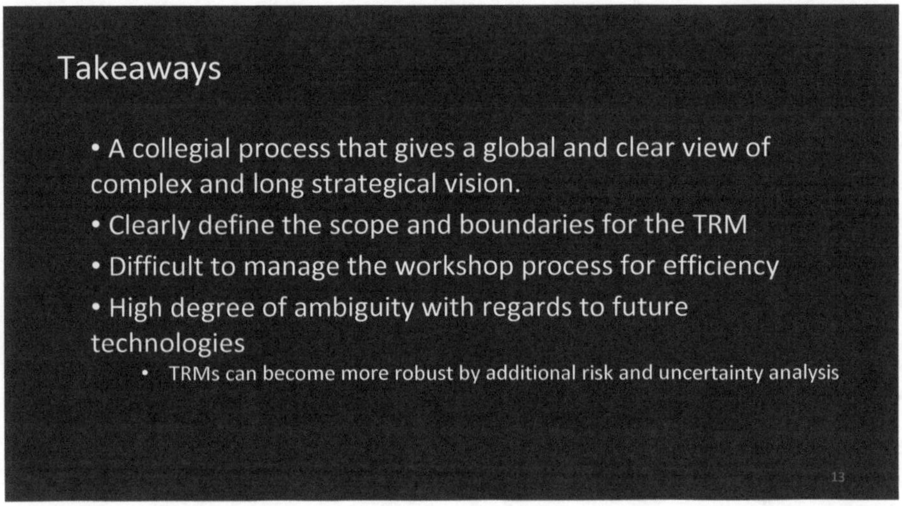
Takeaways
• A collegial process that gives a global and clear view of complex and long strategical vision.
• Clearly define the scope and boundaries for the TRM
• Difficult to manage the workshop process for efficiency
• High degree of ambiguity with regards to future technologies
• TRMs can become more robust by additional risk and uncertainty analysis
13

Questions?
14

References and Research Slides

Data Hardware:

1 - Townsend, Anthony M. *Smart cities: Big data, civic hackers, and the quest for a new utopia*. WW Norton & Company, 2013.

2 - Kitchin, Rob. "The real-time city? Big data and smart urbanism." *GeoJournal* 79.1 (2014): 1-14.

3 - Zanella, Andrea, et al. "Internet of things for smart cities." *IEEE Internet of Things Journal* 1.1 (2014): 22-32.

4 - Schaffers, Hans, et al. "Smart cities and the future internet: Towards cooperation frameworks for open innovation." *The Future Internet Assembly*. Springer Berlin Heidelberg, 2011.

5 - Mitton, Nathalie, et al. "Combining Cloud and sensors in a smart city environment." *EURASIP journal on Wireless Communications and Networking* 2012.1 (2012): 1.

6 - Suciu, George, et al. "Smart cities built on resilient cloud computing and secure internet of things." *2013 19th International Conference on Control Systems and Computer Science* . IEEE, 2013.

7 - Ali-Eldin, Ahmed, Johan Tordsson, and Erik Elmroth. "An adaptive hybrid elasticity controller for cloud infrastructures." 2012 IEEE Network Operations and Management Symposium. IEEE, 2012.

8 - Buyya, Rajkumar, Chee Shin Yeo, and Srikumar Venugopal. "Market-oriented cloud computing: Vision, hype, and reality for delivering it services as computing utilities." *High Performance Computing and Communications, 2008. HPCC'08. 10th IEEE International Conference on*. Ieee, 2008.

Data Software:

9 - U.S. DOT. (2014). *"The Smart/Connected City and its Implications for Connected Transportation"* [Online]. Available: http://www.its.dot.gov/itspac/Dec2014/Smart_Connected_City_FINAL_111314.pdf

10 -A. Adrian. (2013). "Big Data Challenges" [Online]. Available: http://www.dbjournal.ro/archive/13/13_4.pdf

11 - Oracle. (2013). "White Paper: Integrate for Insight" [Online]. Available: http://www.oracle.com/us/technologies/big-data/big-data-strategy-guide-1536569.pdf

12 -J. Treboux et al. (2015). "A Predictive Data-Driven Model for Traffic Jams" [Online]. Available: http://ieeexplore.ieee.org/stamp/stamp.jsp?tp=&arnumber=7122530

13 -"What is sensor? - Definition from WhatIs.com." [Online]. Available: http://whatis.techtarget.com/definition/sensor. [Accessed: 28-Jul-2016].

14 -"Automotive Sensor Applications & Solutions," *TE Connectivity*. [Online]. Available: http://www.te.com/usa-en/industries/sensor-solutions/applications/sensor-solutions-for-automotive-applications.html. [Accessed: 28-Jul-2016].

15 - Yang, Y., Lambert, F. and Divan, D., 2007, June. A survey on technologies for implementing sensor networks for power delivery systems. In *Power Engineering Society General Meeting, 2007. IEEE* (pp. 1-8). IEEE

16 -"Adafruit Ultimate GPS Breakout - 66 channel w/10 Hz updates [Version 3] ID: 746 - $39.95 : Adafruit Industries, Unique & fun DIY electronics and kits." [Online]. Available: https://www.adafruit.com/product/746. [Accessed: 27-Jul-2016].

[1]2016. [Online]. Available: http://smartcitiescouncil.com/sites/default/files/public_resources/Smart%20city%20security.pdf. [Accessed: 28- Jul- 2016].

[2]"IEEE-SA -IEEE Get 802 Program - 802.15: Wireless PANs", Standards.ieee.org, 2016. [Online]. Available: http://standards.ieee.org/about/get/802/802.15.html. [Accessed: 28- Jul- 2016].

General/Product Features:

[P1] V. Kostakos, T. Ojala and T. Juntunen, "Traffic in the Smart City: Exploring City-Wide Sensing for Traffic Control Center Augmentation," in *IEEE Internet Computing*, vol. 17, no. 6, pp. 22-29, Nov.-Dec. 2013.

[P2] 2016. [Online]. Available: http://portal.its.pdx.edu/ [Accessed 28-Jul-2016]

[P3] J. Treboux, A. J. Jara, L. Dufour and D. Genoud, "A predictive data-driven model for traffic-jams forecasting in smart santader city-scale testbed," *Wireless Communications and Networking Conference Workshops (WCNCW), 2015 IEEE* , New Orleans, LA, 2015, pp. 64-68.

15

Product Features

Category	Product Feature	Number
Integration of vehicles and services	Commercial and consumer vehicles	PF1
	Public transportation	PF2
	Emergency vehicles	PF3
	Combine all transp. types	PF4
	Accident/road closure alerts	PF5
	Carpooling/connecting people	PF6
	Adapt to new kinds of technology	PF7
	Parking/EV charging stations	PF8
Usage	Handicap accessible	PF9
	Ease of use	PF10
	Emission preferences	PF11
	Cost preferences	PF12
Analytics	Time/Traffic predictions	PF13
Signal Control	Traffic signal data/control	PF14

16

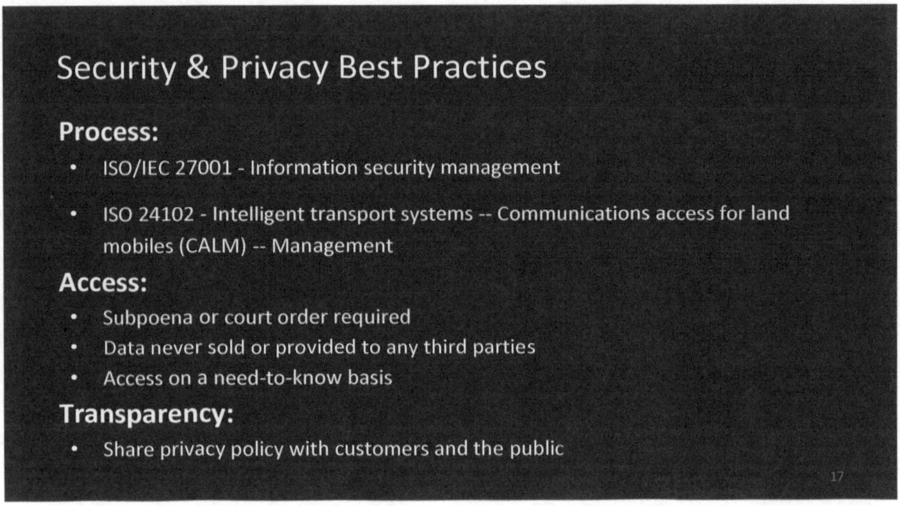
Security & Privacy Best Practices
Process:
• ISO/IEC 27001 - Information security management
• ISO 24102 - Intelligent transport systems -- Communications access for land mobiles (CALM) -- Management
Access:
• Subpoena or court order required
• Data never sold or provided to any third parties
• Access on a need-to-know basis
Transparency:
• Share privacy policy with customers and the public
17

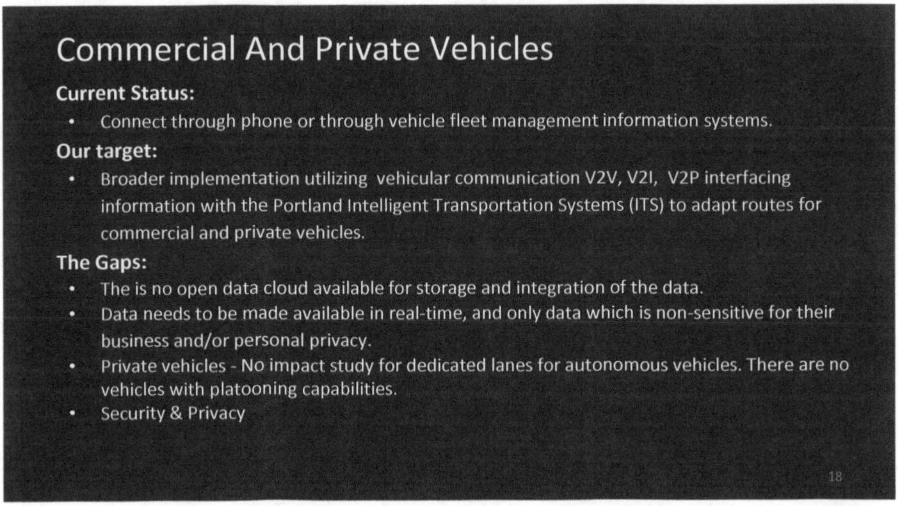
Commercial And Private Vehicles
Current Status:
• Connect through phone or through vehicle fleet management information systems.
Our target:
• Broader implementation utilizing vehicular communication V2V, V2I, V2P interfacing information with the Portland Intelligent Transportation Systems (ITS) to adapt routes for commercial and private vehicles.
The Gaps:
• The is no open data cloud available for storage and integration of the data.
• Data needs to be made available in real-time, and only data which is non-sensitive for their business and/or personal privacy.
• Private vehicles - No impact study for dedicated lanes for autonomous vehicles. There are no vehicles with platooning capabilities.
• Security & Privacy
18

Emergency Vehicles

Simulator

Current Status:

- Police Bureau, Fire, American Medical Response (AMR), have their information standardized and their emergency vehicles tracked by separate entities.
- Lack of real time integrated data between information transportation systems and those entities.

Our target:

- Collect and integrate real time data from a variety of public and private sources making available through an Open Data Cloud and accessible through an app, website or kiosks providing real-time information that will increase the efficiency and quality of transportation services provided to the population.

The Gaps:

- The is no open data cloud available for storage and integration of the data.
- Current systems were not originally developed to be publicly accessible real-time databases.
- The information is not stored or organized around transportation
- Infrastructure
- Security & Privacy

19

Gaps Technological & Social

The Gaps:

- The privacy and security of the people using the smart city transportation system [portland document]
- Establish a dialog between the city and its partners and the people using the smart city transportation system
- The lack of consistent utilization of security standards and process available even today by all parties (public, private, device makers, software makers etc.)
- Fast mobile internet affordable for the majority

20

Gaps & Solutions

Technology Categories	Code	Current Standards	> 2 years	> 5- years	> 15- years
Security Protocols	T5	T5.1.1 - Cryptography (IEEE P1363) T5.1.2 - Access Control (IEEE 802.1: 802.1X) T5.1.3 - WPA2 (IEEE 802.11i-2004) T5.1.4 - IPV6 (IETF) T5.1.5 - Payment Card Industry Data Security Standard (PCI DSS) T5.1.6 - LR-WPANs (IEEE 802.15.4) T5.2.7 - Intrusion Detection Systems	T5.2.1 - Consistent use of encryption between all network nodes and in all network communication. T5.2.2 - Rollout start of 5G mobile networks, designed for much larger number of devices	T5.3.1 - Consistent use of Intrusion Detection Systems and Deep Packet Inspection at large "City Scale" networks. T.5.3.2 - Wide accessibility to new 5G networks, and cheaper access to 4G networks.	T5.4.1 -- Adaptive and quasi-autonomous IDS and DPI T5.4.2 - V2V,V2I, V2P (IEEE 802.11p) T5.4.3 - CEREBRE protocol

21

Technologies & Resources

Technology Categories	Code	Current Standards	> 2 years	> 5- years	> 15- years
Security Protocols	T5	T5.1.1 - Cryptography (IEEE P1363) T5.1.2 - Access Control (IEEE 802.1: 802.1X) T5.1.3 - WPA2 (IEEE 802.11i-2004) T5.1.4 - IPV6 (IETF) T5.1.5 - Payment Card Industry Data Security Standard (PCI DSS) T5.1.6 - LR-WPANs (IEEE 802.15.4) T5.2.7 - Intrusion Detection Systems	T5.2.1 - Consistent use of encryption between all network node and in all network communication. T5.2.2 - Rollout start of 5G mobile networks, designed for much larger number of devices	T5.3.1 - Consistent use of Intrusion Detection Systems and Deep Packet Inspection at large "City Scale" networks. T.5.3.2 - Wide accessibility to new 5G networks, and cheaper access to 4G networks.	T5.4.1 -- Adaptive and quasi-autonomous IDS and DPI T5.4.2 - V2V,V2I, V2P (IEEE 802.11p)
Resources for Security	R5	R5.1 Software engineers Hardware engineers Security engineers Project manager	R5.2 - Software engineers Hardware engineers Security engineers Project manager	R5.3 - Software engineers Hardware engineers Security engineers Project manager Network engineers Data Scientists	R5.4 - Software engineers Hardware engineers Security engineers Project manager Network engineers Data Scientists Partnership with Universities

Identification of Experts using Social Network Analysis (SNA)

Edwin Garces
Momtaj Khanam

Portland State University

Appendix 14

OUTLINE

ETM

1) Introduction
2) Theory and definitions
   a) Social Network Analysis (SNA):
      a) Definition
      b) Metrics
   b) Methodology
      a) Data Sources
      b) Software
3) Application:
   - Agriculture – Energy Efficiency
   - TRM
   - Advanced Commercial Refrigeration
   - Lighting
4) Conclusions

Portland State UNIVERSITY

2

T. U. Daim (ed.), *Roadmapping Future*, Applied Innovation and Technology Management, https://doi.org/10.1007/978-3-030-50502-8

Introduction

- Many methods, tools, and techniques require the opinion of experts or debates of a certain topics.
- Finding these experts is not simple and more if the knowledge and experience on a determined field will determine the quality of the results.
- Research collaboration among experts has gained importance during the last decades, especially in the area of technology.
- Social Network Analysis is a common method for finding who is the knowledgeable person in a community.
- At BPA, **SNA**, **Bibliometric**, and **patent Analysis** are used to determine the **lead authors**, co-authors, institutions, contact information in a determined period.
- Main objectives:
  - To identify experts in a specific area
  - To present a methodology to identify experts by using **Social Network Analysis** and **Bibliographic Analysis**.

Portland State UNIVERSITY

Theory and Definitions

Portland State UNIVERSITY

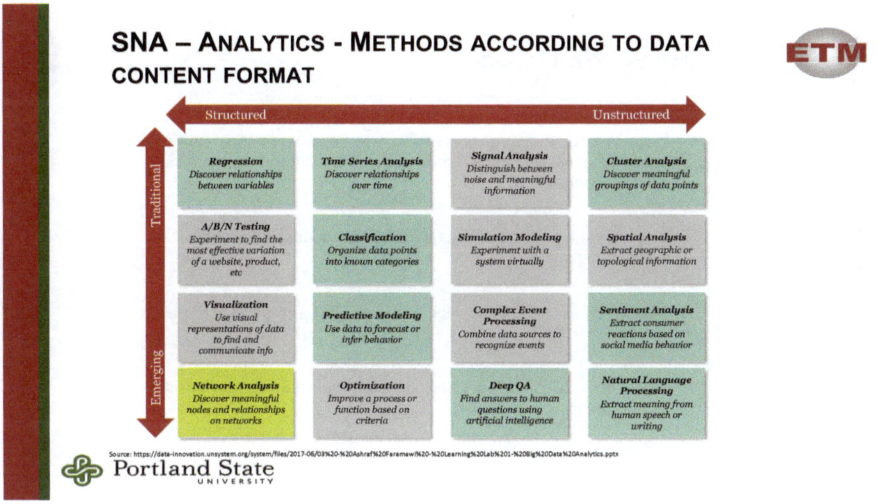

https://data-innovation.unsystem.org/system/files/2017-06/03%20-%20Ashraf%20 Faramawi%20-%20Learning%20Lab%201-%20Big%20Data%20Analytics.pptx

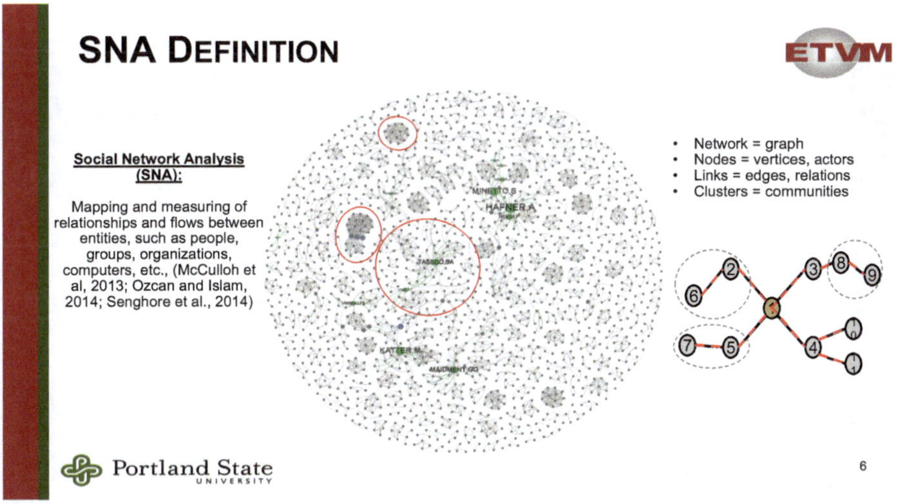

SNA Metrics

- **Interpreting Centrality**
  - **Degree:** A node with the most direct connections is not necessarily better. What really matters is the position in the network and who the node is connected to.
  - **Closeness:** A node close to all other nodes can monitor network information flows.
  - **Betweenness:** A node with high betweenness has significant influence on network flows.
  - **Eigenvector:** Chain effect when one node influences another, which in turn influences many others.

Portland State UNIVERSITY

9

Closeness: Metric is only meaningful for connected networks.

Power Law distribution

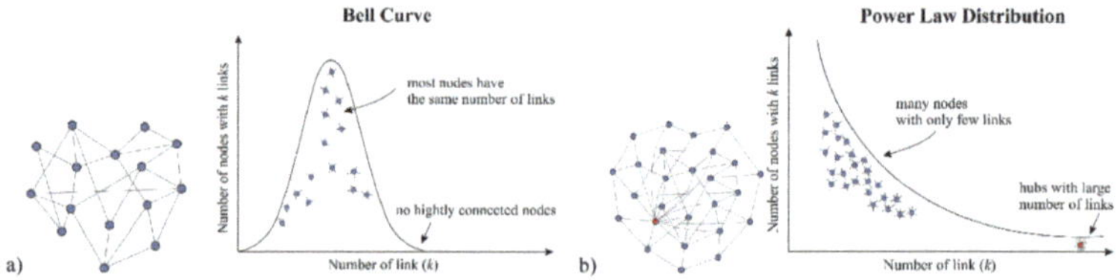

Source: Anna Maria Kowalczyk, The Analysis of Networks Space Structures as Important Elements of Sustainable Space Development. 2017

Node degree distribution

- Node degree - number of nearest neighbours, $k_i = 1, 2, \ldots k_{\max}$
- Degree distribution $P(k) \equiv P(k_i = k)$

$$P(k) = \frac{n_k}{n} = \frac{n_k}{\sum_k n_k}$$

- Power law

$$P(k) = Ck^{-\gamma} = \frac{C}{k^{\gamma}}$$

Portland State UNIVERSITY

FORECASTING – GROW OF NETWORKS

ETM

The example of temporal social network consisting of five time frames

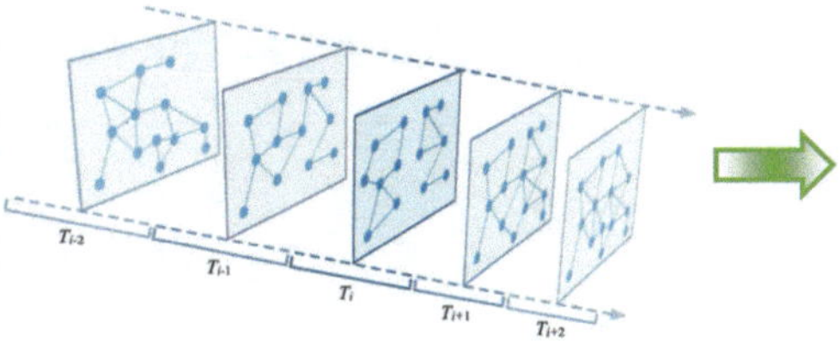

Seven possible types of events in the group evolution

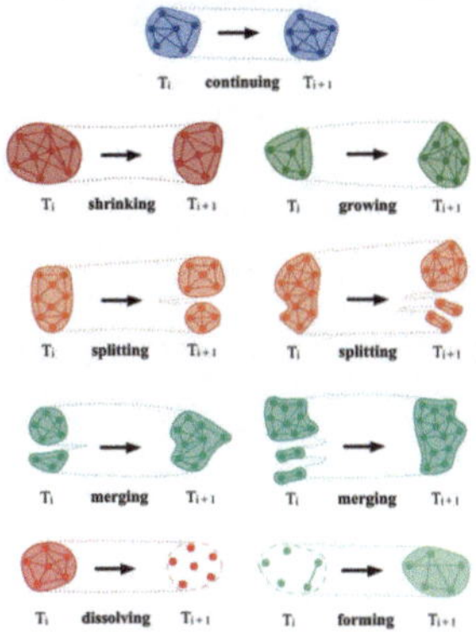

Portland State UNIVERSITY

Source: Piotr Brodka, Przemysaw Kazienk,Stanisaw Saganowski, "GED: the method for group evolution discovery in social networks", 2012

ETM

1978–1993 (Start of merger)

1978–1994 (Start of major merger)

1978–1998 (Major merger)

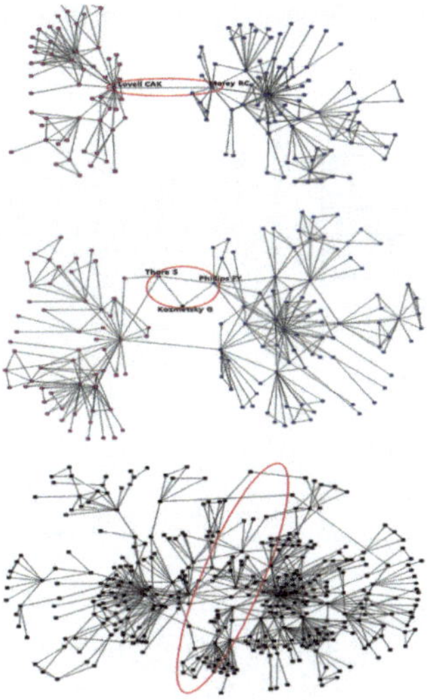

Portland Sta UNIVER

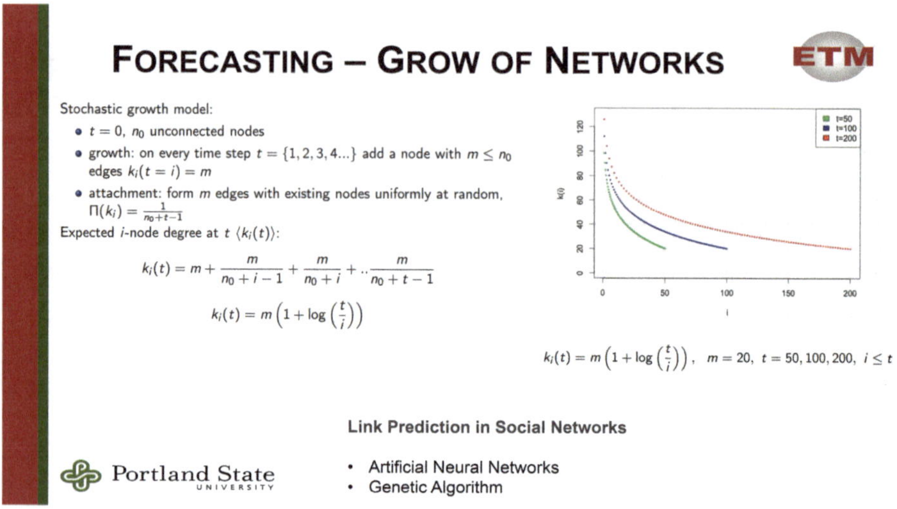

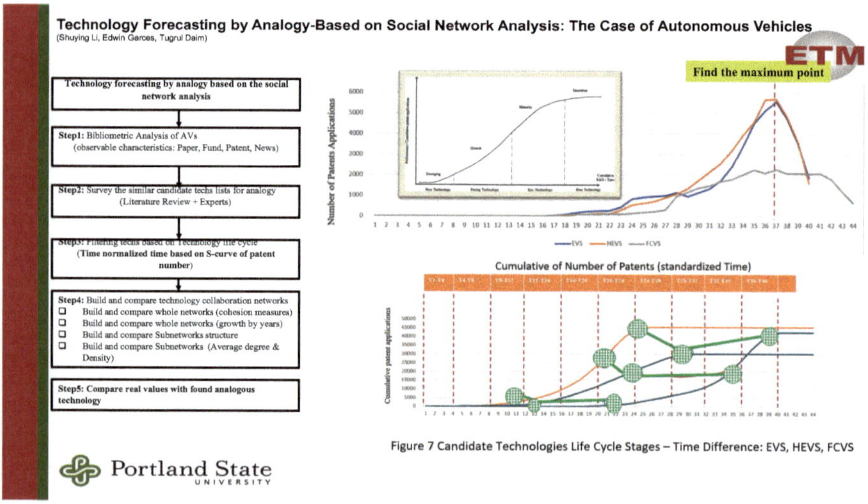

Technology Forecasting by Analogy-Based on Social Network Analysis: The Case of Autonomous Vehicles [Shuying Li, Edwin Garces, Tugrul Daim]

Technology Forecasting by Analogy-Based on Social Network Analysis: The Case of Autonomous Vehicles

(Shuying Li, Edwin Garces, Tugrul Daim)

ETM

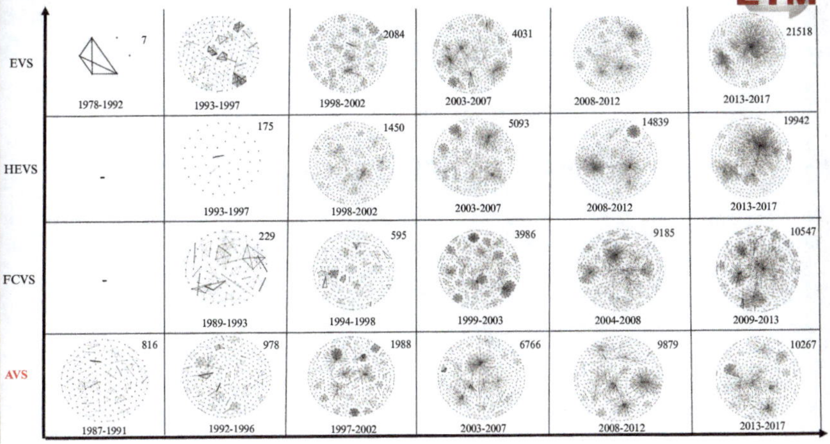

Portland State UNIVERSITY

Technology Forecasting by Analogy-Based on Social Network Analysis: The Case of Autonomous Vehicles

(Shuying Li, Edwin Garces, Tugrul Daim)

ETM

$$\text{Variance} = (\sigma x)^2 = \frac{1}{n}\sum_{i=1}^{n}(x_i - \bar{x})^2$$

	Connection variance	Density variance
EVS	0.57	0.04
HEVS	0.60	0.05
FCVS	0.51	0.02

$$y = \frac{L}{1 + \alpha e^{-bt}}$$

$$y = \frac{L}{1 + \alpha 10^{A-Bt}}$$

$$Y = LOG\left[\frac{y}{L - y}\right] = -A + Bt$$

L (AVs)	40000
A	314.9616
B	0.156563

Figure 12: Forecasting the cumulative Number of patents for AVs

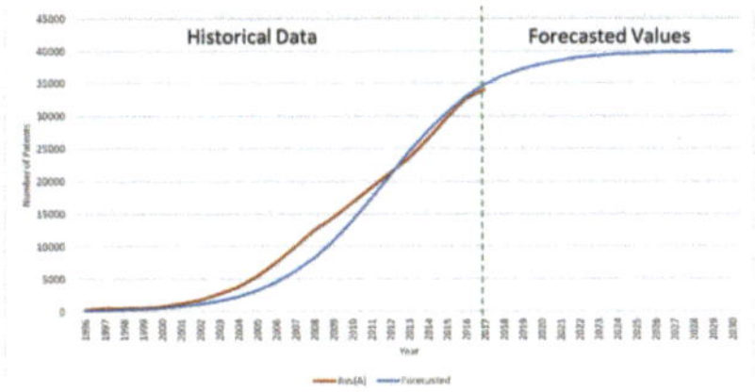

Portland State UNIVERSITY

ETM

SNA – Applications

Portland State University

Network Science – SNA Applications

ETM

- Sociology (SNA)
- Mathematics (Graphs)
- Computer Science (Graphs)
- Statistical Physics (Complex networks)
- Economics (Networks)
- Bioinformatics (Networks)

Portland State University

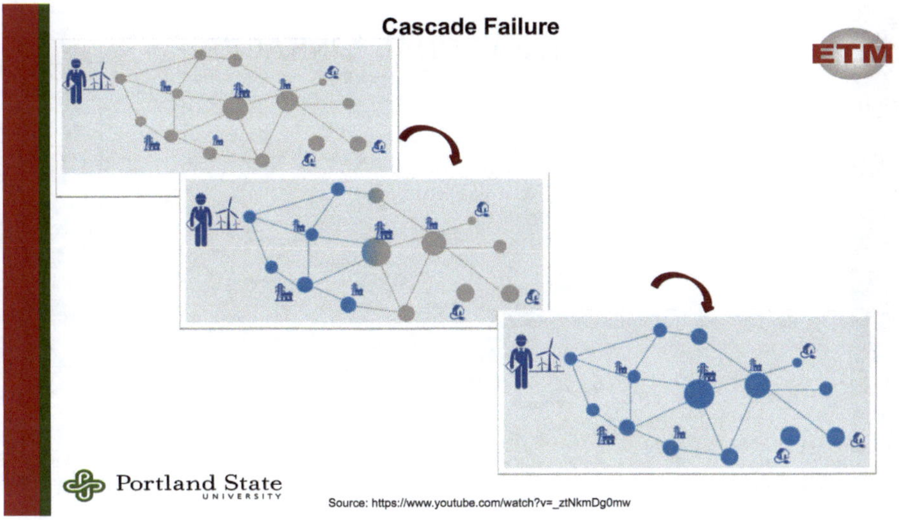
Cascade Failure
ETM
Portland State
UNIVERSITY
Source: https://www.youtube.com/watch?v=_ztNkmDg0mw

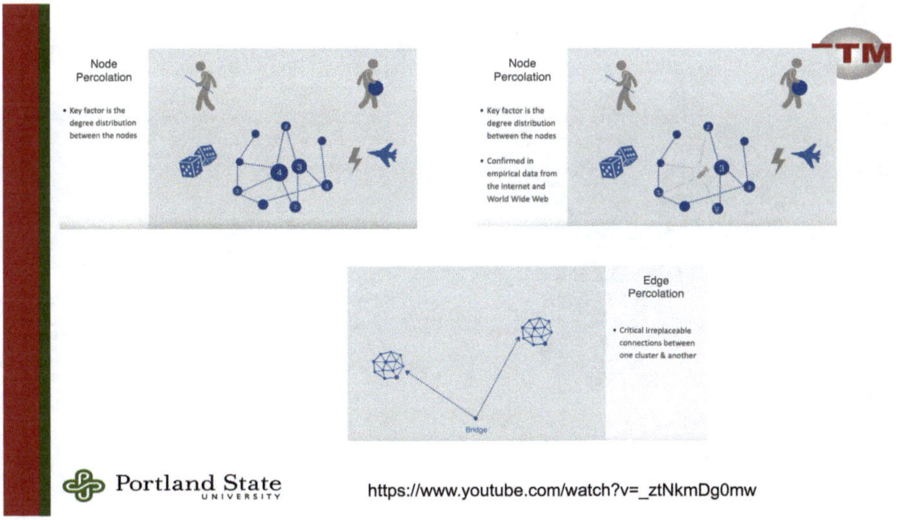
Node
Percolation
• Key factor is the degree distribution between the nodes
Node
Percolation
• Key factor is the degree distribution between the nodes
• Confirmed in empirical data from the internet and World Wide Web
Edge
Percolation
• Critical irreplaceable connections between one cluster & another
Bridge
Portland State
UNIVERSITY
https://www.youtube.com/watch?v=_ztNkmDg0mw

Organizational Linkages

ETM

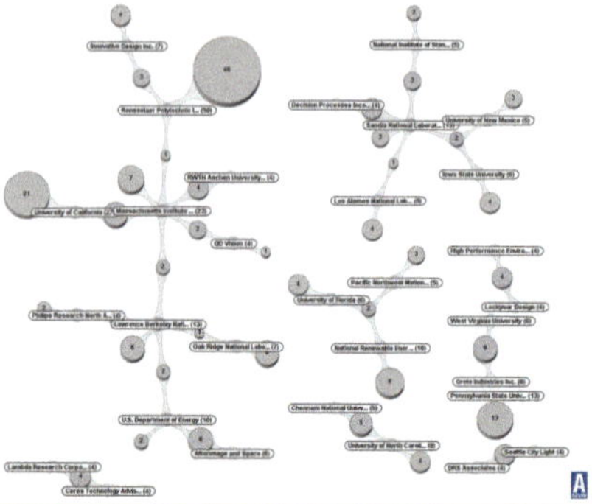

Organizational mapping identified organizations which were not thought about before

Portland State University

Vantage Point SW was used for this analysis

Technology Map

ETM

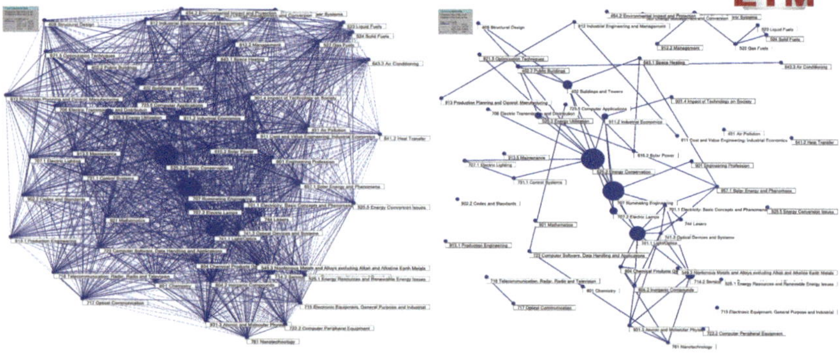

Mapping Technologies Helps Identify Related Competencies

Portland State University

Vantage Point SW was used for this analysis

ETM

IDENTIFICATION OF EXPERTS

METHODOLOGY

Portland State University

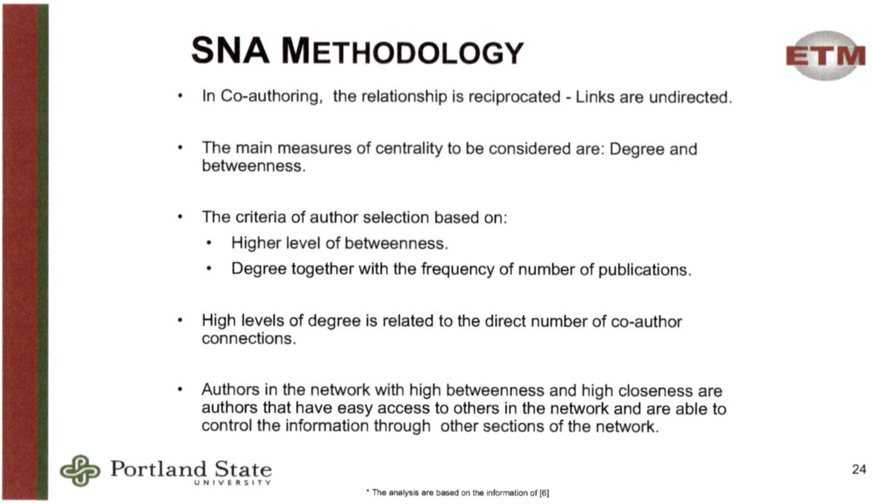

Since the objective is to identify the main researchers about Commercial refrigeration, based on co-authorship, the main measures of centrality to be considered are: Degree and betweenness.

Closeness, and eigenvector can be considered for alternatives objectives of diffusion of knowledge and the strongest connection of the authors.

High levels of degree indicate the direct number of co-author connections.

Higher levels of closeness measures indicates that authors are the closest to all other people in the network. Hence, they can reach all the nodes in the network more efficiently than others. This is useful if information needs to be dispersed quickly across the network.

Higher levels of betweenness as well as degree are indicators of importance; however they can be considered a single point of failure too.

SNA Methodology

ETM

- **For this application:**
  - Nodes are authors.
  - Connections between nodes represents authors of the same publication.
- **For these types of networks:**
  - Closeness centrality is not meaningful since the network mainly disconnected.
  - Degree of centrality is not necessarily meaningful since it only represents the number of co-authors.
  - Betweenness centrality is the key metric because we are looking for subject matter experts.

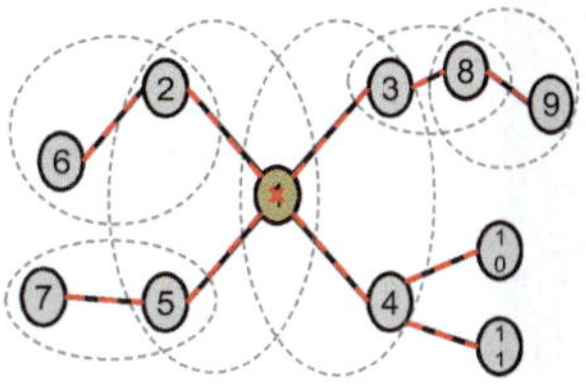

Portland State UNIVERSITY

25

There are two types of data in SNAs: structural and composition. Co-author connections is an example of structural data.

SNA Methodology

ETM

- Case study methodology combines social network and bibliometric analyses; similar processes have been used by other researchers (see [7] and [8]).
- This study's contribution is the systematic use of social network methods, bibliographic analyses, and data mining techniques for a real-world, practical application.
- SNA and data mining methods applied and adjusted iteratively.

Portland State UNIVERSITY

26

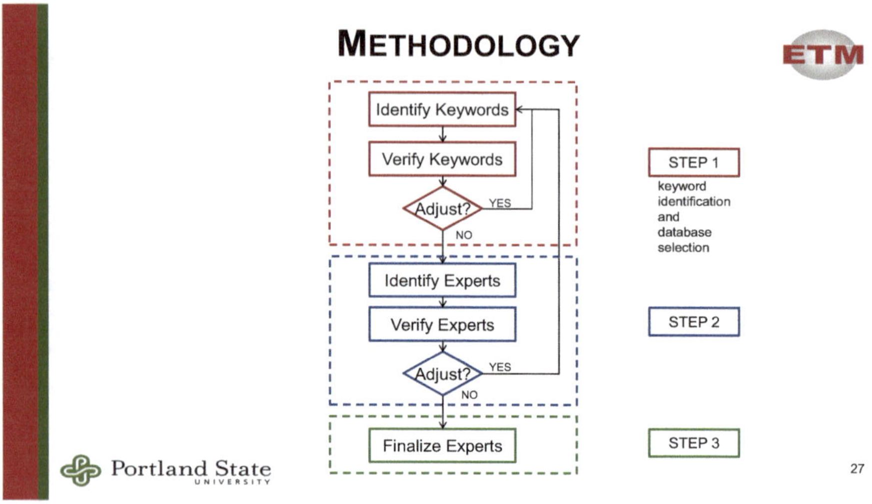

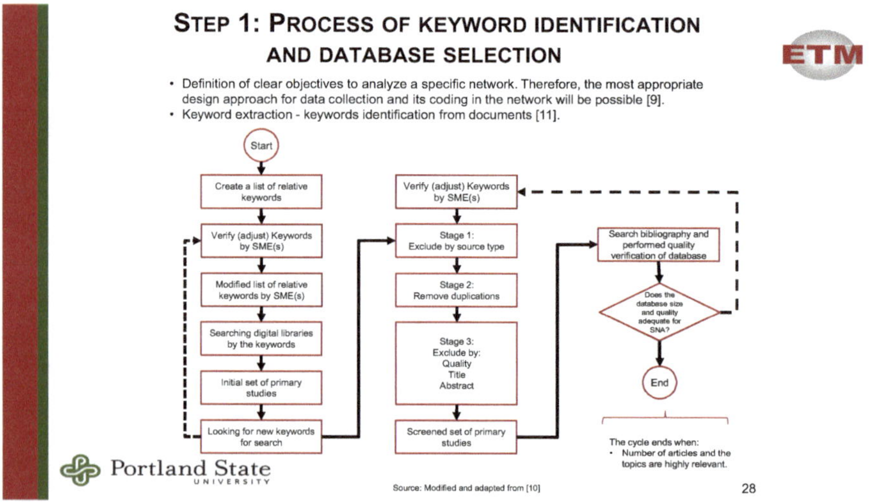

Create a list of relative keywords: Identification of the topic and technology. Literature review. Identification of categories of topics / levels / components. Work directly withSME(s).

Searching digital libraries by the keywords: Keyword extraction using the initial keywords in bibliographic databases. Words identification from documents [11].

Step 2: SNA Process and Methodology

ETM

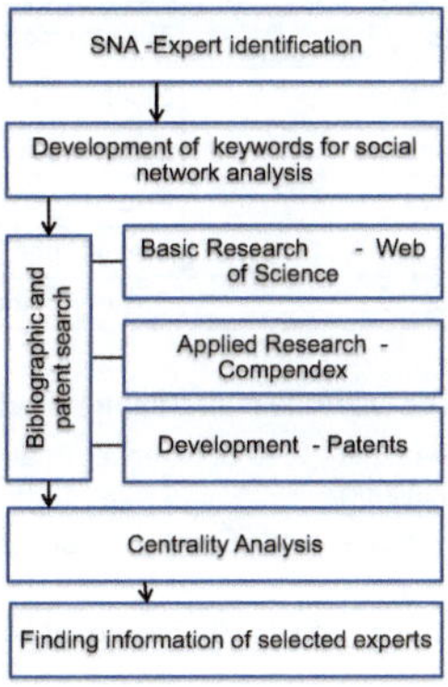

Portland State UNIVERSITY

Step 3: Finalize Experts

ETM

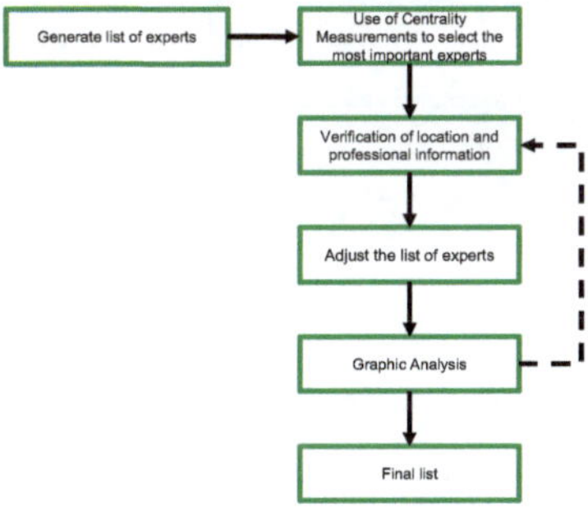

Portland State UNIVERSITY

30

SNA Data Sources

R&D Stage	Typical Source
Basic Research	Science Citation Index
Applied Research	Engineering Index
Development	Patents
Application	Newspaper Abstracts Daily
Social Impacts	Business and Popular Press

- Experts can be identified from different databases and approaches:
  - Basic Research - Web of Science (Citations, publications, and co-author network),
  - Applied Research - Compendex (Publications and co-author network),
  - Development - USPTO (Patent count and co-author network).

Portland State UNIVERSITY

[12] : J. P. Martino, *"A Review of Selected Recent Advances in Technological Forecasting,"* Technol. Forecast. Soc. Change, vol. 70, no. 8, pp. 719–733, Oct. 2003.

31

PSU-ETM - SNA Software

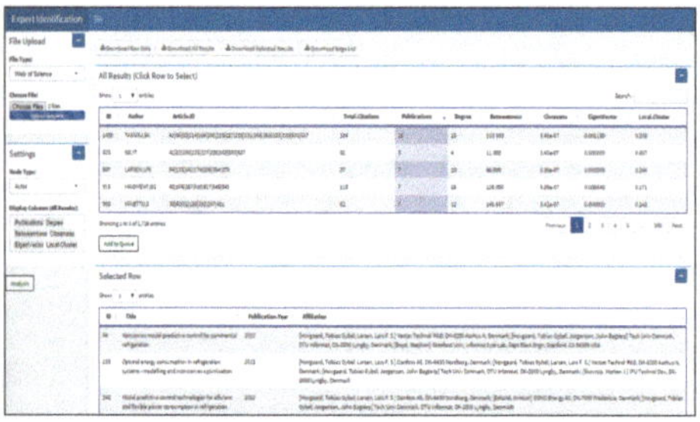

R and Shiny tools used for data extraction and analysis

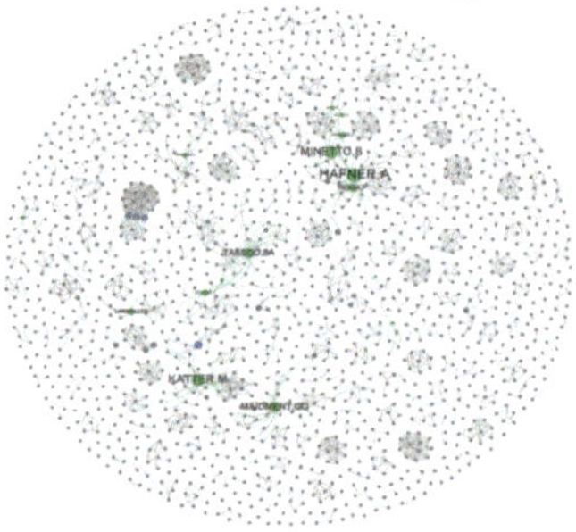

Output from R and Shiny imported into Gephi for visualization

Portland State UNIVERSITY

32

SNA at BPA: Experience About Software Applications

UCINET - NetDraw | **CiteSpace** | **R–Shiny tools -Gephi**

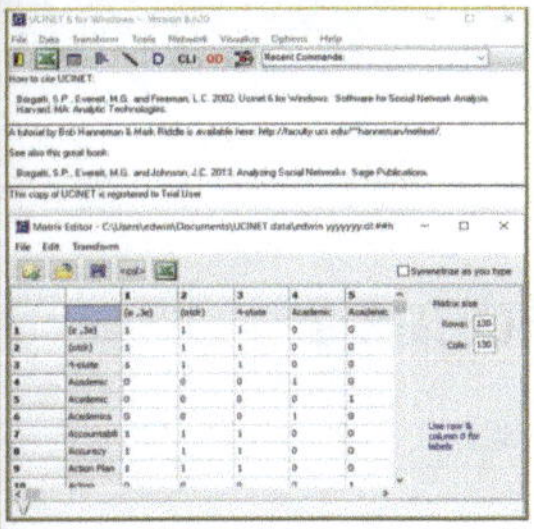

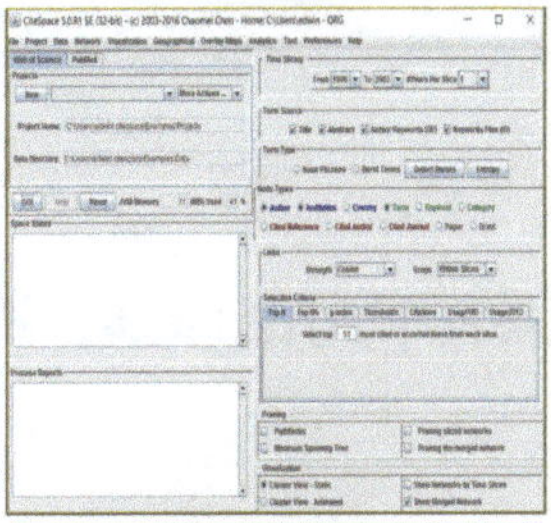

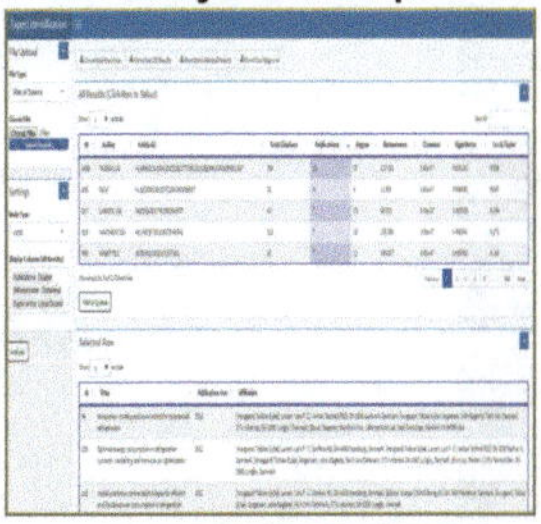

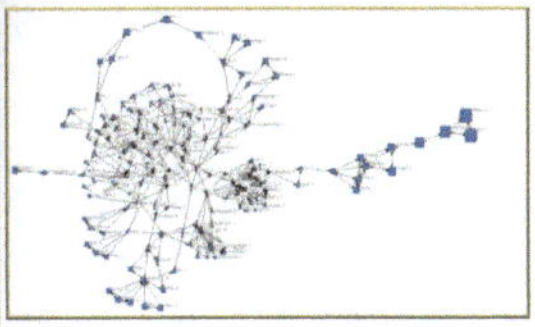

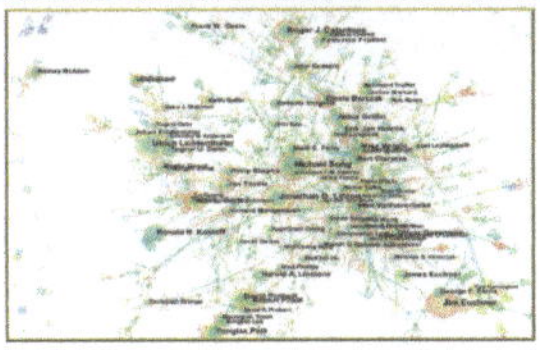

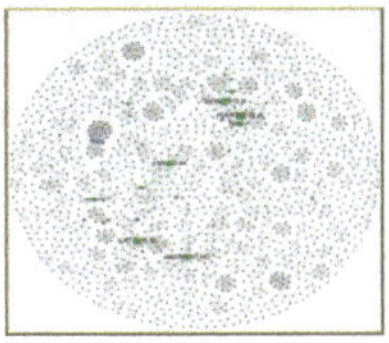

SNA: List of Software and Potential Applications

ETM

- UCINET, Pajek, CiteSpace, SCI2, NodeXL
- Software developed at ETM: R – Shiny - Gephi
- Using multiple software for a specific analysis:
  - SCI2 + NodeXL
  - SCI2 + Gephi + NodeXL
  - SCI2 + Gephi + UCINET
  - ETM + Gephi
  - ETM + Gephi + NodeXL
  - ETM + Gephi + UCINET

Portland State UNIVERSITY

ETM

SNA – Practical Process

Portland State University

ETM

Identification of Keywords

Portland State University

IDENTIFICATION OF KEYWORDS

Example from Power Service RM

ETM

Initial Keywords
Generation Asset Management
power service
power generation
Joint Transmission and Hydropower Modeling
Resource Integration
supply balancing resources
variable energy resources
Outage planning
computational methods
simulation
optimization algorithm
combined power and transmission system
FCRPS
FCRTS
transmission and generation constraints
EIM
Reliability
transmission congestion

AREA: Joint Transmission and Hydropower Modeling	
	(Generation Asset Management OR power service OR power generation)
AND	Transmission AND Hydro power Modeling
AND	(Resource Integration OR supply balancing resources OR variable energy resources OR Outage planning OR computational methods OR simulation OR optimization algorithm OR combined power AND transmission system OR FCRPS OR FCRTS OR transmission AND generation constraints OR EIM OR Reliability OR transmission congestion)

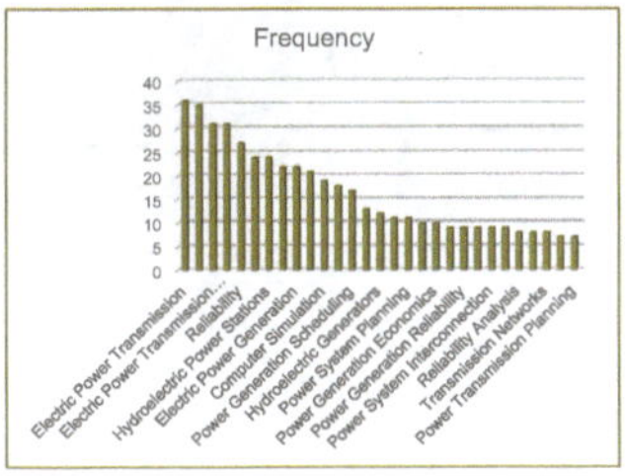

226 articles found in Compendex

New Keywords From bibliographic Search	Frequency
Electric Power Transmission	36
Scheduling	35
Electric Power Transmission Networks	31
Optimization	31
Reliability	27
Electric Power Systems	24
Hydroelectric Power Stations	24
Hydroelectric Power Plants	22
Electric Power Generation	22
Hydrothermal Power Systems	21
Computer Simulation	19
Hydroelectric Power	18
Power Generation Scheduling	17
Load Flow	13
Hydroelectric Generators	12
Power System Reliability	11
Power System Planning	11
Power Generation Dispatch	10
Power Generation Economics	10
Power Generation Planning	9
Power Generation Reliability	9
Integer Programming	9
Power System Interconnection	9
Power Generation	9
Reliability Analysis	8
Power Transmission Economics	8
Transmission Networks	8
Power Transmission Reliability	7
Power Transmission Planning	7

Portland State UNIVERSITY

Structuring the Boolean Search Query

ETM

Field
- Energy
- Irrigation
- Agriculture

Specific Application
- water or irrigat* or AGR* OR CROP OR CULT* or soil

Specific Methods or Topics
- method or manag* or precision or schedu* or precision flat rate or variable speed or control or data or method* or Pumping or PAIL or vsi

Portland State UNIVERSITY

Example from Agr. Irrig.

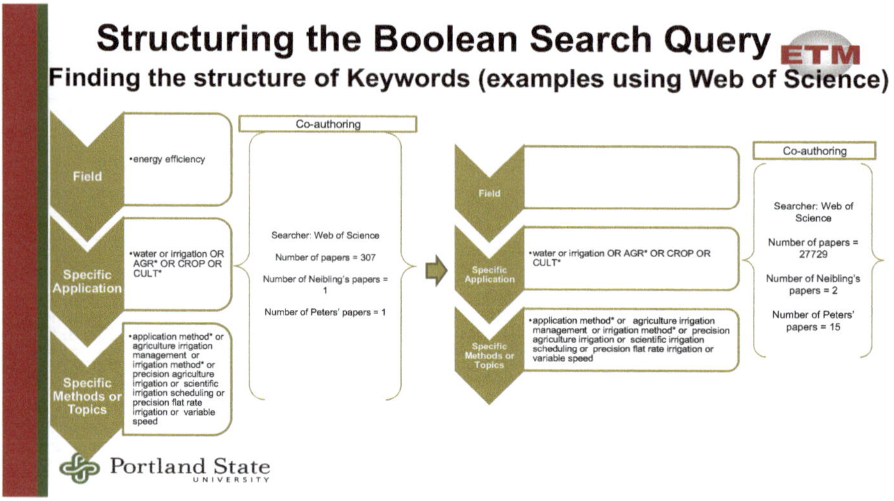
Structuring the Boolean Search Query
ETM
Finding the structure of Keywords (examples using Web of Science)
Field
•energy efficiency
Specific Application
•water or irrigation OR AGR* OR CROP OR CULT*
Specific Methods or Topics
•application method* or agriculture irrigation management or irrigation method* or precision agriculture irrigation or scientific irrigation scheduling or precision flat rate irrigation or variable speed
Co-authoring
Searcher: Web of Science
Number of papers = 307
Number of Neibling's papers = 1
Number of Peters' papers = 1
Field
Specific Application
•water or irrigation OR AGR* OR CROP OR CULT*
Specific Methods or Topics
•application method* or agriculture irrigation management or irrigation method* or precision agriculture irrigation or scientific irrigation scheduling or precision flat rate irrigation or variable speed
Co-authoring
Searcher: Web of Science
Number of papers = 27729
Number of Neibling's papers = 2
Number of Peters' papers = 15
Portland State UNIVERSITY

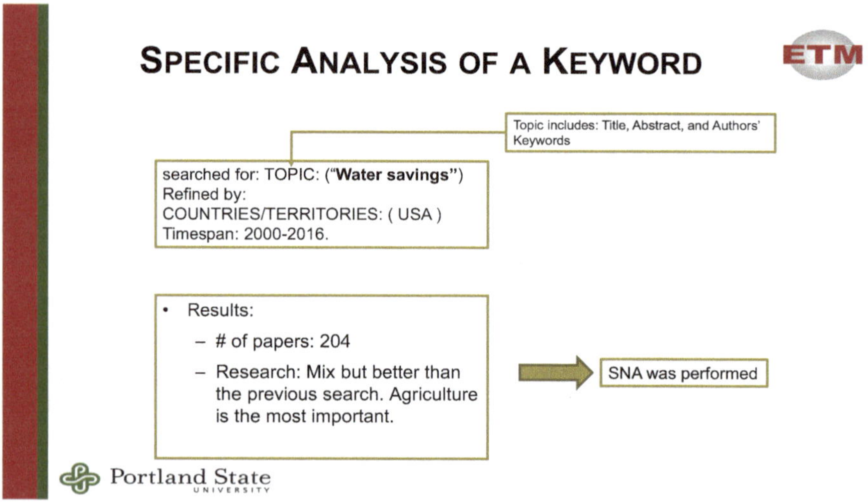
SPECIFIC ANALYSIS OF A KEYWORD
ETM
Topic includes: Title, Abstract, and Authors' Keywords
searched for: TOPIC: ("Water savings")
Refined by:
COUNTRIES/TERRITORIES: (USA)
Timespan: 2000-2016.
• Results:
– # of papers: 204
– Research: Mix but better than the previous search. Agriculture is the most important.
SNA was performed
Portland State UNIVERSITY

Specific Analysis of a Keyword

RESEARCH AREAS

ETM

Field: Research Areas	Number of papers	%
AGRICULTURE	98	48.04%
WATER RESOURCES	92	45.10%
ENGINEERING	66	32.35%
ENVIRONMENTAL SCIENCES ECOLOGY	46	22.55%
GEOLOGY	15	7.35%
ENERGY FUELS	9	4.41%
MARINE FRESHWATER BIOLOGY	9	4.41%
SCIENCE TECHNOLOGY OTHER TOPICS	8	3.92%
BUSINESS ECONOMICS	7	3.43%
METEOROLOGY ATMOSPHERIC SCIENCES	4	1.96%
BIODIVERSITY CONSERVATION	3	1.47%
THERMODYNAMICS	3	1.47%
ARCHITECTURE	2	0.98%
CONSTRUCTION BUILDING TECHNOLOGY	2	0.98%
ENTOMOLOGY	2	0.98%
FOOD SCIENCE TECHNOLOGY	2	0.98%
MECHANICS	2	0.98%
PHYSICS	2	0.98%
PLANT SCIENCES	2	0.98%
URBAN STUDIES	2	0.98%
ZOOLOGY	2	0.98%

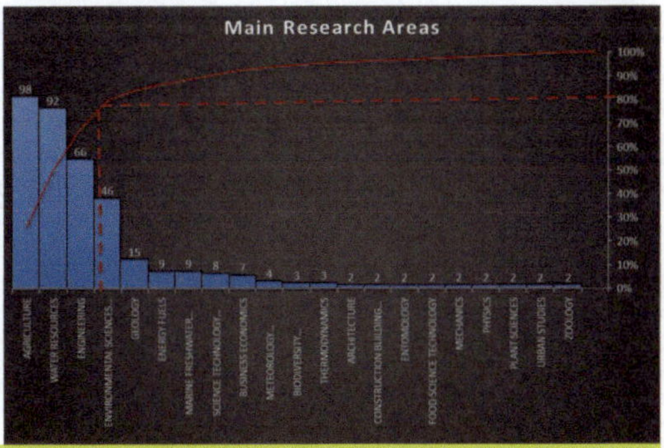

Following the Pareto's criterion, the main research areas associated to the keyword " water savings" are (based on the 80% cumulative percentage): Agriculture, Water Resources, Engineering, Environmental Sciences Ecology.

The term is used in agriculture is significant, 48.04 % of the papers are concentrated on agriculture.

Portland State UNIVERSITY

WEB OF SCIENCE: NETWORK CLUSTERS BASED ON CLAUSET-NEWMAN-MOORE

ETM

Based on the cluster analysis, Neibling is connected only with one author

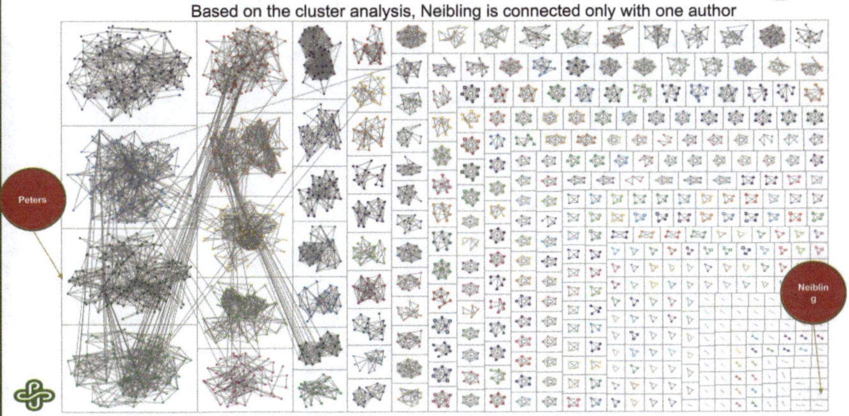

ETM

Running SNA - Instructions

Portland State University

Instructions

Downloading Data:

Web of Science:

a. Search for results under Web of Science,
b. Click on "Save to Other File Formats" -> Select Record Content: "Full Record and Cited References", File Format: **"Tab-delimited (Win, UTF-8)",**
c. Click Send and save to local drive,
d. Repeat until all article entries downloaded.

Compendex:

a. Search for results under Compendex,
b. Click Select -> Maximum (up to 500) and then click "Save to my PC",
c. Select Location: "My PC", Format: "CSV", Output: "Detailed Record" and download,
d. Repeat until all article entries downloaded.

Patents (Sumobrain)

a. Go to www.sumobrain.com and Signup. Once account is created Login,
b. Search for US patents on relevant topic, (choose as date the last 20 years)
c. Save the results to a new portfolio (this only shows when you have logged in),
d. Once the portfolio is created, select the results and then click on the Export button on the right. You can only export 250 results at a time so you have to download them separately and combine the results into one file later,
e. Open the downloaded xls files, delete the first empty row, and then save as a csv file,
f. Go to http://rstudio.cecs.pdx.edu:3838/users/kevin9/Patent/ and upload the csv file,
g. Wait for the download links to show on the right,
h. Download the authors.csv file and select ten authors with the highest number of patents. You will need to search for their contact details through a Google search.

Installing the software:

Download and Install R -> select base option (http://cran.rstudio.com/) ,

1. Download and Install RStudio (http://www.rstudio.com/products/rstudio/download/),
2. In RStudio go to Menu, Tools, Install Packages. Look for "Shiny" package and install it
3. For visualization purposes, Gephi software needs to be installed. Download and install the last version of Gephi (http://gephi.github.io/users/download/).

Analyzing Data:

1. In the Console type "shiny::runGitHub("Identification-of-Experts-by-SNA", "edwingus")". The first time RStudio will install the additional needed packages and then opens a new windows. This new window (called "expert identification") is the SNA interface that will be used for uploading the data, running the SNA, and visualizing the results.
2. In the expert identification window, choose the type of source data to be uploaded. There are three options: Web of Science, Compendex, and Patent.
3. In the expert identification windows, select "choose files" to load your data file. The software allows uploading one or multiple files.
4. The final step is to run "Analyze" option; which is located on the lower left corner of the window

Results

1. The results will be showed in the window "expert identification". The results shows the all used data, main centrality indicators including frequency, betweenness, degree, closeness, and so on.
2. For printing and saving the results, use the alternative browser interface. This can be found on the top of the windows.

For results graph visualization, use the "edges file" and open it in Gephi.

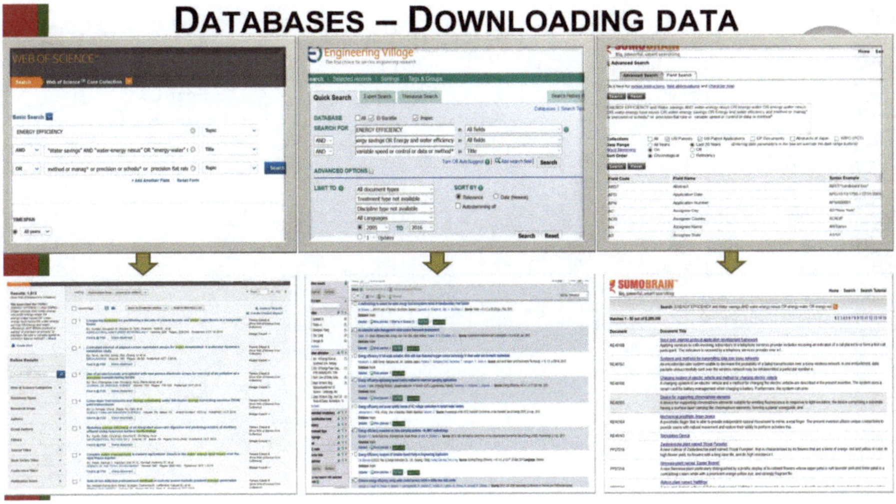
DATABASES – DOWNLOADING DATA
WEB OF SCIENCE
Engineering Village
SUMOBRAIN

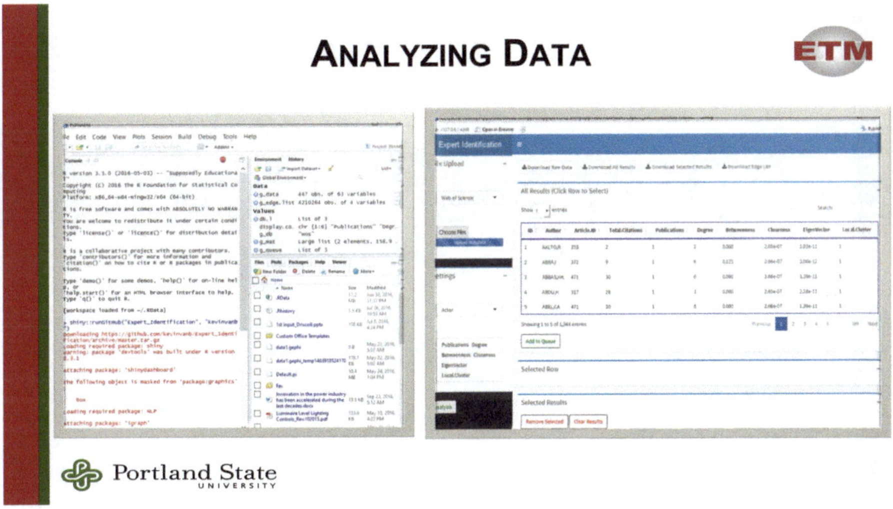
ANALYZING DATA
ETM
Expert Identification
Portland State
UNIVERSITY

RESULTS

Results

ID	Author	Publications	DOI	Article.ID	Title	Total.Citations	Affiliation	Email	Degree	Betweenness	Closeness	EigenVector	Local.Cluster
1	AALTO,R	1	10.1002/esp.	253	Applicatio	2	[Perreault, Lauren M	Perreault.LM@g	2	0	1.17E-07	0	1
2	ABAWI,GS	1	10.2136/sssa	522	Long-term	53	[Moebius-Clune, Bia	bnm5@cornell.e	9	0.083	1.18E-07	6.87E-12	1
3	ABBA,I	1	10.1007/s103	372	Applicatio	9	[Styger, Erika] Corne	eds8@cornell.ed	6	0.125	1.18E-07	0	1
4	ABBAS,HK	1	10.1080/1556	471	Ecology of	30	[Abbas, H. K.; Bruns,	Hamed.Abbas@	6	0	1.18E-07	0	1
5	ABDU,H	1	10.1016/j.jhy	317	Imaging of	28	[Robinson, David A.;	darearthscience	3	0.2	1.18E-07	0	1
6	ABEL,CA	1	10.1080/1556	471	Ecology of	30	[Abbas, H. K.; Bruns,	Hamed.Abbas@	6	0	1.18E-07	0	1
7	ABLER,D	1	10.1023/A:10	743	US agricul	119	MIT, Cambridge, MA	02139 USA; God	16	0	1.19E-07	0.193	1
8	DU KHEIRA	1	10.1007/s112	468	Comparat	0	[Abou Kheira, Abdra	abdo@ksu.edu	0	0	1.17E-07	0	NA
9	DUBACRIN	1	10.3763/ijas.2	375	The syster	9	[Styger, Erika; Upho	eds8@cornell.ed	3	0.125	1.18E-07	0	1
10	BRAHAM,	1	10.1016/j.jen	564	Applicatio	12	Univ Autonoma Cha	raquels85@yahc	3	0	1.17E-07	0	1
11	CCINELLI,	1	10.1080/1556	471	Ecology of	30	[Abbas, H. K.; Bruns,	Hamed.Abbas@	6	0	1.18E-07	0	1
12	TA-MARTI	2	10.1007/s111	42|538	Temporal	20	[Booker, J. D.; Zartn	robert.lascano@	10	27281.488	1.56E-07	0	0.467
13	ACUTIS,M	1	10.1016/j.age	19	Dynamics	0	[Ferrara, Rossana M	marco.carozzi@	8	0	1.18E-07	2.23E-12	1

Edge List

Source	Target	Weight	Type
VAKALIS,S	HEIMANN,R	1	undirected
VAKALIS,S	TALLEY,A	1	undirected
VAKALIS,S	HEIMANN,N	1	undirected
VAKALIS,S	BARATIERI,M	1	undirected
HEIMANN,R	TALLEY,A	1	undirected
HEIMANN,R	HEIMANN,N	1	undirected
HEIMANN,R	BARATIERI,M	1	undirected
TALLEY,A	HEIMANN,N	1	undirected
TALLEY,A	BARATIERI,M	1	undirected
HEIMANN,N	BARATIERI,M	1	undirected

Data

ID	PT	AU	BA	AF	BF	CA	TI	SO	LA	DT	DE	ID.1	AB	C1	RP	EM	RI	FU	FX	CR	NR
1	J	Vakalis, S;	NA	Vakali	NA	NA	Introduction	FUEL	English	Article	Biocoal; C	STEAM-EX	In the bior	[Vakalis, S	Vakalis, S (	stergios.vakalis@natec.unibz.it; bob@engin				Abbas Pou	59
2	J	Yu, CL; Hui	NA	Yu, Ch	NA	NA	Responses of	SCIENTIF	English	Article		WATER-U	Different a	[Yu, Chih-L	Hui, DF (re	dhui@tnstate.edu		USDA- Ca	We gratef	Lehmann J	59
3	J	Anandhi, A	NA	Anand	NA	NA	A system's ap	CLIMATIC	English	Article		IPCC SRES	Estimating	[Anandhi, A	Anandhi, A	anandhi@famu.edu		National S	This work	Adger WN	62
4	J	Jones, MA	NA	Jones,	NA	NA	Economic an	DESALINA	English	Article	Modeling;	REVERSE-(	Global gro	[Jones, M.	Quinn, JC (	jason.quinn@usu.edu		United Sta	The work w	Allen R.G.	39
5	J	Gaxiola, R	NA	Gaxiol	NA	NA	Moving On U	TRENDS I	English	Editorial Material		PROTON-I	Upregulat	[Gaxiola, R	Gaxiola, R	Roberto.gaxiola@asu.edu				Anjaneyulu	12
6	J	Rajib, MA;	NA	Rajib,	NA	NA	Multi-objecti	JOURNAL	English	Article	AMSR-E; C	LAND-SUR	The object	[Rajib, Mo	Merwade,	vmerwade@purdue.e		National S	This work	Abbaspour	91
7	J	Marek, G;	NA	Marek	NA	NA	Estimating pr	AGRICUL	English	Article	Effective p	VARIABLE-	Irrigation f	[Marek, G	Marek, G (	gary.marek@ars.usda		USDA-ARS	We gratef	Moorhead	22
8	J	Huang, LH;	NA	Huang	NA	NA	Impact of cu	JOURNAL	English	Article		SALT-AFFE	Saline-sod	[Huang, L.	Huang, LH	huanglihua@neigae.a		Project of	This work	Hu YC, 200	56
9	J	Hruby, CE;	NA	Hruby	NA	NA	Effects of till	JOURNAL	English	Article	Fecal indic	ENVIRONM	Applicatio	[Hruby, C.	Soupir, ML	msoupir@iastate.edu		Iowa State	This proje	BEVEN K,	59
10	J	Zaccaria, D	NA	Zacca	NA	NA	Risk Assessm	CLEAN-SO	English	Article	Environme	MAPPING	This paper	[Zaccaria,	Zaccaria, D	Dzaccaria@ucdavis.edu				Allen R. G.	40
11	J	Shaw, J; K	NA	Shaw,	NA	NA	Benefits of th	IRRIGATI	English	Article; Pr	irrigation programme		The South	[Shaw, Jef	Shaw, J (re	jeff.shaw@stantec.com				Campbell	3

Portland State UNIVERSITY

GRAPH VISUALIZATION

Gephi

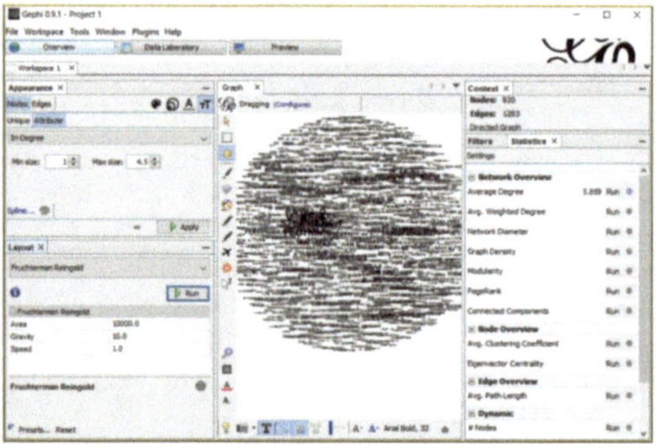

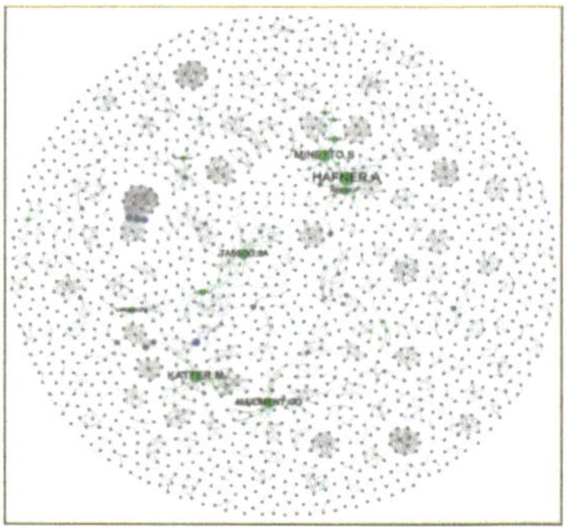

Portland State UNIVERSITY

ETM

APPLICATION CASE

Portland State UNIVERSITY

ETM

Case 1: Social Network Analysis (SNA) for Energy Efficiency in Agriculture Sector

Portland State UNIVERSITY

Case 1: Agriculture Sector Technologies for Energy Efficiency

Organization:

Bonneville Power Administration (BPA) Energy Efficiency Emerging Technology (E3T)

Objective:

- BPA EE Ag Sector SME point-of-contact
- Participants in Technical Advisory Groups (TAGs)
- Experts for developing Technology Roadmaps

Deliverables:

- Listing authors in the field of agriculture
- Research groups and networks

Portland State University

METHODOLOGY

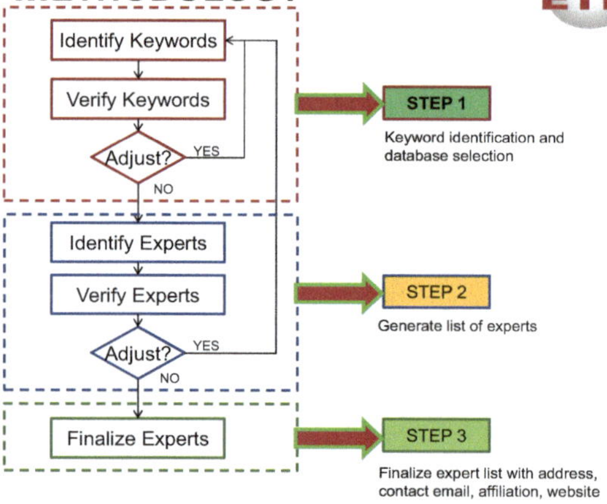

Portland State University

JOURNAL PAPERS

1. Irrigation scheduling performance by evapotranspiration-based controllers S.L. Davis, M.D. Dukes, Agricultural Water Management, Volume 98, Issue 1, 1 December 2010, Pages 19-28.
2. Precision of soil moisture sensor irrigation controllers under field conditions B. Cardenas-Lailhacar, M.D. Dukes, Agricultural Water Management 97 (2010) 666–672.
3. Strip Tillage and High-Efficiency Irrigation Applied to a Sugarbeet–Barley Rotation, William B. Stevens, Robert G. Evans, W.M. Iversen, Jay Jabro, Upendra M. Sainju, Brett L. Allen, Agronomy journal · July 2015.
4. Adoption of site-specific variable rate sprinkler irrigation systems Robert G. Evans • Jake LaRue • Kenneth C. Stone • Bradley A. King, Irrig Sci (2013) 31:871–887.
5. Emerging Technologies for Sustainable Irrigation: Selected Papers from the 2015 ASABE and IA Irrigation Symposium, F. R. Lamm, K. C. Stone, M. D. Dukes, T. A. Howell, Sr., J. W. D. Robbins, Jr., B. Q. Mecham, 2016 American Society of Agricultural and Biological Engineering, Vol. 59(1): 155-161.
6. Surface runoff due to LEPA and spray irrigation of a slowly permeable soil, A.D. Schneider and Terry A. Howell, Transaction of the ASAE, pp. 1089m – 1095, Vol. 43, No. 5, 2000.
7. Adoption of site-specific variable rate sprinkler irrigation systems Robert G. Evans • Jake LaRue • Kenneth C. Stone • Bradley A. King, Irrig Sci (2013) 31:871–887.
8. Opportunities for conservation with precision irrigation E.J. Sadler, R.G. Evans, K.C. Stone, and C.R. Camp, Journal of Soil and Water Conservation Volume 60, Number 6, 2005.
9. Satellite Irrigation Management Support With the Terrestrial Observation and Prediction System: A Framework for Integration of Satellite and Surface Observations to Support Improvements in Agricultural Water Resource Management Forrest S. Melton, Lee F. Johnson, Christopher P. Lund, Lars L. Pierce, Andrew R. Michaelis, Samuel H. Hiatt, Alberto Guzman, Diganta D. Adhikari, Adam J. Purdy, Carolyn Rosevelt, Petr Votava, Thomas J. Trout, Bekele Temesgen, Kent Frame, Edwin J. Sheffner, and Ramakrishna R. Neman, IEEE JOURNAL OF SELECTED TOPICS IN APPLIED EARTH OBSERVATIONS AND REMOTE SENSING, VOL. 5, NO. 6, DECEMBER 2012.
10. Studying uniform and variable rate center pivot irrigation strategies with the aid of site-specific water production functions, Amir Haghverdi, Brian G. Leib, Michael J. Buschermohle, Robert A. Washington-Allen, and Paul D. Ayers, Computers and Electronics in Agriculture · April 2016.

Portland State UNIVERSITY

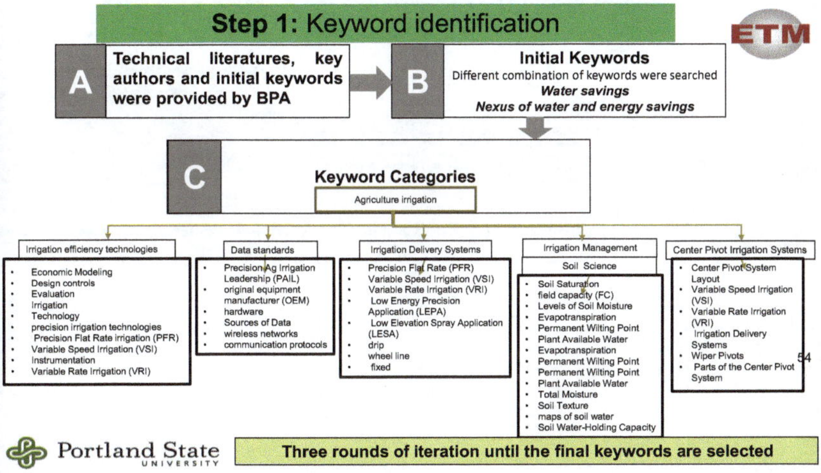

Portland State UNIVERSITY

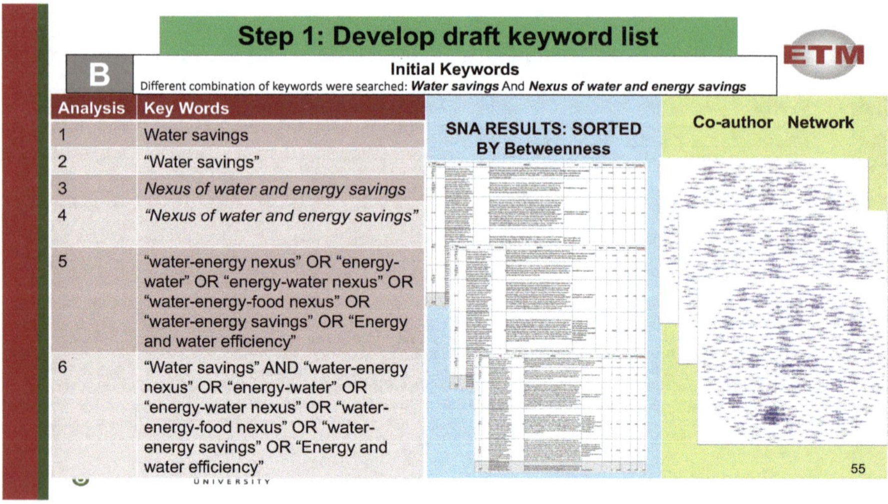
Step 1: Develop draft keyword list
ETM
B
Initial Keywords
Different combination of keywords were searched: Water savings And Nexus of water and energy savings
Analysis
Key Words
1
Water savings
2
"Water savings"
3
Nexus of water and energy savings
4
"Nexus of water and energy savings"
5
"water-energy nexus" OR "energy-water" OR "energy-water nexus" OR "water-energy-food nexus" OR "water-energy savings" OR "Energy and water efficiency"
6
"Water savings" AND "water-energy nexus" OR "energy-water" OR "energy-water nexus" OR "water-energy-food nexus" OR "water-energy savings" OR "Energy and water efficiency"
SNA RESULTS: SORTED BY Betweenness
Co-author Network
55
UNIVERSITY

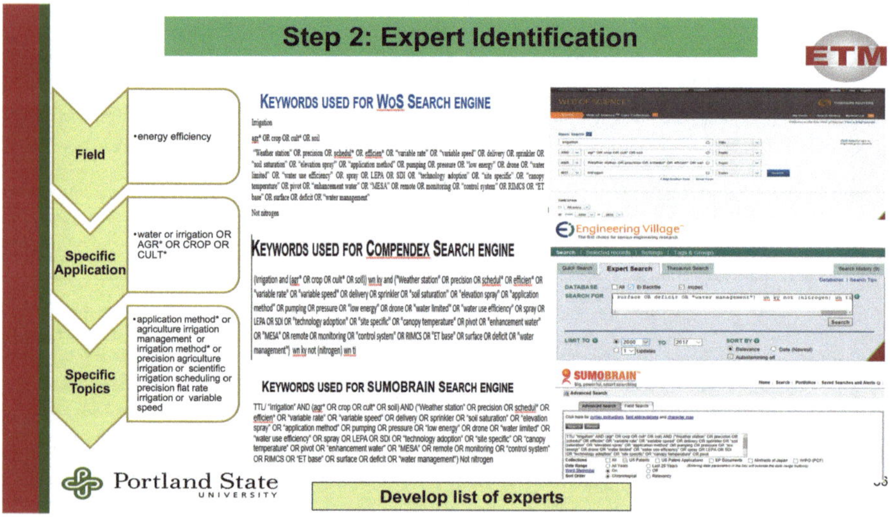
Step 2: Expert Identification
ETM
Field
•energy efficiency
Specific Application
•water or irrigation OR AGR* OR CROP OR CULT*
Specific Topics
•application method* or agriculture irrigation management or irrigation method* or precision agriculture irrigation or scientific irrigation scheduling or precision flat rate irrigation or variable speed
KEYWORDS USED FOR WoS SEARCH ENGINE
Irrigation
agr* OR crop OR cult* OR soil
"Weather station" OR precision OR schedul* OR efficien* OR "variable rate" OR "variable speed" OR delivery OR sprinkler OR "soil saturation" OR "elevation spray" OR "application method" OR pumping OR pressure OR "low energy" OR drone OR "water limited" OR "water use efficiency" OR spray OR LEPA OR SDI OR "technology adoption" OR "site specific" OR "canopy temperature" OR pivot OR "enhancement water" OR "MESA" OR remote OR monitoring OR "control system" OR RIMCS OR "ET base" OR surface OR deficit OR "water management"
Not nitrogen
KEYWORDS USED FOR COMPENDEX SEARCH ENGINE
(Irrigation and (agr* OR crop OR cult* OR soil)) wn ky and ("Weather station" OR precision OR schedul* OR efficien* OR "variable rate" OR "variable speed" OR delivery OR sprinkler OR "soil saturation" OR "elevation spray" OR "application method" OR pumping OR pressure OR "low energy" OR drone OR "water limited" OR "water use efficiency" OR spray OR LEPA OR SDI OR "technology adoption" OR "site specific" OR "canopy temperature" OR pivot OR "enhancement water" OR "MESA" OR remote OR monitoring OR "control system" OR RIMCS OR "ET base" OR surface OR deficit OR "water management") wn ky not (nitrogen) wn ti
KEYWORDS USED FOR SUMOBRAIN SEARCH ENGINE
TTL/ "Irrigation" AND (agr* OR crop OR cult* OR soil) AND ("Weather station" OR precision OR schedul* OR efficien* OR "variable rate" OR "variable speed" OR delivery OR sprinkler OR "soil saturation" OR "elevation spray" OR "application method" OR pumping OR pressure OR "low energy" OR drone OR "water limited" OR "water use efficiency" OR spray OR LEPA OR SDI OR "technology adoption" OR "site specific" OR "canopy temperature" OR pivot OR "enhancement water" OR "MESA" OR remote OR monitoring OR "control system" OR RIMCS OR "ET base" OR surface OR deficit OR "water management") Not nitrogen
Engineering Village
SUMOBRAIN
Portland State
UNIVERSITY
Develop list of experts

Experts Identified From WoS Search Engine Using Keywords

ETM

Author	Title	Affiliation	Email	Website	Phone Number	Database	Measure
Robert G Evans 	1. Strip tillage and high-efficiency irrigation applied to a sugarbeet-barley rotation (2015) 2. Adoption of site-specific variable rate sprinkler irrigation systems (2013) 3. Water use and water productivity of sugarbeet, malt barley, and potato as affected by irrigation frequency (2012) 4. Integrated decision support, sensor networks, and adaptive control for wireless site-specific sprinkler irrigation (2012) 5. A review of mechanical move sprinkler irrigation control and automation technologies (2012) 6. Site-specific sprinkler irrigation in a water-limited future (2012) 7. Development of combined site-specific MESA and LEPA methods on a linear move sprinkler irrigation system (2010) 8. A remote irrigation monitoring and control system for continuous move systems. part a: description and development (2010) 9. Evaluation of closed-loop site-specific irrigation with wireless sensor network (2009) 10. Evaluating the surface irrigation soil loss (SISL) model (2007) 11. Opportunities for conservation with precision irrigation (2005) 12. Apple tree and fruit responses to early termination of irrigation in a semi-arid environment (2001)	United States Department of Agriculture (USDA), Bushland, TX USA	robert.evans@ars.usda.gov	https://www.ars.usda.gov/plains-area/sidney-mt	(406) 433-9496	WOS	Betweenness Degree
James R. Mahan 	1. Comparison of deficit irrigation scheduling methods that use canopy temperature measurements (2013) 2. Determining the Optimum Plant Temperature of Cotton Physiology and Yield to Improve Plant-Based Irrigation Scheduling (2012) 3. Deficit irrigation in a production setting: canopy temperature as an adjunct to ET estimates (2012) 4. Behavior of temperature-based water stress indicators in BIOTIC-controlled irrigation (2006) 5. Determination of temperature and time thresholds for BIOTIC irrigation of peanut on the Southern High Plains of Texas (2005) 6. Establishing differential irrigation levels using temperature-time thresholds (2004)	ARS, USDA, Plant Stress & Water Conservat Lab, Lubbock, TX 79415 US	james.mahan@ars.usda.gov.j mahan@lbk.ars.usda.gov	https://www.ars.usda.gov/people-locations/person/?person-id=3538	(806) 749-5560 ext. 5221	WOS	Betweenness Degree
Gerrit Hoogenboom 	1. Simulating water content, crop yield and nitrate-N loss under free and controlled tile drainage with subsurface irrigation using the DSSAT model (2011) 2. Response of soybean genotypes to different irrigation regimes in a humid region of the southeastern USA(2010) 3. Estimating the demand for irrigation water in a humid climate: A case study from the southeastern United States (2009) 4. Development and application of crop water stress index for scheduling irrigation in cotton (Gossypium hirsutum L) under semiarid environment (2009) 5. Evaluation of FAO-56 crop coefficient procedures for deficit irrigation management of cotton in a humid climate (2007) 6. Irrigation water use estimates based on crop simulation models and kriging (2007) 7. Evaluation of on-farm irrigation applications using the simulation model EPIC (2005) 8. Optimizing irrigation management for a spatially variable soybean field (2003) 9. Center pivot irrigation management optimization of dry beans in humid areas (2000)	Univ Georgia, Dept Biol & Agr Engn, Griffin, GA 30223 USA	gerrit@uga.edu	https://scholar.google.com/citations?user=6t7Zm4oAAAAJ&hl=en	770-228-7261	WOS	Betweenness Degree
Jose O. Payero 	1. Large-scale on-farm implementation of soil moisture-based irrigation management strategies for increasing maize water productivity. 2. Effect of timing of a deficit-irrigation allocation on corn evapotranspiration, yield, water use efficiency and dry mass (2009) 3. Effect of irrigation amounts applied with subsurface drip irrigation on corn evapotranspiration, yield, water use efficiency, and dry matter production in a semiarid climate (2008) 4. Long-term response of corn to limited irrigation and crop rotations (2007) 5. Yield response of corn to deficit irrigation in a semiarid climate (2006) 6. Furrow irrigation management with limited water (2006) 7. Comparison of irrigation strategies for surface-irrigated corn in west central Nebraska (2006) 8. Response of soybean to deficit irrigation in the semi-arid environment of west-central Nebraska (2005) 9. Field scale limited irrigation scenarios for water policy strategies (2004)	Clemson Univ, Clemson, SC USA	jpayero@clemson.edu	http://www.clemson.edu/cafls/faculty_staff/profiles/jpayero	803-284-3343x229	WOS	Degree Betweenness

15

Graph Showing Betweenness Centrality of Experts (WoS)

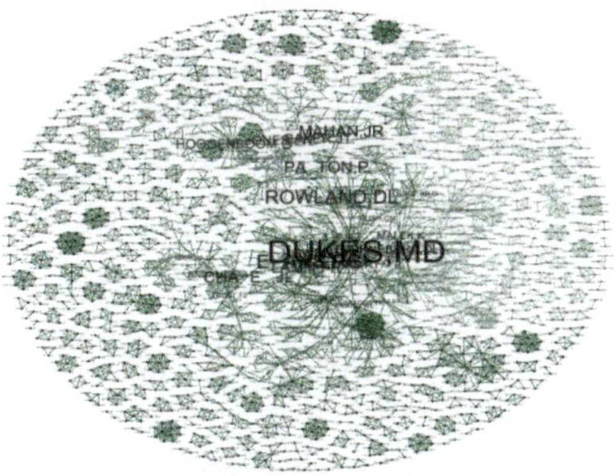

21

EXPERTS IDENTIFIED FROM COMPENDEX SEARCH ENGINE USING KEYWORDS

ETM

Author [First name (s), Surname]	Title	Affiliation	Email	Website	Phone Number	Database	Measure
Christopher Michael Usher Neale 	1. Estimating soil heat flux for alfalfa and clipped tall fescue grass (2005) 2. Vegetation index-based crop coefficients to estimate evapotranspiration by remote sensing in agricultural and natural ecosystems (2011) 3. Water balance of irrigated areas: a remote sensing approach (2011) 4. Combining a water balance model with evapotranspiration measurements to estimate total available soil water in irrigated and rainfed vineyards (2016) 5. Irrigation evaluation based on performance analysis and water accounting at the Bear River Irrigation Project (U.S.A.) (2011)	Daugherty Water for Food Institute, Department Biological Systems Engineering University of Nebraska	cneale@nebraska.edu	http://engineering.unl.edu/faculty/cv/Neale_Biosketch_2014-revised.pdf	(402) 4725359	Compendex	Betweenness Degree
Martha Anderson 	1. Utility of thermal sharpening over Texas high plains irrigated agricultural fields (2007) 2. An Integrated Hydrological and Water Management Study of the Entire Nile River System - Lake Victoria to Nile Delta (2011) 3. A Remote-Sensing Driven Tool for Estimating Crop Stress and Yields (2013) 4. Utility of thermal image sharpening for monitoring field-scale evapotranspiration over rainfed and irrigated agricultural regions (2008) 5. Evaluation of Drought Indices Based on Thermal Remote Sensing of Evapotranspiration over the Continental United States (2011) 6. Monitoring daily evapotranspiration over two California vineyards using Landsat 8 in a multi-sensor data fusion approach (2016)	U.S. Dept. ot Agric., Hydrology & Remote Sensing Lab., Beltsville, MD, United States	Anderson@ars.usda.gov	https://www.ars.usda.gov/northeast-area/beltsville-md/beltsville-agricultural-research-center/hydrology-and-remote-sensing-laboratory/people/martha-anderson/	(301) 504-6616	Compendex	Degree Betweenness
Dennis L. Corwin 	1. Spatio-temporal impacts of dairy lagoon water reuse on soil: heavy metals and salinity (2015) 2. Evaluation of soil salinity leaching requirement guidelines (2011) 3. Site-specific management in salt-affected sugar beet fields using electromagnetic induction (2005) 4. Apparent soil electrical conductivity measurements in agriculture (2005) 5. Analytical steady-state solutions for water-limited cropping systems using saline irrigation water 92014)	U.S. Salinity Lab., Riverside, CA, United States	Dennis.Corwin@ars.usda.gov	https://www.ars.usda.gov/pacific-west-area/riverside-ca/us-salinity-laboratory/water-reuse-and-remediation-research/people/dennis-corwin/	951-369-4819	Compendex	Betweenness Degree
Thomas Marek 	1. The ASCE Standardized Equation Based Bushland Reference ET Calculator (2012) 2. Use of crop-specific drought indices for determining irrigation demand in the texas high plains (2013) 3. Soil water extraction, water use, and grain yield by drought-tolerant maize on the Texas High Plains (2015) 4. Estimating preseason irrigation losses by characterizing evaporation of effective precipitation under bare soil conditions using large weighing lysimeters (2016) 5. Weighing lysimeters for the determination of crop water requirements and crop coefficients (2006)	Texas A&M AgriLife Res. & Extension Center, Amarillo, TX, United States	tmarek@ag.tamu.edu	http://amarillo.tamu.edu/facultystaff/thomas-h-marek-ms-pe/	(806)677-5600	Compendex	Degree Betweenness

26

GRAPH SHOWING BETWEENNESS CENTRALITY OF EXPERTS (COMPENDEX)

ETM

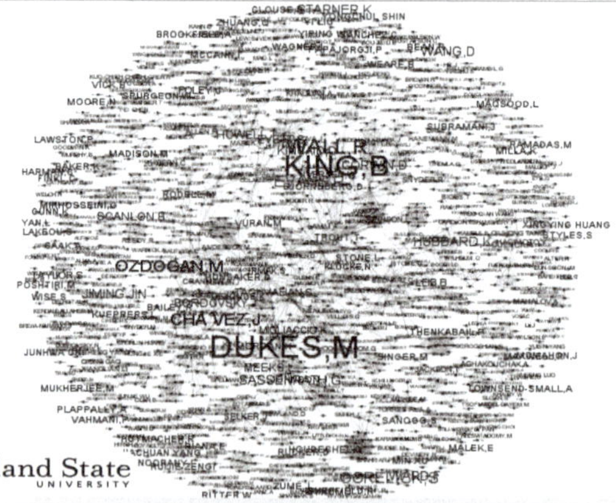

Portland State UNIVERSITY

34

Experts Identified From Sumobrain Search Engine Using Keywords

Author [First name (s), Surname]	Patent Assignee	Article Titles	Email	Website	Address/ Phone Number	Present Affiliation/ Position	Database	Measure
Harvey J. Nickerson 	Rain Bird Corporation Rain Bird Corporation 970 West Sierra Madre Ave. Azusa, CA 91702 (626) 812-3400 http://www.rainbird.com/corporate/contact.htm	1. Modular and expandable irrigation controller (10/28/2008) 2. Modular and expandable irrigation controller (12/29/2009) 3. Automatically adjusting irrigation controller (9/28/2010) 4. Code replacement for irrigation controllers 11/30/2010 5. Modular and expandable irrigation controller (11/30/2010) 6. Modular and expandable irrigation controller (8/9/2011) 7. Automatically adjusting irrigation controller (5/1/2012) 8. Automatically adjusting irrigation controller with temperature and rainfall sensor (6/12/2012) 9. Modular and expandable irrigation controller (8/14/2012) 10. Modular and expandable irrigation controller (9/11/2012) 11. Code replacement for irrigation controllers (4/9/2013) 12. Automatically adjusting irrigation controller (2/11/2014) 13. Code replacement for irrigation controllers (5/27/2014) 14. Method and apparatus for programming a decoder-based irrigation controller (11/25/2014) 15. Automatically adjusting irrigation controller (6/2/2015) 16. Volumetric budget based irrigation control (2/16/2016) 17. Modular and expandable irrigation controller (5/24/2016) 18. Irrigation controller wireless network adapter and networked remote service (10/18/2016)		https://www.linkedin.com/in/harvey-nickerson-67a508 http://www.rainbird.com/ag/index.htm	970 West Sierra Madre Ave. Azusa, CA 91702Agricultural Irrigation (800) 435-5624	Engineering Manager (New Product Development) **Rain Bird Corp**	Smobrain	Betweenness Degree
David M. Redmond 	Rain Bird Corporation Rain Bird Corporation 970 West Sierra Madre Ave. Azusa, CA 91702 (626) 812-3400 http://www.rainbird.com/corporate/contact.htm	1. Sensor device for interrupting irrigation (5/24/2011) 2. Sensor device for use in controlling irrigation (5/27/2014) 3. User interface for a sensor-based interface device for interrupting an irrigation controller (9/29/2015) 4. Irrigation controller wireless network adapter and networked remote service (10/18/2016) 5. Sensor device for use in controlling irrigation (11/22/2016)	davidredmond@cox.net	https://www.linkedin.com/in/davidmredmond?authType=NAME_SEARCH&authToken=D1rm	(520) 906-3627.	Founder & Director **Engineering, Program** Management	Sumobrain	Betweenness Degree
David G Fern 	Rain Bird Corporation Rain Bird Corporation 970 West Sierra Madre Ave. Azusa, CA 91702 (626) 812-3400 http://www.rainbird.com/corporate/contact.htm	1. Sensor device for interrupting irrigation (5/24/2011) 2. Wireless extension to an irrigation control system and related methods (5/22/2012) 3. Wireless extension to an irrigation control system and related methods (8/6/2013) 4. Sensor device for use in controlling irrigation (5/27/2014) 5. Sensor device for use in controlling irrigation (11/22/2016)	info@goldspiredesign.com	https://www.linkedin.com/in/davidgfern http://www.goldspiredesign.com/	9053 Inverness Road Santee, California 92071 USA +1 619-368-7390	President GoldSpire Design LLC	Sumobrain	Betweenness Degree
.Ryan L. Walker	Rain Bird Corporation (Asuza, CA, US)|	1. Same code base in irrigation control devices and related methods (7/17/2012) 2. Same code base in irrigation control devices and related methods (4/30/2013) 3. Volumetric budget based irrigation control (2/16/2016)		https://www.linkedin.com/in/ryan-walker-18390979	6991 E Southpoint Rd, Tucson, AZ 85756 (800) 435-5624	Director Rain Bird Corporation	Sumobrain	Degree Betweenne

37

Graph Showing Betweenness Centrality of Experts (SUMOBRAIN)

Por

41

ETM

Case 2: Social Network Analysis (SNA) for Technology Roadmap (TRM) Co-authorship Network Analysis

Portland State UNIVERSITY

Case 2: Technology Roadmap (TRM) Co-authorship Network Analysis

ETM

Organization:

Institute of Manufacturing (IfM) at University of Cambridge.

Objective:

- Knowledge creation
- Knowledge transfer
- Knowledge contribution

Deliverables:

- Listing authors in the field of TRM based on centrality metrics
- Research groups and networks

Portland State UNIVERSITY

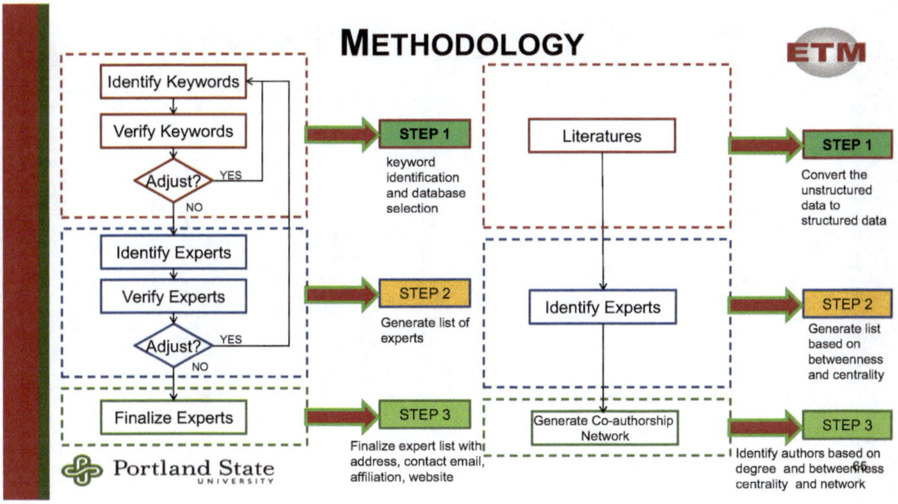
METHODOLOGY
ETM
Identify Keywords
Verify Keywords
Adjust?
YES
NO
Identify Experts
Verify Experts
Adjust?
YES
NO
Finalize Experts
STEP 1
keyword identification and database selection
STEP 2
Generate list of experts
STEP 3
Finalize expert list with address, contact email, affiliation, website
Literatures
Identify Experts
Generate Co-authorship Network
STEP 1
Convert the unstructured data to structured data
STEP 2
Generate list based on betweenness and centrality
STEP 3
Identify authors based on degree and betweenness centrality and network
66
Portland State UNIVERSITY

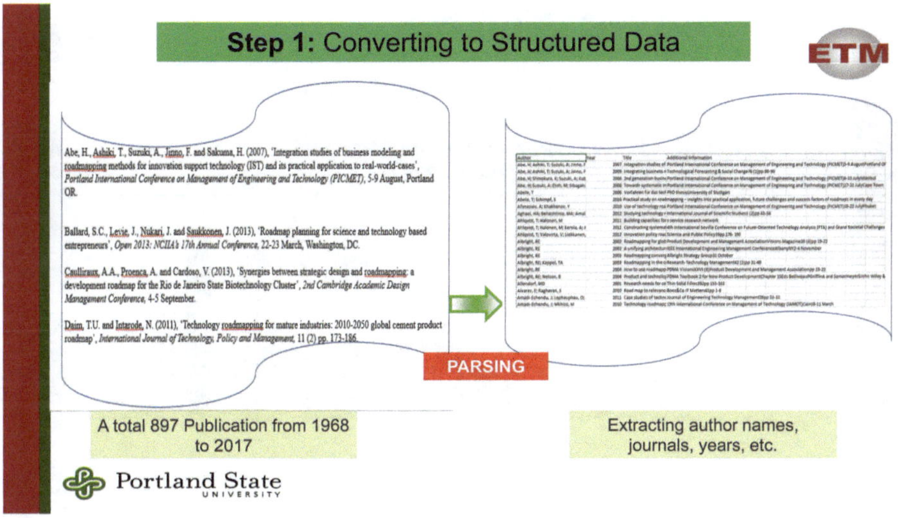
Step 1: Converting to Structured Data
ETM
Abe, H., Ashiki, T., Suzuki, A., Jinno, F. and Sakuma, H. (2007), 'Integration studies of business modeling and roadmapping methods for innovation support technology (IST) and its practical application to real-world-cases', Portland International Conference on Management of Engineering and Technology (PICMET), 5-9 August, Portland OR.
Ballard, S.C., Levie, J., Nukari, J. and Saukkonen, J. (2013), 'Roadmap planning for science and technology based entrepreneurs', Open 2013: NCIIA's 17th Annual Conference, 22-23 March, Washington, DC.
Caulliraux, A.A., Proenca, A. and Cardoso, V. (2013), 'Synergies between strategic design and roadmapping: a development roadmap for the Rio de Janeiro State Biotechnology Cluster', 2nd Cambridge Academic Design Management Conference, 4-5 September.
Daim, T.U. and Intarode, N. (2011), 'Technology roadmapping for mature industries: 2010-2050 global cement product roadmap', International Journal of Technology, Policy and Management, 11 (2) pp. 173-186.
PARSING
A total 897 Publication from 1968 to 2017
Extracting author names, journals, years, etc.
Portland State UNIVERSITY

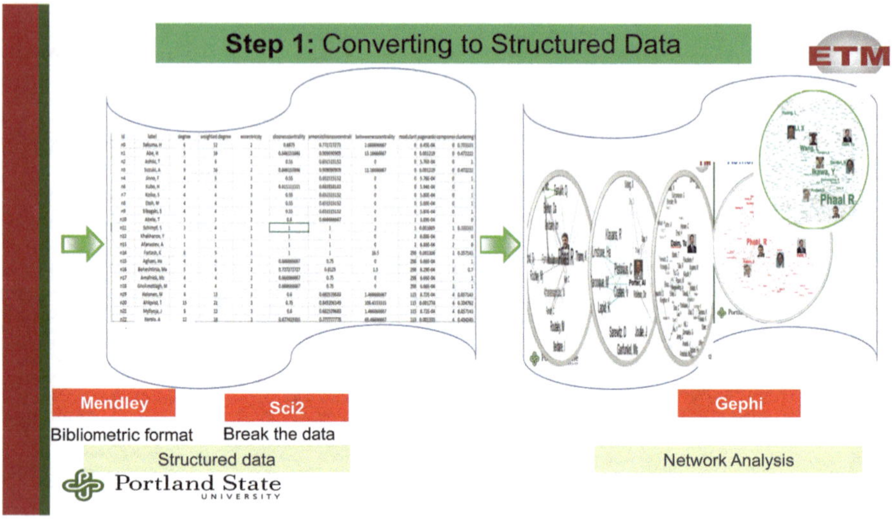
Step 1: Converting to Structured Data
ETM
Mendley
Sci2
Gephi
Bibliometric format
Break the data
Structured data
Network Analysis
Portland State
UNIVERSITY

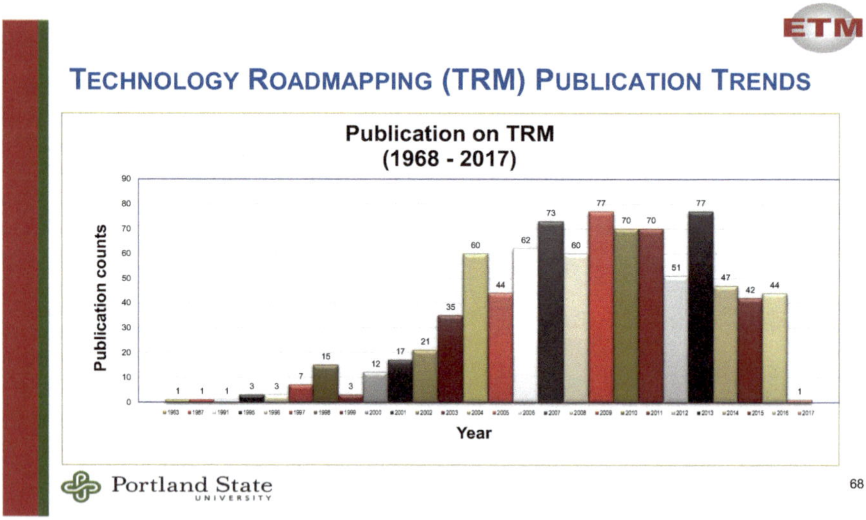
ETM
TECHNOLOGY ROADMAPPING (TRM) PUBLICATION TRENDS
Publication on TRM
(1968 - 2017)
Publication counts
0
10
20
30
40
50
60
70
80
90
1
1
1
3
3
7
15
3
12
17
21
35
60
44
62
73
60
77
70
70
51
77
47
42
44
1
1968
1987
1991
1995
1996
1997
1998
1999
2000
2001
2002
2003
2004
2005
2006
2007
2008
2009
2010
2011
2012
2013
2014
2015
2016
2017
Year
Portland State
UNIVERSITY
68

Words Frequency based on the Publications' Titles

ETM

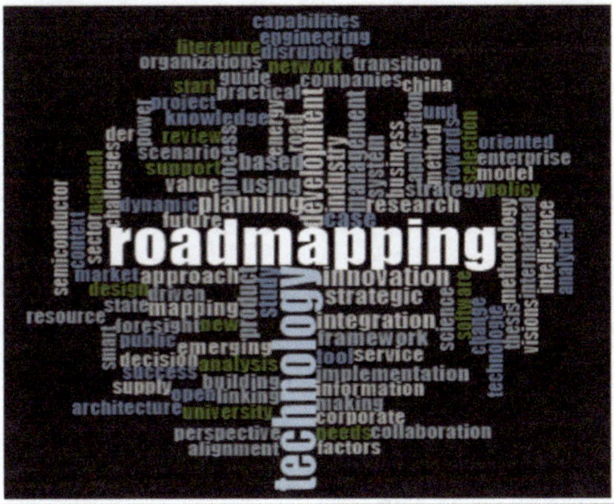

Portland State UNIVERSITY

Words Frequency based on the Publications' Titles

Word	Count	Similar Words
Roadmapping	742	roadmap, roadmapping, roadmaps
Technology	560	engineering, engineers, technical, technologi, technological, technologies, technology
Innovation	165	advance, advanced, advances, advancing, concept, conception, concepts, creation, design, designing, foundation, initial, initiating, initiative, [illegible] innovation, innovations, innovative, innovators, institute, institutions, introducing, introduction
Development	154	acquiring, arising, develop, developer, developing, development, developments, education, educational, evolution, evolutional, evolving, exploitation, formulation, getting, growth, mature, producing, training
Planning	154	design, designing, plan, planning, plans, plotting, program, programme, programs, project, projects
Using	117	applied, applying, employment, exercise, exercises, exploitation, function, functional, functionalities, practical, practice,practices, purpose, purposes, role, usage, use, using, utility, utilization, utilizing
Study	112	analyse, analysing, canvas, considering, field, fields, learned, learning, read, reading, report, studies, study, studying, survey, takes, work, working
Industry	90	industrial, industrialisation, industrialization, industrially, industries, industry, manufacturing
Case	89	case, cases, event, example
Management	88	achievements, care, cope, dealing, direction, directions, handling, manage, managed, management, managing
Based	87	base, based, bases, establish, establishing, establishment, foundation, fundamentals, ground, grounded, home, infrastructure, infrastructures, means, radical, stem, themes
System	87	order, organic, organisation, organisational, organisations, organise, organization, organization', organizations, system, systematic, systematization, systeme, systemic, systems
Product	85	generation, generative, outputs, product, production, products, profitable
Integration	85	incorporating, integrate, integrated, integrates, integrating, integration, integrative, integrator, structuration, structure, structured, structuring, unifying
Process	82	action, actions, advance, advanced, advances, advancing, operating, operation, operational, process, processes, processing, progress, work, working
Strategic	79	strategic, strategize, strategizing
Approach	78	advance, advanced, advances, advancing, approach, approaches, attack, come, near
Research	77	exploration, exploring, investigating, investigation, research, search
Framework	74	framework, frameworks, model, modeling, modelling, models
Mapping	66	function, functional, functionalities, map, mapping, maps
Practical	61	executing, execution, executives, exercise, exercises, expert, experts, feasibility, good, goods, operating, operation, operational, pattern, patterns, practical, practice, practices, virtual
Value	56	appraisal, assess, assessing, assessment, assessments, evaluating, evaluation, evaluations, measured, measurement, measures, measuring, value
Emerging	55	emerged, emergence, emergency, emergent, emerging, growth, issues
Business	54	business, businesses, job, line
Support	54	assist, assisted, assistive, back, documentation, help, support, supported, supporting, supports, sustainability, sustainable
Knowledge	51	cognitive, initial, initiating, initiative, initiatives, knowledge, learned, learning
Road	50	road, route
Implementation	48	applied, applying, executing, execution, executives, implement, implementation, implementing, instrument, instruments
Building	48	build, building, constructing, construction, establish, establishing, establishment, form, frames, make, making, progress, shaping
Future	46	future, futures, next
Tool	46	instrument, instruments, tool, tools
Organizations	44	administration, coordinate, coordinating, coordination, establish, establishing, establishment, form, format, formation, government, governments, organic, organisation, organisational, organisations, organise, organization, organization', organizations
Eng[illegible]ring	43	direction, directions, engineering, engineers, organic, organisation, organisational, organisations, organise, organization, organization', organizations
[illegible]	[illegible]	strategies, strategy
Model	39	example, model, modeling, modelling, models, pattern, patterns, simulation

ETM

Portland State UNIVERSITY

AUTHORS BASED ON DEGREE CENTRALITY

ETM

The top five authors are located in the network, based on the highest number of co-authoring connections.

The size of authors' names corresponds to the level of degree.

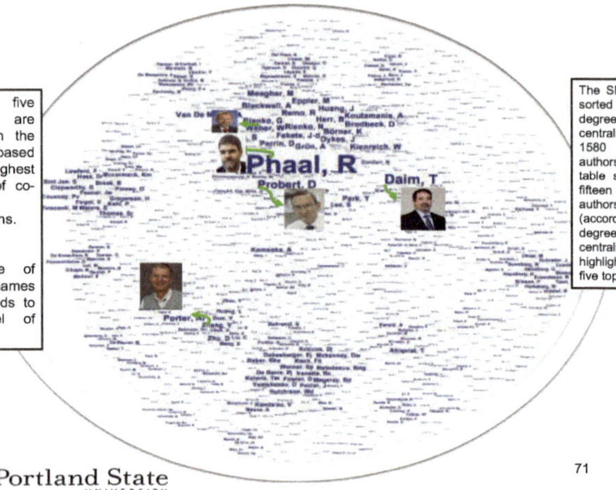

The SNA results sorted by the degree of centrality shows 1580 different authors. The table shows the fifteen top authors (according to degree centrality) and highlights the fist five top authors..

No.	Label	Degree
1	Phaal, R	66
2	Daim, T	34
3	Probert, D	28
4	Porter, Al	21
5	Weber, W	19
6	Remo, R	19
7	Rienko, G	19
8	Rienko, N	19
9	Blackwell, A	19
10	Borner, K	19
11	Brodbeck, D	19
12	Dykes, J	19
13	Eppler, M	19
14	Fekete, J-d	19
15	Grun, A	19

Portland State UNIVERSITY

71

AUTHORS BASED ON BETWEENNESS CENTRALITY

ETM

The top five authors are located in the network based on the highest levels of betweenness.

The size of authors' names corresponds to the level of Betweenness.

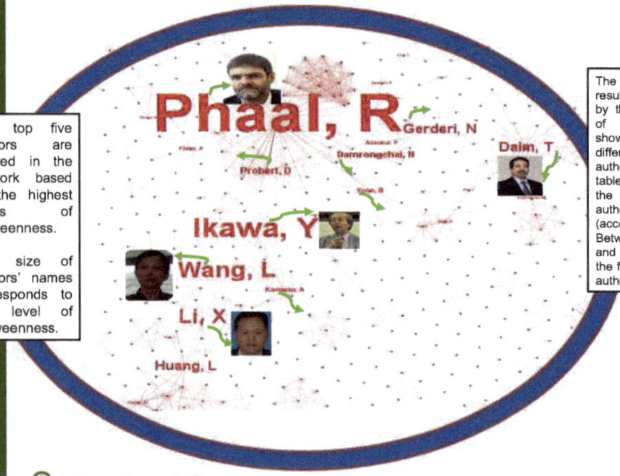

The SNA results sorted by the degree of centrality shows 1580 different authors. The table shows the top fifteen authors (according to Betweenness) and highlights the fist five top authors.

No.	Label	Betweeness
1	Phaal, R	30909.177
2	Ikawa, Y	18745.000
3	Wang, L	15447.000
4	Li, X	14245.667
5	Daim, T	10613.807
6	Gerdsri, N	10408.047
7	Huang, L	9926.8333
8	Probert, D	7607.3631
9	Damrongchai, N	6809.4052
10	Kameoka, A	5445.000
11	Yoon, B	4933.5095
12	Assakul, P	4544.5992
13	Fleury, A	4005.500
14	Kim, H	3854.5897
15	Porter, Al	2906.1833

72

Portland State UNIVERSITY

RESULTS

ID	Author	Publicatio	DOI	Article.ID	Title	Total.Citat	Affiliation	Email	Degree	Betweenn	Closeness	EigenVect	Local.Cluster
1	ABBOTT,M	1	10.1175/1!	467	Analysis o	5	Oregon State Univ, C		3	0	3.32E-07	3.15E-12	1
2	ABDOU,W	1	10.1029/2	243	Mars Clim	65	[Kleinboe	armin.klei	11	0	3.34E-07	0	1
3	ABER,JD	1	10.1046/j.	481	Measuren	143	Oregon State Univ, D		3	0	3.60E-07	7.01E-12	1
4	ABRAHAM	1	10.1063/1.	183	Simulatio	1	[Gorman,	jpabrahan	3	0	3.32E-07	3.15E-12	1
5	ABRAHAM	1	10.1016/j.	316	Applicatic	13	Univ Auto	raquels85	3	0	3.32E-07	3.15E-12	1
6	ABSAR,A	1	10.1016/j.	119	Hot spring	0	[Craig, J.;	Jonathan.	10	0	3.34E-07	0	1
7	ADAM,R	1	10.1046/j.	449	Elevated a	74	USDA ARS	gwall@us	12	0	3.34E-07	0	1
8	ADDAMS,(	2	10.1029/2(	336|339	Reliable c	34	Stanford L	gerrit.sch	3	0	3.32E-07	0	1
9	AFSHAR,R	1	10.2135/cr	48	Potential	1	[Afshar, R	rekeshava	5	0	3.33E-07	0	1
10	AGARWAL	1		176	NUMERICA	0	[Zhang, Zheming; Ag		1	0	3.32E-07	3.53E-12	NA
11	AGDAG,M	1	10.1081/C	466	Row spaci	3	Univ Nebraska, Dept		4	0	3.32E-07	1.41E-13	1
12	AGUE,JJ	1	10.1007/s(	34	Analysis o	3	[Chu, Xu;	xu.chu@y	1	0	3.32E-07	3.53E-12	NA
13	AGUIRRE,(	1		425	Stochastic	0	Purdue Univ, Dept A		1	0	3.32E-07	3.53E-12	NA
14	AHMAD,A	1		252	Developm	7	[Ahmad, S	shakeelah	10	0	3.35E-07	0	1
15	AHMAD,S	1		252	Developm	7	[Ahmad, S	shakeelah	10	0	3.35E-07	0	1
16	AHSAN,SS	1	10.1021/e	71	In Situ UV	1	[Doud, De	la249@co	4	0	3.32E-07	1.42E-13	1
17	AJO-FRAN	1	10.1016/j.	200	Monitorin	47	[Hovorka, Susan D.; I		22	0	3.36E-07	7.92E-10	1
18	ALBRECHT	1		416	Polymer c	16	Short Elliott Hendric		2	0	3.32E-07	4.47E-12	1
19	ALDERFAS	1	10.1016/S(	462	Use of cro	38	King Saud Univ, Dept		1	0	3.32E-07	2.25E-12	NA
20	ALDERSON	1	10.1021/e	458	Applicatic	31	CALTECH, WM Keck L		2	0	3.32E-07	4.47E-12	1
21	ALEXANDE	1	10.1016/j.	181	Modifying	21	[Alexande	Anthony.A	4	0	3.32E-07	1.42E-13	1
22	ALHALABI	1	10.1016/j.	35	Solar-ther	5	[Reiff, Joh	reif@cs.d	1	0	3.32E-07	3.53E-12	NA

Portland State UNIVERSITY 73

TOP 5 AUTHORS NETWORK BASED ON DEGREE METRICS

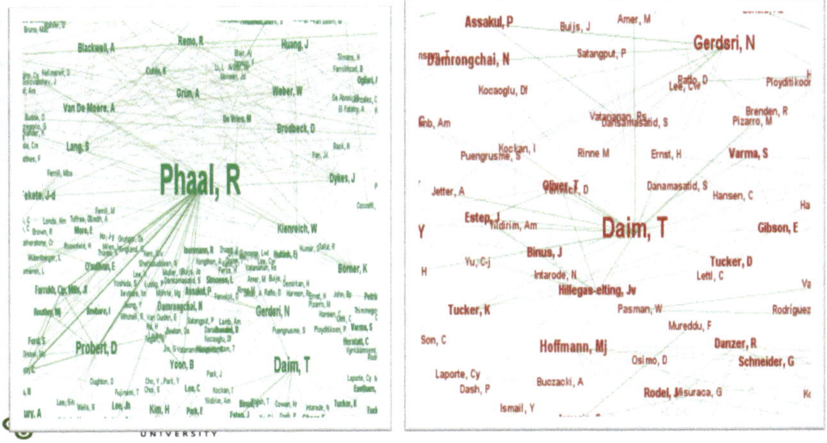

Portland State UNIVERSITY

TOP 5 AUTHORS NETWORK BASED ON DEGREE METRICS

ETM

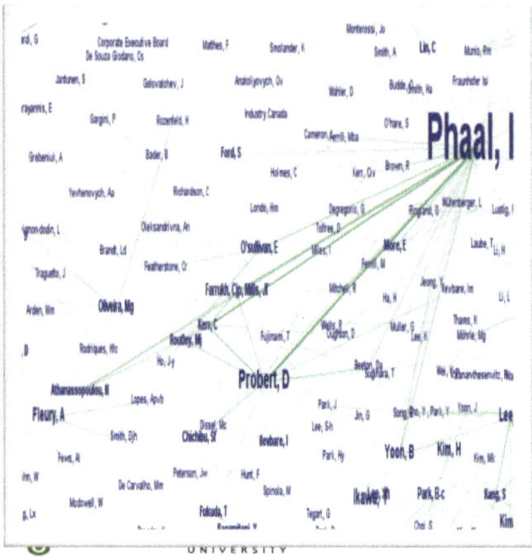

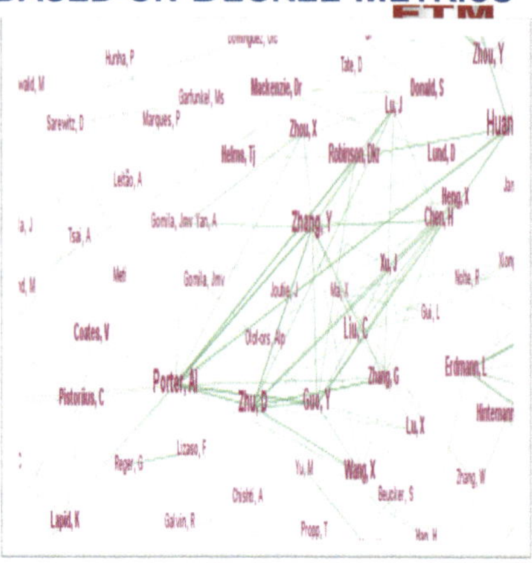

TOP 5 AUTHORS NETWORK BASED ON DEGREE METRICS

ETM

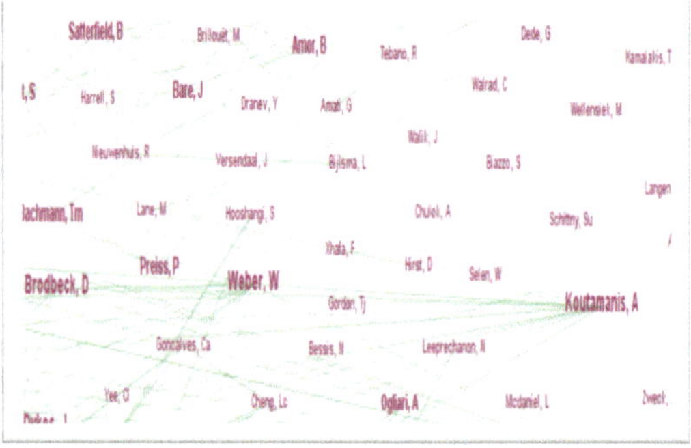

ETM

Visualization of Sub-Networks based on the Five Top Authors According to Betweenness

Portland State University

77

5 Top Authors Network Based on Betweenness Metrics

ETM

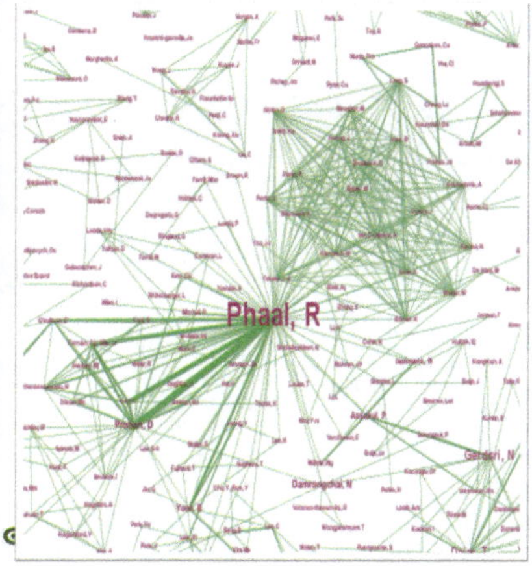

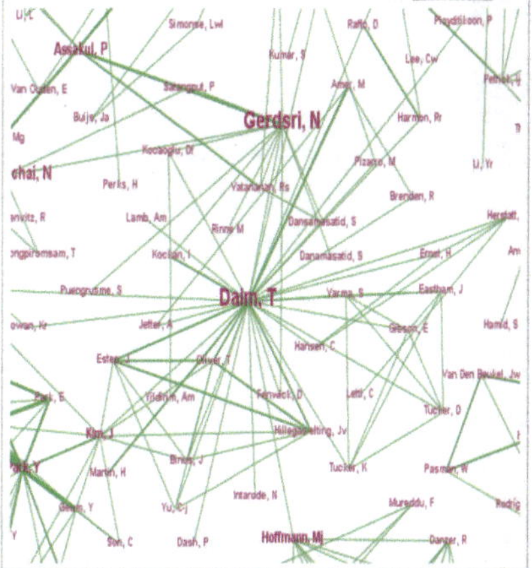

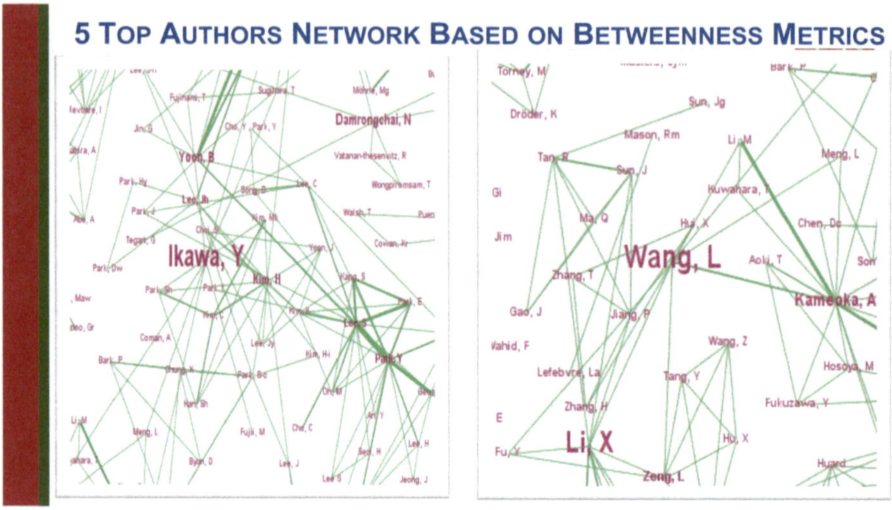
5 Top Authors Network Based on Betweenness Metrics
Ikawa, Y
Damrongchai, N
Yoon, B
Kim, H
Park, Y
Wang, L
Li, X
Kameoka, A
Sun, Jg
Mason, Rm
Meng, L
Kuwahara, T
Chen, Dc
Aoki, T
Zhang, T
Gao, J
Jiang, P
Wang, Z
Lefebvre, La
Tang, Y
Hosoya, M
Zhang, H
Fukuzawa, Y
Hu, X
Zeng, L
Dröder, K

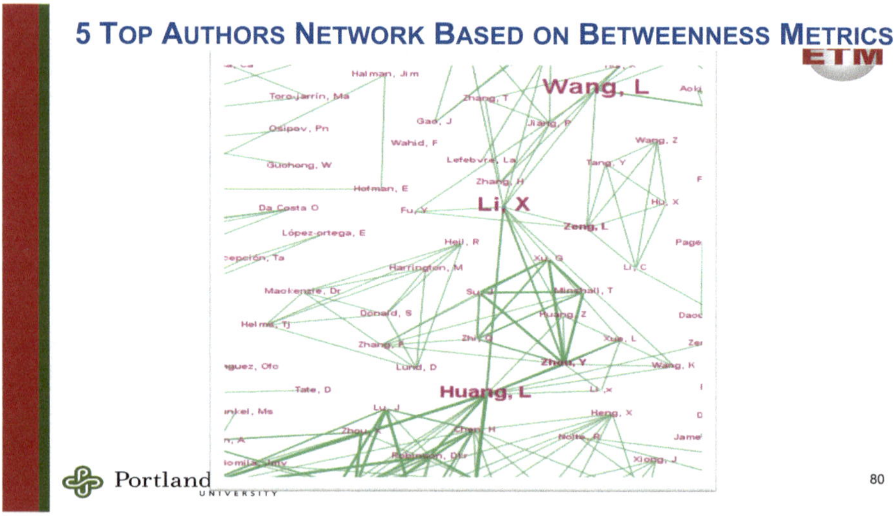
5 Top Authors Network Based on Betweenness Metrics
ETM
Wang, L
Li, X
Huang, L
Zeng, L
Halman, Jim
Toro-jarrin, Ma
Osipov, Pn
Guohong, W
Da Costa O
López-ortega, E
Harrington, M
Mackenzie, Dr
Donald, S
Zhang, T
Gao, J
Jiang, P
Wahid, F
Lefebvre, La
Zhang, H
Wang, Z
Tang, Y
Hu, X
Portland State University
80

ETM

Conclusions

Portland State University

Conclusions

ETM

- Significantly reduces the time required for identifying experts; however, there are some important caveats:
  - Accurate keywords are necessary or results will be meaningless ("garbage in, garbage out").
  - Expert input is required to validate results.
  - Only experts who publish will be identified.
- Analysis can go far beyond identifying experts:
  - Research areas of focus can be identified from titles and abstracts; this can help align roadmaps with emerging research topics.
  - Changes in research over time can be analyzed to estimate future technological changes.
- Integration of Bibliometrics, Patent Analysis and Social Network Analysis provided further intelligence to building technology roadmaps
  - Experts and organizations which were not thought of before were identified and will be invited to the next series of workshops when the roadmap will be updated.
  - Identification of technology trends were found to be beneficial to populating the roadmap documents prior to the roadmapping workshops
- Potential uses of results
  - Recruit other collaborators for mutually-beneficial R&D projects.
  - Identify gaps between important research communities and envision ways to bridge those gaps.
  - Prepare for future technology roadmapping projects.

Portland State University

82

ETM

More Cases

Portland State
UNIVERSITY

ETM

Case 1

Portland State
UNIVERSITY

84

APPLICATION: ADVANCED COMMERCIAL REFRIGERATION

- In early 2015, BPA Energy Efficiency engineer sought knowledge on international experts in refrigeration technologies for commercial applications including grocery stores, supermarkets, and restaurants.
- Key topics & criteria:
  - Refrigerants: low global-warming potential (GWP); natural; hydrocarbon; halocarbon; ammonia; carbon dioxide (CO_2).
  - Technologies: magnetocaloric; electrochemical; thermoelectric; thermoacoustic; thermionic.
  - Affiliated institutions: research universities; national laboratories; manufacturers; design firms.

Portland State UNIVERSITY 85

APPLICATION: ADVANCED COMMERCIAL REFRIGERATION

SNA Project Steps

Step	Summary
1. Develop draft keyword list	BPA engineer provided technical literature to PSU ETM student
2. Finalize keyword list	Iterative processes between BPA engineer and PSU ETM student
3. Develop draft expert lists and diagrams	
4. Finalize expert lists and diagrams	

Portland State UNIVERSITY 86

Application: Advanced Commercial Refrigeration

ETM

Step 1: Develop draft keyword list

A Technical literature consulted to establish first-draft keyword list → **B** Technology Applications: Grocery stores, supermarkets, restaurants, etc.

C Keyword categories within the technology applications

Refrigerants

1. Hydrocarbon (Propane; Isobutene; Propylene; Ethane)
2. Halocarbon (R22; R134a; R404a; R507)
3. Ammonia
4. Carbon Dioxide
5. Natural
6. Low Global-Warming Potential

Energy Efficiency

1. Vacuum Insulated Panel
2. Phase Change Material
3. Defrost Control
4. Electronically Commutated Motor
5. Variable Speed Drive
6. Heat Recovery
7. Heat Transfer
8. Microchannel Heat Exchanger
9. Microchannel Evaporator
10. HVAC Integration

Technologies

1. Magnetocaloric Cooling
2. Electrochemical compression
3. Thermoelectric
4. Thermionic
5. Thermo-tunneling
6. Thermoacoustic
7. Magnetic

Portland State UNIVERSITY

87

Application: Advanced Commercial Refrigeration

ETM

Steps 2–4: Finalize keyword list, develop & then finalize draft expert lists and diagrams

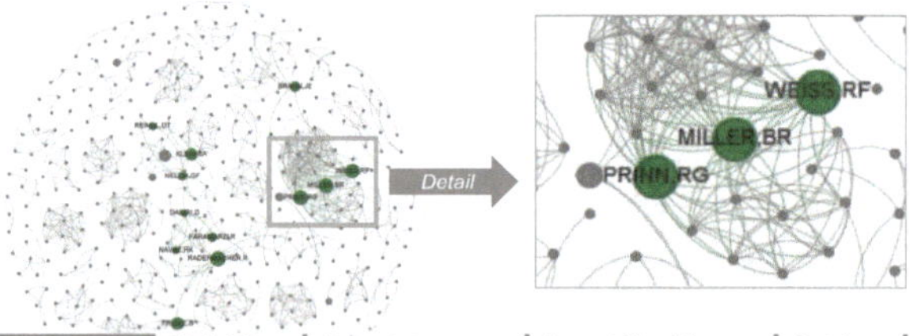

U.S.-based Experts

Circle Size:	Node Colors:	Layout Algorithm:	Database:
Betweenness Centrality	Green: Selected Grey: Not Selected	Fruchterman-Reingold	World of Science

Portland State UNIVERSITY

88

Application: Advanced Commercial Refrigeration

Steps 2–4: Finalize keyword list, develop & then finalize draft expert lists and diagrams

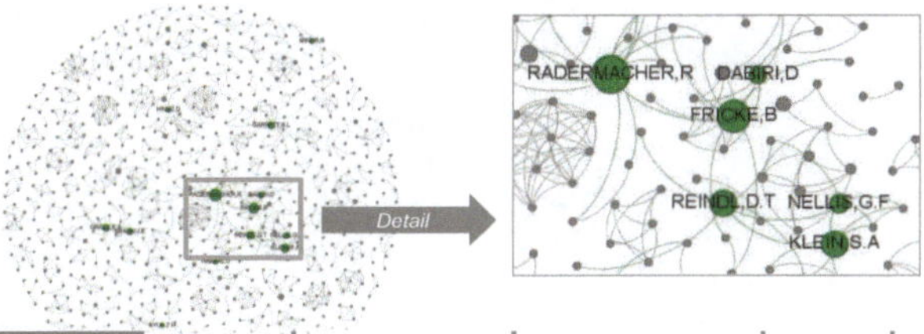

U.S.-based Experts	*Circle Size:* Betweenness Centrality	*Node Colors:* Green: Selected Grey: Not Selected	*Layout Algorithm:* Fruchterman-Reingold	*Database:* Compendex

Portland State UNIVERSITY

Application: Advanced Commercial Refrigeration

Steps 2–4: Finalize keyword list, develop & then finalize draft expert lists and diagrams

U.S.-based Experts

Author	Affiliation	Selected Article Titles
RADERMACHER, REINHARD	University of Maryland	1. An evaluation of the environmental impact of commercial refrigeration systems using alternative refrigerants 2. An experimental study of the performance of a dual-loop refrigerator freezer system 3. Review of secondary loop refrigeration systems
FRICKE, BRIAN	Oak Ridge National Laboratory	1. Comparative analysis of various CO2 configurations in supermarket refrigeration systems 2. An evaluation of the environmental impact of commercial refrigeration systems using alternative refrigerants 3. Comparison of Vertical Display Cases: Energy and Productivity Impacts of Glass Doors Versus Open Vertical Display Cases

Portland State UNIVERSITY

Application: Advanced Commercial Refrigeration

Steps 2–4: Finalize keyword list, develop & then finalize draft expert lists and diagrams

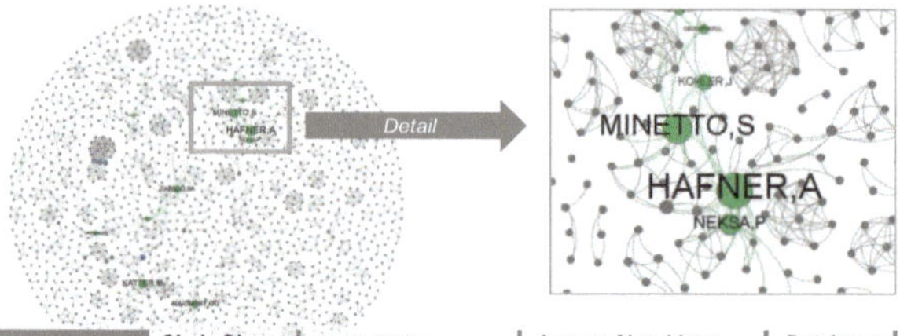

International Experts

Circle Size: Betweenness Centrality

Node Colors: Green: Selected Blue: U.S. Selected Grey: Not Selected

Layout Algorithm: Fruchterman-Reingold

Database: World of Science

Portland State UNIVERSITY

91

Application: Advanced Commercial Refrigeration

Steps 2–4: Finalize keyword list, develop & then finalize draft expert lists and diagrams

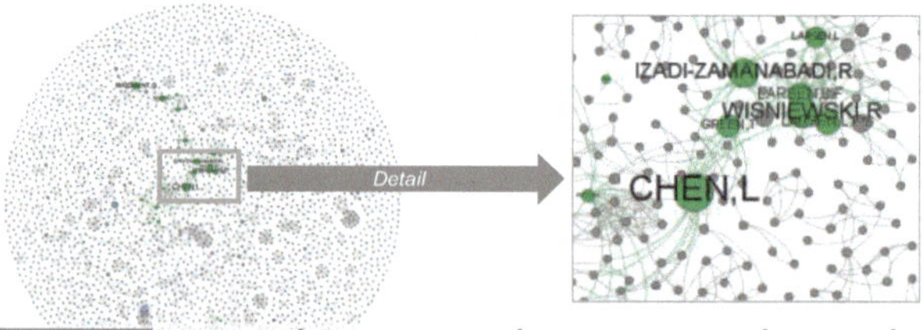

International Experts

Circle Size: Betweenness Centrality

Node Colors: Green: Selected Blue: U.S. Selected Grey: Not Selected

Layout Algorithm: Fruchterman-Reingold

Database: Compendex

Portland State UNIVERSITY

92

Application: Advanced Commercial Refrigeration

Steps 2–4: Finalize keyword list, develop & then finalize draft expert lists and diagrams

International Experts

Author	Affiliation	Selected Article Titles
HAFNER, ARMIN	SINTEF Energy Research, Norway	1. Multi-ejector concept for R-744 supermarket refrigeration 2. Simulation models in the supersmart supermarket energy-benchmark tool 3. R744 refrigeration system configurations for supermarkets in warm climates
MINETTO, SILVA	National Research Council, Construct Technologies Institute, Italy	1. Experimental analysis of a new method for overfeeding multiple evaporators in refrigeration systems 2. Simulation models in the supersmart supermarket energy-benchmark tool 3. Recent installations of CO2 supermarket refrigeration system for warm climates: data from the field

Portland State University

93

Case 2

Portland State University

94

Case Study: Lighting Roadmap

ETM

I. Building Design/Envelope Roadmaps
- I.1 Deep Retrofits for Residential and Commercial
- I.2 Retrofit and New Construction Labeling
- I.3 Solar/Smart Roofing
- I.4 Retrofit Insulation
- I.5 New Construction Insulation
- I.6 Retrofit and New Construction Air / Water Management
- I.7 Zero Net Energy Buildings
- I.8 Manufactured Housing / Modular / Pre Manufactured Systems / Offices
- I.9 Fenestration & Daylighting

II. Lighting Roadmaps
- II.1 General Lighting
- II.2 Solid State Lighting
- II.3 Task/Ambient Lighting
- II.4 Lighting Controls
- II.5 Luminaires
- II.6 Daylighting

III. Electronics Roadmaps
- III.1 Direct Current (DC) Power
- III.2 Use and Virtualization
- III.3 Component-Level Efficiency
- III.4 Complete Electronic System
- III.5 Power Management Control and Communication

IV. Heating, Ventilation, and Air Conditioning roadmaps
- IV.1 Commercial and Residential Water Heating
- IV.2 Fault Detection, Predictive Maintenance, and Controls
- IV.3 Heat Recovery and Economizer Optimization
- IV.4 Heating & Cooling Production and Delivery
- IV.5 HVAC Motor-driven Systems
- IV.6 Commercial Integrated Buildings
- IV.7 Residential HVAC
- IV.8 Modeling, Lab and Field Testing

V. Sensors, Meters, and Energy Management System Roadmaps
- V.1 Smart Device-Level Controls Responsive Environment
- V.2 Easy / Simple User Interface Controls
- V.3 Energy Management Services
- V.4 Low-Cost Savings Verification Techniqu
- V.5 Real-Time Smart Electric Power Measur facilities
- V.6 Enterprise Energy and Maintenance Ma Systems

VI. Industrial Food Processing Roadmaps
- VI.1 Heating
- VI.2 Cooling
- VI.3 Mechanical
- VI.4 Infrastructure

VII. Combined Heat and Power Roadmaps
- VII.1 Production
- VII.2 Resources
- VII.3 Delivery

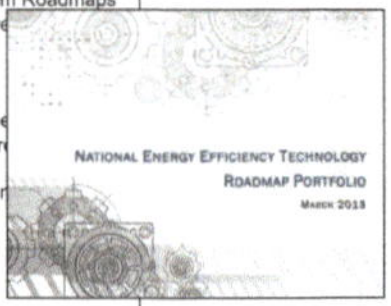

Portland State University

95

Data Sources

ETM

R&D Stage	Typical Source
Basic Research	**Science Citation Index**
Applied Research	**Engineering Index**
Development	**Patents**
Application	Newspaper Abstracts Daily
Social Impacts	Business and Popular Press

- Experts identified from three separate databases and approaches:
  - Basic Research (Citations and publications) – these experts would be ideal to know about the ongoing R&D and to be invited to the workshop focusing on identification of R&D efforts
  - Applied Research (Co-author network) – these experts would be ideal to know about the available technologies and be invited to the workshop focusing on identification of technologies
  - Development (Patent count) - these experts would be ideal to know about the available products and be invited to the workshop focusing on identification of products

Portland State University

96

Basic Research

Author Initials	Author Affiliation	Selected By
Shur, M.S	Rensselaer Polytechnic Institute	Cit
Setlur, A.A	GE Global Research	Cit
Phillips, J.M	Sandia National Lab.	Cit
Krames, M.R	Philips Lumileds Lighting Co	Cit
Ohno, Y	National Institute of Standards and Technology	Cit
Kim, J.K	Rensselaer Polytech Institute	Cit
Dai, Q.Q	Oak Ridge National Lab.	Cit
Dupuis, RD	Georgia Inst Technology	Cit

Basic Research is done at academic, government and industry labs – showing technology maturity

Portland State UNIVERSITY

97

Basic Research

Author/Initials	Article Title	Focus
Shur, MS	1. Solid-state lighting: Toward superior illumination	1. Solid-state lighting
Setlur, A.A	1. Energy-Efficient, High-Color-Rendering LED Lamps Using Oxyfluoride and Fluoride Phosphors	1. Solid-state lighting
Phillips, JM	1. Research challenges to ultra-efficient inorganic solid-state lighting	1. Solid-state lighting
Krames, MR	1. Research challenges to ultra-efficient inorganic solid-state lighting	1. Solid-state lighting
Ohno, Y	1. Research challenges to ultra-efficient inorganic solid-state lighting	1. Solid-state lighting
Kim, JK	1. Transcending the replacement paradigm of solid-state lighting	1. Solid-state lighting
Dai, QQ	1. Semiconductor-Nanocrystals-Based White Light-Emitting Diodes	1. Solid-state lighting
Dupuis, RD	**1. History, development, and applications of high-brightness visible light-emitting**	1. Solid-state lighting

Portlar

Solid State Lighting is the focus in Basic Research

98

Basic Research

ETM

- Academic Institutes
  - Rensselaer Polytechnic Institute was identified as the key academic institute leading the basic research in lighting technology area
  - Georgia Tech also came out as a critical player. In another analysis on transmission technologies, Georgia Tech came out as the major player. Further analysis showed that they have an established competence in power engineering
- Government
  - Sandia and Oak Ridge came out as the leaders among the National Labs
  - National Institute of Standards and Technology also came out as a key contributor
- Industry
  - GE Global Research and Philips Lumileds Lighting Co were listed as the industrial leaders contributing to basic research
- Topic
  - Solid State Lighting is the focus in Basic Research

Portland State UNIVERSITY

99

Applied Research

ETM

Author Initials	Author Affiliation	Selected By
Brodrick,J.	U.S. Department of Energy	Pub
Agogino,A.M.	University of California, Berkeley	Pub
Wetzel,C.	Rensselaer Polytechnic Institute	Pub
Granderson,J.	Lawrence Berkeley Nat. Lab	Deg
Narendran,N.	Rensselaer Polytechnic Institute	Pub, Deg, Bet
Andersen,M.	Massachusetts Institute of Technology	Pub, Deg, Bet
Matthews,D.H.	Carnegie Mellon University	Deg
Freyssinier,J.P.	Rensselaer Polytechnic Institute	Pub, Deg, Bet
Swensen,J.S.	Pacific Northwest National Laboratory	Deg
Yutao Zhou	Rensselaer Polytechnic Institute	Deg, Bet
Tu,J.	Philips Lighting,	Bet
Brown,J.J.	Universal Display Corporation	Bet
Nicklas,M.	Innovative Design	Bet
Meek,C.	University of Washington	Deg

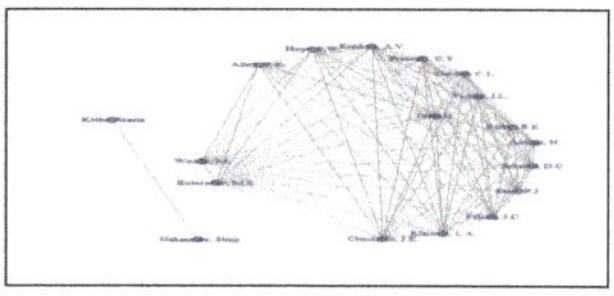

Applied Research Leaders are similar to Basic Research Leaders – National Labs, Industry and Key Universities such as RPI

Portland State UNIVERSITY

100

Applied Research

ETM

Author/Initials	Article Titles	Focus
Brodrick,J.	1. Choosing the right light 2. **Control systems** & LEED 3. Squeezing the Watts out of fluorescent lighting 4. **Solid-State Lighting,** Part 1 5. Lighting and **Standard** 90.1 6. **Inorganic LEDs**	1. General 2. Solid-state lighting
Agogino,A.M.	1. Personalized dynamic design of **networked lighting** for energy-efficiency in open-plan offices 2. Control of **wireless-networked lighting** in open-plan offices 3. **Wireless networked lighting** systems for optimizing energy savings and user satisfaction 4. Towards embedded **wireless-networked intelligent daylighting** systems for commercial buildings	1. Control 2. Daylighting
Wetzel,C.	1. The quantum efficiency of green GaInN/GaN **light emitting diodes** 2. Wavelength-stable rare earth-free green **light-emitting diodes** for energy efficiency 3. Development of high efficiency green and deep green **light emitters** in piezoelectric group-III nitrides	1. Solid-state lighting
Granderson,J.	1. Standardization of **user interfaces** for lighting controls 2. Towards embedded **wireless-networked intelligent daylighting** systems for commercial buildings	1. Control 2. Daylighting
Detchprohm,T.	1. The quantum efficiency of green GaInN/GaN **light emitting diodes** 2. Phosphor-free white 3. Wavelength-stable rare earth-free green **light-emitting diodes** for energy efficiency	1. Solid-state lighting

Portland State University

Solid State Lighting and Control are the leading topics in Applied Research

101

Applied Research

ETM

- Rensselaer was a clear leader here as well
- Other academic institutes leading in applied research were MIT, Berkeley, Carnegie Mellon and University of Washington. These did not come up in the basic research area
- National labs leading in this are LBNL and PNNL
- Phillips was one of the three industry leaders in applied research
- Key conclusion here is that while academy, government and industry are all taking part in both basic and applied research, there are very established institutes which are leaders in one or the other while Rensselaer has a key competency in both.
- Solid State Lighting and Control are the leading topics in Applied Research

Portland State University

102

DEVELOPMENT

ETM

Author/Initials	Author Affiliation	Selected By
Negley, G.H	Cree LED Lighting Solutions	Count
Morgan, F.M	Philips Solid-State Lighting Solutions, Inc.	Count
Dowling, K.J	Philips Solid-State Lighting Solutions, Inc.	Count
Mueller, G.G	Philips Solid-State Lighting Solutions, Inc.	Count
Pickard, P.K	Cree LED Lighting Solutions	Count
Lys, I.A	Philips Solid-State Lighting Solutions, Inc.	Count
Briggs, GE	Arkalumen Inc.	Count
Verfuerth, N.R	Orion Energy Systems, Inc.	Count
Gordin, M.K	Musco Corporation	Count
Rains, J.C	ABL IP Holding LLC	Count
Macadam, R.L	Lutron Electronics, Co., Inc	Count

Development leaders are clearly in the Industry

Portland State UNIVERSITY

103

DEVELOPMENT

ETM

Author/Initials	Patent Title	Focus
Negley, G.H	1. **Light emitting diode** (LED) lighting systems including low absorption, controlled reflectance enclosures 2. Lighting device which includes one or more **solid state** light emitting device 3. Lighting device with multi-chip light emitters, **solid state** light emitter support members and lighting elements 4. ...	1. Solid-state lighting
• Morgan, F.M • Dowling, K.J • Mueller, G.G • Lys, I.A	1. **Controlled** lighting methods and apparatus 2. Methods and apparatus for **controlled** lighting based on a reference gamut 3. Methods and apparatus for **controlling** devices in a networked lighting system 4. Universal lighting network methods and systems 5. Wireless lighting **control** methods and apparatus	1. Lighting Controls
Pickard, P.K	1. Active **thermal management** systems for enclosed lighting and modular lighting systems 2. Conversion kit for lighting assemblies 3. Lighting assemblies and components for lighting assemblies 4. Lighting device with multi-chip light emitters, **solid state** light emitter support members and lighting elements	1. Solid-state lighting 2. General Lighting

Solid State Lighting is the leading topic in Development as well

Portland State UNIVERSITY

104

Development

- As expected, industrial corporations are dominant in this area.
- Phillips is the key leader. Its presence in both basic and applied research as well may indicate that presence in research will lead to leadership in development. This may be an interesting hypothesis to explore further.
- Solid State Lighting seems to be of interest at all phases

References

[1] M. Pascaru, "Methodological aspects of the identification of the community leaders m. pascaru "1," pp. 10–18, 1976.

[2] C. Campbell, P. Maglio, a Cozzi, and B. Dom, "Expertise identification using email communications," Cikm 2003, no. January, pp. 528–531, 2003.

[3] G. Rueda, P. Gerdsri, and D. F. Kocaoglu, "Bibliometrics and social network analysis of the nanotechnology field," Portl. Int. Conf. Manag. Eng. Technol., pp. 2905–2911, 2007.

[4] The R Foundation, "The R Project for Statistical Computing." [Online]. Available: https://www.r-project.org/.

[5] J. M. Bastian M., Heymann S., "Gephi," 2009. [Online]. Available: http://gephi.github.io/.

[6] I. McCulloh, H. Armstrong, and A. Johnson, *Social Network Analysis with Applications.* John Wiley & Sons, 2013.

[7] M. Coscia, F. Giannotti, and R. Pensa, "Social Network Analysis as Knowledge Discovery Process: A Case Study on Digital Bibliography," in *2009 International Conference on Advances in Social Network Analysis and Mining,* 2009, pp. 279–283.

[8] G. Rueda, P. Gerdsri, and D. F. Kocaoglu, "Bibliometrics and social network analysis of the nanotechnology field," *Portl. Int. Conf. Manag. Eng. Technol.*, pp. 2905–2911, 2007.

[9] I. McCulloh, H. Armstrong, and A. Johnson, *Social Network Analysis with Applications.* John Wiley & Sons, 2013.

[10] H. Jafarzadeh and A. Aurum, "A Systematic Review on Search Engine Advertising," *Pacific Asia J. Assoc. Inf. Syst.*, vol. 7, no. 3, pp. 1–32, 2015.

[11] Beliga, Slobodan, "Keyword extraction: a review of methods and approaches." Croatia.

[12] J. P. Martino, "*A Review of Selected Recent Advances in Technological Forecasting*," Technol. Forecast. Soc. Change, vol. 70, no. 8, pp. 719–733, Oct. 2003.

References

- Amer M, Daim T, "Application of Technology Roadmaps for Renewable Energy Sector" Technological Forecasting and Social Change, Vol 77, No 8, 2010, pp 1355-1370
- Daim T, Oliver T, "Implementing Technology Roadmap Process in the Energy Services Sector: A Case Study of a Government Agency", Technology Forecasting and Social Change, Vol 75, No 5, June 2008, pp 687-720
- Fenwick D, Daim T, Gerdsri N "Value Driven Technology Road Mapping (VTRM) Process Integrating Decision Making and Marketing Tools: Case of Internet Security Technologies" Technological Forecasting and Social Change, Vol 76, No 8, 2009, pp 1055-1077
- McCulloh I, Armstrong H, Johnson A, Social Network Analysis with Application, 2013, John Wiley
- Ozcan S, Islam N, Collaborative networks and technology clusters — The case of nanowire, Technological Forecasting and Social Change, Volume 82, February 2014, Pages 115-131
- Phaal, R., Technology roadmapping - A planning framework for evolution and revolution. Technological Forecasting and Social Change, 2004. 71(1-2): p. 5-26.
- Senghore F, Campos-Nanez E, Fomin P, Wasek J S, Using Social Network Analysis to Investigate the Potential of Innovation Networks: Lessons Learned from NASA's International Space Apps Challenge, Procedia Computer Science, Volume 28, 2014, Pages 380-388
- Willyard, C.H. and McClees, C.W., "Motorola's technology roadmapping process," Research Management, Sept.-Oct., pp. 13-19, 1987.

Appendix 15

Luminaire Level Lighting Controls

Luminaire Level Lighting Controls

- "Embedding lighting controls into luminaires reduces costs and simplifies installation and commissioning compared with traditional controls that are separate from the luminaires. And, luminaire-level controls allow each luminaire to function independently – allowing for individual controllability and the possibility of increased energy savings".

- [NEEA - http://neea.org/initiatives/commercial/luminaire-level-lighting-controls]

Objectives

- **(First Stage)To identify competing technologies with automatic luminaire controls**

- **(Next Stage)**To forecast LLLC technologies. Find the patterns of the technologies associated with automatic luminaire controls.

T. U. Daim (ed.), *Roadmapping Future*, Applied Innovation and Technology Management, https://doi.org/10.1007/978-3-030-50502-8

Methodology

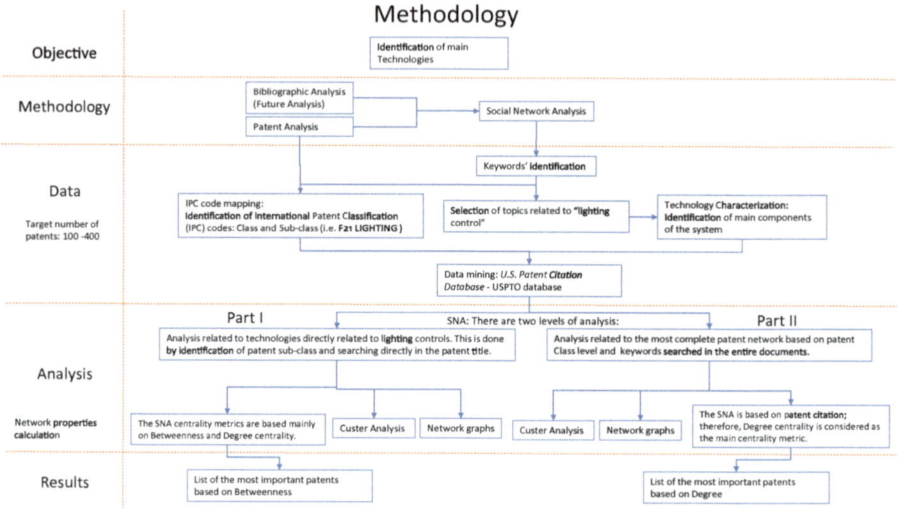

Keyword Identification

Initial Keywords

Original Keywords – Provided by SMEs

- Wireless **lighting control**
  - Add: energy efficiency
  - Add: LED
- Luminaire mesh network
  - **Add: Led lighting**
- Programmable **lighting control** system
- Mesh network of intelligent devices
  - **Add: lighting**
- **Lighting** zone control
  - Add: wireless mesh network
- Programmable **lighting control** system
  - Add: wireless network
  - Add: led
  - Add: energy efficiency
  - Add: dimming
- Wireless **lighting control** wide area network
  - Add: ipv6
- Wireless **lighting control**
  - Add: 6lowpan
- 802.15.4 **lighting** mesh network
  - Add: personal
  - Add: Wifi
- **Wireless lighting** control
  - **Add: energy reporting**
  - Add: POE

Different Options to combine keywords (This can be used for future analyses)

Options	Keywords
Option 1	Wireless **lighting** control OR Luminaire mesh network OR Programmable **lighting** control system OR **Lighting zone** control OR Programmable **lighting** control system OR Mesh network intelligent devices OR Wireless **lighting** control wide area network AND energy efficiency OR LED OR **Led lighting** OR **lighting** OR wireless network OR dimming OR ipv6 OR 6lowpan OR personal OR Wifi OR **energy reporting** OR POE
Option 2	Wireless **lighting** OR Programmable **lighting** OR Luminaire mesh network OR **Lighting zone** OR Mesh network intelligent device AND control wide area network OR control OR system
Option 3	TITLE (**lighting** OR luminaire OR "intelligent device") AND TITLE (wireless OR programmable OR zone OR "mesh network" or wire) AND TITLE ("control wide area network" OR control OR system or LED or dimming **or sensor or software** or lamp) sensor OR Thermal camera OR Micro radar
Optional	energy efficiency OR LED OR **Led lighting** OR **lighting** OR wireless network OR dimming OR ipv6 OR 6lowpan OR personal OR Wifi OR **energy reporting** OR POE OR Occupancy OR Infrared OR light-level OR LED OR Bluetooth OR Voltage **regulation** OR Wifi OR Energy metering OR Networking OR Algorithm OR Trim OR Ultrasonic OR **Motion-detecting** OR **Heating-sensing** OR Sound-sensing OR **optical** camera OR infrared motion OR **Optical** trip wire OR Door contact

Final Configuration of Keywords

Objective	Keywords	Place	Operator
Direct Competing Products / Technologies / Systems	**"Lighting Control"**	Title	AND
	Energy efficiency	Full Text	AND
	h05b33 OR h05b37	CPC Main Group (IPC-R8)	
	vehicle OR camera OR tv OR car OR street	Full Text	NOT
Competing Products / Technologies / Systems / Components / Parts / Elements	**lighting** OR luminaire OR "intelligent device"	Title	AND
	wireless OR programmable OR zone OR "mesh network" or wire	Title	AND
	"control wide area network" OR control OR system or LED OR dimming OR sensor OR **software** OR lamp	Abs + Claim	AND
	H05B OR H01L OR F21V OR G02F OR G01J	CPC Main Group (IPC-R8)	
	vehicle OR camera OR tv OR car OR street	Abs + Claim	NOT

Technology Characterization

Lighting Control Techniques

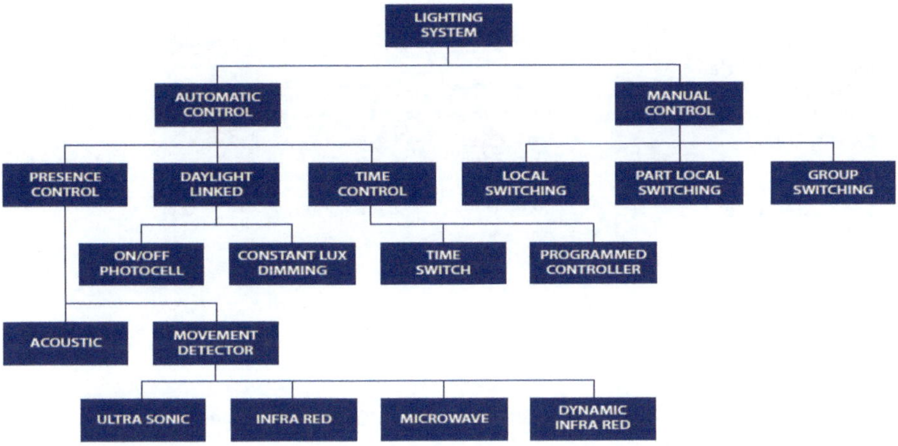

[1] C. Ehrlich, K. Papamichael, J. Lai, and K. Revzan, "Lighting Controls," *Energy*, vol. 1444, pp. 1–7, 2002.

Lighting_Controls.26792.shortcut.pdf

Lighting Control: Components – Functions

Lamps	Software	Localized Processing / Distributed Inteligence	Commitioning Tool	Switch	Connection/ Interface	Light Fixtures	Control	Dimming	sensor	AREA / zoning
					wireless radio		Occupancy Control	Dimming ballast	Occupancy Sensors	Office
					infrared signals.		light-level Control	LED drivers	light-level sensor	Retail
					Bluetooth		Daylight Control	Voltage regulation	Daylight Sensor	K-12
					Wifi					University
							Robust Control Algorithms	Task Tuning w/ High-end Trim	user tuning	Warehouse
							Personal Lighting Controls		air temperature	Grocery
							Standarized wireless control		Ultrasonic	Restaurant
							Institutional Tuning		Motion-detecting (microwave),	Hotel
									Heating-sensing (infrared),	Hospital
									Sound-sensing	Assembly
									optical cameras	Residential Care
									Infrared motion	Other
									Optical trip wires	
									Door contact sensors	
									Thermal cameras	
									Micro radars	

[1] C. Ehrlich, K. Papamichael, J. Lai, and K. Revzan, "Lighting Controls," *Energy*, vol. 1444, pp. 1–7, 2002.
[2] L. Nock, J. Wilson, L. C. Radetsky, and T. Lighting, "Easily Commissioned Lighting Controls Phase 2 Report," pp. 1–45, 2015.
[3] J. Phelan, "Lighting Control Technology for Digital and Dynamic Luminaires Learning Objectives :," 2014.

IPC - CPC Code Mapping

Identification of Patent Section / Class / Sub-Class

- The data were mining by the title of "lighting control": sample patents 1000.
- The cluster were obtained based on the class

IPC-R | US CLS | cls. exp

Section ▾ | None ▾ | CLEAR

cl.	cl.1	cl.2	cl.3	cl.4	cl.5	cl.6	cl.7	cl.8	cl.9	t▾
H	165	11	14	40	1				5	236
G	80	7		2	5					94
F	38	3						6		47
B	14			2	1	1	1	1		20
E	4				2	2	4	2		14
A	3			1						4
D	1									1
C	1									1

Patent Section

The identified Sections related to "lighting control" are: H, F, G

Identification of Patent Class and Sub Class

- The data were mined by the title "lighting control" and considered as data sample:
- Patent Sample = 1000.
- Cluster were obtained based on the class and sub-class classification.
- Based on the 1000 patents, the classes and sub-classes were reviewed to select the ones directly associated with "lighting control"

Patent Network – 100 patents
– Colors group the nodes according keywords (not relevant for the objective of identifying patent classes, but it is an indicator of clear cluster patterns)

Class ▾ | None ▾ | CLEAR

cl.	cl.	cl.	cl.	cl.	cl.	cl	cl.	cl	cl.	cl.	total
H05	43	1...	1...	13	58	92	2...	5	22	14	804
F21	11	10	23	31	32	3	9	8			127
H04	14	8	38	2	2	1	38	3			106
G09	2	5	35	2	2	4	10	9	21		90
G06	1	2	27		3	4	44	4	3		88
G05	9	15	8		5	9	34				80
G02		3	11	15	1		2	39			71
H01	2	4	11	13	9	4	6	21	1		71
B60	4	5	19	1	23	3	4	2	2		63

Patent Class

Sub Class ▾ | None ▾ | CLEAR

cl.	cl.	cl.	cl.	cl.	cl.	cl	cl.	cl	cl.	cl.	total
H05B	43	1...	1...	13	58	92	2...	5	22	14	799
F21V	9	9	17	28	27	2	7	8			107
G09G	2	4	33	1	2	4	9	7	21		83
G06F	1	2	16		3	4	37	4	3		70
B60Q	4	5	16	1	22	3	3	1	1		56
H04N	3	3	30	2			10	3			51
G05B	9	3	5		4	1	24				46
G02B		1	5	12	1		2	24			45
C02F		3	6	5			2	19			35

Patent Sub-class

Cooperative Patent Classification for “Lighting Control”

Section		Class		Subclass		
Symbol	Title	Symbol	Title	Symbol	Title	Description
H	Electricity	H05	Electric Techniques Not Otherwise	H05B	ELECTRIC HEATING; ELECTRIC LIGHTING NOT OTHERWISE PROVIDED FOR	This subclass covers: Electric heating: Heat sources utilising ohmic resistance, electric, magnetic or electromagnetic fields, electric discharge, or combinations thereof; Light sources specially adapted for heating, e.g. infrared sources as used in light ovens. Not only the electric elements and circuitry designs are covered by this subclass but also the electric aspects of their arrangement, where these concern cases of general application. Where heating elements are used in specific applications, please see the relevant subclasses. Electric lighting: Electric arc lamps, electroluminescent light sources and light sources using a combination of different types of light generation; Circuit arrangements for electric light sources. The primary circuits of the above light sources and the elements themselves of the circuits are covered by this subclass. The above light sources when combined with light sources of a different kind are also covered by this subclass
		H01	Basic Electric Elements	H01L	SEMICONDUCTOR DEVICES; ELECTRIC SOLID STATE DEVICES NOT OTHERWISE PROVIDED FOR	This subclass covers: In general discrete and integrated semiconductor devices and other electric solid state devices (as far as not provided for in another subclass) and details thereof. This includes the following kind of devices: Integrated circuit devices, e.g. CMOS integrated devices, DRAM, EPROM, CCD; semiconductor devices (e.g. field-effect, bipolar) adapted for rectifying, amplifying, oscillating or switching, e.g. diodes, transistors, thyristors; semiconductor devices sensitive to radiation, e.g. photo diodes, photo transistors, solar cells; incoherent light emitting diodes, e.g. LED; solid state devices using organic materials as the active part or using a combination of organic materials with other materials as the active part, e.g. organic LED or polymer LED; electric solid state devices using thermoelectric, superconductive, piezo-electric, electrostrictive, magnetostrictive, galvano-magnetic or bulk negative resistance effects, e.g. thermo couples, Peltier elements, Josephson elements, piezo elements; photo-resistors, magnetic field dependent resistors, field effect resistors; capacitors with potential-jump barrier, resistors with potential-jump barrier or surface barrier; thin-film or thick-film circuits; processes and apparatus adapted for the manufacture or treatment of such devices, except where such processes relate to single step processes for which provision exists elsewhere.
F	Mechanical Engineering; Lighting; Heating; weapons; Blasting	F21	Lighting; Heating	F21V	DETAILS OF LIGHTING DEVICES, OF GENERAL APPLICATION	This subclass covers: Details of parts involved in light emission or distribution Details of those parts not involved in light emission or distribution, e.g. fittings Special adaptation related to the use of a non-electric light sources
G	Physics	G02	Optics	G02F	DEVICES OR ARRANGEMENTS, THE OPTICAL OPERATION OF WHICH IS MODIFIED BY CHANGING THE OPTICAL PROPERTIES OF THE MEDIUM OF THE DEVICES OR ARRANGEMENTS FOR THE CONTROL OF THE INTENSITY, COLOUR, PHASE, POLARISATION OR DIRECTION OF LIGHT, e.g. SWITCHING, GATING, MODULATING OR DEMODULATING; TECHNIQUES OR PROCEDURES FOR THE OPERATION THEREOF; FREQUENCY-CHANGING; NON-LINEAR OPTICS; OPTICAL LOGIC ELEMENTS; OPTICAL ANALOGUE/DIGITAL CONVERTERS	This subclass covers: Devices, the optical operation of which is modified by changing the optical properties (refraction, birefringence, absorption, nonlinear susceptibility) of the medium of the devices. The term "optical" applies not only to visible light but also to ultra-violet, infra-red radiations or Terahertz (G02F 1/3534). The following optical elements are therefore covered, the list being not exhaustive: thermo-optic elements. electro-optic elements. magneto-optic elements. elasto-optic elements. acousto-optic elements. liquid crystal devices. electrochromic elements. electrophoretic elements. non-linear optics, i.e. devices or arrangements in which the electric or magnetic field component of the light beam influences the optical properties of the medium. Demodulating light: Transferring the modulation of modulated light, i.e. transferring the information from one optical carrier of a first wavelength to a second optical carrier of a second wavelength, insofar these demodulators are based in substantial manner on elements which are provided for under the bullets above. Optical logic elements: Optical bistable devices, i.e. devices exhibiting two different optical output states for a same optical input value, Optical logic elements, insofar these demodulators are based in substantial manner on elements which are provided for under the bullets above. Optical analogue/digital converters: Optical bistable devices, i.e. devices exhibiting two different optical output states for a same optical input value, Optical logic elements, insofar these devices are based in substantial manner on elements which are provided for under the bullets above.
		G01	Instruments	G01J	MEASUREMENT OF INTENSITY, VELOCITY, SPECTRAL CONTENT, POLARISATION, PHASE OR PULSE CHARACTERISTICS OF INFRA-RED, VISIBLE OR ULTRA-VIOLET LIGHT; COLORIMETRY; RADIATION PYROMETRY	This subclass covers: Apparatus or methods for measuring properties of infra-red, visible, or ultra-violet light, for the purpose of: Photometry not otherwise provided for (G01J 1/00) Spectrometers and spectroscopy per se : measurement of the spectral content of incident light and spectroscopic methods used in this process (G01J 3/00) Colorimetry per se (G01J 3/00) Measurement of temperature by optical means (remote detection of infrared radiation, non-contact) (G01J 5/00) Polarimeters and polarimetry per se : passive measurement of the state of polarisation of incident light (G01J 4/00) Measurement of the velocity of light (G01J 7/00) Measurement of optical phase differences : wavefront and coherence measurements (G01J 9/00) Measurement of the wavelength of incident light (G01J 9/00) Measurement of the characteristics of ultrashort pulses (G01J 11/00)

Patent classification in the Class level - this is used for the complete patent citation analysis

Cooperative Patent Classification for Direct Related Lighting Control Technologies

Section		Class		Subclass		Group		
Symbol	Title	Symbol	Title	Symbol	Title	Symbol	Title	Description
H	Electricity	H05	Electric Techniques Not Otherwise	H05B	ELECTRIC HEATING; ELECTRIC LIGHTING NOT OTHERWISE PROVIDED FOR	H05B33	Electroluminescent light sources	This group covers: Elements, compositions, chemical compounds capable of emitting light and circuit arrangements for driving said sources.
						H05B37	Circuit arrangements for electric light sources in general	This group covers: Application circuits for control of light sources. Where the focus of the control is in the application and not in the light source, controlling the light source when some external condition occurs (light dependent, occupancy, data communication, etc).

The purpose of the identification of Patent sub-class level is to use as keywords for direct related technologies

Analysis / Results

Part I

Direct Related Lighting Control Technologies

Part I

Direct Related Lighting Control Technologies

- Criteria:
  - The search for patents correspond to one that is focused on the title. Therefore, the found patents can be considered as direct competing technologies in the field of "lighting control".
  - In order to identify the most important patents, the number of cited patents has not been considered as a single indicator of importance, since the older patents tend to have higher citation and newest patents have low citation rates even they could be important.
- In order to analyze the data, the following analysis has been perform:
  - The complete list of patents is provided in the next couple slides.
  - Identification and categorization of patents according TIME.
  - SNA citation analysis has been performed. The SNA is focused on the centrality measures of Degree and Betweenness.
  - A Cluster analysis is performed in order to identify and verify the most important patents and the connections among groups. This cluster analysis is based on density analysis directly linked to Degree and uses the Clauset-Newman-Moore algorithm that utilized betweenness levels as a part of the algorithm.
  - The final list is gotten according the Betweenness levels, since this centrality measure indicates the importance of conection with diferent groups in the network.

Data Query

Data 1: Direct Related Competing Technologies

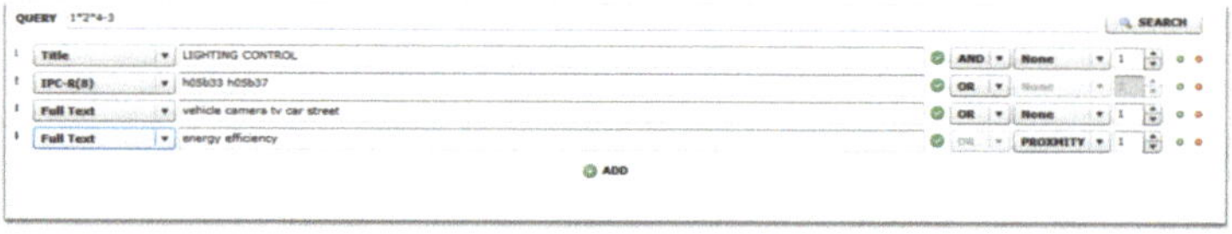

Results: 113 patents

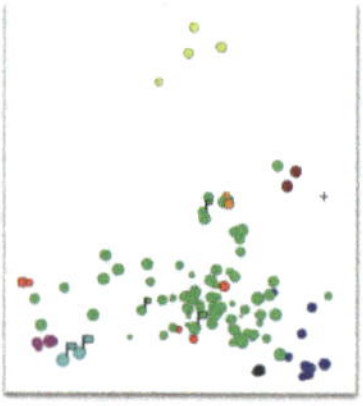

-	-		-	KEYWRD
1	☑		5	algorithm, light, communicate, connect, power, communication, correspond, converter, base, sensor, ballast, transceiver, plurality, signal, control
2	☑		81	associate, circuit, response, source, generate, intensity, frequency, detect, output, set, dimmer, supply, component, path, value
3	☑		9	produce, magnitude, cycle, light-emitting, converter, circuitry, end, threshold, dimmer, input, generate, supply, set, plurality, intensity
4	☑		4	target, store, command, illumination, color, sense, communication, sensor, communicate, transmit, produce, adjust, light-emitting, base, group
5	☑		4	building, setting, regulator, area, LED, state, fixture, response, transmit, output, housing, sensor, controller, receive, light
6	☑		2	element, generate, intensity, communication, LED, fixture, plurality, controller, base, light, lighting, control, glare, shield, angle
7	☑		2	setting, target, period, set, value, module, light, lighting, control, rendering, classify, render, compute, atmosphere, index
8	☑		2	temperature, number, pulse, brightness, group, value, regulator, color, output, plurality, base, light, source, control, bit
9	☑		2	activate, resistor, transistor, path, driver, current, series, threshold, LED, turn, connect, couple, circuit, voltage, base

Graph Metric
Graph Type
Vertices
Unique Edges
Edges With Duplicates
Total Edges

List of Patents

Doc No.	Filing Dt	Title
USA2016/0014872	2015-09-22	Method and Apparatus for Controlling Light Levels to Save Energy
USP9316379	2015-08-26	Landscape lighting systems having light diverting elements and light intensity controllers
USA2015/0338118	2015-07-31	INTEGRATION OF LED LIGHTING WITH BUILDING CONTROLS
USA2015/0327339	2015-07-16	LED (LIGHT-EMITTING DIODE) STRING DERIVED CONTROLLER POWER SUPPLY
USA2015/0353004	2015-06-04	LIGHTING APPARATUS COMPRISING A CONTROL DEVICE AND AIRCRAFT COMPRISING THE LIGHTING APPARATUS
USP9140414	2015-06-02	LED directional lighting system with light intensity controller
USP9307598	2015-03-17	LED lighting device with replaceable driver-control module
USA2015/0237699	2015-02-17	LIGHTING SYSTEM AND CONTROL CIRCUIT FOR THE SAME
USP9247603	2015-02-11	Shunt regulator for spectral shift controlled light source
USA2015/0145431	2014-11-26	VOLTAGE-CONTROLLED DIMMING OF LED-BASED LIGHT MODULES COUPLED IN PARALLEL TO A POWER SUPPLY
USA2015/0035450	2014-07-24	SOLID STATE LIGHTING CONTROL
USP9185758	2014-07-15	Controlling current flowing through LEDs in a LED light fixture
USP9326356	2014-07-03	Method and apparatus for determining a target light intensity from a phase-control signal
USA2014/0239833	2014-05-08	Power Conversion and Control Systems and Methods for Solid-State Lighting
USA2016/0029459	2014-04-16	LIGHTING CONTROL SYSTEM AND LIGHTING CONTROL METHOD
USA2014/0300284	2014-04-03	COLOR TUNABLE LIGHT SOURCE MODULE WITH BRIGHTNESS AND DIMMING CONTROL
USA2014/0300283	2014-04-03	COLOR TUNABLE LIGHT SOURCE MODULE WITH BRIGHTNESS CONTROL
USP9161401	2014-03-20	LED (light-emitting diode) string derived controller power supply
USP9131581	2014-03-14	Solid-state lighting control with dimmability and color temperature tunability
USA2014/0225528	2014-03-14	Devices, Systems, Architectures, and Methods for Lighting and other Building Control applications
USP9215780	2014-03-03	System and method for reducing peak and off-peak electricity demand by monitoring, controlling and metering lighting in a facility
USP9089031	2014-02-10	Add-on smart controller for LED lighting device
USA2015/0351198	2014-01-03	LIGHTING CONTROL ANALYZER
USP8941307	2013-11-27	Light control system using statement of personal computer which is applied to a building for energy saving and method thereof
USP9101026	2013-10-28	Integration of LED lighting with building controls
USA2015/0234369	2013-09-10	UNIFIED CONTROLLER FOR INTEGRATED LIGHTING, SHADING AND THERMOSTAT CONTROL
USA2015/0271899	2013-09-06	CONTROLLABLE LIGHTING DEVICES
USP9144139	2013-08-27	Method and apparatus for controlling light levels to save energy
USA2014/0063857	2013-08-23	METHOD AND APPARATUS FOR CONTROLLING A LIGHTING DEVICE
USP9013467	2013-07-19	Controlled operation of a LED lighting system at a target output color
USP8842009	2013-06-07	Multiple light sensor multiple light fixture control
USA2013/0261768	2013-05-22	Light control system and method for automatically rendering a lighting atmosphere
USP9220136	2013-05-21	Method and apparatus for controlling a lighting device
USP9301374	2013-05-10	Control of lighting devices
USA2013/0270998	2013-04-17	SOLID-STATE LIGHTING SYSTEMS HAVING INTELLIGENT CONTROLS
USP9265105	2013-04-10	Power conversion and control systems and methods for solid-state lighting
USP9326339	2013-03-27	LED lighting device with control means and method for operating the LED lighting device
USP9143697	2013-03-21	Lighting and control systems and methods
USA2014/0265877	2013-03-15	LIGHTING CONTROL DEVICE AND METHOD
USA2013/0200809	2013-03-13	Integrated Ambient and Task Lighting Control
USA2015/0189712	2013-02-20	LED LIGHTING ARRANGEMENT AND METHOD OF CONTROLLING A LED LIGHTING ARRANGEMENT
USP8926121	2013-02-15	Portable light with spectrum control means
USP9220153	2013-02-04	Illuminating apparatus and control method for keeping a constant total brightness of ambient light and light produced by the illuminating apparatus
USP8922125	2013-01-22	Occupancy sensor for controlling an LED light
USP8981655	2013-01-04	Power conversion and control systems and methods for solid-state lighting
USA2013/0229119	2012-10-30	Dimmable Solid State Lighting System, Apparatus and Method, with Distributed Control and Intelligent Remote Control
USP8692479	2012-10-24	Method of controlling a ballast, a ballast, a lighting controller, and a digital signal processor

List of Patents

Doc No.	Filing Dt	Title
USP8787765	2012-10-19	Methods for communication between a lighting node and a controller
USA2013/0033183	2012-10-10	SYSTEM AND METHOD FOR CONTROLLING LIGHTING
USA2014/0320022	2012-10-09	LED LIGHTING CONTROL SYSTEM
USP9000406	2012-09-27	Method and apparatus of drive currents control over a solid state light source
USP8594505	2012-09-26	Lighting and control systems and methods
USP8816604	2012-08-03	Dimming control method and apparatus for LED light source
USP8604713	2012-07-30	Method, apparatus and computer-readable media for controlling lighting devices
USP8571716	2012-06-08	Integration of LED lighting with building controls
USP8803432	2012-05-04	Method and apparatus for determining a target light intensity from a phase-control signal
USP9215779	2012-03-30	Light switch and control device having a touch screen interface
USP9119263	2012-03-19	Multi-zone light controller
USP8872432	2012-03-15	Solution for dynamic lighting control
USA2013/0241420	2012-03-15	DYNAMIC LIGHTING CONTROL
USP9057491	2012-03-14	Network controlled interior lighting system
USP8820867	2012-02-20	Control circuit for light emitting diode indicator
USP8779675	2011-12-16	Controlling current flowing through LEDs in a LED lighting fixture
USP8653750	2011-11-15	Method of controlling an electronic ballast, an electronic ballast and a lighting controller
USA2013/0113366	2011-11-07	COLOR CONTROL OF SOLID STATE LIGHT SOURCES
USA2012/0112638	2011-11-04	Thermal Management In A Lighting System Using Multiple, Controlled Power Dissipation Circuits
USA2014/0292223	2011-08-31	LIGHTING APPARATUS AND LIGHT EMISSION CONTROL METHOD
USP8587212	2011-05-27	Lighting system, dimming control apparatus and dimming control method
USA2012/0299486	2011-02-07	Light level control for building illumination
USA2012/0268027	2010-11-23	ENERGY SAVING LED HAVING DIMMING FUNCTION AND MOOD-LIGHTING CONTROL FUNCTION
USP8446099	2010-10-04	Power conversion and control systems and methods for solid-state lighting
USP8760078	2010-10-04	Power conversion and control systems and methods for solid-state lighting
USP8587225	2010-09-02	Floor plan deduction using lighting control and sensing
USP8110996	2010-08-13	Modular wireless lighting control system using a common ballast control interface
USP9078306	2010-06-18	Lighting control system
USP8492987	2010-06-11	Load control device for a light-emitting diode light source
USP8492988	2010-06-11	Configurable load control device for light-emitting diode light sources
USA2010/0289428	2010-05-11	Controllable Retroffited LED Panel Lighting
USP8633653	2010-03-02	Lighting control system with improved efficiency

Number of Patents/Year for Direct Related Technologies

Year	Number of Patents (Filing Date)
1981	1
1982	1
1991	1
1992	1
1994	2
1995	1
1996	1
1997	1
1999	1
2000	1
2002	1
2003	1
2004	2
2005	1
2006	3
2007	1
2008	6
2009	6
2010	10
2011	7
2012	17
2013	22
2014	14
2015	9
Grand Total	111

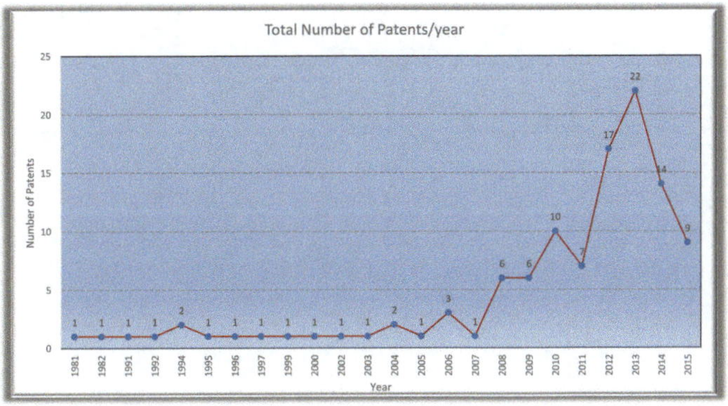

The number of patents has significatively increased since 2008. The filling date was taking as reference. The publication date is usually 2 to 3 years from the application date.

Patent Citation Network

Graph Metric	Value
Graph Type	Undirected
Vertices	961
Unique Edges	1171
Edges With Duplicates	0
Total Edges	1171

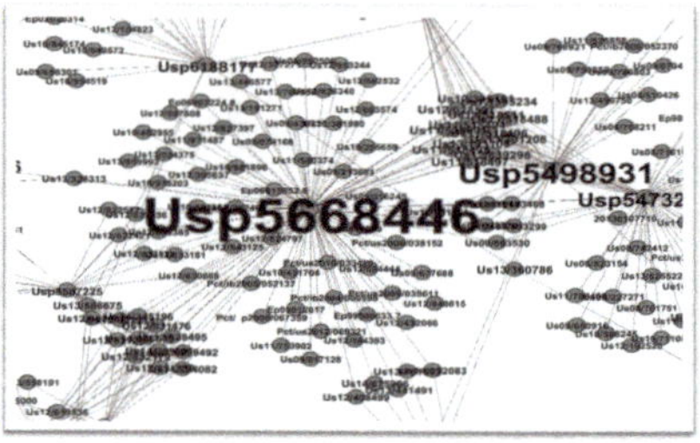

Network layout: Fuchterman Reingold algorithm

The figure shows how patents are linked each other based on their citations. The initial number of patents was 111. The cited patent database is 961 cited patents (vertex – nodes) , corresponding 1171 connections (edges).

Cluster Analysis

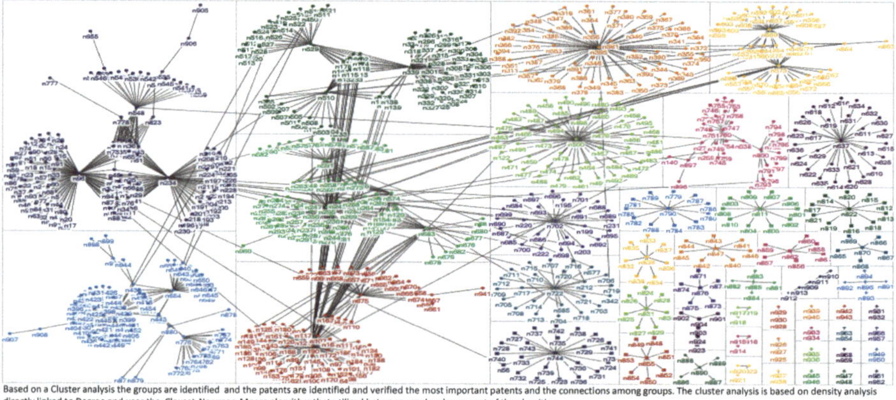

Based on a Cluster analysis the groups are identified and the patents are identified and verified the most important patents and the connections among groups. The cluster analysis is based on density analysis directly linked to Degree and uses the Clauset-Newman-Moore algorithm that utilized betweenness levels as a part of the algorithm.

There were identified 46 Clusters

10 clusters are considered with highest sub-networs and connections.

6 clusters are considered with middle number on connections.

NOTE: Considering future comments, a more dtailed analysis can be performed: like "connections among clusters" to see for example how certain (the most important) technologies play an important roles in other type or cluster of technologies.

List of the Top Most Important Patents Based on Betweenness

			Centrality Metrics				Patent Information				
Network Vertex	Patent Number / Appl. Number	Title	Degree	Betweenness Centrality	Eigenvector Centrality	Comment	Pub Dt	Filing Dt	Assignee,Patentee	Inventor	Information Link
N190*	Usp5668446	Energy management control system for fluorescent lighting	95	145803.673	0.001	Title,Cit, Deg, Bet	9/16/1997	9/23/1996	NEGAWATT TECHNOLOGIES INC	BAKER JAMES ALLAN	http://patentsearch.jp/?no=US5668446
N529*	Usp7075254	Lighting ballast having boost converter with on/off control and method of ballast operation	33	81765.833	0.000	Title,Cit, Deg, Bet	7/11/2006	12/14/2004	LUTRON ELECTRONICS CO INC	CHITTA VENKATESH | GAWRYS BRENT M | CHITTA VENKATESH | GAWRYS BRENT M	http://patentsearch.jp/?no=US7075254
N96*	Usp6683419	Electrical control for an LED light source, including dimming control	96	81456.632	0.009	Title,Cit, Deg, Bet	1/27/2004	6/24/2002	DIALIGHT CORPORATION	KRIPAROS DANIEL J | KRIPAROS DANIEL J	http://patentsearch.jp/?no=US6683419
N510*	Usp5498931	Method for automatic switching and control of lighting	33	71731.581	0.000	Title,Cit, Deg, Bet	3/12/1996	10/3/1994	TLG PLC	BEDOCS LOU	http://patentsearch.jp/?no=US5498931
N234*	Usp5581158	Lamp brightness control circuit with ambient light compensation	83	65974.409	0.008	Title,Cit, Deg, Bet	12/3/1996	4/3/1995	ETTA INDUSTRIES INC	QUAZI FAZLE	http://patentsearch.jp/?no=US5581158
N295*	Usp8344665	System and method for controlling lighting	71	63396.145	0.000	Title,Cit, Deg, Bet	1/1/2013	9/29/2008	ORION ENERGY SYSTEMS INC	VERFUERTH NEAL R | POTTS MICHAEL J | WANG JUN | BARTOL ANTHONY J | VERFUERTH NEAL R | POTTS MICHAEL J | WANG JUN | BARTOL ANTHONY J	http://patentsearch.jp/?no=US8344665
N455*	Usp8492987	Load control device for a light-emitting diode light source	59	49118.000	0.000	Title,Cit, Deg, Bet	7/23/2013	6/11/2010	LUTRON ELECTRONICS CO INC	NUHFER MATTHEW W | SHEARER THOMAS M | CHITTA VENKATESH | IYER KARTIK | NUHFER MATTHEW W | SHEARER THOMAS M | CHITTA VENKATESH | IYER KARTIK	http://patentsearch.jp/?no=US8492987
N395*	Usp5293097	Fully automatic energy efficient lighting control and method of making same	59	45263.604	0.000	Title,Cit, Deg, Bet	3/8/1994	11/27/1991	NOVITAS INC	ELWELL BRIAN E	http://patentsearch.jp/?no=US5293097
N339*	Usp5473202	Control unit for occupancy sensor switching of high efficiency lighting	61	39124.490	0.000	Title,Cit, Deg, Bet	12/5/1995	6/5/1992	PLATNER BRIAN	MUDGE PHILIP H | PLATNER BRIAN	http://patentsearch.jp/?no=US5473202
n416**	Us13/212556 US 20120043900	Method and Apparatus for Measuring Operating Characteristics in a Load Control Device	3	34317.818	0.000	Cit, Deg, Bet	2/23/2012	8/18/2011	Lutron Electronics Co., Inc.	Venkatesh Chitta, Jonathan Robert Quayle, Alexander J. Rovnan, Mark S. Taipale, Dragan Veskovic	http://www.google.st/patents/US20120043900
N450**	Us13/212773 US 8593076	Electronic dimming ballast having advanced boost converter control	3	34317.818	0.000	Cit, Deg, Bet	11/26/2013	8/18/2011	Lutron Electronics Co., Inc.	Dragan Veskovic	https://www.google.st/patents/US8593076?dq=Us+13/212773&hl=pt-PT&sa=X&ved=0ahUKEwiY1dTgmyDMAhUGHGMKHRm6D5kQ6AEIGjAA
N500*	Usp4434388	Electrical lighting controller	45	33330.000	0.000	Title,Cit, Deg, Bet	2/28/1984	9/3/1981	N/A	CARVER LEROY J | PENROD JOHN K	http://patentsearch.jp/?no=US4434388
N122**	Us11/536356 US 7679222	Power and signal distribution system for use in interior building spaces	2	33075.000	0.001	Cit, Deg, Bet	3/16/2010	9/28/2006	Worthington Armstrong Venture	Inventores Brian T. Patterson, Sean D. Browne, William E. Beakes, Sandor A. Frecska, Jere W. Myers	https://www.google.st/patents/US7679222?dq=Us+11/536356&hl=pt-PT&sa=X&ved=0ahUKEwiEI-OFnPDMAhVUSWMKHTWICpUQ6AEIGjAA
N89**	Us12/694353 US 7986103	Method for driving a lamp in a lighting system based on a goal energizing level of the lamp and a control apparatus therefor	2	29805.828	0.005	Cit, Deg, Bet	7/26/2011	1/27/2010	Koninklijke Philips Electronics N.V.	Renatus Bernardinus Maria Geerts, Jacob Stegeman	https://www.google.st/patents/US7986103?dq=Us+12/694353&hl=pt-PT&sa=X&ved=0ahUKEwjHgeOcnfDMAhVkImMKHUKLBSUQ6AEIGjAA
N58**	Us11/570944 US 7667409	Method for driving a lamp in a lighting system based on a goal energizing level of the lamp and a control apparatus therefor	2	29805.828	0.005	Cit, Deg, Bet	2/23/2010	6/28/2005	Koninklijke Philips Electronics, N.V.	Renatus Bernardinus Maria Geerts, Jacobu Stegeman	https://www.google.st/patents/US7667409?dq=Us+11/570944&hl=pt-PT&sa=X&ved=0ahUKEwjirBSOnvDMAhUBR2MKHVBwB7MQ6AEIGjAA

* Patents found directly by the search and has high cited levels and Betweenness and Degree

** Patents found in the citation analysis with highest levels of Betweenness and Degree.

Part II

Related Technologies / components – Citation Analysis

Data Query

Data 2: Related Technologies / components

Results: 327 patents

QUERY 1*2*3*4-5 SEARCH

1	Title	lighting luminaire "intelligent device"	OR	None	1
2	Title	wireless programmable zone wire "mesh network"	OR	None	1
3	Abs+Claim	control system LED dimming sensor software lamp "control wide area network"	OR	None	1
4	IPC-R(8)	H05B H01L F21V G02F G01J	OR	None	
5	Abs+Claim	vehicle camera tv car street	OR	None	1

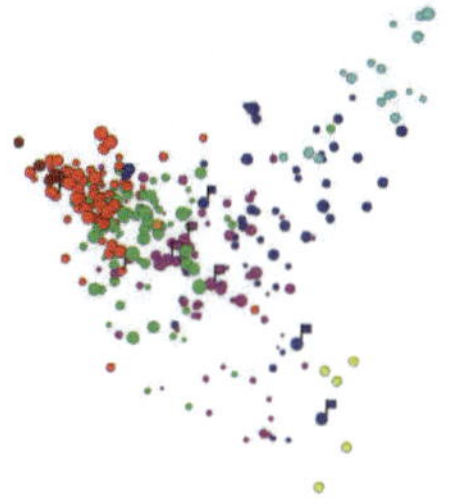

-	-	-	KEYWRD
1	✓	89	LED, source, predetermine, module, memory, plurality, occupancy, transceiver, security, network, couple, output, set, detect, value
2	✓	98	medium, string, pulse, current, plug, voltage, connector, pin, drive, lamp, dim, port, parameter, ballast, transformer
3	✓	50	insulate, electrode, extend, chip, conductor, layer, substrate, bus, socket, bond, wire, contact, light-emitting, dispose, frame
4	✓	5	lead, member, frame, assembly, center, conductor, define, pair, chip, series, surface, connection, contact, end, housing
5	✓	57	interior, lens, aperture, tube, box, side, shape, distance, install, panel, assembly, function, board, member, chamber
6	✓	17	film, insulate, wiring, heat, metal, electrode, region, layer, semiconductor, board, distance, pattern, make, LEDs, light-emitting
7	✓	4	associate, deliver, energy, transfer, frequency, load, distance, operation, receiver, battery, mode, couple, supply, adapt, wireless
8	✓	7	strip, cycle, converter, recite, LEDs, turn, execute, processor, color, produce, send, LED, couple, group, controller

Number of Patents/Year

Row Labels	Count of Filing Dt
1972	1
1973	1
1974	2
1977	1
1980	1
1981	1
1982	2
1983	1
1985	2
1988	1
1989	2
1990	2
1991	1
1992	2
1993	4
1994	1
1995	4
1996	4
1997	2
1998	8
1999	4
2000	11
2001	3
2002	9
2003	6
2004	10
2005	9
2006	13
2007	18
2008	11
2009	28
2010	20
2011	34
2012	25
2013	28
2014	34
2015	21

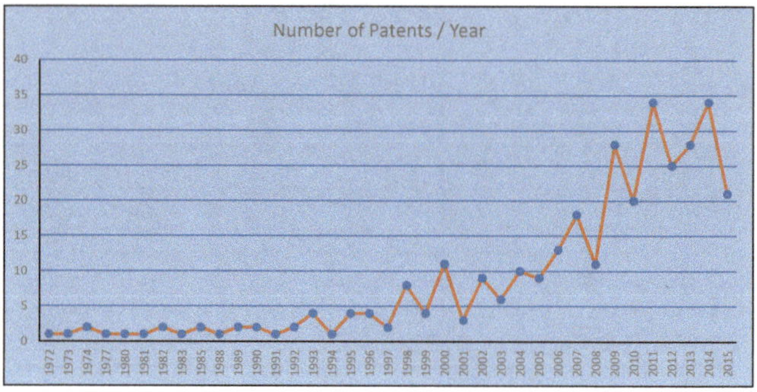

Patent Citation Network

Graph Metric	Value
Graph Type	Directed
Vertices	3285
Unique Edges	9373
Edges With Duplicates	0
Total Edges	9373

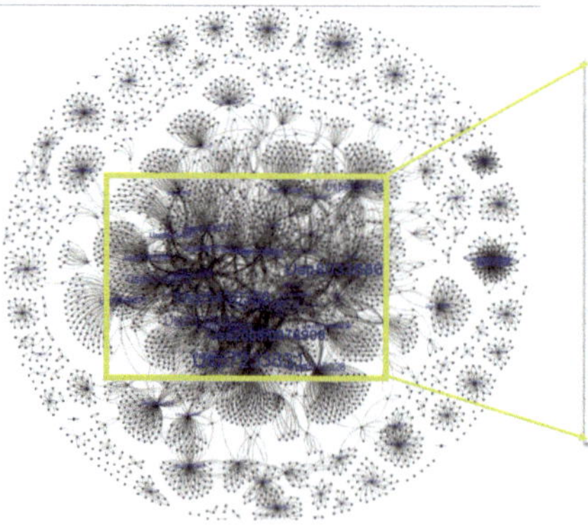

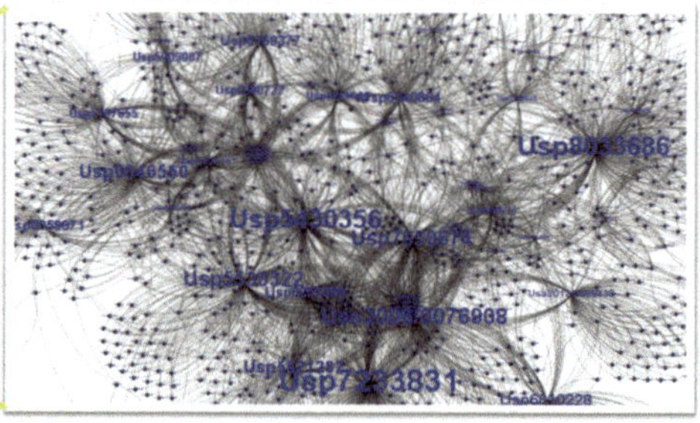

Cluster Analysis

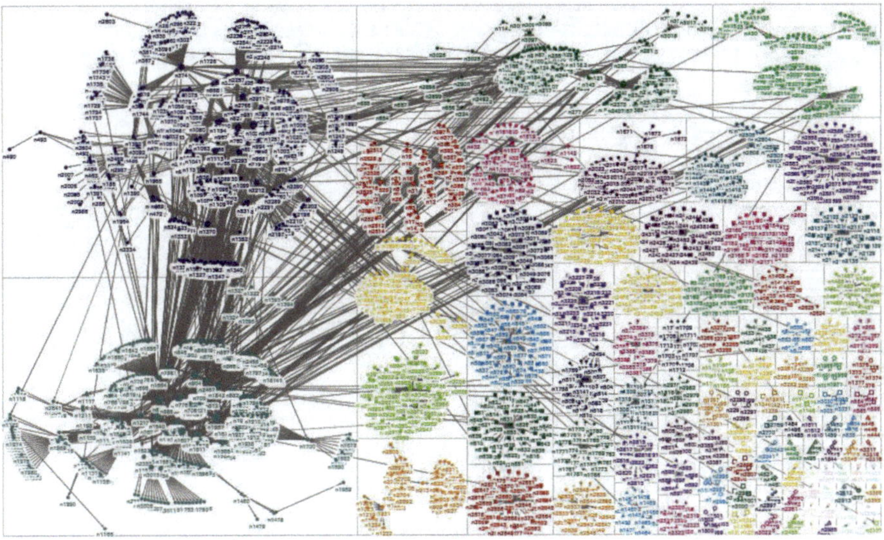

Cluster Analysis

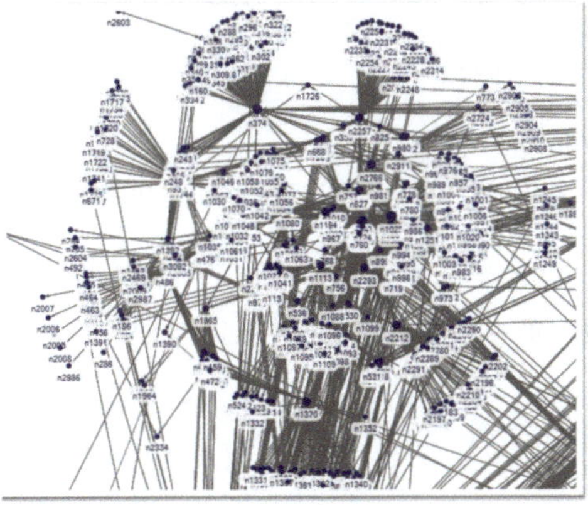

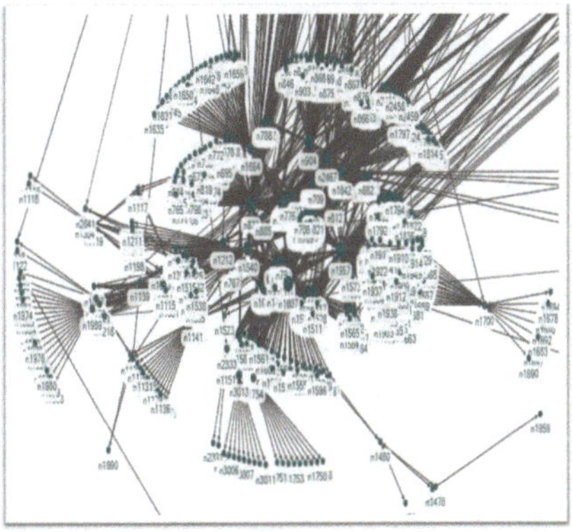

List of the Top Most Important Patents

			Centrality Metrics						Patent Information			
Network Vertex	Patent Number / Appl. Number	Title	Indegree	outdegree	degree	Closness centrality	Betweeness centrality	Eigen centrality	Filing Dt	Assignee,Patentee (newest)	Inventor	Information Link
n1957	Usp7233831	Systems and methods for controlling programmable lighting systems	223	223	446	0.282	980803.400	1.000	6/5/2002	COLOR KINETICS INCORPORATED	BLACKWELL MICHAEL K | BLACKWELL MICHAEL K	http://patentsearch.jp/?no=US7233831
n833	Usp5430356	Programmable lighting control system with normalized dimming for different light sources	190	190	380	0.277	599295.200	0.943	10/5/1993	LUTRON ELECTRONICS CO INC	FERENCE JONATHAN H | LIND III FREDERICK J	http://patentsearch.jp/?no=US5430356
n265	Usp8033686	Wireless lighting devices and applications	188	188	376	0.270	729176.500	0.487	11/26/2009	WIRELESS ENVIRONMENT LLC	RECKER MICHAEL V | LEVINE DAVID B | RECKER MICHAEL V | LEVINE DAVID B	http://patentsearch.jp/?no=US8033686
n1607	Usa2006/0076908	Lighting zone control methods and apparatus	160	160	320	0.270	374452.300	0.785	9/12/2005	COLOR KINETICS INCORPORATED	MORGAN FREDERICK M | CHEMEL BRIAN | PIEPGRAS COLIN | MOLLNOW TOMAS | BLACKWELL MICHAEL K	http://patentsearch.jp/?no=US2006076908
n1540	Usp7659674	Wireless lighting control methods and apparatus	151	151	302	0.287	579867.100	0.900	5/1/2007	COLOR KINETICS INCORPORATED | PHILIPS SOLID-STATE LIGHTING SOLUTIONS INC	DOWLING KEVIN J | MORGAN FREDERICK M | LYS IHOR A | BLACKWELL MICHAEL K | DUCHARME ALFRED | OSTERHOUT RALPH | PIEPGRAS COLIN | MUELLER GEORGE G | GEARY DAWN | HOLMES TIMOTHY | MUELLER GEORGE G | LYS IHOR A | DOWLING KEVIN J | MORGAN FREDERICK M | BLACKWELL MICHAEL K | DUCHARME ALFRED D | OSTERHOUT RALPH | PIEPGRAS COLIN | GEARY DAWN | HOLMES TIMOTHY	http://patentsearch.jp/?no=US7659674
n904	Usp5530322	Multi-zone lighting control system	150	150	300	0.260	384369.200	0.745	4/11/1994	LUTRON ELECTRONICS CO INC	FERENCE JONATHAN H | HAUSMAN DONALD F | LOAR JOHN F | SPEHALSKI ROBERT S | ZAHARCHUK WALTER S	http://patentsearch.jp/?no=US5530322
n1025	Usp6046550	Multi-zone lighting control system	127	127	254	0.260	342745.700	0.335	6/16/1999	LUTRON ELECTRONICS CO INC	FERENCE JONATHAN H | HAUSMAN DONALD F | LOAR JOHN F | SPEHALSKI ROBERT S | ZAHARCHUK WALTER S	http://patentsearch.jp/?no=US6046550
n1842	Usp5621282	Programmable distributively controlled lighting system	118	118	236	0.245	240132.900	0.593	4/10/1995	N/A	HASKELL WALTER	http://patentsearch.jp/?no=US5621282
n2420	Usp6010228	Wireless emergency safety light with sensing means for conventional light switch or plug receptacle	112	112	224	0.237	355376.300	0.309	11/13/1997	BLACKMAN STEPHEN E	BLACKMAN STEPHEN E | WOLFGANG THOMAS	http://patentsearch.jp/?no=US6010228
n374	Usp8491159	Wireless emergency lighting system	103	103	206	0.235	538720.100	0.121	6/30/2010	WIRELESS ENVIRONMENT LLC	RECKER MICHAEL V | LEVINE DAVID B | RECKER MICHAEL V | LEVINE DAVID B	http://patentsearch.jp/?no=US8491159
n2467	Usp6252358	Wireless lighting control	102	102	204	0.275	218315.200	0.569	8/14/1998	N/A	XYDIS THOMAS G | ANGOTT PAUL G	http://patentsearch.jp/?no=US6252358
n1080	Usp6169377	Lighting control with wireless remote control and programmability	99	99	198	0.261	310375.600	0.276	5/24/1999	LUTRON ELECTRONICS CO INC	BRYDE GARY W | WOLBERT III DONALD J | HAKKARAINEN SIMO PEKKA | SPIRA JOEL S	http://patentsearch.jp/?no=US6169377
n1896	Usp6340864	Lighting control system including a wireless remote sensor	99	99	198	0.262	366386.800	0.290	8/10/1999	PHILIPS ELECTRONICS NORTH AMERICA CORPORATION	WACYK IHOR T	http://patentsearch.jp/?no=US6340864
n2766	Usp5059871	Programmable lighting control system linked by a local area network	99	0	99	0.000	0.000	0.146	7/9/1990	LIGHTOLIER INCORPORATED	PEARLMAN GORDON W | CARLSON STEVEN B	http://patentsearch.jp/?no=US5059871
n2257	Usp5187655	Portable programmer for a lighting control	88	88	176	0.257	502421.900	0.231	1/16/1990	LUTRON ELECTRONIC CO INC	POST ROBERT P | D'ALEO MICHAEL J | KEEPORTS DOUG W | FERENCE JONATHAN H | LUCHACO DAVID G | SPIRA JOEL S	http://patentsearch.jp/?no=US5187655
n1194	Usp5909087	Lighting control with wireless remote control and programmability	83	0	83	0.000	0.000	0.230	3/13/1996	LUTRON ELECTRONICS CO INC	BRYDE GARY W | WOLBERT III DONALD J | HAKKARAINEN SIMO PEKKA | SPIRA JOEL S	http://patentsearch.jp/?no=US5909087
n1113	Usp6300727	Lighting control with wireless remote control and programmability	74	74	148	0.252	168306.300	0.250	6/26/2000	LUTRON ELECTRONICS CO INC	BRYDE GARY W | WOLBERT III DONALD J | HAKKARAINEN SIMO PEKKA | SPIRA JOEL S	http://patentsearch.jp/?no=US6300727
n2212	Usp6859644	Initialization of wireless-controlled lighting systems	72	72	144	0.247	173892.000	0.227	12/19/2002	KONINKLIJKE PHILIPS ELECTRONICS N V	WANG LING | WANG LING	http://patentsearch.jp/?no=US6859644
n1370	Usp8731689	Networked, wireless lighting control system with distributed intelligence	69	69	138	0.238	208540.900	0.104	5/6/2008	ABL IP HOLDING LLC	PLATNER BRIAN P | FASSBENDER WILLIAM J | ZAVERUHA RYAN A | RAMIREZ FRANK J | GROSS PHILLIP S | FRIGON RAYMOND A | PLATNER BRIAN P | FASSBENDER WILLIAM J | ZAVERUHA RYAN A | RAMIREZ FRANK J | GROSS PHILIP S | FRIGON RAYMOND A | PLATNER BRIAN P | FASSBENDER WILLIAM J | ZAVERUHA RYAN A | RAMIREZ FRANK J | GROSS PHILIP S | FRIGON RAYMOND A	http://patentsearch.jp/?no=US8731689
n2293	Usp4924151	Multi-zone, multi-scene lighting control system	65	65	130	0.242	98852.770	0.254	9/30/1988	LUTRON ELECTRONICS CO INC	D'ALEO MICHAEL | DARRAGH DENIS | FERENCE JONATHAN | LUCHACO DAVID | ROWEN MICHAEL J | SPIRA JOEL S	http://patentsearch.jp/?no=US4924151
n1664	Usa2012/0098439	Coordinated System of Battery Powered Wireless Lights	63	63	126	0.245	194983.200	0.192	12/22/2011	WIRELESS ENVIRONMENT LLC	RECKER MICHAEL V | LEVINE DAVID B	http://patentsearch.jp/?no=US2012098439

For future Analysis

- Next STEPS:
  - Verify if the patent level is what is needed
  - More detailed analysis of each patent
  - Categorization of the main patents according the technology characterization

- For future analysis:
  - We can analyzed more the relationship among:
    - Clusters – to know how the main clusters focused on
    - Class or Subclass or group to know what area is pushing the development of the technology
    - Forecasting

Luminaire Level Lighting Controls

Report II: Additional Patents

Criteria for the additional list

- The following 6 slides contains the list of patents sorted by betweenness centrality.
- It follows the same criteria as the list that was provided already. These are patents that follows.
- The importance is based on the betweenness; therefore, the most important patents are in slides 4 and 5.

Additional patent list

Vertex	Label	Degree	Betweenness Centrality	Closeness Centrality	Eigenvector Centrality	Title	Pub Dt	Filing Dt	Assignee,Patentee	Inventor	Information Link
n654	Usp8492988	27	23740.00000	0.00021	0.00001	Configurable load control device for light-emitting diode light sources	7/23/2013	6/11/2010	LUTRON ELECTRONICS CO INC	NUHFER MATTHEW W | SHEARER THOMAS M | CHITTA VENKATESH | BIERY ETHAN CHARLES | IYER KARTIK | NUHFER MATTHEW W | SHEARER THOMAS M | CHITTA VENKATESH | BIERY ETHAN CHARLES | IYER KARTIK	http://patentsearch.jp/?no=US8492988
n722	Usp6188177	23	21254.93625	0.00023	0.00008	Light sensing dimming control system for gas discharge lamps	2/13/2001	5/20/1999	POWER CIRCUIT INNOVATIONS INC	ADAMSON HUGH PATRICK | LANGER GEORGE O	http://patentsearch.jp/?no=US6188177
n548	Usp8214084	32	20928.07886	0.00023	0.00266	Integration of LED lighting with building controls	7/3/2012	10/2/2009	ALTAIR ENGINEERING INC | ILUMISYS INC	SIMON DAVID L | IVEY JOHN | IVEY JOHN | SIMON DAVID L	http://patentsearch.jp/?no=US8214084
n591	Usp7812543	29	18542.50764	0.00023	0.00007	Modular wireless lighting control system using a common ballast control interface	10/12/2010	11/15/2006	N/A	BUDIKE LOTHAR E S | BUDIKE JR LOTHAR E S	http://patentsearch.jp/?no=US7812543
n573	Usp8442403	31	17822.21010	0.00025	0.00005	Lighting and control systems and methods	5/14/2013	3/2/2009	MPJ LIGHTING LLC | LUMENETIX INC	WEAVER MATTHEW | WEAVER MATTHEW	http://patentsearch.jp/?no=US8442403
n675	Usp6867558	26	16907.17736	0.00024	0.00005	Method and apparatus for networked lighting system control	3/15/2005	5/12/2003	GENERAL ELECTRIC COMPANY	GAUS RICHARD C JR | WELLES KENNETH | SARKOZI JANOS | GAUS JR RICHARD C | WELLES KENNETH | SARKOZI JANOS	http://patentsearch.jp/?no=US6867558
n609	Usp8698607	28	16151.36622	0.00025	0.00002	Lighting system and remote control method therefor	4/15/2014	12/1/2008	KONINKLIJKE PHILIPS ELECTRONICS N V | KONINKLIJKE PHILIPS N V	VAN DE SLUIS BARTEL MARINUS | FERI LORENZO | VAN DE SLUIS BARTEL MARINUS | FERI LORENZO	http://patentsearch.jp/?no=US8698607
n744	Usa2006/0220571	22	16149.00000	0.00020	0.00008	Light emitting diode current control method and system	10/5/2006	11/23/2005	SUPER VISION INTERNATIONAL INC	HOWELL WAYNE | HOBDAY SIMON | BRADFORD ALLAN | PULLEY ADAM | KVIST CRISTOFFER | FABER STEPHEN M | KINGSTONE BRETT	http://patentsearch.jp/?no=US2006220571
n702	Usp5404080	25	14051.79125	0.00021	0.00050	Lamp brightness control circuit with ambient light compensation	4/4/1995	7/5/1994	ETTA INDUSTRIES INC	QUAZI FAZLE	http://patentsearch.jp/?no=US5404080
n760	Usp7994723	18	13327.22802	0.00023	0.00003	Lighting system and method for controlling a plurality of light sources	8/9/2011	7/18/2006	KONINKLIJKE PHILIPS ELECTRONICS N V	BUDDE WOLFGANG OTTO | ERDMANN BOZENA | BUDDE WOLFGANG OTTO | ERDMANN BOZENA	http://patentsearch.jp/?no=US7994723
n776	Usp7812544	17	11580.33333	0.00016	0.00000	Fluorescent light control	10/12/2010	4/14/2008	ISINE INC	MORALES LOUIS J | STEVENS GARY W | UNGER JOHN B | MORALES LOUIS J | STEVENS GARY W | UNGER JOHN B	http://patentsearch.jp/?no=US7812544
n574	Usa2013/0033183	29	7985.14292	0.00026	0.00025	SYSTEM AND METHOD FOR CONTROLLING LIGHTING	2/7/2013	10/10/2012	ORION ENERGY SYSTEMS INC	VERFUERTH NEAL R | POTTS MICHAEL J | WANG JUN | BARTOL ANTHONY J	http://patentsearch.jp/?no=US2013033183

Additional patent list

Vertex	Label	Degree	Betweenness Centrality	Closeness Centrality	Eigenvector Centrality	Title	Pub Dt	Filing Dt	Assignee,Patentee	Inventor	Information Link
n800	Usp8773030	11	7798.73653	0.00023	0.00002	Low voltage outdoor lighting power source and control system	7/8/2014	9/22/2009	HUNTER INDUSTRIES INC	WOYTOWITZ PETER J | WOYTOWITZ PETER J	http://patentsearch.jp/?no=US8773030
n683	Usp8587225	25	6087.46547	0.00025	0.00025	Floor plan deduction using lighting control and sensing	11/19/2013	9/2/2010	ENLIGHTED INC	ASHAR PREMAL | MOHAN TANUJ | PERKINS DAVID | GOLDING JONATHAN | ASHAR PREMAL | MOHAN TANUJ | PERKINS DAVID | GOLDING JONATHAN	http://patentsearch.jp/?no=US8587225
n880	Usp8803432	5	2334.33333	0.00016	0.00000	Method and apparatus for determining a target light intensity from a phase-control signal	8/12/2014	5/4/2012	LUTRON ELECTRONICS CO INC	HAUSMAN JR DONALD F | HAUSMAN JR DONALD F	http://patentsearch.jp/?no=US8803432
n900	Usp9220136	3	1557.00000	0.00016	0.00000	Method and apparatus for controlling a lighting device	12/22/2015	5/21/2013	MARVELL WORLD TRADE LTD	ZHANG WANFENG | SUTARDJA PANTAS | SONG YONGHUA | ZHANG WANFENG | SUTARDJA PANTAS | SONG YONGHUA	http://patentsearch.jp/?no=US9220136
n897	Usp8110996	4	1256.67255	0.00025	0.00002	Modular wireless lighting control system using a common ballast control interface	2/7/2012	8/13/2010	N/A	BUDIKE JR LOTHAR E S | BUDIKE JR LOTHAR E S	http://patentsearch.jp/?no=US8110996
n778	Usa2008/0258631	15	819.89198	0.00022	0.00232	Light-adjusting and current-limiting control circuit	10/23/2008	1/31/2008	FUMONG TECHNOLOGY LTD	WU AI-MIN | HUANG HAI-TING	http://patentsearch.jp/?no=US2008258631
n864	Usp8594505	5	781.00000	0.00018	0.00000	Lighting and control systems and methods	11/26/2013	9/26/2012	LUMENETIX INC	WEAVER MATTHEW D | WEAVER MATTHEW D	http://patentsearch.jp/?no=US8594505
n906	Usp9143697	2	779.00000	0.00017	0.00002	Lighting and control systems and methods	9/22/2015	3/21/2013	LUMENETIX INC	WEAVER MATTHEW D | WEAVER MATTHEW D	http://patentsearch.jp/?no=US9143697
n908	Usp9131581	2	779.00000	0.00016	0.00000	Solid-state lighting control with dimmability and color temperature tunability	9/8/2015	3/14/2014	LIGHTEL TECHNOLOGIES INC	HSIA CHUNGHO | SHEN PAI-SHENG | HSIA CHUNGHO | SHEN PAI-SHENG	http://patentsearch.jp/?no=US9131581
n637	Usp4495446	27	351.00000	0.03704	0.00000	Lighting unit with improved control sequence	1/22/1985	12/27/1982	GENERAL ELECTRIC COMPANY	BROWN THOMAS A | DISSOSWAY MARC A | PEIL WILLIAM | VAMVAKAS SPIRO	http://patentsearch.jp/?no=US4495446
n823	Usp8571716	7	197.76097	0.00022	0.00105	Integration of LED lighting with building controls	10/29/2013	6/8/2012	ILUMISYS INC	IVEY JOHN | SIMON DAVID L | IVEY JOHN | SIMON DAVID L	http://patentsearch.jp/?no=US8571716
n790	Usp7973495	11	55.00000	0.09091	0.00000	Adaptive control apparatus and method for a solid state lighting system	7/5/2011	3/12/2007	TIR TECHNOLOGY LP | KONINKLIJKE PHILIPS ELECTRONICS N V	TOMA ION | MOHAMMADI SHAHRIAR | BJELJAC BOJANA | JUNGWIRTH PAUL | ION TOMA | MOHAMMADI SHAHRIAR | BJELJAC BOJANA | JUNGWIRTH PAUL	http://patentsearch.jp/?no=US7973495
n811	Usa2010/0289428	10	45.00000	0.10000	0.00000	Controllable Retroffited LED Panel Lighting	11/18/2010	5/11/2010	ADVANCED CONTROL TECHNOLOGIES INC	FRAZIER JOSEPH E | SHELOW KEVIN J | COLIP GARY D	http://patentsearch.jp/?no=US2010289428

Additional patent list

Vertex	Label	Degree	Betweenness Centrality	Closeness Centrality	Eigenvector Centrality	Title	Pub Dt	Filing Dt	Assignee,Patentee	Inventor	Information Link
n822	Usa2012/0112638	10	45.00000	0.10000	0.00000	Thermal Management In A Lighting System Using Multiple, Controlled Power Dissipation Circuits	5/10/2012	11/4/2011	N/A	MELANSON JOHN L | KING ERIC J	http://patentsearch.jp/?no=US2012112638
n855	Usp8779675	7	21.00000	0.14286	0.00000	Controlling current flowing through LEDs in a LED lighting fixture	7/15/2014	12/16/2011	COOPER TECHNOLOGIES COMPANY	MIKANI VASKE | LI LI | BOHLER CHRISTOPHER LEE | RHODES SCOTT EDWARD | MIKANI VASKE | LI LI | BOHLER CHRISTOPHER LEE | RHODES SCOTT EDWARD	http://patentsearch.jp/?no=US8779675
n839	Usp8587212	7	21.00000	0.14286	0.00000	Lighting system, dimming control apparatus and dimming control method	11/19/2013	5/27/2011	INDUSTRIAL TECHNOLOGY RESEARCH INSTITUTE	LI HUNG-CHUN | YU WUN-LONG | LI HUNG-CHUN | YU WUN-LONG	http://patentsearch.jp/?no=US8587212
n831	Usp8692479	7	21.00000	0.14286	0.00000	Method of controlling a ballast, a ballast, a lighting controller, and a digital signal processor	4/8/2014	10/24/2012	NXP B V	MERCIER FREDERIC | MAUGARS PHILIPPE | MERCIER FREDERIC | MAUGARS PHILIPPE	http://patentsearch.jp/?no=US8692479
n847	Usp5894200	7	21.00000	0.14286	0.00000	Two terminal method of and apparatus for improving electrical and light producing efficiency in low voltage direct current incandescent lamp intensity control	4/13/1999	3/26/1997	N/A	GOODALE JR GAROLD JOSEPH | BEHLOW JR HERBERT WALLACE	http://patentsearch.jp/?no=US5894200
n862	Usp9013467	6	15.00000	0.16667	0.00000	Controlled operation of a LED lighting system at a target output color	4/21/2015	7/19/2013	INSTITUT NATIONAL D'OPTIQUE	SISTO MARCO MICHELE | SISTO MARCO MICHELE	http://patentsearch.jp/?no=US9013467
n870	Usa2007/0159107	5	10.00000	0.20000	0.00000	Apparatus and method for controlling discharge lights	7/12/2007	12/24/2004	N/A	POWELL DAVID JOHN	http://patentsearch.jp/?no=US2007159107
n876	Usp7804252	5	10.00000	0.20000	0.00000	Two way lighting control system with dual illumination sources	9/28/2010	6/23/2008	N/A	CHEN CHIA-TEH | CHEN CHIA-TEH	http://patentsearch.jp/?no=US7804252
n904	Usp8604713	3	7.00000	0.12500	0.00000	Method, apparatus and computer-readable media for controlling lighting devices	12/10/2013	7/30/2012	ARKALUMEN INC	BRIGGS GERALD EDWARD | BRIGGS GERALD EDWARD	http://patentsearch.jp/?no=US8604713
n895	Usp8463408	4	6.00000	0.25000	0.00000	Light control system and method for automatically rendering a lighting atmosphere	6/11/2013	11/4/2008	KONINKLIJKE PHILIPS ELECTRONICS N V	BOLEKO RIBAS SALVADOR EXPEDITO | BOLEKO RIBAS SALVADOR EXPEDITO	http://patentsearch.jp/?no=US8463408

Additional patent list

Vertex	Label	Degree	Betweenness Centrality	Closeness Centrality	Eigenvector Centrality	Title	Pub Dt	Filing Dt	Assignee,Patentee	Inventor	Information Link
n890	Usp8842009	4	6.00000	0.25000	0.00000	Multiple light sensor multiple light fixture control	9/23/2014	6/7/2013	MOJO LABS INC	JONES MORGAN | JONES MORGAN	http://patentsearch.jp/?no=US8842009
n885	Usp8872432	4	6.00000	0.25000	0.00000	Solution for dynamic lighting control	10/28/2014	3/15/2012	KERCSO BERTALAN | BALAZS LASZLO | MAROS ISTVAN | GENERAL ELECTRIC COMPANY	KERCSO BERTALAN | BALAZS LASZLO | MAROS ISTVAN	http://patentsearch.jp/?no=US8872432
n924	Usp8232742	2	4.00000	0.10000	0.00000	Method, apparatus and computer-readable media for controlling lighting devices	7/31/2012	11/24/2009	ARKALUMEN INC	BRIGGS GERALD EDWARD | BRIGGS GERALD EDWARD	http://patentsearch.jp/?no=US8232742
n911	Usp9185758	2	3.00000	0.14286	0.00000	Controlling current flowing through LEDs in a LED light fixture	11/10/2015	7/15/2014	COOPER TECHNOLOGIES COMPANY	BOHLER CHRISTOPHER LEE | RHODES SCOTT EDWARD | LI LI | MIKANI VASKE | MIKANI VASKE | LI LI | RHODES SCOTT EDWARD | BOHLER CHRISTOPHER LEE	http://patentsearch.jp/?no=US9185758
n913	Usa2015/0035450	2	3.00000	0.14286	0.00000	SOLID STATE LIGHTING CONTROL	2/5/2015	7/24/2014	CAMBRIDGE SEMICONDUCTOR LIMITED	WERNER ANTONIUS JACOBUS JOHANNES	http://patentsearch.jp/?no=US2015035450
n930	Usa2009/0160365	2	1.00000	0.50000	0.00000	Apparatus Having Supply Voltage Adaptive Light Emitting Component Circuitry And Method Of Controlling	6/25/2009	5/10/2006	N/A	NIEMITALO PAAVO	http://patentsearch.jp/?no=US2009160365
n916	Usp8373360	2	1.00000	0.50000	0.00000	Lighting control system and LED lamp	2/12/2013	9/1/2009	HONG KONG APPLIED SCIENCE AND TECHNOLOGY RESEARCH INSTITUTE CO LTD	LEUNG CHUN KAI | WU KAI CHIU | LEUNG TIK SHUN | LI KIN CHEONG | LU MING | LEUNG CHUN KAI | WU KAI CHIU | LEUNG TIK SHUN | LI KIN CHEONG | LU MING	http://patentsearch.jp/?no=US8373360
n919	Usp8446099	2	1.00000	0.50000	0.00000	Power conversion and control systems and methods for solid-state lighting	5/21/2013	10/4/2010	MCCUNE JR EARL W	MCCUNE JR EARL W | MCCUNE JR EARL W	http://patentsearch.jp/?no=US8446099
n922	Usp8760078	2	1.00000	0.50000	0.00000	Power conversion and control systems and methods for solid-state lighting	6/24/2014	10/4/2010	MCCUNE JR EARL W	MCCUNE JR EARL W | MCCUNE JR EARL W	http://patentsearch.jp/?no=US8760078
n927	Usp8894741	2	1.00000	0.50000	0.00000	Water, light and airflow control system and configuration for a plant air purifier	11/25/2014	9/18/2009	MITTELMARK MARC ANTHONY	MITTELMARK MARTIN | MITTELMARK MARTIN	http://patentsearch.jp/?no=US8894741
n938	Usp9089031	1	0.00000	1.00000	0.00000	Add-on smart controller for LED lighting device	7/21/2015	2/10/2014	LIGHTEL TECHNOLOGIES INC	MAA CHIA-YIU | LIN CHING-FENG | YU CHUN-TE | SHEN PAI-SHENG | MAA CHIA-YIU | LIN CHING-FENG | YU CHUN-TE | SHEN PAI-SHENG	http://patentsearch.jp/?no=US9089031

Additional patent list

Vertex	Label	Degree	Betweenness Centrality	Closeness Centrality	Eigenvector Centrality	Title	Pub Dt	Filing Dt	Assignee,Patentee	Inventor	Information Link
n954	Usa2014/0300284	1	0.00000	1.00000	0.00000	COLOR TUNABLE LIGHT SOURCE MODULE WITH BRIGHTNESS AND DIMMING CONTROL	10/9/2014	4/3/2014	LEDENGIN INC	LEE KACHUN | YAN XIANTAO | MEI ZEQUN	http://patentsearch.jp/?no=US2014300284
n955	Usa2014/0225528	1	0.00000	0.00017	0.00002	Devices, Systems, Architectures, and Methods for Lighting and other Building Control applications	8/14/2014	3/14/2014	CLAYTON RANDY	CLAYTON RANDY	http://patentsearch.jp/?no=US2014225528
n959	Usa2013/0229119	1	0.00000	1.00000	0.00000	Dimmable Solid State Lighting System, Apparatus and Method, with Distributed Control and Intelligent Remote Control	9/5/2013	10/30/2012	LUXERA INC	KOROBOV VLADIMIR | LIVSCHITZ LEONARD SIMON | ORLOV NKOLAI | NIKITIN ANTON | SHTEYNBERG ANATOLY	http://patentsearch.jp/?no=US2013229119
n941	Usa2013/0241420	1	0.00000	0.00017	0.00000	DYNAMIC LIGHTING CONTROL	9/19/2013	3/15/2012	BAL ZS LASZLO | KERCSO BERTALAN | MAROS ISTVAN	BAL ZS LASZLO | KERCSO BERTALAN | MAROS ISTVAN	http://patentsearch.jp/?no=US2013241420
n957	Usp9220153	1	0.00000	1.00000	0.00000	Illuminating apparatus and control method for keeping a constant total brightness of ambient light and light produced by the illuminating apparatus	12/22/2015	2/4/2013	LUXUL TECHNOLOGY INCORPORATION	PAN CHENG-HUNG | CHEN ERIC WEN-CHIEN | PAN CHENG-HUNG | CHEN ERIC WEN-CHIEN	http://patentsearch.jp/?no=US9220153
n934	Usp9140414	1	0.00000	1.00000	0.00000	LED directional lighting system with light intensity controller	9/22/2015	6/2/2015	MIND HEAD LLC | BEAUSOLEIL DAVID M	BEAUSOLEIL DAVID M | BEAUSOLEIL DAVID M	http://patentsearch.jp/?no=US9140414
n943	Usp9326339	1	0.00000	1.00000	0.00000	LED lighting device with control means and method for operating the LED lighting device	4/26/2016	3/27/2013	DIEHL AEROSPACE GMBH	NIEBERLEIN UWE | DOBLER STEFAN | KOENINGER GUENTHER | NIEBERLEIN UWE | DOBLER STEFAN | KOENINGER GUENTHER	http://patentsearch.jp/?no=US9326339
n936	Usp9307598	1	0.00000	1.00000	0.00000	LED lighting device with replaceable driver-control module	4/5/2016	3/17/2015	LIGHTEL TECHNOLOGIES INC	MAA CHIA-YIU | LIN CHING-FENG | SHEN PAI-SHENG | MAA CHIA-YIU | LIN CHING-FENG | SHEN PAI-SHENG	http://patentsearch.jp/?no=US9307598
n952	Usp9215779	1	0.00000	1.00000	0.00000	Light switch and control device having a touch screen interface	12/15/2015	3/30/2012	RESTREPO CARLOS EDUARDO | LACEY DARRON KIRBY | PABST JR RICHARD A | LANDRUM THOMAS | ZHANG SHELDON | DU LILY | COOPER TECHNOLOGIES COMPANY	RESTREPO CARLOS EDUARDO | LACEY DARRON KIRBY | PABST JR RICHARD A | LANDRUM THOMAS | ZHANG SHELDON | DU LILY | RESTREPO CARLOS EDUARDO | LACEY DARRON KIRBY | PABST JR RICHARD A | LANDRUM THOMAS | ZHANG SHELDON | DU LILY	http://patentsearch.jp/?no=US9215779

Additional patent list

Vertex	Label	Degree	Betweenness Centrality	Closeness Centrality	Eigenvector Centrality	Title	Pub Dt	Filing Dt	Assignee,Patentee	Inventor	Information Link
n948	Usp8633653	1	0.00000	1.00000	0.00000	Lighting control system with improved efficiency	1/21/2014	3/2/2010	GENERAL ELECTRIC COMPANY	DUFFY MARK | ROBERTS BRUCE | PLASKON JEFFREY | BALAZS LAZLO | MAROS ISTVAN | SHERTOK AVI | DUFFY MARK | ROBERTS BRUCE | PLASKON JEFFREY | BALAZS LAZLO | MAROS ISTVAN | SHERTOK AVI	http://patentsearch.jp/?no=US8633653
n944	Usa2014/0063857	1	0.00000	0.00016	0.00000	METHOD AND APPARATUS FOR CONTROLLING A LIGHTING DEVICE	3/6/2014	8/23/2013	MARVELL WORLD TRADE LTD	PENG HAO | ZHANG WANFENG | CHOI JINHO | DOAN TUYEN	http://patentsearch.jp/?no=US2014063857
n932	Usp9119263	1	0.00000	1.00000	0.00000	Multi-zone light controller	8/25/2015	3/19/2012	KONINKLIJKE PHILIPS N V	ZNAMENSKIY DMITRY NIKOLAYEVICH | ABBO ANTENEH ALEMU | ZNAMENSKIY DMITRY NIKOLAYEVICH | ABBO ANTENEH ALEMU	http://patentsearch.jp/?no=US9119263
n946	Usp8922125	1	0.00000	1.00000	0.00000	Occupancy sensor for controlling an LED light	12/30/2014	1/22/2013	IR-TEC INTERNATIONAL LTD	LIN SHIH-TSUNG | LIN SHIH-TSUNG	http://patentsearch.jp/?no=US8922125
n940	Usp9247603	1	0.00000	1.00000	0.00000	Shunt regulator for spectral shift controlled light source	1/26/2016	2/11/2015	ONCE INNOVATIONS INC	GRAJCAR ZDENKO | ERICKSON LEIF | GRAJCAR ZDENKO | ERICKSON LEIF	http://patentsearch.jp/?no=US9247603
n950	Usa2013/0270998	1	0.00000	1.00000	0.00000	SOLID-STATE LIGHTING SYSTEMS HAVING INTELLIGENT CONTROLS	10/17/2013	4/17/2013	AXLEN INC	PI BO	http://patentsearch.jp/?no=US2013270998
n960	Usp9215780	1	0.00000	0.00020	0.00001	System and method for reducing peak and off-peak electricity demand by monitoring, controlling and metering lighting in a facility	12/15/2015	3/3/2014	ORION ENERGY SYSTEMS INC	VERFUERTH NEAL R | POTTS MICHAEL J | WANG JUN | VERFUERTH NEAL R | POTTS MICHAEL J | WANG JUN	http://patentsearch.jp/?no=US9215780

Analysis of patents based on all centrality measures

- As it was required, the analysis is focused on finding the most important patents in each clusters and the patents located on the peripheries.
- The criteria based on looking for the patents that plays an important role in the cluster (based on degree and betweenness).
- The patents located on the peripheries of the clusters are considered usually as isolated agents that do not play an important role in the networks. However, due to the importance of new inventions and the possibility of finding very new patent that will perform better in the future, these patents were selected based on the following criteria:
  - 5 clusters were selected as the most important in the network
  - These patents have a low Degree of centrality and low betweenness.
  - Only patents that have been cited and found in the search in the "title of the patents" are considered in the list.
  - The patents are differentiated by colors and match the graph highlights

Patent Citation Ne

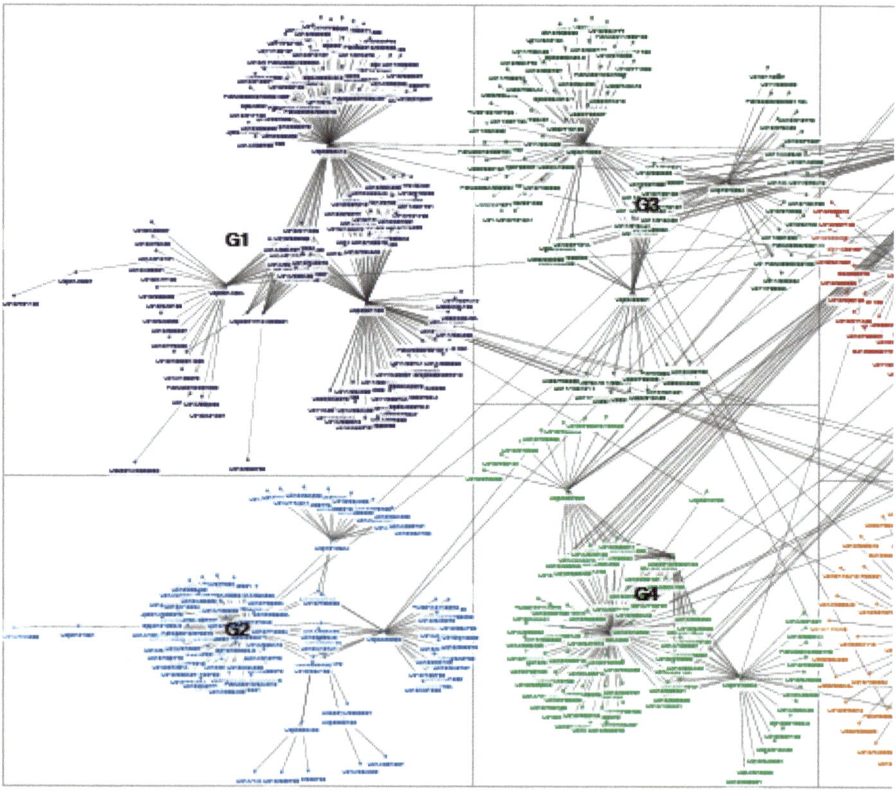

G1, G2,

:twork by Clusters

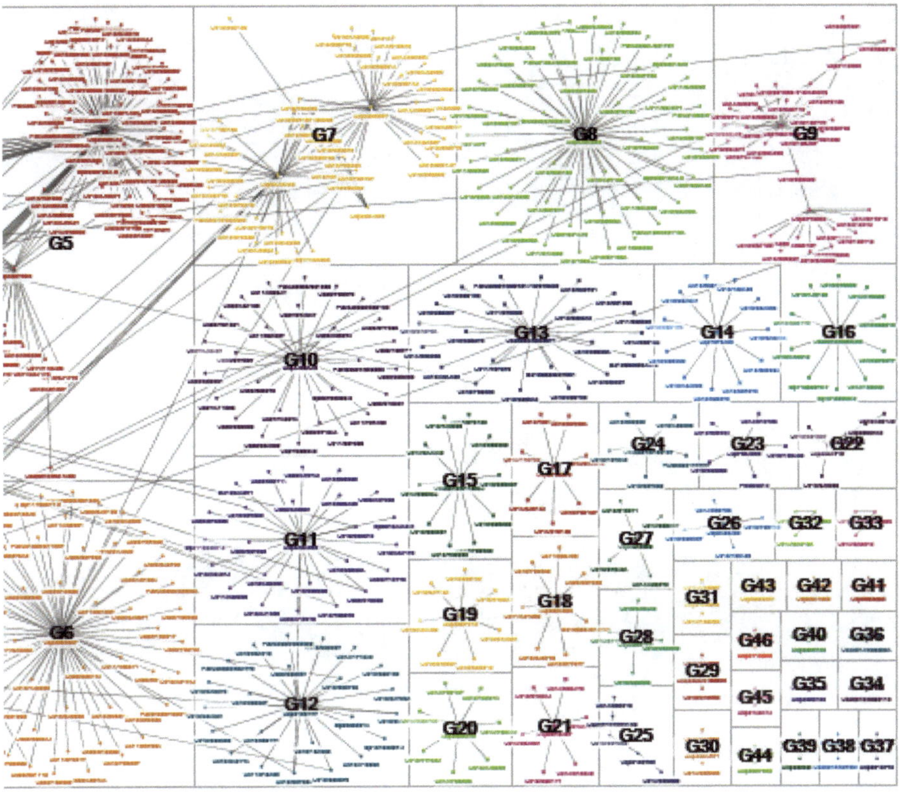

.....G44 = Clusters

Custers G1

Vertex	Label	Degree	Betweenness Centrality	Closeness Centrality	Eigenvector Centrality	Title	Pub Dt	Filing Dt	Assignee,Patentee	Inventor	Information Link
n96	Usp6683419	96	81456.6315	0.000277	0.009434	Electrical control for an LED light source, including dimming control	1/27/2004	6/24/2002	DIALIGHT CORPORATION	KRIPAROS DANIEL J | KRIPAROS DANIEL J	http://patentsearch.jp/?no=US6683419
n234	Usp5581158	83	65974.409	0.000274	0.008178	Lamp brightness control circuit with ambient light compensation	12/3/1996	4/3/1995	ETTA INDUSTRIES INC	QUAZI FAZLE	http://patentsearch.jp/?no=US5581158
n548	Usp8214084	32	20928.0789	0.000228	0.002656	Integration of LED lighting with building controls	7/3/2012	10/2/2009	ALTAIR ENGINEERING INC | ILUMISYS INC	SIMON DAVID L | IVEY JOHN | IVEY JOHN | SIMON DAVID L	http://patentsearch.jp/?no=US8214084
n778	Usa2008/0258631	15	819.891981	0.000217	0.002318	Light-adjusting and current-limiting control circuit	10/23/2008	1/31/2008	FUMONG TECHNOLOGY LTD	WU AI-MIN | HUANG HAI-TING	http://patentsearch.jp/?no=US2008258631
n823	Usp8571716	7	197.760969	0.000216	0.001047	Integration of LED lighting with building controls	10/29/2013	6/8/2012	ILUMISYS INC	IVEY JOHN | SIMON DAVID L | IVEY JOHN | SIMON DAVID L	http://patentsearch.jp/?no=US8571716
n806	Usp9143697	2	779	0.000168	0.000019	Lighting and control systems and methods	9/22/2015	3/21/2013	LUMENETIX INC	WEAVER MATTHEW D | WEAVER MATTHEW D	http://patentsearch.jp/?no=US9143697
n855	Usa2014/0225528	1	0	0.000168	0.000019	Devices, Systems, Architectures, and Methods for Lighting and other Building Control applications	8/14/2014	3/14/2014	CLAYTON RANDY	CLAYTON RANDY	http://patentsearch.jp/?no=US2014225528

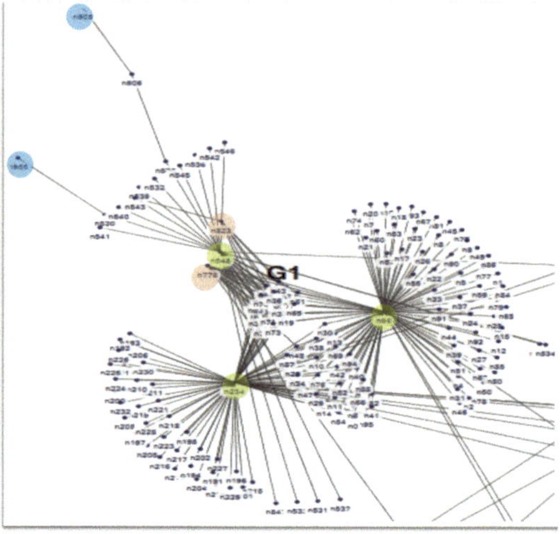

Cluster G2

Vertex	Label	Degree	Betweenness Centrality	Closeness Centrality	Eigenvector Centrality	Title	Pub Dt	Filing Dt	Assignee,Patentee	Inventor	Information Link
n455	Usp8492987	59	49118	0.00021	0.000007	Load control device for a light-emitting diode light source	7/23/2013	6/11/2010	LUTRON ELECTRONICS CO INC	NUHFER MATTHEW W | SHEARER THOMAS M | CHITTA VENKATESH | IYER KARTIK | NUHFER MATTHEW W | SHEARER THOMAS M | CHITTA VENKATESH | IYER KARTIK	http://patentsearch.jp/?no=US8492987
n654	Usp8492988	27	23740	0.000207	0.000005	Configurable load control device for light-emitting diode light sources	7/23/2013	6/11/2010	LUTRON ELECTRONICS CO INC	NUHFER MATTHEW W | SHEARER THOMAS M | CHITTA VENKATESH | BIERY ETHAN CHARLES | IYER KARTIK | NUHFER MATTHEW W | SHEARER THOMAS M | CHITTA VENKATESH | BIERY ETHAN CHARLES | IYER KARTIK	http://patentsearch.jp/?no=US8492988
n776	Usp7812544	17	11580.33333	0.000161	0	Fluorescent light control	10/12/2010	4/14/2008	ISINE INC	MORALES LOUIS J | STEVENS GARY W | UNGER JOHN B | MORALES LOUIS J | STEVENS GARY W | UNGER JOHN B	http://patentsearch.jp/?no=US7812544
n880	Usp8803432	5	2334.333333	0.00016	0	Method and apparatus for determining a target light intensity from a phase-control signal	8/12/2014	5/4/2012	LUTRON ELECTRONICS CO INC	HAUSMAN JR DONALD F | HAUSMAN JR DONALD F	http://patentsearch.jp/?no=US8803432
n900	Usp9220136	3	1557	0.00016	0	Method and apparatus for controlling a lighting device	12/22/2015	5/21/2013	MARVELL WORLD TRADE LTD	ZHANG WANFENG | SUTARDJA PANTAS | SONG YONGHUA | ZHANG WANFENG | SUTARDJA PANTAS | SONG YONGHUA	http://patentsearch.jp/?no=US9220136
n908	Usp9131581	2	779	0.000159	0	Solid-state lighting control with dimmability and color temperature tunability	9/8/2015	3/14/2014	LIGHTEL TECHNOLOGIES INC	HSIA CHUNGHO | SHEN PAI-SHENG | HSIA CHUNGHO | SHEN PAI-SHENG	http://patentsearch.jp/?no=US9131581
n944	Usa2014/0063857	1	0	0.000159	0	METHOD AND APPARATUS FOR CONTROLLING A LIGHTING DEVICE	3/6/2014	8/23/2013	MARVELL WORLD TRADE LTD	PENG HAO | ZHANG WANFENG | CHOI JINHO | DOAN TUYEN	http://patentsearch.jp/?no=US2014063857

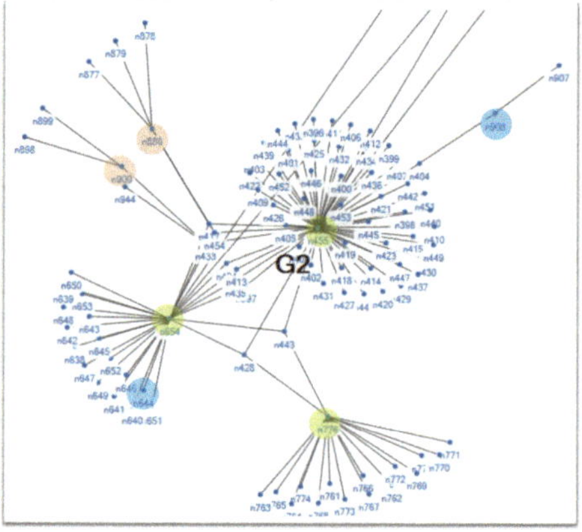

Cluster G3

Vertex	Label	Degree	Betweenness Centrality	Closeness Centrality	Eigenvector Centrality	Title	Pub Dt	Filing Dt	Assignee, Patentee	Inventor	Information Link
n339	Usp5473202	61	39124.49	0.000281	0.000431	Control unit for occupancy sensor switching of high efficiency lighting	12/5/1995	6/5/1992	PLATNER BRIAN	MUDGE PHILIP H | PLATNER BRIAN	http://patentsearch.jp/?no=US5473202
n529	Usp7075254	33	81765.83	0.00028	0.000258	Lighting ballast having boost converter with on/off control and method of ballast operation	7/11/2006	12/14/2004	LUTRON ELECTRONICS CO INC	CHITTA VENKATESH | GAWRYS BRENT M | CHITTA VENKATESH | GAWRYS BRENT M	http://patentsearch.jp/?no=US7075254
n510	Usp5498931	33	71731.58	0.000325	0.000482	Method for automatic switching and control of lighting	3/12/1996	10/3/1994	TLG PLC	BEDOCS LOU	http://patentsearch.jp/?no=US5498931

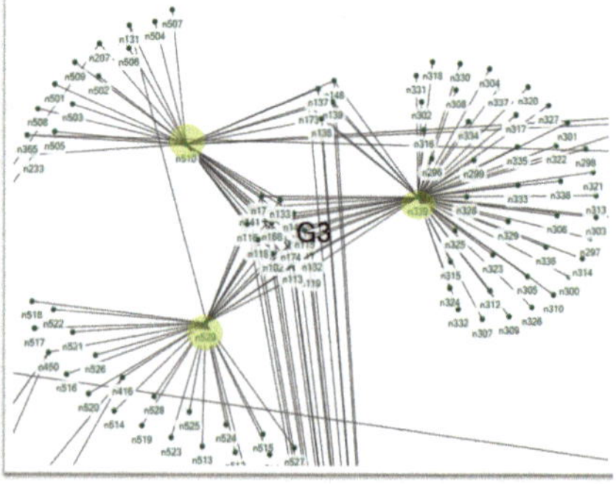

Cluster G4

Vertex	Label	Degree	Betweenness Centrality	Closeness Centrality	Eigenvector Centrality	Title	Pub Dt	Filing Dt	Assignee, Patentee	Inventor	Information Link
n295	Usp8344665	71	63396.15	0.000292	0.000379	System and method for controlling lighting	1/1/2013	9/29/2008	ORION ENERGY SYSTEMS INC	VERFUERTH NEAL R | POTTS MICHAEL J | WANG JUN | BARTOL ANTHONY J | VERFUERTH NEAL R | POTTS MICHAEL J | WANG JUN | BARTOL ANTHONY J	http://patentsearch.jp/?no=US8344665
n591	Usp7812543	29	18542.51	0.000227	0.00007	Modular wireless lighting control system using a common ballast control interface	10/12/2010	11/15/2006	N/A	BUDIKE LOTHAR E S | BUDIKE JR LOTHAR E S	http://patentsearch.jp/?no=US7812543
n574	Usa2013/0033183	29	7985.143	0.000259	0.000251	SYSTEM AND METHOD FOR CONTROLLING LIGHTING	2/7/2013	10/10/2012	ORION ENERGY SYSTEMS INC	VERFUERTH NEAL R | POTTS MICHAEL J | WANG JUN | BARTOL ANTHONY J	http://patentsearch.jp/?no=US2013033183
n683	Usp8587225	25	6087.465	0.000247	0.000248	Floor plan deduction using lighting control and sensing	11/19/2013	9/2/2010	ENLIGHTED INC	ASHAR PREMAL | MOHAN TANUJ | PERKINS DAVID | GOLDING JONATHAN | ASHAR PREMAL | MOHAN TANUJ | PERKINS DAVID | GOLDING JONATHAN	http://patentsearch.jp/?no=US8587225
n960	Usp9215780	1	0	0.000201	0.000005	System and method for reducing peak and off-peak electricity demand by monitoring, controlling and metering lighting in a facility	12/15/2015	3/3/2014	ORION ENERGY SYSTEMS INC	VERFUERTH NEAL R | POTTS MICHAEL J | WANG JUN | VERFUERTH NEAL R | POTTS MICHAEL J | WANG JUN	http://patentsearch.jp/?no=US9215780

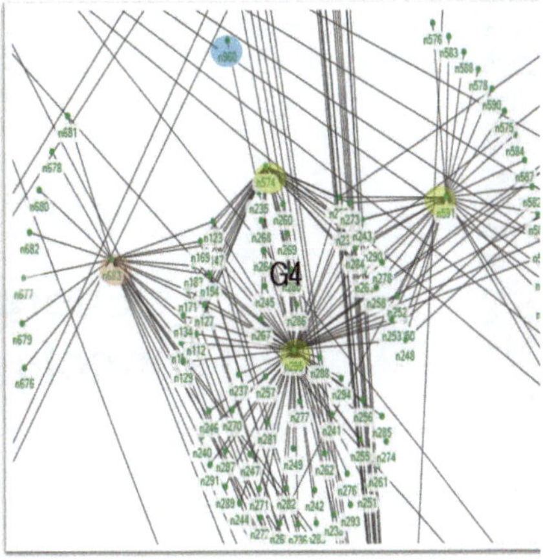

Cluster G5

Vertex	Label	Degree	Betweenness Centrality	Closeness Centrality	Eigenvector Centrality	Title	Pub Dt	Filing Dt	Assignee, Patentee	Inventor	Information Link
N190	Usp5668446	95	145803.7	0.000333	0.001175	Energy management control system for fluorescent lighting	9/16/1997	9/23/1996	NEGAWATT TECHNOLOGIES INC	BAKER JAMES ALLAN	http://patentsearch.jp/?no=US5668446
n675	Usp6867558	26	16907.18	0.000238	0.000053	Method and apparatus for networked lighting system control	3/15/2005	5/12/2003	GENERAL ELECTRIC COMPANY	GAUS RICHARD C JR | WELLES KENNETH | SARKOZI JANOS | GAUS JR RICHARD C | WELLES KENNETH | SARKOZI JANOS	http://patentsearch.jp/?no=US6867558
n941	Usa2013/0241420	1	0	0.000174	0	DYNAMIC LIGHTING CONTROL	9/19/2013	3/15/2012	BAL ZS LASZLO | KERCSO BERTALAN | MAROS ISTVAN	BAL ZS LASZLO | KERCSO BERTALAN | MAROS ISTVAN	http://patentsearch.jp/?no=US2013241420

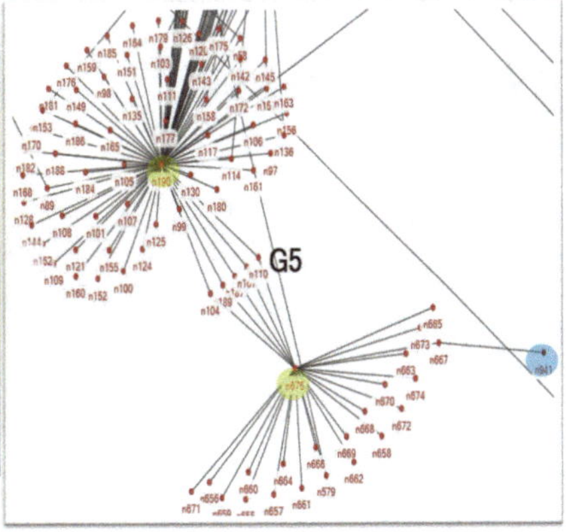

ETM

Energy Efficient Technology Data Mining

Portland State
UNIVERSITY

Appendix 16

Aerogel Technology

ETM

- Aerogel: synthetic porous ultra-light material derived from a gel, in which the liquid component of the gel has been replaced with a gas; used as a catalyst
- Process technologies: supercritical drying, freeze drying, spray drying
- Characteristics: a solid with extremely low density and low thermal conductivity
- In 1931, Samuel Stephens Kistler first invented it, but wasn't used much until new applications were developed

Portland State UNIVERSITY

1

T. U. Daim (ed.), *Roadmapping Future*, Applied Innovation and Technology Management, https://doi.org/10.1007/978-3-030-50502-8

Industry application

- Thermal Insulation: industrial and building insulation(Potential energy efficiency application)
- Pharmacology: controlled drug delivery systems
- Effluents treatments: oils removal, gases capture

Portland State UNIVERSITY

2

Research Questions

- Research requirements:
  - Identify application areas and needs
  - Is the cost likely to drop any time soon so broader applications are feasible?
  - Who is doing it?
  - What is being done globally and Where is it being done?
  - What are the trends identified through data mining?

Portland State UNIVERSITY

3

Searching Keywords

ETM

- "Aerogel" limited in Title
- 3379 articles found in Compendex for 1884-2013
- Filtered data: 3320 (59 duplicates removed by title)
- CAGR: 16% since 1990

1934-2013

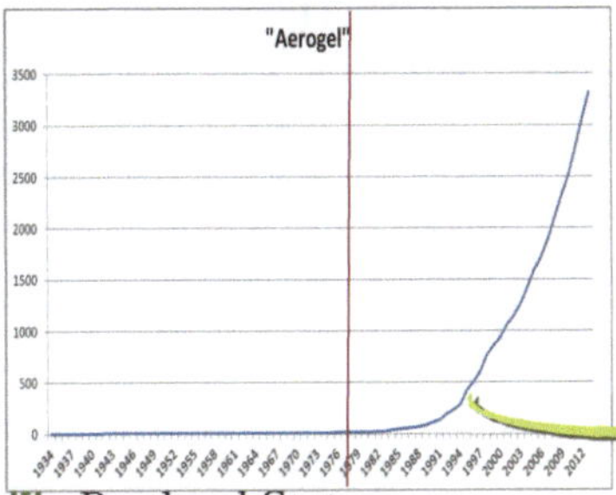

1978-2013

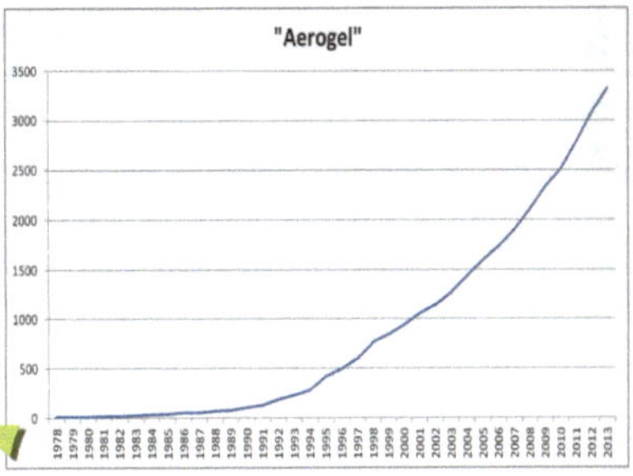

Portland State UNIVERSITY

Filtering Data

ETM

- Period: last 10 years (2003 – 2013)
- DB: Compendex
- Filtered data: 2176
- Process technologies (All fields, Title)
  - Supercritical drying (2031, 230)
  - Freeze drying (6480, 1108)
  - Spray drying (7588,1284)
- Set the Maximum is 8000

Portland State UNIVERSITY

Technology Forecasting(Fisher-Pry Curve

ETM

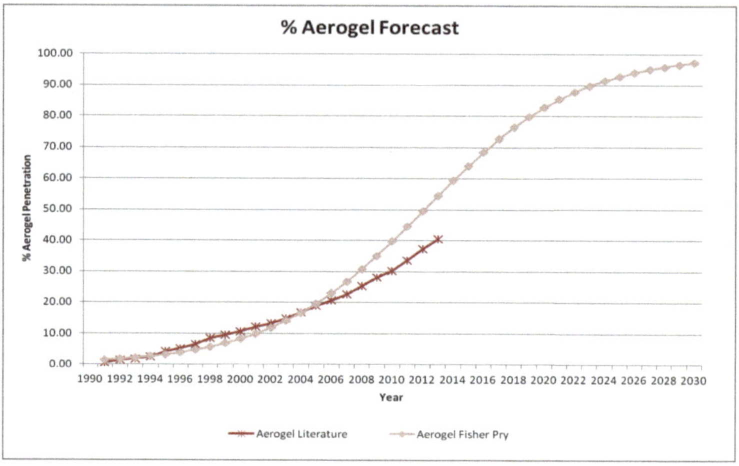

Portland State UNIVERSITY 6

Identified Experts by Citation Index

ETM

Author/Initials	Author Affiliation	Selected By
Leventis, N.	NASA Gleen Research Center, Cleveland, OH, United States	Cit, Pub
Sotiriou-Leventis, Chariklia	University of Missouri-Rolla, Rolla, MO, United States	Cit, Pub
Rolison, D.R.	Naval Research Lab, Washington, United States	Cit, Pub
Mulders, Norbert	Cornell Univ, Ithaca, United States	Pub, Cit
Baumann, T.	Lawrence Livermore Natl. Laboratory, Livermore, CA, United States	Cit, Pub
Lu, H.	Oklahoma State University, Stillwater, OK, United States	Cit, Pub
Meador, Mary Ann	NASA Glenn Research Center, Cleveland, OH, United States	Cit, Pub
Dunn, Bruce	Univ. of California, Los Angeles, CA, United States	Pub, Cit
Pekala, Richard W.	Lawrence Livermore Natl Lab, Livermore, United States	Pub, Cit
Brock, Stephanie L.	Wayne State University, Detroit, MI, United States	Cit, Pub
Worsley, M.A.	Lawrence Livermore National Laboratory, Livermore, CA, United States	Cit, Pub
Satcher Jr., J.H.	Lawrence Livermore National Laboratory, Livermore, CA, United States	Cit, Pub

Portland State UNIVERSITY 7

SNA Result

ETM

- Minimum Degree of Centrality>=18

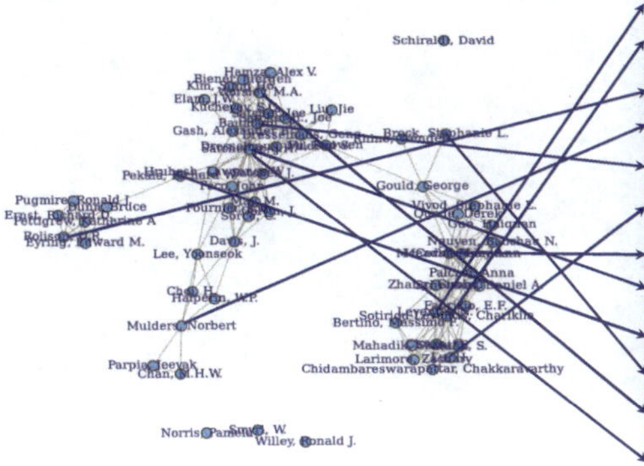

Author/Initials
Leventis, N.
Sotiriou-Leventis, Chariklia
Rolison, D.R.
Mulders, Norbert
Baumann, T.
Lu, H.
Meador, Mary Ann
Dunn, Bruce
Pekala, Richard W.
Brock, Stephanie L.
Worsley, M.A.
Satcher Jr., J.H.

Portland State UNIVERSITY

Filtering Data

ETM

- Period: last 10 years (2003 – 2012)
- DB: USPTO
- Filtered data: 215

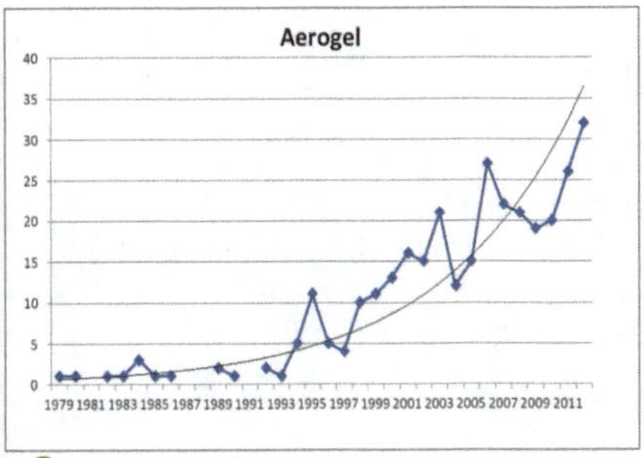

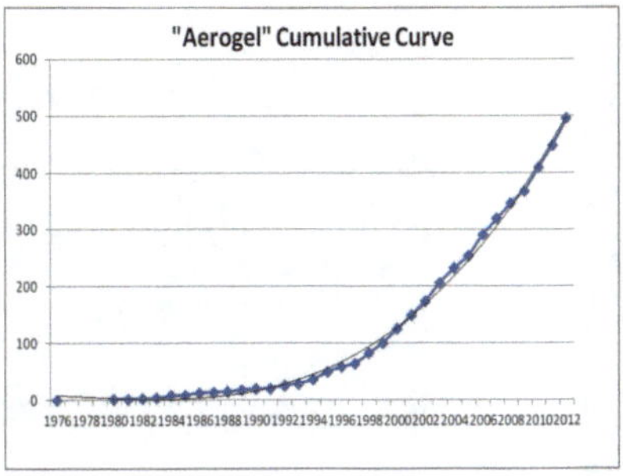

Portland State UNIVERSITY

Source: AcclaimIP.com

Patent Analysis

ETM

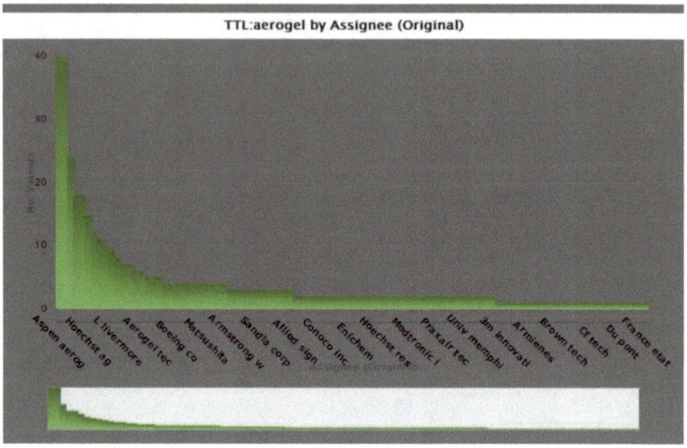

Source: AcclaimIP.com

Portland State UNIVERSITY

10

Patent Analysis

ETM

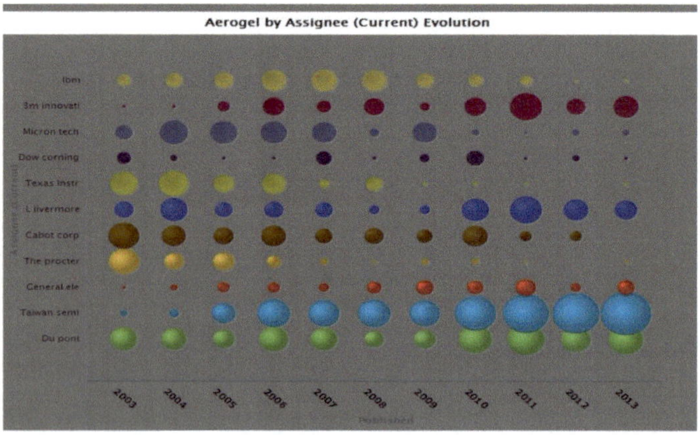

Source: AcclaimIP.com

Portland State UNIVERSITY

11

References

- http://www.aerogel.org/

Portland State UNIVERSITY

12

Back up

Portland State UNIVERSITY

13

Processing Technologies Trends

ETM

- "Spray Drying" in all fields, CAGR (1989-2012): 7.4%
- "Freeze drying " in all fields, CAGR (1989-2012): 8.6%
- "Supercritical drying " in all fields, CAGR (1989-2012): 18.4%

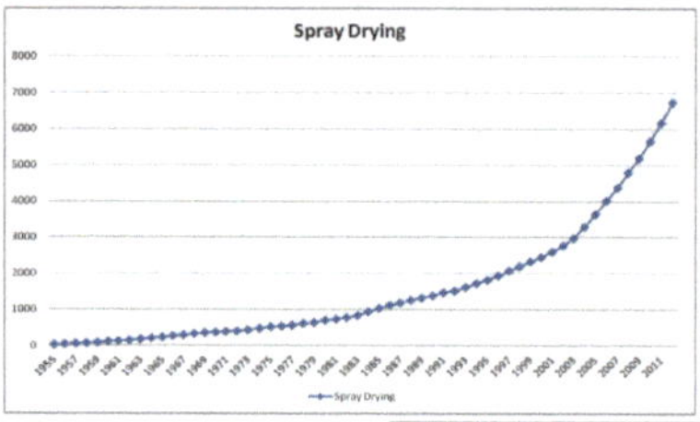

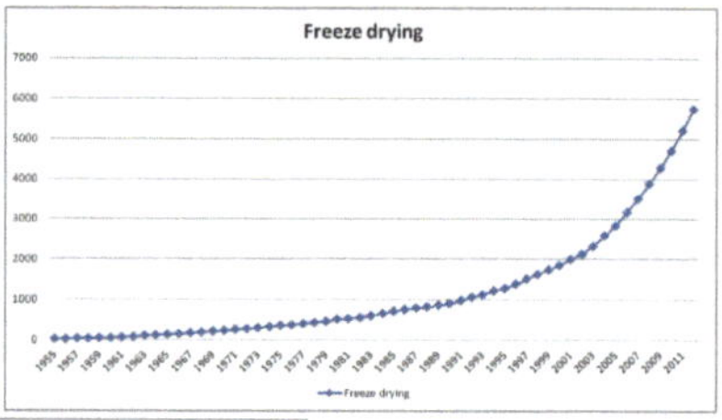

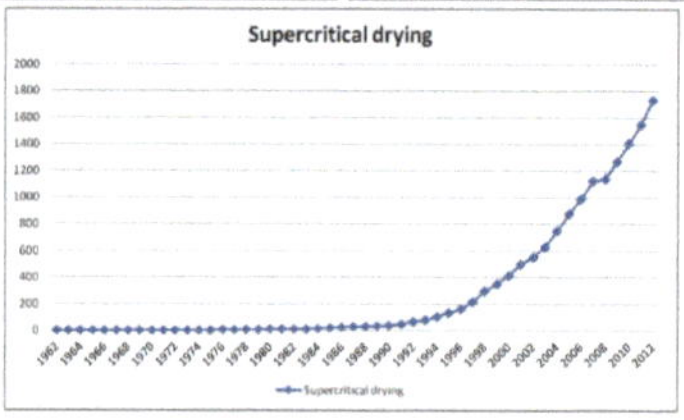

Portland State UNIVERSITY

14

Literature Analysis

ETM

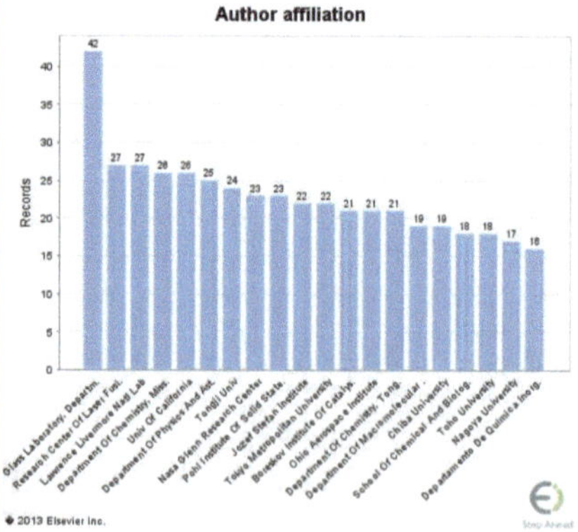

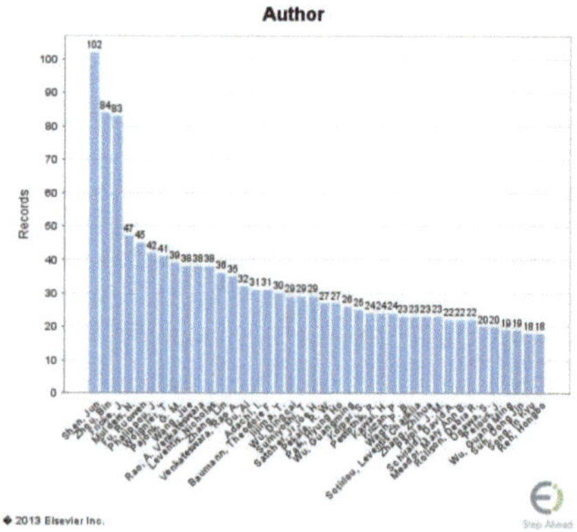

Portland State UNIVERSITY

15

Searching Keywords

ETM

- "Aerogel & Insulation" all limited in Title
- 97 articles(26 US) found in Compendex for 1884-2013
- CAGR: 13.8%(1990-2013)

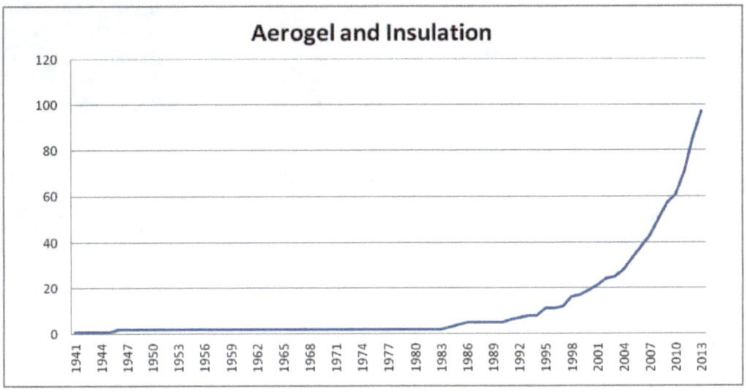

Portland State UNIVERSITY

16

Identified Experts by Citation Index

ETM

Author/Initials	Author Affiliation	Selected By
Gould, George L.	Aspen Aerogels, Inc., MA, United States	Cit, Pub
Lee, Je Kyun	Aspen Aerogels, Inc., MA, United States	Cit, Pub
Agonafer, Damena	Carnegie Mellon University, PA, United States	Cit
Amon, Cristina H.	Carnegie Mellon University, PA, United States	Cit
Gu, Jason	Carnegie Mellon University, PA, United States	Cit
Romero, David	Carnegie Mellon University, PA, United States	Cit
Smith, Brian	Carnegie Mellon University, PA, United States	Cit
Rhine, Wendell	Aspen Aerogels, Inc., MA, United States	Cit, Pub
Bardy, Erik	State University of New York at Buffalo, NY, United States	Cit, Pub
Mollendorf, Joseph	State University of New York at Buffalo, NY, United States	Cit, Pub
Pendergast, David	State University of New York at Buffalo, NY, United States	Cit, Pub
Rouanet, S.	Cabot Corp., Tuscola, IL, United States	Cit, Pub

Portland State UNIVERSITY

17

SNA Result

ETM

- Minimum Degree of Centrality>=1

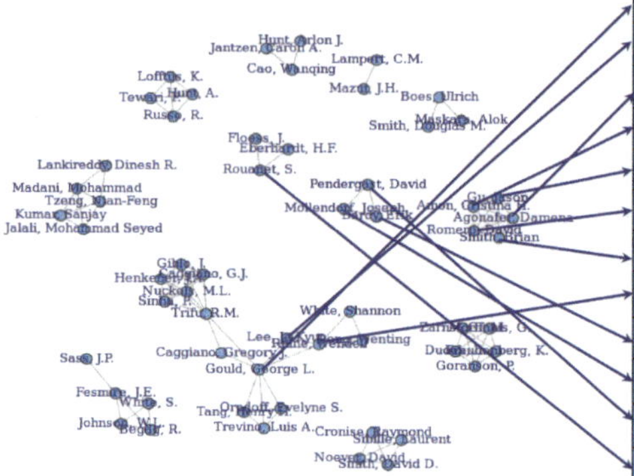

Author/Initials
Gould, George L.
Lee, Je Kyun
Agonafer, Damena
Amon, Cristina H.
Gu, Jason
Romero, David
Smith, Brian
Rhine, Wendell
Bardy, Erik
Mollendorf, Joseph
Pendergast, David
Rouanet, S.

Portland State UNIVERSITY

ETM

PAX Technology Technology Data Mining

Portland State UNIVERSITY

PAX Fans, Blowers, Mixers

ETM

- Liquids and gases naturally move in a swirling non-linear path, unlike the linear path traditional fans, mixers, etc. try create.
- PAX technology – biomimicry* designed fans, mixers, etc., taking advantage of nature's efficiencies (Streamlining Principle).
- Developed by PAX Scientific (www.http://paxscientific.com/)

Portland State UNIVERSITY *biomimicry - imitation of systems, elements, and models of nature. 20

PAX Fans (cont.)

ETM

- Benefits of PAX fans [2]:
  - Increased efficiency,
  - Significantly quieter,
  - Optimized Flow Pattern
  - Decreased Manufacturing and Maintenance Cost
- Research requirements:
  - Technology landscape analysis,
  - Identify products in development and initiatives in place,
  - Identify potential applications.

Portland State UNIVERSITY

[2] PAX Scientific, "PAX Fan: Innovative Air Handling." [Online]. Available: http://paxscientific.com/fansblowers. [Accessed: 03 -Oct-2013].

21

Industry application

ETM

- Broad industrial and commercial application
  - Most fluid-handling devices (mixers, pumps, turbines, propellers, heat exchangers, ducts, and other applications)
  - Fans, appliances, refrigeration
  - Computer cooling, automotive and transportation fans
- Petroleum, Pharmaceutical, and Beverage industries
  - Rotary and static (in-line) mixer technology
- Water Purification and Desalination

Portland State UNIVERSITY

22

Research Questions

ETM

- Research requirements:
  - Identify application areas and needs
  - Who is doing it?
  - What is being done globally and where is it being done?
  - What are the trends identified through data mining?

Portland State UNIVERSITY

23

Data Mining Application

ETM

R&D Stage	Typical Source
Basic Research	Science Citation Index
Applied Research	**Engineering Index**
Development	**US Patents**
Application	Newspaper Abstracts Daily
Social Impacts	Business and Popular Press

- Martino [1] describes the above relationship between the R&D stage of a technology and the typical source of information,
- To gain a further understanding on the progress the PAX technology, data mining was done using articles, conference proceedings and patents.

Portland State UNIVERSITY

[1] J. P. Martino, "A review of selected recent advances in technological forecasting," *Technological Forecasting and Social Change*, vol. 70, no. 8, pp. 719–733, Oct. 2003.

24

Data Mining (Academic Journals and Conferences)

ETM

- Search String:
  - (***Fan OR Blower OR Mixer***) in title, **AND**
  - ((***Natural OR Energy***) **AND (Efficien*)**) in all.
- 23 articles found in US for 2003-2013,
- CAGR: 50.2% since 2006

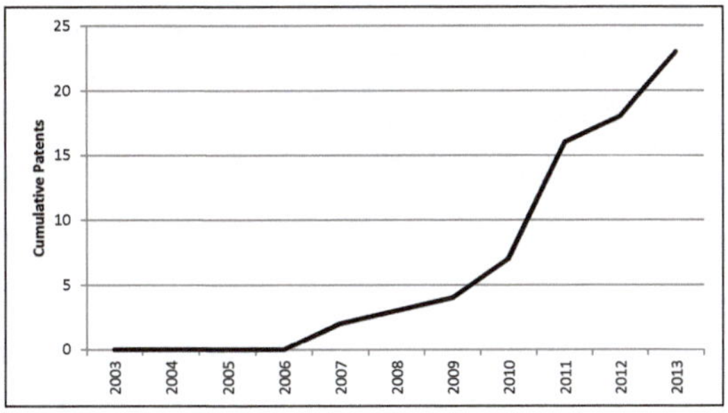

Portland State UNIVERSITY

Expert Identification

- Experts in the field of energy efficient fans, blowers and mixers were identified from the literature using Social Network Analysis (SNA).

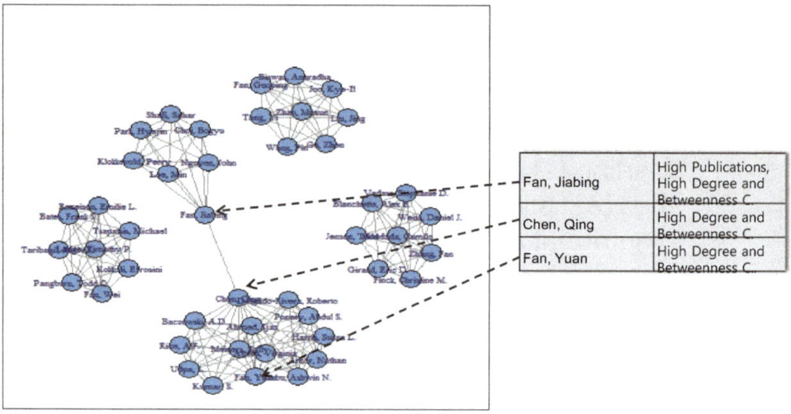

Portland State UNIVERSITY

Identified Experts by SNA

Author/Initials	Author Affiliation	Selected By
Chen, Qing	Department of Electrical and Computer Engineering, Michigan State University	Pub, Deg, Bet
Fan, Yuan	Department of Electrical and Computer Engineering, Michigan State University	Pub, Deg, Bet
Fan, Jiabing	Department of Fiber Science and Apparel Design College of Human Ecology, Cornell University	Pub, Deg, Bet
Ayres, Virginia	Department of Electrical and Computer Engineering, Michigan State University	Deg
Fan, Wei	Department of Chemical Engineering and Materials Science, University of Minnesota	Deg

Portland State UNIVERSITY

Data Mining (Patents)

- Identify technology development
- DB: USPTO
- Filtered data: 94 patents in the US

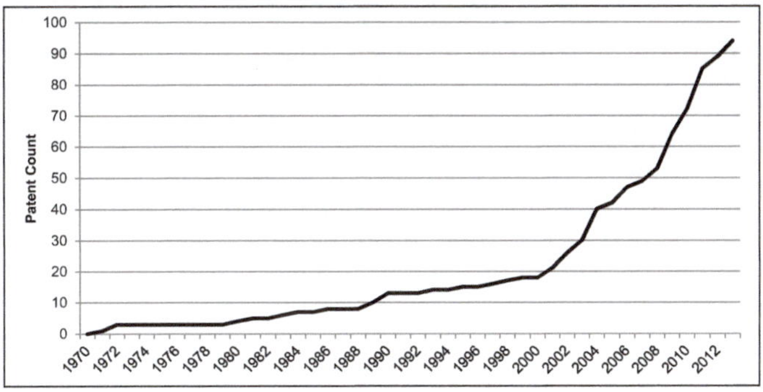

Source: AcclaimIP.com

Portland State University

Data Mining (Patents)

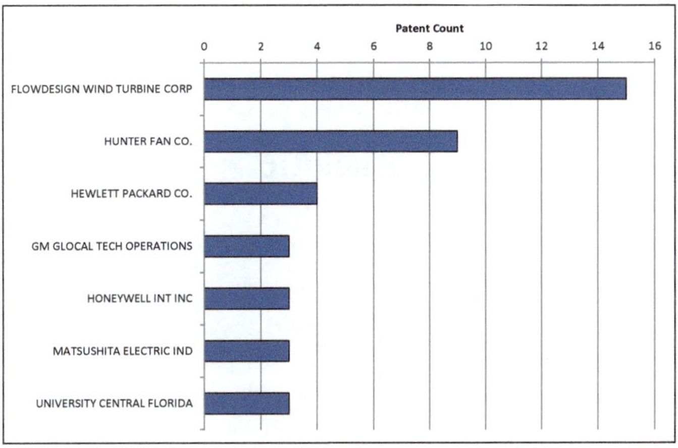

Source: AcclaimIP.com

Portland State University

References

ETM

- http://paxscientific.com/
- http://www.paxwater.com/

Portland State UNIVERSITY

30

ETM

Back Up

Portland State UNIVERSITY

31

Searching Keywords

ETM

- "Fluid-handling" in all field (limited in title: just 64)
- 498 articles found in Compendex for 1884-2012
- CAGR: 5% since 1990

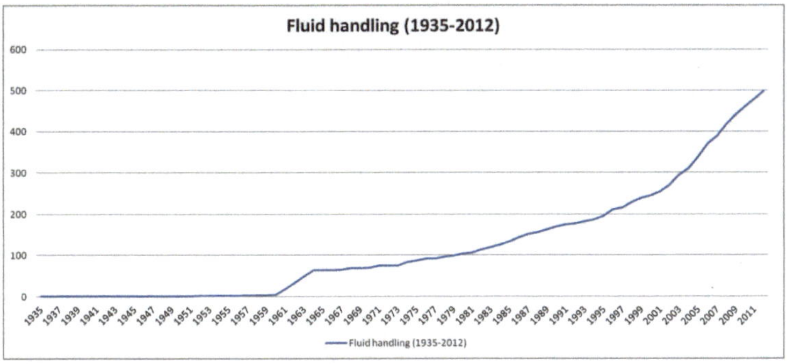

Portland State UNIVERSITY

Searching Keywords

ETM

- "Blower" limited in title
- 828 articles found in Compendex for 1884-2012
- CAGR: 3.9% since 1990

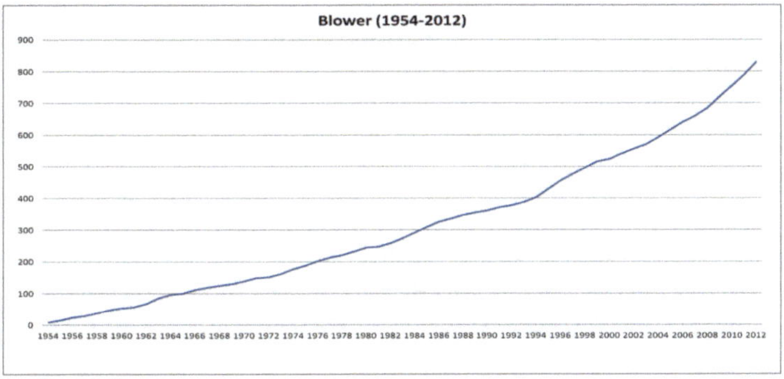

Portland State UNIVERSITY

Searching Keywords

- "Fan" limited in title
- 8704 articles found in Compendex for 1884-2012
- CAGR: 5.7% since 1990

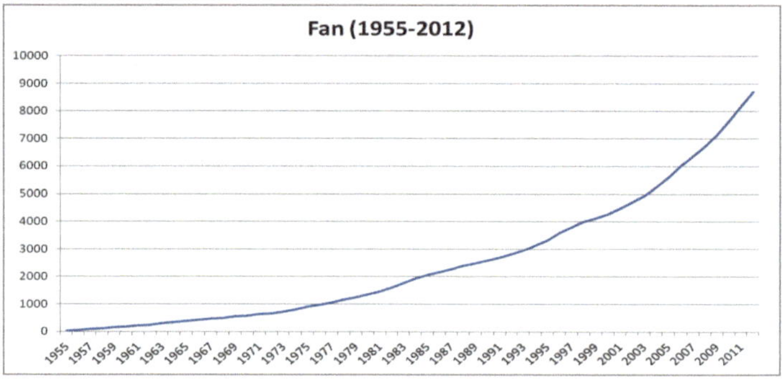

Portland State UNIVERSITY

Searching Keywords

- "Impeller" limited in title
- 3660 articles found in Compendex for 1884-2012
- CAGR: 6.2% since 1990

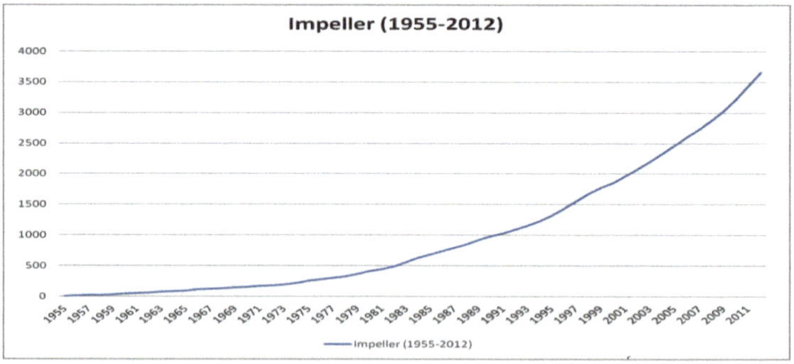

Portland State UNIVERSITY

Searching Keywords

ETM

- "*Biomimetics and Fan*" in all field (limited in Title: 59)
- 142 articles found in Compendex for 1884-2013
- CAGR: 53.2% since 2002

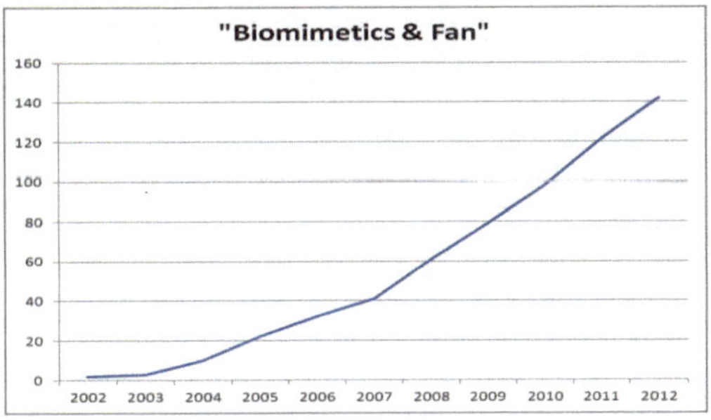

Portland State UNIVERSITY

SNA Result

ETM

- Minimum Degree of Centrality>=2

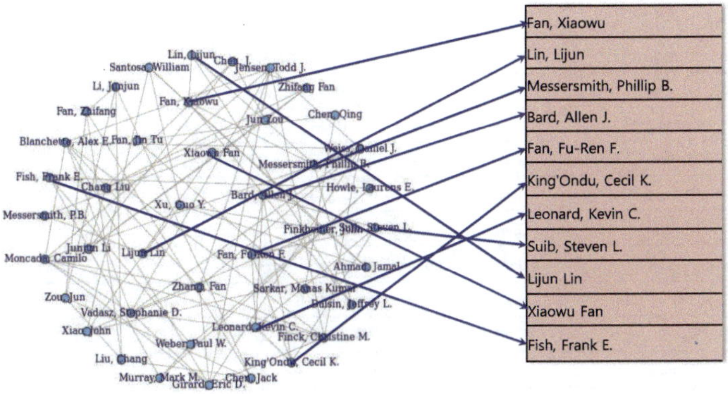

Portland State UNIVERSITY

Identified Experts by Citation Index

ETM

Author/Initials	Author Affiliation	Selected By
Fan, Xiaowu	Northwestern University, IL ,United States	Cit, Pub
Lin, Lijun	Northwestern University, IL ,United States	Cit, Pub
Messersmith, Phillip B.	Northwestern University, IL ,United States	Cit, Pub
Bard, Allen J.	University of Texas at Austin, Austin, TX, United States	Cit
Fan, Fu-Ren F.	University of Texas at Austin, Austin, TX, United States	Cit
King'Ondu, Cecil K.	University of Connecticut, CT, United States	Cit
Leonard, Kevin C.	University of Texas at Austin, Austin, TX, United States	Cit
Suib, Steven L.	University of Connecticut, CT, United States	Cit
Lijun Lin	Northwestern Univ., Evanston, IL, USA	Cit
Xiaowu Fan	Northwestern Univ., Evanston, IL, USA	Cit
Fish, Frank E.	West Chester University, United States	Cit

Portland State UNIVERSITY

38

Identified Experts by Citation Index

ETM

Author/Initials	Author Affiliation	Selected By
Fan, Jintu	Northwestern University, IL ,United States	Cit, Pub
Fan, Yichen	Northwestern University, IL ,United States	Cit, Pub
Messersmith, Phillip B.	Northwestern University, IL ,United States	Cit, Pub
Bard, Allen J.	University of Texas at Austin, Austin, TX, United States	Cit
Fan, Fu-Ren F.	University of Texas at Austin, Austin, TX, United States	Cit
King'Ondu, Cecil K.	University of Connecticut, CT, United States	Cit
Leonard, Kevin C.	University of Texas at Austin, Austin, TX, United States	Cit
Suib, Steven L.	University of Connecticut, CT, United States	Cit
Lijun Lin	Northwestern Univ., Evanston, IL, USA	Cit
Xiaowu Fan	Northwestern Univ., Evanston, IL, USA	Cit
Fish, Frank E.	West Chester University, United States	Cit

Portland State UNIVERSITY

39